Handbook of Water-Soluble Gums and Resins

ROBERT L. DAVIDSON

Editor in Chief

McGRAW-HILL BOOK COMPANY

New York   St. Louis   San Francisco   Auckland   Bogotá
Hamburg   Johannesburg   London   Madrid   Mexico
Montreal   New Delhi   Panama   Paris   São Paulo
Singapore   Sydney   Tokyo   Toronto

Library of Congress Cataloging in Publication Data

Main entry under title:

Handbook of water-soluble gums and resins.

   Includes index.
   1. Gums and resins—Handbooks, manuals, etc.
I. Davidson, Robert L.
TP978.H26   668'.37'0202   79–24007
ISBN 0–07–015471–6

1234567890   KPKP   89876543210

The editors for this book were Harold B. Crawford and
Joseph Williams, and the production supervisor was
Thomas G. Kowalczyk. It was set in Gael by The
Kingsport Press.

Printed and bound by The Kingsport Press.

Contents

Preface, v
Editorial Advisers and Contributors, vii

1	Introduction	1-1
2	Alginates	2-1
3	Alkyl and Hydroxyalkylalkylcellulose	3-1
4	Carboxymethylcellulose	4-1
5	Carrageenan	5-1
6	Guar Gum	6-1
7	Gum Agar	7-1
8	Gum Arabic	8-1
9	Gum Ghatti	9-1
10	Gum Karaya	10-1
11	Gum Tragacanth	11-1
12	Hydroxyethylcellulose	12-1
13	Hydroxypropylcellulose	13-1
14	Locust Bean Gum	14-1
15	Pectins	15-1
16	Polyacrylamide	16-1
17	Poly(Acrylic Acid) and Its Homologs	17-1
18	Polyethylene Glycol	18-1
19	Poly(Ethylene Oxide)	19-1
20	Polyvinyl Alcohol	20-1
21	Polyvinylpyrrolidone	21-1
22	Starch and Its Modifications	22-1
23	Tamarind Gum	23-1
24	Xanthan Gum	24-1

Index follows Chapter 24.

Preface

One cannot do justice in a few words to the omnipresence of water-soluble gums and resins. It is virtually impossible for a day to pass without each of us having benefitted from one of these materials, either in the clothing we wear, the foods we eat, the beverages we drink, the toiletries or cosmetics we use, the medicines we take, and more. Much more. A scan of the index will convert the most obdurate of skeptics.

This book is designed to be used. Both organization and writing are tailored for speedy, accurate referencing. Each chapter is divided by category and subcategory with a degree of built-in redundancy so that users can enter it according to their different needs, for example:

- by gum or resin
- by basic function
- by end product
- by industry
- by property or characteristic

The text is supported by a detailed index. With it, one can locate both major and minor uses, applications, interactions, compatibilities, system components, or whatever other data is required. Together, these data-search aids implement the editor's conviction that the utility of a reference work, regardless of its content, is no better than the speed and accuracy with which a user can enter and search the work.

The cumulative qualifications of the authors is gratifying. Their biographies and, where available, photographs are presented in the "Editorial Advisors and Contributors" section, beginning on page ix. In particular, I owe much to John Glavis, formerly of Rohm and Haas, and now deceased; Kenneth Guiseley of FMC's Marine Colloids; Morton Rutenberg of National Starch and Chemical; and George Greminger of Dow Chemical. Their generous consultation and expert advice was invaluable in the planning of the book.

One final note, a caveat. The information in this book is offered solely for your consideration. It is not a warranty or recommendation for the use of any product, nor for the practice of any procedure or patented invention without prior investigation and verification of your own.

Robert L. Davidson
Editor in Chief

Editorial Advisors and Contributors

EDITORIAL ADVISORS

FRANK JOHNSON GLAVIS

Research Chemist, Rohm and Haas Company, Philadelphia, Pennsylvania, from 1938 to 1974; retired 1974; deceased June 11, 1976. He received his A.B. degree from Dartmouth College in 1935, and his Ph.D. degree from the University of Illinois in 1938, when he joined Rohm and Haas. He was the author of twenty-five U.S. patents, and the holder of many foreign patents in the fields of synthetic lubricants, additives for lubricants, and monomer preparation, as well as for the polymerization and use of acrylic and methacrylic acids and their derivatives, especially the esters. Among his many contributions to polymer science, he developed a method for the hydrolysis of methacrylic polymers, used as a chemical means of distinguishing the isotactic and syndiotactic forms of such polymers.

Dr. Glavis was a man of indominable will who, despite diabetes since the age of 10 and blindness since the age of 35, continued a fruitful scientific career in chemistry. His interests were broad ranging, and included not only polymer chemistry, but ice skating, sailing and boating, music, and reading. He was a family man, married in 1940 to Doris Ashworth who, with their daughter, Wendy Jane, survives. His courage will be remembered, his kindness will be missed.

MORTON W. RUTENBERG
Starch and Its Derivatives (author)

Associate Director, Starch Modification Research, Starch Division, National Starch and Chemical Corporation, Bridgewater, New Jersey; received his B.S. degree (1942) and M.S. degree (1947) in chemistry, and his Ph.D. degree (1949) in organic chemistry from the University of Pennsylvania. He joined National Starch as Chemist in 1949, advancing through Supervisor, Protein Research, and Section Leader in By-Products Research to his present position in 1973. He has worked in starch chemistry and technology, polysaccharides, hemicelluloses, and natural gums. He is co-inventor on 19 patents on starch and hemicelluloses. He is a member of the American Chemical Society, the American Association of Cereal Chemists, Sigma Xi, and the AAAS.

KENNETH B. GUISELEY
Carrageenans (author)

Director of Polymer Research, FMC Corporation, Marine Colloids Division, Rockland, Maine; received his B.A. degree in chemistry, *summa cum laude*, from Hartwick College, Oneonta, New York, in 1955, and received his Ph.D. degree in organic chemistry from Syracuse University, New York, in 1960. From 1960 to 1961, he was a Post-Doctoral Fellow at Syracuse University. In 1961, he joined Marine Colloids as Senior Chemist, and in 1974 he was appointed to his present position. He is a member of the Carbohydrate Division of the American Chemical Society, and is the author of seven publications and three patents on carbohydrate sulfates and various seaweed polysaccharides.

GEORGE K. GREMINGER
Alkyl and Hydroxyalkylalkylcelluloses (author)

Associate Scientist, The Dow Chemical Company, Midland, Michigan; received his B.S. degree from the New York State College of Forestry at Syracuse University. He joined Dow in 1940, and has been involved in production, research, patent liaison, and product development with ethylcellulose, carboxymethylcellulose, methylcellulose, and methylcellulose derivatives. He has been involved in projects in Europe, Latin America, and Canada, which provided additional expertise with global markets and problems. He holds 33 patents in the area of cellulose ether technology. He is a member of the American Chemical Society, the Technical Associations of the Pulp and Paper Industry, and the AAAS.

CONTRIBUTORS

L. BLECHER
Polyvinylpyrrolidone

Manager, Food & Drug Industries, Chemicals Division, Acetylene Chemicals Department, GAF Corporation; received his B.S. degree in chemistry from Morningside College in 1949. He joined GAF in 1967 as Marketing Specialist. He has specialized professionally in the market development of pharmaceutical excipients and food additives. Mr. Blecher has published articles on alcoholic beverage stabilization. His memberships include the American Chemical Society, the Master Brewers Association of the American Society of Brewing Chemists, the Institute of Food Technologists, the Society for Cryobiology, and the Academy of Pharmaceutical Sciences.

DAVID B. BRAUN
Poly(Ethylene Oxide)

Development Scientist and former Technology Manager, Polyox® Water-Soluble Resins, Union Carbide Corporation, Tarrytown, New York; received his B.S. degree in chemistry from the State University of New York at Buffalo in 1956. Since then he has been employed by Union Carbide in various phases of research and development. He spent a number of years involved with the development of silicone products, particularly silicone elastomers. He then worked in the field of application development of vinyl chloride resins and plasticizers. For the past several years, Mr. Braun has been responsible for a major portion of the laboratory development program for Polyox poly(ethylene oxide) water-soluble resins.

ROBERT W. BUTLER
Hydroxypropylcellulose

Manager, Technical Services, Hercules, Inc., Wilmington, Delaware; graduated from the Imperial College of Tropical Agriculture, Trinidad, West Indies, in 1943 (Sugar Technology), and received his Ph.D. degree in chemistry from McGill University in 1950. He joined Hercules in 1952 after two years at Industrial Cellulose Research, Hawkesbury, Ontario, Canada. He has spent most of his career at Hercules in the development of cellulose and starch derivatives and water-soluble polymers. For five years, he was involved with product development and market introduction of hydroxypropylcellulose. He has also held other positions at Hercules in research, market and product development, and sales.

IAN W. COTTRELL
Alginates, Xanthan Gum

Assistant Technical Director for New Process and Product Research, Kelco Division of Merck & Company, Inc., San Diego, California; received his B.Sc. from the University of Edinburgh in chemistry, 1965, and his Ph.D. degree in polysaccharide chemistry from the University of Edinburgh, 1968. Following a post-doctoral fellowship at Trent University, Ontario, Canada, Dr. Cottrell joined the research staff of Stein Hall & Compay, Inc., in 1970. In 1972, he joined Kelco as a Senior Research Chemist, and later became Section Head of Organic Development before being named to his present position. Dr. Cottrell is the author of several publications on pectic substances, alginates, and xanthans and other fermentation gums. He is a member of the American Chemical Society and the Society for Industrial Microbiology.

DONALD F. DURSO
Introduction

Director, Absorbent Technology, Johnson & Johnson, New Brunswick, New Jersey; received his B.S. degree in Chemical Engineering from Case Institute of Technology (1947), and his M.S. degree (1949) and Ph.D. degree (1951) in biochemistry from Purdue University. He worked for Buckeye Cellulose Corp. from 1951 to 1971, and was director of research; was professor at Texas A & M University, Department of Forest Science, from 1971 to 1975; and joined Johnson & Johnson in 1975 in his present position. His professional career as a polysaccharide industrial chemist has included work in the preparation and use of purified cellulose as a chemical raw material. He has authored a number of articles and patents. He is a member of the American Chemical Society, Sigma Xi, the Canadian Pulp & Paper Association, and the Technical Association of the Pulp and Paper Industry.

RALPH E. FRIEDRICH
Polyacrylamides

Senior Research Specialist, Central Research, Physical Research Laboratory, The Dow Chemical Company, Midland, Michigan; graduate of the University of Michigan, B.Sc. in chemistry, 1947. His work has been concerned with the polymerization of vinyl monomers, particularly acrylamide, and with electroconductive resins. He is the coauthor of five publications and the author or coauthor of fifteen patents, many of which deal with acrylamide manufacture or use, including his studies on the mechanism of flocculation of silica and clays by high-molecular-weight polymers. Another of his areas of interest is electrostatics as it concerns xerographic printing and the prevention of vapor and dust explosions.

THOMAS W. GERARD
Tamarind Gum

Senior Research Chemist, Eastern Research Center, Stauffer Chemical Company, Dobbs Ferry, New York; received his B.S. degree in chemistry from St. Louis University in 1947. He joined Stauffer in 1976. Prior to then, he was Consultant for Denard Associates, 1975–1976; Director of Plant Hydrocolloids, Meer Corporation, 1959–1975; Chemist, Stein, Hall & Co., Inc., 1956–1958; and Senior Research Chemist and Project Leader in Corn Products Research, Anheuser-Busch, Inc., 1949–1956. For 27 years, he has been active in research, development, and application of carbohydrates in a variety of industries, including food, pharmaceutical, cosmetic, paper, textile and mining. He has published a number of technical papers, and has starch patents to his name. He is a member of the American Chemical Society, the Institute of Food Technologists, and the Technical Association of the Pulp and Paper Industry.

HAROLD L. GREENWALD
Poly(Acrylic Acid) and Its Homologs

Patent Agent, Patent Department, Rohm and Haas Company, Philadelphia, Pennsylvania; received his B.S. degree in chemistry from Pennsylvania State College in 1939, and his Ph.D. degree in chemistry from Columbia University in 1952. He became a registered patent agent (United States) in 1975. From 1941 to 1945, he was engaged in rocket propellant research for Division 8 of the National Defense Research Committee. He joined Rohm and Haas in 1950, and has worked on rocket propellants, surface chemistry, organic coatings, and aqueous polymer dispersions. His work has ranged from process development in oil recovery and emulsion polymerization to the physical properties of surfactants and polymer dispersions. He has contributed a number of scientific papers, patents, and book chapters in the above-named fields.

KENNETH S. KANG
Xanthan Gum

Head and Manager of Biochemical Development, Kelco Division of Merck & Company, Inc., San Diego, California; a native of Korea, he was educated at Yonsei University in Seoul. He came to the United States in 1961, and studied microbiology and biochemistry at the University of Delaware, receiving his M.S. degree in 1963 and his Ph.D. degree in 1965. After post-doctoral research at Pennsylvania State University, he joined Kelco in 1966. Dr. Kang has been instrumental in the development of food-grade xanthan gum, and in the improvement of the manufacturing process. He also led the research for new microbial polysaccharides, and has authored publications about them. He is a member of the American Society of Microbiologists, the American Chemical Society, Sigma Xi, and the American Association for the Advancement of Science.

EUGENE D. KLUG
Carboxymethylcellulose, Hydroxypropylcellulose

Consultant (Europe), Hercules, Inc., Wilmington, Delaware; was Research Associate with Hercules until his retirement in 1976; received his A.B. degree from Oberlin College, 1936, and joined Hercules. Most of his career at Hercules was spent in the research and development of derivatives of cellulose, starch, and other polysaccharides. The patents he has authored have played a major role in the development of Hercules sodium carboxymethylcellulose (CMC), hydroxyethylcellulose (Natrosol®), and hydroxypropylcellulose (Klucel®). He was Secretary (1952–1954) and Chairman (1955) of the Cellulose Chemistry Division of the American Chemical Society and has served on committees of the ASTM relating to cellulose and cellulose derivatives.

PETER KOVACS
Alginates, Xanthan Gum

General Manager (Europe, Middle East, Africa), International Division, Kelco Division of Merck & Company, Inc., San Diego, California; received his B.Sc. degree in chemistry from San Diego State University in 1955. He worked as a Research Chemist for Narmco R & D Corporation, then joined Kelco in 1967. His prior positions with Kelco have included Senior Research Chemist, Technical Manager, International Division, and Section Head of Food Development and Regulatory Compliance. Mr. Kovacs is the author of several publications on the applications and toxicology of alginates and xanthan gum. He is a member of the Institute of Food Technologists, and has served as vice president of MARINALG International, the world association of seaweed processors.

KARL L. KRUMEL
Alkyl and Hydroxyalkylalkylcellulose

Senior Research Specialist in Process and Applications Research, The Dow Chemical Company, Midland, Michigan; received his B.A. degree in chemistry from Grinnell College in 1960, his M.A. degree in chemistry from DePauw University in 1962, and his Ph.D. degree in organic chemistry from Michigan State University in 1965. He joined Dow Chemical in 1965. From 1969 to 1975, he was active in cellulose ether technology as related to the production support and product development of these materials. Since 1975, he has been involved in process and product research in agricultural chemicals. He has several patents and publications in these areas, and is a member of the American Chemical Society.

D. H. LORENZ
Polyvinylpyrrolidone

Manager, Polymer Research, Chemicals R&D, GAF Corporation; received his B.S. degree in organic chemistry in 1958 and his Ph.D. degree in organic chemistry in 1963 from the Polytechnic Institute of New York. He joined GAF in 1965 as a Research Chemist. In 1970, he was named Group Leader, and in 1975 he moved to his present position. Dr. Lorenz has specialized in water-soluble polymers, cosmetic and pharmaceutical polymers, urethanes, and radiation curing. He holds 20 U.S. patents and 40 foreign patents. He is a member of the American Chemical Society.

H. L. LOWD
Polyvinylpyrrolidone

Product Manager, Polyvinylpyrrolidone Monomers and Polymers, Acetylene Chemicals Group, GAF Corporation; received his B.S. degree in chemical engineering from Tufts University in 1970, and his M.B.A. degree in marketing from the University of Bridgeport in 1973. He joined GAF in 1971 as Sales Trainee. Other positions he has held are Acetylene Specialist and Chemical Regional Sales Manager. He moved to his present position in 1977. He is a member of the American Institute of Chemical Engineers, the American Pharmaceutical Association, and the MBAA.

LEO S. LUSKIN
Poly(Acrylic Acid) and Its Homologs

Senior Technical Writer, Advertising Department, Rohm and Haas Company, Philadelphia, Pennsylvania; received his B.S. degree in chemistry (1936) and his M.S. degree (1942) from the University of Michigan, and a Bachelor of Music degree (1941) from the Curtis Institute of Music. He joined Rohm and Haas in 1944 as Senior Research Chemist in organic chemical synthesis. Since 1962, he has held writing and promotional positions in Special Products, Industrial Chemicals, and Plastics Intermediates Departments of the company. He has published many papers and has patents in organic synthetic applications, such as textile finishing agents, coatings, specialty monomers, and bactericides.

TETSUJIRO MATSUHASHI
Agar Notes

Section Chief, Nagano State Laboratory of Food Technology, Nagano, Japan; responsible for processed foods (frozen and dried), soybean products, general food technology, and wastewater treatment; educated at Tokyo College of Physics (1948–1949), Tokyo University of Fisheries (graduated 1953), Food Science Department of the University of Georgia (M.S. degree in 1966), and the Agricultural and Biological Chemistry Department of the Tokyo University of Agriculture (Ph.D. degree in 1978; dissertation on "Fundamental Studies on Agar Manufacture"). He was with the Chiba State Fisheries Experimental Station, 1953–1955; the Fisheries Administration Division at Chiba, 1955–1961; the Nagano State Laboratory for Agar-Agar Industry, Chino, 1961–1969. He received the Academic Award from the Japanese Society of Food Science & Technology in 1973.

GEORGE MEER
Gum Ghatti

President, Meer Corporation, North Bergen, New Jersey; received his B.S. degree in chemistry from Rensselaer Polytechnic Institute, his M.S. degree in biochemistry from New York University, and his M.B.A. degree from the New York University Graduate School of Business Administration. He joined Meer in 1955 as a Product Control Chemist, became Technical Director and Vice President in 1959, and President in 1971. He is active in the water-soluble gums field, has coauthored a number of articles, and has presented many technical papers. He belongs to the American Chemical Society, the American Pharmaceutical Association, the Institute of Food Technologists, the Essential Oil Association, where he is on the Board of Directors, and is on the Board of Directors of the Drug, Chemical and Allied Trades Association.

WILLIAM A. MEER
Gum Agar, Gum Arabic, Gum Karaya

Vice President, Meer Corporation, North Bergen, New Jersey; received his B.S. degree in pharmacy from the University of Florida, his M.S. degree from Columbia University College of Pharmacy, and his Ph.D. degree from the University of Connecticut. He joined Meer Corporation in 1959 as Director of Quality Control, and was named Vice President in 1961. His background is in phyto chemistry, but he has worked in pharmacognosy, industrial pharmacy, and food technology. He has coauthored a number of articles and papers. He belongs to the American Chemical Society, the American Pharmaceutical Association, the Institute of Food Technologists, and the American Society of Pharmacognosy.

THEODORE W. MODI
Polyvinyl Alcohol

Technical Consultant, Ethylene Polymers Division, Plastic Products and Resins Department, E. I. du Pont de Nemours & Co., Inc., Chestnut Run, Wilmington, Delaware; received his B.S. degree in chemical engineering from Michigan State University in 1951. He has been employed by Du Pont since graduation, during which more than 20 years were associated with Elvanol® polyvinyl alcohol. His assignments have been in development, research, and marketing, with major emphasis on customer service for applications in textiles, paper, adhesives, automotive, film, building products, and other industries. He has also given general technical support and consultation on polyvinyl alcohol to field personnel and others.

JENS KRISTIAN PEDERSEN
Pectins

Research and Development Manager, A/S Kovenhavens Pektinfabrik (since 1972 a subsidiary of Hercules, Inc., Wilmington, Delaware); received his M.Sc. degree in chemical engineering from the Royal Technical University of Denmark in 1958. He joined A/S Kovenhavens Pektinfabrik in 1960 as a research chemist, working on methods of manufacturing carrageenans. He worked on all aspects of both pectin and carrageenans, from raw materials through manufacturing development of quality control methods, application development, and technical service. He has been a speaker at numerous meetings, courses, and seminars in Europe, Japan, Canada, and the United States, and has authored several articles on the same subjects.

GEORGE M. POWELL, III
Hydroxyethylcellulose, Polyethylene Glycol

Retired from Union Carbide Corporation in 1969; from 1970 through 1976, engaged in consulting and technical writing activities, principally for Union Carbide. He received his professional Chemical Engineer degree from Columbia University in 1933, then joined the Research & Development Department of Union Carbide's chemical subsidiary in South Charleston, West Virginia, progressing to Associate Director. He has broad experience in industrial applications for plastics and chemicals, with emphasis on vinyl resins and the organic coatings fields. In 1950, Mr. Powell received the John Wesley Hyatt award for pioneering work on vinyl resin dispersions.

JAMES K. SEAMAN
Guar Gum, Locust Bean Gum

Section Leader, Water Soluble Polymers, Natural Products Laboratory, Celanese Polymer Specialties Company, Jeffersontown, Kentucky; received his undergraduate degree from New York University in 1951, and his M.S. degree in organic chemistry from Brooklyn Polytechnic Institute in 1960. He joined Stein, Hall & Co. in 1954 as a laboratory assistant, advancing to Chemist, Chief Chemist, and Group Leader. He remained with Celanese after its purchase of Stein, Hall & Co., advancing to his present position in 1979. He has published articles on guar gum and locust bean gum. He is a member of the American Chemical Society and the Society of Plastics Engineers.

NORMAN F. STANLEY
Carrageenans

Senior Scientist, FMC Corporation, Marine Colloids Division; largely self-educated in chemistry and mathematics. He began his career in seaweed hydrocolloids in 1940 as a laboratory technician with the Algin Corporation of America, and was promoted to Laboratory Manager in 1941, then to Research Director in 1953. From 1959 to 1964, he was Assistant Technical Director for Marine Colloids, and in 1964 he became Senior Chemist, receiving his present appointment in 1974. He developed processes to extract and modify carrageenan, holds a patent, has authored or coauthored seven publications on carrageenan chemistry, and is a member of the Society of Rheology, the American Chemical Society, the AAAS, and the AIAA.

KENNETH R. STAUFFER
Gum Tragacanth

Consultant to the food industry involved with emulsions, stabilizers, and other gum-related technology; also teaching and conducting research as Assistant Professor at the University of Rhode Island. Received his A.A.S. degree from the State University of New York at Morrisville in food processing (1963), his B.S. degree in food science from the University of Georgia (1965), and his Ph.D. degree from Rutgers University in the Department of Food Science (1979). He has been Technical Director for Tragacanth Importing Corporation, a Food Scientist for R. T. French, and an analytical researcher for Hooker Chemical Research Center. He is a member of the Carbohydrate Division of the Institute of Food Technologists, and of Sigma Xi. He has authored several papers on the industrial use of gums.

GLENN I. STELZER
Carboxymethylcellulose

Regional Director of Marketing, Western Region, Hercules, Inc., San Francisco, California; received his B.S. degree in chemical engineering in 1951 from the University of Missouri at Columbia. He joined Hercules in 1951, and has held positions as Development Engineer, Sales Representative, District Manager for Water Soluble Polymers (Mid Central States; New York and New England), Sales Manager for Sodium Carboxymethylcellulose, and Director of Sales for the Food & Fragrance Development Department. He was appointed to his present position in 1978, and is in charge of all Hercules product sales in the eleven western states, plus Alaska and Hawaii. He is a licensed professional engineer in Missouri and is a member of the Institute of Food Technologists.

JOHN C. TINKER
Gum Arabic

General Manager, Water Soluble Gum Division, Meer Corporation, North Bergen, New Jersey; received his B.A. degree in chemistry from Syracuse University. He joined Meer in 1973. Prior to then, he was General Manager of Gelatin, Gums, and Stabilizers for eight years at Swift & Company, preceded by eight years as Vice President of Clarence Morgan, Inc., and as President of its subsidiary, Certified Proteins, and Vice President of a predecessor company, Paul D. Dunkel, Inc. Before then, he was a technical sales representative for the Sheffield Chemical Division of Kraft. His memberships include the American Chemical Society and the Institute of Food Technologists.

HENRY VOLK
Polyacrylamides

Research Specialist, Dow Chemical Company, Midland, Michigan. He joined Dow in 1943, and has been active in the research and development of water-soluble polymers for many years. His interests have included the synthesis and product development of polymers. He is the author or coauthor of seventeen patents, the majority of which are related to the preparation and/or utility of water-soluble polymers. He has also been involved with monomers synthesis; e.g., styrene sulfonic acid, tertiary butyl styrene sulfonic acid. His work on the flocculation properties of water-soluble polymers has led to an increased knowledge of the proper techniques for maximizing polymer performance.

PHILIP A. WHITEHOUSE, JR.
Carrageenans

Received his B.Sc. degree in chemistry from the University of Maine in 1962. For the next five years, he researched antineoplastic agents and natural products, coauthoring many papers on these subjects with Dr. G. R. Pettit. He joined Marine Colloids Division of FMC Corporation in 1967, and for 10 years was closely involved in all phases of carrageenan manufacture and applications, holding titles from Research Chemist to Assistant Production Manager. He established a system linking the finished application through the manufacturing process to the raw material. He coauthored one patent relating to the sulfation of polysaccharides and is a member of the Sigma Xi. He left Marine Colloids in 1977 to pursue an independent career in marketing.

ANDREW W. WOOD
Polyvinylpyrrolidone

Plant Manager, Calvert City Plant, Chemical Group, GAF Corporation, Calvert City Kentucky; received his B.S. degree in chemistry from Rutgers University in 1944, and his M.S. degree (1962) and Ph.D. degree (1968) in chemistry from the Polytechnic Institute of New York. He joined GAF in 1947 as a Chemist and has held positions as Plant Chemist and Research Chemist, moving to his present position in 1976. Dr. Wood holds 10 patents, and is a member of Sigma Xi, Lambda Chi Epsilon, and the American Chemical Society. He has taught organic chemistry in the Paducah Community College extension program for a number of years.

DONALD P. WYMAN
Polyvinylpyrrolidone

Technical Director, Polymers, Plastics, Reprographic Materials R&D, Research and Development Department, GAF Corporation; received his B.S. degree in chemistry from Ohio University in 1953 and his Ph.D. degree in organic chemistry from Michigan State University in 1957. He joined GAF in 1974 as Technical Director, Reprographic Materials R&D. His previous professional positions have included Director of Marketing for Scott Graphics from 1968 to 1974, and Manager of Plastics Research for Borg Warner Chemicals from 1964 to 1968. Dr. Wyman has some 30 publications and holds several U.S. and foreign patents. He is a member of the American Chemical Society and the American Association for the Advancement of Science.

Chapter 1

Introduction

D. F. Durso
Johnson & Johnson

Not many years ago, a comprehensive report on water-soluble gums and resins would have treated mainly natural materials. The few synthetic compounds were based on cellulose. With the advent of practical synthetic organic polymer chemistry, a revolution took place which greatly increased the variety of chemical compounds and physical reactions available. Users could almost specify independently each of the solution attributes they wished to obtain and confidently expect to purchase some stock or custom-made molecule to suit their needs. More recently, we are seeing another revolution in this field of "water modifiers." Today, the chemist is producing superabsorbents which differ from the past materials in that they do not dissolve but merely make the water part of their internal structure.

The most descriptive single term for all of these materials is *hydrogels*. These are substances which associate with water molecules in such a way that the behavior of the water is modified, allowing us to perform functions not normally possible. This change in water behavior is usually accomplished by minor amounts of the hydrogel, commonly less than 10% of the weight of the water.

There are many ways in which one could choose to group the water-soluble gums and resins so that the user of the Handbook might quickly acquire the desired information. After much deliberation and several false starts, the reader will find a combined chronological-source format which introduces this complicated but exciting and dynamic field. It should be noted that many "old" materials compete successfully today after almost a century of efforts to replace them. Thus, it is the usual balance of economics and performance which determines the commercial realities.

Exudates

Originally the term *gum* was applied to the exudates from various plants. These substances were usually sticky and deformable masses produced in response to some injury. Centuries ago it was learned that these substances could be dissolved in water and then used in special ways. Perhaps lubrication and gluing would represent the extremes of the actions available. From some obscure beginning, procedures evolved for stimulating the production of gum by the plant, for purification of the raw material, for reduction to a shippable form, and for use in end products.

Despite the evolutionary indications in this report, it should be noted that a large commerce exists today in many of the gums known even in ancient times. For many of them, the basic processes for their production and collection are unchanged from those used centuries ago. It is a testimony to the synthesis ability of plants that, in many cases, the natural exudates provide functions which cannot be equaled economically by man.

As will be seen in more detail later in this Handbook, the plant exudates are polysaccharides containing various sugars other than glucose, and having significant quantities of oxidized groups as an adjunct to their normal polyhydroxy format. In all cases, a large proportion of the groups are carboxyl, and these are found as salts of calcium, magnesium, or potassium rather than in the free form.

Extractives

In many cases, water-soluble polysaccharides in general similar to the exudates are components of land and marine plants and their seeds. These materials result from normal metabolic and life processes, and many times they represent the reserve carbohydrate in that system. In the latter case, the commercial value of the extract is such that the plant is cultivated for the express purpose of providing the gum. Unlike the rudimentary systems for collection and processing of the exudates, the extractive-type gums result from quite sophisticated ventures, including seed selection, plant placement, seed collection, and seed processing. In many cases, it will be seen that seed processing requires fractionation to obtain the desired commercial gum quality.

This class of gums includes those obtained from seaweeds and from trees. Here the entire plant is harvested and comminuted to increase accessibility; then the gum is removed by water leaching. The entire cost of the operations, including planting, can be borne by the gum product because of the inimitable properties of the natural material.

Like the exudates, these materials are salts of acidic polysaccharides. The polymer structures may be more varied and more complex than those of the previous group.

Synthetic, Natural-Based

Using purified cellulose as a starting material, organic chemists have learned to imitate or replace nature in many applications. The chief commercially important materials are those known as ethers, wherein the hydroxyl groups have been derivatized by a reaction which replaces the hydrogen atom with a simple or complex group.

The first and still largest-volume material is carboxymethylcellulose (CMC), made by the addition of chloroacetic acid to soda cellulose. In many cases, this material replaced the natural gums when problems arose because of economic, supply, or chaotic conditions. Originally needed to replace a long-standing natural material, this first synthetic gum developed a technology of use and production which has stood the test of time. Production of CMC has been increasing almost continually since its first commercial use and, like the natural gums, it has carved certain niches which cannot be filled by other similar hydrogels.

Unlike nature, man is not restricted to the use of carboxylic salt groups in order to accomplish the purpose of modification of water. Neutral (non-ionic) and basic groups can also be introduced. In some cases, the materials can pass through a water-soluble stage and become soluble in organic solvents. In other cases, the derivative group will impart water and solvent miscibility simultaneously. The possibilities here are limited only by the imagination of the chemist and the ability of the user to pay for the transformation of insoluble cellulose into a new form.

Among the present commercially important materials are these neutral ethers of cellulose: methyl-, ethyl-, hydroxyethyl-, and hydroxypropyl-. Details concerning their preparation and applications will be found elsewhere in this Handbook.

While starch can also be converted into almost all the same derivatives as cellulose, commercially there are no important neutral ethers. In many cases, starch itself serves as a hydrogel after enough processing to disrupt the natural structure of the starch granule. Huge amounts of corn and potatoes are processed each year in order to fill the needs of the industrial, technical applications where other "gums" cannot compete. It is quite literally true that the cellulose industry as we know it today, where the cellulose is used in solid form for its physical attributes, could not exist without the almost equal-volume industry which supplies starch in many grades.

Besides the uses for "natural" starch, there are many applications where the starch has been made cationic by the incorporation of nitrogen-based groups. Here again, the hydroxyl hydrogen has been chemically removed and replaced by a substituent which provides a different "gum" reaction.

It has been known for many years that derivatives of cellulose and starch can be prepared by grafting. In these reactions, free radicals are generated within the substrate by some suitable catalyst system. Usually the hydrogen of the hydroxyl is eliminated and thus the oxygen atom serves as the site of the free radical. In the presence of suitable vinyl-type monomers, these active species will then cause a closely associated synthetic chain molecule to be produced. Whether the new molecule is attached covalently or by weaker bonds, the result is a major modification of the water-solubility of the natural substrate. The components of the final product can be further modified by chemical reactions generally known as "crosslinking." These can be as simple as internal ester formation (between hydroxyl and carboxyl groups), or they can be as complicated as those obtained by chemical additions to double bonds.

It should be obvious that with these tools available, a very large number of water-interacting "gums and resins" can now be prepared. In 1976, laboratory and commercial quantities of several materials were introduced into products where minimal water-solubility of the hydrogel is desired. The performance value of these new materials is yet to be proven, since they command a premium price over all previous "derivatives." Strangely, all of the current products are based on carboxylic ethers of cellulose or starch.

A relatively new group of cellulose-based gums are those containing sulfonic acid groups. These are being utilized where shortages of seaweed gums have provided the necessary economic basis.

Another large group of uses has arisen in recent years for gums produced by the action of fungi or bacteria (specifically, their enzymes) on natural materials, usually starch. These biologically produced polymers result from the almost total decomposition of the substrate into its monosaccharides followed by re-synthesis into new molecules utilizing some or all of these changes: points of attachment other than C_4, partial oxidation of the hydroxyl groups, interchain crosslinking, or introduction of new chemical groups to replace the hydroxyl hydrogen.

Synthetic

In this group are those hydrogel materials prepared from simple molecules, usually petroleum-based. In general, these are based on polymerization of vinyl-type monomers, and usually the monomer has been modified such that it contains an acidic (or acid-based) function. The materials were first synthesized as part of fundamental research projects, then tested for the possibility of replacing some of the gums mentioned previously, and finally accepted into some commercial uses because of unique properties.

Here again it is strange to note that with the myriad chemical moieties available to the organic chemist, the materials of interest contain hydroxyl and carboxyl groups, just like the "original" gums and resins.

These totally synthetic products have found applications in those fields where resistance to normal biological decomposition is needed. They can be structured so as to be impervious to the organisms (and their "natural" enzymes) requiring monosaccharide-based polymers. In addition, these products also can be prepared so as to permit usage at elevated temperatures where all other types discussed previously would decompose and/or lose their ability to interact with water molecules.

Summary

Depending on source, four major groups of water-soluble gums and resins have been identified. When examined chemically, we find that they are polymers with acidic, neutral, or basic groups scattered among the linear, branched, or crosslinked chain molecules. The materials are all used for the simple reason that, in rather minute amounts, they so alter the normal properties of water as to result in a valuable function. This can be in nutrition, in medicine, in graphic arts, in boxboard manufacture, etc.

The varied needs are such that a rather unbelievably large family of materials is available, and many of these are the basis for multimillion-pound annual commercial enterprises. Many of the suppliers can provide more than one material, and where required, they also provide applied technology in the form of special formulations prepared by them or instructions as to how the users can prepare their own custom gums.

Chapter 2

Alginates

Ian W. Cottrell and Peter Kovacs
Kelco, div. of Merck & Co., Inc.

General Information 2-2
  Chemical Structure 2-2
  Manufacture 2-3
  Physical Properties 2-5
  Solution Properties 2-6
  Compatibilities 2-15
  Toxicology/Environment 2-17
  Application Procedures 2-20
  Film Forming 2-23
Commercial Uses 2-25
  Food Applications 2-25
  Industrial Applications 2-26
Formulations 2-26
  Stabilizing Frozen Foods 2-26
  Food Gels 2-28
  Dessert Soufflés 2-31
  Dressings 2-32
  Fabricated Fruit 2-33
  Pie Fillings 2-34
  Industrial Applications 2-35
Laboratory Techniques 2-37
  Viscosity Measurement 2-37
  Moisture Determination 2-37
  Powder Color Determination 2-37
  Alginates in Mixtures (Detection) 2-38
  Alginates in Mixtures (Determination) 2-38
Product Tradename Directory 2-38
Further Reading 2-42

GENERAL INFORMATION

As early as 600 B.C., seaweed was used as a food for man, but algin, a component of seaweed, was first discovered by British chemist E. C. C. Stanford in 1880. In 1896, A. Krefting prepared a pure alginic acid. In 1929, Kelco Company began commercial production of alginates and introduced milk-soluble algin as an ice cream stabilizer in 1934. In 1944, propylene glycol alginate was developed.

Algin is a polysaccharide found in all brown seaweeds, Phaeophyceae,* which grow on rocky shores or in ocean areas that have clean, rocky bottoms. Although some species can be found at the high-tide line, others exist along the shore where depths are less than about 40 m (125 ft), the maximum depth to which sunlight will penetrate. (Since algae do not have true roots, stems, or leaves, nourishment comes directly from sunlight and the mineral nutrients in ocean water.)

Only a few species of brown seaweeds are used for commercial production of algin. The principal source of the world's supply of algin is the giant kelp, *Macrocystis pyrifera*, found along the coasts of North and South America, New Zealand, Australia, and Africa. Other seaweeds used for algin manufacture are *Ascophyllum nodosum* and species of *Laminaria* and *Ecklonia*.

Algin exists in the kelp cell wall as the insoluble mixed salt (calcium, magnesium, sodium, potassium) of alginic acid. Alginic acid is a high-molecular-weight linear glycuronan comprising solely D-mannuronic acid and L-guluronic acid.

Algin is used in foods and general industrial applications because of its unique colloidal behavior and its ability to thicken, stabilize, emulsify, suspend, form films, and produce gels. These properties are discussed in greater detail in later sections of this chapter, such as Solution Properties and Commercial Uses.

Chemical Structure

It has been in recent years only that the composition of alginic acid has become understood. Table 2.1 shows the composition of alginic acid, whereas Table 2.2 shows the

TABLE 2.1
Composition of Alginic Acid Obtained from Commercial Brown Algae

Species	Mannuronic acid content, %	Guluronic acid content, %	Ratio of mannuronic to guluronic acid*	Range of mannuronic to guluronic acid ratios†
Macrocystis pyrifera	61	39	1.56	—
Ascophyllum nodosum	65	35	1.85(1.1)	1.40–1.95
Laminaria digitata	59	41	1.45	1.40–1.60
Laminaria hyperborea (stipes)	31	69	0.45	0.40–1.00
Ecklonia cava and *Eisenia bicyclis*	62	38	1.60	—

* Data of Haug (1964) and Haug and Larsen (1962) for commercial algin samples. Of the two ratios shown for *Ascophyllum nodosum*, the algin sample manufactured in Canada has the higher M/G value; the lower ratio corresponds to a European sample.
† Data of Haug (1964) showing the range in composition for mature algae collected at different times at each of several locations.
Note: See Further Reading for citations to above references.

proportions of polymannuronic acid segments, polyguluronic acid segments, and alternating segments of these two uronic acids in three commercial samples of alginic acid. Figures 2.1 to 2.3 illustrate the structures of mannuronic and guluronic acids; the apparent discrepancies between the data of Tables 2.2 and 2.3 are accounted for by variations between alginates derived from different species of brown algae.

Chemical Derivatives   The propylene glycol ester of alginic acid is the only organic derivative of alginic acid currently on the market. Propylene glycol alginate has im-

* Other seaweed hydrocolloids, e.g., carrageenan (Chap. 5), are extracted from various types of red algae, Rhodophyceae.

proved acid stability and resists precipitation by calcium and other polyvalent metal ions.

Amine alginates can be made by reacting alginic acid with organic amines. Suitable amines are triethanolamine, triisopropanolamine, butylamine, dibutylamine, and diamyl-

TABLE 2.2
Proportions of Polymannuronic Acid, Polyguluronic Acid, and Alternating Segments of the Two in Alginic Acid Isolated from Brown Algae*

Source	Polymannuronic acid segment, %	Polyguluronic acid segment, %	Alternating segment, %
Macrocystis pyrifera	40.6	17.7	41.7
Ascophyllum nodosum	38.4	20.7	41.0
Laminaria hyperborea	12.7	60.5	26.8

* Data of Penman and Sanderson (1972); see Further Reading for citation.

amine. Algin acetate and algin sulfate esters have been prepared but have no known applications. Carboxymethyl alginate can be made by treating sodium alginate with chloroacetic acid and alkali. A number of alkylene glycol esters of alginic acid have been prepared and evaluated.

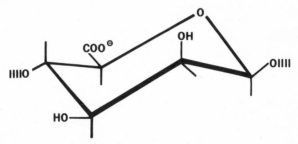

Figure 2.1 Conformation of mannuronic acid.

Ethylene oxide can be reacted with alginic acid to form 2-hydroxyethyl alginate. Alginamides can be prepared by reacting propylene glycol alginate with primary amines, such as ammonia, ethanolamine, ethylenediamine, ethylamine, propylamine, isopropylamine, and butylamine. Very little reaction occurs with secondary amines.

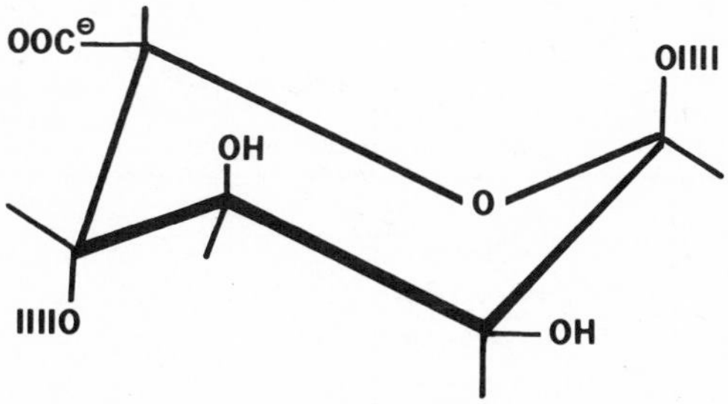

Figure 2.2 Conformation of guluronic acid.

Manufacture

Macrocystis pyrifera, the brown seaweed that is the main source of algin, grows in relatively calm waters and in large, dense beds. The plant is a perennial and can be harvested on a continuing basis. Its rapid growth permits up to four cuttings per year.

Only mature beds are cut. At the time of harvesting, a dense mat of fronds floats on the ocean surface. Cutting the dense mat on the surface allows light to penetrate the water and reach the immature fronds; this stimulates their growth. Harvesting is actually a massive pruning of the kelp bed. Underwater blades mow the kelp approxi-

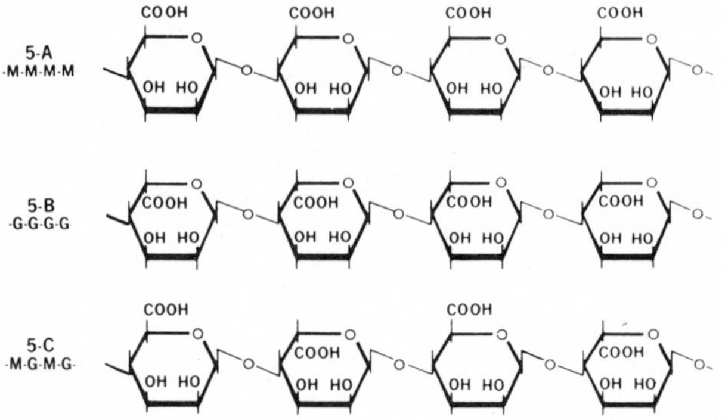

Figure 2.3   Structure of the polymer segments contained in alginic acid.

mately 3 ft below the water surface; then the cut kelp is automatically conveyed into the hold of the barge by a moving belt.

Although commercial methods of producing sodium alginate from seaweed are proprietary, the fundamental steps in a typical process, essentially one of ion exchange,

TABLE 2.3
Typical Physical Properties of Alginates

	Alginic acid	Refined sodium alginate	Specially clarified sodium alginate	Ammonium alginate	Propylene glycol alginate
Moisture content, %	7	13	9	13	13 max
Ash, %	2	23	23	2	10 max
Color	White	Ivory	Cream	Tan	Cream
Specific gravity	—	1.59	1.64	1.73	1.46
Bulk density, kg/m^3 (lb/ft^3)	—	54.62 (875)	43.38 (695)	56.62 (907)	33.71 (540)
Browning temperature, °C	160	150	130	140	155
Charring temperature, °C	250	340,460	410	200	220
Ashing temperature, °C	450	480	570	320,470	400
Ignition temperature, °C	*	*	*	*	*
Heat of combustion, Cal/g	2.80	2.50	2.44	3.04	4.44
As a 1% solution, dist. water:					
Heat of solution, Cal/g soln.	0.090	0.080	0.115	0.045	0.090
Refractive index, 20°C	—	1.3343	1.3342	1.3347	1.3343
pH	2.9	7.5	7.2	5.5	4.3
Surface tension, dyn/cm	53	62	70	62	58
Freezing point depression, °C	0.010	0.035	0.020	0.060	0.030

* Spontaneous combustion did not occur in an air environment.

are shown in Fig. 2.4. In the seaweed, the algin is apparently present as a mixed salt of sodium and/or potassium, calcium, and magnesium and is a high-molecular-weight polymer. The exact composition varies considerably with the type of seaweed but does not affect processing.

It is possible to extract sodium alginate from seaweed with a strong solution of a sodium salt; however, for the production of purified alginates, the commercial processes are much more efficient. Alginic acid may also be neutralized with bases to give salts and reacted with propylene oxide to make propylene glycol alginate.

Physical Properties

Commercially available, water-soluble alginates include the sodium, potassium, ammonium, calcium, and mixed ammonium-calcium salts of alginic acid, propylene glycol

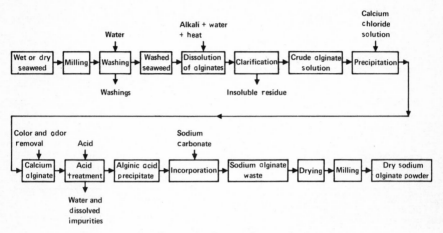

Figure 2.4 Flow diagram for the manufacture of sodium alginate.

alginate, and alginic acid itself. The physical properties of several of these alginates are given in Table 2.3.

   Powdered Alginates   Alginate, as a hydrophilic polysaccharide, absorbs moisture from the atmosphere; therefore, equilibrium moisture content is related to relative humidity, as shown in Fig. 2.5. The dry storage stability of alginates is excellent at

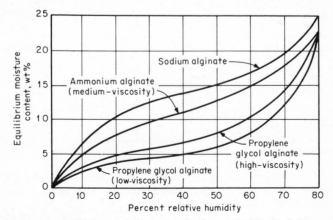

Figure 2.5 Equilibrium moisture curves for three typical kinds of alginates.

moderate temperatures, 25°C (77°F) or less; however, they should be stored in a cool, dry place. Table 2.4 gives data showing the effects of storage for 1 year at 24.9°C (75°F) on typical alginates. Table 2.5 shows the effects of various storage temperatures on the stabilities of alginates.

Solution Properties

Pure alginates dissolved in distilled water form smooth solutions with long-flow characteristics. The physical variables that affect the flow properties of alginate solutions are temperature, shear rate, polymer size, concentration, and the presence of solvents miscible with the distilled water. The chemical variables that affect algin solutions are pH and the presence of sequestrants, monovalent salts, polyvalent cations, and quaternary ammonium compounds.

Rheological Properties   The flow properties of sodium alginate solutions are concentration-dependent. A 2.5% medium-viscosity sodium alginate solution is pseudoplastic

TABLE 2.4
Loss in Viscosity by Dry Alginate Powder After 1 Year at 23.9°C (75°F)

	Initial*		After 1 year*	
	Pa · s	cP	Pa · s	cP
Ammonium alginate	1.5	1,500	0.675	675
Potassium alginate	0.27	270	0.248	248
Propylene glycol alginate	0.115	115	0.067	67
Propylene glycol alginate	0.4	400	0.236	236
Sodium alginate	0.027	27	0.026	26
Sodium alginate	0.4	400	0.330	330
Sodium alginate	0.42	420	0.380	380

* Brookfield Model LVF viscometer at 60 rpm, appropriate spindles; 1% solutions.
NOTES:
(1) High-viscosity alginates usually have a more rapid viscosity decrease than do low-viscosity alginates.
(2) Ammonium alginate is generally less stable than are sodium, potassium, or propylene glycol alginates.
(3) Propylene glycol alginates gradually become insoluble when stored at elevated temperatures for extended periods of time.

TABLE 2.5
Effect of Temperature on Viscosity of Dry Alginate Powders After 1 Year of Storage

		Viscosity of 1% solution			
		Initial*		After 1 year*	
	Storage temperature	Pa · s	cP	Pa · s	cP
Sodium alginate	1.7°C (35°F)	0.027	27	0.026	26
(low viscosity)	23.9°C (75°F)	0.027	27	0.026	26
	32.2°C (90°F)	0.027	27	0.022	22
Sodium alginate	1.7°C (35°F)	0.42	420	0.41	410
(medium viscosity)	23.9°C (75°F)	0.42	420	0.38	380
	32.2°C (90°F)	0.42	420	0.23	230
Ammonium alginate	1.7°C (35°F)	1.5	1500	1.35	1350
(high viscosity)	23.9°C (75°F)	1.5	1500	0.625	625
	32.2°C (90°F)	1.5	1500	0.3	300

* Brookfield Model LVF viscometer at 60 rpm, appropriate spindle.

over a wide range of shear rates (10 to 10,000 s^{-1}), whereas a 0.5% solution is Newtonian at low shear rates (1 to 100 s^{-1}) and pseudoplastic only at high shear rates (1000 to 10,000 s^{-1}), as shown in Fig. 2.6.

Because of high molecular weight and molecular rigidity, sodium alginate forms solutions of unusually high apparent viscosity, even at low concentrations. Propylene glycol alginate solutions are shear thinning over a wide range of shear rates at 3% concentrations. However, at 1% or lower concentrations, solutions have almost constant viscosity below shear rates of 100 s^{-1} (Fig. 2.7).

Figure 2.8 shows that viscosity-shear curves of medium-viscosity sodium and potassium alginates are virtually the same over the entire shear range. On the other hand, in comparing low-viscosity propylene glycol and sodium alginates, the curves are identical at shear rates greater than 10,000 s^{-1} but diverge at low shear rates.

The effects of solution solids on shear thinning are illustrated in Fig. 2.9. The viscosity-shear curves of a 2% solution of medium-viscosity sodium alginate were the same as those of a 9% solution of a low-viscosity sodium alginate. Measurements were taken using shear rates in the Brookfield-viscometer range (1 to 10,000 s^{-1}). At high shear rates, such as those experienced at 100,000 s^{-1}, measured with a capillary viscometer, the curves diverge.

Figure 2.10 illustrates the effect of temperature on the flow of a high-viscosity propylene glycol alginate. Addition of a sequestrant, sodium hexametaphosphate, to a medium-viscosity sodium alginate (Fig. 2.11) gives a viscosity-shear curve comparable to that of a low-calcium sodium alginate.

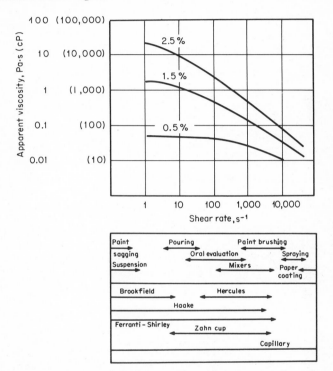

Figure 2.6 Apparent-viscosity and shear-rate curves for medium-viscosity sodium alginate solutions with process and viscometer shear-rate changes.

Xanthan gum can be used to modify the rheological behavior of sodium alginate solutions (Fig. 2.12). As shown, the curves for the 0.5% solutions of sodium alginate and xanthan differ greatly. A combination of the two gums produces flow properties intermediate between the two materials.

Latex paints illustrate the importance of rheological properties to the design of product performance. If the paint is highly pseudoplastic, application will be easy and sagging will be prevented, but flow and leveling will be minimal and brush marks will be left on the dried paint film. Elimination of the yield value will result in settling out of the pigments in the can. If dilatancy occurs, stirring will be difficult, and brush drag will be excessive.

Effect of Temperature The viscosities of algin solutions decrease as temperatures increase, approximately 12% for each 5.6°C (10°F) increase in temperature. The decrease is reversible if the high temperatures are not held for long periods. Table 2.6

shows the effect of time and temperature on solution viscosity. It is apparent that the heating of sodium alginate results in some thermal depolymerization, the amount being related to both temperature and time.

Although a reduction in temperature of an alginate solution results in an increase

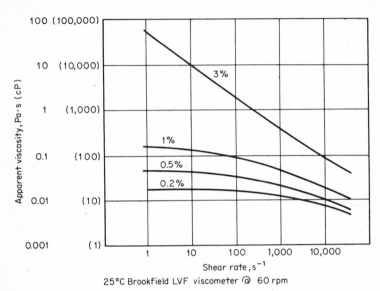

Figure 2.7   Apparent-viscosity and shear-rate curves for low-viscosity propylene glycol alginate solutions.

in viscosity, it does not produce a gel. A sodium alginate solution can be frozen and thawed without any change in its appearance or viscosity after remelting. It is possible to form a freeze-dried, sodium-calcium alginate gel with an absorptive capacity of more than 5000%.

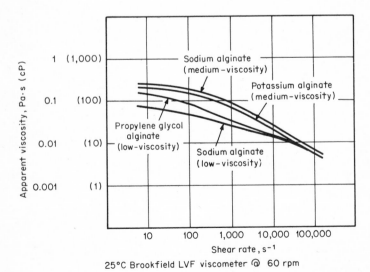

Figure 2.8   Apparent-viscosity and shear-rate curves for 1% dispersions of various alginates.

Effect of Solvents Addition of increasing amounts of nonaqueous, water-miscible solvents, such as alcohols, glycols, or acetone, or an aqueous alginate solution increases solution viscosity and eventually causes precipitation of the alginate. Tolerance of the alginate solution to such solvents is influenced by the source of the alginate, the degree of polymerization, the cation type present, and the solution concentration. Table 2.7 gives data on solvent tolerances of various types of alginates in solution.

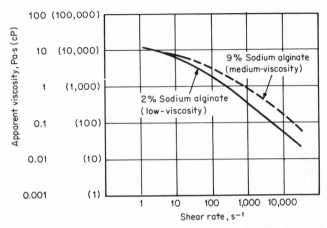

Figure 2.9 Effect of solution solids concentration on solution shear thinning.

Effect of Concentration Figure 2.13 shows the effect of solution concentration on selected grades of sodium, ammonium, potassium, and propylene glycol alginates.

Effect of pH Sodium alginates with some residual calcium content increase in viscosity at a pH of 5.0 and are unstable at pH levels of about 11.0. Sodium alginates with minimal calcium content do not show the viscosity increase until the pH reaches 3.0 to 4.0. Lower-molecular-weight sodium alginates are stable at a pH as low as 3.0 if calcium is completely sequestered.

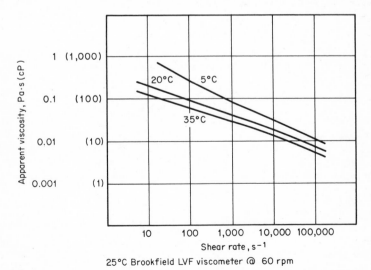

Figure 2.10 Apparent-viscosity and shear-rate curves for 1.5% solutions of high-viscosity propylene glycol alginate.

Propylene glycol alginates do not gel until the pH is below 3.0, but they do saponify at pH levels above 7.0. The long-term stability of sodium alginate solutions is poor when the pH reaches 10.0. At even higher pH values there is depolymerization with an accompanying viscosity loss. Figure 2.14 illustrates the effect of pH on viscosity for several types of alginates in solution.

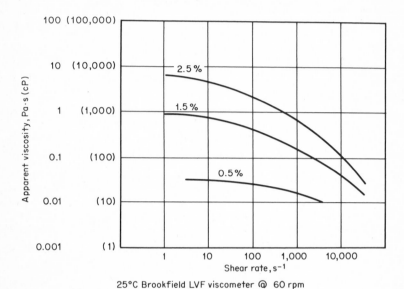

25°C Brookfield LVF viscometer @ 60 rpm

Figure 2.11   Apparent-viscosity and shear-rate curves for various concentrations of medium-viscosity sodium alginate with sodium hexametaphosphate sequestrant.

Gelation   Algin polymers will react with most polyvalent cations (magnesium excepted) to form crosslinkages. As the content of polyvalent ion increases, the algin solution thickens, then gels, and finally there is precipitation. The proposed structure of an alginate gel in which the calcium ions are bound between the associated segments of the polymer chain is shown in Fig. 2.15.

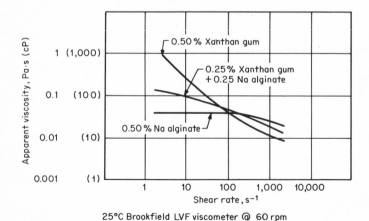

25°C Brookfield LVF viscometer @ 60 rpm

Figure 2.12   Apparent-viscosity and shear-rate curves for dispersions containing both sodium alginate and xanthan gum.

TABLE 2.6
Viscosity vs. Time and Temperature for Three Viscosity Grades of Sodium Alginate
(1% solutions in distilled water)

Storage at:	25°C (77°F)		37.8°C (100°F)				54.4°C (130°F)				71.1°C (160°F)				82.2°C (180°F)			
	(room temp.)		(hot)		(room temp.)		(hot)		(room temp.)		(hot)		(room temp.)		(hot)		(room temp.)	
Viscosity at: Hours	Pa·s	cP	Pa·s	cP	Pa·s	cP	Pa·s	cP	Pa·s	cP	Pa·s	cP	Pa·s	cP	Pa·s	cP	Pa·s	cP
High-viscosity sodium alginate																		
0	0.900	900	0.600	600	0.800	800	0.420	420	0.740	740	0.360	360	0.680	680	0.300	300	0.600	600
1	0.900	900	0.540	540	0.720	720	0.400	400	0.660	660	0.310	310	0.600	600	0.220	220	0.510	510
2	0.900	900	0.520	520	0.700	700	0.380	380	0.640	640	0.290	290	0.570	570	0.200	200	0.480	480
3	0.900	900	0.500	500	0.680	680	0.360	360	0.620	620	0.270	270	0.550	550	0.180	180	0.460	460
4	0.900	900	0.490	490	0.670	670	0.350	350	0.610	610	0.260	260	0.530	530	0.170	170	0.440	440
Medium-viscosity sodium alginate																		
0	0.530	530	0.330	330	0.400	400	0.215	215	0.330	330	0.185	185	0.295	295	0.180	180	0.290	290
1	0.530	530	0.290	290	0.385	385	0.200	200	0.310	310	0.170	170	0.285	285	0.160	160	0.280	280
2	0.530	530	0.280	280	0.380	380	0.195	195	0.280	280	0.165	165	0.275	275	0.155	155	0.260	260
3	0.530	530	0.275	275	0.370	370	0.190	190	0.270	270	0.160	160	0.260	260	0.150	150	0.250	250
4	0.530	530	0.260	260	0.360	360	0.185	185	0.265	265	0.155	155	0.250	250	0.145	145	0.240	240
Low-viscosity sodium alginate																		
0	0.062	62	0.040	40	0.054	54	0.026	26	0.051	51	0.022	22	0.044	44	0.019	19	0.040	40
1	0.062	62	0.038	38	0.054	54	0.025	25	0.049	49	0.022	22	0.043	43	0.019	19	0.040	40
2	0.062	62	0.037	37	0.052	52	0.025	25	0.047	47	0.021	21	0.042	42	0.019	19	0.039	39
3	0.062	62	0.037	37	0.052	52	0.024	24	0.047	47	0.020	20	0.041	41	0.018	18	0.038	38
4	0.062	62	0.036	36	0.051	51	0.024	24	0.046	46	0.020	20	0.040	40	0.018	18	0.037	37

TABLE 2.7
Maximum Solvent Tolerance of Alginate Solutions

Type of alginate solution	Methanol, %	Ethanol, %	Isopropanol, %	t-Butanol, %	Glycerol, %	Ethylene glycol, %	Propylene glycol, %	Butyl cellosolve, %	Acetone, %
2% Sodium (low-viscosity)	20	20	10	20	70+	70+	40	20	10
1% Sodium (medium-viscosity)	20	20	10	10	70+	70+	40	30	10
1% Sodium (high-viscosity)	20	20	10	10	70+	70+	70+	20	10
1% Potassium	20	20	20	20	70+	70+	50	40	20
1% Ammonium	30	30	20	20	70+	70+	70+	20	20
1% Propylene glycol (low-viscosity)	40	30	30	30	60	70+	50	30	30
1% Propylene glycol (high-viscosity)	30	20	20	20	60	70+	40	30	20

COMMENTS ON DATA:

(1) The alginate was dissolved in water first, then diluted with solvent to a given water/solvent ratio. Concentration of alginate in the final solution was 1% except where shown. Solvent levels were varied in 10% increments.

(2) The *maximum solvent tolerance* is the solvent percentage next below the percentage at which alginate separation was evident.

(3) With most solvents, the compatibility end point is quite evident due to alginate precipitation and/or viscosity loss. With glycerol and ethylene glycol, however, the end point is not sharp, and apparent compatibility may extend to as high as 90% solvent. High solvent combinations are difficult to prepare because the alginate must be hydrated prior to addition of the solvent.

All alginate gels are the result of interactions between the alginate molecules, which produce a three-dimensional structure controlling the mobility of the water molecules. They are not thermally reversible. By the proper selection of gelling agent, gel structure and rigidity are controlled. Loss of water to the atmosphere, and resultant shrinkage, is very slow in algin gels.

Metallic polyvalent ions, e.g., zinc, aluminum, copper, and silver, form complexes with alginates in the presence of excess ammonium hydroxide. When the ammonia

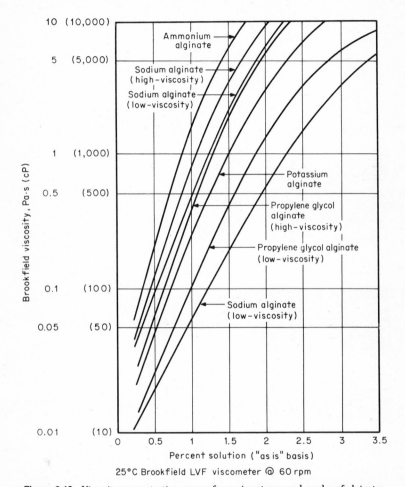

Figure 2.13 Viscosity-concentration curves for various types and grades of alginates.

is driven from the system, the insoluble metal alginate is formed. Calcium is the polyvalent cation most often used to change the rheological properties and gel characteristics of algin solutions. Calcium is also used to form insoluble alginate filaments and films.

The method of calcium addition to an alginate system greatly influences the properties of the final gel. If calcium is added too rapidly, the result is spot gelation and a discontinuous gel structure. The rate of calcium addition can be controlled by use of a slow-dissolving calcium salt or by the addition of a sequestrant, such as tetrasodium pyrophosphate or sodium hexametaphosphate.

Following are some general rules that are useful for changing the firmness or setting time of a gel formula:

1. An increase in calcium sequestrant weakens the gel and vice versa; however, too little sequestrant gives a grainy gel.

2. A decrease in calcium gives a softer gel; an increase gives a firmer gel. Too much calcium, however, gives a grainy gel or causes precipitation of calcium alginate.

3. In an acid-gel system, an increase of a slowly soluble acid accelerates gel setting time but can give a grainy gel.

4. An increase of soluble alginate will give a firmer gel, but the texture may be objectionable. The higher the alginate viscosity, the crisper the gel and the sharper the cut.

5. The tendency toward syneresis increases as the calcium approaches the stoichiometric amount needed for complete reaction with the alginate.

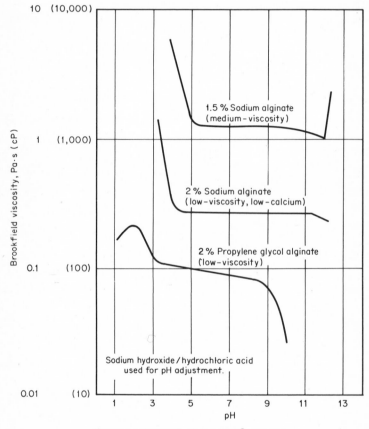

25°C Brookfield LVF viscometer @ 60 rpm

Figure 2.14 Effect of pH on alginate solutions.

Effect of Sequestrants The purpose of sequestrants in alginate solutions can be either to prevent the alginate from reacting with polyvalent ions present in the solution or to sequester the calcium inherent in the alginate. Polyvalent ion contaminants can come from water, chemicals, pigments, or various natural-origin materials. Figures 2.16 and 2.17 show viscosity-concentration relationships for two types of alginates with and without sodium hexametaphosphate as the sequestrant.

In Fig. 2.16, a low-calcium sodium alginate shows a very small viscosity change upon addition of the polyphosphate sequestrant to the solution. In contrast, Fig. 2.17 shows

that a sodium-calcium alginate solution has a major change in viscosity when the sequestrant is added. Sequestered alginate solutions are more Newtonian in behavior than
are those with some available calcium.

Effect of Monovalent Salts   Monovalent salts depress the viscosities of dilute sodium
alginate solutions. The maximum effect on viscosity is attained at a salt level of $0.1N$
in the solution. Except for alginates high in calcium, an increase in alginate concentration decreases the effect of the monovalent electrolyte.

Figure 2.18 shows the effect of sodium chloride on the viscosity of several kinds of
alginate solutions, whereas Table 2.8 shows the effects of 1% and 5% sodium chloride
concentrations over a 210-day period at temperatures of 4.4, 23.9, and 48.9°C (40, 75,
and 120°F). The effect of a salt on an alginate solution will vary with the source of
the alginate, as well as with its degree of polymerization, the concentration of alginate
in the solution, and the type of salt.

Insolubilization   Normally, insoluble adducts result when sodium alginate reacts with
cationic organic ammonium compounds. This insolubilization can be prevented by
adding an electrolyte, e.g., NaCl, to suppress the activity of the cation. Salt concentrations needed to solubilize insoluble alginate adducts are listed in Table 2.9.

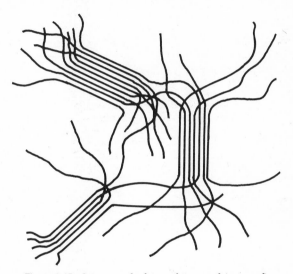

Figure 2.15   Structure of calcium alginate gel (proposed).

Compatibilities

Alginates in solution have compatibility with a wide variety of materials, including other
thickeners, synthetic resins, latices, sugar, oils, fats, waxes, pigments, various surfactants,
and alkali metal solutions. Incompatibilities are generally the result of a reaction with
divalent cations (except magnesium) or other heavy metal ions, cationic quaternary
amines, or chemicals that cause alkaline degradation or acid precipitation. In many
cases the incompatibility can be avoided by sequestration of the metal ion or by careful
control of the solution pH.

Table 2.10 lists materials that were tested for compatibility with a solution of a medium-
viscosity purified sodium alginate.

Preservatives   Alginates have compatibility with most commonly used preservatives,
except quaternary ammonium compounds. The polysaccharide is quite resistant to
the common enzyme systems produced by bacteria; however, since the solutions will
support microbiological growth, a preservative should be used if alginate solutions are
to be stored for any considerable period of time. The preservatives listed in Table
2.10 exhibit good compatibility.

Sodium benzoate can be used to protect against bacterial action in acid systems. For additional protection against yeast and mold, potassium sorbate or calcium or sodium propionate can be effective.

Thickeners   The alginates show compatibility with most commercially available thickeners, both synthetic and natural. With some thickeners a synergistic viscosity increase may be noticed. If the residual polyvalent ion content of a natural gum causes gelation of an algin solution, the gelation can be controlled by the proper use of a sequestrant.

Water-Soluble Resins   The compatibility of the alginates with most water-soluble resins is excellent. Polyvinyl alcohol exhibits definite synergism with sodium alginate in the formulation of grease-resistant films.

Latices   Those latices normally used in the formulation of paints, paper coatings, and adhesives have compatibility with the alginates. However, latex emulsions with

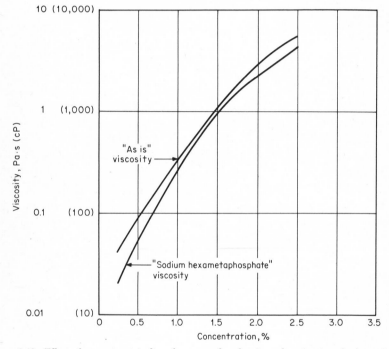

Figure 2.16   Effect of sequestrant (sodium hexametaphosphate) on the viscosity of a low-calcium sodium alginate solution.

pH of 4.0 or less will cause gelation of the alginate. This apparent incompatibility may be overcome by proper buffering. High-viscosity ammonium alginate may be used as a creaming agent for natural rubber latex and for several types of synthetic latex.

Organic Solvents   As shown in Table 2.10, alginate solutions will tolerate up to 30% water-miscible solutions. However, viscosity increases may occur with long-term storage. To prevent localized gelation, it is necessary that there be good agitation of the solution at the time the organic solvent is added.

Enzymes   Enzymes commonly encountered as by-products or as commercially available products, e.g., protease, cellulase, amylase, galactomannanase, have no effect on the alginate molecule. Storage test data is given in Table 2.11 for representative enzymes.

Surfactants   Although alginate solutions have compatibility with anionic, non-ionic, and amphoteric surfactants, high concentrations of surfactants will result in a loss of viscosity, and eventually the alginate will salt out of solution.

Non-ionic surfactants can be used at concentrations higher than those allowable for the anionics or amphoterics. Some cationic surfactants may be used if approximately 2.5% of a soluble salt, such as sodium chloride, is added to the system. The exact salt level required depends upon the particular cationic material in the system.

Plasticizers Plasticizers, such as glycols or glycerol, may be used to improve the flexibility of alginate films. Data for a number of plasticizers and their effects on alginate solutions are given in Table 2.10.

Inorganic Salts The compatibility of alginate solutions with inorganic salts is limited to ammonium, magnesium, or the alkali metal salts. Divalent or higher-valence cationic

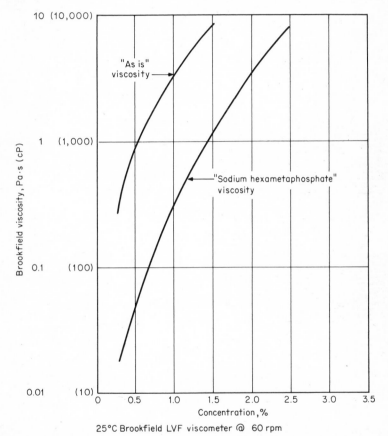

25°C Brookfield LVF viscometer @ 60 rpm

Figure 2.17 Effect of sequestrant (sodium hexametaphosphate) on the viscosity of a sodium-calcium alginate solution.

salts will, unless sequestered, cause gelation or precipitation of the alginate. The alginates will also be precipitated by molar solutions of monovalent salts.

Salts which are slightly acidic may produce large viscosity increases after prolonged storage. Sequestrants in many cases will improve the salt compatibility and stability of the sodium alginate. Mixed alginate salts (sodium/calcium alginate) are much more salt-sensitive than are the alkali metal alginates.

Toxicology/Environment

Numerous studies have verified the high safety level of alginates in foods. Ammonium alginate, calcium alginate, potassium alginate, and sodium alginate are included in a list of food stabilizers that are generally recognized as safe (GRAS) under 21 CFR 182.

Propylene glycol alginate is approved as a food additive under 21 CFR 172.858 for use as an emulsifier, stabilizer, or thickener in foods in accordance with that regulation.

The use of the edible salts of alginic acid, as well as propylene glycol alginate, is approved in all the appropriate standard of identification regulations for the United States and the European Economic Council (EEC). In these standards alginates are either specifically mentioned by their common and usual name, or they are included under the provision of "safe and suitable optional ingredients." For maximum allowable usage levels, each regulation must be consulted separately.

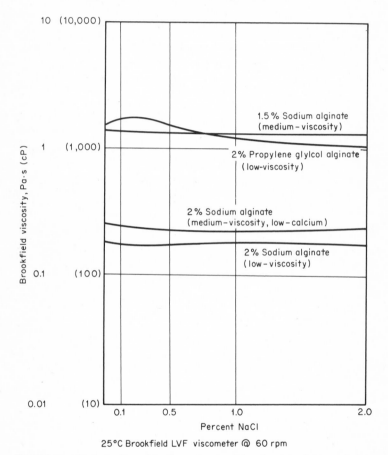

25°C Brookfield LVF viscometer @ 60 rpm

Figure 2.18 Effect of sodium chloride on the viscosities of various alginate solutions.

Propylene glycol alginate is approved for use in defoaming agents (21 CFR 173.340), in coatings for fresh citrus fruits (21 CFR 172.210), as an inert pesticide adjuvant (40 CFR 180.1001), and as a component of paper and paperboard products that come into contact with aqueous and fatty foods (21 CFR 176.170).

Ammonium alginate and sodium alginate are approved for use as boiler water additives under 21 CFR 173.310. The edible salts of alginic acid are also approved for use as components of paper and paperboard in contact with aqueous and fatty foods, under 21 CFR 176.170, which specifies that substances generally recognized as safe (GRAS) can be used.

The *Food Chemicals Codex* contains monographs on alginic acid, ammonium alginate, calcium alginate, potassium alginate, sodium alginate, and propylene glycol alginate.

TABLE 2.8
Effect of Sodium Chloride on Viscosity of Stored Alginate Solutions

		Solutions containing 1% sodium chloride								Solutions containing 5% sodium chloride							
		Initial viscosity		Viscosity at 210 days†						Initial viscosity		Viscosity at 210 days†					
				4.4°C (40°F)		23.9°C (75°F)		48.9°C (120°F)				4.4°C (40°F)		23.9°C (75°F)		48.9°C (120°F)	
Alginate	Solution strength, %	Pa·s	cP	Pa·s	cP	Pa·s	cP	Pa·s	cP	Pa·s	cP	Pa·s	cP	Pa·s	cP	Pa·s	cP
Propylene glycol	2.0	0.079	79	0.171	171	0.017	17			0.460	460	0.230	230	0.037	37		
Propylene glycol	2.0	0.380	380	4.200	4200	0.790	790			0.720	720	1.050	1050	0.570	570		
Calcium-sodium	1.0	0.895	895	6.900	6900	4.900	4900	0.020	20	0.840	840	4.150	4150	3.700	3700	0.345	345
Sodium	1.5	1.280	1280	gel	gel	8.000*	8000*	2.000*	2000*	1.270	1270	3.300	3300	5.750	5750	1.800*	1800*
Sodium	2.0	0.260	260	gel	gel	1.600	1600	0.215	215	0.265	265	gel	gel	1.650	1650	1.000	1000
Sodium	3.0	0.124	124	gel	gel	1.060	1060	0.200	200	0.101	101	gel	gel	0.760	760	0.225	225
Sodium (low calcium)	1.5	0.950	950	1.100	1100	1.120	1120	0.115	115	1.140	1140	1.850	1850	1.320	1320	0.300	300
Sodium (low calcium)	2.0	0.266	266	0.225	225	0.268	268	0.035	35	0.293	293	0.950	950	0.363	363	0.053	53
Ammonium	1.0	0.930	930	7.600	7600	2.250	2250	0.006	6	0.910	910	3.000	3000	2.400	2400	0.023	23

* Gelatinous character.
† Brookfield Model LVF viscometer at 60 rpm, appropriate spindle.

Monographs on sodium alginate and alginic acid are also included in the *National Formulary*.

Alginic acid and its edible salts, as well as propylene glycol alginate, are included in the approved emulsifier/stabilizer lists published by the European Economic Council and the Council of Europe. The FAO/WHO Joint Expert Committee established an acceptable daily intake (ADI) limit of 25 mg/kg for alginic acid and its edible salts and 25 mg/kg for propylene glycol alginate.

Biological Oxygen Demand   Biological and chemical oxygen demand data for a medium-viscosity sodium alginate, a high-viscosity ammonium alginate, and a low-viscosity propylene glycol alginate have been determined and are presented in Table 2.11.

Application Procedures

Alginates are hydrophilic colloids. A granule will wet immediately when put into water and then will dissolve rapidly. Without agitation, however, masses of granules put into water will form lumps, the interiors of which will not be wetted. On the other hand, the dry alginate powder will dissolve quickly in water if one of the following methods is used:

High-Shear Mixing   Use of a high-shear mixer that produces a good vortex will achieve good dispersion of the dry alginate in water solutions. The mixing agitator is placed off center to produce maximum turbulence at the lower portion of the vortex. The agitator blades should be submerged to prevent excessive aeration, and the powdered alginate should be sifted slowly onto the upper wall of the vortex so that the individual granules are wetted out. Addition of the alginate should be completed before its progressive thickening action destroys the vortex.

TABLE 2.9
Concentration of Electrolyte Needed to Solubilize Adducts of Alginates and Trimethyldodecylammonium Chloride

Alginate (1% solution)	Minimum sodium chloride level, %
Sodium	2.5
Potassium	2.5
Propylene glycol	1.5

Eductor (Aspirator) Mixing   Possibly the best dispersion of alginates can be obtained with a funnel and mixing eductor as shown in Fig. 2.19. The use of this aspirator may eliminate the need for high-shear mixing equipment.

Before adding the alginate, the dissolving tank should be filled with enough water to cover the mixer blades when a vortex is developed. In small tanks this water may be run through the aspirator. Larger tanks may be filled faster by bypassing the aspirator. When sufficient water has been added, the agitator is turned on and water flow is started in the aspirator. Water flow must be started before the alginate is put into the funnel.

After water flow has begun, the dry alginate granules are poured into the funnel, which is attached to the top of the aspirator. While the alginate is being added, and until it is completely dissolved, the dissolving tank must be agitated to prevent settling out of any of the dispersed granules. This agitation can be accomplished with a mechanical mixer, as shown in Fig. 2.19, or with an air sparger.

After all of the powder has been dispersed into the mixer, make-up water should be added to achieve the desired solution concentration. A minimum water pressure of at least 345 kPa (50 psi) is required. If water pressure exceeds 689 kPa (100 psi), flow will be too fast, and the granules will lump.

An alternate procedure employs a pump to circulate the water from the tank and through the eductor. If this method is used, it may be possible to eliminate the mixer shown in Fig. 2.19.

Dry-Mix Dispersion   Quite often a formulation that includes an alginate will also contain ingredients such as sugar or starch that can be blended with the alginate in dry form to facilitate later dispersion in solution. A physical separation of the alginate

TABLE 2.10
Compability of Medium-Viscosity Sodium Alginate
(In distilled water, with 0.2% formaldehyde preservative, stored in glass bottles)

Material	Percent material	Percent alginate	Viscosity*			
			Initial		After 90 days	
			Pa · s	cP	Pa · s	cP
Preservatives:						
Dowicide® A	0.005	1.0	0.395	395	0.265	265
	0.1	1.0	0.350	350	0.260	260
Formaldehyde	0.1	1.0	0.315	315	0.275	275
	1.0	1.0	0.325	325	0.310	310
Methyl Parasept®	0.5	1.0	0.345	345	0.285	285
Vancide® TH	0.1	1.0	0.275	275	0.207	207
	0.5	1.0	0.280	280	0.202	202
Sindar® G-4	0.1	1.0	0.285	285	0.222	222
	0.5	1.0	0.286	286	0.115	115
Advacide® 340-A	0.1	1.0	0.392	392	0.257	257
	0.5	1.0	0.390	390	0.268	268
Omacide® 24	0.1	1.0	0.375	375	0.420	420
	0.5	1.0	0.370	370	0.650	650
Surflo® B-17	0.1	1.0	0.377	377	0.320	320
	0.5	1.0	0.361	361	0.325	325
Nalco® 248	0.1	1.0	0.345	345	0.192	192
	0.5	1.0	0.295	295	0.125	125
Nalco 243	0.1	1.0	0.346	346	0.257	257
	0.5	1.0	0.322	322	0.227	227
Metasol® D3T	0.1	1.0	0.360	360	0.217	217
	0.5	1.0	0.347	347	0.147	147
Metasol DX3-S	0.1	1.0	0.332	332	0.167	167
	0.5	1.0	0.325	325	0.125	125
Thickeners:						
Xanthan gum	0.5	0.5	0.760	760	0.730	730
	1.0	1.0	3.740	3740	4.000	4000
Guar gum	1.0	1.0	8.320	8320	2.950	2950
Gum tragacanth	0.5	0.5	0.350	350	1.050	1050
	1.0	1.0	5.170	5170	9.150	9150
Methocel® 90HG	0.5	0.5	0.290	290	0.160	160
Locust bean gum	0.5	0.5	0.340	340	0.220	220
	1.0	1.0	3.950	3950	3.000	3000
Water-soluble resins:						
Vinol® PA-20	1.0	1.0	0.550	550	0.350	350
	5.0	1.0	1.420	1420	0.230	230
Carboset® 525	1.0	1.0	0.480	480	0.395	395
	10.0	1.0	0.370	370	0.290	290
Carbopol® 934	0.5	0.5	0.540	540	0.100	100
	1.0	0.5	1.250	1250	0.650	650
Latex emulsions:						
Rhoplex® AC490	60.0	0.5	0.170	170	0.150	150
Ucar® 360	60.0	0.5	0.215	215	0.030	30
Dow Latex 460	60.0	0.5	0.330	330	0.300	300
Dow Latex 307	60.0	0.5	0.320	320	0.280	280
Geon® 652	60.0	0.5	0.280	280	0.010	10
Airflex® 500	60.0	0.5	0.470	470	0.260	260
Genflo® 355	60.0	0.5	0.380	380	0.405	405
Genflo 67	60.0	0.5	0.980	980	0.850	850
Organic solvents:						
Acetone	10.0	1.0	0.590	590	0.610	610
	20.0	1.0	1.710	1710	3.150	3150
Methanol	10.0	1.0	0.660	660	0.680	680
	20.0	1.0	1.590	1590	3.000	3000
Isopropanol	10.0	1.0	0.840	840	0.970	970
	20.0	1.0	2.720	2720	4.250	4250
Benzyl alcohol	10.0	1.0	0.660	660	0.480	480

* Brookfield Model LVF viscometer at 60 rpm, appropriate spindle.

TABLE 2.10
Compability of Medium-Viscosity Sodium Alginate (*continued*)
(In distilled water, with 0.2% formaldehyde preservative, stored in glass bottles)

Material	Percent material	Percent alginate	Viscosity* Initial Pa · s	Initial cP	After 90 days Pa · s	After 90 days cP
Enzymes:						
Alkalase®	1.0	1.0	0.190	190	0.155	155
Cellulase® 4000	1.0	1.0	0.315	315	0.195	195
Papain	1.0	1.0	0.323	323	0.223	223
Rhozyme® A-4	1.0	1.0	0.310	310	0.210	210
Gumase® HP 150	1.0	1.0	0.290	290	0.165	165
Surfactants:						
Stepanol® WAT	10.0	0.5	0.043	43	0.037	37
Igepal® CO 630	20.0	0.5	0.560	560	0.250	250
Tween® NPX	20.0	0.5	1.290	1290	0.630	630
Tergitol® NPX	20.0	0.5	0.340	340	0.220	220
Miranol® 2 MCA	10.0	0.5	0.028	28	0.038	38
Plasticizers (glycols):						
Glycerol	50.0	0.5	1.240	1240	3.750	3750
Propylene glycol	50.0	0.5	0.325	325	0.305	305
Triethanolamine	50.0	0.5	0.320	320	0.305	305
Hexylene glycol	50.0	0.5	0.990	990	0.480	480
Kromfax®	50.0	0.5	1.010	1010	1.450	1450
Ethylene glycol	50.0	0.5	1.000	1000	1.120	1120
Inorganic salts:						
Ammonium chloride	1.0	1.0	0.250	250	1.100	1100
	5.0	1.0	0.250	250	0.510	510
	10.0	1.0	0.230	230	0.145	145
Diammonium phosphate	1.0	1.0	0.240	240	0.210	210
	5.0	1.0	0.210	210	0.150	150
	10.0	1.0	0.280	280	0.180	180
Ammonium sulfate	1.0	1.0	0.235	235	0.150	150
	5.0	1.0	0.210	210	0.060	60
Magnesium chloride	1.0	1.0	0.290	290	0.580	580
	5.0	1.0	0.130	130	0.120	120
Potassium chloride	1.0	1.0	0.210	210	1.480	1480
	5.0	1.0	0.210	210	0.350	350
Potassium phosphate, dibasic	1.0	1.0	0.190	190	0.170	170
	5.0	1.0	0.290	290	0.230	230
Potassium sulfate	1.0	1.0	0.290	290	0.430	430
	5.0	1.0	0.290	290	0.230	230
Sodium chloride	1.0	1.0	0.310	310	0.880	880
	5.0	1.0	0.340	340	1.090	1090
Sodium phosphate, dibasic	1.0	1.0	0.230	230	0.160	160
	5.0	1.0	0.215	215	0.215	215
Sodium sulfate	1.0	1.0	0.260	260	0.360	360
	5.0	1.0	0.265	265	0.255	255
Sodium tetraborate	1.0	1.0	0.155	155	0.175	175
Sodium citrate	1.0	1.0	0.215	215	0.170	170
	5.0	1.0	0.240	240	0.245	245

* Brookfield Model LVF viscometer at 60 rpm, appropriate spindle.

particles can be attained by preblending the alginate with these other ingredients. The blend of alginate and the other dry ingredients is added slowly to the vortex created by an agitator. Solution is completed in a matter of minutes.

Liquid-Mix Dispersion Physical separation of algin granules can be obtained with nonsolvents, such as miscible nonaqueous liquids (alcohols, glycols), or with nonmiscible liquids (vegetable oil, mineral oil).

The alginate is slurried in the nonaqueous liquid, then poured into water that is being agitated. Time required for complete solution will be governed by the time needed for the nonsolvent to diffuse and allow the alginate particle to solvate in water.

Stock Paste   In some instances, the use of a stock paste may be desirable. Alginate pastes of high solids content may be prepared and stored indefinitely. At the time of use the appropriate amount of stock paste is weighed into a container equipped with an agitator. Dilution water should be added slowly to the stock paste under agitation. The paste will not disperse or dissolve readily if it is added to water. The paste should be prepared under agitation by a high-shear mixer and held for 24 h before use to ensure uniform composition.

TABLE 2.11
Biological and Chemical Oxygen Demand
(Data reported as mg oxygen/g substrate)

	Biological oxygen demand			Chemical oxygen demand
Alginate	5-day	10-day	20-day	
Sodium (medium viscosity)	285	394	408	665
Ammonium	334	469	499	720
Propylene glycol	124	184	266	1238

Film Forming

A number of alginate applications, such as paints, paper coatings, glazes, and icings, depend upon the film-forming ability of the alginates. Film properties vary according to the method of film preparation; the five basic procedures are listed below:
1. Evaporate the water from a cast film of soluble alginate solution.
2. Extrude a solution of a soluble alginate into a precipitating bath. This in turn yields an insoluble alginate.
3. Treat a soluble film of alginate with a di- or trivalent metal salt solution or an acidic solution to form an insoluble film.
4. Dry a cast film of alginate to drive off the ammonia from a solution of a di- or trivalent metal alginate that is soluble in excess ammonium hydroxide.
5. Extrude a soluble alginate into a water-miscible nonsolvent, such as acetone or isopropanol.

It is possible to produce insoluble films by treatment with polyvalent ions or acids as detailed in procedures (2), (3), and (4) or by crosslinking with formaldehyde, glyoxal, or a formaldehyde-donor resin such as urea-formaldehyde. Insoluble films are not water-

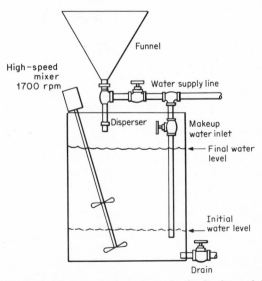

Figure 2.19 Typical funnel-and-mixer eductor for the dissolving of alginates.

repellent, and they swell in water. Zinc alginate films swell the least of the alginate films made by polyvalent ion treatment.

Plasticization of sodium alginate films can be produced by reaction with glycerol, sorbitol, urea, or urea-sodium nitrate. These films are impervious to greases, oils, fats, waxes, and most organic solvents but will transmit water vapor.

Propylene glycol alginate can be reacted with gelatin or a polyfunctional amine to make permeable water-insoluble films. Combinations of polyvalent metal ions may be used to control or to vary permeability of algin films and to control hydrophilic and oleophilic film characteristics.

TABLE 2.12
Food Applications of Alginates

Property	Product	Performance
Water holding	Frozen foods	Maintains texture during freeze-thaw cycle.
	Pastry fillings	Produces smooth, soft texture and body.
	Syrups	Suspends solids, controls pouring consistency.
	Bakery icings	Counteracts stickiness and cracking.
	Dry mixes	Quickly absorbs water or milk in reconstitution.
	Meringues	Stabilizes meringue bodies.
	Frozen desserts	Provides heat-shock protection, improved flavor release, and superior meltdown.
	Relish	Stabilizes brine, allowing uniform filling.
Gelling	Instant puddings	Produces firm pudding with excellent body and texture, and better flavor release.
	Cooked puddings	Stabilizes pudding system, firms body, and reduces weeping.
	Chiffons	Provides tender gel body that stabilizes instant (cold make-up) chiffons.
	Pie and pastry fillings	Acts as cold-water gel base for instant bakery jellies and instant lemon pie fillings. Develops soft gel body with broad temperature tolerance; gives improved flavor release.
	Dessert gels	Produces clear, firm quick-setting gels with hot or cold water.
	Fabricated foods	Provides a unique binding system that gels rapidly under a wide range of conditions.
Emulsifying	Salad dressings	Emulsifies and stabilizes various types.
	Meat and flavor sauces	Emulsifies oil and suspends solids.
Stabilizing	Beer	Maintains beer foam under adverse conditions.
	Fruit juice	Stabilizes pulp in concentrates and finished drinks.
	Fountain syrups, toppings	Suspends solids; produces uniform body.
	Whipped toppings	Aids in developing overrun, stabilizes fat dispersion, and prevents freeze-thaw breakdown.
	Sauces and gravies	Thickens and stabilizes for a broad range of applications.
	Milkshakes	Controls overrun and provides smooth, creamy body.

It should be noted that alginate films are not thermoplastic, and they do not adhere to polished metal surfaces.

COMMERCIAL USES

The ability of alginates to hold water, to gel, to emulsify, and to stabilize have led to numerous food and industrial applications, as summarized in Tables 2.12 and 2.13.

TABLE 2.13
Industrial Applications of Alginates

Property	Product	Performance
Water holding	Paper coating	Controls rheology of coatings; prevents dilatancy at high shear.
	Paper sizings	Improves surface properties, ink acceptance, and smoothness.
	Adhesives	Controls penetration to improve adhesion and application.
	Textile printing	Produces very fine line prints with good definition and excellent washout.
	Textile dyeing	Prevents migration of dyestuffs in pad dyeing operations. (Algin is also compatible with most fiber-reactive dyes).
Gelling	Air-freshener gel	Firm, stable gels are produced from cold-water systems.
	Explosives	Rubbery, elastic gels are formed by reaction with borates.
	Toys	Safe, nontoxic materials are made for impressions or puttylike compounds.
	Hydro-mulching	Holds mulch to inclined surfaces; promotes seed germination.
	Boiler compounds	Produces soft, voluminous flocs easily separated from boiler water.
Emulsifying	Polishes	Emulsifies oils and suspends solids.
	Antifoams	Emulsifies and stabilizes various types.
	Latices	Stabilizes latex emulsions; provides viscosity.
Stabilizing	Ceramics	Imparts plasticity and suspends solids.
	Welding rods	Improves extrusion characteristics and green strength.
	Cleaners	Suspends and stabilizes insoluble solids.

Food Applications

The primary food products in which alginates are used include frozen desserts, dairy products, bakery products, salad dressings and sauces, fabricated foods, beverages, and dessert gels.

In ice cream, algin ensures a smooth texture by regulating the formation of ice crystals. Alginates can also be used to stabilize water ices, sherbets, milk shakes, and ice milks.

A number of dairy products are also benefited by alginate stabilization, including processed and cream cheeses, whipped cream, and other cultured dairy products. Alginates are also used to stabilize bakery toppings and icings, as well as to prevent the sticking of products to wrapping paper. Other bakery products in which alginates are used include meringues and fruit or chiffon pie fillings.

Propylene glycol alginate is an excellent thickener and stabilizer for bottled salad dressing emulsions. Gravies and meat and barbecue sauces are also stabilized by propylene glycol alginate. The fruit pulp in noncarbonated fruit beverages is suspended by this alginate, and the foam of beer is stabilized.

Sodium alginate finds applications in dessert gels that can be made to set at room temperature, in quick-setting dessert puddings and in custards containing milk or milk solids that remain soft without forming a hard skin. In fabricated foods, such as onion rings from fresh diced onions, or meat pieces, sodium alginate provides the matrix gel. A sodium alginate, calcium chloride treatment of quick-frozen fish creates a strong, freeze-thaw stable coating that keeps air away from the fish and lengthens shelf life.

Industrial Applications

The alginates are also of value in various industrial procedures that can benefit from water-holding, gelling, emulsifying, and stabilizing actions of these materials, as detailed in Table 2.13.

In the manufacture of paper, alginates are used to control the flow properties of coatings and to prevent dilatancy at high shear rates. The alginates improve surface properties and ink acceptance and enhance smoothness of the final product.

For the manufacture of textiles the alginates are used to prevent the migration of dyestuffs in pad-dyeing operations. They produce very fine line prints with good definition and excellent washout. It should be noted that alginates also are compatible with most fiber-reactive dyes.

The ability to gel is useful in the formation of firm, stable gels from cold-water systems for use in air fresheners and for the production of elastic gels by reaction with borates for explosives formulations. For the production of toys the alginates in gel form are safe, nontoxic materials that can be used as mold impressions or as safe puttylike compounds. And in boiler compounds the alginates produce soft, voluminous flocs that are easy to separate from the treated boiler water.

Emulsification properties of the alginates are valuable in the emulsification of oils and suspension of solids necessary for the manufacture of polishes. Alginates provide viscosity required for the stabilization of latex emulsions. The alginates impart plasticity to and suspend solids in ceramic glazes, giving green strength but burning off upon firing. In the production of welding rods the alginates improve extrusion characteristics as well as green strength. And in cleaners of various types the alginates suspend and stabilize the insoluble salt content.

FORMULATIONS

Stabilizing Frozen Foods

Algin products can be used to build quality into frozen foods. Some of the advantages of algins are:

■ They do not mask or hide fine flavors, and are themselves free from odor or flavor.

■ They discourage weeping or syneresis during the thawing and heating periods. Viscosities of algin solutions decrease very slowly on heating, and increase slowly on cooling.

■ They retard ice crystal formation, helping to maintain smooth body and texture.

■ Their emulsifying qualities help prevent fat or oil separation during freezing, thawing, or heating.

■ They do not cause discoloration or color changes.

■ They are easy to use, being soluble in hot or cold water.

■ They may be added directly to cold mixes, or they may be preblended with other dry ingredients.

Fruit Pie Filling   Flavor release is improved in frozen fruit pies by replacing a part of the pregelatinized starch with algin, allowing the elimination of some so-called flavor aids or boosters. Also, objectionable gelation may be avoided upon cooling.

Because of the different types of fruit and starches used for fruit pie fillings, it is not possible to offer exact formulas. However, suggested steps for trial mixtures are as follows:

1. In a batch using 45.4 kg (100 lb) of pregelatinized starch, reduce the starch to 36.3 kg (80 lb), a 20% reduction.
2. Replace the 9.1 kg (20 lb) of starch with 1.36 to 2.27 kg (3 to 5 lb) of algin, depending on the particular fruit, starch, and formula. This is a 15 to 25% replacement of the omitted pregelatinized starch.

Frozen Gel   The following formulation will withstand a complete freeze-thaw cycle, and after thawing it maintains excellent gel body with pleasing eating quality. The gel does not weep or shrink, and is appropriate for frozen desserts and frozen salads. It may be used to coat fruits, vegetables, poultry, meats, and similar products.

	Weight percent
Granulated sugar	91.74
Refined sodium alginate	3.81
Citric acid, anhydrous, USP fine granulated (Pfizer, Inc.)	2.75
Sodium hexametaphosphate, powdered, unadjusted (Calgon Corp.)	1.10
Dicalcium phosphate, monohydrate, powdered (Stauffer Chemical)	0.37
Flavor and color, if desired	0.23

For a small package batch, use 109.3 g (3.85 oz) of the powder. Heat 474 mL (2 cups) of water to boiling in a saucepan. Remove from the flame, and pour into a mixing bowl. Add the powder to the water, and mix immediately with an eggbeater for 2 min. Pour into molds, cover, and place in a freezer. Fast freezing produces best results; therefore sections of approximately 2.54-cm (1-in) thickness are recommended for uniform processing.

Frozen Fruit Algin can be used to provide smooth body and texture consistency to frozen fruits because of the effective stabilization of the water content of the fruit with added sugar. Bright fruit color and natural flavor release are obtained. Strawberries, blueberries, peaches, raspberries, and other fruits can be used.

A typical unit will be a 13.6-kg (30-lb) tin of fruit with 0.17 to 0.28 kg (6 to 10 oz) of algin. The algin is blended with all or part of the sugar to be used in the frozen pack, usually with a standard sugar blender or simple baffled barrel rotation. The algin-sugar blend is then added to the fruit in the usual manner with which the sugar itself is added.

Cream Sauce Algin will improve the flavor release of cream sauces that are used with frozen foods. It also helps to emulsify the fat during preparation, freezing, thawing, and heating. A typical formulation is as follows:

	Weight percent
Water	66.7
Margarine	18.1
Whole milk powder	12.0
Starch	2.4
Propylene glycol alginate, low-viscosity	0.8
Flavor	As desired
Salt	As desired

Dry-mix the starch, milk powder, and algin, and add the mix to water while stirring vigorously. Heat to 82°C (180°F) to hydrate the starch, meanwhile continuing to stir. Cool until the starch is completely gelatinized. Add the margarine, and blend intimately. Add salt and other flavors last.

Barbecue Sauce A barbecue sauce for frozen barbecue beef may be processed without heat treatment if it is subsequently maintained in cold storage, or if it is used in a frozen food product. With incorporation of algin in the formulation, there is complete release of fine flavors, weeping is prevented, and there is consistent body in the final product. A typical formulation is:

	Weight percent
Cane sugar	31.00
Water	23.60
Tomato paste, 30% solids	21.90
Malt vinegar, 50 grain	15.20
Salt	3.00
Barbecue beef seasoning (W. J. Strange Co.)	1.50
Vegetable oil	1.46
Worcestershire sauce, 32 grain	1.14
Bead molasses, thick soy	0.70
Propylene glycol alginate, high-viscosity	0.50

To cold-process, dry-blend the algin with five times its weight of sugar and with the seasoning. Slowly add the dry blend to the total amount of water and vinegar under good agitation, and dissolve completely.   Add the Worcestershire sauce, tomato paste, bead molasses, salt, and finally the vegetable oil.   Continue the agitation of the mixture until the ingredients are dissolved and completely dispersed.

Frozen Shortcake Berry Filling   This formulation uses previously unstabilized berries which have been defrosted.

	Weight percent
Strawberries, 4 plus 1	93.76
Sugar	4.68
Propylene glycol alginate, locust bean gum composition	1.56

First dry-blend the algin with the sugar.   Then add the defrosted berries to a vat with complete but slow-moving agitation to produce thorough blending without excess crushing.   Add the berries stepwise, at about 68 kg (150 lb) per step; also add about 4.5 kg (10 lb) of the algin-sugar mixture.   This will assist in obtaining uniform blending and solution.   About 1 h should be allowed to complete the mixing and solution.

Tomato Sauce (Pizza and Spaghetti)   The use of refined sodium alginate in tomato sauce for frozen pizza and frozen spaghetti helps hold the juice and water in the sauce itself so that it will not soak into the porous pizza or spaghetti.   The concentration of algin will vary depending upon the desired viscosity for the sauce, with a range of 0.3 to 0.5% algin based on total weight being recommended.   The sauce itself may be prepared with a tomato paste, water, spices, and flavorings, or it may be prepared using tomato purée or tomatoes.

The procedure for preparing the sauce is:

1. Dry-blend the algin with five times its weight of sugar or the combination of sugar and spices.   Omit salt.
2. Add the dry blend to cold water, using 3.8 L (1 gal) for each 71 g (2.5 oz) of algin.   Stir with maximum agitation (as with a Lightnin' or Hobart mixer) until the algin is completely dissolved.
3. Add the tomato paste (or purée) and salt.   Blend until smooth.   To obtain a 0.3 to 0.5% algin concentration, use from 3.8 to 7.6 L (1 to 2 gal) of 2% algin solution for each 38 L (10 gal) of tomato paste or purée.   If the sauce is heat-processed, add the algin as a straight 2% solution after the heating is completed.

Macaroni and Cheese   Prepare a 1.5% solution of refined sodium alginate by dissolving it in cold water with rapid agitation.   The concentration of dry algin in the melted cheese should be 0.30 to 0.50%.   A blend of 80 parts of melted cheese with 20 parts of a 1.5% algin solution will give a concentration of 0.30% algin in the finished cheese sauce, and will retard or prevent bleeding.

Chop Suey   Prepare a 1.5% solution of refined sodium alginate as above, and blend with gravy and/or solids.   Because of the added water in chop suey, it is usually advantageous to use 0.50 to 1.0% dry algin based on the weight of the gravy.

Food Gels

Algin can be used to make a wide variety of food gels, from tender dessert gels to firm, strong ones.   The common method for making an algin gel is to introduce calcium ions into an algin solution, forming a calcium alginate gel which is nonreversible upon heating.   Gelling or setting time may be controlled by addition of phosphate buffers or sequestering agents.   Another method of control is to adjust the pH to effect the release of calcium ions, or by combination of phosphate buffer with an acid.

Dessert Gel   The following formulation contains no protein or any other animal-derived material, yet the use of algin gives a very fast setting gel that will gel at either room temperature or under refrigeration.   The gel will not melt or bleed when warm, nor will it become "livery" upon standing.

	Weight percent
Baker's special sugar	90.447
Refined sodium alginate	3.529
Adipic acid, fine granular (Monsanto)	3.235

Sodium tripolyphosphate, anhydrous powder	
(Stauffer Chemical)	1.765
Calcium carbonate, food-grade powder	0.475
Artificial strawberry flavoring,	
59.389/AP (Firmenich)	0.400
F.D.&C. Red 40	0.035
Magnesium carbonate, USP powder	0.118

Mix the ingredients with boiling water (472 mL, or 2 cups, of water for 85 g, or 3 oz, of dry mix). Stir for 1 min, then pour into molds and chill in a refrigerator. Gelling time is about 30 min, or as soon as chilled.

Cold-Water Gel   This formulation may be used to prepare a dessert gel using room temperature water and sodium alginate.

	Weight percent
Baker's special sugar	89.95
Refined sodium alginate	3.42
Adipic acid, food-grade powder (Monsanto)	3.23
Sodium citrate, fine granular (Pfizer)	2.82
Dicalcium carbonate, anhydrous powder	
(Stauffer Chemical)	0.35
Artificial strawberry flavoring,	
59.389/AP (Firmenich)	0.20
F.D.&C. Red 40	0.03

To prepare, add a dry blend of the ingredients to cold water (85 g, or 3 oz, to 472 mL, or 2 cups, of water), and stir briskly for 1 min. Pour into gel dishes. The gel will set at room temperature or in a refrigerator in a few minutes.

Cold-Milk Gel   The following formulation is for instant vanilla pudding or bakery filling.

	Weight percent
Baker's special sugar	85.1648
Sodium alginate, sodium phosphate combination	7.0588
Tetrasodium pyrophosphate, pudding grade	
(Stauffer Chemical)	3.8235
Calcium sulfate dihydrate, powdered (U.S. Gypsum)	1.8118
Beatreme 3458 (Beatrice Foods)	1.1765
Fine salt	0.5882
Artificial vanilla flavoring, 59.290/AP	
(Firmenich)	0.3529
F.D.&C. Yellow 5	0.0188
F.D.&C. Yellow 6	0.0047

To prepare instant vanilla pudding, add 472 mL (2 cups) of cold whole milk to a bowl. Next, add a dry blend of 85 g (3 oz) of the above ingredients while stirring at slowest speed on electric mixer, or slow speed with a hand mixer. Continue mixing at slow speed for 3 min. Pour the mixture into molds, pie or pastry shell, and allow to set either at room temperature or in a refrigerator.

To prepare instant vanilla bakery filling, add 3785 mL (1 gal) of cold milk to a bowl. Add 0.68 kg (1.5 lb) of the dry blend while stirring at slowest speed with a wire whip or paddle, or in an electric mixer of the Hobart type. Stir for 3 min and allow to set.

Instant Chiffon Pie Filling   The following formulations produce rich and tasty chiffons which can be served as puddings, frozen desserts, or pie fillings.

For a whole-milk type of lemon chiffon, the formulation is:

	Weight percent
Baker's special sugar	63.706
Wip-Treme 2320 (Beatrice Foods)	31.216
Sodium alginate, sodium phosphate	
combination	2.166
Adipic acid, fine granular (Monsanto)	1.529
Lemon juice powder 1626 (Borden)	0.637

Tetrasodium pyrophosphate, pudding grade (Stauffer Chemical)	0.637
Felcofix Lemon 5-fold N.O. Flavoring (Felton International)	0.096
F.D.&C. Yellow 5	0.013

To prepare, add a dry blend of 156.97 g of the above to 236 mL (1 cup) of cold milk in a small mixing bowl. Blend at lowest speed, then beat at high speed for 4 min.

For a whole-milk type of chocolate chiffon, the formulation is:

	Weight percent
Baker's special sugar	60.17
Wip-Treme 2320 (Beatrice Foods)	27.13
Cocoa (Cokay 35, R. A. Johnston Co.)	8.14
Sodium alginate, sodium phosphate combination	2.17
Calcium gluconate, USP powder (Pfizer)	1.63
Tetrasodium pyrophosphate, pudding grade (Stauffer Chemical)	0.38
Fine salt	0.27
Natural and artificial vanilla sugar 15 (D. Michael & Co.)	0.11

To prepare, add a dry blend of 184.3 g of the preceding ingredients to 236 mL (1 cup) of cold milk in a small mixing bowl. Blend at lowest speed, then beat at high speed for 4 min.

Instant Cheese Cake Mix   The following formulation will produce a rich, high-quality cheese cake when blended with cold milk.

	Weight percent
Baker's special sugar	38.3067
E-Z Cheez, Type S (Borden)	24.5399
Wip-Treme 2320 (Beatrice Foods)	24.5399
Beatreme 1038 (Beatrice Foods)	6.1350
Sodium alginate, sodium phosphate combination	3.6810
Calcium gluconate, USP powder (Pfizer)	1.5357
Tetrasodium pyrophosphate, pudding grade (Stauffer Chemical)	1.2270
Aromalok Artificial Cheese Flavor 180460 (Fritzsche)	0.0307
F.D.&C. Yellow 5	0.0049
F.D.&C. Yellow 6	0.0012

To prepare, dry-blend the above ingredients (163 g) with cold milk (354 mL, or 1.5 cups) in a small mixing bowl, and beat at slow speed with an electric mixer or rotary beater until blended. Beat at medium speed for another 3 min. Pour into graham cracker crust and chill 1 h before serving.

Instant Imitation Bakery Jelly   This formulation produces a fast-setting and firm gel that has good heat stability and a slight plastic consistency that provides for easy workability.

	Weight percent
Baker's special sugar	58.80
Instant Clearjel Starch (National Starch)	14.89
Adipic acid, fine granular (Pfizer)	8.13
Refined sodium alginate	6.30
Calgon, unadjusted powder (Calgon Corp.)	3.52
Guar gum, 200 mesh	2.96
Dicalcium phosphate dihydrate, USP (Stauffer Chemical)	2.22
Artificial strawberry flavoring 59.389/AP (Firmenich)	1.41
Potassium sorbate, N.F. powder	0.67
Sodium benzoate, USP powder	0.67
F.D.&C. Red 40	0.42

To prepare, dry-blend the above ingredients (28.4 g) with 113.5 g (¼ lb) sugar. Add 236 mL (1 cup) of tap water, and mix vigorously with a wire whip for 1 min. The jelly will be ready for use in 30 to 45 min.

Banana Gel Base   This gel base acts as an excellent carrier for most kinds of fruit purée, and can be cut or molded into various sizes and shapes.

	Weight percent
Fine granulated sugar	91.00
Refined sodium alginate	4.88
Fumaric acid, purified powder	1.76
Sodium hexametaphosphate, unadjusted powder	1.41
Sodium benzoate powder	0.48
Dicalcium phosphate dihydrate, USP (Stauffer Chemical)	0.47

To prepare, place 205 g (7.2 oz) of banana purée and 114 g (4 oz) of cold water in a mixing bowl, and mix briefly. Add 85 g of a dry blend of the formulation above, stir with an egg beater for about 45 s, and pour into gel dishes. The gel will set in a refrigerator or at room temperature within 30 min.

Meringue Powder with "Dried Egg Whites"   A high-quality meringue for institutional use with "dried egg whites" can be made using the following formulation:

	Weight percent
P-20 powdered egg white (Henningson Foods)	50.20
Baker's special sugar	30.11
Amaizo 721-A starch (American Maize)	8.78
Cerelose 2001 (Corn Products)	6.27
Propylene glycol alginate, low-viscosity	4.39
Artificial vanilla powder 674 (Felton International)	0.25

To prepare, blend dry ingredients (79.7 g) and water (354 mL) at lowest speed on a Kitchen-Aid K-45 mixer with the whip attachment. Mix for 2 min at medium speed, then whip 1 min at highest speed. Adjust to medium speed and add 227 g Baker's special sugar; then mix for 3 min. Bake at 190°C (375°F) for 8 to 10 min.

Dessert Soufflés

By adding sodium alginate to dessert soufflé mixes, the user has the advantages of convenience in preparation, stability during service, and a fat content and total caloric value lower than in home-prepared recipes. In addition, by the omission of egg yolks in the formula, the cholesterol level is reduced.

Vanilla Soufflé

	Weight, g	Percent
Baker's sugar	35.25	66.6
Redisol No. 4 starch (A. E. Staley Corp.)	7.5	14.2
Clearjel starch (National Starch & Chemical Corp.)	7.5	14.2
Sodium alginate, sodium phosphate combination	1.5	2.8
Tetrasodium pyrophosphate, anhydrous powder, pudding grade (Stauffer Chemical Co.)	0.75	1.4
Table salt	0.25	0.47
Artificial vanilla flavor, 59.290 (Firmenich)	0.15	0.28
Calcium sulfate dihydrate	0.05	0.09
F.D.&C. Yellow 5	0.0025	0.005
F.D.&C. Yellow 6	0.00075	0.001
	53.00	100.00

Dry-blend the ingredients, then add to 180 mL (6 oz) of cold milk in a small mixing bowl. Blend at a slow speed (to prevent air bubbles) for 3 min with an electric mixer, then set aside.

To prepare the egg white mix, add 15.1 g of egg white solids (P-20, Henningsen Foods, Inc.) to 120 mL (4 oz) of cold water. Whip at a high speed until soft peaks are formed (about 2 min). Sprinkle in 14 g (1 tbsp) of sugar and continue to whip until stiff peaks are formed (approximately 1 min).

The soufflé mixing and baking procedure is as follows:

1. Stir one-fourth of the beaten egg white mix into the soufflé mix. Fold this mixture into the rest of the beaten egg whites.
2. Place the product from (1) into a 6-cup ceramic soufflé dish which has been greased with butter and coated with sugar. The product should be 2.5 cm (1 in) below the rim for best results. Bake at 177°C (350°F) for about 30 min. If a metal container is used, place it into a slightly larger second container with 2.5 cm (1 in) of water.
3. Serve the soufflé with a favorite sweet sauce.

Chocolate Soufflé

	Weight, g	Percent
Baker's sugar	39.0	59.5
Cocoa (Cokay No. 35, R. A. Johnston Co.)	9.0	13.7
Redisol No. 4 starch (A. E. Staley Corp.)	7.5	11.4
Clearjel starch (National Starch & Chemical Corp.)	7.5	11.4
Sodium alginate, sodium phosphate combination	1.5	2.3
Tetrasodium pyrophosphate, anhydrous powder, pudding grade (Stauffer Chemical Co.)	0.75	1.1
Table salt	0.25	0.38
Calcium sulfate dihydrate	0.05	0.08
	65.55	100.00

The procedure for dry blending and mixing of the chocolate soufflé is the same as for the vanilla soufflé, with the following exceptions: sprinkle in 30 g (2 tbsp) of sugar during the whipping of the egg whites, and increase baking time from 30 min to about 1 h.

Lemon Soufflé

	Weight, g	Percent
Baker's sugar	36.00	64.2
Redisol No. 4 starch (A. E. Staley Corp.)	7.50	13.4
Clearjel starch (National Starch & Chemical Corp.)	7.50	13.4
Sodium alginate, sodium phosphate combination	1.50	2.7
Lemon Powder 14110 (Borden, Inc.)	1.50	2.7
Adipic acid	1.00	1.8
Tetrasodium pyrophosphate, anhydrous powder, pudding grade (Stauffer Chemical Co.)	0.75	1.3
Table salt	0.25	0.44
Calcium sulfate dihydrate	0.05	0.09
F.D.&C. Yellow 5	0.01	0.02
	56.00	100.00

The procedure for mixing and baking is the same as for the vanilla soufflé.

Dressings

Propylene glycol alginate has excellent emulsifying properties for use in food emulsions such as French dressings. The alginate dissolves readily in water or a mixture of water and vinegar. This allows rapid and thorough incorporation of oil during the emulsifying process. In addition, this alginate produces a nonseparating French dressing with long shelf life and excellent resistance to separation even when exposed to large changes in temperature.

Propylene glycol alginate is an approved food additive for unrestricted use as a stabilizer, emulsifer, or thickener in all nonstandardized foods and confectionary products. It is not an allergen, and it is essentially odorless, tasteless, and colorless. In addition to the formula which follows, there are other formulations using propylene glycol alginate for products where its emulsifying, suspending, thickening, and stabilizing properties are desired in addition to higher viscosity. Examples are Roquefort and romano cheese dressing, Italian dressing (for partial stabilization), mustard dressing, cole slaw dressing, and pickle relish.

The average use levels for propylene glycol alginate vary from 0.40 to 0.50%, based on total weight of dressing. The higher amount is used for the lower oil content (35%), and the lower amount for higher oil content (45%). To determine the correct level for each formulation, trial is recommended. A typical 35% oil formula follows:

	Liquid	or	Weight, %
Vegetable oil	17.0 L		35.00
Water	15.1 L		32.35
Tomato purée			9.00
Vinegar, 100 grain	3.8 L		8.35
Cider vinegar	2.5 L		5.50
Sugar			5.50
Salt			2.00
Paprika			1.50
Propylene glycol alginate			0.50
Mustard			0.30
			100.00

To prepare, place water in the mixing tank, then add tomato purée or juice. Dry-blend the propylene glycol alginate with the dry spices (except the salt), and add to liquid in the tank under vigorous agitation. Continue the agitation for about 2 min for a 95-L (25-gal) batch, and about 8 min for a 380-L (100-gal) batch.

Add vinegar, sugar, salt, and liquid spices, if used. Continue the agitation for another 2 to 4 min, until dissolved. Then add oil while mixing. After mixing thoroughly, pump the mixture through a colloid mill, prior to bottling.

Fabricated Fruit

A gel is readily formed when a soluble calcium salt is added to a sodium alginate solution. This gel has properties of stability over a wide range of temperatures, has excellent syneresis control, and is irreversible to heat. Possible uses for this fabrication concept include imitation cocktail cherries; imitation glacé fruit pieces for cakes, bread, cookies, ice cream, and candy products; icings; and gelled products containing puréed fruit. With variations on the technique, it is possible to produce vegetable products and spaghetti, as well as exotic foods, such as imitation caviar.

A formula and procedure for fabricated fruit pieces is as follows:

	Weight, g	Percent
Sodium alginate (low viscosity, low calcium)	10.0	1.00
Anhydrous sodium carbonate	0.5	0.05
Dicalcium phosphate dihydrate	2.0	0.20
Sugar	435.0	43.50
Corn syrup, 42 DE	291.0	29.10
Water	261.5	26.15
Color and flavor	add to suit	
Total weight of mixture	1000.0	100.00
Total weight after boiling	907.0	(75% solids)

To prepare the preceding syrup formulation, dry-blend the alginate with the sodium carbonate, dicalcium phosphate, color, and 50 g of the sugar. Dissolve the mixture in the water, then add the solution to warmed corn syrup and heat to nearly boiling. Then add the remainder of the sugar and boil the mixture (at approximately 100°C, or 230°F) until the desired solids concentration is reached. Add flavor, and mix well. The syrup is now ready for extrusion into an acid bath.

The composition of the acid bath is:

	Weight, g	Percent
Anhydrous citric acid	600	15.0
Dicalcium phosphate dihydrate	40	1.0
Water	3360	84.0
Total	4000	100.0

To prepare the acid bath, dissolve the citric acid in the water, then add the calcium salt and stir until dissolved.

The extrusion procedure is as follows:

1. Pour hot syrup at 65 to 104°C (150 to 200°F) into an extrusion device, such as a Berry mill, and extrude just beneath the surface of the citric acid bath. It is best to use a tall column, at least a foot high, for the citric acid bath. This helps to produce more spherical particles. [An even more spherical particle is obtained if the syrup is cooled to 65°C (150°F) before extrusion.]
2. Leave the gelled pieces immersed in the acid bath with mild agitation for 15 to 30 min. Firmness and tartness can be controlled by the length of time held in the bath. For example, less time is required if a softer body is desired, more time if a firmer body is desired.
3. Rinse, drain, and soak the pieces in a 70° Brix sugar solution for 24 h, and then rinse and drain again. Pieces may then be air-dried for several hours, or can be dried by heated forced air for an hour or so. Soaking in the 70° Brix sugar solution tends to give more uniformly spherical–shaped jellies with clearer, brighter appearances.

Pie Fillings

Alginates (sodium, propylene glycol, ammonium/calcium) can be used to stabilize the starch system in pie fillings. They do not mask the flavor, and they produce full body without pastiness. The short, nongummy texture has a pleasant mouth-feel. And these alginates may be used in canned, freshly baked, or frozen fillings. They are equally effective in cooked and cold-mix types.

Sodium alginate is used for fruit fillings. It produces high viscosity, good freeze-thaw stability, exceptional clarity, high gloss, maximum flavor release (even with delicate flavors), and excellent resistance to weeping. The flow properties of the filling, as well as the gel body, are easily regulated with calcium ions.

Propylene glycol alginate is used for cream and whipped fillings, high-acidity fruit fillings, and for prestabilization of strawberries, raspberries, etc., prior to the makeup of the filling. This alginate is a combination emulsion stabilizer and viscosity producer. Fillings made with propylene glycol alginates have good foam cell stability, excellent freeze-thaw stability, and especially good flavor release.

Ammonium/calcium alginate is used for fruit and cream fillings, soft fillings such as lemon, and to enhance remix-reset properties. Fillings made with this alginate are unusually resistant to high-temperature breakdown. Other filling properties are excellent freeze-thaw stability, maximum flavor release, special types of short gel structures with "dead flow" or "cutability," and thixotropic viscosity that gives clean-breaking mouth-feel.

When thickening with alginates, best results are obtained by replacing a significant portion of the starch in the formula with one part by weight of the alginate for four to seven parts by weight of starch. Too low a starch replacement produces insignificant improvement, while excessive starch reduction sometimes will cause oven boil-out, an inferior filling being the result. It is best to test a formula to determine the optimum level of starch replacement.

Cooked Fillings For starch which requires cooking, here is a suggested procedure:
1. Place water, sugar, and liquid color in the kettle.
2. Heat to a light boil.
3. Add a slurry consisting of water, starch, salt, and preservative to the boiling liquid.
4. Cook to 85 to 90°C (185 to 195°F). Hold at this temperature for 5 min.
5. Adjust soluble solids. Fill into containers.

To work an alginate into this method of preparation, mix it with the liquid or dry sugar before combining it with the water, or with the water and fruit juice. Heat the batch up to temperature, adding the starch slurry and so on as in the procedure above. It is possible to add the alginate after the starch has been cooked, but handling is easier when the alginate is added at the start.

To reduce lumping and to give a smooth hydration, it is strongly recommended that the user blend the alginate with 5 to 10 times its weight in liquid sugar or dry granulated sugar before adding it to a batch. Some users predisperse the alginate in three to four parts glycerin or propylene glycol before adding it to the liquid sugar.

Exposure of the filling to elevated temperature for an extended period of time can cause a loss of the alginate and/or starch viscosity. The result will be runniness. On the other hand, slow cooking can also be detrimental to quality.

Cold-Mix Fillings When using pregelatinized starch, the following procedure can be used:
1. Place liquid sugar into the kettle or mixing bowl.
2. Add starch, then mix until uniform.
3. Add water and all available liquids, then mix until smooth.
4. Add fruit and the remaining ingredients, then blend until uniform.

If only dry granulated sugar is used, the incorporation of algin into a formula with pregelatinized starch is very simple. It is added at the same time as the starch, and in the same manner as the starch.

Industrial Applications

Kraft Linerboard Sizing The following formulation and procedure are recommended for six press applications on kraft linerboard where it is desired to obtain higher vanceometer tests:

10 parts
  1 part sodium alginate (medium viscosity)
  0.1 to 0.3 parts Calgon or TSPP (trisodium pyrophosphate)
  0.2 part sodium aluminate (optional)

To make up the sizing formulation:
1. Add water to the make-up tank, and turn on the agitator.
2. Add the starch to the agitated water.
3. Add the alginate slowly with agitation. If possible, use an aspirator to prevent lumping.
4. Add the Calgon or TSPP.
5. Run the mixture through a normal starch cook cycle.
6. Add the sodium aluminate, if used, in water solution.

If the starch is to be enzyme-converted, inactivate the enzyme with heat only; do not use copper sulfate, etc. The usual operating temperature is 71 to 82°C (160 to 180°F). Run at the highest viscosity possible. If higher solids are desired, use a thinner-boiling starch.

The Calgon or TSPP is used for two reasons. First, it facilitates solution of the alginate by sequestering calcium or other polyvalent ions present in the water. Second, it eliminates thickening of the alginate which could occur as alum leaches into the size press solution from the sheet. The amount of Calgon or TSPP to be used will depend upon the hardness of the raw water.

In those situations for which the alginate results in higher Cobb test values, it has been found that the addition of sodium aluminate may prevent this increase. It is necessary that the aluminate be completely in solution before the sodium alginate is added, and it is necessary that the aluminate be predissolved.

When the alginate is used at percentages less than 5% on the weight of the starch, it is possible that it will show little or no effect.

Corrugating Adhesives When using starch adhesives, alginate can be added to control the rate of penetration of the liquid phase and stabilize its viscosity. This use of sodium alginate improves bonding, reduces weep and washboarding, increases machine speed, cuts scrap loss, and produces a more uniform, stiffer, smoother corrugated board.

Single-Starch System. When using a single-starch system, since the alginate controls great quantities of water and builds viscosity, it is necessary to reduce the amount of starch used in the cooking to arrive at the normal finished-batch viscosity. This is illustrated in the following corn starch formula for a 2.52 m³ (666 gal) mix:

	Unmodified	Alginate modified
Top mixer:		
Pearl corn starch	90.72 kg (200 lb)	68.04 kg (150 lb)
Caustic soda	14.51 kg (32 lb)	14.51 kg (32 lb)
Bottom mixer:		
Borax, 10 mol	14.51 kg (32 lb)	14.51 kg (32 lb)
Pearl corn starch	453.6 kg (1000 lb)	430.92 kg (950 lb)
Formaldehyde	3.79 L (1 qt)	3.79 L (1 qt)
Sodium alginate		5.44 kg (12 lb)
Total volume	2.52 m³ (666 gal)	2.52 m³ (666 gal)
Viscosity, 37.8°C(100°F)	35 s	40 s

To incorporate the alginate into the adhesive, simply add it to the secondary water while agitating. Allow 5 to 10 min for the alginate to hydrate before adding borax or starch, then drop the primary mix in the normal manner. When preparing adhesive in a single tank system, add the alginate after the cooling water and prior to borax and slurried starch. The alginate may not fully hydrate unless this procedure is followed, and as a result its efficiency will be reduced.

Concentration of alginate should be varied to accommodate local operating conditions, as well as the proportion of starch solids used and the characteristics of the paper. Generally, the alginate is used as 0.15 to 0.25% of the total wet weight of the adhesive mixture. In areas where water hardness is more than 250 ppm, it is desirable to add a polyphosphate to the water prior to the addition of the alginate. Usually 200 to 450 g (approximately 0.5 to 1 lb) is sufficient.

Two-Starch System. For plants requiring or desiring the two-starch system, the following formulation is suggested:

	Single facer	Double backer
Top mixer:		
Pearl corn starch	68.04 kg (150 lb)	72.58 kg (160 lb)
Caustic soda	14.06 kg (31 lb)	15.88 kg (35 lb)
Lower mixer:		
Alginate	4.54 kg (10 lb)	5.44 kg (12 lb)
Borax, 10 mol	14.51 kg (32 lb)	15.42 kg (34 lb)
Pearl corn starch	408.24 kg (900 lb)	449.06 kg (990 lb)
Formaldehyde	3.79 L (1 qt)	3.79 L (1 qt)
Total volume	2.52 m³ (666 gal)	2.52 m³ (666 gal)
Viscosity, 37.8°C(100°F)	35 s	50 s

Preparation techniques are the same as for the single-starch system.

Fiber-Reactive Dyes Fiber-reactive dyes are unique in that they link chemically with cellulosic fibers in the presence of alkali at elevated temperatures to produce brilliant shades having excellent fastness. The fiber reacts or links chemically with the dyestuff, locking the dye to the fiber. The thickener should not interfere with this function.

If the thickener reacts with both the dyestuff and the fiber, it is then locked or joined chemically with both of them, and will be difficult to wash out. The result will be a dyed fabric with very stiff, brittle hand and a loss of color value. Such does not happen when sodium alginate is the thickener.

Both the medium- and low-viscosity alginates are suitable for average printing conditions. The concentrations recommended in the following formulations are in the middle range. They result in viscosities between that required for screen printing and roller printing. In practice, the concentration of alginate in a formulation must be adjusted to work best with plant conditions and print specifications.

Print pastes which are prepared with low-viscosity alginates are inherently the most stable. They make possible preparation of higher-solids print pastes which in turn produce denser films on drying and give increased color yield.

Stock Thickener Formulation. Typical formulations for a 378.5-L (100-gal) batch of stock thickener are given below.

	Screen printing	Roller printing
In Water or Aqueous Systems:		
Cold water	302.8 L (80 gal)	302.8 L (80 gal)
Preservative (e.g., sodium benzoate)	624 g (22 oz)	567 g (20 oz)
Ammonia, 28%	454 g (16 oz)	397 g (14 oz)
Sodium alginate	15 to 45 kg (33 to 100 lb)	
Bulk to 378.5 L (100 gal)		
with cold water	As needed	As needed
In "Rapid-made" Emulsion Systems:		
Cold water	181.7 L (48 gal)	181.7 L (48 gal)
Mineral spirits	181.7 L (48 gal)	181.7 L (48 gal)
Emulsifier, 1% total weight	3632 g (8 lb)	
Preservative, 0.1% total weight	454 g (1 lb)	

To prepare, all ingredients are added to the vat with no mixing as follows:
1. Put all required water, preservative (0.1%), and emulsifier (1.0%) into the mixing drum.
2. Add mineral spirits, but never more than the total water content.
3. Add the alginate. Sift it into the liquid slowly with no mixing. By using this method of addition, the alginate passes through the top layer of mineral spirits and is thoroughly wetted out and dispersed with no lumping.
4. Start the mixer. An emulsion is produced with a maximum mixing time of ½ h.

Pad Dyeing. Any of the standard sodium alginate products may be used in pad dyeing, either in dry form or from a prepared stock solution. If the dry, granular grade of alginate is used, care must be exercised in the mixing to ensure proper dispersion and hydration. A typical pad-dyeing formulation for a 378.5-L (100-gal) batch follows:

Reactive dye	737 g (26 oz)
Urea	71 g (2.5 oz)
Sodium alginate	340 g (12 oz)
Sodium bicarbonate	71–170 g (2.5–6 oz)
Ludigol (GAF Corp.) or	
Albatex BD (Ciba Corp.)	85 g (3 oz)
Water to bulk to 378.5 L (100 gal)	As needed

The practical upper limit for alginate content is 1%.

LABORATORY TECHNIQUES

Methods for determination of trace elements and an assay procedure for alginates may be found in the alginate monographs in the *Food Chemicals Codex*, National Research Council, 1972. Other standard procedures follow:

Viscosity Measurement

1. Add 250 mL of distilled water to a pint jar or beaker.
2. Stir at approximately 800 rpm with a propeller-type stirrer.
3. Weigh out the appropriate amount of alginate to yield 300 g of solution of the desired concentration, that is, 3 g for a 1% solution.
4. Disperse the alginate into the stirred solution by means of a high-shear mixing technique (see Application Procedures).
5. After the alginate is dispersed, add enough distilled water to bring the total solution weight to 300 g, that is, 47 mL for a 1% solution.
6. Continue to stir for 2 h.
7. Remove the solution, adjust the temperature to 25°C, stir vigorously by hand to eliminate any possible thixotropic effects, and read the viscosity immediately. (Use Brookfield Model LVF viscometer at 60 rpm with appropriate spindle.)

Moisture Determination

The equilibrium moisture content of any alginate is a function of the relative humidity of the surrounding environment.

1. Weigh 3 to 5 g of alginate into a wide (50 mm inside diameter) low-form weighing bottle. Weigh bottle and contents.
2. Dry the open bottle and contents in an oven at 105°C for 3 h.
3. Replace bottle top, cool in a desiccator, and reweigh.
4. Calculate moisture content as the percentage of weight lost during the drying step.

Powder Color Determination

1. Use a Photovolt reflectometer for this analysis.
2. Place a flat-bottomed, 90-mm petri dish or optically flat glass container on top of the reflectometer search unit.
3. Spread the powder into the glass container to give a layer of powder at least 6 mm (¼ in) deep over the entire search unit. Do not shake the glass container; erroneous readings may result.

4. Standardize the search unit against a white enamel working standard of about 75% reflectance. Record the reflectometer reading while using the green tristimulus filter in the search unit.
5. Read the color.

Alginates in Mixtures (Detection)

Alginates may be precipitated by adding one volume of a 1% solution to three volumes of isopropyl alcohol. After drying, the precipitated alginate will respond to the identification tests for the appropriate alginate in the *Food Chemicals Codex.*

Another procedure is to precipitate alginate by adding one volume of a 0.2 to 1.0% solution to an equal volume of a 1.0% calcium chloride solution. In the case of propylene glycol alginate it should be saponified with dilute alkali and neutralized before precipitating with the calcium.

Alginate precipitated from solution with isopropanol or calcium chloride may be identified by its infrared spectrum. Low levels of alginate in solution may be detected by precipitation with calcium chloride combined with absorption of Night Blue dye (see McDowall, 1956, in Further Reading). Alginate may be detected in various food products by methods given in *Official Methods of Analysis,* A.O.A.C., 1975.

Alginates in Mixtures (Determination)

The following quantitative methods are suggested, if there are no interfering substances present in the solution to be analyzed.

Gravimetric Algin may be determined gravimetrically by weighing the isopropanol alcohol precipitate obtained in either of the first two detection methods, preceding. The precipitate may be assayed by the decarboxylation method given on p. 863 of the *Food Chemicals Codex,* 2d ed., 1972.

Colorimetric For the determination of alginates in various food products, this method involves papain digestion of protein, separation of the alginate as the insoluble calcium salt, and a final colorimetric determination with ferric oxide–sulfuric acid reagent (see Graham, 1970, in Further Reading).

Spectrophotometric This method will work only if alginate is known to be the only carbohydrate present. High salt levels will cause slight changes in the standard curve; therefore, the standard solvent used should have a salt content similar to that of the unknown. Also, the accuracy is improved if standard and unknown both contain the same type of alginate.

The procedure:

1. Pipette a 1.0-mL sample solution (gives 5 to 250 ppm alginate) into a 10-mL Erlenmeyer flask.
2. Add 1.0 mL of 4% resorcinol solution (prepared fresh daily) and mix.
3. From a Mohr pipette, rapidly add 6 mL of concentrated sulfuric acid to the mixture. *Warning:* Heat is generated in this step with possible spattering. Safety goggles and gloves are recommended.
4. Place the flask in an ice-water bath until the solution is near room temperature.
5. Remove the flask from the bath; let it stand at room temperature for 10 to 25 min.
6. Transfer the solution to a 10-mm spectrophotometer cell. Avoid the formation of air bubbles.
7. Read the absorbence of the peak at about 494 mm; use a water blank.
8. Run a reagent blank.
9. Determine the concentration of alginate by comparison of the corrected absorbence with a standard curve prepared with solutions of known concentration ranging from 5 to 250 ppm alginate.

PRODUCT TRADENAME DIRECTORY

(Reference taken from TSCA 1977 Candidate List, provided by Chemical Abstracts: Kelco products supplemented)

Algin product	Chemical abstracts registry number	Supplemental information from chemical abstracts	Synonyms
Alginic acid, ammonium calcium salt	9005-31-6	Formula: Unknown Code designation: R246-5507 CA name (1): HP=9CI Alginic acid, NM=ammonium calcium salt	Keltose® (food, technical)
Alginic acid	9005-32-7	Formula: Unknown Code designation: R246-5617 CA name (1): HP=8CI Alginic acid	Kelacid® (pharmaceutical) Norgine® Landalgine®
Alginic acid, ammonium salt	9005-34-9	Formula: Unknown Replaces CAS registry number(s): 9036-51-5 Code designation: R246-5728 CA name (1): NP=9CI Alginic acid, NM=ammonium salt	Amoloid® LV (technical) Superloid® (technical) Ammonium alginate Callatex® Digamon® Protomon® Analgine® Collatex Arm Extra
Alginic acid, calcium salt	9005-35-0	Formula: Unknown Replaces CAS registry number(s): 9010-42-5-9019-43-6-9060-2- 0-2-37228-92-5 Code designation: R246-5834 CA name (1) HP=9CI Alginic acid, NM=calcium salt	Calcium alginate Calginate® (pharmaceutical) Combinace CA 33
Alginic acid, potassium salt	9005-36-1	Formula: Unknown Code designation: R246-5958 CA name (1): HP=8CI Alginic acid, NM=potassium salt	Kelmar® (food) Improved Kelmar® (food) Potassium alginate
Alginic acid, ester with 1,2-propanediol	9005-37-2	Formula: Unknown Replaces CAS registry number(s): 39306-87-1-51374-11-9-5244- 1-26-6-59125-52-9 Code designation: R246-6042 CA name (1): HP=9CI Alginic acid NM=ester with 1,2=propanediol	Dariloid® K, KB (food) Dricoid® KB (food) Kelcoloid® D (food) Kelcoloid® DH (food) Kelcoloid® O, DO (food) Kelcoloid® HVF, LVF (food, technical) Kelcoloid® DSF (food) Kelcoloid® S (food) Sherbelizer® (food)

Algin product	Chemical abstracts registry number	Supplemental information from chemical abstracts	Synonyms
			Propylene glycol alginate
			Propylene glycol ester of alginic acid
			Propylene glycol alginate ester
			Alginic acid propylene glycol ester
			Propylene alginate
Alginic acid, sodium salt	9005–38–3	Formula: Unknown Replaces CAS registry number(s): 12772–46–2–32129–82–1–3219–7–42–5–37332–19–7–50643–02–2–56940–21–7 Code designation: R246–6152 CA name (1): HP=8CI Alginic acid, NM=sodium salt	Cocoloid® (food) Dariloid® (food) Dariloid® XL (food) Dariloid® Q, QH (food) Dricoid® (food) Kelco Gel® LV (food) Kelco Gel® HV (food) Kelco Pac® (food) Kelcosol® (food, technical) Kelgin® F (food, technical) Kelgin® HV (technical) Kelgin® LV (food, technical)
Algin	9005–40–7	Formula: Unknown Code designation: R246–6262 CA name (1): HP=9CI Algin	Kelgin® MV (food, technical) Kelgin® Q, QH, QM, QL (technical) Kelgin® RL (technical) Kelgin® XL (technical) Kelset® (food, technical) Keltex® (technical) Keltex® S (technical) Keltex® P (technical) Keltone® (food, technical) Margel® (food)

L'-Algiline
Cohasal-IH
Lamitex
Manucol
Manutex
Manutex SA/KP
Minus
Proctin
Protanal
Protatek
Stipine
Tagat
Sodium alginate
Halltex
OG 1
Meypralgin R/LV
Snow Algin L
Manugel F 331
Antimigrant C 45
Manutex F
Manucol SS/LD 2
Amnucol
Nouralgine
Protacell 8
Cecalgine TBV
Algipon L-1168
Pectalgine
Tragaya

FURTHER READING

The following list of references will give the reader of this chapter a broad choice for further specialized reading on the subject of alginates.

Anonymous, *The Seaweed Story,* California Dept. of Fish and Game, Sacramento, 1954.

Anonymous, *Report on Coagulant Aids for Water Treatment,* U.S. Public Health Service, Bureau of Water Hygiene, Cincinnati, Ohio, 1970.

Anonymous, *GRAS (Generally Recognized as Safe) Food Ingredients—Alginates,* PB-221 226. N.T.I.S., 1972.

Anonymous, *The National Formulary,* 14th ed., American Pharmaceutical Association, Washington, D.C., 1975.

A.O.A.C., *Official Methods of Analysis of the Association of Official Analytical Chemists,* 12th ed., including supplements, Washington, D.C., 1975.

Atkins, E. D. T., Mackie, W., and Smolko, E. E., "Crystalline Structure of Alginic Acids," *Nature (London),* **225,** 626–28 (1970).

Atkins, E. D. T., Mackie, W., Parker, K. D., and Smolko, E. E., "Crystalline Structures of Poly-D-mannuronic and Poly-L-guluronic Acids," Part B. Polymer Letters, 9, *J. Polym. Sci.,* London, 1970.

Chapman, V. J., *Seaweeds and Their Uses,* 2d ed., Methuen & Co., Ltd., London, 1970.

Charm, S. E., and McComis, W., "Physical Measurements of Gums," *Food Technol.,* 19, 948–953 (1965).

Clendenning, K. A., "Photosynthesis and growth in *Macrocystis pyrifera,*" in *Proceedings of the Fourth International Seaweed Symposium,* Pergamon Press, Oxford, 1964, pp. 55–65.

Engel, L., *The Sea,* Time, Inc., New York, 1961, p. 40.

Ensink, A. L., *Planographic Printing Plate,* U.S. Patent 2,806,424 (1957).

Ensink, A. L., *Light-Sensitive Polyvalent Metal Alginate Photolithographic Element,* U.S. Patent 2,835,576 (1958).

Epstein, S. S., Fujii, K., Andrea, J., and N. Mantel, "Carcinogenicity Testing of Selected Food Additives by Parenteral Administration to Infant Swiss Mice," *Toxicol. Appl. Pharmacol.,* 16, 321–324 (1970).

Fischer, F. G., and Dörfel, H., "Die Polyuronsauren der braunalgen Kolenhydrate der Algen. Part I.," *Z. Physiol. Chem.,* 302, 186–203 (1955).

Graham, H. D., "Specificity of the Ferric-H_2SO_4 Reagent for Alginates," *J. Food Sci.,* 35, 494–498 (1970).

Grant, G. T., Morris, E. R., Rees, D. A., Smith, P. J. C., and Thom, D., "Biological Interactions Between Polysaccharides and Divalent Cations: The Egg-Box Model," *FEBS Lett.,* 32, 195–198 (1973).

Haug, A., *Composition and Properties of Alginates,* Rept. No. 30, Norwegian Institute Seaweed Research, Trondheim, Norway, 1964.

Haug, A., and Larsen, B., "Quantitative Determination of the Uronic Acid Composition of Alginates," *Acta. Chem. Scand.,* 16, 1908–1918 (1962).

Haug, A., Larsen, B., and Smidsrod, O., "A Study of the Constitution of Alginic Acid by Partial Acid Hydrolysis," *Acta. Chem. Scand.,* 20, 183–190 (1966).

Haug, A., Larsen, B., and Smidsrod, O., "Studies on the Sequence of Uronic Acid Residues in Alginic Acid," *Acta. Chem. Scand.,* 21, 691–704 (1967).

Haug, A., Myklestad, S., Larsen, B., and Smidsrod, O., "Correlation between Chemical Structure and Physical Properties of Alginates," *Acta. Chem. Scand.,* 21, 768–778 (1967).

Hermans, P. H., "Gels," in H. R. Kruyt (ed.), *Colloid Science,* vol. 2, Elsevier Publishing Company, Amsterdam, pp. 483–651 (1949).

Hirst, E. L., Jones, J. K. N., and Jones, W. O., "Structure of Alginic Acid. Part I," *J. Chem. Soc.,* 1880–1885 (1939).

Hirst, E. L., Percival, E., and Wold, J. K., "The Structure of Alginic Acid. Part 4. Partial Hydrolysis of the Reduced Polysaccharide," *J. Chem. Soc.,* 1493–1499 (1964).

Krefting, A., *An Improved Method of Treating Seaweed to Obtain Valuable Products Therefrom,* British Patent 11,538 (1896).

McDowell, R. H., "The Detection of Small Quantities of Alginates," in *Proceedings of the Second International Seaweed Symposium,* Pergamon Press, London, pp. 131–137 (1956).

McNeely, W. H., and Kovacs, P. "The Physiological Effects of Alginates and Xanthan Gum," in A. Jeanes and J. Hodges (eds.), *Physiological Effects of Food Carbohydrates,* ACS Symp. Ser. 15, American Chemical Society, Washington, D.C. pp. 269–291 (1975).

Morris, E. R., Rees, D. A., and Thom, D., "Characterization of Polysaccharide Structure and Interactions by Circular Dichroism: Order-Disorder Transition in the Calcium Alginate System," *J. Chem. Soc., Chem. Commun.,* 245–246 (1973).

Mozes, G., and Vamos, E., "Generalized Newtonian Fluids and Structural Viscosity: Thixotropic Phenomena," *Intern. Chem. Eng.,* 6(1), 150–159 (1966).

National Research Council, *Food Chemicals Codex,* 2nd ed., with supplements, National Academy of Sciences, Washington, D.C., 1972.

Nelson, W. L., and Cretcher, L. H., "The Isolation and Identification of D-Mannuronic Acid Lactone from the *Macrocystis pyrifera,*" *J. Am. Chem. Soc.,* 52, 2130–2132 (1930).

Newberger, S. H., Jones, J. H., and Clark, G. R., "A Technique for Obtaining Infrared Spectra of Water-Soluble Gums," *Toilet Goods Assoc., Proc. Sci. Sect.* (18), 38–39 (1952).

Newberger, S. H., Jones, J. H., and Clark, G. R., "A technique for obtaining spectra of water-soluble gums. Part II. Identification of Gums," *Toilet Goods Assoc., Proc. Sci. Sect.,* (19), 25–29 (1953).

North, W. J., "Giant Kelp: Sequoias of the Sea," *Nat. Geogr.* **142**, 251–269 (1972).

Papenfuss, G. F., "Studies of South African Phaeophyceae. I. *Ecklonia maxima, Laminaria pallida, Macrocystis pyrifera,"Am. J. Bot.,* **29**, 15–24 (1942).

Parkes, S., *Chemical Catechism,* 6th ed., London, 1814, p. 151.

Penman, A., and Sanderson, G. R., "A Method for the Determination of Uronic Acid Sequence in Alginates," *Carbohydr. Res.,* **25**, 273–282 (1972).

Pourade, R. F., *Time of the Bells,* Union Tribune Publishing Company, San Diego, 1961, p. 137.

Rees, D. A., "Structure Conformation and Mechanism in the Formation of Polysaccharide Gels and Networks," in M. L. Wolfrom and R. S. Tipson (eds.), *Advances in Carbohydrate Chemistry and Biochemistry,* vol. 24, Academic Press, Inc., New York, 1969, pp. 267–332.

Scott Blair, G. W., *Elementary Rheology,* Academic Press, Inc., New York, 1969.

Sharratt, M., and Dearn, P., "An Autoradiographic Study of Propylene Glycol Alginate in the Mouse," *Food Cosmet. Toxicol.,* 10, 35–40 (1972).

Steiner, A. B., and McNeely, W. H., "Organic Derivatives of Alginic Acid," *Ind. Eng. Chem.,* **43**, 2073–2077 (1951).

Steiner, A. B., and McNeely, W. H., "Algin in Review," in *Natural Plant Hydrocolloids, Advan. Chem. Ser. No. 11,* American Chemical Society, Washington, D.C., 1954, pp. 68–82.

Van Wazer, J. R., Lyons, J. W., Kim, K. Y., and Colwell, R. E., *Viscosity and Flow Measurement,* Interscience Publishers, Inc., New York, 1963.

Vincent, D. L., "Oligosaccharides from Alginic Acid," *Chem. Ind.,* 1109–1111 (1960).

Wise, R. G., *Algin Sponge and Process Therefor,* U.S. Patent 3,653,383 (1972).

Chapter **3**

Alkyl and Hydroxyalkylalkylcellulose

G. K. Greminger, Jr., and K. L. Krumel
Dow Chemical U.S.A., Midland, Michigan

Cellulosic Ethers 3-2
General Information 3-3
   Chemistry 3-4
   Manufacture 3-5
   Toxicity and Handling 3-6
   Solution Properties 3-6
   Powder and Film Properties 3-11
   Physical and Chemical Properties 3-13
Commercial Uses: Compounding and Formulating 3-14
   Adhesives 3-14
   Agricultural Chemicals 3-14
   Chemical Specialties 3-14
   Construction Industry Products 3-14
   Cosmetics 3-14
   Food Products 3-14
   Latex Paint 3-15
   Paint Removers 3-15
   Paper Products 3-15
   Pharmaceuticals 3-15
   Printing Inks 3-15
   Resins 3-15
   Elastomers 3-15
   Textiles 3-15
   Tobacco Sheet 3-15
Commercial Uses: Processing Aids 3-15
   Ceramics 3-15
   Leather 3-16
   Polyvinyl Chloride 3-16
Industries Using Alkyl and Hydroxyalkylalkylcellulose 3-16
Formulations 3-16
. . . . 3-16

Paint Remover 3-17
Construction Industry Products 3-21
Food Products 3-21
Pharmaceutical Products 3-22
Tobacco 3-22
Leather 3-23
Agricultural Uses 3-24
Laboratory Techniques 3-24
Product/Tradename Glossary 3-25

Table of Abbreviations

DS   Degree of substitution
EHEC   Ethylhydroxyethylcellulose
HBMC   Hydroxybutylmethylcellulose
HEMC   Hydroxyethylmethylcellulose
HPMC   Hydroxypropylmethylcellulose
MC   Methylcellulose
MS   Molar substitution

CELLULOSIC ETHERS

Cellulose, a large-volume, renewable agricultural raw material, is transformed into hundreds of products affecting every phase of daily life. Its use and versatility are exploited by the chemical industry much as the meat industry exploits its raw materials: using "everything but the squeal."

The production of water-soluble cellulose derivatives, in contrast to that of polymers based on petrochemical resources, starts with a preformed polymer backbone of either wood or cotton cellulose instead of a monomer. Cellulose is a linear polymer of anhydroglucose with the β-O-glucopyranosyl structure shown below:

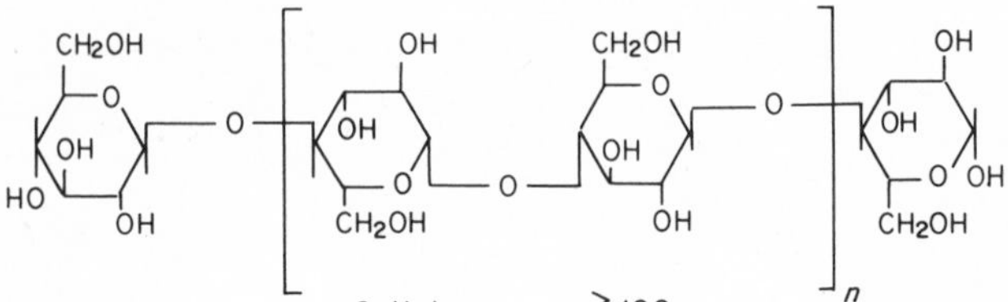

Cellulose $n = \gtrsim 100$

The properties of a specific cellulose ether depend on the type, distribution, and uniformity of the substituent groups. For each β-O-glucopyranosyl ring, there are three hydroxyl groups available for the nucleophilic substitution reaction. Reactions at these sites can occur either on a one-to-one basis or with formation of side chains depending on choice of reagent employed to modify the cellulose. In the former case, the term *degree of substitution* (DS) is used to identify the average number of sites reacted per ring. The maximum value is 3, corresponding to the number of hydroxyls available for reaction. When side-chain formation is possible, the term *molar substitution* (MS) is used and the value can exceed 3.

The water-soluble cellulose ethers possess a range of multifunctional properties resulting in a broad spectrum of end uses. For the purposes of this Handbook, this group of materials includes the alkyl and hydroxyalkylalkylcelluloses (specifically methylcellulose, hydroxypropylmethylcellulose, hydroxybutylmethylcellulose, hydroxyethylmethylcellulose, and ethylhydroxyethylcellulose), discussed in this chapter; hydroxyethylcellulose, discussed in Chap. 12; hydroxypropylcellulose, discussed in Chap. 13; and carboxymethylcellulose, discussed in Chap. 4.

GENERAL INFORMATION*

This family of commercial water-soluble cellulose ethers comprises methylcellulose (MC) and the methylcellulose derivatives hydroxypropylmethylcellulose (HPMC), hydroxyethylmethylcellulose (HEMC), and hydroxybutylmethylcellulose (HBMC). Water-soluble ethylhydroxyethylcellulose (EHEC) also possesses many properties and end uses in common with the methylcellulose products and is included in this section.

Methylcellulose and hydroxypropylmethylcellulose are two examples of this versatile class of water-soluble hydrocolloids derived from the etherification of cellulose. MC and HPMC are polymers having the useful properties of thickening, thermal gelation, surfactancy, film formation, and adhesion. Those characteristics earn them application in areas such as foods, cosmetics, paints, construction, pharmaceuticals, tobacco products, agriculture, adhesives, textiles, and paper. Additionally, to tailor a product for a specific end use, the properties of MC and HPMC may be modified by changing the molecular weight or the relative amounts of etherifying reagents.

Commercial MC products have an average degree of substitution (DS) ranging from 1.5 to 2.0; hence one-half to two-thirds of the available hydroxyl units are substituted with methyl groups (Table 3.1). In commercial HPMC products, the DS for methyl groups ranges from 0.9 to 1.8 and the molar substitution (MS) of hydroxypropyl groups ranges from 0.1 to 1.0.

MC and HPMC possess the rather unusual property of solubility in cold water and insolubility in hot water, so that when a solution is heated, a three-dimensional gel structure is formed. By modifying production techniques and by altering the ratios of methyl and hydroxypropyl substitutions, it is possible to produce products whose thermal gelation temperature ranges from 50 to 90°C (122 to 194°F) and whose gel texture ranges from firm to rather mushy.

Altering the amounts of methyl and hydroxypropyl substitution also affects the solubility properties of the cellulose ether. Decreasing the substituent groups below a DS of 1.4 gives products whose solubility in water decreases. Concentrations of 2 to 8% sodium hydroxide are required for solubility as the level of substitution decreases. Increasing the substitution above an MS of 2.0 improves solubility in polar organic solvents.

Kalle & Co. A.-G. of West Germany also produces hydroxyethylmethylcellulose. The small amount of hydroxyethyl substitution increases the solubility of the polymer and raises the thermal gel point from about 55°C to about 70°C. The more polar nature of the hydroxyethyl group versus the hydroxypropyl group allows for the formation of a slightly stiffer gel than is possible with an HPMC material of comparable gelation temperature.

A third product, ethylhydroxyethylcellulose, is similar in many properties to MC and HPMC. The small amount of hydroxyethyl substitution raises the thermal gel point from about 55°C (131°F) to about 70°C (158°F). The more polar nature of the hydroxyethyl group allows for the formation of a slightly stiffer gel than is possible with an HPMC material of comparable gelation temperature.

In many respects the properties and uses of EHEC are also very similar to those of methylcellulose and hydroxypropylmethylcellulose. The product has the characteristic properties of thickening, surfactancy, film forming, binding, solubility in cold water, and insolubility in hot water, plus a broad range of solubility in many organic solvents.

The properties of EHEC are quite dependent upon the relative amounts of ethyl and hydroxyethyl substitution. By varying the ratio of substituents, the gelation temperature, the gel characteristics, the solubility properties in different solvents, and the surfactancy can be modified. Increasing the amount of ethyl substitution increases the solubility in organic media and the tendency to form a firm gel, while increasing the hydroxyethyl substitution improves the water solubility, reduces the tendency to form a gel on heating, and improves the brine tolerance of the polymer in various salt solutions.

Methylcellulose was first produced commercially in the United States in 1938 by

---

* While this chapter concentrates on the methylcellulose and hydroxypropylmethylcellulose products, much of the data and end-use information is pertinent to all in this class of cellulose ethers. Specific comments on hydroxyethylmethylcellulose, hydroxybutylmethylcellulose, and ethylhydroxyethylcellulose will be found at appropriate points in the text.

The Dow Chemical Co. under the registered trademark of Methocel.   Hydroxypropyl-methylcellulose achieved commercial significance in the early fifties.   In addition to The Dow Chemical Co., other suppliers of these products are Shin-Etsu Chemical Products Ltd. (Metolose) of Japan; British Celanese Ltd. (Celacol) of Great Britain; and Kalle & Co. A.-G. (Tylose), Henkel and Cie GmbH (Culminal), and Wolff A.-G. of Germany.
 The worldwide capacity of MC and HPMC in 1979 is estimated to be about 159 million pounds per year and is growing.   Water-soluble ethylhydroxyethylcellulose is produced by Berol Kemi AB (formerly Modokemi AB) of Sweden.

TABLE 3.1
Commercial Methylcellulose and Modified Methylcellulose Products

| | Substitution | | | | | |
	Percent methoxyl	Percent hydroxy-propoxyl	Gel point, °C	Viscosity range, cP* 2% soln, 20°C	Tradename†	Manufacturer
MC	26–33		50–55	15–4,000	Methocel A	The Dow Chemical Co.
				5–8,000	Metolose SM	Shin Etsu Chemical Products Ltd.
				20–100,000	Celacol M&MM	British Cela-nese Ltd.
				25–4,000	Cuminal	F. Henkel and Cie GmbH
			~65	30–15,000	Tylose‡	Kalle & Co. A.-G.
HPMC	28–30	7–12	60	15–4,000	Methocel E	The Dow Chemical Co.
					Metolose 60SH	Shin Etsu Chemical Products Ltd.
	27–30	4–7.5	65	50–4,000	Methocel F	The Dow Chemical Co.
					Metolose 65SH	Shin Etsu Chemical Products Ltd.
	23	10	65	10–400	Celacol HPM	British Cela-nese Ltd.
	19–24	7–12	75	35–15,000	Methocel K	The Dow Chemical Co.
					Metolose 90SH	Shin Etsu Chemical Products Ltd.
	16.5–20	23–32	~70	5,000–75,000	Methocel J	The Dow Chemical Co.
HBMC	29.5–33.5	2.0–5.0§	45–50	8,000–20,000	Methocel HB	The Dow Chemical Co.

 * To convert to Pa·s, multiply by 0.001.
 † The tradenames are all registered trademarks of the respective manufacturers.
 ‡ The Tylose products are basically methylcelluloses that contain a small amount of hydroxyethyl substitution.
 § As hydroxybutoxyl.

Chemistry

MC is prepared by reacting purified wood pulp or cotton linters having a high α-cellulose content with aqueous sodium hydroxide and then with methyl chloride according to the following simplified reaction scheme where R_{cell} is the cellulose radical:

Main Reactions:

$$R_{cell}OH + NaOH \rightarrow R_{cell}OH \cdot NaOH \text{ (complex)}$$
$$R_{cell}OH \cdot NaOH \rightleftarrows R_{cell}ONa + H_2O$$
$$R_{cell}ONa + CH_3Cl \rightarrow R_{cell}OCH_3 + NaCl$$

Side Reactions:

$$CH_3Cl + NaOH \rightarrow CH_3OH + NaCl$$
$$CH_3Cl + H_2O \rightarrow CH_3OH + HCl$$
$$CH_3OH + NaOH \leftrightharpoons CH_3ONa + H_2O$$
$$CH_3ONa + CH_3Cl \rightarrow CH_3OCH_3 + NaCl$$

For the production of HPMC, propylene oxide is also added to the mixture and reacts as follows:

Main Reaction:

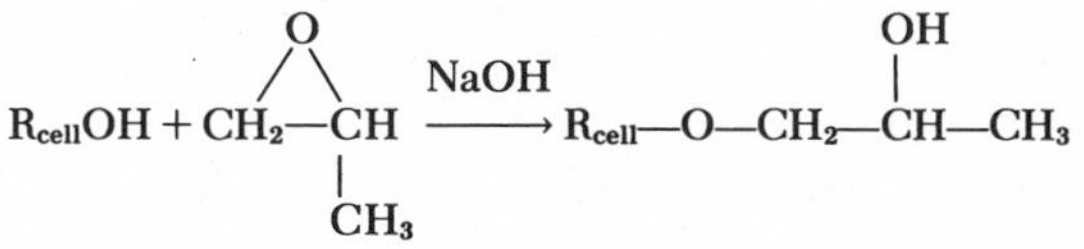

$$R_{cell}OH + CH_2\overset{O}{\overbrace{}}CH \xrightarrow{NaOH} R_{cell}\!-\!O\!-\!CH_2\!-\!\overset{OH}{\overset{|}{CH}}\!-\!CH_3$$
$$\underset{CH_3}{}$$

Side Reaction:

$$CH_2\overset{O}{\overbrace{}}CH\!-\!CH_3 \xrightarrow[CH_3Cl]{NaOH} \text{glycols} + \text{glycol ethers}$$

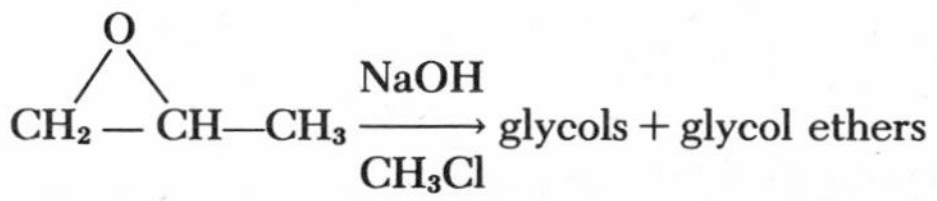

The relative amounts of methyl and hydroxypropyl substitution are controlled by the weight ratio and concentration of sodium hydroxide and the weight ratios of methyl chloride and propylene oxide per unit weight of cellulose.

EHEC is prepared by reacting dissolving-grade wood pulp with aqueous sodium hydroxide and then with ethyl chloride and ethylene oxide as schematically illustrated below:

$$R_{cell}OH + NaOH \rightarrow R_{cell}ONa + H_2O$$
$$R_{cell}ONa + CH_2\overset{}{\underset{O}{\overbrace{}}}CH_2 \rightarrow R_{cell}\!\!-\!\!(O\!-\!CH_2CH_2)_n OH$$

$$R_{cell}OCH_2CH_2OH + CH_3CH_2Cl \rightarrow R_{cell}\!\!\overset{\diagup OCH_2CH_2OH}{\underset{\diagdown OCH_2CH_3}{}}$$

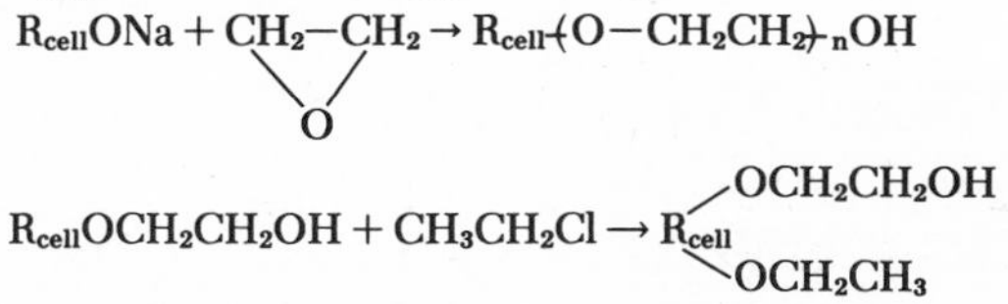

The amount of ethylation is controlled by the amount of caustic used in the formation of alkali cellulose and the amount of hydroxyethylation is controlled mainly by the amount of ethylene oxide added to the reactor. The addition is stepwise since ethylene oxide is far more reactive than ethylene chloride and hence reacts with the cellulose first.

Manufacture

There are three main steps used in the manufacture of MC and HPMC.

Preparation of Alkali Cellulose   Alkali cellulose is prepared by contacting cellulose and 35 to 60% aqueous caustic according to several procedures that include dipping a cellulose sheet in a caustic solution, spraying the caustic onto agitated cellulose flock, slurrying the cellulose in aqueous caustic and removing the excess, or mixing the cellulose and aqueous caustic in an inert diluent.

Viscosity control of the final product is obtained by choice of pulp, by aging the alkali cellulose in warm air, and by controlling the amount of oxygen left in the reactor during methylation. For high-viscosity products the higher-molecular-weight cotton linters are used with minimum aging. Since alkali cellulose is susceptible to oxidative degradation, exposing it to air for varying time periods is an effective method for viscosity reduction.

Reaction   The alkali cellulose, methyl chloride, and (if required) propylene oxide are loaded into a jacketed, nickel-clad, agitated vessel and heated under controlled conditions to a maximum pressure of 1.38 MPa (200 psig.) The heat of reaction is removed by condensation of the solvents. In addition to controlling the substitution levels, variations in the amounts of methyl chloride and changes in the reaction profile will affect the properties of the final product.

Purification Since MC and HPMC are insoluble in hot water, the reaction by-products are removed by slurrying the crude product in water heated to above 90°C (194°F) and then filtering. The purified wet product is then dried, ground to >95% through 40-mesh screen, and commonly packaged in 22.68-kg (50-lb) bags.

Toxicity and Handling

Commercial MC and HPMC products have been used by the food, pharmaceutical, and cosmetic industries for many years.* They are odorless, tasteless powders and are considered to be physiologically inert.

MC products are listed in the *United States Pharmacopeia XIX* and *Food Chemicals Codex* and are listed by the FDA as Generally Recognized as Safe (GRAS).

HPMC compounds whose methoxyl substitution ranges from 19 to 30% and hydroxy-propyl substitution ranges from 4 to 12% are also listed in the *United States Pharmacopeia XIX* and *Food Chemicals Codex*. Both products can meet the requirements of Food Additive Regulations 182.1480 and 172.874 as a miscellaneous and/or general purpose food additive for nonstandardized foods.

While a gross exposure to MC or HPMC can conceivably cause temporary mechanical irritation to skin and eyes, exposure to normal amounts presents no significant health hazards from either contact or inhalation.

In storage, good housekeeping is suggested to prevent dusts from building up to possibly explosive levels. All the cellulose ethers are organic materials that will burn under the right conditions of heat and oxygen supply. Fires can be extinguished by conventional means. Gross powder spills should be swept up to avoid accidents caused by slippery floors or equipment, and the trace residual product can be flushed to a sewer. These products showed no biochemical oxygen demand (BOD) with the standard 5-day test. However, radioassay tests with activated sludge showed breakdown over a 15- to 20-day period. These products should provide no ecological hazard. The products may be disposed of by either landfill or incineration.

Solution Properties

Thickening
MC and HPMC can be dissolved in cold water to yield smooth, clear solutions (Table 3.2). Commercial products are available in a wide range of viscosities varying from

---

* Most manufacturers provide premium or other special grades suggested for use in foods, pharmaceuticals, cosmetics, etc.

TABLE 3.2
Properties of Methylcellulose and Hydroxypropylmethylcellulose Aqueous Solutions

	MC	HPMC
Surface tension 25°C: mN/m (dyn/cm)	47–53	44–56
Interfacial tension 25°C: mN/m (dyn/cm) (vs. paraffin oil)	19–23	18–26
pH	Neutral	
Freezing point	0.0°C (32°F)	
Solution stability	Stable from pH = 3 to pH = 11	
Refractive index	$n_D^{20} = 1.336$	
Specific gravity 20°/4°	1% = 1.0012 5% = 1.0117 10% = 1.0245	

0.005 to 75 Pa·s (5 to 75,000 cP) as measured by ASTM Method D1347-72 and D2363-72 on aqueous 2% wt/wt concentrations at 20°C (68°F). Under these conditions, the MC and HPMC viscosity/concentration relationship is approximated by Philippoff's equation [*Cellul. Chem.*, **17**(57) (1936)] shown below in Eq. (1), which simplifies to Eq. (2) when the measurements are made at 20°C in water where $\eta_0 = 1.0$.

$$\frac{\eta}{\eta_0} = \left(1 + \frac{[\eta]\, c}{8}\right)^8 \tag{1}$$

where η = apparent viscosity
η_0 = solvent viscosity
$[\eta]$ = intrinsic viscosity, dL/g
c = concentration, g/dL

$$\eta^{1/8} = 1 + Kc \tag{2}$$

where

$$K = \frac{[\eta]}{8}$$

When the viscosity is known at one concentration, the viscosity can be calculated for any other concentration by using Eq. (2) to first calculate K for the sample at the concentration for which the viscosity is known, and then using Eq. (2) again to calculate the viscosity at the new concentration knowing the value of K.

EHEC may be dissolved in cold water to yield clear, smooth solutions. Commercial products range in viscosity from 0.050 to 12.000 Pa·s (50 to 12,000 cP) at 2% concentration. The solutions are pseudoplastic in that the apparent viscosity decreases with increasing rate of shear; the solutions are not thixotropic unless they are gelled. (See Fig. 3.1.)

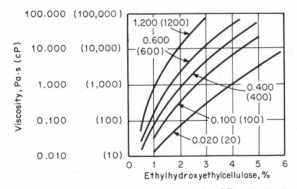

Figure 3.1 The relationship of viscosity vs. concentration for different viscosity grades of ethylhydroxyethylcellulose from Berol Kemi AB. (Similar and additional data can be found in *The Modocoll Manual*, Mo och Domsjö of Sweden, 1960.)

Rheology Solutions of MC and HPMC generally show pseudoplastic nonthixotropic flow properties at 20°C (68°F) that are not a function of substitution within the range of available commercial products, and whose deviation from Newtonian character increases with increasing molecular weight (Figs. 3.2 and 3.3). Dilute solutions of low-viscosity products (Fig. 3.4) do closely approach Newtonian flow, but increasing the concentration of the gum to over 5% may give a solution showing some thixotropy due to weak chain-to-chain interactions. Since flow properties are dependent on the molecular weight and the molecular weight distribution of the polymer, a blend of high- and low-molecular-weight polymers can have different flow properties than a polymer having the same solution viscosity as the blend but having a narrow molecular weight distribution (Fig. 3.5). This effect is generally not important for dilute solutions of higher-viscosity materials (Fig. 3.6) but can be significant when applied to solutions of over 5% of the low-viscosity derivatives.

Heating a solution of MC or HPMC shows the normal effect of lower viscosity until

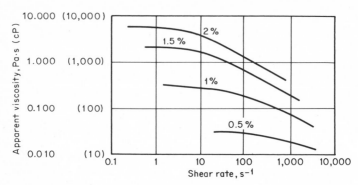

Figure 3.2 Apparent viscosity vs. shear rate relation for aqueous solutions of methylcellulose [4 Pa·s (4000 cps)] at different concentrations at 25°C. *(Data courtesy of The Dow Chemical Co.)*

the gelation temperature is reached; at that point, the viscosity of the solution increases rapidly and highly thixotropic flow is observed.

The normal effect of temperature in the range of 0 to 45°C (32 to 113°F) is roughly a 3% reduction in viscosity for every degree Celsius increase in the temperature of the solution (when applied to aqueous solutions containing no added solutes and showing no evidence of gelation).

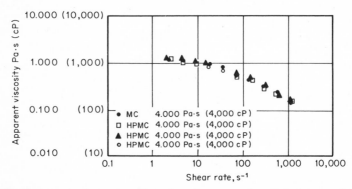

Figure 3.3 Viscosity vs. shear rate: effect of product type of methylcellulose and hydroxy-propylmethylcellulose on flow curve. *(From data on Dow brands of methylcellulose and hydroxy-propylmethylcellulose.)*

Thermal Gelation   MC and HPMC solutions show the unusual property of forming a structured gel when heated.

In solution, these polymers exist as aggregates of long colloidal molecules. These molecules are highly hydrated with the solvent water in layers that are held through hydrogen bonding, thereby giving the chains some lubricity and smooth flow. As the temperature is raised, the hydrogen bonding between the water molecules weakens and the interactions between chains become significant, eventually leading to the formation of a structured gel. Unlike many chemical gels, those made from MC and HPMC are primarily a result of phase separation and are susceptible to shear thinning (a mechanical breaking up of the gel without affecting the molecular weight). With cooling, this process is reversible and the gel reverts back to a solution whose flow properties are not changed.

The gelation temperature is dependent on the relative amounts of methyl and hydroxypropyl substitution and may be used as an indication of the relative hydrophilicity of

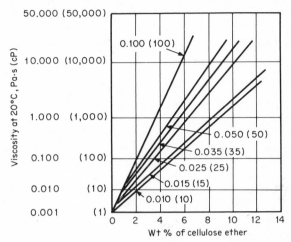

Figure 3.4 Relationship of concentration vs. viscosity for methylcellulose and hydroxypropylmethylcellulose low-viscosity products. *(Data courtesy of The Dow Chemical Co.)*

the derivative. In general, the more highly substituted derivatives have lower gelation temperatures and will be less compatible with added solutes or electrolytes. The gelation temperature of products currently produced varies from about 50 to 85°C (122 to 185°F) with the resultant gels ranging from firm to rather mushy in consistency (when determined by heating a 2% solution of the gum in pure water). The gelation temperature

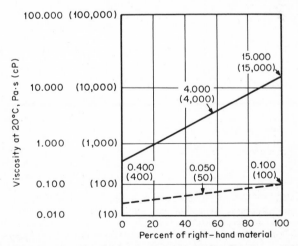

Figure 3.5 Blending chart for determining the viscosity of a mixture of two methylcellulose or hydroxypropylmethylcellulose components. To determine the viscosity of a mixture, locate the viscosity of one component on the left side and the viscosity of the second component on the right side (*y* axis). Connect the points with a straight line and read the resulting viscosity at the desired concentration. The lines shown serve as examples. *(Data courtesy of The Dow Chemical Co.)*

of a product is affected by the concentration of the gum and, more importantly, by other dissolved solutes. Presence of salts (Table 3.3) will lower the gel point; addition of ethanol or propylene glycol can raise the gel point as much as 20°C (36°F).

Upon reaching the thermal gelation temperature, EHEC will separate out of solution as either a floc or a gel depending upon molecular weight and the concentration (see Table 3.4).

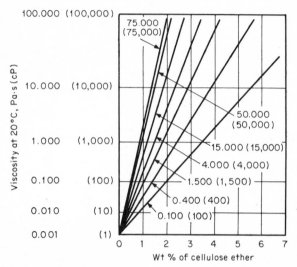

Figure 3.6 The relationship of concentration vs. viscosity for methylcellulose and hydroxy-propylmethylcellulose high-viscosity products. *(Data courtesy of The Dow Chemical Co.)*

TABLE 3.3
Tolerance of 2% Solutions of MC and HPMC to Additives (The number of grams that 100 g of solution will tolerate before the gel point is lowered to room temperature or below)*

Additive	MC		HPMC (Gel Point 65°C)		HPMC (Gel Point 75°C)	
	15 cP†	4000 cP	50 cP	4000 cP	100 cP	4000 cP
NaCl	11	7	17	11	19	12
MgCl$_2$	11	8	35	25	40	39
Na$_2$SO$_4$	6	4	6	4	6	4
Al$_2$(SO$_4$)$_3$	3.1	2.5	4.1	3.6	4.1	3.6
Na$_2$CO$_3$	4	3	5	4	4	4
Na$_3$PO$_4$	2.9	2.6	3.9	3.5	4.7	4.3
Sucrose	100	65	120	80	160	115

* Data courtesy of The Dow Chemical Co.
† To convert to Pa·s, multiply by 0.001.

Surface Activity   MC and HPMC reduce the surface tension and interfacial tension of aqueous systems to values of 41 to 55 dyn/cm and 18 to 28 dyn/cm, respectively (depending on chemical structure), thereby functioning as moderate emulsifiers for two-phase mixtures. Since they are polymeric materials, they are active surfactants at very low use levels ranging from 0.001 to 1.0%. Their status as approved additives in foods makes them useful as edible surfactants. EHEC also behaves as a moderate surfactant, lowering the surface tension of water to 47 to 52 dyn/cm.

Moderate foaming is usually encountered. This can be controlled, if desired, by use of commercially available defoamers: Polyglycol P-1200 (The Dow Chemical Co.), Anti-foam A, AF, B, or FG (Dow Corning Corp.), Nopco KFS (Nopco Chemical Co.), or tri-*n*-butylphosphates.

Solution Stability   MC and HPMC solutions are generally stable in the pH range of 3 to 11. Below pH = 3, acid-catalyzed hydrolysis of the glucose-glucose linkage becomes significant, and above pH = 11, oxidative degradation takes place. Although generally resistant to microorganisms, these solutions are subject to very slow degradation catalyzed by enzymes, especially in the case of the lower-substituted products.

Solution Makeup Attempts to dissolve MC or HPMC in cold water with mild stirring result in formation of gel-coated agglomerates of dry powder that are extremely difficult to dissolve. Therefore, solutions should be made up by one of the following techniques:

1. The powder is dispersed in water heated to above the gel point, and the solution is then cooled while stirring.
2. The powder is dispersed in cold water using a high-shear mixer. Subjecting the polymer or polymer solution to moderately high shear for periods of up to several minutes should not lead to appreciable degradation.
3. The powder is dispersed in a small amount of nonsolvent, and the slurry then added to the water.

The solubility and clarity of these solutions are often improved by chilling to below 10°C (50°F) during solution makeup.

Dispersible Powders MC and HPMC can be made temporarily water-insoluble by reaction with a small amount of glyoxal to form a hemiacetal crosslink. Such product will be insoluble in cold water whose pH is less than 7, but raising the pH to above 8.5 causes the crosslink to be immediately hydrolyzed and the product then dissolves normally. These surface-treated products are commercially available but generally are not recommended for use in food, cosmetic, or pharmaceutical applications.

TABLE 3.4
Tolerance of EHEC* to Added Salts [The number of grams of added salt that can be added to 100 g of 2% solution before the flocculation temperature is lowered to 20°C (68°F)]†

Additive	Modocoll E 600	Modocoll E 100
NaCl	8	10
MgCl₂	5	5
Na₂SO₄	2.5	4.5
Na₂CO₃	3.0	3.0
Na₃PO₄	3.0	4.5
Sucrose.	70	70

* Modocoll E.
† Additional data can be found in *The Modocoll Manual,* Mo och Domsjö of Sweden, 1960.
NOTE: Product line has recently been changed.

Compatibility MC and HPMC are neutral, non-ionic polysaccharides that are not susceptible to chemical gelation or precipitation with di- or trivalent metals, with borates, or by interaction with other polymers to form complexes or coacervates. As highly hydrated colloids, however, MC and HPMC can be gelled or salted out of solution when the concentration of added solutes or electrolytes exceeds certain limits (Table 3.3). HPMC is in general more hydrophilic than MC, and hence is more compatible with added solutes.

Though primarily used in aqueous solutions, certain of the more highly substituted MC and HPMC products are soluble in aqueous alcohol mixtures, in alcohol/chloroalkane mixtures, and in hot polar organic compounds including, for example, ethylene glycol, glycerin, glyceryl diacetate, and diethanolamine. These solubility properties have been utilized in pharmaceutical tablet coatings and in the formulation of materials to be used in thermoplastic fabrication techniques.

Powder and Film Properties

MC and HPMC are available as powders whose particle size is >95% through 40 mesh and which, when properly dissolved in cold water, will develop maximum viscosity in 30 min or less (Table 3.5).

These products yield high-strength, clear, water-soluble films (Table 3.6) that are impervious to many organic and petroleum-based solvents. They are compatible with a wide range of materials that include other gums, glues, and soaps. The mechanical properties of the films can be modified by a number of plasticizers (Table 3.7). The water resistance of these films may be improved by crosslinking the cellulose chains through the remaining hydroxyl units using polybasic acids, dialdehydes, melamine-formaldehyde or urea-formuladehyde resins, or polyphenolic compounds, thereby rendering the cellulose derivative water-insoluble (Table 3.8).

A typical example of this technique is treatment of the cellulose ether with 15% of Uformite 700 (Rohm & Haas Co.) in the presence of citric acid (pH = 4.5) at 120°C (248°F) for 20 min, giving a product that is 88 to 89% water-insoluble.

TABLE 3.5
Properties of MC and HPMC Powders

Physical appearance	White to off white
Apparent density	0.3–0.7 g/mL
Particle size	>95% through 40 mesh
Browning temperature	190–200°C (374–392°F)
Charring temperature	225–230°C (237–446°F)
Percent moisture	3% maximum

The more highly substituted HPMC derivatives are somewhat thermoplastic and may be used for extruded sheeting and injection or compression molding. The HPMC having a gel point of 60°C (140°F) (when properly blended with propylene glycol and suitable plasticizers) may be molded or extruded as a finished water-soluble product at 120 to 190°C (248 to 374°F). Properly plasticized MC or HPMC sheeting or tubing products can be sealed at about 130°C (266°F) using standard heat-seal equipment.

TABLE 3.6
Properties of Unplasticized Films of Methylcellulose

Specific gravity	1.39
Refractive index	1.49
Area factor	341 m²/kg/25.4μm (24,000 in²/lb per mil)
Equilibrium moisture	6.5%
Melting point	290–305°C (554–561°F)
Tensile strength	58.6–78.6 MPa (8,500–11,400 lb/in²)
Elongation	10–15%
MIT folds	>10,000
Mullen burst	164 kPa (23.9 psi)
Moisture vapor transmission rate 100°F, 90–100% RH	10.5 g/100 cm² per 24 h per 25.4μm (67.5 g/100 in² per 24 h per mil)
Oxygen transmission rate 75°F	3.9 mL/100 cm² per 24 h per 25.4μm (25 mL/100 in² per 24 h per mil)
Ultraviolet transmission	
400 mμ	54.6%
290 mμ	49.0%
210 mμ	25.7%

These data are courtesy of The Dow Chemical Co. and are for MC only. Films may also be made from HPMC.

TABLE 3.7
A Partial List of Plasticizers, Nonaqueous Solvents, and Nonsolvents for MC or HPMC

Plasticizer	Nonaqueous solvent	Nonsolvent
Glycerine	Furfuryl alcohol	Peanut oil
Propylene glycol	Dimethyl sulfoxide	Machine oil
D-Sorbitol	Dimethyl formamide	Vegetable oil
Triethylene glycol	Methyl salicylate	Castor oil
Triethanolamine	Pyridine	Mineral oil
N-acetyl ethanolamine	Glacial acetic acid	Benzene
	Formic acid	Toluene
		Cyclohexane

Physical and Chemical Properties

The various manufacturers of these products provide a range of products designed for specific end uses. The physical and chemical properties will vary with the type and amount of substituents reacted with the cellulose polymer backbone. This section provides a general overview of chemical and physical properties. For further details

TABLE 3.8
Effect of Additives on Water Resistance of MC and HPMC Films*

Crosslinking agent	Percent in film	Catalyst	Amount of catalyst	Temper- ature, °C	Time, min	Percent insolu- bilized
Uformite 700†	15	Citric acid, pH = 4.5		120	20	88–89
Kymene 234‡	20	HCl HCl		120	20	96
Rhonite R-2†	15	NH₄Cl	pH = 2.0 0.75%	120	20	95
Glyoxal	15	Citric acid	4.0%	120	20	95

* Data courtesy of The Dow Chemical Co.
† Rohm & Haas Co.
‡ Hercules Inc.

regarding specific products, the commercial literature available from the individual manufacturer should be consulted.

General information on the identity and substitution of MC and MC derivative products are given in Table 3.1. Water-soluble EHEC is available in two basic product types and several viscosity grades (Table 3.9). The more hydrophilic EHEC product

TABLE 3.9
Physical Properties of Commercially Available Water-Soluble Ethylhydroxyethylcellulose from Berol, Sweden*

	Modocoll E	Modocoll M
Substitution:		
Ethyl DS	0.9	1.4
Hydroxyethyl MS	0.8	0.5
Powder properties:		
Percent sodium chloride	<2	<0.5
Percent water	<8	—
Particle appearance	Granules/powder	Fibrous/granules
Particle size		
+ 35 mesh	<5%	—
− 200 mesh	<5%	—
Bulk density	0.3–0.6 g/mL	0.1–0.6 g/mL
Film Properties:		
Specific gravity	1.33	1.49
Tensile strength at	44.1–53.8 MPa	24.1–34.5 MPa
20°C and 65% R.H.	(6400–7800 psi)	(3500–5000 psi)
% Elongation at 20°C		
and 65% R.H.	25–35%	5–15%
Solution properties of a 1–3% solution:		
Freezing point	0°C (32°F)	
pH of aqueous solution	~6	
Specific gravity (2%)	1.006	
Surface tension at 20°C	40–45 mN/m (dyn/cm)	
Interfacial tension at 20°C		
vs. paraffin oil	15–20 mN/m (dyn/cm)	

* Additional data on these products can be found in R. Whistler (ed.), *"Industrial Gums,"* Academic Press, Inc. N.Y., 1973; *The Modocoll Manual,* Mo och Domsjö of Sweden, 1960, chap. 30.
NOTE: Product time has recently been changed.

having a higher gel point and better salt compatibility is called Modocoll E and has an ethyl DS of 0.9 and a hydroxyethyl MS of 0.8. The more highly substituted derivative having a similar gel point of about 40°C (104°F) but improved solubility in organic media is called Modocoll M and has an ethyl DS of 1.4 and a hydroxyethyl MS of 0.5.

COMMERCIAL USES: Compounding and Formulating

The range of multifunctional properties of methylcellulose and methylcellulose derivatives earns these products an extensive spectrum of end uses. New products are evolved constantly in response to the need for specific and unique combinations of properties in the marketplace. Cellulosic gums are a classic example of how the utility of a basic raw material can be expanded through chemical modifications.

The following end uses are evidence of the commercial utility of methylcellulose and modified methylcelluloses. These products are used as:

Adhesives

For their innate adhesive properties and to provide thickening to adhesive formulations. Also, the thermal gel point provides setting instead of thinning upon heating. This allows control of the penetration and placement of the adhesive, factors that affect cost and performance. End uses include plywood adhesives, industrial adhesives, wallpaper paste, library paste, and latex adhesives.

Agricultural Chemicals

For use as spray adherents, fungicide stickers, protective films, and dispersing agents for wettable powders. The wet-tack and surface wetting properties reduce run-off, particularly on waxy plant surfaces. Residues of these cellulose derivatives are not a problem since they are permitted as food additives under most circumstances.

Chemical Specialties

For use as thickeners, suspending agents, binders, film formers, and emulsion stabilizers in a wide range of consumer products such as polishes, cleaning compounds, protective hand creams, coatings, emulsions, and water-soluble film end uses.

Construction Industry Products

For use as workability and water-retention aids in the construction industry. In addition to use in drywall finishing compounds, other applications include refractories; "dryset" ceramic adhesive and grouting systems, cement-based paints, stuccos, and patching and repair formulations; mortars for cellular concrete blocks, and formulations for sound insulating systems.

Increased water retention in portland cement compositions results in increased bond strength, longer board life, and economies in application costs. Reduced cracking and better bonding have also been observed in latex plaster formulas. Additive levels range from 0.05% in masonry mortars to 1.0% in tile grouts (based on cement = B.O.C.).

Cosmetics

For use in providing viscosity control, emulsification and stabilization, lubricity and feel, clarity, foam stability, and surfactant compatibility in lotions, hand and face creams, hair dressings, deodorants, depilatory creams, shampoos, and toothpastes. Methylcellulose and its modifications can function as polymeric surfactants and show activity at very low concentrations. These products are nonallergenic and should present no problems upon accidental ingestion (see food additive section, below).

Food Products

For use as thickeners, binders, emulsifiers, stabilizers, and colloidal suspending agents in a variety of products including salad dressings, fruit pie fillings, baked goods, dietetic foods, breading batters, fried foods, milkshake drinks, and convenience snack foods. The products that meet Federal Drug Administration (FDA) requirements can function as polymeric surfactants. The thermal gel point is particularly useful in the processing of fried foods where binding action and film forming to reduce oil absorption are required. Gels of methylcellulose and its derivatives do not synerese upon freezing. This property is necessary in frozen food items.

Latex Paint

For use as a protective colloid, thickener, and pigment-suspension aid in latex paints. These products offer viscosity stability, wet-edge retention, and ease and flexibility of incorporation, and also contribute to paint film integrity. The high level of substitution in the methylcellulose products results in excellent resistance to enzyme contamination in latex paints, assuring excellent product shelf life.

Paint Removers

For a unique combination of organic and water solubility, allowing the use of hydroxypropylmethylcellulose as a thickener for flush-off and scrape-off paint removers. Commercial use includes military, industrial, and consumer product–type formulations.

Paper Products

For excellent barrier to oily materials of mineral, vegetable, or animal origin. Also used in packaging products, surface sizings, and release coatings. Food additive clearance and adhesive properties are also factors in end uses for the paper industry.

Pharmaceuticals

For use as binders, granulating agents, and film coatings for tablets; in bulk laxatives; as film formers for burn therapy; as stabilizers for ointments and creams; as thickeners for lotions and jellies; as suspending agents in liquid products; and as troche bases. Several diagnostic tests and separation procedures also utilize methylcellulose and hydroxypropylmethylcellulose. The lubricity property of these products also finds application in eye medication for conjunctivitis and in jellies for cystoscopes, etc.

Printing Inks

For use as thickeners and suspending agents for water-base inks.

Resins

For use in mold-release agents for fiber-reinforced plastics and as thickeners and stabilizers for water-based coating systems. The protective colloid properties are used to stabilize formaldehyde solutions and in manufacture of resins.

Elastomers

For thickening and stabilization of latex formulations. Also used for viscosity control in the dipping process for production of rubber goods.

Textiles

For use as binders in textile printing pastes, in latex coatings, and as sizing agents. Also used as thickeners for carpet backsizing formulations providing faster drying speeds, wider effective coating viscosity range, more uniform coating, and better holdout. The thermal gel point provides quick grab, and the surfactant properties are used in the production of latex foams for garment fabrication and carpet backing.

Tobacco Sheet

For use in binders and film formers for reconstituted tobacco sheet, and as adhesives in cigar manufacture.

COMMERCIAL USES: Processing Aids

Methylcellulose and its derivatives are used as processing aids in a number of industries. Uses of EHEC as a processing aid are similar to those of methylcellulose and its modifications. Also, EHEC products are used in foundry practice as an ingredient of cores to improve green strength and to achieve improved collapsibility.

Ceramics

They provide green strength, water retention, and lubricity in refractory mortars and cements and in glaze slips. These cellulose ethers, being non-ionic in structure, leave minimal ash when fired. These properties also are important in the fabrication of porcelain items for the electrical industry. The thermal gel point provides quick set and minimizes sagging during glazing operations and during placement of refractory mortars in repair of hot furnace walls, etc.

Leather

Their thermal gelation property is used in pasting adhesives for the drying operation where the hides are pasted to large plates which pass through a tunnel drier. The quick set minimizes drop-off of hides during drying. Thickening and stabilization of leather finishes are other end uses.

Polyvinyl Chloride

They are used as protective colloids in suspension polymerization to provide improved resin porosity, higher plasticizer absorption rates, and control of resin particle size distribution. Easier cleanup of reactors also reduces exposure to monomer. The quantity used generally varies from 500 to 1000 ppm, based on monomer.

INDUSTRIES USING ALKYL AND HYDROXYALKYLCELLULOSE

Methylcellulose and its modifications are used in hundreds of commercial applications. The following tables provide general guidelines for selecting the optimum product for a specific formulation. The multiplicity of process and raw material variables will always necessitate end-use testing and adjustment to ensure the desired performance in the end product. The methylcellulose products provide a multifunctional combination of properties that can result in savings in raw material costs and in improved technical performance in the final product.

The end uses of HEMC and EHEC are, in general, similar in scope to those of HPMC, and the viscosities shown for HPMC can serve as guides to the selection of an HEMC or EHEC product.

In Table 3.10, products that are surface-treated to allow dispersion directly in cold water without lumping are designated with a suffix "S." The viscosity shown for each end use is the value for a 2% aqueous solution at 20°C (68°F) as determined by ASTM D1347-72 or D2363-72.

FORMULATIONS

The following formulations and recipes provide examples of how the methylcellulose products are used commercially. Modifications should be made to fit specific raw material and end-use requirements.

Latex Paint

Interior Flat Polyvinyl Acetate

	lb/100 gal	kg/100 L
a. *Pigment Grind*		
Place in mixing container:		
water	300	36
Then add:		
Methocel J12MS (Dow brand HPMC)	5	0.600
Pigment dispersant (25%)	8	0.960
Dow Polyglycol P1200 antifoam	3	0.360
Mix for 5 to 10 min; check		
to see that the HPMC is fully		
solubilized before continuing.		
Then add:		
Titanium dioxide, rutile	200	24
Calcium carbonate	250	30
Mix for 15 to 20 min; check		
for proper dispersion of pigments.		

	lb/100 gal	kg/100 L
b. *Paint Let-down*		
To this pigment grind, add directly:		
Ethylene glycol	20	2.4
Dalpad* A coalescing agent	7.5	0.900
Polyvinyl acetate copolymer latex		
(55% solids)	270	0.0325
Defoamer	6	0.720
Dowicil* 100 preservative	1	0.120
Mix at blending speed for		
approximately 10 min.		
Pigment volume concentration.	50%	
Nonvolatile content	52%	
pH	adjust to 6–8	
Viscosity	80–90 KU	

* Registered trademark of The Dow Chemical Co.

Exterior High-Solids Acrylic

	lb/100 gal	wgt/100 L
a. *Pigment Grind*		
Place in mixing container: water	250 lb	30 kg
Then add:		
Pigment dispersant (25%)	9.2 lb	1.1 kg
Dow Polyglycol P1200 antifoamer	2 lb	240 g
Titanium dioxide, rutile	185 lb	22.2 kg
Titanium dioxide, anatase	45 lb	5.4 kg
Mica	30 lb	3.6 kg
Calcium carbonate	100 lb	12 kg
Clay	55 lb	6.6 kg
Dowicil S-13 preservative	6 lb	720 g

Grind in high-speed mill for 15 to 20 min, or pass pigment slurry through high-shear mill to achieve proper dispersion. Slurry together the following, then add to pigment grind:

	lb/100 gal	wgt/100 L
Ethylene glycol	17.5 lb	2.1 kg
Methocel K15MS (Dow brand HPMC) powder	3.5 lb	420 g
Mix for 5 to 10 min.		
b. *Paint Let-down*		
(Blending should be done at a slower speed on the high-speed equipment.)		
Acrylic copolymer latex (47%)	450 lb	54.2 kg
Defoamer	5 lb	600 g
Mix for approximately 10 min.		
Pigment volume concentration	40%	
Nonvolatile content	54.9%	
pH	adjust to 8.5–9.0	
Viscosity	85–90 KU	

Several methods of addition can be used depending upon plant layout and process.

Direct Addition to Grind HPMC or HEMC in dispersible powder form is added directly to the pigment grind on a high-speed mixer of the Cowles, Hockmeyer, or similar type. This method should not be used on mills where high viscosity in the pigment grind will interfere with proper processing.

Glycol Slurry The HPMC product in powder form can be added as a glycol slurry to the paint let-down. Any type of either low- or high-speed blending equipment may be used for this addition. Maximum ratio of powder to glycol for ease of handling is 1:5.

Paint Remover

For home use on wood surfaces, a scrape-off paint remover formulation is preferred because water flushing is impractical and may cause grain raising. For industrial use, which is usually removal of paint from metal, a flush-off formulation is preferred because

TABLE 3.10
Applications for Various Cellulose Derivatives*

	MC	HPMC	HPMC	HPMC	HPMC
Percent methoxyl:	26–33	28–30	27–30	16.5–20	19–24
Percent hydroxypropoxyl:	—	7–12	4–7.5	23–32	7–12
Adhesives:					
Leather pasting	4,000	—	4,000	—	—
Temporary binding agent for glass fibers	15	—	—	—	—
Thickener in phenolformaldehyde adhesives	4,000	—	—	—	—
Stationery adhesives	25	—	—	—	—
	400				
	1,500				
Wallpaper	1,500	—	4,000	12,000S	15,000
	4,000			75,000S	—
General adhesives	4,000		4,000	75,000S	—
Agriculture:					
Dispersing agent for wetting powders	15	50	—	—	—
Dust stickers	—	—	4,000	—	—
Seed stickers	15	—	—	—	—
Spray drift control	25	—	—	75,000S	—
Spray stickers	—	—	4,000	12,000S	—
Weed killers	1,500	—	—	—	—
Asphalt:					
Asphalt emulsion	—	—	—	12,000S	4,000
				75,000S	
Release coating	15	—	—	—	35
Caulking compounds	—	—	4,000	5,000S	4,000
				12,000S	15,000
Ceramics:					
Refractory mortars	4,000	—	4,000	—	—
Glaze slips	15–25	—	—	—	—
Hi-temp glaze slips	4,000	—	4,000	5,000S	4,000
Porcelain enamels	15–25	—	—	—	—
Cements	4,000	—	4,000	—	—
Tile mortars	4,000	—	4,000	5,000S	4,000
Plastic mixes	4,000	—	4,000	—	—
Chemical specialties:					
Aerosols	25	50	50	—	—
Cleaning and polishing compounds	1,500	—	—	5,000S	4,000
Insecticides	15	—	—	—	—
Sanitizers	4,000	—	4,000	—	—
Construction products:					
Drywall joint cements	—	—	4,000	5,000S-	4,000-
				75,000S	15,000
Masonry mortars	—	—	4,000	5,000S-	4,000-
				12,000S	15,000
Pumpability aids	—	—	—	75,000S	15,000
Release coatings	15	—	—	—	—
Stuccos	—	—	4,000	5,000S-	4,000-
				12,000S	15,000
Tile grouts and adhesives	—	—	4,000	5,000S	4,000
				12,000S	15,000
Cosmetics:					
Creams and lotions	—	4,000	—	†	4,000
Deodorants	—			†	15,000
Hair dressings	—	4,000	—	†	—

* These numbers refer to proximate viscosity in cP of 2% solutions (20°C) of the product types shown. To convert to Pa·s, multiply by 0.001.
† Viscosities dependent on application need. Consult supplier.

Shampoos	—	4,000	—	—	—
Toothpastes	—	—	4,000	‡	4,000
Foods:					
Baked goods	4,000	—	4,000	—	4,000
Breading batters	25	—	50	—	100
Dietetic foods	‡	‡	—	—	‡
Milkshake drinks	15	50	50	—	—
Pie fillings	150	—	4,000	—	4,000
Salad dressings	—	—	—	—	4,000
Snack foods	‡	‡	‡	—	‡
Whipped toppings	—	50	50	—	100
Latexes:					
Creaming of natural rubbers	4,000	—	4,000	5,000S 12,000S	4,000 15,000
Protective colloids	25	50	50	5,000S 12,000S	100
Thickeners	—	—	—	75,000S	—
Leather:					
Finishings	4,000	—	—	5,000S 12,000S	—
Pasting adhesives	4,000	—	4,000	—	—
Paints:					
Acrylics, polyvinyl acetate, styrene-butadiene	—	—	—	5,000S 12,000S 20,000S 75,000S	4,000 15,000S
Cement paints	—	—	4,000	5,000S 12,000S 20,000S 75,000S	4,000 15,000S
Multicolor lacquers	25–4,000	—	50–4,000	—	—
Texture paints	—	—	4,000	5,000S 12,000S	4,000 15,000
Paint removers§	—	4,000	4,000	—	15,000
Paper:					
Adhesives	15–400	—	50–400	—	100–400
Barrier coatings	15–400	—	50–400	—	100–400
Dielectric papers	15–400	—	—	—	—
Release coatings	15–400	—	50–400	—	100–400
Pencils and crayons	25–400	—	50–400	—	—
Pharmaceuticals:					
Bulk laxatives	—	—	—	—	15,000
Creams and ointments	—	4,000	4,000	—	4,000
Ophthalmic preparations	1,500 4,000	—	4,000	—	4,000
Suspensions	1,500 4,000	4,000	4,000	—	4,000
Tablet binders	15–25	50	50	—	100
Tablet film coats	—	15–50	—	—	—
Plywood control of glue viscosity	4,000	—	4,000	5,000S 12,000S	15,000
Printing inks (Water-based inks)	25–4,000	15–4,000	50–4,000	5,000S 12,000S	100–4,000
Polyvinyl chloride	15–25	15–50	50	—	35–100
Resins:					
Emulsion coatings	—	4,000	4,000	5,000 12,000	4,000 15,000
Mold-release agents	15–25	—	50	—	35–100

‡ Viscosities dependent on application need.   Consult supplier.
§ Evaluate HB products.

TABLE 3.10
Applications for Various Cellulose Derivatives *(Continued)*

	MC 26–33	HPMC 28–30 7–12	HPMC 27–30 4–7.5	HPMC 16.5–20 23–32	HPMC 19–24 7–12
Percent methoxyl:					
percent hydroxypropoxyl:	—				
Rubber:					
Latex stabilizers and thickeners	4,000	—	4,000	5,000S 12,000S	4,000
Mold release	15–25	—	50	—	35–100
Textiles:					
Adhesives	400	—	50	—	35–100
Carpet backsizing	—	—	—	75,000S	15,000S
Dye thickening	1,500	—	4,000	—	4,000
	4,000	—	—	—	15,000
Flocking adhesives	—	—	4,000	5,000S 12,000S	4,000 15,000
Latex coatings	15–25	—	4,000	12,000S	15,000
	4,000				
Printing pastes	1,500	—	4,000	5,000S	—
	4,000	—	—	12,000S	—
	10,000	—	—	75,000S	
Warp sizes	15	—	50	—	100
Tobacco:					
Reconstituted sheet	400	—	4,000	—	—
	15–4,000	—	—	—	—
Viscosity control	—	—	—	—	15,000

of the time saving that results from water rinsing. This accelerated method of paint removal can contribute to large savings from a labor standpoint.

The following formulations are typical of those used to remove most conventional lacquers, varnishes, enamels (drying-oil type, alkyds, or drying-oil alkyd-modified types), many of the epoxy esters, epoxy amides, amine-catalyzed epoxies, acrylics, etc. Many of the paints mentioned are stripped within a matter of seconds; others may require 10 to 15 min or more.

Scrape-off Paint and Varnish Remover

	Quantity when using:	
Component	Methocel F4M,* 4000 cP	Methocel HB†
1. Methylene chloride, technical grade	75 gal (284 L)	75 gal (284 L)
2. Toluene	3 gal (11.4 L)	3 gal (11.4 L)
3. Paraffin (ASTM 122–127° F.M.R.)	16 lb (7.3 kg)	16 lb (7.3 kg)
4. Hydroxypropylmethylcellulose or hydroxy-butylmethylcellulose	12 lb (5.4 kg)	9.6 lb (4.4 kg)
5. Methanol	12 gal (45.4 L)	12 gal (45.4 kg)
6. Mineral spirits	10 gal (37.9 L)	10 gal (37.9 L)

* Dow brand of HPMC.
† Dow brand of HBMC.

Mixing:
1. Add toluene to methylene chloride.
2. Melt paraffin and dissolve with agitation in above solvent mixture.
3. Add the cellulose ether and stir until it is completely dispersed.
4. Add methanol slowly under constant agitation.
5. Add mineral spirits and blend ingredients until uniform.

Flush-off Paint Remover Formulation

Component	Quantity
1. Methylene chloride, technical grade	75.0 gal (284 L)
2. Toluene	3.5 gal (13.2 L)
3. Paraffin (ASTM 122–127° F.M.R.)	18.0 lb (8.2 kg)
4. Hydroxypropylmethylcellulose, 4 Pa·s (4000 cP)	12.0 lb (5.4 kg)
5. Methanol	12.0 gal (45.4 L)
6. Ammonia 26°Bé	5.0 gal (19 L)
7. Potassium oleate (80% active)	75.0 lb (34 kg)

NOTE: If desired, 4.1 kg (9 lb) of 12.000 Pa·s (12,000 cP) Methocel HB (Dow brand of hydroxybutyl-methylcellulose) may be substituted.

Mixing. This composition can be blended rapidly and efficiently by combining the ingredients in the order indicated in the above formulation. Constant agitation should be employed. Paraffin will go into solution more easily if melted prior to addition.

This formulation can be sprayed or brushed on. Complete blistering usually occurs within a few minutes, after which the surface is flushed with water. A second application is made on areas not completely stripped by the first coat of remover.

Although primarily for fast, economical removal of paint from metal, this formulation is usable on wood if slight staining or grain raising is not important.

Construction Industry Products

Table 3.11 provides information on use levels for various cellulose ether applications in the construction industry.

TABLE 3.11
Functions of MC and HPMC in Construction Applications

End use	Function	Use level	Type	°C Gel point	2% viscosity, cP at 20°C*
Tile grout	Water retention	0.5–1.0%	HPMC	65	4,000
	Improved bond	Based on total wt.	HPMC	85	4,000
Masonry cement	Water retention	0.05–0.1%	HPMC	65	4,000
	Air entrainment	Based on total wt.			
Portland cement plasters	Water retention	0.1–0.25%	HPMC	65	4,000
	Sprayability	Based on mortar	Surface-	65	4,000
	Increased yield	cement	treated HPMC		
Portland cement mortars	Improved bond	0.25–0.50%	HPMC	65	4,000
	Water retention	Based on cement			
Concrete	Pumping aid	¼ lb/yd³	HPMC	—	12,000
		0.15 kg/m³	Dow "J12MS" type		
Tape joint Compounds	Workability	0.50% on total wt.	HPMC	65	4,000
	Binding		Dow "228" type		
Refractory Mortars	Green strength	0.5–1.0% on	MC	—	4,000
	Low ash	total weight			

* To convert to Pa·s, multiply by 0.001.

Food Products

The methylcellulose products possess two properties that are particularly useful to the processor of foods: thermal gelation and surface activity. Thermal gelation generates a hydrophilic gel which reduces penetration of oil during frying of items such as potatoes. This property is also useful in strengthening the structure of baked goods prepared with low-gluten flours. A range of gel points is available. This effect is augmented by improved gas retention (in bakery items such as breads, cakes, etc.) resulting from the film properties of the methylcelluloses.

MC products can function as polymeric surfactants where whipping action and foam stabilization are required in whipped toppings and dairy products.

MC and HPMC have received clearance as food additives in the United States. These products are listed in one or more of the following compendia: *United States Pharmacopeia, National Formulary,* and the *Food Chemicals Codex.* Monographs on these products also appear in the *Codex Alementarius.*

Examples of end uses and related functional properties are found in Table 3.12.

TABLE 3.12
Food Uses of Methylcellulose and Hydroxypropylmethylcellulose Products

End Use	Functions
Frozen pastries	No syneresis on freezing, thermal gelation on baking
Frozen meat patties	No syneresis on freezing, thermal gelation on baking
TV dinners, gravies	No syneresis on freezing
Frozen fish products	No syneresis on freezing, binding
Breading batters	Improved stability to spoilage
Additive to low-gluten flours	Gas retention, stronger cell structure, thermal gelation, water retention
Food dressings	Surface activity, thickening
Whipped toppings	Surface activity, thickening, foam stabilization
Frozen desserts	Surface activity, crystal modification
Dairy mixes	Whipping action, thickening
Flavor emulsions	Surface activity
Glazes	Binding, whipping action
Doughnut mixes	Gas and moisture retention, reduced oil uptake
French fried potatoes	Reduced oil uptake
Food coatings	Binding, reduced oxygen transmission
Pie fillings	Water retention on baking via thermal gelation
Specialty sauces	Thickening, inhibition of phase separation
Cake mixes	Gas and moisture retention, improved volume
Dietetic syrups	Viscosity control, physiologically inert
Condiment carrier	Water retention
Meringue	Whipping action, foam stabilization
Extruded potato shapes	Reduced oil uptake, binding, lubricity

Pharmaceutical Products

Methylcellulose and hydroxypropylmethylcellulose are used in many pharmaceutical and medical products. The choice of a particular material depends on the function required and the ingredients of the formula. Each manufacturer of MC and HPMC has types for specific end uses and will provide assistance in choosing the optimum product.

Some general use guidelines can be furnished. The low-viscosity products are used where film build is required, and the higher-viscosity products are used when thickening and bulking action is required. Compatibility with organic products and solubility in organic solvents improve as substitution increases. The surface active and protective colloid properties also increase as hydroxypropoxyl substitution increases.

Examples of end uses and associated functional properties are given in Table 3.13.

Tobacco

Methylcellulose is used as a film-former and binder in the manufacture of reconstituted tobacco sheet for cigars. A typical formula is:

	Parts by weight
Tobacco powder (200–300 mesh)	100
Methylcellulose [1.500 Pa·s (1500 cP)]	11
Propylene glycol	3
Water to give castable slurry	Variable

Leather

The thermal gel point of methylcellulose provides improved adhesion to the plates used for the drying of hides in leather processing. A suggested starting formulation is:

	Weight percent
Methylcellulose [4.000 Pa·s (4000 cP)]	2.4
Casein, 15% solution (ammonia or borax cut)	8.1
N-acetyl ethanolamine	0.4
Water	89.1
Total	100.0

Preparation of Leather Paste   Several procedures for the preparation of pastes of methylcellulose are available. When limited time for preparation of the paste is available, the following procedure for a 208-L (55-gal) batch is recommended:

Add 5.0 kg (11 lb) of methylcellulose [4.000 Pa·s (4000 cP)] powder, 16.8 kg 37 lb of a 15% (ammonia or borax cut) casein solution, and 0.9 kg (2 lb) of N-acetyl ethanolamine

TABLE 3.13
Pharmaceutical and Medical Uses of Methylcellulose and Hydroxypropylmethylcellulose

End Use	Functions
Control of diarrhea	Nonmetabolized bulk, film forming, water retention
Control of constipation	Nonmetabolized bulk, water retention
Surgical jellies	Lubricity, thickening
Ointments and lotion	Thickening, lowered surface tension
Suspensions	Thickening, dispersing activity
Burn therapy	Film forming, hydrophilic gel
X-ray contrast media:	
Gastrointestinal	Filming, pigment dispersing, bulking, water retention
Bronchography	Filming, hydrophilic coating
Ophthalmic medicinals	Non-ionic, bland, ointment, good wetting of cornea
Tissue culture	Reduces coating of cells on glassware, inert
Microscopic diagnostic techniques	Thickening, physiologically inert, slows movement of protozoa
Tablet binder	Adhesiveners, non-ionic
Tablet coatings	Filming, soluble in both water and organic solvents
Nose drops	Thickening, hydrophilic film spreads easily
Dental medicinals	Inert to calcium salts
Crystal modification	Absorption on surfaces
Vitamin emulsions	Surface activity, thickening
Nonglycogenetic medicinals	Not metabolized, thickening
Dietetic foods for phenyl ketone via therapy	Improved gas retention in doughs, thermal gelation, filming
Control of bleeding	Filming, thickening, hydrophilic gel
Surgical casts	Controls set of plaster of paris

to 56.7 kg (125 lb) of hot [88°C (190°F+)] water; stir until the cellulose derivative is completely wetted out; then dilute with 458.6 kg (280 lb) of cold water to give 208 L (55 gal) of paste. This procedure should take no longer than ½ h to complete. The choice of alkali used to solubilize the casein can have some effect on the washability of the paste from the platens and hides. Ammonia is volatile and leaves an insoluble form of casein that is difficult to remove. Borax or caustic soda is preferred if paste removal is a problem.

Some producers offer methylcellulose in granular form. These granules should be dispersed in water first, followed by addition of the casein solution and the plasticizer immediately thereafter. Thorough agitation with minimal air entrapment should be continued until dissolution is complete.

The formulations given are only starting-point suggestions. To meet the specific requirements of many tanneries and to satisfy individual situations and needs, certain changes in the formulations can be made.

Agricultural Uses

Methylcellulose polymers are used in agricultural products as seed stickers, dust stickers, dispersing agents for wettable powders, spray spreaders and stickers, and protective films.

Seed Stickers   In the slurry method of seed treatment to adhere fungicides, etc., to the seeds, methylcellulose [0.015 Pa·s (15 cP)] is commonly used. From 25 to 50% of MC (based on weight of dry protectant) is suggested; coarse-grained protectants may require a higher concentration of methylcellulose.

Spray Spreader and Sticker   For spray spreaders and stickers 0.153 to 0.306 g of MC or HPMC [4.000 Pa·s (4000 cP) per liter of water (2 to 4 oz/100 gal) provides the necessary wetting action.

Dust Stickers   Methylcellulose in powder form is used in dust stickers; the usual addition is 62.6 to 125.2 g of MC or HPMC [4.000 Pa·s (4000 cP)] per kilogram of agricultural dust (1 to 2 oz/lb).

Wettable Powders   MC [0.015 Pa·s (15 cP)] at a level of 0.5% is recommended.

Protective Films   Aqueous dips made with MC [0.015 Pa·s (15 cP)] at concentrations of at least 5% are successful in this application.

LABORATORY TECHNIQUES

Analytical tests and procedures for methylcellulose and hydroxypropylmethylcellulose are found in the following mongraphs:

Methylcellulose:
  American Society for Testing and Materials
  United States Pharmacopoeia, 19th ed.
Hydroxypropylmethylcellulose:
  American Society for Testing and Materials
  United States Pharmacopoeia, 19th ed.
Test procedures are also listed in:
  Food Chemicals Codex
  FAO/WHO Codex Alementarius

Additional information on analysis is found in the following references:

Aldrich, J. C., and E. P. Samsel, "Application of Anthrone Test to Determination of Cellulose Derivatives in Nonaqueous Media," *Anal. Chem.,* 29, 574–576 (1957).
"Hydroxypropylmethylcellulose," *Food Chemicals Codex,* 2d ed., National Academy of Sciences and National Research Council, Washington, D.C., 1972.
"Hydroxypropylmethylcellulose," *The National Formulary,* 13th ed., American Pharmaceutical Association, Washington, D.C., 1970.
Kanzaki, G., and E. Y. Berger, "Colorimetric Determination of Methylcellulose with Diphenylamine," *Anal. Chem.,* 31, 1383–1385 (1959).
Methods of Testing Methylcellulose, ASTM D 1347–72, American Society for Testing and Materials, Philadelphia.
Methods of Testing Hydroxypropylmethylcellulose, ASTM D 2363–72, American Society for Testing and Materials, Philadelphia.
"Methylcellulose," *Food Chemicals Codex,* 2d ed., National Academy of Sciences and National Research Council, Washington, D.C., 1972.
"Methylcellulose," *United States Pharmacopoeia,* 19th ed., The United States Pharmacopeial Convention, Inc., Bethesda, Md., 1970.
Samsel, E. P., and J. A. McHard, "Determination of Alkoxyl Groups in Cellulose Ethers," *Ind. Eng. Chem. Anal. Ed.,* 14, 750–754 (1942).
Crössman, F., W. Klaus, E. Mergenthaler, and S. W. Souci, "Zur physicalis chem und chemischem Charakterisiering der als Lebeusmittel-Zusatzstoffe verwendeten Celluloseäther," II Miteilung "Methoden zur Prüfung von Celluloseäthern auf Identität und Reinheit," *Z. Lebensm. Unters. Forsh.,* 125 (5), 413–427 (1964).

PRODUCT/TRADENAME GLOSSARY

Ethylhydroxyethylcellulose: A cellulose ether containing both ethyl and hydroxyethyl substitution.

Hydroxybutylmethylcellulose: A modified methylcellulose containing hydroxybutyl substitution.

Hydroxyethylmethylcellulose: A modified methylcellulose containing hydroxyethyl substitution.

Hydroxypropylmethylcellulose: A modified methylcellulose containing hydroxypropyl substitution.

Celacol: Brand name for British Celanese, methylcellulose products.

Culminal: Brand name for Henkel methylcellulose products.

Marpolose: Brand name for Mitsubishi Yushi methylcellulose products.

Metolose: Brand name for Shin-Etsu methylcellulose products.

Methocel: Brand name for Dow methylcellulose products.

Tylose: Brand name for Kalle methylcellulose products.

Bermocoll: Brand name for Berol Kemi AB ethylhydroxyethylcellulose products.

Chapter **4**

Carboxymethylcellulose

Glenn I. Stelzer and E. D. Klug

Hercules Inc.

General Information 4-1
   Chemical Nature 4-2
   Physical Properties 4-5
   Manufacture 4-6
   Biological Properties 4-7
   Toxicological Properties 4-8
   Rheology 4-9
   Storage and Handling 4-11
   Applications 4-17
   Specialties 4-22
   Future Developments 4-22
   World Production 4-23
Tradename Glossary 4-23
Further Reading 4-24

GENERAL INFORMATION

Sodium carboxymethylcellulose is a water-soluble anionic linear polymer. It is universally known as CMC and will sometimes be so designated here. In the food, pharmaceutical, and cosmetic industries, the highly purified types required are referred to as *cellulose gum*. The United States Food and Drug Administration (FDA) has defined cellulose gum (see section on toxicological properties); also the *Food Chemicals Codex* and the Food and Agriculture Organization (FAO) of the United Nations have established specifications for identity and purity of sodium carboxymethylcellulose for food uses worldwide.

Purified sodium carboxymethylcellulose is a white to buff-colored, tasteless, odorless, free-flowing powder. Less-purified grades contain the reaction salts (sodium chloride and sodium glycolate) and can be off-white to a light brown for the low-assay types (50% purity).

Sodium carboxymethylcellulose is probably used in more varied applications worldwide than any other water-soluble polymer known today. Applications vary from the large worldwide detergent use to the specialized barium sulfate suspension for medical diagnosis.

Worldwide applications of CMC in order of size of estimated end use are given in Table 4.1. These estimated usages demonstrate the versatility throughout the world for this modified natural long-chain water-soluble polymer.

TABLE 4.1
Worldwide Use Patterns for
Carboxymethylcellulose

End use	Estimated world usage, %
Detergents	27
Drilling fluids	10
Paper	7
Mining	7
Textiles	6
Food	5
Coatings	4
Cosmetics	1
Miscellaneous	33

It is estimated that over 250 types of sodium carboxymethylcellulose are manufactured throughout the world by over 50 producers with outputs ranging from as little as 200 metric tons to over 35,000 metric tons per year.

The growth of CMC was accelerated by the world conflict in the early 1940s when fatty acids usage was drastically shifted from civilian soap manufacture to wartime manufacture of explosives. Even though CMC was developed shortly after World War I as a possible replacement for some gelatin uses, the major growth in the use of CMC began after it was discovered that it improved the efficiency of synthetic detergents. Usage during the early 1940s was primarily for detergent systems, although many new applications were developed on a laboratory scale where control of water movement was important.

With the end of the world conflict in 1945 and with the huge demand for consumer products, CMC, backed with several years of laboratory studies, began finding uses in all types of areas requiring water control in systems with various levels of soluble and insoluble solids.

In the United States, a landmark in the growth of purified sodium carboxymethylcellulose (cellulose gum) was the approval by the FDA for its use as an intentional food additive (see section on toxicological properties for details). Following this was the definition in the *United States Pharmacopoeia* for subsequent use in pharmaceutical applications.

Development of specialized types of CMC followed rapidly as new-property demands were defined by widening uses:

■ Expansion of CMC use was rapid in paper manufacture as a result of its film-forming ability and oil resistance.

■ Pollution abatement regulations in the United States accelerated the use of CMC (technical grades) as a warp size in the textile industry.

■ The rapid growth of popularity of water-based latex paints which required a leveling control agent (again, water control) broadened the usage spectrum of CMC.

Chemical Nature

Cellulose is a linear polymer of β-anhydroglucose units. Each anhydroglucose unit contains three hydroxyl groups. CMC is prepared by the reaction of the cellulose hydroxyls with sodium monochloroacetate as follows:

$$R_{cell}OH + NaOH + ClCH_2COONa \rightarrow ROCH_2COONa + NaCl + H_2O$$

The extent of the reaction of cellulose hydroxyls to form a derivative is called the *degree of substitution* (DS) and is defined as the average number of the three hydroxyl groups in the anhydroglucose unit which have reacted. Thus, if only one of the three hydroxyl groups has been carboxymethylated, the DS is 1.0. Commercial products have DS values ranging from 0.4 to about 1.4. The most common grade has a DS of 0.7 to 0.8, and if the DS is not specifically mentioned, it can be assumed to be in this range. CMC is commercially available in several different viscosity grades ranging from 4.5 Pa·s (4500 cP) in 1% solution to 0.010 Pa·s (10 cP) in 2% solution. The various viscosity grades correspond to products having molecular weights from about 1,000,000 to 40,000. Table 4.2 shows that 19 different DS-viscosity combinations are available from one producer.

TABLE 4.2
Types of Cellulose Gum Commercially Available*

Viscosity range at 25°C		Types			
		DS 4	DS 7	DS 9	DS 12
High, at 1% concentration					
Pa · s	cP				
0.4–1.0	400–1000	4H1			
1.0–2.8	1000–2800		7H3S, 7HO		
1.5–2.5	1500–2500		7H		
2.5–4.5	2500–4500		7H4		
Medium, at 2% concentration					
Pa · s	cP				
0.05–0.1	50–100		7M1		
0.1 –0.2	100–200		7M2		
0.2 –0.8	200–800		7M8S		
0.3 –0.6	300–600	4M6, 4M6S	7M		
0.4 –0.8	400–800			9M8	12M8
0.8 –3.1	800–3100			9M31	12M31
0.8 –2.7	800–2700		7M27S		
Low, at 2% concentration					
Pa · s	cP				
0.018 max.	18 max.		7L1, 7L2		
0.025–0.05	25–50		7L		

* Sodium carboxymethylcellulose as commercially available from Hercules Inc., Wilmington, Del.

CMC is a salt of a carboxylic acid having approximately the same acid strength as acetic acid. The pK varies somewhat with degree of substitution. The pure commercial product of DS 0.8 has a pK value of 4.4; the corresponding value of K, the ionization constant, is 4×10^{-5}. A dilute solution of such a product has a pH of about 7 and has over 99% of its carboxylic acid groups in the sodium salt form and very few in the free acid form.

CMC forms soluble salts with alkali metal and ammonium ions. Calcium ion, present in concentrations normally found in hard water, prevents CMC from developing its full viscosity, and thus its dispersions are hazy. At much higher concentrations, calcium ions precipitate CMC from solution.

Magnesium and ferrous ion have a similar effect on CMC dispersions. Heavy metal ions like silver, barium, chromium, lead, and zirconium precipitate CMC from solution. Quaternary salts attached to a long hydrocarbon chain, such as dimethylbenzylcetylammonium chloride, also precipitate CMC from solution.

CMC is precipitated from solution by the polyvalent cations Al^{3+}, Cr^{3+}, or Fe^{3+}. If the ion concentration is carefully controlled, for example, by the presence of a chelating agent such as citric acid, it is possible to form more viscous solutions, soft gels, or very rigid gels. In these instances the polyvalent ion functions as a crosslinking agent.

CMC reacts with certain proteins. For example, soy protein, which is insoluble in its isoelectric range, can be solubilized by CMC. Thus the solubility can be extended over a wider pH range. CMC has a similar solubilizing effect on casein; but in the case of gelatin, which is a more soluble protein, the reaction with CMC manifests itself as a rise in solution viscosity.

Like all polymers, CMC may be salted out of solution. However, CMC, being a very hydrophilic polymer, is more tolerant of alkali metal salts than many other water-soluble polymers. Its salts compatibility is much greater if the salt is dissolved in the CMC solution than if the CMC is dissolved in the salt solution (see Fig. 4.1). Such behavior relates to the fact that CMC of DS 0.7 is aggregated in solution. This is discussed later in connection with its rheological properties.

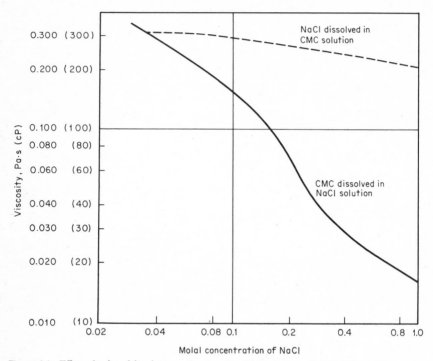

Figure 4.1   Effect of order of dissolving on the viscosity of 1% solutions of CMC (DS of 0.7) containing sodium chloride.

TABLE 4.3
Typical Physical Properties of CMC of DS 0.7

Solid powder	
Moisture content as packed, %	8.0
Browning temperature, °C	227
Charring temperature, °C	252
Bulk density, g/mL	0.75
Biological oxygen demand*	
DS 0.8, high viscosity	11,000 ppm
DS 0.8, low viscosity	17,300 ppm
Solutions	
Specific gravity, 2% solution at 25°C	1.0068
Refractive index, 2% solution at 25°C	1.3355
pH, 2% solution	7.0
Surface tension, 1% solution at 25°C	71 dyn/cm
Bulking value in solution, L/kg (gal/lb)	0.544 (0.0652)
Films 02	
Specific gravity, g/mL	1.59
Refractive index	1.515

* After 5-day incubation. Under these conditions cornstarch has a BOD of 800,000 ppm.

Commercial grades of CMC have most of their carboxyl groups in the sodium salt form. Such products may be converted into the free acid form, e.g., by passing a solution through a suitable ion exchange resin. If the resulting free acid is freed from water by drying a film or precipitating with alcohol and drying, the product is no longer water-soluble. It may be dissolved, however, in aqueous NaOH, that is, by re-forming a soluble salt.

Physical Properties

The general physical properties of CMC are summarized in Table 4.3. Other physical properties follow.

Equilibrium Moisture Content   CMC is a very hydrophilic polymer whose equilibrium moisture increases with DS. Figure 4.2 gives the equilibrium moisture content for products of DS 0.4, 0.7, and 1.2 at different humidities. These data were obtained on dry commercial samples by conditioning the powder to constant weight at 25°C.

Molecular Weights   The molecular weights shown in Table 4.4 were calculated from intrinsic viscosity measurements in 0.1% NaCl at 25°C using the relationship $[M] = 2.9 \times 10^{-4}\ M^{0.78}$, where M is the weight-average molecular weight.

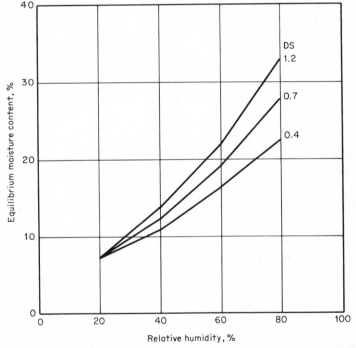

Figure 4.2 Effect of relative humidity on the equilibrium moisture content of CMC (DS of 0.4, 0.7, and 1.2) at 25°C.

TABLE 4.4
Molecular Weights and Solution Viscosities of CMC of DS 0.8

Commercial grade*	Concentration, %	Viscosity range Pa·s	cP	Molecular weight
7H	1	1.5–2.5	1,500–2,500	700,000
7M	2	0.3–0.6	300–600	250,000
7L	2	0.025–0.05	25–50	100,000
7L2	2	0.018 or less	18 or less	50,000

* Hercules, Inc., commercial grades used for illustrative purposes.

Solubility   The only good common solvent for CMC is water.   The degree of dispersion in water varies with the DS and the molecular weight.   CMC with a DS of 0.7 may be dissolved in glycerin, particularly in the presence of a slight amount of water, by heating with good agitation.   Aqueous solutions of CMC will tolerate considerable quantities of water-miscible organic solvents such as methanol, ethanol, and acetone. For example, a 1% solution of the high-viscosity grade will tolerate 1.6 volumes of ethanol per volume of CMC solution before it becomes hazy and precipitates.   Low-viscosity grades will tolerate as much as 3.5 volumes.   Aqueous solutions of CMC will tolerate large amounts of alkali metal salts and small amounts of calcium and magnesium salts.   Heavy metals and multivalent salts precipitate CMC, as discussed earlier under Chemical Nature.

Film Properties   Table 4.5 gives mechanical properties of 0.508-mm (2-mil) films containing about 18% moisture for three different viscosity grades of CMC with a DS of 0.7.   It is evident that the strength and flexibility are greater for the types which have high viscosities or molecular weights.   Films may be insolubilized by crosslinking at the hydroxyl groups using suitable water-soluble resins such as Hercules Kymene 917 and, Kymene 754 Resin, or Aerotex M-3.   The crosslinks are formed by reaction of cellulosic hydroxyls with the aldehyde functionality of the resins.   A film is cast from an aqueous solution of the resin and CMC.   Upon drying and further curing, the film becomes insoluble.   The degree of insolubilization depends on the extent of the curing treatment.   Dry CMC films may be insolubilized by treatment with aqueous solutions of aluminum salts.

TABLE 4.5*
Mechanical Properties of CMC Films

Commercial grade†	7L	7M	7H
Viscosity conc. 1%, Pa · s (cP)			1.5–2.5 (1,500–2,500)
conc. 2%, Pa · s (cP)	0.025–0.05 (25–50)	0.3–0.6 (300–600)	
Molecular weight	100,000	250,000	700,000
Tensile strength, MPa (psi)	55.16 (8,000)	89.63 (13,000)	103.42 (15,000)
Elongation, % at break	8.3	14.3	14.3
Flexibility, MIT double folds	93	131	513
Electrostatic charge	Negative	Negative	Negative
Refractive index	1.515	1.515	1.515
Specific gravity	1.59	1.59	1.59

* The films, which were cast from aqueous solution, were 2 mils (0.508 mm) thick and contained about 18% moisture.
† Hercules, Inc., grade designations.

Manufacture

The manufacture of CMC involves treatment of cellulose with aqueous sodium hydroxide followed by reaction with sodium chloroacetate:

$$R_{cell}OH + NaOH + ClCH_2COONa \rightarrow R_{cell}OCH_2COONa + NaCl + H_2O$$

A side reaction, the formation of sodium glycolate, also occurs:

$$ClCH_2COONa + NaOH \rightarrow HOCH_2COONa + NaCl$$

Cellulose is a fibrous solid.   Chemical cellulose which is used for the manufacture of CMC is derived from cotton linters or wood pulp.   To obtain uniform reaction, it is essential that all the fibers be wetted out with the aqueous NaOH.   One process for accomplishing this is to steep sheeted cellulose in aqueous NaOH, and then press out the excess.   The sheets are then shredded and the sodium chloroacetate is added. Reactions are generally conducted at 50 to 70°C.   In some cases, a greater amount of NaOH is added, and the monochloroacetic acid is added as such, the sodium salt being formed in the presence of the cellulose.

In an alternate process, the steeping and pressing steps are eliminated by conducting the reaction in the presence of an inert water-miscible diluent such as tertiary butyl alcohol or isopropanol.   At the end of the reaction, the excess alkali is neutralized and the crude product which contains sodium chloride and sodium glycolate is purified or partially purified (see Table 4.6).   There are many variations of these processes, depending on the DS level and the quality of the product desired.

Biological Properties

Water-soluble cellulose derivatives are all subject to microbiological attack under certain conditions. The magnitude of the microbiological degradation is influenced by a number of factors, which include contaminants present, temperature, pH of system, oxygen available, and concentration. The first sign of biological degradation is usually loss of viscosity (i.e., chain length). This loss can be rapid under extreme conditions, or very slow under much less severe conditions. Biological attack is greater in solution systems than on the dry form of sodium carboxymethylcellulose. Usually, the moisture content of sodium carboxymethylcellulose is from 5 to 10%, and biological degradation is generally not severe under normal dry storage conditions at this moisture level.

TABLE 4.6
Classification of Purity Range of CMC

Class	Purity range, %NaCMC	General use
Low assay	50–60	Detergent, mining, petroleum
Crude	60–70	Detergent, mining, petroleum
Semitechnical	70–90	Detergent, mining, petroleum
Semirefined	90–95	Detergent, mining, petroleum, textile or warp size
Technical*	95–98	Industrial uses
Refined	98–99.5	Industrial uses
Purified (standard and premium)†	99.5+	Food, pharmaceutical, cosmetic, toothpaste

* In Europe "technical grade" contains 60–95% CMC.
† In Europe "purified grade" contains 95+% CMC.

Manufacturing conditions and subsequent packaging systems can greatly influence the microbiological stability of the final product. In some commercial manufacturing systems using solvents, the final packaged sodium carboxymethylcellulose is essentially aseptic. In the dry process production of CMC, there is a greater possibility of residual biological contamination. Generally, however, the caustic present in the system necessary for the alkali cellulose stage is detrimental to any microorganisms present.

Packaging environment is also extremely important. Clean containers and air free of microbiological organisms are necessary in the packaging of CMC. Generally speaking, the product as produced and packaged is relatively free of microbiological organisms which would promote degradation of the cellulose chain. It has been found in actual application that most biological organisms causing CMC degradation have been introduced from outside sources (other than in the CMC). Each manufacturer of sodium

CMC usually runs periodic bacteriological examinations. An example of these analyses is as follows:

	Organisms/0.02 g		Estimated no./g*	
Class	Bacteria	Molds	Bacteria	Molds
Purified (99.5 + % NaCMC)	0	0	<50	<50

* Information for sodium carboxymethylcellulose produced in a solvent medium.

Other specific bacteriological testing has been done on the example purified sodium carboxymethylcellulose, testing for the presence of coliforms, thermophilic anaerobic spores, pathogenic staphylococci, beta-hemolytic streptococci, *Salmonella* species, and *Pseudomonas aeruginosa.*

It is stressed that most individual manufacturers of CMC throughout the world have had bacteriological examination of their product and will have data comparable to the preceding. Varying manufacturing processes (i.e., different solvents and dry process techniques) will yield different biological analyses, but generally speaking, most known processing does not promote or support bacterial growth.

The major bacteriological problems are generally caused when the sodium carboxymethylcellulose is used in solution. Contaminating spores can be introduced in the system from influent water used for solution preparation, as well as from the air surrounding the mixing and makeup vessels. This is especially true in warm, humid climates which are supportive of bacteriological spores. Care should always be taken in handling containers of sodium carboxymethylcellulose that are stored in locations exposed to warm, humid air.

Introduction of bacteriological contamination has been observed in plant operations where the vessels, piping, pumps, etc., have not been cleaned after each use. In many fluid-handling systems, the possibility exists for areas which could retain residual product. In these areas, under the proper conditions, microbiological growth can take place at a rapid rate, thus contaminating subsequent batches pumped or handled in the same system. Industrial experience has shown that this is a major source of microbiological contamination. Simple clean-out measures in any solution-handling system where CMC is used are all that are necessary to prevent subsequent contamination of other batches of product.

Toxicological Properties

In the United States toxicological information on sodium carboxymethylcellulose has primarily been developed on the food-additive grade, or cellulose gum, which is of 99.5% purity. Extensive testing has been performed on this purified grade of sodium carboxymethylcellulose to assess its safety in foods as an additive. Details on testing and results follow. However, a definition of food-grade sodium carboxymethylcellulose is necessary to assure safe use. The United States Food and Drug Administration defines cellulose gum as the sodium salt of carboxymethylcellulose, not less than 99.5% on a dry weight basis, with a maximum substitution of 0.95 carboxymethyl groups per anhydroglucose unit, and with a minimum viscosity of 0.025 Pa·s (25 cP) in a 2% (dry weight) aqueous solution at 25°C.

Sodium carboxymethylcellulose (cellulose gum) is classified under "Substances That Are Generally Recognized As Safe" (GRAS) by Title 21, Section 182.1745 (formerly 121.101) of the *Code of Federal Regulations* (U.S.A.).

Both the *Food Chemicals Codex* and the Food and Agriculture Organization of the United Nations World Health Organization (FAO/WHO) have established specifications for identity and purity of sodium carboxymethylcellulose which include the same properties as defined by the United States Food and Drug Administration. For detailed specifications and analyses methods, consult the latest edition of the *Food Chemicals Codex.*

The toxicological background for purified sodium carboxymethylcellulose has been well established by a number of investigators. The most significant findings of all this work are summarized in the following.

Six-Month Oral Toxicity In tests, 100 rats, 100 guinea pigs, and 10 dogs were fed dietary levels of 1% and 2% (0.5 g and 1.0 g of food-grade CMC per kilogram of body weight added to diet daily) for 6 months.   Normal growth, fertility, urinalysis, and hematology were observed during the course of the experiment, and no gross or microscopic pathology was detected upon termination.

Three dogs were also fed dietary levels of 5% and 10% for 6 months.   Food-grade sodium alginate and karaya gum were fed for comparison at a level of 10%.   At the 5% level, all observations during the experiment and at termination were normal. At the 10% level, growth was retarded with NaCMC, as well as with the sodium alginate and karaya gum.   Loose stools were observed with all three products.   No other changes (urinalysis, hematology, gross and microscopic pathology) were detected.   Attempts to feed dogs dietary levels of 20% CMC were unsuccessful due to food refusal.

One-Year Studies   Twenty guinea pigs were fed dietary levels of 1% and 2% (0.5 g/kg and 1.0 g/kg added to the diet daily) for 1 year.   No mortality occurred, growth was normal, and upon termination no gross or microscopic pathology was detected.

Chronic Oral Toxicity   In tests, 25 rats were fed dietary levels of 0.2%, 1%, and 2% (0.1 g/kg, 0.5 g/kg, and 1.0 g/kg added to their diet) for 2 years.   Mortality, growth, monthly urinalysis, and hematology were normal.   Gross and microscopic examination after 25 months' feeding revealed no pathology other than the senility changes present in controls.   No increase in the number of neoplasms was found in the rats which were fed CMC.

Reproduction   Rats fed dietary levels of 0.2%, 1%, and 2% CMC were carried through three-generation reproduction studies, with offspring being maintained on these same dietary levels.   No alterations in fertility or reproduction of these animals were detected.

Gastrointestinal Absorption   Radioactive CMC, manufactured with ^{14}C-tagged sodium chloroacetate, was administered orally to rats as an aqueous solution at a dosage of 1.3 g/kg.   Urine was collected in special metabolism cages to prevent crosscontamination with feces.   The animals were sacrificed 48 h after dosaging, and the livers and kidneys were analyzed for radioactivity.   None was found at a sensitivity equivalent to less than 0.02% of the administered dose.   Urine specimens showed an average activity equivalent to 0.14% of the administered dose.   This activity corresponded to the amount of radioactive salt (sodium glycolate) formed in the synthesis, and was established by chromatography of the urine solid not to be NaCMC or other saccharide polymer.

Clinical Study   Eleven human volunteers ingested 10 g of NaCMC each day for 6 months.   Complete hematology and urinalysis on all subjects revealed no alterations during NaCMC administration.   Bone marrow studies were made on three subjects, and at completion all were within normal limits.   Some additional volunteers experienced a laxative effect at 10 g daily.   No other physiological manifestations were detected.

Skin Irritation and Sensitization   Two hundred human volunteers have been patch-tested by the Schwartz technique with no evidence of primary irritation or sensitization from NaCMC.

Getting Information   The toxicological information available for other grades of CMC should be requested from the respective manufacturers.   When requesting information on other than purified CMC (99.5+%), the intended use or application is extremely important and should be specified along with the intended usage level of CMC in the product.   From this information the manufacturer will be able to assist the user in evaluating the safety for use in such a system.

Over the years several applications have been developed in which semipurified and technical grades of CMC have been used in products which have had contact with foods and products which have been used to clean and otherwise treat food-containing vessels.   Each case should be considered individually, weighing use, concentration levels, and specific conditions to determine if any toxicological aspects are important in the use of the product coming in contact with the public, either as a food or in a food, cosmetic, or pharmaceutical product.

Rheology

Probably the most useful property of CMC is its ability to impart viscosity and other special rheological properties to its aqueous solutions.   CMC is a linear polymer, and its solutions exhibit many typical rheological properties of linear polymers.   Figure 4.3

shows the relationship between viscosity and concentration, and Fig. 4.4 shows the relationship between viscosity and temperature for commercial grades of CMC. Most solutions of CMC are pseudoplastic, that is, the measured viscosity decreases with increase in shear rate. Most products having DS values below about 1.0 are also thixotropic. This is believed to be because of the presence of many aggregates in the aqueous dispersions. Thus, the viscosity not only depends on the shear history, but also on the time after shearing when the viscosity is measured.

Thixotropy, or the lack of it, is a function not only of degree of substitution, but also of uniformity of substitution. Solution appearance can be changed from thixotropic (applesauce consistency) to very smooth (syruplike) with no change in DS by special

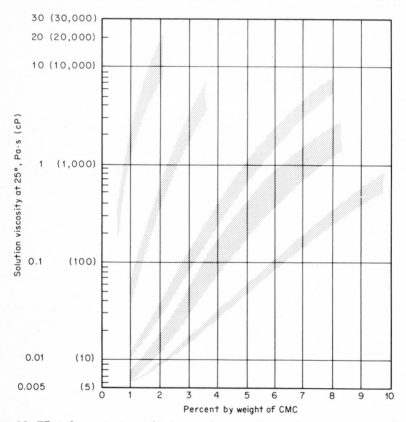

Figure 4.3 Effect of concentration on the viscosity of aqueous solutions of some commercial grades of CMC (DS of 0.7). The bands show the range of viscosities likely to be encountered in a given commercial grade.

reaction schedules and raw materials selections. Uniformity of substitution also increases tolerance to acid systems and dissolved ions. Special "smooth" types of CMC are available for highly demanding applications.

The presence of salts in solution represses the disaggregation of CMC and therefore affects the viscosity. Table 4.7 gives specific examples of the effect of mild shear (Anchor stirrer) and vigorous shear (Waring blender) on the viscosity of solutions of CMC of different substitution levels in water and in aqueous salt solutions.

The viscosity of CMC in a glycerin-water mixture is greater than the viscosity in pure water; the viscosities in the two solvents appear to be proportional to the viscosities of the solvents. This is shown in Fig. 4.5 for a 1% solution of CMC of DS 0.7 in glycerin-water (60:40) at various shear rates. Another way of achieving very high viscosity with CMC is to blend it with a non-ionic cellulose derivative such as water-soluble hydroxyethyl- or hydroxypropylcellulose. Table 4.8 shows that such blends in 1% solution give about twice the expected viscosity based on that of the individual ingredients.

The effect of pH on the viscosity of concentrated solutions of CMC in the pH range from 5 to 9 is very slight. At pH levels below 3, precipitation of the CMC may occur. At pH of 10 or above, there is a slight decrease in viscosity. CMC solutions containing

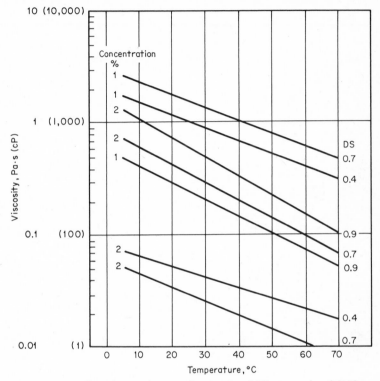

Figure 4.4 Effect of temperature on the viscosity of different grades of CMC.

TABLE 4.7
Effect of Agitation on the Viscosity of CMC in Water and Aqueous Sodium Chloride

| | | After agitation with Anchor stirrer | | | | | | After agitation with Waring blender | | | | | |
| | | Dist. H_2O | | 4% NaCl | | 26% NaCl | | Dist. H_2O | | 4% NaCl | | 26% NaCl | |
DS	% Conc.	Pa·s	cP	Pa·s	cP	Pa·s	cP	Pa·s	cP	Pa·s	cP	Pa·s	cP
0.4	1	0.900	900	0.011	11	0.006	6	4.000	4000	0.065	65	0.016	16
0.75	1	1.680	1680	0.140	140	0.045	45	0.760	760	1.040	1040	2.240	2240
0.9	2	0.215	215	0.160	160	0.225	225	0.125	125	0.095	95	0.235	235
1.3	2	0.175	175	0.080	80	0.180	180	0.100	100	0.055	55	0.140	140

1.0% citric or lactic acid and 5.0% acetic acid may be stored for months at room temperature without any significant change in viscosity (see Fig. 4.6).

Storage and Handling

The handling of sodium carboxymethylcellulose starts with the manufacturer through to the preparation of the final product, either a solution system or a dry compound. In each step, care and precautions are necessary to ensure gaining full benefits of the desired properties from sodium carboxymethylcellulose.

Packaging Standard packaging worldwide for sodium carboxymethylcellulose is in multilayer paper bags. In the United States, the packaging is 22.7 kg (50 lb) net weight, while in other countries CMC is either packaged in 20- or 25-kg net weight bags. The multiwall paper bag usually contains an inside liner of polyethylene film or has one ply coated with polyethylene. The number of plies varies from three to five, according to the manufacturer and the grade being packaged. The multiwall bag is, in some cases, extensible paper which can withstand considerable handling—from the manufacturer to final use by the customer.

The moisture barrier in the multiwall paper bag is necessary since CMC is hygroscopic in nature and will pick up moisture very readily. The 7-type (DS = 0.7) CMC will pick up approximately 15% moisture at 25°C when exposed to a relative humidity of 50%. At a constant relative humidity the higher DS CMC will pick up the higher moisture. For example, a 4-type (DS = 0.4) CMC at an 80% relative humidity will

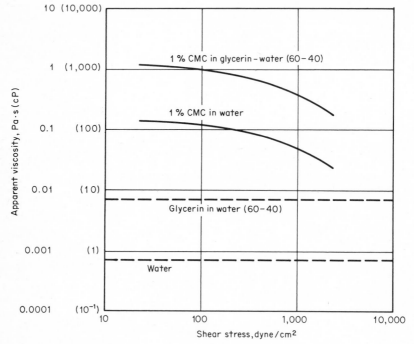

Figure 4.5 Effect of the presence of glycerin on the viscosity of aqueous solutions of CMC (DS of 0.7).

TABLE 4.8
Synergistic Effect on Viscosity of Blending CMC
with a Non-ionic Polymer (1% solutions)

			Viscosity	
Polymer	DS	MS	Pa · s	cP
Sodium carboxymethylcellulose (CMC)	0.8		1.5	1500
Hydroxyethylcellulose (HEC)		2.5	1.8	1800
Hydroxypropylcellulose (HPC)		4.0	1.64	1640
CMC:HEC (1:1) expected*			1.65	1650
CMC:HEC (1:1) measured			3.2	3200
CMC-HPC (1:1) expected*			1.57	1570
CMC:HPC (1:1) measured			3.28	3280

* Calculated by the usual procedure for polymer blends.

pick up approximately 20% moisture, while under the same conditions a 12-type (DS = 1.20) CMC will pick up as much as 35% moisture.

CMC is also packaged in the United States in 45.4- and 90.7-kg (100- and 200-lb) fiberboard drums. These drums can have a sprayed polyethylene coating internally or the CMC can be placed in a free-film polyethylene liner and then placed in the drum. For maximum moisture protection, the polyethylene liner inside a fiberboard drum is recommended. Special containers, such as 227- and 454-kg (500- and 1000-lb) paperboard cartons, can be used for CMC. These cartons should be lined with polyethylene liners to fully protect the CMC from high humidity and moisture during transit and storage.

The packaging of CMC is an extremely important aspect in the handling of this hygroscopic product. One can request covered palletizing from the manufacturer and receive

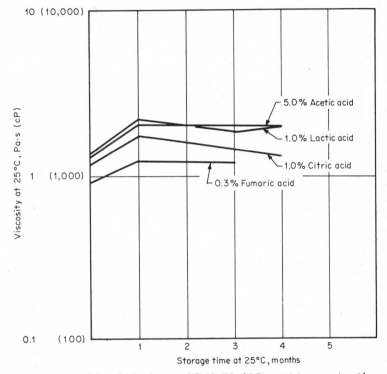

Figure 4.6 Stability of 1% solutions of CMC (DS of 0.7) containing organic acids.

the protection of bags during shipment for critical applications (e.g., food, cosmetics, pharmaceuticals, toothpaste, etc.). This can be accomplished in several ways, but generally by placing a polyethylene film over the top of a pallet [generally 907 kg (2000 lb) net weight per pallet] held in place by strapping and a paperboard cap. A number of variations of this particular protection is available from different manufacturers. This type of protection is recommended in critical uses, both from handling and storage standpoints.

CMC can be shipped in bulk for certain end-use applications. These applications include textile warp sizing, detergent use, and some paper applications. To date, bulk shipments have been confined worldwide to bulk trucks (capacities of approximately 18,100 to 20,400 kg, or 40,000 to 45,000 lb). The acceptable type of system for bulk transporting CMC is what is known as a *differential pressure system*. This simply consists of a positive pressure (13.82 to 35.55 kPa, or 2 to 5 psig) exerted on the bulk load with a return line from the silo to the truck. The air-handling system under this small differential pressure will unload a 18,100-kg (40,000-lb) truck within 60 min.   Other systems of bulk handling CMC have been used where an air-siphoning system can be used to unload large containers (227 to 454 kg, or 500 to 1000 lb).

It is recommended that bulk handling of CMC be considered when it is to be used in technical- and crude-grade applications.   These applications include paper, detergents, textile sizing, mining, etc.   For other uses, where highly purified grades of sodium carboxymethylcellulose are used (e.g., pharmaceutical, cosmetic, toothpaste, food, etc.), it is recommended, for control and quality purposes, that smaller containers [no larger than 454-kg (1000-lb) cartons] be used.   If contamination or a questionable quality is present, then a smaller quantity of material will be affected.

In-Plant Handling   In-plant handling of sodium carboxymethylcellulose, of course, is dictated by the amount used and the type of final product in which it is used.   Generally speaking, CMC is used in small amounts as a part of a final formulation.   Therefore, it is not necessary to bulk handle or use air-conveying systems.   However, in some systems (i.e., detergent, paper, textile, etc.) it is convenient to store CMC in silos and then convey it via screw-feeding mechanisms or air-conveying make-up (either dry-mixing or solution) systems.

Bulk Handling   A suggested in-plant bulk-handling system is shown in Fig. 4.7.

Normal precautions for handling a flammable organic dust in a finely divided and suspended state should be taken in the use of CMC in plant operations.   Each use

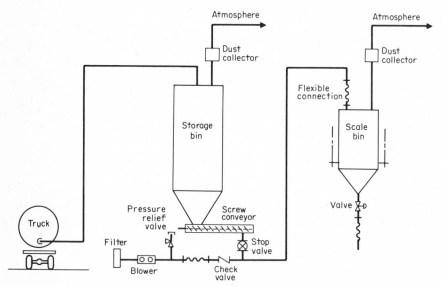

Figure 4.7   Bulk CMC unloading system.

condition must be fully considered, and the above general guideline must be used to ensure proper, safe handling conditions.

Purified sodium carboxymethylcellulose, when tested using the standard Hartmann dust explosion test, revealed the minimum explosive concentration when the 200 United States mesh size dust fully suspended in air is 0.150 kg/m^3 (0.150 oz/ft^3).   The ignition temperature under the above condition is about 370°C.   It is cautioned that these values are only for the carefully defined Hartmann test conditions, and each plant system needs to be defined and studied carefully to ascertain the actual hazard existing, if any.

Bag Handling and Storage   The customary and usual system for handling CMC is in the 20- to 25-kg (or 50-lb) bags.   Here, the bag is sometimes received palletized, and thus in many cases the bags are then handled singularly.   The palletized material can be stored in any cool, dry, out-of-sunlight location used for normal warehousing of powdered materials.   Special care should be taken when CMC is used in certain industries (e.g., food, cosmetic, pharmaceutical, toothpaste, etc.) that it be stored to prevent contamination by spills or accidents with other materials which could cause problems in purified CMC usage. For example, when being stored in a warehouse, all food products should be stored at the highest levels to prevent contamination by gravity of materials of nonfood or nonpharmaceutical quality spilled from above.

The other consideration in storage is the possible contamination of CMC by spores from air. This is especially true in warm, humid climates. It is suggested that CMC and/or other food and pharmaceutical ingredients be covered during storage in a warehouse to prevent any spores, dirt, and/or other contaminants from being on the container, or package, which will be opened in the actual makeup of the final system, whether solution, paste, or dry mixture.

Prior to use it is suggested that the bags, drums, and large containers of CMC be taken into a "ready room" where the containers are decontaminated. Here, vacuuming can be used to remove any spores and other contaminants which may come in contact with the containers during normal storage. Prior to actual removal of contaminants by air in this ready room, removal of shrouds, etc., could be done; then the actual blowing off of the bags is suggested. After this preliminary decontamination the material can be moved into the make-up area for actual use in the final formulation. When making up food and pharmaceutical quality products, it is suggested that the original container not be brought next to the make-up vessel, but that the CMC be removed from the bag or drum and then placed into another clean container for weighing purposes and addition to the final product.

In any use, the above precautions are recommended as good practice.

In-process use of CMC requires that a clean system be used. All cellulosic backbone materials are subject to enzyme attack. Above a threshhold enzyme content of a system, degradation of the cellulose backbone is quite rapid. It is recommended that preservatives be used in liquid systems, and details are given in another section on the type and amount of preservative used. However, it is good practice, whenever CMC solutions have been used in a system, that good, thorough clean-out procedures be used for cleanup. It has been found in numbers of industrial applications where CMC solutions have been used that residual amounts of CMC solution remain in the piping, pumps, mixers, etc., which will promote the growth of bacteria. The next time the equipment is used this bacteria will contaminate the new solution of CMC being processed in the equipment. Thus, it is recommended that the equipment be thoroughly washed out with hot water and a disinfectant and, if at all possible, that steam be used throughout the entire system to clean out and decontaminate all the areas to remove residual solution. This is especially true in systems that have filling devices. It is recommended that, when CMC-containing solutions are filled into smaller containers, the cleanup include every possible area in the make-up and filling system that could have residual CMC-containing solutions remaining.

In handling dry powder CMC, it is always worthwhile to have a "scalping screen" available prior to actually mixing or adding to other dry ingredients. Also, when using a mixing or adding device for powder, it is recommended that this be covered so that miscellaneous and overhead contaminants do not fall unnoticed into this system at the same time that the dry materials are added prior to makeup in a system.

Screw conveyors, automatic weighing bins, and conveyors of all types recommended generally for powders with a bulk density of 0.5 to 1.0 g/mL are satisfactory for use with CMC systems.

Dry flow characteristics of CMC are important in design of in-plant handling equipment. These characteristics vary with the type CMC, moisture content, bulk density, and particle size distribution.

Laboratory tests on a technical-grade (95+% purity) CMC with a moisture content of 6% revealed the angles of slide shown in Table 4.9. *Angle of slide* is the minimum angle at which a material will flow from rest on an inclined surface (galvanized sheet metal used for practical plant testing).

In any design practical tests with actual dry powders being used should be considered. Also, vibrators are worthwhile on storage and feed bins.

Shipping Proper packaging to fit the specific product requirements is the first step in assuring safe shipment. The purified food-grade material needs to be well protected to withstand the normal rigors of shipping and thus ensure being received in acceptable condition for usage.   Most manufacturers have, over the years, developed the necessary packaging as described above.

For shipping in the United States, the multiwall paper bag has been shown to be fully adequate under practically all conditions.   Problems have occurred when the bags have been subjected to direct water (rain), as would be expected.   Thus, precautions

TABLE 4.9
Angles of Slide for Various Screen Sizes of Dry CMC

Cumulative % on U.S. 50-mesh sieve	Angle of slide, degrees
1.5	55–60
2.0	50–55
12.0	50–60
16.6	45–50
25.0	40–50
27.0	40–45

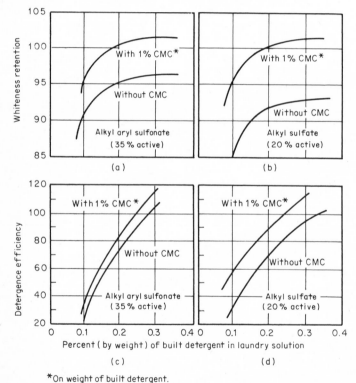

*On weight of built detergent.

Figure 4.8   Effect of CMC on detergence efficiency and whiteness retention of cloth washed in synthetic detergents.   The use of 1% CMC on weight of detergent raises both detergence efficiency and whiteness retention.

to prevent handling under extremely damp or wet conditions must be available.   Most hygroscopic materials have the tendency to absorb odors when exposed in closed areas to a highly contaminated and concentrated atmosphere, even in the acceptable and commonly used multiwalled bags.

Shipment of CMC along with highly volatile materials should always be avoided, and common carriers so advised. Whenever possible, food-grade materials should be shipped together, as well as segregated and stored apart from nonfood (pharmaceutical and cosmetic) materials.

Overseas shipment is more demanding and requires, in many cases, special packages. In critical usages (i.e., food, pharmaceutical, cosmetic, etc.) drums with polyethylene liners are highly recommended. This type container can withstand damp conditions as well as multihandling by various carriers and warehouses. Another commonly used package is a "bag in a bag," where the regular bag is inserted into a larger similar bag, thus giving the material double the normal multiwall paper bag protection.

Containerized overseas shipments are also preferable to eliminate exposure to the immediate danger of physical handling as well as contamination by spills, breakage, etc., of nearby stored materials.

Applications

Detergents   The world's largest single use for sodium carboxymethylcellulose (CMC) is in detergent systems to prevent soil redeposition after the soil has been removed

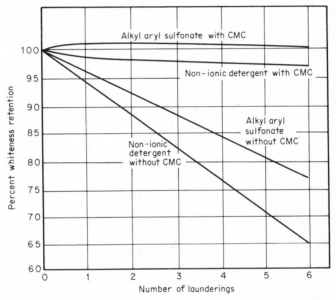

Figure 4.9 CMC prevents soil redeposition. When cloth is washed several times in the presence of soiled cloth, the use of CMC in the detergent keeps the cloth from becoming "tattletale gray." Detergent mix concentration for data shown here was 0.2%.

from the fabrics by the synthetic detergent. This significant use was discovered in the mid-1930s in Europe. The major clothes cleaner prior to World War II was fatty-acid soap. The world turmoil from 1939 through 1945 removed the major fatty-acid supply from the commercial market and channeled it to military use.

Synthetic detergents were found to be efficient soil removers from fabrics; however, in the normal washing and rinsing cycles, a portion of the removed soil was reabsorbed on the fabric, yielding an undesirable appearance sometimes referred to in advertisements as "tattletale gray." The addition of a small amount of CMC to the built detergent system eliminated this undesirable effect. Low-assay and crude types of CMC have been found suitable in detergent systems. Concentrations (based on the total system) of 0.2 to 2.0% CMC have been found satisfactory. The effect of CMC (technical detergent grade) on the whiteness of clothes and detergence of typical detergents is shown in Fig. 4.8.

The addition of CMC improves whiteness retention of unsoiled fabrics after washing several times in the presence of soiled material, as is shown in Fig. 4.9.

The antisoil redeposition properties of CMC are believed to be caused by electrostatic repulsion between the negatively charged dirt particles and the negativity induced by the CMC absorbed on the fabrics. Another theory advanced is that the colloid system developed by the CMC polymer is sufficient to keep the dirt particles from reestablishing on the surface of the fabric.

The advent of fabric blends (e.g., polyester-cotton) has not changed the need for an antisoil redeposition agent in detergent systems. The latter, newly developed and highly accepted fabric blends lend some validity to the colloid theory on the functionality of CMC in a detergent system.

Petroleum In the petroleum industry, CMC has long been established as a water-soluble colloid in drilling-fluid systems. Drilling muds are used to lubricate the bit and develop a medium to facilitate removal of cuttings from the bore hole. In these specialty petroleum fluid systems, water-loss inhibitors are used to control viscosity and reduce the loss of water into porous strata. The fluids commonly used contain various aqueous clay dispersions, surfactants, weighting agents, and other necessary ingredients needed according to the type of earth constituents at the particular drilling site.

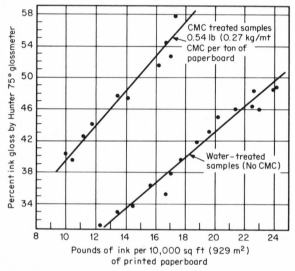

Figure 4.10 CMC-ink relationship for the gloss of CMC-treated and water-treated paperboard at various add-ons.

The suspending properties, along with the compatibility and economics, make CMC especially suitable for petroleum-field fluids. Another very important facet of the utilization of CMC is the ability to build into CMC the properties necessary to meet specific petroleum-producing fluids requirements which vary within a particular producing area as well as throughout different parts of the world.

The development of secondary and tertiary recovery methods for oil recovery may further expand the need for special property types of CMC.

Paper The film-forming and oil-resistance properties of CMC have made this water-soluble polymer a valuable and much-used additive in the paper industry. CMC is used in surface sizing, coating colors, and pulp furnish preparation in paper manufacture.

After the development of a highly purified CMC, solutions of CMC were used on calendar stacks as a surface sizing on paperboard to give gloss-ink holdout, food-grease resistances, wax holdout, fiber bonding, increased wax pick, and other surface characteristics where a highly efficient film-former is required. The highly purified type of CMC is necessary to eliminate the possibility of salt corrosion on the highly polished steel rolls of a paperboard calendar stock.

An accepted paper usage for many years for CMC has been gloss-ink holdout on paperboard. Figure 4.10 shows the improvement effected with a calendar treatment of as little as 0.25 kg of CMC per metric ton of paperboard (0.5 lb/ton).

CMC is also used for the sizing of paper. Here the solution is applied at a size press (tub sizing) and gives the paper higher densometer, improved ink resistance, higher wax pick, and improved smoothness.

In coating colors, which are used both on paper and paperboard, CMC aids in controlling the flow properties of a coating-color solution. The rheology of the coating color at the instant of application is critical to the final properties of the paper product. Proper rheology, aided by a water-soluble colloid, will promote improved final surface characteristics. One theory suggests that this is a result of the delay in pigment dewatering at the point of application, in turn promoting smoothness of the final coating.

Internal bonding of the paper fibers is improved by addition of special types of CMC, when used under proper conditions. This yields greater dry strengths in Fourdrinier products, and improved ply bonding in cylinder machine products. CMC is a highly efficient film-former, and is used alone or in conjunction with plain or modified starches, depending upon the desired properties and economics involved.

Textiles The textile industry utilizes CMC in cloth manufacture and in the finishing of the fibers.

The use of CMC as a low add-on warp sizing began in the early 1950s, prompted primarily by the increasing demand on the textile manufacturers in the United States to reduce the pollution load in plant effluents. The lesser amount of CMC required

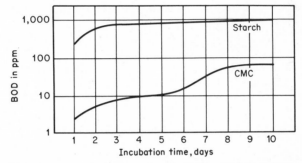

Figure 4.11 BOD of CMC compared with starch. After 10-day incubation, CMC exerts only one-tenth as much BOD as starch does.

to effect proper sizing, as well as the far lower biological oxygen demands (BOD) of CMC as compared with traditionally used starch accelerated its use in warp sizing through the late 1950s and the 1960s.

BOD and Desizing Wastes Compared to starch, a major advantage in the use of CMC warp size is a reduction in treatment costs in cases where stream pollution is a problem. Figure 4.11 compares the BOD of CMC with that of starch. The inherently low BOD of CMC warp size often permits discharging of desizing wastes directly into a nearby stream, a practice which may be prohibited when using starch sizes. CMC warp size is more soluble than starch, and desizing is simplified, since enzymes are not required. As a consequence of the research stimulated by the pollution problem that arose when blends (polyester-cottons primarily) became accepted by the trade, it was discovered that CMC was a more efficient size, especially in the high-humidity weave rooms. Through the 1960s, CMC lowered textile mill pollution loads and became more and more accepted as the major warp size, on blends primarily, although it was found that CMC performed quite well on all cottons.

Another major factor which had been overlooked before arose in the 1970s. The warp sizing is removed after weaving, prior to finishing. This is accomplished traditionally on starch with enzymes and water washing with volumes of hot water, producing the previously mentioned pollution load. However, the cost of energy and water has increased dramatically, and this has increased desizing costs. CMC can be removed easily with much less water and at lower temperatures than any other normally used size (either starch-based or low add-on types).

CMC is usually used as a 6% solution, with small amounts of textile wax added. Sizing add-on with CMC is 5 to 10%, as compared with 15 to 25% for starches. Operations are relatively trouble-free. The film forming of CMC is the key property in this use, although CMC's moisture affinity is important in the weave room since moisture will plasticize the CMC film and thus prevent "shedding" on the high-speed looms. This "shedding" reduction, a cleanliness factor, is only one part of the overall mill efficiency improvement realized by using CMC.

The use of CMC in finishing systems (e.g., various printing pastes, etc.) is also an important textile usage. Compatibility in varying pH ranges and easy removability are two advantages in the utilization of CMC in textile finishing. The rheology of a finishing system under today's high-speed printing conditions demands special water colloids to fit special needs. The types of CMC used in finishing vary widely.

Food   The use of sodium carboxymethylcellulose is growing in food applications, especially in the developed countries where convenience and fabricated foods have been growing rapidly since the early 1950s. The highest purified form (99.5+%) is used in food applications and, as mentioned previously, is referred to in the United States as *cellulose gum.*

Cellulose gum has been introduced into food systems primarily to maintain an original property and/or properties for as long a period of time as practical. Also, cellulose gum has been used in fabricated and convenience foods to give a property and/or properties desirable to make the product acceptable. In ice cream (frozen desserts), cellulose gum inhibits the formation of ice crystals, thus enabling ice cream to withstand thermal shock. Also, "mouth-feel" is improved in ice cream. Instant-type dry beverages contain cellulose gum to improve the mouth-feel, suspend pulp solids, give body, and stabilize the system. Rapid viscosity buildup is necessary in instant-type dry beverages, and cellulose gum is utilized for this application.

Dehydrated foods, such as fruit and vegetable juices and dry soup powders, are examples of uses in which cellulose gum functions to facilitate hydration and also adds to the texture of the final product.

In bakery products, moisture retention, batter viscosity control, and improved keeping qualities of baked products are properties enhanced by the addition of cellulose gum. Pet food (semimoist and canned dry-gravy types) is a product area where the cellulose gum properties are utilized to give a rapid gravy viscosity build-up, as a processing aid for extruded and canned products, for moisture retention, and for appearance enhancement.

Cellulose gum is physiologically inert and has no caloric value. Its use in low-calorie foods to give desirable properties is evident. Examples of the properties of cellulose gum in various food applications are given in Table 4.10.

TABLE 4.10
Cellulose Gum Properties and Their Applications in Food Products

Cellulose gum action	Products benefiting from action
Texturizing	Frozen desserts, desserts, low-calorie products
Thickening	Bakery batters, pet foods, gravies, low-calorie foods, soups, sauces, snack foods, instant-makeup beverages, desserts
Bodying	Beverages, desserts, syrups
Mouth-feel	Beverages, frozen desserts, desserts, syrups, low-calorie products, instant-makeup beverages,
Stabilizing	Desserts, toppings, bakery goods, snack foods, frozen eggs
Moisture retention	Baked goods, cakes, meringues, pie fillings, confections, pet foods
Suspending	Low-calorie products, beverages, fruit drinks, protein beverages, instant-makeup beverages, desserts
Binding	Extruded foods, pet foods, snacks, specialty foods

Incorporating cellulose gum into processed foods, convenience foods, and specialty-type foods varies considerably according to the necessary conditions. Dry-mixing with other ingredients, prewetting with suitable organic humectants, and the use of various mixing techniques (as described under Mixing) are all used today.

Cellulose gum will withstand boiling, freezing, and subfreezing temperatures for a short time—useful properties for the processing of foods and the fabrication of specialty foods.

The United States Food and Drug Administration, the *Food Chemicals Codex,* and the Food and Agriculture Organization of the United Nations World Health Organization (FAO/WHO) all recognize the use of sodium carboxymethylcellulose (99.5+% purity) as a direct food additive. In the United States, establishment of cellulose gum as an accepted common name for high-purity sodium carboxymethylcellulose resulted from a Hercules petition granted by order of the United States Deputy Commissioner of Food and Drugs, effective June 26, 1963.

Future uses will include specialty cellulose gum types now under development. These include ultra-high-viscosity types, improved acid- and base-tolerance types, and a special potassium salt variety to give the desirable cellulose properties without addition of the sodium ion.

Coatings   The advent of the water-based latex paints presented a need for specialty CMC types to meet the rheology requirements of the point of flow under the roller or brush. The availability of property variation in manufacture of CMC to meet the specific coating-system requirement has been greeted with acceptance by the coatings industry. Viscosity control, suspension properties, and in-use rheology control are all needed in water-based paint systems. This application is not unlike the use of CMC in coating colors used in the paper industry. Both systems usually contain pigments and latex binders in a high-solids system. Both depend upon the flow characteristics at the point of application for final coating properties. The use of a water-soluble colloid is vitally important in careful control of the in-use flow characteristics.

Types and concentrations of CMC used vary considerably according to the system, although the usable concentration would be 1 to 3% based on the total system weight. The growth of CMC use for latex paints has been primarily because it has (1) the necessary properties for rheology control and (2) the ability to vary these properties to fill specific system requirements.

Cosmetics and Pharmaceuticals   CMC has many specialty applications in cosmetics, where water binding, viscosity control, suspension of solids, film forming, and physiological inertness of the water-soluble gum are important. Specialty types of highly purified CMC are used to give the property and/or properties desired in the final cosmetic products, such as hand lotions, shampoos, hair treatment compounds, and many other beauty aid products.

Toothpaste is a system with extremely demanding requirements on the necessary water-soluble colloid. In a toothpaste system, specialty types of CMC give stand-up qualities for the paste on the brush, the property of consistency so necessary from a mouth-feel standpoint. CMC also retains moisture to prevent drying and, due to its thixotropy, controls the "at rest" flow characteristics of a paste. Each toothpaste system presents special needs which have to be met by a highly controlled CMC. Toothpaste systems are very lean in water (one of the best solvents for CMC); thus CMC properties have to be carefully controlled to be compatible in the toothpaste system and yet give the final above-mentioned properties economically.

Pharmaceutical uses of CMC include calamine lotions, salt-containing systems, delayed-release tablet systems, ointment bases, bulk-laxative products, and many other systems where water control is necessary.

Miscellaneous Applications   The binding, film-forming, thickening, suspending, and water-holding properties of CMC are useful properties in many widely varied applications not covered previously. Again, the control of water is the major concern.

Ceramic glazes require incorporation of a flow-control agent prior to firing. This agent preferably is a burn-out material, its purpose being to protect the glaze from "sag" prior to final firing. CMC has successfully fulfilled this function, utilizing all levels of viscosity types according to the particular glaze formulation.

CMC has been used to provide viscosity control in aerial-drop fire-fighting systems. The CMC prevents dispersion of the phosphate-containing fluid upon impact. Rapid makeup, compatability, and storage stability are all required properties in this unique system.

Another successful use of CMC is as a slip agent in ceramic-type coatings for welding rods. Again, the burn-out characteristic of CMC is an advantage in this ceramic-type application.

Nonstaining wallpaper adhesives are a major application for CMC in some areas of the world. Characteristics of CMC which are important for this application are its ease of slip, nonspoiling property, high adhesive efficiency, and ease of makeup.

Its suspending property and its compatibility with many ion-containing systems facilitate the use of CMC in auto and other industrial polishes.

Specialties

The common salt of carboxymethylcellulose is sodium carboxymethylcellulose. The solubility characteristics, manufacturing procedures, and general uses all quite conveniently accommodate the sodium salt. Other salts of carboxymethylcellulose are possible; e.g., the monovalent alkali metals all can be used to make salts of carboxymethylcellulose. Manufacturing reaction and purification techniques vary considerably, based on the particular hydroxide form of the salt used.

The potassium salt of carboxymethylcellulose can be prepared using potassium hydroxide in place of sodium hydroxide. Potassium hydroxide is a common reactant and is readily available. The manufacturing conditions and purification system are similar to those used by most manufacturers in preparation of the sodium salt utilizing sodium hydroxide.

The major advantage of the potassium salt of carboxymethylcellulose is the exclusion of the sodium ions in the resultant product system. In some cases, especially in convenience and/or fabricated foods, using a building block based on potassium rather than sodium would be quite helpful to the formulator, since one of the goals in fabricating any food system is to keep the ion balance as near as possible to that of the similar product occurring in nature.

The potassium salt of carboxymethylcellulose has been produced commercially, and toxicological studies are under way in the United States on this unique and quite convenient salt of carboxymethylcellulose. From preliminary data, it does not appear that the physiological inertness of potassium carboxymethylcellulose will differ from that of the already proven and approved sodium salt of carboxymethylcellulose.

The divalent calcium salt of carboxymethylcellulose has been prepared for use in highly specialized applications, primarily in the pharmaceutical industry. This salt is insoluble in water, although it can be solubilized by a weak ammonium hydroxide solution. The advantage of this type carboxymethylcellulose is again the exclusion of sodium ions in systems where this cation could be harmful or detrimental to the final use of the product. Production of the calcium salt is by use of calcium hydroxide with CMC-free acid, with subsequent purification with large excesses of water to remove the reactant soluble salts. To date, this has not been a major product in the world market, although for special applications it could be a quite useful specialty.

The free-acid form of carboxymethylcellulose has been used in a number of applications where, again, the sodium salt is detrimental. This form of carboxymethylcellulose is prepared by acidifying a solution of sodium carboxymethylcellulose with a mineral acid. At a low pH, the free-acid form of CMC is precipitated and then can be purified with large volumes of water to remove the soluble salts. The resultant free-acid CMC can be solubilized by ammonium hydroxide. Free-acid CMC can be used as a binder in systems where the sodium salt is not desired, or in any system where an alkaline salt would be detrimental to the final product. Manufacturing the free-acid salt requires reaction and purification beyond that required for the normal manufacture of sodium carboxymethylcellulose.

Because of advancing CMC technology specialty ultra-high-viscosity types of CMC heretofore unavailable are now being presented to the market. These 1% solution grades of CMC (4.5 to 6.0 Pa·s, or 4500 to 6000 cP) have been commercially produced in both DS-0.7 and DS-0.9 types.

Future Developments

The future of sodium carboxymethylcellulose is quite bright, especially in view of the growing environmental awareness throughout the world. It is evident that considerable effort is being taken in a number of areas to change from solvent-based systems to water-based systems to alleviate environmental problems. In a number of these systems, a water-soluble film-former, protective colloid, thickening agent, stabilizer, and suspending agent is necessary. In the future, controlling the movement of vast quantities of water appears to be quite important. In controlling water, hydrophilic colloids are going to be needed, and one of the most readily available from the economic and property versatility standpoints is sodium carboxymethylcellulose.

In the laboratory, extremely high DS (approximately 1.6) types have been produced which develop higher viscosities in high pH systems than any commercially available DS types today. These high DS types would yield almost gel-free solutions and would exhibit less viscosity variability in salt systems. In many fabricated-food systems, as well as in industrial systems, these controlled properties would be useful.

In convenience and fabricated foods, the use of sodium carboxymethylcellulose and other salts of carboxymethylcellulose will be needed. The production, processing, and distribution of a number of food products to the ultimate consumer require that in the future the preparation of "fabricated" foods incorporating hydrophilic colloids will gain importance because they can be readily produced and easily stored and distributed.

The increased use of hydrophilic colloids in foods is thought to be growing at a faster rate than the production of many of the natural gums used in the food industry. Control of the properties and production of sodium carboxymethylcellulose is not influenced greatly by weather or local economic influences.

Sodium carboxymethylcellulose is used in textile, mining, and petroleum industries because of the large volumes of water necessary in processing products from these industries, and its uses are expanding with the growth of these industries. The ability of the various sodium carboxymethylcellulose manufacturers to tailor the properties of the final colloid to meet specific needs is an advantage to the various users, and will further expand CMC usage and the number of grades of CMC available.

World Production

World production figures for CMC are at the very best only approximate as a result of many factors. These include lack of data from several parts of the world (especially in the Eastern bloc countries), varying bases for reporting production (some producers use total production regardless of purity, while others report on 100% CMC basis), and known producers who do not report production.

The current world production of all grades of CMC (50% to 99.5+% purity) is estimated at 150,000 metric tons. Again, this is only an estimated production; however, the size alone demonstrates that in the relatively short period of time since the mid-1940s CMC usage has grown tremendously. This growth is attributed to the wide variety of uses for CMC, the ability of producers to manufacture at a reasonable cost a wide variety of types to fit the expanding needs, and the increasing world need to control the flow of water.

CMC is produced in most areas of the world in one form or another. Over the years, primarily since the world conflict of the 1940s, producers have commenced operations, discontinued operations, merged, changed names, and expanded. No listing would be complete or be fully accurate at any single time; however, in the next section the authors have attempted to list the world producers and tradenames as an aid to the reader. Omissions and obsolete tradenames are regretted, but as pointed out above, the CMC-producing industry is ever changing.

TRADENAME GLOSSARY

AKU-CMC	Akzo Plastics bv	Holland
Biscon	Kyoto Sosii	Japan
Blanose	Novacel	France
Bomud	Bononic	Brazil
Boniadril	Bononic	Brazil
Boniasol	Bononic	Brazil
Bonracel	Bononic	Brazil
Buckeye CMC	Buckeye Cellulose Corp.	North America
Carbocel	Lamberti	Italy
Carbose	BASF-Wyandotte Corp.	North America
Carboxal	Chemical Development of Canada	North America
Cekol	Uddeholm	Sweden
Cellofas B	ICI Ltd.	United Kingdom
Cellofix	Svenska Cellulosa	Sweden
Cellogen	Daiichi Koggo Seiyuku Co. Ltd.	Japan
Cellucol	Adashi Koryo	Japan
Cellujel	Svenska Cellulosa	Sweden
Cel-Pro	Cellulose Products of India Ltd.	India

CMC	Industries of India Ltd. Gujchem	India
CMC-T	Hercules Inc.	North America
CMC-Warp Size	Hercules Inc.	North America
Collowell	Svenska Cellulosa	Sweden
Copagen	Cooppal	Belgium
Courlose	British Celanese Ltd.	United Kingdom
Daicel-CMC	Daicel Co. Ltd.	Japan
Edifas B	ICI Ltd.	United Kingdom
Extropol	Quimica Amtex	Colombia
Finnfix	Aanekoski Chemical Works	Finland
HK Sodium CMC	H. Kohnstamm Co., Inc.	North America
Horsil	Henkel GmbH	Germany
Kaserose	Shikoku Kasei	Japan
Kiccolate	Nechirin Kasei	Japan
Lucel	Milan Blogojevic	Yugoslavia
Majol	Uddeholm	Sweden
Methylan	Henkel GmbH	Germany
Nymcel	Nijma	Holland
Phrikolat	Phrix	Germany
Proger Sodio CMC	Color Prager S.A.	Spain
Purified CMC	Hercules Inc.	North America
Relatin	Henkel GmbH	Germany
Sodium carboxymethylcellulose	Essex International	North America
Sunrose	Sanyo Kokusaku Pulp	Japan
Tylose C	Kalle/Hoechst	Germany
Walsroder CMC	Wolff	Germany
Zellin	Finnowtal	Finland

FURTHER READING

This section contains some of the more pertinent sources of information relating to use of sodium carboxymethylcellulose in various industries.

Adhesives

1. Casey, J. P., and E. R. Lehman (to A. E. Staley Mfg. Co.), *Dry Potential Adhesive Compositions,* U.S. Patent 3,015,572 (1962); "Electrostatic Flocking Makes Gains," *Chem. Eng. News,* **43**(2) (January 11, 1965).

Analysis

2. Hercules, Inc., *Analysis Procedures for Assay of CMC and Its Determination in Formulations,* Bulletin VC-472A, 1971; *Sodium Carboxymethylcellulose Testing,* ASTM D 1439-65, American Society for Testing and Materials, Philadelphia.
3. Black, H. C., "Determination of Sodium Carboxymethylcellulose in Detergent Mixtures by the Anthrone Method," *Anal. Chem.,* **23**, 1792–1795 (1951).
4. Conner, A. Z., and R. W. Eyler, "Analysis of Sodium Carboxymethylcellulose," *Anal. Chem.,* **22**, 1129–1132 (1950).
5. Eyler, R. W., and R. T. Hall, "Determination of Carboxymethylcellulose in Paper," *Paper Trade J.,* **125**(15), 59–62 (1947).
6. Eyler, R. W., E. D. Klug, and F. Diephuis, "Determination of Degree of Substitution of Sodium Carboxymethylcellulose," *Anal. Chem.,* **19**, 24–7 (1947).

Cement

7. Ludwig, N. C. (to Universal Atlas Cement Co.), *Retarded Cement,* U.S. Patent 2,673,810 (1954).
8. Starnes, P. E., *Cellular Concrete,* U.S. Patent 2,635,052 (1953).

Ceramic

9. Danielson, R. R., "How to Glaze Structural Clay Products," *Brick Clay Rec.,* **121**(3), 53–55, 1952.
10. Stawitz, J., "Water-Soluble Cellulose Ethers as Auxiliary Materials in the Ceramic Industry," *Tonind. Ztg.,* **77**, 14–15, 1953.

Cosmetic

11. Knechtel, A. H., "Cellulose Ethers: What They Are and Their Use in Cosmetics," *Am. Perfum.,* **78**, 95–97 (October 1963).
12. Rufe, R. G., "Cellulose Polymers in Cosmetics and Toiletries," *Cosmet. Perfum.,* **90**, 93–100 (March 1975).

13. Lesser, M. A., "Cellulose Derivatives in Drugs and Cosmetics," Parts I and II, *Drug & Cosmet. Ind.*, **62**(5), 612–614, 670–671, 692–694; **62**(6), 750–752, 830–832 (1948).
14. Mehoffey, R. J. (to Colgate-Palmolive Co.), *Hand Lotion*, U.S. Patent 2,678,902 (1954).
15. Scapparino, G., "Carboxymethylcellulose in Cosmetics," *Riv. Itali. Essenze Profumi Piante Off. Olii Veg. Saponi*, 34, 343–344 (1952).

Emulsion

16. Matreyek, W., and F. H. Winslow, "Particle Size in Suspension Polymerization," *Ind. Eng. Chem.*, **43**, 1108–1112 (1951).
17. Morrison, R. I., and B. Campbell, "Water-Soluble Cellulose Ethers as Emulsifying Agents," *J. Soc. Chem. Ind.*, **68**, 336–6 (1949).
18. Neil, C. E., and P. E. McCoy (to American Bitumuls and Asphalt Co.), *Oil-in-Water, Coal Tar Emulsions*, U.S. Patent 2,670,332 (1954).

Food

19. Batdorf, J. B., "How Cellulose Gum Can Work for You," *Food Eng.*, **36**, 66–68 (August 1964).
20. Batdorf, J. B., and J. B. Klis, "Makes Cakes Easier to Prepare," *Food Process.*, (May 1963).
21. Bayfield, E. G., "Improving White Layer Cake Quality by Adding CMC," *Baker's Dig.*, **36**(2), 50–52 (April 1962).
22. Burt, L. H. (to Hercules Inc.), *Carboxymethylcellulose in Ice Cream*, U.S. Patent 2,548,865 (1951).
23. Burt, L. H. (to Hercules Inc.): *Fruit and Vegetable Processing with CMC*, U.S. Patent 2,728,676 (1955).
24. "CMC: Versatile Gel Former," *Food Eng.*, **33**, 92 (May 1961).
25. Desmarais, A. J., and A. J. Ganz, "Effect of Cellulose Gum on Sugar Crystallization and Its Utility in Confections," *Manuf. Confect.*, **42**(10), 33–36 (1962).
26. Ganz, A. J. (to Hercules Inc.), *Stabilizing Icings with CMC and Gluten*, U.S. Patent 3,009,812 (1960).
27. Ganz, A. J. (to Hercules Inc.), *Making Stable Foams by Codrying CMC and Egg Whites*, U.S. Patent 3,287,139 (1966).
28. Ganz, A. J. (to Hercules Inc.), *Peptization of Protein with CMC*, U.S. Patent 3,407,076 (1968).
29. A. J. Ganz (to Hercules Inc.), *Prehydrated CMC for Instant Solubility*, U.S. Patent 3,485,651 (1969).
30. Ganz, A. J., "CMC and Hydroxypropyl Cellulose: Versatile Gums for Food Use," *Food Prod. Dev.*, 3(6) 65–71 (1969).
31. Levin, H. M. (to HCA Corp.), *CMC in Salad Dressings*, U.S. Patent 3,414,413 (1968).
32. Seas, S. W., and K. R. Spurgeon, "New Spread-Type Dairy Products," *S.D. Agric.* Exp. Stn., Tech, *Bull. 543* (April 1968).
33. Young, W. E., and E. G. Bayfield, "Hydrophilic Colloids as Additives in White Layer Cakes," *Cereal Chem.*, **40**(3) (May 1963).
34. Gould, A. A., "Recent Developments in Ice Cream," *Ice Cream Rev.* (June 1949).
35. Josephson, D. V., and C. D. Dahle, "A New Cellulose Gum Stabilizer for Ice Cream," *Ice Cream Rev.* (June 1945).
36. Onderzoekinginstituut "Research" N.V., *Milk Products Containing Sodium Carboxymethylcellulose*, Dutch Patent 73,644 (1953).
37. Perech, R., *Vegetable-or-Fruit-Juice Concentrate*, U.S. Patent 2,393,561 (1946).
38. Pompa, A., "Recipes for Ice Cream Utilizing Sodium Carboxymethylcellulose as the Stabilizer," *Food*, 14(231) (1945).

General

39. Baird, G. S., and J. K. Speicher, "Carboxymethylcellulose," in R. L. Davidson and M. Sittig (eds.), *Water-Soluble Resins*, Reinhold Publishing Corporation, New York, 1962, chap. 4.
40. Batdorf, J. B., and J. M. Rossman, "Sodium Carboxymethylcellulose," in R. L. Whistler (ed.), *Industrial Gums*, rev. ed., Academic Press, Inc., New York, 1971, chap. 25.
41. Batdorf, J. B., and P. S. Francis, "The Physical Behavior of Water-Soluble Cellulose Polymers," *J. Soc. Cosmet. Chem.*, **XIV**(3) (March 1963).
42. Brown, W., and D. Henley, "The Configuration of the Polyelectrolyte Sodium Carboxymethylcellulose in Aqueous Sodium Chloride Solutions," *Makromol. Chem.*, **79**, 68–88, 1964.
43. Butler, R. W. (to Hercules Inc.), *Free Acid Cellulose Ether Film*, U.S. Patent 3,064,313 (1962).
44. Butler, R. W., and G. I. Keim (to Hercules Inc.), *Insolubilizing CMC with Cationic Epichlorohydrin Modified Polyamide Resin*, U.S. Patent 3,224,986 (1965).
45. DeButts, E. H., J. A. Hudy, and J. H. Elliott, "Rheology of Sodium Carboxymethylcellulose Solutions," *Ind. Eng. Chem.*, **49**, 94–98 (1957).
46. Francis, P. S., "Solution Properties of Water-Soluble Polymers," *J. Appl. Polym. Sci.*, **5**(15), 261–270 (May–June 1961).

47. Gloor, W. E., "A Continuum of Molecular Weight Distribution Applicable to Linear Homopolymers," *J. Appl. Polym. Sci.*, 19(1), 273–279, 1975.
48. Klug, E. D., "Sodium Carboxymethylcellulose," in *Encyclopedia of Polymer Science and Technology*, vol. III, Interscience Publishers, Inc., New York, 1965, pp. 520–539.
49. Ott, E., H. M. Spurlin, and M. W. Grafflin, *Cellulose and Cellulose Derivatives*, 2d ed., part II, Interscience Publishers, New York, 937–945, specifically, and other references throughout the book.
50. Samuels, R. M., "Quantitative Structural Analysis of Mechanical Behavior in Polycrystalline Polymers," *Appl. Polym. Symp.*, (24), 37–43 (1974).
51. Shelanski, H. A., and A. M. Clark, "Physiological Action of Sodium Carboxymethylcellulose on Laboratory Animals and Humans," *Food Res.*, 13(1), 29–35 (1948).
52. Wirick, M. G., "A Study of the Enzymic Degradation of CMC and Other Cellulose Ethers," *J. Polym. Sci.*, A1, 6, 1965–1974 (1968).

Paint

53. Floyd, J. D., J. W. Gill, and M. G. Wirick, "Viscosity Stability of Latex Paints Containing Water-Soluble Cellulose Polymers," *J. Paint Technol.*, 38(498), 398–401 (1966).
54. Jaffee, H. L., "Water-Soluble Thickeners as Flow Control Additives," *J. Paint Technol.*, 32, 706–721 (May 1960).

Paper

55. Afonchikov, N. A., and E. A. Terent'ev, "Effect of Sodium Carboxymethylcellulose on Pulp Beating and the Properties of Paper," *Bum. Prom.*, 38, 1963.
56. Barber, E. J.: "CMC as a Paper Machine Additive," *Tappi*, 44, 179A–183A (February 1961).
57. Casey, J. P., *Pulp and Paper*, 2d ed., Interscience Publishers, Inc., New York, 1960, vols. I–III.
58. Hurst, A. R., "Surface Application to Improve Printability of Paperboard," paper given at Annual Convention, Technical Association of the Pulp and Paper Industry, February 1957.
59. Kohne, H., Jr., "Use of CMC as a Coating Adhesive," *Tappi*, 42, 294–298 (April 1959).
60. Ware, H. O., and E. J. Barber, "Water- and Alkali-Soluble Cellulose Derivatives in Paper and Paperboard," *Tappi*, 5(40), 365–373 (1957).
61. Honma, N. J., "Carboxymethylcellulose Sizing," *Jpn. Technol. Assoc. Pulp Pap. Ind.*, 4(2), 10–14 (1950).
62. Horsey, E. F., "Sodium Carboxymethylcellulose for Paper-making," *Pap. Trade J.*, 125(4), 52–56 (1947).
63. Horsey, E. F., and D. Price (to Hercules Powder Company), *Sizing of Pulp with Rosin and Carboxymethylcellulose*, U.S. Patent 2,572,932 (1951).
64. Wurz, O., "Experiences with Carboxymethylcellulose in the Paper Industry," *Das Papier*, I, 377–381 (1953).

Pharmaceutical

65. Allen, B. F., "Pharmaceutical Applications of Sodium Carboxymethylcellulose," *Md. Pharm.*, 37, 612–616, 1961.
66. Dekay, H. G., "A Review of Some Dispersing and Suspending Agents," *Am. J. Hosp. Pharm.*, 9, 520–523 (1952).
67. Ishii, K., H. Nakatani, and F. Imai (to Takeda Chem. Ind. Ltd.), *Stabilization of Barium Sulfate Suspension*, Japanese Patent 3297, 1964).
68. Inone, H., and S. Imaza, *Vaginal Antiseptic*, Japanese Patent 2696 (1950).
69. Lowry, M. L., "A Third Report: A New Hemostatic Agent," *Arch. Surg.*, 60, 793–805 (1950).
70. Necheles, H., and H. Kroll (to Michael Reese Research Foundation), *Antacid Preparation*, U.S. Patent 2,477,080 (1949).
71. Schultz, J., "Carboxymethylcellulose as a Colloid Laxative," *Am. J. Dig. Dis.*, 16(9), 319–322 (1949).
72. Stawitz, J., "The Water-Soluble Cellulose Ethers in Pharmacy and Medicine," *Pharm. Ind.*, 12, 39–44, 71–75, 90–93 (1950).
73. Yalcindag, O. N., "A Sodium Carboxymethylcellulose, Water and Glycerol Mixture as a New Ointment Base," *Am. J. Pharm.*, 124, 386–389 (1952).

Rubber

74. Dennstedt, I., W. Becker, and G. Fromandi (to Farbenfabriken Bayer A. G.), *Lubricants for Rubber Molds*, German Patent 824,553 (1951).
75. Livinston, H. K., "Creaming Neoprene Latex," *Ind. Eng. Chem.*, 39, 550–554 (1947).

Soaps and Detergents

76. Batdorf, J. B., "CMC in Liquid Detergents," *Soap Chem. Spec.*, 38, 58–61 (January 1962).
77. Jarrell, J. G., and H. B. Trost, "Carboxymethylcellulose, an Aid to Washability," *Soap Sanit. Chem.*, 28(7), 40–43, 155; (8), 50–52, 163 (1952).

78. Trost, H. B., "Soil Redeposition," *J. Am. Oil Chem. Soc.*, **40**(11), 669–674 (November 1963).
79. Niewenhuis, K. J., Carboxymethylcellulose as Soap-Forming Washing Agent, *Meded. Proefstn. Wasind.* (Netherlands), **66**, 10 pp. (1947).
80. Pollok, F. J., "Sodium Carboxymethylcellulose as a Detergent Aid," *Perfum. Cosmet.*, **24**, 991–995 (1951).
81. Suter, H. R., and M. G. Kramer, "Detergency Properties of Systems Containing a Solid Nonionic Detergent," *Soap Sanit. Chem.*, **27**(8), 33–36, 149 (1951).
82. Weatherburn, A. S., "The Influence of Sodium Carboxymethylcellulose on the Suspending Power of Built Soap Solutions," *Text. Res. J.*, **20**, 510–513 (1950).
83. Wilkinson, C. R., "Modern Wool Scouring," *Dyer, Text. Printer, Bleacher, Finisher*, **105**, 627–629, 691–692 (1951).

Textile Sizing

84. Baird, G. S., and A. L. Griffiths, "The Advantages of CMC as a Warp Size," *Mod. Text.* (June 1965).
85. "CMC Kayoed Stream Pollution," *Text. Ind.*, **124**, 161–162 (October 1960).
86. Gardner, R. T., Jr., "CMC as Aerosol Fabric Size," *Soap Chem. Spec.*, **38**, 98–101, 122 (1962).
87. "Preventing Soil Redeposition with CMC," *Soap Chem. Spec.*, **35**, 135 (April 1959).
88. Trost, H. B., and H. B. Bush: "Sizing Up War Sizes," *Text. Ind.*, **134**, 127–134 (June 1970).
89. "CMC Warp Size—Abates Stream Pollution," *Hercules Chemist*, **26**, 14–17 (1956).
90. Dreyfus, H., W. A. Dickey, and P. F. C. Sawter (to Celanese Corp. of America), *Warp Sizing of Yarns with Aqueous Solutions of Sodium Carboxymethylcellulose*, U.S. Patent 1,950,664 (1934).
91. Elliott, J. H. (to Hercules Powder Company), *Textile Sizing with Zirconium Salts of Water-Soluble Carboxymethylcellulose*, U.S. Patent 2,650,887 (1953).
92. Ferdinand School A. G. *Size Solutions and Size Compositions Containing Salts of Cellulose Ether Carboxylic Acids as Their Essential Constituents*, British Patent 496,351 (1938).
93. Hart, R., "An Outline on Warp Sizing of Spun Rayon," *Rayon Text. Mon.*, **27**, 192–195 (1946).
94. Kitazawa, F., "Sorption of Carboxymethylcellulose (CMC) on Synthetic and Rayon Stable Yarns," *Bull. Nagoya Inst. Technol. (Anniversary Issue)*, **4**, 155–158 (1952).
95. Seydel, P., "Cotton Warp Sizing IV," *Cotton (Atlanta)*, **110**(12) 82–84, 149 (1946).
96. Seymour, R. B., and G. M. Schroeder (to Henry H. Frede Co.), *Sodium Carboxymethylcellulose in Textiles*, U.S. Patent 2,486,803 (1949).
97. Sponsel, "Cellulose Ether Sizing and Finishing Agents," *Melliand Textilber.*, **19**, 738–739 (1938).
98. Thomas, E. B., and H. F. Oxley (to Celanese Corp. of America), *Sizes and Coating Compositions*, U.S. Patent 2,135,128 (1938).
99. Wiegerink, J. F., *Ironing Aid and Textile Refinishing Composition*, U.S. Patent 2,645,584 (1953).
100. Wurz, E., and O. Wurz, "Applications of Carboxymethylcellulose in the Textile Industry," *Text. Rundsch.*, **8**, 557–560 (1953).

Toxicological/Biological

101. Rowe, V. K., H. C. Spencer, E. M. Adams, and D. D. Irish, *Food Res.*, **9**, 175 (1944).
102. Shelanski, H. A., and A. M. Clark, *Food Res.*, **13**, 29 (1948).
103. Brown, C. J., and A. A. Houghton, *J. Soc. Chem. Ind.*, **60**, 254 (1941).
104. Werle, E., *Chem. Ztg.*, **65**, 320 (1941); *Chem. Abstr.*, **37**, 2807 (1943).
105. Ziegelmayer, W., A. Columbus, W. Klausch, and R. Wieske, *Arch. Tierernaehr.*, **2**, 35 (1951); *Chem. Abstr.*, **46**, 5680 (1952).
106. Schwartz, L., L. W. Spolyar, F. U. Gastineau, J. E. Dalton, A. B. Loveman, M. B. Sulzberger, E. P. Cope, and R. L. Baer, *J. Am. Med. Assoc.*, **115**, 906 (1940).
107. Heuper, W. C., *Am. J. Pathol.*, **21**, 1021 (1945).
108. Lusky, L. M., and A. A. Nelson, *Fed. Proc. Am. Soc. Exp. Biol.*, **16**, 318 (1957).
109. Jasmine, G., *Rev. Can. Biol.*, **20**, 701–707 (1961).
110. Jasmine, G., and R. Gaveau, *Laval Med.*, **37**, 547–550 (1966).
111. Walpole, A. L., *Imperial Chem. Industries*, Proc. Intern. Conf. Perugia, Italy, 83–88 (1961).
112. Frawley, J. P., et al., *Food Cosmet. Toxicol.*, **2**, 539–543 (1964).
113. Y., Nishiyawa, *Gann*, **29**, 1–9 (1935); Y., Nishiyawa, *Gann*, **30**, 419–420 (1936).
114. Nishiyawa, Y., *Gann*, **31**, 223–225 (1937).
115. Nishiyawa, Y., *Gann*, **32**, 85–98 (1938).
116. Nonaka, T., *Gann*, **32**, 234–235 (1938).
117. *Problems in Evaluation of Carcinogenic Hazard from Use of Food Additives*, National Academy of Sciences, NRC, Publications 749 (December 1959).
118. Takizawa, N., *Gann*, **32**, 236–237 (1938).
119. Takizawa, N., *Gann*, **34**, 1–5 (1940).
120. Takizawa, N., *Gann*, **33**, 193–195 (1939).
121. Tokoro, Y., *Gann*, **34**, 149–155 (1940).
122. Petroff, N., and N. Krotkina, *Bull. Assoc. Fr. Etude Cancer*, **17**, 566–589 (1923).
123. Lorenz, E., M. Shimkin, and H. Stewart, *J. Nat. Cancer Inst.*, **1**, 355 (1941).
124. Bech, S., et al., *Cancer Res.*, **5**, 135–139 (1945).

125. Truhart, R., *Ann. Pharm. Fr.*, **5**, 619–623 (1947).
126. Peacock, P. R., *Br. J. Nutr.*, **2**, 201–204 (1948).
127. Hieger, I., *Br. J. Cancer*, **3**, 123 (1949).
128. Hieger, I., *Br. Med. J.*, **14**, 159–160 (1958).
129. White, C. P., *J. Pathol. Bacteriol.*, **14**, 145 (1909).
130. White, C. P., *J. Pathol. Bacteriol.*, **14**, 450–462 (1910).
131. Burrowe, H., I. Hieger, and E. L. Kennaway, *J. Pathol. Bacteriol.*, **43**, 419–426 (1936).

Miscellaneous Applications

132. Batdorf, J. B. and K. J. Fletcher, "Fighting Fires With Viscous Water," *Hercules Chem.*, (47), 11–15, 1963.
133. Davidson, S. H. (to Imperial Chemical Industries, Ltd.), *Explosives Containing Polysaccharide Ethers*, U.S. Patent 2,680,067 (1954).
134. Pearson, S. C., and R. G. Soltis, "Chemical Treatment as a Means of Maintaining Effluent Quality in an Overloaded Activated Sludge Plant: The Western Branch Experience," presented at Industrial Water and Pollution Conference, Detroit, April, 1974.
135. *Spray Drift Control of Biologically Active Compositions*, British Patent 948,185 (1964).
136. Wirick, M. G., "Aerobic Biodegradation of Carboxymethylcellulose," *J. Water Pollut. Control Fed.*, **46** (3), 512–521, 1974.

Chapter **5**

Carrageenan

Kenneth B. Guiseley,
Norman F. Stanley, and
Philip A. Whitehouse
FMC Corporation, Marine Colloids Division

General Information 5-2
   Chemical Nature 5-2
   Physical Properties 5-9
   Manufacture 5-10
   Biological/Toxicological Properties 5-11
   Rheological Properties 5-11
   Additives/Extenders 5-17
   Handling 5-17
   Applications 5-17
   Application Procedures 5-18
   Specialties 5-18
   Future Developments 5-18
Commercial Uses: Compounding and Formulating 5-19
   Milk Applications 5-19
   Water Applications 5-21
   Nonfood Applications 5-22
Commercial Uses: Processing Aids 5-22
   Beverage Clarification 5-22
   Abrasive Suspensions 5-23
   Ceramic Glazes and Core Washes 5-23
Industries Using Carrageenans 5-23
   Food 5-23
   Pharmaceuticals and Toilet Goods 5-23
   Metal Fabrication 5-23
   Ceramics 5-23
   Coatings 5-23
   Agriculture 5-24
   Household Products 5-24
Formulations 5-24
   Chocolate Milk 5-24

Canned Water-Dessert Gel 5-24
Air-Treatment Gel 5-25
Toothpaste 5-25
Milk Puddings 5-25
Antacid Gel 5-26
Laboratory Techniques 5-27
Water Viscosity Measurement 5-27
Water Gel Strength Measurement 5-27
Milk Gel Strength Measurement 5-28
Product/Tradename Glossary 5-28
Further Useful Reading/References 5-29
General References 5-29
References 5-29

GENERAL INFORMATION

Carrageenans are water-soluble gums which occur in certain species of red seaweeds of the Gigartinaceae, Solieriaceae, Phyllophoraceae, and Hypneaceae families. Chemically, they are sulfated linear polysaccharides of D-galactose and 3,6-anhydro-D-galactose (3,6-AG). By virtue of these half-ester sulfate groups, carrageenans are anionic polyelectrolytes.

The charged nature of the sugar units and their structural arrangement within the macromolecule render the carrageenans highly reactive chemically and account for physical properties such as the ability to form gels.

Carrageenans are used commercially as thickening, suspending, and gelling agents. Typical applications are as a thickener or "binder" in toothpaste, a suspending agent for cocoa in chocolate milk, and a gelling agent for milk puddings, water-gel desserts, and air-freshener gels. Selected properties are given in Table 5.1.

Chemical Nature

Structure All carrageenans have the common structural feature of being linear polysaccharides built up of alternating 1,3-linked β-D-galactopyranosyl and 1,4-linked α-D-galactopyranosyl units (Fig. 5.1). The 1,3-linked units occur as the 2- and 4-sulfates, or occasionally unsulfated. The 1,4-linked units occur as the 2- and 6-sulfates, the 2,6-disulfate, the 3,6-anhydride, and the 3,6-anhydride-2-sulfate.[10] Sulfation at C_3 apparently never occurs. This wealth of possibilities for substitution on the basic copolymer admits the possibility of a continuous spectrum of carrageenan types. However, these exist as variants and hybrids of a small number of ideal or limit polysaccharides of definite chemical structure (Fig. 5.2), which for convenience are named by Greek-letter prefixes: *mu-, kappa-, nu-, iota-, lambda-, theta-*, and *xi*-carrageenans.[8,15] Commercially available *kappa-, iota-*, and *lambda*-carrageenans are mixtures which approach the respective ideal types in chemical composition.

In *kappa*-carrageenan, the 1,3- and 1,4-linked units are respectively D-galactose-4-sulfate and 3,6-anhydro-D-galactose. *Mu-* differs from *kappa-* in that the anhydride is replaced by D-galactose-6-sulfate. It is considered to be the biological precursor of *kappa-*. In the seaweed, the change from *mu-* to *kappa-* is catalyzed by the enzyme *dekinkase*.[5] Sulfate at C_6 is eliminated from the 1,4-linked units with concomitant ring closure to form the anhydride. While *kappa-* forms gels with water in the presence of certain cations, notably potassium, its precursor, *mu-*, is nongelling.[1]

Similarly, *nu-* is believed to be the precursor of *iota-*. Chemically they differ from their respective counterparts, *mu-* and *kappa-*, only in having a sulfate group at C_2 on the 1,4-linked units. Again, *nu-* is a nongelling fraction and *iota-* a gelling fraction.[14]

Lambda- is a nongelling carrageenan which differs from *nu-* in that about 70% of the 1,3-linked units are sulfated at C_2 rather than C_4, the remainder being unsulfated. Base-catalyzed S_N2 elimination of 6-sulfate converts the 1,4-linked units to 3,6-anhydro-D-galactose 2-sulfate, as in *iota-*; however, unlike either *iota-* or *kappa-*, the alkali-modified *lambda-*, which has been named *theta*-carrageenan, is nongelling.

Lambda- and *kappa*-carrageenans occur together in the carrageenan extracted from

TABLE 5.1
Selected Properties of Carrageenan

	Kappa-	Iota	Lambda
Solubility:			
Hot water	Soluble above 70°C	Soluble above 70°C	Soluble
Cold water	Na$^+$ salt soluble. From limited to high swelling of K$^+$, Ca^{2+} and NH$_4^+$ salt	Na$^+$ salt soluble. Ca^{2+} salt gives thixotropic dispersions	All salts soluble
Hot milk	Soluble	Soluble	Soluble
Cold milk	Insoluble	Insoluble	Disperses with thickening
Cold milk (TSPP added)	Thickens or gels	Thickens or gels	Increased thickening or gelling
Concentrated sugar solutions	Soluble hot	Difficultly soluble	Soluble hot
Concentrated salt solutions	Insoluble cold and hot	Soluble hot	Soluble hot
Water-miscible solvents	Up to about 30% solvent; see text*	Same	Same
Organic solvents	Insoluble	Insoluble	Insoluble
Gelation:			
Effect of cations	Gels most strongly with K$^+$	Gels most strongly with Ca^{2+}	Nongelling
Type of gel	Brittle with syneresis	Elastic with no syneresis	Nongelling
Locust bean gum effect	Synergistic	None	None
Stability:			
Neutral and alkaline pH	Stable	Stable	Stable
Acid (pH 3.5)	Hydrolysis of solution, accelerated by heat. Gelled state stable.	Stable	Hydrolysis
Compatibility	Generally compatible with non-ionic and anionics, but not with cationics.		

* See "Incompatibilities," page 5-18.

Chondrus crispus and some *Gigartina* species, although they do not occur together in the same plant. *Kappa-* has been shown to occur only in the haploid gametophytic forms of *Chondrus crispus,* and *lambda-* only in the diploid tetrasporophytic form.[7] Since the various forms grow and are harvested together, extraction yields a mixture of the two carrageenans. Their ratio varies, but averages about 70% *kappa-* and 30% *lambda-*.

Xi-carrageenan, which replaces *lambda-* in some *Gigartina* species (*G. chamissoi* and *G. canaliculata*), has not been completely characterized, but seems to differ from *lambda-* in that the 1,3-linked units are completely sulfated at C_2, while at least some of the 1,4-linked units are unsubstituted at C_6. The limit polysaccharide, which so far has not been isolated, would be completely devoid of 6-sulfate.

Carrageenans from different sources may be regarded as varying mixtures of the limit polysaccharides and intermediate species ranging in degree of 3,6-anhydrization and 2-sulfation of the 1,4-linked units. This is shown graphically in Fig. 5.3. This diagram divides the carrageenans into two general groups. One consists of the *mu-*, *nu-*, *kappa-*, and *iota-* types and their hybrids. Carrageenans of this group gel with potassium ions or can be made to gel by alkali treatment; they are characterized by

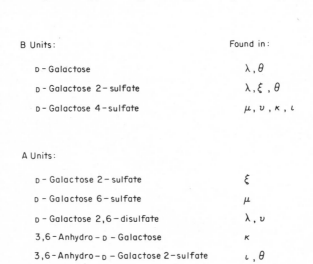

Figure 5.1   Repeating structure of carrageenans.

having their 1,3-linked units sulfated at C_4. The other group consists of the *lambda-* and *xi-* types and their alkali modification products; these do not gel either before or after alkali treatment, and both their 1,4- and 1,3-linked units are partially or completely sulfated at C_2.

Molecular Weight   As is true of both synthetic polymers and naturally occurring polysaccharides in general, carrageenans do not have sharply defined molecular weights, but rather have average molecular weights representing a distribution of molecular species identical in structure but of varying chain length. Various types of average, such as number-average $(\overline{M}_n)$ or weight-average $(\overline{M}_w)$, are defined. Figure 5.4 shows cumulative molecular weight distributions for carrageenans of various average molecular weights.

Commercial food-grade carrageenans typically have number-average molecular weights in the region of 200,000 daltons. The functionality of carrageenans in most food and industrial applications depends on molecular weight, and is largely lost at molecular weights under 100,000 daltons. As Fig. 5.4 shows, food-grade carrageenans have very little material in this region.

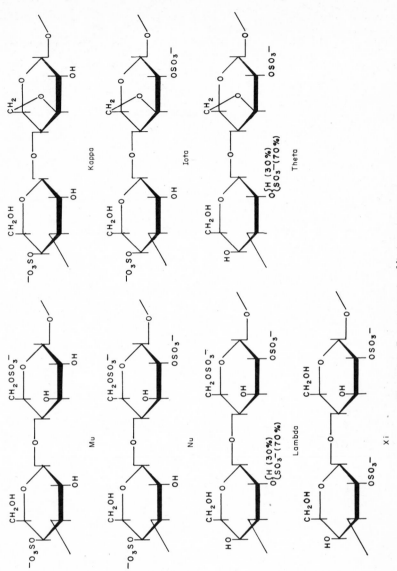

Figure 5.2 Repeating units of limit carrageenans.

Reactivities   The chemical reactivity of carrageenans is primarily due to half-ester sulfate groups, $R-O-SO_3^{\ominus}$, which are strongly anionic, being comparable to sulfuric acid in this respect.   The free acid is unstable, and commercial carrageenans are available as stable sodium, potassium, and calcium salts or, most generally, as a mixture of these. The associated cations together with the conformation of the sugar units in the polymer chain determine the physical properties of the carrageenans.   For example, *kappa*- and *iota*-carrageenans form gels in the presence of potassium or calcium ions.

Reactivity with proteins is exhibited by both gelling and nongelling carrageenans, although regularity of the polymer is an important factor.   In most, if not all, cases

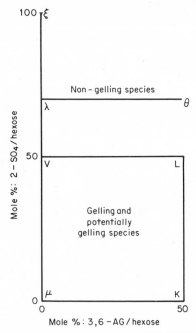

Figure 5.3   Composition of carrageenans.

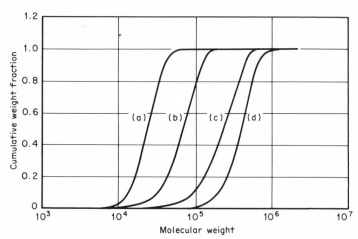

Figure 5.4   Cumulative molecular weight distributions of carrageenans.   *(a) Iota-,* $\bar{M}_n = 14,300$; *(b) Iota-,* $\bar{M}_n = 48,700$; *(c) Iota-,* $\bar{M}_n = 160,000$; *(d) Kappa-,* $\bar{M}_n = 266,000$.

ion-ion interactions between the sulfate groups of the carrageenan and the charged groups of the protein are involved. Reaction depends on protein/carrageenan net charge ratio, and thus is a function of the isoelectric point of the protein, the pH of the system, and the weight ratio of carrageenan to protein.

The extent of reaction is greatest when the net charge ratio is equal to 1. Usually this results in the precipitation of an insoluble protein-carrageenan complex. Figure 5.5 is a somewhat stylized representation of this type of interaction. Below the isoelectric point of the protein, direct polyanion-polycation (carrageenan-protein) interaction can occur (Figs. 5.5III and IV).

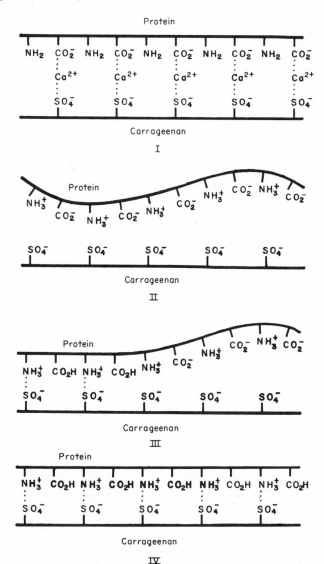

Figure 5.5 Ionic interactions of carrageenan and protein. (I) pH > I.E.P. Interaction mediated by divalent cation. (II) pH = I.E.P. Protein/carrageenan net-charge ratio = 0. No direct interaction. (III) pH < I.E.P. Protein/carrageenan net-charge ratio < 1. Partial interaction; no precipitation. (IV) pH < I.E.P. Protein/carrageenan net-charge ratio = 1. Complete interaction; precipitation.

For some proteins, interaction also occurs above the isoelectric point even though both the carrageenan and protein then have net negative charges. A common explanation for this behavior invokes mediation by calcium or other polyvalent cations (Fig. 5.5I). However, direct interaction can also occur if the protein molecule has exposed regions of contiguous positively charged amino acid residues, even though its net charge may be negative (Fig. 5.6).[12] The *kappa*-casein component of milk has been shown to react with both *lambda*- and *kappa*-carrageenan by this mechanism. The calcium-sensitive milk proteins α_s- and β-casein do not react directly with carrageenans, but do form weak complexes in the presence of calcium ions.[4] It is probable that both types of interaction occur in the commercially important reaction of carrageenans with milk to form gels or to stabilize the suspension of cocoa in chocolate

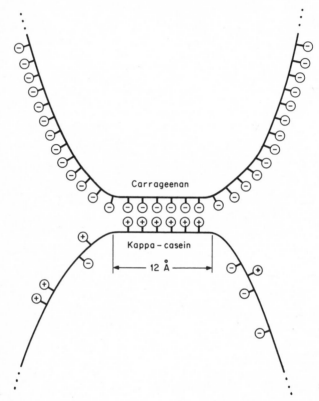

Figure 5.6 Ionic interaction of carrageenan and *kappa*-casein. pH $\geq$ I.E.P.

milk. This reaction is commonly referred to in the carrageenan and dairy industries as *milk reactivity*.

A rather remarkable property of carrageenans is their ability to stabilize casein micelles. In natural milk, the calcium-sensitive α_s- and β-caseins are stabilized in micellar form by a third protein, *kappa*-casein. On a weight-for-weight basis, *kappa*-carrageenan is as effective a micelle-builder as *kappa*-casein, while *lambda*-carrageenan is less effective. *Theta*-carrageenan is more effective than *lambda*-, though less so than *kappa*-.[6] Evidently the presence of 3,6-anhydride, or possibly the concomitant absence of 6-sulfate, is helpful here, though not essential. This property of micelle stabilization may account for the effectiveness of carrageenans as stabilizers for evaporated milk and infant feeding formulas. Proteins other than casein can be stabilized in this manner; an example is the protein in soy-based infant formulas.

A less desirable property of carrageenans is their susceptibility to depolymerization through acid-catalyzed hydrolysis. This is related to 3,6-anhydride content; Fig. 5.7 shows some typical data. Cleavage occurs preferentially at the 1 → 3 glycosidic linkages, and is promoted by the presence of the strained ring system of the anhydride. Sulfation at C_2 of the 1,4-linked units appears to mitigate attack by acid, degradation rates for *iota*- being roughly one-half those for *kappa*-. Carrageenans in the gel state are more stable to acid than those in the sol state. Depolymerization follows pseudo-zero-order kinetics in either state, but the rate for the gel state is about one-sixth of that for the sol state. The secondary and tertiary structures developed on gelation may exert a shielding effect on the glycosidic bonds. This effect permits the use of carrageenans in acid systems, such as relishes, if enough potassium salt is present to develop a gel structure.

The rate of acid-catalyzed degradation is proportional to hydrogen ion activity, with maximum stability at a pH of about 9. Carrageenans are relatively resistant to alkaline degradation, though this does occur through "peeling" reactions at high pH. The rate of acid-catalyzed degradation increases with temperature in accordance with the Arrhenius equation, with activation energies in the range of 84 to 126 kJ/mol (20 to 30 kcal/mol). Carrageenans with high 3,6-AG content tend to have activation energies at the low end of this range. Relations have been developed which afford good estimates of degradation, as measured by viscosity or molecular weight loss, for a given set of processing conditions defined by pH, temperature, and time.

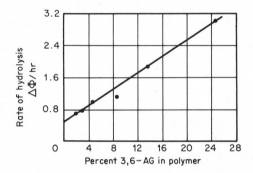

Figure 5.7   Rate of hydrolysis of carrageenans at pH 1.8, 60°C, as a function of 3,6-AG content.

Physical Properties

Appearance   Commercial carrageenans are cream-colored to light brown powders. Under low-power magnification, individual particles are seen to be short-fiber segments (alcohol-precipitated products), or thin flakes (roll-dried products).

Particle Size   Commercial carrageenans are usually ground to pass (99.5%) through a No. 40 United States standard sieve. For certain special applications, fine-mesh products, passing through a No. 270 United States standard sieve, are available.

Density   Particle density averages about 1.7 g/cm³. Nominal bulk density is about 0.6 g/cm³ (39 lb/ft³) for roll-dried products and 1 g/cm³ (64 lb/ft³) for alcohol-precipitated products.

Solubilities   *Hot Water:* Carrageenans are soluble in hot (>75°C) water. Solubility is limited only by the viscosity of the solution. For commercial carrageenans, solutions containing up to 10% carrageenan can be prepared and handled with conventional mixing equipment.

Cold Water. Sodium salts of *kappa*- and *iota*- are soluble in cold water, while salts of other cations such as K^+ and Ca^{2+} do not dissolve completely, but exhibit limited to very high swelling depending on such factors as type and level of cations present, particle density and porosity of the carrageenan, etc. *Lambda*- is fully soluble in cold water, regardless of the cations with which it is associated. *Iota*- is particularly sensitive to calcium ions with which it forms highly thixotropic cold dispersions, thus lending itself very well for use as a suspending agent. Similar effects are obtained with potassium ions when used at sufficient concentration.

The question as to the degree of solubility of a product which swells on hydration often arises. In practice, if the viscosity of an aqueous carrageenan solution that has been heated and cooled is essentially the same as that of a cold dispersion, it is presumed that the product is cold-water soluble. This measurement should be supplemented by visual observations to ensure that the cold mix is really a solution, as evidenced by its smoothness, instead of a swollen dispersion.

Another question that frequently arises concerns the temperature of solubility. For *kappa-* and *iota-*carrageenans, which require heat for solution, this temperature is closely related to the gel melting temperature (q.v.). It therefore depends strongly on the cations associated with the carrageenan and those present in the system into which the carrageenan is to be incorporated, but is relatively insensitive to carrageenan concentration. In most systems, heating to 75°C should assure solution.

Hot Milk. All carrageenans are soluble in hot milk.

*Cold Milk. Lambda-*carrageenan has the greatest ability to disperse in and thicken milk without the need of solubilizing salts. Its insensitivity to potassium and calcium ions, along with its high ester sulfate content, may account for this cold-milk action. The higher the molecular weight of the *lambda-*, the lower the concentration required for a given degree of thickening. Also, the state of dispersion and rate of hydration are much dependent on both the applied shear and the particle size of the carrageenan. High shear and/or use of fine-mesh *lambda-* will serve to hasten the thickening of cold milk.

With regard to *kappa-* and *iota-*carrageenans, the higher the 3,6-AG content, and the lower the ester sulfate content, the more insoluble they are in cold milk. This is attributable in part to the increased sensitivity of these carrageenans to K^+ and Ca^{2+}, which are constituents of the milk. However, even *kappa-* and *iota-* types, which are practically insoluble in cold milk, may be used effectively for thickening and gelling if tetrasodium pyrophosphate (TSPP) is used. This salt has the ability to form gels in cold milk and is commonly used in conjunction with pregelatinized starches for the preparation of cold-set milk puddings. While its action in the carrageenan–milk protein system is not known, it is possible to obtain much firmer gels when carrageenan is present than could be obtained with phosphate alone.

Concentrated Sugar and Salt Solutions. Kappa- and *lambda-* carrageenans are soluble in hot sucrose solutions with concentrations as high as 65%. *Iota-*carrageenan, however, is sparingly soluble under these conditions. On the other hand, *iota-* and *lambda-* solutions will tolerate high concentrations of strong electrolytes (20 to 25% of NaCl), while *kappa-* will be salted out.

Water-miscible Solvents. Water-miscible solvents (e.g., alcohol, propylene glycol, glycerin, dimethyl sulfoxide) may be incorporated into carrageenan solutions. The concentration of solvent which can be tolerated depends upon the molecular weight of carrageenan and the type of carrageenan and cations present, as well as the method of incorporation of solvent. The carrageenan is held in solution by the water in the system, so that those factors which tend to make the carrageenan more hydrophilic somewhat paradoxically act to increase the solvent tolerance. Thus, the higher the ester sulfate content of the carrageenan and the lower its molecular weight, the higher the solvent tolerance. Also, the lower the concentration of salts present, the higher the tolerance.

Organic Solvents. All carrageenan products are insoluble in organic solvents, even the most polar ones.

Manufacture

The production of carrageenan is a straightforward operation consisting of extraction, purification, and isolation. The major raw materials used are the red seaweeds *Chondrus crispus, Eucheuma spinosum, Eucheuma cottonii,* and related types. These grow along the coasts of North and South America, Europe, Korea, and islands in the Pacific.

Though simple in principle, the manufacture of carrageenan incorporates a number of proprietary technical refinements and control measures which are generally well guarded by the few producers; however, the basic process is the same for all. The seaweed, either mechanically dried or sun-dried, is first washed to remove sand, salt, and other debris picked up from the beaches and rocks where it grows. Once cleaned, it is mixed with fresh water and made alkaline. This facilitates extraction and prevents hydrolysis of the carrageenan.

Completion of the extraction requires from 1 h to over a day in many cases, after which the seaweed residue is removed by filtration. An inert filter-aid is used, as the filtration is rather difficult because of finely dispersed solids and very high viscosity. The filtrate contains about 1% carrageenan.

Single- or multistage evaporators concentrate the stream to about 2 to 3% gum, which is then recovered by admixture with isopropyl alcohol, in which carrageenan is insoluble. Separation, drying, and grinding yield the pure alcohol-precipitated extract.

Additional methods of purification and recovery involve treating the filtrate with activated carbon, refiltering to separate the carbon and adsorbed impurities, then roll-drying on large steam-heated drum dryers which evaporate the water, yielding carrageenan diluted with some natural sea salts.

The milled carrageenans are separated into batches and set aside for testing. Based on the results of this testing, formulas are calculated for optimum properties of the finished products. The batches are then blended accordingly. The quality control laboratory tests each blend and retains samples. When approval comes from the lab, the finished product is sealed in polyethylene bags placed inside of fiber drums. This is the shipping package, and each container is marked with the blend number.

Biological/Toxicological Properties

Carrageenan is listed by the FDA as Generally Recognized As Safe (GRAS). Following reports of cecal and colonic ulceration in guinea pigs and rabbits induced by a highly degraded carrageenan produced, ironically, for symptomatic relief and cure of peptic and duodenal ulcers in man, intensive investigations into carrageenan's safety were carried out by the FDA and other groups sponsored by the carrageenan industry. By late 1976, food-grade carrageenan [defined as having a water viscosity of no less than 0.005 Pa $\cdot$ s (5 cP) at 1.5% concentration and 75°C, which corresponds to a molecular weight of 100,000], had been demonstrated to be safe.

Gastrointestinal Ulceration There is no evidence that carrageenan is ulcerogenic in humans. Food-grade carrageenan is not absorbed. Tests have been carried out on rats, guinea pigs, monkeys, baboons, pigs, gerbils, and hamsters. In guinea pigs, some results were inconclusive, but in all the other animals carrageenan has never shown any deleterious effects on the gut.

Teratogenicity Carrageenan is not teratogenic. A three-generation rat study by Collins[2] at the FDA laboratories in Washington, D.C., has shown this conclusively. Rats were given doses of carrageenan ranging up to 5% of the diet (3000 to 5000 mg/kg), with no effect on subsequent generations.

Carcinogenicity In a study conducted at the Eppley Institute for Cancer Research, rats and hamsters were allowed to complete their normal life span on diets containing 0% (100 controls), 0.5%, 2.5%, and 5.0% carrageenan (30 in each of the latter groups). The results demonstrated that carrageenan is not a carcinogen.

A review of the physiological effects of carrageenan was presented at the American Chemical Society Symposium on the Physiological Effects of Food Carbohydrates in 1974.[13]

Rheological Properties

Viscosity Carrageenans typically form highly viscous solutions. This is due to their unbranched, linear macromolecular structure and polyelectrolytic nature. The mutual repulsion of the many negatively charged ester sulfate groups along the polymer chain causes the molecule to be rigid and highly extended, while their hydrophilic nature causes the molecule to be surrounded by a sheath of immobilized water molecules. All of these factors contribute to resistance to flow.

Viscosity depends on concentration, temperature, the presence of other solutes, and the type of carrageenan and its molecular weight. The viscosity increases nearly exponentially with concentration. This behavior is typical of linear polymers carrying charged groups and is a result of the increase with concentration of interaction between polymer chains. Salts lower the viscosity of carrageenan solutions by reducing the electrostatic repulsion among the sulfate groups. This behavior is also normal for ionic macromolecules.

The viscosity decreases with increase in temperature. Again, the change is exponential. The change is reversible, provided that heating is done at or near the stability optimum at pH 9 and is not prolonged to the point where significant thermal degradation occurs. Both gelling *(kappa-, iota-)* and nongelling *(lambda-)* carrageenans behave in this manner at temperatures above the gelling point of the carrageenan. On cooling, however, the gelling types will abruptly increase in apparent viscosity when the gelling point is reached, if the counterions (K^+ or Ca^{2+}) essential for gelation are present. For this reason, viscosity measurements are made at a sufficiently high temperature (for example, 75° C) to avoid gelation. In this way, both gelling and nongelling types of carrageenan can be compared with each other and with other water-soluble gums. The carrageenan concentration employed is generally 1.5% by weight of the aqueous solution, although cold water–soluble carrageenans are frequently dissolved at 1.0% concentration in water of room temperature, and the viscosity measured at 25°C.

Viscosity increases with molecular weight, in accordance with the well-known Mark-Houwink equation:

$$[\eta] = KM^\alpha$$

where $[\eta]$, the intrinsic viscosity, is defined as the limit of the ratio of specific viscosity to concentration on extrapolation to zero concentration. Intrinsic viscosities correlate closely with practical viscosity measurements taken at 1.5% concentration at 75°C. Since carrageenans are polydisperse, the molecular weight term M in the equation is an average value. Strictly speaking, a special type of average, the viscosity-average molecular weight, applies here. As a practical matter, however, good Mark-Houwink relations are found between intrinsic viscosity and the more readily determined number-average and weight-average molecular weights. The values of the constants K and α vary with the type of average used. The exponent α is close to unity, indicative of a rather rigid, rodlike molecule, as expected for a highly charged linear polyanion.

In industrial practice, viscosity measurements are often made with easily operated rotational viscometers such as the Brookfield viscometer. Commercial carrageenans are generally available in viscosities ranging from about 0.005 to 0.800 Pa · s (5 to 800 cP) when measured at 1.5% concentration at 75°C. Carrageenans having viscosities less than 0.100 Pa · s (100 cP) have flow properties very close to Newtonian. The degree of deviation from Newtonian flow increases with concentration and molecular weight of the carrageenan. The solutions exhibit pseudoplastic (shear-thinning) flow properties and are sufficiently shear-dependent that it is necessary to specify the shear rate used for making the viscosity measurement. This is recognized in industry, and the specifying of spindle and speed is well accepted. Where non-Newtonian behavior is expected, viscosity measurements should be made at a shear rate comparable to that encountered in the process being used.

The pseudoplastic behavior of carrageenans corresponds closely to the power law equation:

$$\log \eta_a = \log a + (b-1) \log \dot{\gamma}$$

where η_a is the apparent viscosity, $\dot{\gamma}$ the shear rate, a the consistency index, and b the flow behavior index. For carrageenan solutions, b is either unity (Newtonian flow) or less than unity (shear thinning). For a wide range of shear rates (two or more decades) a better fit to experimental data can be had by adding a quadratic term:

$$\log \eta_a = \log a + (b-1) \log \dot{\gamma} + c \log^2 \dot{\gamma}$$

The coefficient c in this equation is either zero (strict power-law behavior) or negative.

Carrageenans specifically tailored for water-thickening applications are usually *lambda-* types or the sodium salt of unmodified mixed *lambda-* and *kappa-*. Here high water viscosities are desirable, and the high molecular weight and hydrophilicity of *lambda-* contribute to this. For gelling applications, a low viscosity in hot solution is usually desirable for ease of handling, and fortunately high-gel-strength carrageenans (mixed calcium and potassium salts of modified *kappa-* or *iota-*) fulfill this requirement because of their greater hydrophobicity and the effect of the calcium ions.

Gelation   *Water Gels.   Kappa-* and *iota-*carrageenans have the ability to form gels on cooling of a hot solution. These gels are thermally reversible, i.e., they melt on heating and gel again on cooling. According to Rees, carrageenans which form aqueous gels do so because of double-helix formation, as shown in Fig. 5.8.[11] At temperatures higher than the melting point of the gel, thermal agitation overcomes the tendency

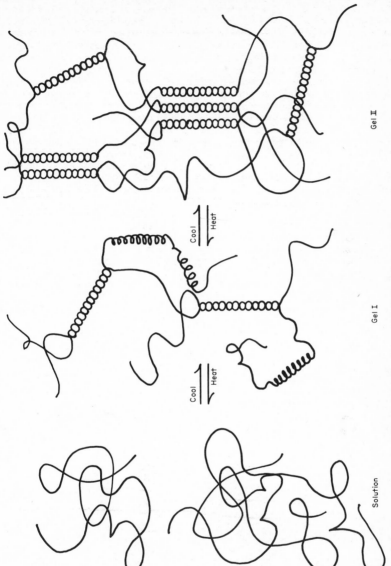

Figure 5.8 Sol-gel transformation of *kappa*- and *iota*-carrageenans.

Gel II

Gel I

Solution

Cool / Heat

to form helixes, and the polymer exists in solution as a random coil. On cooling, a three-dimensional polymer network builds up wherein double helixes form the junction zones of the polymer chains (gel I). Further cooling leads to aggregation of these junction zones (gel II). The inhibitory effects of kinks in the polymer chain can be seen. Anderson[1] has shown that the presence of even one kink in 200 residues has a dramatic effect in lowering gel strength.

The effect of sulfation on gelling properties likewise can be explained sterically on the basis of the double-helix secondary structure. Examination of molecular models shows that sulfate at C_2 of the 1,3-linked units, such as occurs in *lambda-*, acts as a wedging group to prevent the double helix from forming. Sulfate at C_2 on the 3,6-AG units, as occurs in *iota-*, projects outward from the double helix, and so does not sterically interfere with its formation. Sulfate at C_4 on the 1,3-linked galactoside, as occurs in *kappa-* and *iota-*, similarly projects outward and does not interfere with double-helix formation. Sulfation at C_6 of the 1,4-linked galactoside forms kinks in the chains which tend to inhibit double-helix formation. This is understandable if we consider the effect of 3,6-anhydride on the conformation of the carrageenan chain. Where a

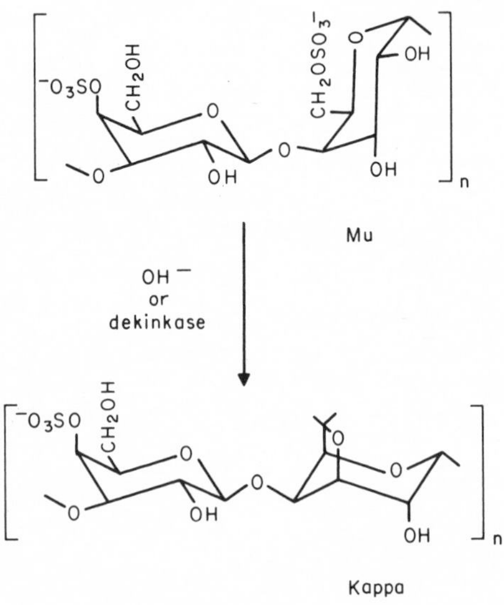

Figure 5.9   Conformation of repeating units of *mu-* and *kappa-*carrageenans.

1,4-linked unit is 6-sulfated, it tends to exist in the C1 chair conformation, as do all the 1,3-linked units. As can be seen (Fig. 5.9), this introduces a kink into the polymer chain. Ring closure to form the 3,6-anhydride constrains the 1,4-linked pyranose unit to the 1C form with resultant removal of the kink. The introduction of 3,6-anhydride is thus a chain-straightening process leading to greater regularity in the polymer, which results in enhanced gelling potential as a result of increased capability of forming double helixes.

Thus, the more the idealized structures of *kappa-* and *iota-* can be approached, the higher the gelling potential. Native *kappa-* and *iota-* contain some 6-sulfated units and are less strongly gelling than the completely anhydrized limit polysaccharides. These units can be converted to the anhydride by base-catalyzed S_N2 elimination of the 6-sulfate. This reaction is commercially important in carrageenan production, where it is used to enhance the gelling properties of the native carrageenans. On the other hand, 3,6-AG formation in going from *lambda-* to *theta-* has no effect, since gelation is still precluded by the 2-sulfate on the 1,3-linked galactoside.

In order for gelation of *kappa-* and *iota-* to take place, it is essential that certain cations be present. These may be associated with the carrageenan or may be constituents of the system in which the carrageenan is used. *Kappa*-carrageenan will not gel in the Na⁺ form.   Gel structure does develop in the presence of excess Na⁺, but a useful, coherent gel is not produced.   For applications, K^+, Ca^{2+}, or NH_4^+ ions may be added to induce gelation, with K^+ producing the strongest gel.   Apparently the size of the hydrated ion is important, as cations of ionic radius similar to that of K^+ (for example, NH_4^+, Rb^+, Cs^+, Tl^+) behave similarly.

The higher the 3,6-AG content of the *kappa-*, the greater the enhancement of gel strength by gelling cations.   This may be explained by the increased hydrophobicity imparted to the polymer by the 3,6-AG units, plus (by analogy to monosaccharide sulfates) the lower solubility of the potassium salt, since, according to one theory of gelation, gel formation is looked upon as a type of precipitation.   Other contributing factors are the increased regularity of the polymer, leading to increased helical content and improved ion packing.   Pure potassium *kappa-* produces a gel that is clear and compliant. Excess K^+ affords a gel which is stiffer, but still highly transparent.   In commercial practice, however, these products are not realized, and the amount of Ca^{2+} normally present in the carrageenan produces a gel which is rigid, brittle, and somewhat cloudy. Figure 5.10 shows the effect of K^+ ion concentration on the breaking strength of high 3,6-AG *kappa-* gels.

Kappa-carrageenan gels are subject to syneresis, which may be described as the appearance of fluid on the surface of the gel.   This fluid arises from shrinkage of the gel (as a result of increasing aggregation of junction zones) and contains a proportionate

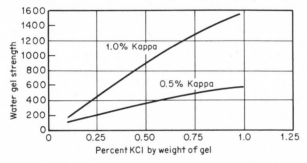

Figure 5.10   Effect of K^+ ion concentration on the breaking strength of high 3,6-AG *kappa*-carrageenan gels.

part of the soluble substances of the gel: salts, sugars, *lambda-* and *theta*-carrageenan (but not the *kappa-*).   This syneresis depends on the concentration of gelling cations, so that care must be taken to avoid excessive levels of these.   Carrageenan manufacturers will often include the required cations in their products, or make specific recommendations regarding further additions.

Iota-carrageenan, like *kappa-*, does not form gels when it is in the sodium form. Although *iota-* will gel with K^+ and NH_4^+ ions, it reacts most strongly with Ca^{2+} forming very compliant and transparent gels that are not subject to syneresis.   In appearance, these very closely resemble gelatin gels.   The much higher gelling temperature of *iota-* gels, however, allows this product to be used in preparation of dessert gels that do not require refrigeration.   Moreover, on storage, *iota-* gels do not develop the undesirable toughness characteristic of an aged gelatin gel.

The gelling temperature of a specific type of carrageenan is relatively insensitive to carrageenan concentration and is primarily a function of the concentration of gelling cation present.   Figure 5.11 shows the effect of concentration of gelling cations on the gelling temperatures of *kappa-* and *iota-*.   Gelling temperatures have been determined by a variety of rheometric methods, as well as by optical measurements such as optical rotation and light scattering.   Values obtained by different methods vary somewhat, but in general the data are well fitted by functions of the form $\theta_g = a + b\sqrt{c}$, where θ_g is the gelling temperature and c is the concentration of gelling cation.   This linear dependence on the square root of ion concentration indicates that coulombic effects, rather than salting-out effects, predominate the sol-gel transition.

The melting temperature of a carrageenan gel is higher than its setting temperature. The extent of this hysteresis is dependent only on the type of carrageenan; for *kappa-* it is 10 to 15°C, for *iota-*, about 5°C. That is, when the gelling temperature is changed by varying the concentration of gelling cation, the melting temperature is similarly altered, and the two parallel each other as further adjustments are made. Thus, melting temperature, like gelling temperature, depends linearly on the square root of cation concentration.

The melting and gelling temperatures of *kappa*-carrageenan are increased by the presence of sucrose. This non-ionic additive causes these temperatures to increase more rapidly than is predicted by using the square root law which is applicable to cations. Hysteresis is independent of sucrose concentration.

Aqueous *kappa-* gels do not normally exhibit good freeze-thaw stability. Considerable change in gel texture, as well as substantial water release, may take place. The more hydrophilic *iota-* shows better stability to the freeze-thaw cycle and compares favorably with gelatin in this respect.

Milk Gels. All carrageenan products have the ability to form gels by cooling a solution of the carrageenan in hot milk. Even *lambda-*, which does not gel in water regardless of the cations present, will form milk gels at levels of 0.2% by weight of the milk. This gelation is attributed to the formation of carrageenan-protein bonds as previously described.

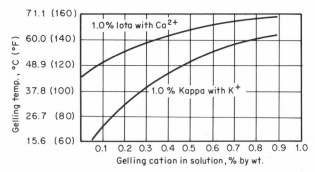

Figure 5.11 Effect of concentration of gelling cations on the gelling temperatures of *kappa-* and *iota*-carrageenans.

With *kappa-* and *iota-* there is, in addition to the carrageenan-casein bonding, a water-gelling effect from the cations present in the carrageenan as well as the Ca^{2+} and K^+ present in the milk. These cations appear to be required for milk-gel formation, as milk which has been ion-exchanged to remove the Ca^{2+} and K^+ does not gel with the sodium salts of *Lambda-, kappa-,* or *iota-*. On the other hand, the strength of milk gels is enhanced by addition of soluble Ca^{2+} and K^+ salts in a manner similar to that which occurs in water gels.

The presence of fat also influences the behavior of carrageenans in milk. Strongly gelling *kappa*-carrageenans can be used in high-fat systems, but they are not well tolerated in low-fat systems wherein they may exert a destabilizing action resulting in whey separation. For the latter, weak milk-gelling *kappa*-carrageenans with high ester sulfate and moderate to high 3,6-AG are more suitable. The reason that strong gelling types can be employed in high-fat but not in low-fat systems is due in part to the presence of the dispersed fat phase. This apparently tempers the carrageenan–milk protein complex, serving to interrupt it to some extent. Interaction may also occur between carrageenan and the phospholipid which is present as a monomolecular layer covering the dispersed globules of butterfat in the milk. Since the phospholipid contains basic amino groups with which the ester sulfate groups of the carrageenan can react, it is very likely that carrageenan-phospholipid bonds are formed. This may account for the extraordinary efficiency with which very low levels (approximately 50 ppm) of carrageenan stabilize evaporated milk against fat separation.

Kappa- produces gels in milk which have the same brittle nature that they have in water. Moreover, these gels are very prone to syneresis. The addition of salts such as orthophosphates, polyphosphates, carbonates, citrates, etc., which have the ability to chelate or precipitate Ca^{2+}, may be used to modify the texture of carrageenan-milk gels (presumably by suppressing the water-gel contribution to the structure) to ameliorate these undesirable properties.

Iota-carrageenan does not produce the same syneresis-free gels in milk that it does in water. If TSPP is included, however, syneresis is markedly reduced and the gel becomes more compliant. Similar, but not as dramatic, results are achieved with polyphosphates, orthophosphates, citrates, and carbonates.

In cold-milk systems, soft gels can be produced by *lambda-* or *theta-* when used at sufficient concentration. Moderate to high 3,6-AG *kappa-* and *iota-*, by themselves, are virtually insoluble in cold milk, but can be employed to produce a gel if TSPP is present.

Additives/Extenders

Because of ion sensitivities, salts of gel-inducing cations must be controlled in formulation. An excess of potassium or ammonium ion in a *kappa*-carrageenan system will lead to copious syneresis. Calcium and other divalent cations will raise the gelling and melting temperatures of *iota*-carrageenan gels so high that it may not be possible to prepare them properly.

Locust bean gum is frequently used with *kappa*-carrageenan for enhancement and modification of gel strength and texture. The most generally employed ratios vary from 1:3 to 3:1, although satisfactory gels can be prepared outside this range. The synergistic interaction between the gums can be explained by structural considerations of the gums and the gels: Locust bean gum is a galactomannan, a polysaccharide having a mannose backbone with single-unit branches of galactose. Although the average galactose/mannose ratio is about 1:4, the galactose side groups tend to occur on series of contiguous mannose residues, forming "hairy" regions and leaving "smooth" unsubstituted stretches of mannose backbone between the hairy regions. The smooth regions have been shown to bind with the helixes in the carrageenan gel, thus providing additional crosslinkages which enhance gel strength. In addition, the hairy regions between these bonds cannot bind, and so contribute to the flexibility of the crosslinkages, producing a compliant gel. Such gels, in addition to being stronger and more compliant, are far less prone to syneresis.[11] Guar gum, another galactomannan, shows no such interactions with *kappa*-carrageenan. Its mannose backbone is too highly substituted (1:2) to provide the necessary binding sites. *Iota*-carrageenan shows no interaction with either galactomannan.

Handling

The shipping package usually consists of 90.8 kg (200 lb) net of carrageenan sealed inside a polyethylene bag placed inside of a fiber drum. The drum is sealed and bears the lot number of the blend from which it was made. The tradename, net weight, customer address, and any special markings to facilitate shipping or by customer request are then added.

All common shipping methods can be used, provided the drums are transported in an upright position and are not subjected to extremes in temperature or humidity. In-plant storage should meet the same criteria. When a container has been opened and partially consumed, it should be resealed to prevent absorption of moisture from the atmosphere. Moisture will cause the carrageenan to cake, and will make further handling difficult.

Applications

By Result Carrageenans are used to gel, thicken, or suspend; therefore, they are used in emulsion stabilization, for syneresis control, and for bodying, binding, and dispersion.

By End Product Carrageenans are used in chocolate milk, milk puddings, ice cream, infant formulas, evaporated milk, real and artificial whipped cream, yogurt, process cheeses and cheese dips, water dessert gels, low-calorie jellies, fish gels, relishes, tomato sauces, pet foods, meats, toothpaste, lotions and creams, pharmaceutical suspensions, abrasive suspensions, barium sulfate and pigment dispersions, agricultural suspensions, and air-freshener gels.

By Industry   Carrageenans are used to produce dairy and convenience foods, pharmaceuticals, cosmetics, household products, and in metals fabrication and coatings.

By Process   Carrageenans are used as processing aids in the fining of beer and wine.

Application Procedures

Use Levels   Most water applications require carrageenan concentrations of about 1%. In milk or other protein systems, the usual concentrations range from a low of 0.005% (50 ppm) in evaporated milk, through 0.03% in chocolate milk, to 0.2 to 0.8% for different types of milk puddings.

Dispersion   Because of extremely rapid water pickup, carrageenan tends to clump. The resulting viscous coating of the particles then impedes dissolution. Therefore, proper dispersion of the carrageenan into a liquid system is very important. If possible, the carrageenan can be mixed with a diluent, such as a part of the sugar or other dry ingredient of the final product (preferably, a minimum of three parts diluent to one of gum). A "retardant," such as liquid sugar, alcohol, or glycerin, can be used for the initial dispersion of the carrageenan, since the carrageenan will not dissolve in it. Thus, when one of these is a part of a formulation, it can be used as a retardant. If dispersed without any such aids, the carrageenan should be added to the liquid slowly with sufficient agitation to draw it down into the vortex rapidly. Cold water is preferred over hot water, which merely compounds the dispersion problem.

Stability   Carrageenan is most stable at neutral to alkaline pH, but can be used in acid systems if not subjected to prolonged heating. This can generally be avoided by mixing in the acid as close to the packaging operation as possible and then cooling, by using the high-temperature, short-time (HTST) process, or by any other measure which will minimize contact time of acid and carrageenan at elevated temperatures. One notable exception to this need for caution is the use of carrageenan in tomato sauces of all kinds; here its functionality is retained, despite the acidity of the system.

Incompatibilities   Although carrageenan is used in conjunction with proteins in numerous applications, under certain circumstances the two may react to form insoluble complexes, for example, at a pH below the protein's isoelectric point. Thus, gelatin forms a precipitate with carrageenan at a pH of 4.0. Quaternary ammonium salts, often used as germicides, will quantitatively precipitate carrageenan. *Kappa-* and *lambda-*carrageenans will tolerate high sugar concentrations, but *iota-* is sparingly soluble. On the other hand, *iota-* is soluble in strong brines, as is *lambda-*, whereas *kappa-* is salted out, i.e., is insoluble. Alcohol concentrations in excess of 30 to 40% will cause precipitation of carrageenan, except for the sodium salt of *lambda-*carrageenan, which has a limited solubility in even higher than 50% alcohol.

Specialties

Carrageenan has been reported to be beneficial in the treatment of peptic and duodenal ulcers. However, its generally viscous nature makes the preparation of a convenient dosage form difficult. A highly depolymerized carrageenan, available in Europe, circumvents this problem, but its safety has been questioned as a result of adverse side effects in some test animals. Carrageenan's mode of action in such therapy is postulated to be one of protecting the ulcerated area by a coating action akin to that of the chemically similar mucopolysaccharides normally present in the stomach lining.

Future Developments

The use of carrageenan in the food industry has been steadily increasing, generally following the economy, since many of its applications are in convenience foods. Because of the increasing interest in new protein sources and carrageenan's rather unique ability to interact with proteins in a variety of stabilization functions, increased usage is predicted in this area. Recent successes in aquaculture technology virtually assure a stable, continuing supply of carrageenan-bearing seaweeds.[3,9] In the past, the threat of natural-harvest shortfalls has limited applications and market development, and kept prices at a premium. With ready availability of raw material, broadened markets can be expected, along with the price stability such expansion brings. The cost-effectiveness of carrageenan, coupled with its unique water-gelling properties, can be expected to lead to new uses of carrageenan in nonfood applications, as, for example, in the relatively recent innovation, the room-deodorizer gel. Other areas of potential interest include hand lotions, gelled antacids, pigment dispersions, emulsion stabilization, and specific cation reactivities.

COMMERCIAL USES: Compounding and Formulating

In food applications, carrageenan is used primarily to gel, thicken, or stabilize (the latter, a general term meaning to prevent phase separation). Secondary advantages include improved palatability and appearance. The manufacture of some products, principally involving milk or imitation milk, would not be possible without it. In nonfood uses, its viscosity and gelling properties give it a wide range of applicability: emulsion stabilization, suspension of abrasives and other insolubles, air-treatment gel formulation, and replacement of solvent or oil-based systems by aqueous ones. The most successful application of carrageenan technology is often accomplished through user-manufacturer collaboration, since the manufacturer maintains a strong technical sales staff for that purpose.

In the following listing, carrageenan usage is divided into food and nonfood applications, with the former further separated into milk and water systems. This is a natural division based on the carrageenan's mode of action.

Milk Applications

Carrageenan is employed in many consumer-prepared milk systems; in others, full utility and economy are best achieved by incorporation into manufactured milk products. Table 5.2 gives typical milk (dairy) applications of carrageenan.

Uses in Dry Mixes *Milkshake and Instant Breakfast Powders.* Carrageenan is used to suspend the ingredients and to impart a richness and body to these drinks when the powders are mixed with liquids. Fine-mesh (<250) *lambda-* is very effective at levels of 0.10 to 0.20%, based on the finished drink.

Cooked Flans and Custards. Light-bodied custard desserts (blanc mange) may be prepared by incorporating *kappa*-carrageenan with the other ingredients in a dry mix. Levels of *kappa*-carrageenan from 1.05 to 2.1 g/L (0.5 to 1.0 g/pt) of milk produce a delicate, brittle gel; a creamier product can be produced by combinations of *kappa*- and *iota*-carrageenans together with TSPP. A wide variety of textures is available depending upon the ratios and levels used.

Cooked Puddings and Pie Fillings. Starch has traditionally been used to stabilize these products. *Kappa*-carrageenan at levels of 0.42 to 1.05 g/L (0.2 to 0.5 g/pt) of milk provides a more uniform set to these products.

Cold Prepared Flans and Custards. *Lambda*-carrageenan at levels of 0.2 to 1.0% produces "instant" gelling in cold-milk systems. Puddings of this type have particularly good flavor release. *Lambda-* or *iota*-carrageenan in combination with TSPP functions in an equivalent manner, the mixture providing syneresis control and texture modification.

Uses in Manufactured Products *Chocolate Milk.* A typical chocolate milk contains 1% cocoa, 6% sugar and from 0.025 to 0.035% carrageenan. The carrageenan is used to keep the cocoa in suspension, and gives the drink a rich mouth-feel. *Kappa*-carrageenan is used when the drink is pasteurized. *Iota-* and *lambda-* are functional, but not as economical to use.

Chocolate Syrups. Chocolate milk is sometimes prepared from a syrup concentrate. This is blended into the cold milk at the dairy before packaging. One part syrup usually is added to 10 to 12 parts milk. *Lambda*-carrageenan at a level of 0.04 to 0.05% based on the finished drink keeps the cocoa in suspension and adds richness and body to it.

Ice Cream and Sherbet. Used at a concentration of 0.01 to 0.05% in combination with a primary stabilizer such as guar, locust bean gum, and/or carboxymethylcellulose (CMC), *kappa*-carrageenan adds creaminess, controls ice crystal formation, and prevents syneresis under freeze-thaw conditions. It also prevents whey separation in the unfrozen ice cream mix.

Filled and Skim Milk. *Iota-* and *kappa*-carrageenans are very effective at levels of

0.02 to 0.04% (by weight of the finished product) in stabilizing the fat emulsion and in improving the appearance and mouth-feel of these products.

Cottage and Cream Cheese Products. *Kappa*-carrageenan at levels of 0.02 to 0.03% in combination with locust bean gum at 0.10 to 0.20% stabilizes the creaming mixture, induces curd formation, imparts shape retention, and prevents syneresis.

Evaporated (Canned) Milk. As little as 50 ppm (0.005%) *kappa*-carrageenan is commonly used to prevent fat separation in evaporated milk. Prior to carrageenan's inclu-

TABLE 5.2
Typical Milk (Dairy) Applications of Carrageenan

Use	Function	Product	Approx. use level, %
Frozen desserts:			
Ice cream, ice milk	Whey prevention Control meltdown	*Kappa-*	0.010–0.030
Pasteurized milk products:			
Chocolate, eggnog, fruit-flavored	Suspension, bodying	*Kappa-*	0.025–0.035
Fluid skim milk	Bodying	*Kappa-, iota-*	0.025–0.035
Filled milk	Emulsion stabilization, bodying	*Kappa-, iota-*	0.025–0.035
Creaming mixture for cottage cheese	Cling	*Kappa-*	0.020–0.035
Sterilized milk products:			
Chocolate, etc.	Suspension, bodying	*Kappa-*	0.010–0.035
Controlled calorie	Suspension, bodying	*Kappa-*	0.010–0.035
Evaporated	Emulsion stabilization	*Kappa-*	0.005–0.015
Infant formulations	Fat and protein stabilization	*Kappa-*	0.020–0.040
Milk gels:			
Cooked flans or custards	Gelation	*Kappa-, kappa- + iota-*	0.20 –0.30
Cold prepared custards (with added TSPP)	Thickening, gelation	*Kappa-, iota-, lambda-*	0.20 –0.50
Pudding and pie fillings (starch base)			
Dry mix cooked with milk	Level starch gelatinization	*Kappa-*	0.10 –0.20
Ready-to-eat	Syneresis control, bodying	*Iota-*	0.10 –0.20
Whipped products:			
Whipped cream	Stabilize overrun	*Lambda-*	0.05 –0.15
Aerosol whipped cream	Stabilize overrun, stabilize emulsion	*Kappa-*	0.02 –0.05
Cold prepared milks:			
Instant breakfast	Suspension, bodying	*Lambda-*	0.10 –0.20
Shakes	Suspension, bodying, stabilize overrun	*Lambda-*	0.10 –0.20
Acidified milks:			
Yogurt	Bodying, fruit suspension	*Kappa-* + locust bean gum	0.20 –0.50

sion, it was necessary to turn over the cases of milk in storage to prevent separation of the milk.

Infant Formulations. Carrageenan is required for fat and protein stabilization in food formulations for infants. Medium-strength *kappa-* is used at 0.02 to 0.04% in both milk and soy milk products.

Canned Ready-to-eat Milk Puddings. *Iota*-carrageenan at a level of 0.10 to 0.20% may be used to replace part of the starch in canned, ready-to-eat milk products, offering advantages during processing (reduced viscosity and thus better heat transfer) and in the finished product (controlled degree of set, minimized syneresis development, and improved flavor release).

Whipping Cream. Natural cream to which carrageenan is added will show improved stabilization of the whip in whipping cream. If added to the cold cream, *lambda-* at 0.05 to 0.15% is best. If added during pasteurization, *kappa-* at only 0.02 to 0.05% is adequate.

Aerosol Spray Cream Toppings. *Kappa-*carrageenan stabilizes both natural and artificial aerosol-propelled cream toppings in the can without developing excessive viscosity. *Kappa-*carrageenan is used at about 0.03% for natural cream and about 0.05 to 0.10% for the artificial type. Often, locust bean gum at concentrations of about 0.10% is used in conjunction with carrageenan.

Yogurt. *Kappa-*carrageenan can be used to stabilize yogurt to which fruit is added. A typical system includes about 0.25% carrageenan and 0.75% locust bean gum, based on the yogurt. Dissolution of the gums is achieved by incorporating them prior to pasteurization and inoculation.

Frozen Whipped Toppings. Combinations of *kappa-* and *lambda-*carrageenans at concentrations of 0.03 to 0.05% improve the body of frozen whipped toppings and reduce syneresis under freeze-thaw conditions.

Imitation Milk. In imitation milk products, sodium caseinate and/or soy protein are used in place of milk solids, and vegetable fat replaces the butterfat. *Iota-* and *lambda-* are effective at concentrations of about 0.05% in stabilizing the fat emulsion and providing body to the product.

Water Applications

Typical water applications for carrageenans, both for food and nonfood products, are shown in Table 5.3, which shows not only the end products, but also what levels of which types of carrageenans are used and the specific function that the gum serves in each product.

TABLE 5.3
Typical Water Applications of Carrageenan

Use	Function	Carrageenan type	Approx. use level, %
Dessert gels	Gelation	*Kappa-* + *iota-* *Kappa-* + *iota-* + locust bean gum	} 0.5 –1.0
Low-calorie jellies	Gelation	*Kappa-* + *iota-* *Kappa-* + galactomannans	} 0.5 –1.0
Pet-foods (canned)	Fat stabilization, thickening, suspending, gelation	*Kappa-* + locust bean gum	0.2 –1.0
Fish gels	Gelation	*Kappa-* + locust bean gum *Kappa-* + *iota-*	} 0.5 –1.0
Syrups	Suspension, bodying	*Kappa-*, *lambda-*	0.3 –0.5
Fruit drink powders and frozen concentrates	Bodying Pulping effects	Sodium *kappa-*, *lambda-* Potassium/calcium *kappa-*	0.1 –0.2 0.1 –0.2
Relishes, pizza, barbecue sauces	Bodying	*Kappa-*	0.2 –0.5
Imitation milk	Bodying, fat stabilization	*Iota-*, *lambda-*	0.03–0.06
Imitation coffee creams	Emulsion stabilization	*Lambda-*	0.1 –0.2
Whipped toppings (artificial)	Stabilize emulsion, overrun	*Kappa-*, *iota-*	0.1 –0.3
Puddings (nondairy)	Emulsion stabilization	*Kappa-*	0.1 –0.3
Air-treatment gels	Gelation	*Kappa-* + *iota-* *Kappa-* + galactomannans	} 2.0 –3.5
Toothpastes	Binder	Sodium *kappa-*, *lambda-*, *iota-*	0.8 –1.2
Lotions	Bodying, emollient	Sodium *kappa-*, *lambda-*, *iota-*	0.2 –1.0
Suspensions (industrial)	Suspension	*Iota-*	0.2 –1.0
Dispersions	Suspension, dispersion	Hydrolyzed *kappa-*, *lambda*, *iota-*	0.2 –0.5
Water-based paints	Suspension, flow control emulsion stabilization	*Kappa-* + galactomannans, *iota-*	0.15–0.5

Uses in Dry Mixes   *Dessert Gels.* Mixtures of *kappa*- and *iota*-carrageenans at a concentration of 0.10 to 1.0%, either alone or in combination with a refined locust bean gum, are ideally suited for gelled desserts.   Unlike those made with gelatin, these gels do not melt at room temperature.   *Iota-* gels will withstand freeze-thaw cycling.

Fruit Drinks.   *Lambda-* or the sodium salt of *kappa*-carrageenan can be used with the necessary sugars, acid, and flavor in forming a fruit-drink mix.   Concentrations of 0.1 to 0.2% carrageenan, based on the prepared drink, are effective in providing body and a pleasing mouth-feel.

Uses in Manufactured Products   *Low-Calorie Jellies.*   A jelly containing either artificial sweeteners or a low level of sugar in combination with artificial sweeteners can be made using combinations of either *kappa*- and *iota*- or *kappa*- and clarified locust bean gum at levels of 0.5 to 1.0%.

Pet Foods.   *Kappa*-carrageenan at a concentration of about 0.20 to 0.50% in combination with a similar amount of locust bean gum prevents fat separation during processing and imparts a richness to the gravy that accompanies the canned pet foods.   Particles of gel can be obtained in the product, if desired.

Fish Gels.   Combinations of *kappa*- and *iota*- at concentrations of 0.5 to 1.0% gel the broth and preserve the flavor of fish packed in cans or jars.   Seasonings visible in the broth remain suspended.

Frozen Fish Coating.   A solution of *kappa*-carrageenan, locust bean gum, and potassium chloride produces a gelled film that coats frozen fish, protecting it from freezer burn and mechanical distintegration during processing.   A solution of about 0.4% of the mixture is used.   The processed frozen fish is passed through the solution in a one-step operation, and is then returned to freezer storage.

Relishes, Pizza, Barbecue Sauces.   *Kappa-* or *iota*-carrageenan can be used in concentrations up to 0.5% to provide texture, sheen, and improved adhesion properties in relishes, pizza, and barbeque sauces.

Nonfood Applications

Pharmaceuticals and Toilet Goods   *Toothpastes.*   Carrageenan at concentrations of 0.8 to 1.2% is used to prevent separation of the elixir (liquid portion) and abrasive, and to impart short texture and good rinseability characteristics to toothpaste.   Stability of the paste in storage and during use in areas where high temperature and humidity prevail is improved by using carrageenan relative to cellulosic binders.

Mineral Oil and Insoluble Drug Preparations.   *Iota*-carrageenan at concentrations of 0.10 to 0.50% gives stable emulsions and suspensions for mineral oil and insoluble drug preparations.

Lotions and Creams.   *Lambda*-carrageenan at concentrations of 0.1 to 1.0% may be used in hand lotions and creams to provide slip and improved rub-out.   Therapeutic value cannot be claimed for the carrageenan, but it is frequently noted that fishermen who collect Irish moss (from which the carrageenan is obtained) have surprisingly soft skin on their hands.

Antacid Gels.   The chalkiness of antacids can be masked by incorporating them into a glycerin-water mixture, gelled with about 0.7 to 2.0% *kappa*-carrageenan, or a blend of *kappa*-carrageenan and locust bean gum.

Other Applications   *Air-Freshener Gels.*   Mixtures of *kappa*-carrageenan, other gums, and a gelling salt such as potassium chloride are used to prepare air-freshener gels.   Total gum levels may go up to 2.5%.   Volatile odor-absorbing compounds and fragrance oils incorporated in the gel are released uniformly from the gel surface as the gel dries down.

Water-based Paints.   *Iota*-carrageenan or blends of *kappa*-carrageenan and locust bean gum may be used at concentrations of 0.15 to 0.25% (based on the total weight) to thicken latex emulsion paints.

COMMERCIAL USES: Processing Aids

Beverage Clarification

Carrageenan added to beer or wine serves as a fining agent.   Its action is believed to be one of removing proteins through its ability to interact with them.   Traditionally, bleached, ground seaweed has been the form of choice here.

Abrasive Suspensions

Iota-carrageenan (partial calcium salt) at concentrations of 0.25 to 0.80% will suspend polishing and cutting abrasives, such as ceric oxide, aluminum oxide, and silicon carbide, in liquid media.

Ceramic Glazes and Core Washes

As with abrasive suspensions, a carefully controlled calcium ion level in *iota*-carrageenan provides suspension of insolubles in liquid media. Concentrations of carrageenan ranging from 0.25 to 0.80% are satisfactory.

INDUSTRIES USING CARRAGEENANS

Food

Dairy
- *Chocolate milk.* Suspension of cocoa particles.
- *Ice cream.* Prevention of whey separation prior to freezing when galactomannan gums or carboxymethylcellulose are used as stabilizers; control of melt-down.
- *Liquid diets and instant breakfasts.* Thickening and suspension of solids.
- *Evaporated milk.* Prevention of fat separation following the high-temperature, short-time sterilization process.
- *Yogurt, Swiss style.* Suspension of fruit following breaking of the curd in bulk culture.
- *Processed cheeses and cheese dips.* Syneresis prevention.
- *Whipped cream.* Foam stabilization.
- *Infant formulas.* Prevention of fat and whey separation.

Dairy Substitutes
- *Frozen whipped toppings.* Foam stabilization and prevention of fat separation and syneresis following freeze-thaw cycle.
- *Infant formulas (vegetable protein).* Prevention of fat and whey separation.

Packaged Desserts
- *Milk gels (puddings, flans, custards, and pie fillings).* Primary or secondary (with starch) gelling agent.
- *Water Gels (canned, frozen, dry mixes.)* Gelling agent.

Other Food Uses
- *Relishes.* Bodying agent; provides glaze (sheen).
- *Fish gels.* Gelling agent.
- *Tomato sauces.* Bodying agent.
- *Pet foods (canned).* Gelation and fat stabilization.
- *Meats.* Syneresis prevention; binder.
- *Frozen soups and food sauces.* Prevention of syneresis upon thawing.

Pharmaceuticals and Toilet Goods
- *Toothpaste.* Binder.
- *Lotions and creams.* Bodying agent; provides slip and emollience.
- *Antibiotic suspensions.* Prolonged shelf life.
- *Barium sulfate suspensions.* Dispersant.
- *Antacid gels.* Gelling agent; masks chalkiness.

Metal Fabrication
- *Abrasive suspensions.* Suspending agent.
- *Core washes.* Suspending agent.

Ceramics
- *Glazes.* Suspending agent.

Coatings
- *Pigments.* Dispersion and suspension.
- *Water-based paints.* Thickener.

Agriculture

- *Pesticides and herbicides.* Suspension; sticking agent.

Household Products

- *Air-freshener gels.* Primary gelling agent.

FORMULATIONS

As will be evident from some of the following recipes, it is common for carrageenan manufacturers to have proprietary blends for specific end uses. For example, in the field of air-treatment gels, several different formulations may exist, each tailor-made to suit the needs of a particular customer. Such considerations as gelling temperature, viscosity at the pouring temperature, hardness and dryness of the resulting gel, or even the compatibility of an essential oil may have been worked out on an individual basis.

Controlled seaweed formulas and processing conditions allow the manufacturer to create numerous shades of functionality. Each user will thus have several products available for the performance of a given function. Generally, a great deal of time can be saved by contacting the manufacturer directly for suggestions of products to suit a specific need. The manufacturer usually will supply samples and suggestions for use, and a pilot recipe, if it exists. Thus, the *kappa-, lambda-,* and *iota-*designations in these basic recipes are merely indicative of the type of carrageenan best suited to a particular end use.

Chocolate Milk

Two basic methods have long been used for the incorporation of carrageenan into dairy chocolate milk for the purpose of suspending the cocoa particles. Of these, the "hot process" is far more common today, although the "cold process" is still used in some areas. A basic formula for each method is presented here. The hot process employs a weak-gelling *kappa*-carrageenan, whereas the cold process, to avoid gelling of the syrup, uses a carrageenan which is predominantly *lambda*.

Hot Process   For each pint of 2% butterfat milk, preblend the following ingredients and disperse them with agitation into the cold milk:

Fine sugar	32.3 g
Cocoa	6.9 g
Mixed *kappa*- and *lambda*-carrageenans	0.15 g
Vanillin	0.08 g

Heat the milk to 71°C (160°F). Maintain this temperature for several minutes under agitation, then cool the mixture rapidly with constant agitation (as over a surface cooler) and pack into sterile containers.

Cold Process   Prepare a chocolate syrup by dispersing a well-blended mixture consisting of 2.15 g *lambda*-carrageenan and 30.5 g of sugar into 170 mL of water, then heating to 82°C (180°F) to complete solution. Add this solution to a mixture of:

Cane sugar	265 g
Cocoa	44.6 g
Salt	3.4 g
Vanillin	1.0 g

Mix thoroughly, heat to 93°C (200°F), and stir 30 min at 88 to 93°C (190 to 200°F). Cool with constant agitation to 49°C (120°F), then make up to 500 g with water. When 1 part of this syrup is added with agitation to 11 parts of cold milk, a stable chocolate milk will result.

Canned Water-Dessert Gel

A basic formulation consists of the following relative proportions:

Cane sugar	91.9%	
Adipic acid	3.0%	
Calcium *iota*-carrageenan	3.0%	(imparts elasticity and controls syneresis)
Potassium *kappa*-carrageenan	0.9%	(controls rigidity and unmolding properties)
Tripotassium citrate	1.2%	
Color and flavor	q.s.	
	100.0%	

This dry blend is used at a 15% level in the final dessert gel. Because of the susceptibility of carrageenan to hydrolysis at high temperature and acid pH, certain precautions should be taken in large-scale preparation of these gels. In an HTST process, all ingredients may be mixed cold into the water and sterilized on a continuous basis, then canned and quickly cooled. For aseptic filling, the solution may be quickly cooled to a few degrees above the gelling temperature, then filled and further cooled.

A modified batch method employs separately heated solutions of (1) the acidic ingredients and (2) the carrageenan, sugar, and salts dissolved in water. The two solutions are metered into a small-surge mixing tank just prior to filling. Again, it is important to cool the solution as quickly as possible. The approximate pH is 4.1, while the gelling and melting temperatures are 41°C (105°F) and 63°C (145°F) respectively. The gel texture can be varied by changing the ratio of the two carrageenans.

Air-Treatment Gel

Mixed *kappa*- and *iota*-carrageenan (with added potassium chloride)	3.0–3.5%
Color (2% dye solution in water)	3.0%
Perfume oil	7.5%
Water	86.0–86.5% (to 100%)

Disperse the carrageenan in cold water, and heat to 80°C with vigorous agitation until the gum is dissolved. Add the color and perfume, mix in thoroughly, cool to 65°C, then pack into containers of about 114 g (4 oz) capacity. Cool further, rapidly, to prevent oil separation. Emulsifiers and preservatives may be added with the carrageenan.

Toothpaste

Glycerin (99%)	143.0 g
Spearmint oil	5.9 g
Lambda- and *iota*-carrageenan	5–10 g
Water	150 mL
Dicalcium phosphate, dentifrice grade	270 g
Sodium lauryl sulfate	7.5 g

Prepare a slurry of the carrageenan in the glycerin plus oil, add the water, and mix well. Add the other two dry ingredients, and mix well. Heat, with agitation, to a temperature of 82 to 93°C (180 to 200°F) and hold for at least 5 min. Cool to 27 to 32°C (80 to 90°F) with frequent agitation, and replace water lost by evaporation. Mix thoroughly (using, for example, a Read or Day heavy-duty mixer), then deaerate the paste under vacuum. Pack into tubes.

Milk Puddings

Numerous types of milk puddings exist; three are presented here to illustrate the range of applicability of carrageenan.

Creamy Type (Cold Set)

Lambda-carrageenan, 270-mesh	4.0 g
Nonfat milk solids	20.0 g
Sugar	47.0 g
Color and flavor	q.s.
	71.0 g

Blend dry ingredients and place in small electric mixer bowl. Mix in 237 mL (1 cup) cold milk at low speed. Whip at high speed for 3 min. Fold in 237 mL (1 cup) cold milk at low speed for about 1 min. Pour into mold; refrigerate. Pudding may be consumed in 5 to 10 min and unmolded after about 1 h.

Cooked Custard Type (Dessert and Pie Filling)

Iota-carrageenan	1.4 g
Mixed *kappa*- and *lambda*-carrageenan	0.3 g
Tetrasodium pyrophosphate	0.8 g
Salt	0.3 g
Sugar	54.2 g
Color and flavor	q.s.
	57.0 g

For Custard. Blend the dry ingredients, add 474 mL (1 pint) of milk, and stir thoroughly in a saucepan until completely dispersed. Bring to a full boil over medium heat, stirring constantly to prevent sticking. Remove from heat, pour into molds, and refrigerate. An emulsifier such as Myvatex 8–20 (Distillation Products Industries, Rochester, N.Y.) may be incorporated into the dry blend at a level of 0.2 g to minimize foaming.

For Pie Filling (or Custard). Proceed as above, except use only 237 mL (1 cup) of milk initially, then, after cooking, remove from the heat, and stir in another 237 mL (1 cup) of cold milk. Pour into a pie crust or molds. (The hot mix of the first procedure could damage a pie crust; additionally, the second method renders the finished product ready for consumption more quickly.)

Cooked Custard or Flan Flans are far more popular in Europe than in the United States, where the creamy-textured starch-based milk pudding is preferred. However, prepackaged flans and box mixes are now being introduced in the United States. These are far less rich than the usual cookbook flan which relies heavily on cream, sugar, and eggs. The typical flan texture is provided by carrageenan, alone or in combination with other gums. The following recipe can be prepared by either of the two preceding methods given for the cooked custard pie filling:

Cane sugar (fine granulated)	54.2 g
Kappa-carrageenan	1.1 g
Sodium hexametaphosphate	1.0 g
Salt	0.4 g
Color and flavor to suit	q.s.
	56.7 g

Antacid Gel

The chalkiness of antacids is readily masked by inclusion of them in a carrageenan gel. This convenient dosage form may be varied in texture from creamy to chewy by modification of the formula. The following will give a chewy gel:

Mixture A		*Mixture B*	
Kappa-carrageenan	0.6%	Antacid powder	25%
Locust bean gum	0.6%	Distilled water	43.6%
Methyl paraben	0.18%		
Propyl paraben	0.02%		
Cane sugar	5.0%		
Glycerin	25.0%		

Disperse the dry ingredients of each mixture into their respective liquids; then combine the two mixtures. Heat to 77°C (170°F) with agitation, to dissolve the gums; then add color and flavor to suit, make up evaporation losses with water, and pour into molds holding 2 g each (to give an antacid dose of 0.5 g per unit). When the mixture is cool, unmold and wrap.

LABORATORY TECHNIQUES

Water Viscosity Measurement

Equipment Required:
 Boiling-water bath
 Mechanical agitator and propeller-type stirrer blade
 600-mL beaker, Berzelius
 Balance
 Thermometer
 Brookfield Viscometer, LVF, 110v–60 cycle, equipped to run at 12 rpm, with #1
 and #2 spindles and guard
Procedure:
1. Disperse 7.50 g of the extractive into 450 mL of distilled water contained in a
   tared beaker. After the powder is well dispersed (10 to 20 min), bring to final
   weight (500 g + tare) with distilled water and heat in the boiling-water bath under
   continuous mechanical agitation until a temperature of 80°C is reached. Adjust
   for evaporation losses with distilled water, mix well, and cool to 76 to 77°C.
2. Preheat appropriate spindle and guard to approximately 75°C; quickly wipe dry
   and attach to viscometer. Position in hot solution, adjusting spindle to proper
   height. Start viscometer rotating and measure solution temperature.
3. At a solution temperature of 75°C, remove the thermometer and observe the
   scale reading (0 to 100 scale). Take reading after six complete revolutions. Record
   reading and multiply by 5 to obtain viscosity in centipoises when using the #1
   spindle, or by 25 for the #2 spindle (viscosities over 0.005 Pa·s). *Note:* Variations
   of only 1°C in the temperature will cause a 2.5 to 5% change in the viscosity.

Water Gel Strength Measurement

Reagent Required:
 U.S.P. potassium chloride
Equipment Required:
 Boiling-water bath
 Mechanical agitator and propeller-type stirrer blade
 Thermostatically controlled, circulating water baths: 25°C for *kappa*-carrageenan,
   10°C for *iota*-carrageenan
 Crystallizing dishes, 7.0 cm diameter by 5.0 cm deep
 Device for determining force exerted on plunger necessary to break gels, such as
   Cherry-Burrell Curd Tension Meter or Marine Colloids Gel Tester
 Plungers for tester 1.09 cm (0.431 in) diameter for *kappa*-carrageenan, 2.15 cm
   (0.845 in) for *iota*-carrageenan
 Dietary balance, 1000-g capacity, or transducer/recorder (for MC Gel Tester)
Procedure (for *kappa*-carrageenan):
1. Disperse 10.00 g carrageenan and 1.00 g potassium chloride in 490 mL of distilled
   water.
2. Heat with agitation in boiling-water bath until a temperature of 80°C is reached.
   Remove from bath and adjust weight to 500 g plus tare, using distilled water.
3. Divide the hot solution among three crystallizing dishes, and place them in the
   25°C bath. Hold in bath for 1 h.
4. Remove dishes from the bath.
5. Invert the gels in the dishes, using a small spatula between the gel and the dish
   to let air in as the gel slides out, and to help it escape as the gel is returned.
6. Place the dishes on the measuring device so that the plunger will contact the
   gel approximately in the center. Start the plunger and observe the pointer on
   the dietary scale for the reading at which it springs back, or read the break force
   off the recorder chart.
7. Repeat with the other two gels, and average the three results.
Procedure (for *iota-carrageenan*). Same as for *kappa*-carrageenan except:
1 and 2. Normally, sufficient calcium is present in the extractive to produce a good
   gel. If not, add 2.00 g CaCl$_2$ · 2H$_2$O in addition to the KCl.
3. Use 10°C bath.

4. Holding time is 2 h.
5. Do not invert gels, but loosen edges with small spatula.
6. Use large plunger.

Milk Gel Strength Measurement

Equipment Required:
  Boiling-water bath
  Mechanical agitator and propeller-type stirrer
  1-L beaker
  Crystallizing dishes, 7.0 cm diameter by 5.0 cm deep
  Petri dish cover
  Thermostatically controlled circulating-water bath, 10°C
  Device for determining force exerted on plunger necessary to break gels, such as Cherry-Burrell Curd Tension Meter or Marine Colloids Gel Tester
  Plunger for tester, 2.15 cm (0.845 in) diameter
  Dietary balance, 1000-g capacity or transducer/recorder (for MC Gel Tester)
Material Required:
  Homogenized milk
Procedure:
1. Disperse 1.00 g carrageenan into 486 g milk contained in tared beaker, and heat in boiling-water bath with mild agitation (to minimize foaming) to a temperature of 82°C.
2. Remove from boiling-water bath and replace evaporation losses with distilled water (to 487 g plus tare). Mix briefly and pour into three crystallizing dishes, filling them within 3 mm of the top.
3. Place the dishes in the 10°C bath for 1 h, then invert the gels into Petri dish covers. A small spatula may be used to facilitate removal of the gel from the dish by providing an air passage between the gel and the dish.
4. Position the gel in the Petri dish on the measuring device so the plunger will make contact in the center. Start the plunger and observe the pointer on the dietary scale for the reading at which it springs back, or read the break force off the recorder chart.
5. Repeat with the other two gels, and average the three results.

PRODUCT/TRADENAME GLOSSARY

Aubygel: Gelling carrageenans used for milk and water systems, Satia Div., Ceca, S.A.

Aubygum: Carrageenans used to thicken water systems, Satia Div., Ceca, S.A.

Carastay: Carrageenans used for gelling and thickening of milk and water systems, Stauffer Chemical Co.

Carrageenan: Any of several sulfated galactans found in certain red seaweeds and used for thickening, gelling, and stabilizing water and milk systems. For specific types, see *iota-*, *kappa-*, and *lambda-*carrageenans.

Flanogen: Carrageenan-galactomannan mixtures used to gel and thicken milk systems, Satia Div., Ceca, S.A.

Furose: Carrageenan used as a dispersant and suspending agent, Marine Colloids Div., FMC Corp.

Gelcarin: Carrageenans used for gelling or thickening both milk and water systems, Marine Colloids Div., FMC Corp.

Gelloid: Carrageenans standardized with sugars; used for gelling or thickening both milk and water systems, Marine Colloids Div., FMC Corp.

Gelogen: Mixtures of carrageenans and phosphates used to gel and thicken milk systems, Satia Div., Ceca, S.A.

Genugel: Carrageenans used as gelling or thickening agents, some types containing galactomannans, Copenhagen Pectin Factory (Hercules, Inc.)

Genulacta: Carrageenans primarily for use in milk systems, Copenhagen Pectin Factory (Hercules, Inc.)

Genuvisco: Carrageenans used as gelling or thickening agents, Copenhagen Pectin Factory (Hercules, Inc.)

*Iota-*carrageenan: A polysaccharide consisting of alternating 1,3-linked β-D-galactose-4-sulfate and 1,4-linked 3,6-anhydro-α-D-galactose-2-sulfate residues. It is most commonly obtained from the red seaweed *Eucheuma spinosum* and is used for its thickening, gel-forming, and suspending properties in both food and nonfood uses.

Kappa-carrageenan: A polysaccharide comprised of alternating 1,3-linked β-D-galactose-4-sulfate and 1,4-linked 3,6-anhydro-α-D-galactose residues. It is commonly obtained from the red seaweeds *Chondrus crispus* and *Eucheuma cottonii* and is used for its gelling properties.

Lambda-carrageenan: A polysaccharide consisting of alternating 1,3-linked β-D-galactose having about 70% of the C_2's sulfated and 1,4-linked α-D-galactose-2,6-disulfate. It is not generally available free from *kappa*-carrageenan, with which it occurs. The richest sources are the red seaweeds *Gigartina acicularis* and *Gigartina pistillata*, though *Chondrus crispus* is more commonly used in its manufacture. In such products, care is taken to avoid development of the gelling properties of the *kappa* component of the seaweed. *Lambda*-carrageenan is nongelling in water and is used for thickening and binding (as in toothpastes).

Lygomme: Carrageenan blended with other gums; used for thickening and gelling milk and water systems, Satia Div., Ceca, S.A.

Nutricol: Carrageenan, salts, and galactomannan used to gel water systems. Certified Kosher. Marine Colloids Div., FMC Corp.

Satiagel: Carrageenan used to gel milk systems, Satia Div., Ceca, S.A.

Satiagum: Carrageenans used to thicken milk and water systems, Satia Div., Ceca, S.A.

SeaGel: Carrageenan blended with galactomannans; used for gelling milk and water systems, Marine Colloids Div., FMC Corp.

SeaKem: Carrageenans used for gelling or thickening both milk and water systems, Marine Colloids Div., FMC Corp.

SeaSpen: Carrageenan and salts used to form aqueous thixotropic gels capable of suspending high levels of solids, Marine Colloids Div., FMC Corp.

Viscarin: Carrageenans used for thickening and stabilizing milk and water systems, Marine Colloids Div., FMC Corp.

FURTHER USEFUL READING/REFERENCES

General References

Glicksman, M., *Gum Technology in the Food Industry*, Academic Press, Inc., New York, 1969.

Lawrence, A. A., *Edible Gums and Related Substances*, Noyes Data Corporation, Park Ridge, N.J., 1973, pp. 98–127.

Moirano, A. L., in H. D. Graham (ed.), *Food Colloids*, AVI Publishing Co., Westport, Conn., 1976, chap. 8.

Rees, D. A., in M. L. Wolfrom and R. S. Tipson (eds.), *Advances in Carbohydrate Chemistry and Biochemistry*, vol. 24, Academic Press, Inc., New York, 1969, pp. 267–313.

Towle, G. A., in R. L. Whistler and J. N. BeMiller (eds.), *Industrial Gums*, 2d ed., Academic Press, Inc., New York, 1973, chap. 5.

References

1. Anderson, N. S., T. C. S. Dolan, C. J. Lawson, A. Penman, and D. A. Rees, "Carrageenans. Part V. The Masked Repeating Structures of λ- and μ-Carrageenans," *Carbohyd. Res.*, 7, 468–473 (1968).
2. Collins, T., statement at FDA Conference, April 8, 1976; the study was not yet reported in final form at time of this publication.
3. Doty, M. S., *Eucheuma Farming for Carrageenans*, Sea Grant Advisory Report, UNIHI-SEAGRANT-AR-73–02, University of Hawaii, Honolulu, 1973.
4. Grindrod, J., and T. A. Nickerson, "Effect of Various Gums on Skim Milk and Purified Milk Proteins," *J. Dairy Sci.*, 51, 834–841 (1968).
5. Lawson, C. J., and D. A. Rees, "An Enzyme for the Metabolic Control of Polysaccharide Conformation and Function," *Nature*, 227, 390–393 (1970).
6. Lin, C. F., and P. M. T. Hansen, "Stabilization of Casein Micelles by Carrageenan," *Macromolecules*, 3, 269–274 (1970).
7. McCandless, E. L., J. S. Craigie, and J. A. Walter, "Carrageenans in the Gametophytic and Sporophytic Stages of *Chondrus crispus*," *Planta*, 112, 201–212 (1973).
8. Mueller, G. P., and D. A. Rees, "Current Structural Views of Red Seaweed Polysaccharides," in H. D. Freudenthal (ed.), *Drugs from the Sea*, Marine Technology Society, Washington, D.C., 1968, pp. 241–255.
9. Parker, H. S., III, "Seaweed Farming in the Sulu Sea," *Oceans*, 9(2), 12–19 (1976).
10. Percival, E., "Chemistry of Agaroids, Carrageenans and Furcellarans," *J. Sci. Food Agric.*, 23, 933–940 (1972).
11. Rees, D. A., "Mechanism of Gelation in Polysaccharide Systems," in *Gelation and Gelling Agents*, British Food Manufacturing Industries Research Association, Symposium Proceedings No. 13, London, 1972, pp. 7–12.
12. Snoeren, Th. H. M., T. A. J. Payens, J. Jeunink, and P. Both, "Electrostatic Interaction between κ-Carrageenan and κ-Casein," *Milchwissenschaft*, 30, 393–396 (1975).

13. Stancioff, D. J., and D. W. Renn, "Physiological Effects of Carrageenan," in A. Jeanes and J. Hodge (eds.), *Physiological Effects of Food Carbohydrates*, American Chemical Society, Washington, D.C., 1975, pp. 282–295.
14. Stancioff, D. J., and N. F. Stanley, "Infrared and Chemical Studies on Algal Polysaccharides," in R. Margalef (ed.), *Proceedings VI International Seaweed Symposium, 1968*, Subsecretaría de la Marina Mercante, Dirección General de Pesca Marítima, Madrid, 1969, pp. 595–609.
15. Stanley, N. F., "The Properties of Carrageenans as Related to Structure," *Proceedings C.I.C. Conference on the Marine Sciences, Charlottetown, P.E.I., Canada, August 1970.*

Chapter **6**

Guar Gum

James K. Seaman
Celanese Plastics and Specialties Company

General Information 6-1
  Manufacture 6-2
  Chemical and Physical Properties 6-3
  Biological Properties 6-9
  Handling 6-10
  Applications 6-10
Commercial Uses: Compounding and Formulating 6-12
  Food 6-13
  Explosives 6-13
Commercial Uses: Processing Aids 6-13
  Oil and Gas 6-13
  Textile 6-13
  Paper 6-14
  Mining 6-16
Industries Using Guar Gum 6-17
  Oil and Gas 6-17
  Explosives 6-18
  Food 6-18
  Paper 6-18
  Textile 6-18
  Mining 6-18
  Formulations 6-18

GENERAL INFORMATION

Guar gum is an edible carbohydrate polymer which is useful as a thickening agent for water and as a reagent for adsorption and hydrogen bonding with mineral and cellulosic surfaces. Etherification with non-ionic, anionic, and cationic reagents has extended the usefulness of the gum. Within the past 5 years guar derivatives have become an important fraction of total guar production. Intensive development efforts by guar gum producers suggest a continuation and expansion of the trend in the years ahead.

Guar gum is found in the seeds of two annual leguminous plants *(Cyamopsis tetragona-lobus* and *psoraloides)*. The seeds are contained in pods 2.54 to 5.08 cm (1 to 2 in) long. In India and Pakistan, where the crop has been grown for centuries as a food for both humans and animals, hand labor is used in the harvest. In the southwestern United States, however, it has established itself within the past 30 years as a cash crop suitable for modern mechanical farm technology. The need for reliable supplies of guar gum continues to grow rapidly, and efforts are underway to establish it as a cash crop in other parts of the world. In 1976, the worldwide consumption of guar products for food and industrial applications reached an estimated 57 million kg (125 million lb) annually.

Manufacture

The production of guar gum in the broadest sense is a series of crushing, sifting, and grinding steps to separate the seeds from the pod and then to separate the valuable gum from the seeds. Processing techniques influence the properties and economics of guar gum and its derivatives, and are generally held as valuable proprietary information by the various producers.

Seed Structure The gum is contained within a portion of the seed known as the *endosperm*. This is the reserve food supply for the developing embryo during germination. Since the seed is dicotyledenous, two endosperm halves are obtained from each seed. These endosperm halves surround the embryo, and they are in turn surrounded by a hull, which is usually light tan in color. In some of the plant varieties, the hull is normally darker. In all varieties, there is a tendency for the hull to darken if the crop is not harvested promptly following the first frost and before the next rain. In severe cases of weather damage, many of the seeds wither and turn black. The yield of gum is lowered, and the gum can be contaminated with black specks.

Guar seeds range from 0.254 to 0.635 cm (0.1 to 0.25 in) in diameter, and the endosperm is from 35 to 42% of the weight of the seed.

Purification Endosperm is separated from the hull and embryo by taking advantage of the difference in hardness of the various seed components. Multistage grinding and sifting operations are combined with other physical treatments to crack the seeds and separate the parts. There are many types of grinders, such as attrition mills and special types of hammer and roller mills, that can be employed in combination with water or acid soaks and embrittling heat treatments.

Grades The endosperm is ground to powder form and marketed as guar gum.

Food-grade. Substantially but not completely pure endosperm is obtained by the usual commercial purification techniques. There may be a small amount of hull and germ present, owing to imperfect purification. Since the entire seed is edible, this contamination dilutes slightly the amount of available gum, but does not harm its suitability as a food additive. See Tables 6.1 and 6.2 for *Food Chemicals Codex* specifications and typical analysis.

Determination of moisture, protein, and acid-insoluble residue content is generally considered to be the minimum analysis needed to monitor the efficiency of the purification process. The presence of the germ is reflected in high protein content. Hull fragments show up as a high acid-insoluble residue. Moisture contents above 15% or

TABLE 6.1
Food Chemicals Codex Specifications for Guar Gum

Galactomannans: Not less than 66.0%
Limits of impurities:
  Acid-insoluble matter: Not more than 7%
  Arsenic (as is): Not more than 3 ppm (0.0003%)
  Ash (total): Not more than 1.5%
  Heavy metals (as Pb): Not more than 20 ppm (0.0002%)
  Lead: Not more than 10 ppm (0.001%)
  Loss on drying: Not more than 15%
  Protein: Not more than 10%
  Starch: Passes test (negative)

TABLE 6.2
Typical Analyses of Guar Powder

	Food-grade fine mesh	Food-grade intermediate mesh	Food-grade coarse mesh	Industrial-grade fine mesh
Percent moisture	10–12	10–12	10–12	10–12
Percent protein	4–6	4–6	4–6	4–6
Percent acid insoluble residue	2.5–5.5	2.5–5.5	2.5–5.5	2.5–5.5
Viscosity:*				
Pa·s	3–5	2.5–4.5	2–3.5	3–5
cP	3000–5000	2500–4500	2000–3500	3000–5000

* At 25°C as measured on a Brookfield Synchro-Lectric viscometer at 20 rpm.

below 8% exceed normal equilibrium moisture and indicate abnormal treatment of the seed or powder.

A crude fiber determination is sometimes used in place of the acid-insoluble residue. Crude fiber is the acid- and alkali-insoluble residue and typically amounts to about half of the acid-insoluble residue. It is recognized that other natural impurities are present in food-grade guar gum. The natural mineral content of the seeds gives rise to an ash content. Low-molecular-weight soluble sugars are present but not usually assayed, and a small ether extractable fraction reflects the natural presence of fats and oils.

Industrial-grade. Historically, most of the guar gum sold for industrial applications has been made from endosperm as pure as that used for food-grade products. Industrial grades, however, make extensive use of chemical additives to modify and control such properties as rate of swell, viscosity, and solution stability for specific end uses. These additives can lead to controlled gelation and controlled viscosity decrease.

Processing techniques can also control dispersion, rate of hydration, and viscosity. When these techniques are combined with the use of chemical additives, they lead to the historic pattern of tailor-made products which has dominated the growth of guar consumption.

Derivatives. In recent years, anionic (carboxymethyl), non-ionic (hydroxyalkyl), and cationic (quaternary amine) derivatives have become important factors in total guar consumption. These derivatives are not presently approved for food use, although some may be in the future. The development of the derivatives has followed the pattern previously noted for industrial-grade products. That is, they make full use of chemical additives and processing techniques and are designed to satisfy specific customer end-use requirements. Many products are currently available, and their number and importance are growing rapidly.

Chemical and Physical Properties

Guar gum is a carbohydrate polymer containing galactose and mannose as the structural building blocks. The ratio of the two components may vary slightly, depending on the origin of the seed, but the gum is generally considered to contain one galactose unit for every two mannose units.

Structure It has long been accepted that the structure of guar gum is a linear chain of β-D-mannopyranosyl units linked (1 → 4) with single-membered α-D-galactopyranosyl units occurring as side branches. The α-D-galactopyranosyl units are linked (1 → 6) with the main chain. Recent work has confirmed the long-held belief that the side branches are spaced uniformly. Guar gum is a natural alternating copolymer, as shown in Fig. 6.1.

All of the commercially important derivatives of guar are formed by etherification. The primary C_6 hydroxyl position is the most highly reactive, but the secondary hydroxyls are also sites for substitution.

Solubility in Water Guar gum is a cold water swelling polymer. In its powdered commercial form the rate of thickening and the final viscosity reflect the process history

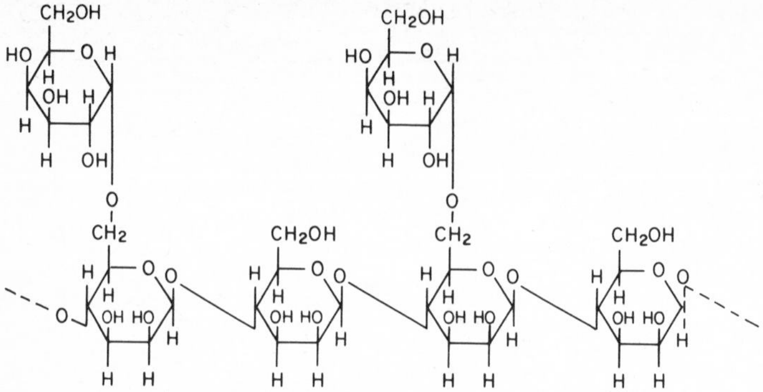

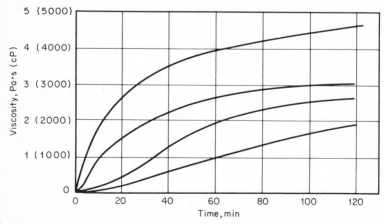

Figure 6.1 Structure of guar gum.

of the product, including the particle size of the powder. See Fig. 6.2 for typical swell curves of a range of available products.

Heating a guar gum solution reduces the time needed to reach its full viscosity potential.

Water is the only common solvent for guar gum, although it will tolerate limited concentrations of water-miscible solvents, such as alcohols. Dimethylformamide and dimethylsulfoxide are solvents as they are for most other polymers.

Commercial guar gum solutions are typically turbid. The turbidity is largely caused by the presence of insoluble portions of the endosperm. Polymer solutions, intensively purified in the laboratory by noncommercial methods, approach water clarity.

Derivatives of guar gum can show dramatically different solubilities and clarities in solution. There are hydroxyalkyl derivatives available, much clearer than guar gum, capable of tolerating major proportions of water-miscible solvents. Carboxymethyl derivatives in general are much clearer than guar gum. Improved clarity results from derivatization and solubilization of insoluble seed impurities. Changes in solubility occur when sufficient hydroxyl groups of the galactomannan are substituted and the new polymer takes on some of the characteristics of the derivatizing reagent.

Rheology Guar gum is one of the most highly efficient water-thickening agents known. Solutions of guar gum and its derivatives are non-Newtonian, classed as pseudoplastic. They thin reversibly when heat is applied and degrade irreversibly with time

Figure 6.2 Viscosity development as a function of time for typical guar products. Solution concentration is 1% in water. (Measured on a Brookfield Synchro-Lectric viscometer at 20 rpm, 25°C.)

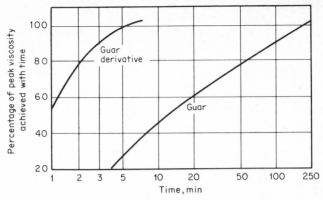

Figure 6.3 Percent viscosity developed as a function of time for typical fast hydrating guar and guar derivative types. Solution concentration is 1% in water. (Measured on a Brookfield Synchro-Lectric viscometer at 20 rpm, 25°C.)

when an elevated temperature is maintained. Some of the recently developed hydroxy-alkyl derivatives resist such degradation to a much greater degree. Solutions resist shear degradation well compared to other water-soluble polymers, but they will degrade with time under high shear. Figures 6.2 to 6.9 show various viscosity and shear properties of guar products.

Viscosity. Guar gum and its derivatives are most commonly used at concentrations below 1%. High-viscosity products are thick solutions at this concentration and look more like gels at 3%. Guar and guar derivatives are available in lower-viscosity form for special applications, for example, where high gum solids are favored, where it is desirable to have charged molecules with controlled thickening power, or where less pseudoplastic, more Newtonian solutions are appropriate. The wide range of viscosities commercially available is consistent with the historic approach of designing guar products to meet the needs of an industry. See Table 6.2 for the range of typical products.

Shear Response. Solutions of guar gum have zero yield value at the most commonly used concentrations. That is, they begin to flow as soon as the slightest shear is applied. The apparent viscosity of the solution decreases sharply as the rate of shear is increased,

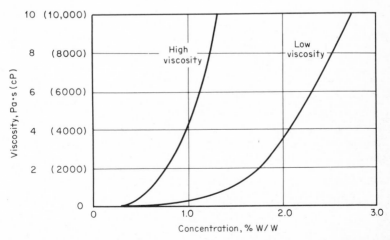

Figure 6.4 Effect of concentration on the viscosity of typical high- and low-viscosity guar products. (Measured on a Brookfield Synchro-Lectric viscometer at 20 rpm after 24 h hydration at 25°C.)

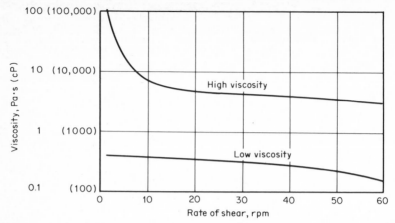

Figure 6.5　Effect of shear rate as measured on a Brookfield Synchro-Lectric viscometer for typical high- and low-viscosity guar products.　Solution concentration is 1% in water, 25°C.

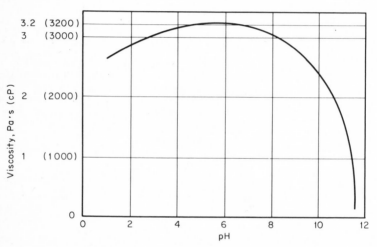

Figure 6.6　Effect of pH on the viscosity of a 1% guar solution at 25°C as measured on a Brookfield Synchro-Lectric viscometer at 20 rpm.　Solution was hydrated at the indicated pH for 24 h.

then levels off and approaches a minimum limiting value that is dependent on the concentration of the solution.　The apparent viscosity of a guar gum solution at any given rate of shear is independent of time and prior shear testing.　It makes no difference whether the final rate of shear is approached from low shear rate to high shear rate or vice versa, provided the shear rates are not high enough to degrade the molecular structure.　The molecular structure can be irreversibly degraded by high rates of shear that would be found, for example, in pumps used to transfer solution from one tank to another.　Such breakdown is related to the degree of shear and the time over which it is applied.　Degraded solutions progressively lose viscosity at a given rate of shear. They also show a change in the shape of their viscosity vs. shear rate curve.　They become less pseudoplastic, the curve flattens out, and they tend toward the linear response of a Newtonian fluid.

A significant variation in shear response has been found in a mixed ether derivative, carboxymethyl hydroxypropyl guar.　This derivative holds insoluble particles in suspen-

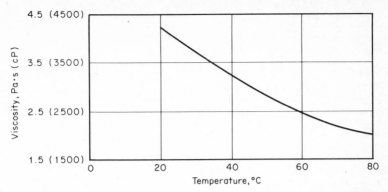

Figure 6.7   Effect of temperature on the viscosity of a 1% guar solution as measured on a Brookfield Synchro-Lectric viscometer at 20 rpm.

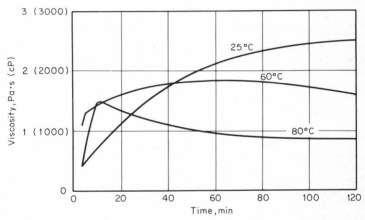

Figure 6.8   Viscosity stability of a 1% guar solution at various temperatures as measured on a Brookfield Synchro-Lectric viscometer at 20 rpm.

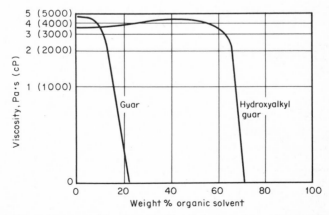

Figure 6.9   Effect of water-miscible solvent (methanol) concentration on the viscosity of 1% guar and hydroxypropyl guar solutions. (Measured on a Brookfield Synchro-Lectric viscometer at 20 rpm.)

sion better than guar or its other commercial derivatives. At the very low shear rates typical of a particle falling under the force of gravity, the solution appears to possess a very high apparent viscosity.

Lower-viscosity, lower-molecular-weight guar and guar derivatives are less pseudoplastic and more Newtonian than their high-viscosity, high-molecular-weight counterparts.

Reactivity   The cyclic neutral sugar structures which make up the polymer contain numerous hydroxyl groups (an average of three per sugar unit). The various reactions of guar gum with other substances are explained by the number and position of these hydroxyl groups.

Derivatives. Etherification and esterification reactions are readily carried out through the hydroxyl functionalities. The commercially important derivatives to date have been confined, however, to the etherification reactions: carboxymethylation with monochloroacetic acid, hydroxyethylation with ethylene oxide, hydroxypropylation with propylene oxide, and quaternization with various quaternary amine compounds containing reactive chloride or epoxide sites.

Significant changes in properties occur when the guar molecule is substituted with these reagents. The degree of change is a function of the type of reagent, the amount of reagent, and the conditions of reaction.

Inclusion of charged sites into the guar molecule is a most obvious source of change. Anionic sites (carboxymethylation) and cationic sites (quaternization) modify the way in which the guar molecule reacts with inorganic salts, hydrated mineral and cellulosic surfaces, and organic dyes. Many products have been designed to take advantage of these changes for specific applications. Depending on the need, the rate and degree of adsorption can be strengthened or lessened, flocculation increased or decreased, gelation made easier, and compatibility improved. Carboxymethylation has the added effect of improving the clarity of guar solutions by solubilizing some of the non-guar seed impurities that result from commercial processing.

Changes in solubility in and compatibility with water-miscible solvents are particularly promoted by hydroxyalkylation. For example, dramatic increases in alcohol tolerance are seen with increasing derivatization.

Hydroxyalkylation is a very useful tool for changing the way in which guar reacts with hydrated mineral and cellulosic surfaces. Increasing degrees of derivatization have the effect of lessening the affinity for such surfaces. This leads to compatibility with high-solids mineral dispersions, modified flocculation rates in complex ore dispersions, and easier removal of gum from dyed fabric by way of illustration.

Electrolyte compatibility is extended by hydroxyalkylation (see Table 6.3). Derivatives have been designed to thicken di- and trivalent salt solutions in which guar will fail to hydrate.

TABLE 6.3
Electrolyte Compatibility of Guar vs. Hydroxypropyl Guar
Viscosity measured on a Brookfield Synchro-Lectric viscometer at 20 rpm. Solutions at 1% concentration based on weight of salt solution, 25°C.

| | Viscosity, Pa·s   (cP) | | | |
	Guar		Hydroxypropyl guar	
26% NaCl	9.1	(9,100)	9.45	(9,450)
38% CaCl$_2$	0.02	(20)	7.7	(7,700)
56% Ca(NO$_3$)$_2$	0.035	(35)	22.55	(22,500)

Biodegradability is progressively changed with increasing hydroxyalkylation. Guar which is easily degraded can be made highly resistant and stable in viscosity with suitable substitution.

The solubility and clarity of guar solutions are markedly improved by hydroxyalkylation. As with carboxymethylation, it is believed that this is a result of derivatizing and solubilizing seed impurities, particularly the cellular components of the endosperm.

Mixed ethers of the various reagents are used to combine the benefits of charged and uncharged substitution for specific applications.

The central conclusion to be drawn from the preceding discussion is that guar gum has proved to be an excellent polymeric raw material for the creation of new and important water-soluble gums. In the process, the range of problems that can be solved has been greatly expanded. The outlook is for a continuing flow of new products in the years ahead. These products will provide increasingly more sophisticated answers to major industrial needs.

Borax Reaction. Each of the sugar residues in the polymer has two hydroxyl groups positioned in the cis- form. This leads to an interesting and valuable reaction with dissociated borate ions that is characteristic of such polymers. The reaction is fully reversible with changes in pH. An aqueous solution of the gum will gel in the presence of borate when the solution is made alkaline, and will liquify again when the pH is dropped below 7. If the dry powdered gum is added to alkaline borate solution, it will not hydrate and thicken until the pH is dropped below 7. The critical pH at which gelation occurs is modified by the concentration of dissolved salts. The effect of dissolved salts is to change the pH at which a sufficient quantity of dissociated borate ions exists in solution to cause gelation.

Derivatives of guar gum tend to react less and less with borate ions as the amount of substituting groups increases in the molecule. This results because the shear bulk of substituting groups changes the regular, alternating, and single-member branched, linear configuration of the molecule and prevents adjacent chains from approaching as closely as before and the substitution of secondary cis- hydroxyl positions decreases the number of such unsubstituted positions available for reaction.

Gels. A considerable number of reactions other than the borax reactions are routinely used to convert guar and guar derivative solutions to gels. These gels may range from pumpable high-viscosity compositions to solid, rubbery products. The common factor is that they are formed by transition metal complexes through the cis- hydroxyl positions. Derivatization alters the strength of the gels in the same way that it affects the borax reaction.

In the special case where a relatively minor amount of charged substituent (e.g., carboxymethyl) has been introduced into the molecule, this can serve as a useful site for ionic crosslinking of adjacent chains.

Salt Reactions. Strong reactions are also obtained with solutions of certain inorganic cations. The addition of a high concentration of calcium salt, for example, will cause a gel to form under alkaline conditions. If dry powder is added to the salt solution, the gum will not hydrate and thicken. In general, the gum will react with polyvalent cations much as it does with borate anion. Significant differences, however, are that many of the reactions lead to insolubilization, precipitation, or unstable, non-useful gels.

Derivatization with non-ionic hydroxyalkyl groups greatly improves the compatibility of guar with most salts.

Biological Properties

Guar gum has the status of a direct food additive. Gum which meets *Food Chemicals Codex* specifications is affirmed as GRAS as a direct food additive under FDA regulation 184.1339.

Solutions of guar gum serve as a food source for the growth of common microorganisms. Where it is necessary to hold solutions for a length of time, common food preservatives, such as benzoic or sorbic acid, may be added to inhibit bacterial growth. In non-food applications, quaternary ammonium preservatives, chlorinated phenolics, or formaldehyde are effective bacterial inhibitors.

The advent of guar derivatives as important commercial products has brought another technique for the preparation of stable solutions. Substitution of the hydroxyl groups alters the structure of the molecule and makes it less suitable as a food for common microorganisms. The more the molecule is altered, the more resistant it becomes to biological attack. The derivatives in wide use today are more resistant than guar to a significant degree. The outlook is for the introduction of derivatives that are stable in solution for long periods of time. They will even resist inoculation with potent cultures that would destroy a guar solution within an hour.

Pending the introduction of completely stable derivatives, in those cases where the bacterial contamination has been allowed to get out of hand and manufacturing equipment contains the residue of preceding batches, it is much easier to thoroughly clean the equipment and start again than to try to control fermentation with preservatives.

Handling

Consideration of the storage of the powdered product and preparation of guar solutions follows:

Dry Storage   The powdered commercial grades of guar gum and its derivatives are all stable in the dry form.   In common with other water-soluble polymers, the one condition that adversely affects the gum is allowing it to get wet.   As long as it is stored in a dry place, there should be no problem from fermentation or lumping of the powder.

Solution Preparation   Many techniques are available to form smooth, uniform solutions from these cold-water-swelling powders.   The principal need is to uniformly disperse and wet each particle.   Particles that are not properly separated will stick together as they begin to swell, leading to lumpy, nonhomogeneous solutions requiring abnormally vigorous agitation and time to come to a uniform, fully hydrated state.

Vigorous agitation is the most common means of dispersing rapidly swelling powder. Additional flexibility can be had by selecting coarser-mesh products that swell less rapidly. If other powders are to be added to a solution, the gum may be preblended with these powders, which will assist in ease of dispersion.   Similarly, liquids that must be added to a formulation can serve as a carrier for the gum, providing the gum does not swell in the liquid.

For the really difficult cases where agitation is poor and the gum must be added directly to water, there are chemical means to briefly retard the hydration of even the finest-mesh, fastest-swelling gum.   These techniques are suitable only for industrial products.   Such products are routinely supplied by the manufacturers to meet the specific needs of a customer.

Applications

Guar gum and its derivatives are among the most important water-soluble polymers. Major uses are in the oil and gas, textile, paper, food, explosives, and mining industries.

Oil and Gas   It is common industry practice to increase the productivity of oil and gas wells by cracking open the hydrocarbon-bearing zones with hydraulic pressure. Guar and its derivatives are used in this process, which is known as *hydraulic fracturing*. High viscosity, at reasonable cost, is required to carry the graded sand which is pumped into the fractured rock.   The sand props open the fracture when the hydraulic pressure is released.   Oil and/or gas is then recovered at an increased rate because the porous rock containing the hydrocarbon has much more surface area exposed and connected to the well bore through the fracture channels just formed.

Guar products provide the viscosity needed for this operation.   They are also compatible with a wide range of field waters and can be formulated to lose viscosity at a controlled rate.   This facilitates removal from the well bore when flow is reversed upon completion of the job.   Guar products are also useful in controlling fluid lost to the porous formations being fractured and in reducing frictional pressure losses while fluid is being pumped.

Explosives   The last 20 years have seen the growth of a sophisticated technology which uses nitrate salts, various organic and inorganic sensitizing ingredients, water, and a water-soluble crosslinkable thickener as a means of making what are called either *slurry explosives* or *water gels*.   These compositions are safer to use than previous explosives and have been formulated to meet such a variety of application needs that they are often the most economical way of formulating an appropriate explosive.

Guar gum and its derivatives are used in the formulation of such products.   It is the ability of guar products to thicken efficiently under a variety of difficult conditions and to be readily crosslinked or gelled that makes them useful.   A properly formulated slurry explosive gel may be anything from a rather loose, cohesive, and pourable gel to a rubbery, almost rigid solid.

Textile   Guar and its derivatives are used as dye solution thickeners in textile printing applications.   The derivatives are primarily carboxymethyl and hydroxylalkyl ethers of guar.   They are often oxidized under controlled conditions so that the thickening power of the product vs. concentration is brought to a predetermined level.   Derivatization promotes solubility. This is useful in preventing the buildup of guar and its derivatives on printing screens and aids in their removal by washing after printing.

Food This industry has made wide use of guar gum's ability to bind large amounts of water.

Ice Cream. Small amounts of guar gum do not greatly affect the viscosity of the mix during manufacture, but impart a smooth, chewy texture to the finished product. A slow meltdown of the product and increased heat shock resistance are other benefits. Ice cream stabilized with guar gum is notably free of the graininess caused by ice crystal formation.

Canned Pet Food. High free-water content is characteristic of this type of product. Guar gum is used to thicken the free water in the product, and the meat and vegetable solids are coated with a thick gravy. Special, slow-swelling grades of guar gum are sometimes used to limit viscosity during the can-filling operation.

Cheese. In soft cheese processing, guar gum controls the consistency and spreading properties of the product. Smoother, more homogeneous cheese spreads with high water content are possible because of the gum's water-binding properties.

Sauces and Salad Dressings. These products make use of the high viscosity at low concentration that is a basic property of guar gum.

Paper The major use of guar gum is as a wet-end additive. This means that the gum is added to the pulp suspension just before the sheet is formed on either a Fourdrinier machine or a cylinder machine. The pulping process, which is designed to remove lignin and thereby produce a fibrous cellulosic pulp, also removes a large part of the hemicelluloses normally present in the wood. These hemicelluloses, which are mostly mannans and xylans, could contribute greatly to the hydration properties of the pulp and the strength of the paper formed from the pulp. Galactomannans replace or supplement the natural hemicelluloses in paper bonding.

It is generally agreed that the hydrogen-bonding effect is one of the major factors affecting fiber-fiber bonding. An examination of the molecular structure of galactomannans reveals a rigid long-chain polymer molecule with primary and secondary hydroxyl groups. Such molecules are capable of bridging across and bonding to adjacent fibers.

It is difficult to compare the efficiency of locust bean gum and guar gum in paper products because too many variables exist, such as type of gum formulation, pulp type, beating time, and machine conditions. However, many persons in the paper industry believe that guar gum is the more efficient additive.

Commercial locust bean gum and guar gum contain small amounts of hull as contaminants. Locust bean hulls are red-brown and in white sheets, and appear as small but visible specks. However, guar hull is almost white. When present in the paper sheet, it is indistinguishable from the bleached pulp.

Advantages gained by the addition of galactomannans to pulp are:

- *Sheet formation:* A more regular distribution of pulp fibers (less fiber bundles) improves sheet formation.
- *Mullen:* Mullen bursting strength is increased.
- *Fold strength:* Fold strength is increased.
- *Tensile strength:* Tensile strength is increased.
- *Pick:* Pick is a measure of the force required to pull a fiber from the surface of a sheet. It applies to printing grades of paper. Pick is increased.
- *Pulp hydration:* Pulp is normally passed through jordans or wet mills which increase the surface of the fiber, thereby allowing it to bind more water. Galactomannans added to a pulp aid in the binding of water and decrease the amount of refining necessary, thereby lowering power consumption.
- *Finish:* Finish is improved, as measured by the smoothness and amount of protruding fibers.
- *Porosity:* Porosity is decreased.
- *Flat crush:* The pressure required to crush a corrugated flute is increased.
- *Machine speed:* Machine speed is increased without losing desirable tested properties.
- *Retention of fines:* Retention of fines is increased. Recently, cationic derivatives of guar have been developed with outstanding ability to increase drainage and fines retention.

Mining In this industry, guar gum and its derivatives are widely used as flocculants to produce liquid-solids separations. These may be broadly classified as filtration, settling, or clarification of mineral slurries. The property used is the ability of the gums to adsorb on the hydrated mineral particles by hydrogen bonding, followed by agglomeration as a result of bridging. Guar gum is also used in flotation to recover base metals. Guar gum acts as a depressant for talc or insoluble gangue mined along with the valuable minerals.

As an illustration of the use of guar products as flocculants, consider the case of an ore ground to a desirable particle size, leached to solubilize a valuable mineral, and then fed to a series of thickeners to separate the pregnant solution from fine solid waste. When dilute guar solutions are added to the leached ore, settling takes place within the thickeners. The rate at which this process takes place is governed by the slowest settling velocity in the thickeners. In turn, this depends on particle size, which is governed by the ability of guar molecules to form large flocs by first adsorbing on the hydrated solids and then causing them to agglomerate by bridging.

Settled solids flowing from the thickeners may be further dewatered by filtration. Guar products increase the permeability of the filter cake, resulting in an increased filtration rate. The floc formed is also incompressible and is readily penetrated by wash water. The cake is readily handled and transported by conveyor belts without decrepitation or dusting. The pregnant solution of this illustration may be further clarified before passing to metal recovery by treating it with dilute guar solution and allowing flocculated solids to settle, leaving behind clarified supernatant.

The terms *settling, filtration,* and *clarification,* then, usually relate to the individual steps of an overall mining recovery or wastewater treatment operation, and the terminology is governed by the intent of the particular step in that operation. Improvement of the operation by flocculation of solids is the common denominator.

In flotation operations the ore is ground to a desirable particle size and slurried in water prior to treatment. When guar or guar derivative is added as a dilute solution to the slurry, the gum adsorbs on the hydrated surfaces of the unwanted talc or gangue. Reactive chemicals called *collectors* are then added to the slurry and they react with the desirable minerals rather than the talc or gangue. The collectors condition the mineral surfaces so that the minerals will rise to the surface when a frother is added and air is bubbled through the slurry. Froth is skimmed off, separating the values from the talc or gangue that has been depressed by the guar product.

COMMERCIAL APPLICATIONS: Compounding and Formulating

The use of guar gum products is primarily in food or explosive applications.

TABLE 6.4
Maximum Usage Levels Permitted

Food (as served)	Percent
Baked goods and baking mixes	0.35
Breakfast cereals	1.2
Cheese	0.8
Dairy products analogs	1.0
Fats and oils	2.0
Gravies and sauces	1.2
Jams and jellies	1.0
Milk products	0.6
Processed vegetables and vegetable juices	2.0
Soups and soup mixes	0.8
Sweet sauces, toppings, and syrups	1.0
All other food categories	0.5

Food

Table 6.4 lists permissible use levels under FDA regulation 184.1339.

No stabilizer is composed entirely of guar gum. Guar derivatives are not presently permitted in food products. Use of the gum is further regulated under Section 170.3 of the Food and Drug Administration as to its function in specific food products.

Explosives

Slurry explosive formulations are best left to those skilled in the art, working under properly controlled conditions.

COMMERCIAL USES: Processing Aids

Oil and Gas

An oil or gas well can be fractured with a solution containing up to about 1% (by weight) of a guar product by supplying sufficient hydraulic horsepower to overcome the overburden pressure and crack the rock. The fracture thus formed is extended outward as far as possible by maintaining pressure and pumping the fluid, which contains graded sand, into the fracture. The sand props open the fracture when the operation is complete.

Textile

As discussed earlier, guar and guar derivatives are widely used in printing and dyeing applications as thickeners to control the mobility of dyestuffs. In some cases the viscosity they develop controls the volume of dye feed. The derivatives are largely hydroxyalkyl and carboxymethyl products, sometimes oxidized to varying degrees to control their thickening power. Locust bean gum products compete directly with guar gum products in this area, and the choice is usually governed by economics since such things as flow, viscosity, and solubility can be altered at will.

The following discussion of applications in the carpet industry illustrates the wide range of techniques in current practice. Similar techniques would apply to flat goods after taking into account the differences in construction of flat goods vs. carpets.

Carpets There are two use areas for gums in the carpet industry; one is dyeing and the other is printing. Some carpet yarns (primarily the spun yarns) are stock-dyed or package-dyed. In package-dyeing, gums are used at very low concentrations (0.1 to 0.2%) as migration control agents to produce a more even distribution of dye in the package.

Most of the carpet yarns are dyed after the carpet is formed, either by the beck method, or where long runs of the same shade are involved, by the newer continuous-dyeing machines. Beck dyeing is done by the exhaustion of dyes onto the fiber and requires no gum, but with one exception all of the newer continuous-dyeing machines use gum as the viscosity control to regulate the pickup of dye on a polished stainless steel roller immersed in the dye feed pan. This film of dye is continuously knifed off and cascaded in a sheet across the width of the carpet just prior to the fixation process, which takes place in a steaming chamber. During the steaming operation, the gum content of the dye solution serves to control the migration of the dye molecules so that a level shade is obtained from side to side of the carpet and also from the surface to the backing of the carpet.

Very-high-viscosity gums are used in this application so that an extremely low percentage of gum can be employed to maintain the critical viscosity level. The actual viscosity of the dye solution is low, in the range of 0.02 to 0.03 Pa·s (20 to 30 cP), and the concentration of gum used is in the area of 0.20 to 0.25%. This low level of gum is an advantage since there is less gum to wash off after the dyeing operation.

Carpet printing is done by various methods and the gum requirements vary with the method. Space printing is a technique originated some years ago. Essentially space printing is the production of a completely random distribution of a number of colors to produce a "pepper and salt" type pattern. The original patented process consisted of printing a geometrical distribution pattern on either flat or tubular knotted fabric and, after fixation of the color, deknitting and winding on cones. This printed yarn was then supplied to tufters to produce the patterned carpet. Other methods have been developed to produce similar effects by warp sheet printing, splatter printing of droplets of color, and computer-programmed distribution of color through nozzles or by pressure bars.

All of these methods require a gum thickener, and the thickener must have very special characteristics for optimum performance. The thickener should be low in insolubles content, high in viscosity, easily dispersible to a lump-free paste, extremely soluble for easy washout, produce good apparent color value, and be compatible with auxiliary chemicals and dyes being used.

Gums for space printing are generally used at 0.35 to 0.5%, and the viscosity will be 0.25 to 0.5 Pa·s (250 to 500 cP), depending on the method, fiber, and pattern.

Pattern designs on carpet are produced by two different screen printing techniques; the first and oldest method is by use of flat screen stencils, and the second is by means of rotary screen stencils which are tubular in shape and made of metal, such as nickel or stainless steel.

The flat screen method is somewhat slower than rotary printing but has a wider range of pattern choices. It also takes up more floor space. But if screen damage or blockage should occur, it is easier and quicker to rectify, and so produces less damaged carpet. Flat-bed machines run at speeds in the range of 4.6 to 6.1 linear meters (15 to 20 feet) per minute for most patterns and types of carpet, while rotary machines are capable of speeds of 18 linear meters (60 feet) per minute, or more. Flat-bed machines use medium-viscosity gums for most patterns and may use high-viscosity gums at lower solids under certain special conditions or on certain pattern or carpet types. Rotary machines can run most patterns on the more economical high-viscosity gums, but do require medium-viscosity gums for finer line patterns.

Flat-bed machines generally operate with dye pastes in the range of 0.8 to 7.5 Pa·s (800 to 7500 cP) viscosity, depending on the pile type and pattern. The gum percentages may be anywhere between 0.75 and 3%, depending on the type gum and viscosity requirements.

Rotary machines operate in a narrower range of viscosities, 6.5 to 10 Pa·s (6500 to 10,000 cP) using gum concentrations of from 1.25 to 2.5%, depending on the gum type and viscosity requirement.

There are several printing machine types which do not fit the description of either the flat-bed or rotary which are also capable of producing pattern effects. These are: (1) the modified roller type which uses raised patterns on a series of rolls arranged radially around a central cylinder, (2) the jet nozzle type which feeds color through jet tubes programmed to produce the pattern required, (3) the reverse dip method machine which feeds carpet with the pile face down and applies color by compartmented reservoirs which rise and fall as the carpet traverses, and (4) the pressure bar segmented roller machine which prints a warp sheet of yarn run between a number of pairs of such rolls using one pair of rolls for each color. These machines all use relatively low-viscosity dye pastes made up with high-viscosity gum at low solids. The gum viscosities will vary from 0.05 to 0.3 Pa·s (50 to 300 cP), and the gum concentration will be in the range of 0.3 to 0.5%.

This area of carpet processing is growing rapidly and innovations are the rule rather than the exception in this particular segment of dye application.

The viscosities mentioned are all by Brookfield RVF Model using the appropriate spindle at 25°C and 20 rpm. Viscosities measured by other means will be considerably different.

Paper

Naturally occurring hemicellulosics which promote hydration and play an important role in interfiber bonding are removed during the pulping process. To supplement and replace hemicellulose, the most suitable polysaccharides have proven to be the galactomannan gums, guar and locust bean.

In general, there are two ways to prepare guar or locust bean gum for use; either by cooking or by dry addition.

For mills that do not have and do not wish to install cooking equipment, specially processed products which can be added dry to the beaters or pulpers are available. These products are formulated for ease of dispersibility in the stock, and designed to suit the specific conditions of stock preparation, i.e., temperature, pH, and retention time. They develop full hydration while in the stock at specific rates determined by individual product design.

However, most mills find it easier and more practical to prepare solutions of guar or locust bean gum before addition to the pulp slurry. This enables them to maintain a finer control on the amount of gum being added to their pulp system and also allows them to quickly increase or decrease the amount of additive as needed.

Guar or locust bean gum solutions are prepared by one of two methods: the batch system or a continuous cooking system. Solutions can be prepared in the batch system by dispersing 0.5 to 1.0% of the dry powder in cold water under mechanical agitation, and then heating the dispersion to 90 to 96°C (195 to 205°F), holding the solution at that temperature for 10 to 20 min. The benefit of a continuous cooker is the preparation of a gum solution on an automatic, continuous basis with lower manpower requirements.

After preparation of guar or locust bean gum solution, it can be added at any convenient point in the stock system that lies after the point of refining and provides good distribution. Since these materials have such a high affinity for adsorption on cellulose, it is recommended that the solution be diluted as much as possible before addition so that a more uniform distribution rate of gum to cellulose can be accomplished.

In most mills, addition in the vicinity of the fan pump offers the best control of the additive process. The rate of delivery of the prepared guar or locust bean gum to the paper machine can best be regulated by use of a rotometer. The normal requirement for attaining measurable improvements is in the range of 1 to 6 kg/metric ton (2 to 12 lb/ton) dry basis, but conditions of the pulp furnish, basis weight, machine speed, stock freeness, etc., will ultimately govern the proper level of addition.

The general benefits gained through the use of guar or locust bean gum vary with the particular paper product being manufactured.

Kraft Papers   A growing use of gums by makers of multiwall bag and gumming grades of kraft paper has shown specific improvements. The rate of addition varies considerably from mill to mill, but is usually 2.5 to 4.5 kg/metric ton (5 to 9 lb/ton). Benefits include:

Improved tensile strength

Improved tear resistance, brought about by reduction of refining made possible by tensile improvement

No adverse effect on porosity

Higher production speeds

Kraft Linerboard   This industry was among the earliest users of guar and locust bean gum and remains so due to the introduction of newer high-speed, high-production paper machines. Guar and locust bean gum products have proved to be the most reliable and effective materials used to improve the Mullen test on this grade of board. Addition rates of 1 to 3 kg/metric ton (2 to 6 lb/ton) are normal in order to accomplish:

Higher speed at equivalent Mullen due to less need for refining, higher freeness, and thus, better drainage

Higher Mullen at reduced basis weights

More uniform Mullen values between top and wire sides

Optimum balance of refining, drainage, drying rates, and Mullen tests in 31.4-kg (69-lb) and heavier-weight boards

Recycled Linerboard   Galactomannans are almost universally used by producers of this grade of paper, largely because of the use of reclaimed paper and board used in the pulp furnish of this grade. Gums provide the necessary upgrading of quality needed in the finished product. The need for guar and locust bean gum is particularly great in paper mills employing high-temperature, high-pressure dispersion units, but in all cases, gums provide these needed benefits at addition rates of 3 to 6 kg/metric ton (6 to 12 lb/ton). Benefits include:

Higher Mullen

Lower basis weight

Higher production speeds

More economical pulp furnish through use of more recycled paper

Improved finish

Corrugating Medium   Almost all medium made from recycled stock requires an additive to meet their strength tests. Galactomannans are commonly used additives.

The addition rate normally varies from 3 to 6 kg/metric ton (6 to 12 lb/ton). The benefits include:

Higher concorra

Higher speed

Better crush resistance in the finished container

Boxboard   Depending on the individual mill's conditions for stock preparation, specific products are recommended for addition to either beaters or hydropulpers. Usually mills making these grades prefer a product designed for dry addition to the stock at addition rates of 3 to 6 kg/metric ton (6 to 12 lb/ton) of liner stock. Ordinarily no gum is used in the filler portion. The resultant production or product benefits include:

Improved ply bonding

Increased board stiffness

Better formation of top liner

Better bonding strengths on coated grades

Offset News Stock   Paper mills have been able to realize through the use of galactomannans marked benefits in the production of stock for continuous web printing of newsprint at a use level of 1 to 1.5 kg/metric ton (2 to 3 lb/ton). Benefits have included:

Reduction of linting and dusting during paper manufacture

Higher pick values

Less build-up in printing on offset press blankets, permitting more impressions between cleanups.

White Papers   Guar and locust bean gum are used in a wide variety of white papers at addition rates of 1.5 to 4 kg/metric ton (3 to 8 lb/ton). By increasing sheet strength, it is possible to reduce the size concentration at the size press, with resultant improvement in tear, opacity, and calendering.

Gums used in high hardwood furnishes reduce surface picking, improve bonding of vessel segments, and increase fold.

Guar or locust bean gum is used in neutral or alkaline furnishes to improve the retention of calcium carbonate.

Galactomannans are added dry to the beater to reduce the refining time required to develop strength characteristics of rag content paper, particularly banknote.

These products are also very effective in decreasing two-sidedness and thus reducing the tendency of papers to curl.

Where formation is uppermost, gums are used to deflocculate the fiber bundles, closing up the sheet and markedly improving the appearance.

Mining

Guar gum is used as a flocculant to perform liquid-solid separations by means of filtration, settling, and clarification. As a typical filtration illustration, consider the acid leaching of a uranium-bearing siliceous ore.

After crushing and grinding to 65% minus 200 mesh, the ore is leached at 55 to 60% solids with sulfuric acid and manganese dioxide at ambient temperature to extract the uranium as a uranyl sulfate complex anion. Some iron and silica also dissolve in the lixiviant, while the bulk of the silicates remain as residue. The pH of the leached slurry is less than 1 and the free acid content is 5 to 15 g/L.

The leached slurry is then distributed to a number of rotary vacuum drum filters using nylon or canvas duck media. Guar gum is added as a 0.1% solution to the leached slurry en route to the filter after hydrating for at least one hour at ambient temperature at a dosage of 0.05 to 0.1 kg/metric ton (0.1 to 0.2 lb/ton) of dry residue solids. This causes the fine residue particles to flocculate by bridging, whence a permeable cake is formed on the filters. The filtrate containing the bulk of the uranium is pumped to the uranium recovery circuit.

The residue from the primary filter is discharged onto a conveyor and dumped into a repulping tank with recycle filtrate from the secondary filter, thereby reducing the concentration of the uranium trapped by the residue. More guar gum is added to the repulped residue as it goes to the secondary filter at a dosage of about 0.025 kg/metric ton (0.05 lb/ton). This causes the fine particles to reflocculate by bridging, whence a cake is formed on the secondary filter. Barren wash water is added to the top of the secondary filter to remove most of the entrapped uranium. Part of the secondary filtrate is recycled to the leach and part is used for repulping. The residue cake is conveyed to a tailing dump or goes to pyrite flotation.

A typical settling/clarification illustration follows the filtration example, except that a series of 5 to 6 thickeners are used in countercurrent decantation (C.C.D.) fashion, instead of filters.

After leaching with sulfuric acid and manganese dioxide or sodium chlorate, the leached slurry passes to the launder of a thickener, where it is diluted with return liquor from the next thickener in the C.C.D. series. Guar gum is added as a less than 0.05% solution (having been diluted from 0.5% with barren recycle liquor) at a series of three points spaced about 3 m (10 ft) from the thickener center well at a dosage of 0.05 to 0.1 kg/metric ton (0.1 to 0.2 lb/ton) of dry solids. The overflow of the first thickener goes to metal recovery if the solution is clear. If not, some cationic guar gum is added to this solution prior to filtration on a diatomaceous earth filter to remove the fine colloidal silica. Another device is to pass this solution to which cationic guar has been added to a large settling tank for clarification prior to metal recovery.

The residue is removed from the first thickener by a diaphragm or centrifugal pump which causes some degradation of floc, and is pumped to the next thickener in the C.C.D. series. Guar gum is added in a similar manner as the first thickener but at a dosage of 0.025 to 0.05 kg/metric ton (0.05 to 0.10 lb/ton) to refloc the residue. The process is repeated with residue passing from the last thickener and pumped to tailing disposal. Fresh water or barrens are added to the last thickener to compensate for the rich uranium-bearing liquor removed from the first thickener.

Guar gum is used as a talc depressant in mineral flotation operations. As an illustration of a typical application, consider the flotation of pentlandite (Ni-Fe-sulfide) from a high-talc ore containing chalcopyrite, pyrrhotite, pyrite, serpentine, and biotite.

After crushing and grinding to a 90% minus 325 mesh a nickel-pyrrhotite concentrate is removed magnetically, followed by a treatment of the nonmagnetic fraction by flotation.

Flotation feed is diluted to 25% solids at a natural pH of 8.5 to 9.0. Guar gum is added at 0.2 to 0.3 kg/metric ton (0.4 to 0.6 lb/ton) to depress talc. The gum is added as a 0.5% solution hydrating at room temperature for 30 min. This is followed by 0.65 to 0.75 kg/metric ton (1.3 to 1.5 lb/ton) sodium ethyl xanthate as a collector and triethoxybutane as a frother. Rougher concentrate is removed with a total float time of 5 min.

Rougher tails is scavenged for 9 min after the addition of 0.05 to 0.075 kg/metric ton (0.1 to 0.15 lb/ton) guar gum and 0.3 to 0.35 kg/metric ton (0.6 to 0.7 lb/ton) sodium ethyl xanthate. Subsequently, the rougher and scavenger concentrates are combined and floated in the cleaners to produce a final concentrate and a middling recycle. No additional guar gum is used in the cleaners, but 0.15 to 0.2 kg/metric ton (0.3 to 0.4 lb/ton) of sodium ethyl xanthate and triethylbutane are added for a float time of 6 min.

The final concentrate assays 12 to 15% Ni, and the final tails assays 0.2 to 0.4% Ni.

In summary, guar and guar derivatives are used at very low concentrations below the level at which they thicken water. They are used as reagents which bond to hydrated mineral surfaces. In their application as depressants in base metal flotation, they function as agents to block the adsorption of other reagents on talc and gangue surfaces. In flocculation, guar and guar derivatives adsorb on hydrated mineral surfaces by hydrogen bonding, and cause agglomeration by bridging.

INDUSTRIES USING GUAR GUM

Oil and Gas

Guar gum and guar derivatives are used in the fracturing of oil and gas wells because of their ability to thicken water efficiently at low concentration. They carry the sand or propping agent needed to keep the fracture from closing when the pressure is released. They are reliable polymers that will hydrate in field waters under many conditions.

Explosives

Guar gum and guar derivatives are used in the explosives industry because of their ability to efficiently thicken the nitrate salt solutions which are the basic components of slurry explosive formulations. The gums are readily crosslinked to form stable water-resistant gels.

Food

Guar gum is used in the food industry for its ability to bond and immobilize large amounts of water, thereby contributing to viscosity, inhibiting ice crystal formation in frozen products, modifying product texture, and stabilizing product consistency to changes in temperature. An illustration of the use of guar gum can be found in the manufacture of ice cream where small amounts of gum promote smooth chewy texture, inhibit ice crystal formation caused by heat shock, and impart slow meltdown qualities to the product.

Paper

In the paper-making process, guar gum adsorbs on hydrated cellulosic surfaces. It is a wet-end additive that promotes dry strength, improves sheet formation, and makes manufacturing easier and faster.

Textile

Guar and guar derivatives thicken the dye baths that are used in the printing and dyeing of fibers, fabrics, and carpets. The gums control the rheological characteristics of the dye formulations and permit complex multiple dye patterns that are sharp, bright, and controlled as to penetration.

Mining

In this industry, guar gum is used principally as a flocculant to produce a liquid-solid separation by filtration, settling, and clarification of mineral slurries. The property used is the ability of the gum to adsorb on the hydrated mineral particles by hydrogen bonding followed by agglomeration as a result of bridging.

Guar gum is also used in flotation to recover base metals. Guar gum acts as a depressant for talc or insoluble gangue minerals mined along with the valuable minerals.

FORMULATIONS

Ice Cream

10.0%	Butterfat
12.0%	Milk, nonfat
11.0%	Cane sugar
5.0%	Corn syrup solids (42DE)
0.3%	Stabilizer/emulsifier
38.3%	Total solids

Ice Milk

3.5%	Butterfat
12.0%	Milk solids, nonfat
11.0%	Cane sugar
8.0%	Corn syrup solids (42DE)
0.45%	Stabilizer/emulsifier
34.95%	Total solids

Sherbet

1.0%	Butterfat
3.0%	Milk solids, nonfat
20.0%	Cane sugar
8.0%	Corn syrup solids (42DE)
0.25%	Stabilizer
32.25%	Total solids

Sour Cream

18.0%	Butterfat
9.0%	Milk solids, nonfat
0.4%	Stabilizer
27.4%	Total solids

Buttermilk

1.0%	Butterfat
10.0%	Milk solids, nonfat
0.4%	Stabilizer
11.4%	Total solids

Yogurt

1.0%	Butterfat
12.0%	Milk solids, nonfat
0.8%	Stabilizer
13.8%	Total solids

Instant Imitation Bakery Jelly

30.00%	Modified food starch
0.50%	Guar gum
1.30%	Sodium hexametaphosphate
1.60%	Dicalcium phosphate
8.20%	Adipic acid
0.30%	Raspberry flavor
0.25%	Raspberry Shade
0.01%	Purple Grade Shade
0.70%	Potassium sorbate
57.14%	Sodium benzoate
100.00%	Total

Whipping Composition for Frozen Deserts

Parts by Weight

50–70	Sucrose
3–11	Modified food starch
3–6	Gelatin (200 Bloom)
0.5–1.5	Soy protein (degraded)
1.0–1.5	Organic food acid
0.1–0.5	Sodium hexametaphosphate
0.2–0.6	Guar gum
65.0	Fructose
10.5	Modified food starch
4.3	Gelatin (200 Bloom)
1.0	Soy protein
0.6	Flavor
1.2	Citric acid
0.25	Salt
0.2	Sodium hexametaphosphate
1.5	Color
0.2	Guar gum

Chapter **7**

Agar*

William Meer
Meer Corporation

General Information 7-2
  Chemical Nature 7-2
  Physical Properties 7-2
  Manufacture 7-4
  Biological/Toxicological Properties 7-4
  Electromechanical Properties 7-4
  Rheological Properties 7-4
  Additives/Extenders 7-5
  Handling 7-5
  Applications 7-6
  Application Procedures 7-6
  Compatibility 7-6
  Specialties 7-6
  Future Developments 7-7
Commercial Uses of Agar 7-7
  Food Industry 7-7
  Microbiology 7-8
  Impression Materials 7-8
  Medicine and Pharmaceuticals 7-8
  Laboratory Applications 7-8
  Miscellaneous and Potential Uses 7-9
Formulations 7-9
  Icing Stabilizers 7-10
  Dental Molds 7-10
  Culture Media 7-10
Laboratory Techniques 7-13
  Gel Strength Method 7-13
References 7-14

* With notes by Dr. Tetsujiro Matsuhashi, Nagano State Laboratory of Food Technology, Japan.

GENERAL INFORMATION

Agar is a complex water-soluble polysaccharide. It is a hydrophilic colloid extracted from certain marine algae of the class Rhodophyceae. Generally, two types of agar are produced, one from the *Gelidium* species, and one from the *Gracilaria* species. Agar is approved for food use and is in the GRAS (Generally Recognized As Safe) list under the Food and Drug Act. It has monographs in the *United States Pharmacopoeia* and in the *Food Chemicals Codex* (see Table 7.1 for United States standards; see Note 1, p. 7-16 for Japanese standards).

Agar has a unique position in its commercial importance because it forms firm gels at concentrations as low as 1%. Gel strength varies in direct relationship to the concentration between 1 to 2%. Its low setting temperature offers another unique property that is extremely useful.

TABLE 7.1
General Specifications of Agar USP[66] and FCC[45]

	USP	FCC
Microbial limit, Salmonella	Negative	(—)
Water, % maximum	20	20
Total ash, % maximum	6.5	6.5
Acid-insoluble ash, % maximum	0.5	0.5
Foreign organic matter, % maximum	1	(—)
Foreign insoluble matter, % maximum	1	1 (boiling water)
Arsenic, ppm	3	3
Lead, ppm	10	10
Heavy metals, ppm	40	40
Foreign starch	Negative	Negative
Gelatin	Negative	Negative
Water absorption, mL water maximum	75	75

USP = *United States Pharmacopoeia*
FCC = *Food Chemicals Codex*

Chemical Nature

Agar occurs as a mixture of at least two polysaccharides: agarose, the gelling agent which is predominantly composed of repeating units of alternating β-D-galactopyranosyl and 3,6-anhydro-α-L-galactopyranosyl units coupled 1 → 3. Agar is relatively insoluble in cold water, although certain agars do swell in water. Agar readily dissolves in boiling water.[2,4,5] Uronic acids are present at levels of less than 1%, or are entirely absent.

Izumi[38] has shown the heterogeneous nature of agar by separating it into four fractions using anion-exchange chromatography. This was confirmed by Duckworth and coworkers,[17,18] who also found certain *Gracilaria* agaroids to contain up to five components. Cellulose acetate electrophoresis of the dyed polysaccharides may facilitate identification.[19] More work is needed in this area.

The agarose gel has a structure consisting of a regular array of micron-sized, spherical, polymer-rich regions thought to be formed by a nucleate-free phase separation. Hysteresis in the optical rotation of the agarose solution indicates the existence of a wide spectrum of junction zones.

Physical Properties

Agar is commercially available in bundles of thin, membranous, agglutinated strips, or in cut, flaked, granulated, or powdered forms. It is white to pale yellow in color. It has a mucilaginous taste, and is either odorless or has a slight characteristic odor. Agar is insoluble in cold water, but it is soluble in boiling water and is slowly soluble in hot water.

A 1.5% by weight solution congeals at between 32 and 39°C to form a firm, resilient gel which does not melt below 85°C (see Note 2). This behavior is characteristic for agar and contrasts with the properties of other seaweed colloids. Many of its applications hinge upon this temperature difference between gelation and liquefaction. Other important characteristics of agar are its ability to swell in water and the high viscosities of its solutions at concentrations of 5 to 10%. This latter property is greatly influenced by the type of raw material used and the method of manufacture.

Pure agar-water gels are fairly stable. Those made with high-gel-strength agar appear to be as stable as dry agar itself if sterilized and stored in airtight containers. Low-gel-strength agars, however, deteriorate more rapidly in the gel form than in the solid form. Few microorganisms metabolize agar or elaborate enzymes that degrade it. This may account for the greater stability of agar over other naturally occurring colloid gels. The metallic and nonmetallic constituents of agar are also important, especially in the bacteriological grades (see Tables 7.2 and 7.3).

TABLE 7.2
Essential Metallic and Nonmetallic Constituents in Bacteriological-Grade Agar and Their Concentrations[55]

S (inorganic)	1.0–1.5%
Na	0.6–1.2%
Ca	0.15–0.25%
Mg	400–1200 ppm
K	100–300 ppm
P	10–80 ppm
Fe	5–20 ppm
Mn	1–5 ppm
Zn	5–20 ppm
Sr	10–50 ppm

TABLE 7.3
Summary of Various Constituents of Agars, Agaroids, and Agaroses[70]

	Domestic	Foreign	Agaroids	Agaroses
Aluminum, ppm	0–50	50–1200	200–8000	0–230
Arsenic, ppm	0	0–2	0–9	—
Barium, ppm	0–3	3–50	3–70	0–13
Bismuth, ppm	0	0–5	0	—
Boron, ppm	20–110	10–300	20–240	—
Cadmium, ppm	0–0.8	0–2.1	0–1.5	0.2–1.3
Calcium, %	0.15–0.3	0.02–0.9	0.7–3.0	0.2
Chromium, ppm	0–2	0–10	0–5	0–2
Cobalt, ppm	0–2	0–10	0–15	0–2
Copper, ppm	0–2	2–66	0–140	0–10
Iron, ppm	5–25	24–800	40–9000	25–140
Lead, ppm	0–0.8	0.3–100	0–60	2–8
Magnesium, %	0.04	0.02–0.62	0.1–2.1	0–0.02
Manganese, ppm	0.1–1	2–150	3–200	1–2
Nickel, ppm	0.1–4	0–10	0–1	0–1
Rubidium, ppm	0	0	0–40	—
Sodium, %	1.5	0.01–4.2	0.1–0.8	—
Strontium, ppm	8–25	0–400	10–200	—
Tin, ppm	0	0–50	0	—
Titanium, ppm	0–10	0–200	20–200	—
Zinc, ppm	2–15	4–2000	20–2000	4–20
Ash, %	3.5–4.4	3.0–7.4	7–15	0.6–2.8
Protein nitrogen, %	0.1–0.2	0–2.0	0.5–3	0–0.2
Reducing substances, %	0–1	0.4–21	0.5–32	0–8
Sulfate, organic, %	0.6–1.3	0.9–4.2	2.5–5	0.1–0.4
Sulfate, inorganic, %	1–3	0.6–2.7	1.5–8	0.1–0.3
Uronic acids, %	0–0.3	0–4	0–10	0–0.2
Moisture, as purchased, %	20	11–24	10–22	9–16
Clarity, 1.6% sol, 45°, mm	900	50–850	20–300	200–600
Gelation temp., 1.6% sol, °C	35–37	36–39	38–63	31–35
Melting temp., 1.6% gel, °C	86–88	60–88	60–76	64–90
Insolubles, %	0–0.1	0.2–2.8	0.3–1.5	0–0.4
Thermoduric spores, no./g	0–1	0–1000	0–2000	0–2
Number of sources tested	1	32	14	4

Note: 118 samples; 51 sources; about 5000 determinations; 1950–1971; dry basis.

Manufacture

Most commercial agar is manufactured by hot-water extraction followed by freezing and filtration. The freezing allows removal of the soluble salts. Manufacture may involve the following steps: cleaning or washing of the raw material, bleaching or chemical treatment, extraction, filtration, gelation, freezing, thawing, filtration, drying, bleaching, washing, and drying. The actual steps followed will depend on the country of origin of the raw agar, as well as the degree of purity desired for the finished product. Newer techniques include the use of countercurrent extraction, centrifugation, artificial freezing, chemical bleaching, alcohol precipitation, and resin-bed decolorization of the extract. (See also Note 3.)

Most modern agar-producing plants no longer utilize freezing of the gel to remove moisture. After gelation, the agar is pressed in hydraulic presses to remove the water. The dried gel is then oven-dried and milled to the desired particle size.

The following operations are employed in this country in the manufacture of agar: (1) cleaning of raw material, (2) chemical pretreatment, (3) extraction under pressure, (4) chemical posttreatment, (5) filtration, (6) gelation, (7) freezing, (8) separation, (9) washing, (10) drying, (11) sterilization, (12) bleaching, (13) washing, and (14) drying.

Several promising areas of technology have received attention in the recent past. For example, caustic pretreatment of *Gracilaria* has enabled agaroid processors to increase the gel strength of their products substantially.[47]

Irradiation of dried species of *Gelidiella, Gelidium, Gracilaria,* and *Hypnea* by 1000 curies (Ci) of cobalt 60 in the dosage range of 0.9 to 6.4×10^4 rd/g has been reported to improve yield, gel strength, and stability of the dry extracts.[56] Similar treatment of agarophytes and agar elsewhere has, however, given negative results. Pretreatment of agarophytes with cellulolytic enzymes has in some cases accelerated extraction rates, improved yields, and increased gel strength.[4,5]

Biological/Toxicological Properties

Agar, as mentioned earlier, is GRAS, and is on the FDA list. It is nontoxic and is generally considered not to be metabolized by humans. Handling recommendations include storage in sealed polyethylene-lined containers while being warehoused in cool, dry places for extended periods.

Biological data is available on the effect of agar in short-term studies on several animals, including humans. This data supports the finding that agar is nontoxic. Tested on chicks, it has no effect on growth rate.[68] It has also been tested on rats.[21,46]

In a feeding study by Ariyama and Takahasi,[6] the growth of male rats weighing from 40 to 45 g and fed 23% agar for 70 days was compared to that of rats maintained on a low-carbohydrate diet. It was concluded that agar contributed no nutritive value to the diet. Other studies have reported similar results for rats and rabbits.[12,23,35,53,54,71,72]

In tests on humans, agar has been used as a laxative in daily doses of from 4 to 15 g since 1905. The efficiency of agar is not great, with the laxative effect usually not noticeable for nearly a week.[57]

Fetotoxicity Investigations by Frohberg[25] and coworkers disclosed that the intraperitoneal injection of 0.2 mL of a 1% agar solution daily from the eleventh to the fifteenth day of gestation in NMRI mice produced no fetotoxic effects.

Electromechanical Properties

Agar is quantitatively flocculated in the presence of electrolytes by 10 volumes of ethanol, 2-propanol, or acetone. It is salted-out by near saturation with sodium sulfate, magnesium sulfate, or ammonium sulfate. Prior to drying, such flocculated agar exists in a metastable state, in which it is dispersible in cold water and in other solvents. In general, the higher the temperature of flocculation and the higher the concentration of electrolytes, the less soluble is the floc.

Many quaternary ammonium compounds cause turbidity and agaropectin precipitation, as does silicotungstic acid. The most sensitive precipitants for agar appear to be tannic, phosphotungstic, and phosphomolybdic acids, when used at pH of 1.5 to 2.5.

Rheological Properties

Viscosity The viscosities of agar and agaroid dispersions are markedly influenced by the type of raw material and the processing conditions employed. The relative viscosities of agar and *Gracilaria* agaroid at 1% and 1.5% concentrations have been reported.[62,69] The viscosity of an agar dispersion at 45°C is relatively constant from pH 4.5 to 9.0, and is not greatly affected by age or ionic strength within the limits of pH 6.0 to 8.0. Once gelation begins, however, viscosity at constant temperature increases with time.

Gelation Temperature   Agar is unique among polysaccharides in that gelation occurs at a temperature relatively far below the gel-melting temperature. Many uses of agar depend upon this high hysteresis. Agars and agaroids from different species have markedly different gelation temperatures, each of which is practically constant, with the exception of the agaroid from *Gracilaria confervoides,* which exhibits seasonal variations of more than 20°C (see Note 4). The gelation temperature of agarose (agaran) sols is correlated with the methoxyl content of the gum.[29]

Agar and agarose (agaran) are among the most potent gel-forming agents known, for gelation is perceptible at concentrations as low as 0.04%. Threshold gels are valuable for their protective action, diffusion prevention, and texture enhancement effects. Stronger gels are of value because of their strength, resilience, elasticity, relative transparency, relative permanence, and reversibility. Agarose gels are firmer but less elastic than are gels from the parent agar.

Melting Temperature   The melting temperature of an agar gel is a function of concentration and of molecular weight. Agar and agaroid gels with 1.5% solids melt at temperatures of 60 to 97°C.

Strength   The threshold gel concentration (TGC) is determined by the concentration of agar solids necessary for the formation of a particular gel under standard conditions. The results correlate well with emulsifying, stabilizing, and protective ability and with gel strength in the 0.2 to 2.0% range.

Rupture at Constant Stress (CS) Method. In Japan, the Nikkansui-Shiki gelometer method has been used in official grading. Ten plungers loaded in an ascending series of mass are simultaneously lowered on ten replicate samples of a 1.5% gel (air-dry basis) that have been aged for 15 h at 20°C. The maximum stress in grams per square centimeter withstood for 20 s without rupture is reported as the solidity of the sample.

Stress-Strain Curve (SSC) Method. This development from Stoloff's method[61] yields information on gel strength, tenacity, and resilience. Screw-topped, 65-mL ointment jars are filled to the 50-mL level with a 1.6% solids solution at 45°C, sealed, and held for 1 h in a stirred water bath maintained at 19.5°C. A 1-cm^2 cylindrical plunger is pressed into the gel without lateral movement at 2 mm/s. Load and depth of gel depression are recorded simultaneously with rupture. Rupture stress is a measure of strength. Depression depth at failure varies with tenacity, and the curve slope is a function of elasticity.

Additives/Extenders

Sodium alginate and starch decrease the strength of agar gels, whereas dextrin and sucrose increase the strength. Locust bean gum has a marked synergistic effect on the strength of agar gels. The incorporation of 0.15% of locust bean gum can increase the rupture strength of an agar gel by 50 to 200%, depending on conditions. Iceland moss extractive (lichenin) and carboxyethylcellulose show similar action to a lesser degree. Gelatin, Russian isinglass, and gum karaya tend to weaken agar gels, but only slightly. When added to warm agar dispersions, most salts, glycerol, sorbitol, the alkanolamines, and 1,2,6-hexanetriol have little effect on the strength of the gels.

Handling

Agar is normally packed in polyethylene-lined fiber drums. The product should be stored at 20 to 25°C to enhance shelf life. Warm storage temperatures tend to reduce gel strength over a long period of time.

The material is nontoxic and free of odors. It is usually available in strips, flake form, or powder form, the latter being free-flowing and easy to handle with conventional equipment. Agar is nonflammable and does not require any special safeguards or precautions.

Applications

The applications of agar are outgrowths of its basic characteristics. As a gelling agent, agar in low concentrations forms firm, resilient, transparent, and stable gels suitable for microbiology. It also forms firm gels suitable for denture and other types of molds. As an emulsifier, suspending agent, and bulking agent at low concentrations, agar forms thick emulsions and absorbs large amounts of water.

Agar is used as a stabilizer for icings, sherbets, ices, cheeses and confectioneries. Its quick-set properties and firmness make it suitable in a variety of stabilized and gelled candies.

The large molecular weight of the agar polysaccharide acts as a flocculant in wines and other related products.

Application Procedures

There are no special procedures to be followed for the dry blending of agar except that the mix should not be allowed to become moist or damp. In preparing gels, agar

TABLE 7.4
American Society of Microbiology Specification for Agar[1]

Nephelos (sol)	<40 (BA)
	<35 (AA)
OD (sol):	
430 nm	<0.125
525 nm	<0.040
655 nm	<0.025
OD (gel) 655 nm	<0.034
Gel strength (g/cm^2)	485–675*
pH	6.3–7.1†
Mesh size	100% >40
Loss on drying, %	>5<10
Ash	>3.5<6.0*
MP, °C	80 ± 5
GP, °C	33–38†
Solution time	6 ± 1 min. (at 100°C)
Flocculation	S
Viable spores	3/gm (max.)
Biological performance	S

* Preferably high.
† Preferably low.
Note: Agar concentration (sol/gel) 1.4%; autoclaved except where noted.

can be allowed to soak and swell before boiling, ensuring full solubility and avoiding scorching.

Compatibility

Near neutrality, agar is compatible with most other polysaccharide gums and with proteins in the sense that flocculation or marked degradation does not occur when their dispersions are mixed. An exception is gum kino.

Near a pH of 3, flocculation occurs when warm agar and gelatin dispersions are mixed. Such mixtures are used to excellent advantage, however, in sherbets and ices by delaying the additions of acids until freezing has begun.

Specialties

Fairbrother and Mastin[22] found that the calcium of a native agar could be replaced by other cations to produce agaroses. Hoffman and Gortner[34] showed agar to behave like a salt of a highly ionized acid capable of salt formation by neutralization with bases. Other exchanges[15] and derivatives, such as the acetate[3,48] and methyl ether[3] derivatives, were investigated. Partial precipitation of agar by quaternary ammonium compounds was found to be a useful separation method.[73]

A method of obtaining agarose by precipitation of agaropectin by a quaternary ammonium salt and its separation from agarose by centrifugation has become generally useful.[11,32] Interest in this fraction is increasing, and other agarose separation methods based on polyethylene glycol,[51] enzymes and sequestrants,[36] ammonium sulfate,[7] aluminum chloride,[10] methyl sulfoxide,[31] EDTA and aluminum hydroxide,[9] and buffer extraction or electrophoretic purification followed by anion exchange[33] have been devised.

Future Developments

Generally speaking, there has been a large increase in the use of agar for microbiological use because of the use of prepared media by industry as well as by research and university laboratories. In addition, it has been possible to prepare specialty media plates that are not easily prepared in individual laboratories.

The commercial preparation of agarose, as well as prepared plates, has enabled the study of specialty areas in the fields of microbiology and biochemistry. (See Table 7.4 for American Society of Microbiology specifications, basic to one of the most important uses for agar.)

COMMERCIAL USES OF AGAR

Agar has a wide variety of uses. It is employed in the bakery, confectionery, and dairy industries; in microbiological culture media; in dentistry; in pharmaceuticals; in meat packing; and in other miscellaneous applications.

Food Industry

In the food industry, agar is used predominantly for its stabilizing and gelling characteristics. It has the unique ability of holding large amounts of moisture. It is employed as a stabilizer in pie fillings, piping gels, meringues, icings, cookies, cream shells, and similar products.

A good use level in icings will range from 0.2 to 0.5%. In addition to stabilizing the icing, agar prevents the adhesion of the sugar coating to the wrapper. Shorter or longer drying times, depending on the type of icing desired, can be regulated by increasing or decreasing the amount of agar. An improved stabilizer for icings has been patented and is prepared by the addition of 0.4 to 0.6% of surface-active agent to a use level of 1.2 to 2.4% agar.[60]

A concentration of 0.5 to 1.0% agar, based on the sugar needed, is used in doughnut-glaze stabilizers. Here the stabilizer increases the viscosity of the glaze, increases its adherence to the doughnut, and provides quicker setting and flexibility with reduced chipping and cracking. The use of agar as a stabilizer also reduces the tendency of the glaze to melt and to be absorbed by the doughnut. Agar is often used in combination with other plant hydrocolloids, such as gum guar and locust bean gum, in these applications.

Agar is useful in low-calorie breads or biscuits since it is nonnutritive.[52] Because it acts as a bulking agent, it can replace starch in prepared cereals, nonstarch breads, and desserts.[26] In concentrations of 0.1 to 1.0%, it is a useful antistaling agent in breads and cakes.

Agar is the preferred additive in jellied candies and in many specialty confectionery products, such as marshmallows and sugared fruit slices. It has been used as a filler in candy bars and for preparing different forms of edible, rigid gels. Agar is commonly employed at a concentration of 0.3 to 1.8% in these applications. Monosodium citrate is preferred as the acidulant for these agar-based jellies.

Agar has been used in sherbets and ices to make them smooth and noncrumbly. Combinations with other gums, such as tragacanth and locust bean, are necessary, however, as agar has a low whipping characteristic. The optimum stabilizer concentrations use 0.12% agar, 0.07% locust bean gum, and 0.2% gelatin. Agar is employed at levels of 0.05 to 0.85% in Neufchâtel and cream cheeses and in fermented milk products, such as yogurt. It helps to reduce wheying-off and improves body and slicing qualities.

Agar is used as a thickening and gelling agent by poultry, fish, and meat canners, especially for the softer types of these products. It is used at levels of 0.5 to 2.0% in the broth weight. Alone or in combination with guar gum, it also finds application in pet foods as well as meat pies. It has also been used as a sausage casing. Agar can be used as a protective colloid in the prevention of color fading in cured meat products. For this purpose, cooked meats, such as ham, are treated with solutions containing gelatin or agar and ascorbic acid. After packaging, when exposed to fluorescent light, such treated material shows less fading than untreated meat, or meat with ascorbic acid alone.[8]

Microbiology

Agar is most valuable in microbiology, although outside of the United States larger quantities are used in other applications.

The ideal agar is low in metabolizable or inhibitory substances, debris, and thermoduric spores; has a gelation temperature of 35 to 40°C, a gel-melting temperature of 75 to 85°C; is readily soluble; and has good gel firmness, resilience, clarity, and stability. Agar concentrations of 1 to 2% are commonly used for this purpose.

In low concentrations, agar prevents the entry of oxygen into liquid media, making the cultivation of anaerobes feasible in air-exposed broths. The usual range of agar concentrations used in liquid media is 0.007 to 0.03%.

Few organisms metabolize agar or elaborate agarolytic enzymes. *Vibrio pupureus,*[74] *Vibrio agarliquefaciens,*[37] a flavobacterium,[30] a pseudomonad,[67] *Cytophaga* species,[67] and certain diatoms[41] are exceptions.

Impression Materials

In prosthetic dentistry, criminology, tool making, and other fields, it is necessary to make accurate casts of intricate undercut objects. Poller[49] and others developed moulage compositions utilizing the low gelation temperature, high gel strength, and superior gel resilience of agar.

Restriction of the use of agar to microbiology, including antibiotic manufacture, by the United States government during World War II caused substitutes to be sought. Some were found and have supplanted agar in the less critical areas. Alginate gels and natural and synthetic elastomers, such as silicone rubbers, are examples. Agar concentrations used for impression materials range from 6 to 14%.

Medicine and Pharmaceuticals

Agar has been used widely as a laxative for several decades. When well hydrated, agar furnishes the smooth nonirritating bulk that appears to be necessary for normal peristalsis. Its use appears to be non-habit-forming, and is said not to be contraindicated, even in cases of uncomplicated peptic ulcer.[27] Medicinal-type agar especially prepared in the form of thin flakes designed to prevent the formation of obstructive masses[24] and to absorb 12 to 15 times its weight in fluid is well received professionally.

Agar is used as a suspending agent for barium sulfate in radiology, as an ingredient of slow-release capsules, in suppositories, in surgical lubricants, in emulsions of many types, and as a carrier of topical medicaments. It is used as a disintegrating agent and an excipient in tablets. Its use as an anticoagulant was patented,[20] and sulfated agar has been shown to have antilipemic activity equal to that of heparin. The ingestion of agar appears to increase the excretion of fats and might promote the intestinal synthesis of niacin. It has been reported to inhibit the aerobic oxidation of vitamin C.

Laboratory Applications

The accuracy of particle-size determinations can be increased with agar, as can turbidimetric determinations involving suspensions of dense solids (high-clarity agar is preferred). Certain dyes, such as methylene blue, toluidine blue, thionine, and pinacyanol, can be reversibly polymerized by agar. In the analytical coagulation of calcium sulfate, arsenic sulfide, ferric hydroxide, and barium sulfate, agar can be employed to advantage.

Microtomy of plant tissues is facilitated when a 5% agar gel is used as the imbedding medium. Agar gels containing 20 to 25% of potassium chloride and 2 to 2.5% of agar have been used for many years as antidiffusion, conductive bridges in connection with calomel electrodes. The electrophoretic migration of proteins through agar gels has been used to resolve ferritin, ovalbumin, hemoglobin, and pepsin.

Sodium and ammonium agar, agarose, and sodium agarose have proved valuable in globulin electrophoresis, immunodiffusion diagnostic techniques, gel filtration, and gel chromatography. The work in these fields is too voluminous to be referred to in detail.

Agar has been found to stabilize cholesterol solutions.[64] Dried agar gel slices are used to obtain infrared absorption spectra of amino acids.[65] Agar and agarose derivatives incorporating epichlorhydrin-linked carboxymethyl and dimethylaminoethyl units[28] have been used in the electrophoresis and chromatography of proteins.

Agar and agarose gel techniques are being used to separate bacterial toxins, to estimate the chain length of nucleic acids, for fractionation of antibiotics and antimycoplasma substances, for classification of viral particle sizes, and for separation and purification of enzymes.

Miscellaneous and Potential Uses

Agar has been found suitable for use in photographic stripping films and papers when esterified with succinic or phthalic anhydride and after enzymic hydrolysis. The moisture content and dimensions of agar films change promptly and reproducibly with ambient humidity. Such films are useful as sensing elements in mechanical and electrical hygrometers. Tough transparent films can be made from agar acetylated in pyridine.

The use of agar in solidified alcohol fuel, dyed coatings for paper, textiles and metals, and in pressure-sensitive tape adhesives in conjunction with a phthaloyl resin has been disclosed. As a flash inhibitor in sulfur-mining explosives, agar–ammonium chloride gels have been used. The foaming properties of dodecyl and hexadecyl detergents are enhanced by agar, and its detergent power alone, as well as in synergistic mixture with other detergents, has been noted.

Better density, brightness, and adhesion are claimed for copper-plate on aluminum when agar is employed. The use of agar in dry cell separators has been suggested. Agar is an ingredient of some cosmetic creams and lotions. As a corrosion inhibitor for aluminum, agar has been used in caustic solutions with some success. The activity of nicotine plant sprays is increased and action is prolonged when agar is incorporated. Agar-shellac-wax mixtures are popular in shoe and leather polishes and dressings. The use of agar media in orchid culture is standard. Advantage can also be taken of the cationic-exchange ability of agar.

Some of the newer uses for agar include its use as a setting inhibitor for deep-well cements,[39] as an adhesive in gloss finishing of paper products,[50] as an ingredient in Fe-Ni-Cr-Ti electrolytes in the alloy plating of steel,[63] as an iron-corrosion inhibitor in citric and malic acid solutions,[13] as a temporary neutron absorber in reactors,[40] and as an inhibitor of iron and lead corrosion by distilled water.

In the past, agar was commonly employed alone or in combination with diatomaceous earths in the clarification of wines, ciders, and juices. In these applications, the range used was 0.1 to 0.5%.

FORMULATIONS

As a gelling agent, agar is very useful in preparing strong water-gel systems for forming in molds, or for creating molds themselves. For emulsification, agar is generally useful in difficult to handle products, as well as where a very thick emulsion is desired. Because agar is a large-molecular-weight polysaccharide, it is useful for the flocculation of large-molecular-weight proteins and other polysaccharides.

Agar is used by the pharmaceutical industry for media preparation and as culture media for microbiology, to make gel molds for denture impressions, and generally as an emulsifying, suspending, and bulking agent. The food industry uses agar as a stabilizer for bakery icings and glazes, confectioneries, ices, sherbets, and cheeses, and as a gelling agent and binder in processed meats. Agar is also valuable as a flocculant for wines and other beverages and liquid-based products.

Typical uses that take advantage of agar's properties are presented in the following formulations.

Icing Stabilizers

Agar is used in the formulation of water-based, quick-set and slow-set bakery icings needed for wrapped and unwrapped baked goods. The flat- or sweet-roll icings are prepared by boiling the agar stabilizer in a sucrose-water solution, then adding this hot mixture to the icing sugar. The icing is then applied hot at 50 to 60°C (120 to 140°F), and the iced item can be wrapped as soon as 60 s later. Shorter or longer drying times for the icing can be regulated by increasing or decreasing the amount of agar in the following formulation:

	Percent
Agar	0.35
Salt	0.3
Emulsifier	0.4
Starch	1.0
Sucrose, granulated	15.0
Icing sugar	73.0
Water	q.s. (for 100%)

A cake icing with a slightly slower setting time may be prepared by first dry-blending agar with salt, emulsifier, dextrose, starch, and sucrose. This dry mix is metered into well-creamed Dex-Kreme (a Corn Products product); the entire mixture is heated to 48°C while mixing, and is then ready for application. The suggested basic formulation for this icing is:

	Percent
Agar	0.5
Salt	0.2
Emulsifier	0.6
Dextrose	0.7
Starch, pregelatinized	1.0
Sucrose	12.0
Dex-Kreme	q.s. (for 100%)

Typical Japanese food products using agar are shown at the end of the chapter (see Note 5).

Dental Molds[44]

	Percent
Agar	13–17
Borates	0.2–0.5
Sulfates	1.0–2.0
Wax, hard	0.5–1.0
Thixotropic materials	0.3–0.5
Water	q.s. (for 100%)

The composition of this commercial, hydrocolloid dental-impression material can be manipulated by changing the gel into a sol with heat. The material is placed in an impression tray in the solution state and impressed against the mouth tissues which are, in turn, to be reproduced in dental stone. The tray is held rigidly in place, and water is circulated through the cooling tubes attached to the outside surface of the tray. When the material has gelled, the tray is removed and the impression is prepared for the pouring of the dental stone.

The gelation temperature of agar is approximately 36 to 38°C. The temperature at which the gel changes to the sol is 60 to 70°C. This is the basis for the use of agar in such a product.

Culture Media

Agar is used for the two types of media employed for the purpose of microbiological assay of vitamins and amino acids: one for carrying the culture in stock or preparing the inoculum, the other for assay. The latter type of media contains all the factors necessary for growth of the test organism except one essential ingredient.

Microassay Culture[16] For the making of a dehydrated agar culture for microassay, the following composition is suggested:

	Weight
Yeast extract	20 g
Proteose peptone No. 3	5 g
Dextrose	10 g
Monopotassium phosphate	2 g
Sorbitan monooleate complex	0.1 g
Agar	10 g

This microassay culture agar is recommended for carrying stock cultures of lactobacilli and other test organisms used for microbiological assay. This medium is also recommended for the general cultivation of lactobacilli and many other microorganisms.

To rehydrate the medium, 47 g of the micro-assay culture agar is suspended in 1000 mL of cold distilled water, then heated to boiling to dissolve the medium completely. The medium is distributed in 10-mL quantities in tubes 16 to 20 mm in diameter, then sterilized in an autoclave for 15 min at 103 kPa (15 psi) and 120°C. The tubes are allowed to cool in an upright position. The pH of the final medium will be 6.7.

Nutrient Agar[16] Dehydrated nutrient agar, as shown in the following formulation, is recommended for the examination of water according to the methods recommended by the American Public Health Association and the American Water Works Association.[59] It is also recommended as a general culture medium for the cultivation of the majority of the less fastidious microorganisms, as well as a base to which a variety of materials are added to give selective, differential, or enriched media. The formulation is:

	Weight
Beef extract	3 g
Peptone	5 g
Agar	15 g

To rehydrate the medium, 23 g of the nutrient agar formulation are suspended in 1000 mL of cold distilled water, then heated to boiling to dissolve the medium completely. To prepare large volumes of the medium, the heating period required to effect complete solution may be reduced by boiling about three-fourths of the distilled water over a free flame and evenly suspending the dehydrated medium to be sure that it will be thoroughly wetted. The suspended material is then slowly added to the boiling water, and the boiling is continued for a minute or two to complete solution. The medium is distributed in tubes or flasks and sterilized in the autoclave for 15 min at 103 kPa (15 psi) and 121°C. The final pH of the medium will be 6.8. One pound (0.454 kg) of the nutrient agar formulation will make 19.7 L of medium.

Desoxycholate Agar[16] Dehydrated desoxycholate agar is used for the direct enumeration of coliform bacteria in dairy products as specified in *Standard Methods for the Examination of Dairy Products.*[58] Organisms other than those of the enteric group are inhibited. Coliform colonies are red in contrast to light, colorless colonies produced by enteric organisms not capable of attacking lactose. The recommended formulation is:

	Weight
Peptone	10 g
Lactose	10 g
Sodium desoxycholate	1 g
Sodium chloride	5 g
Dipotassium phosphate	2 g
Ferric citrate	1 g
Sodium citrate	1 g
Agar	15 g
Neutral Red	0.03 g

To rehydrate the medium, 45 g are suspended in 1000 mL of cold distilled water and heated to boiling to dissolve the medium completely. It should not be sterilized in an autoclave. The dissolved liquid medium is cooled to 45 to 50°C before using it to make poured plates for the coliform count. After the inoculated medium has solidified in the petri dish, the surface is covered with a thin layer of the medium cooled to 45 to 50°C. The final pH of the medium will be 7.3.

Endo Agar[16] Dehydrated endo agar is recommended for the confirmation of the presumptive test for members of the coliform group in the bacteriological examination of water, milk, and other dairy products.[58] The recommended formulation is:

	Weight
Peptone	10 g
Lactose	10 g
Dipotassium phosphate	3.5 g
Agar	15 g
Sodium sulfite	2.5 g
Basic fuchsin	0.4 g

To rehydrate the medium, suspend 41.5 g in 1000 mL of cold distilled water, then heat to boiling to dissolve the medium completely. The solution is then distributed into tubes or flasks and sterilized in an autoclave for 15 min at 103 kPa (15 psi) and 121°C. The characteristic flocculent precipitate present in the medium following autoclaving may be evenly dispersed by twirling or gently shaking the flask just prior to pouring into sterile petri dishes. The final pH of the medium will be 7.5. From 0.45 kg (1 lb) of the endo agar formulation, 10.9 L of medium can be made.

Enterococci Confirmatory Agar[16] This medium is used for the detection and confirmation of enterococci in water supplies, swimming pools, sewage, and other specimens suspected of containing these organisms. The recommended formulation is:

	Weight
Tryptone	5 g
Yeast extract	5 g
Dextrose	5 g
Sodium azide	0.4 g
Agar	15 g
Methylene Blue	0.01 g

To rehydrate the medium, 30.4 g are suspended in 1000 mL of cold distilled water, then heated to boiling to dissolve the medium completely. The solution is then distributed in tubes and sterilized by autoclave for 15 min at 103 kPa (15 psi) and 121°C. The medium is allowed to cool in a slanted position. The final pH of the medium is 8.0.

Sabouraud's Dextrose Agar[16] Dehydrated sabouraud dextrose agar is particularly adapted for the cultivation and identification of molds, especially those infecting the skin. The recommended formulation is:

	Weight
Neopeptone	10 g
Dextrose	40 g
Agar	15 g

To rehydrate the medium, 65 g are suspended in 1000 mL of cold distilled water, then heated to boiling to dissolve the medium completely. The solution is then distributed in tubes or flasks and sterilized by autoclave for 15 min at 103 kPa (15 psi) and 121°C. The final pH of the medium will be 5.6.

Tryptone Glucose Extract Agar[16] Dehydrated tryptone glucose extract agar is recommended for use in determining the standard plate count of milk and other dairy products. The recommended formulation is:

	Weight
Beef extract	3 g
Tryptone	5 g
Dextrose (D-glucose)	1 g
Agar	15 g

To rehydrate the medium, 24 g are suspended in 1000 mL of cold distilled water. Tap water, or unsatisfactory distilled water, may give a precipitate in the final medium. The suspension is then heated to boiling to dissolve the medium completely; then it is distributed in tubes or flasks and sterilized by autoclave for 15 min at 103 kPa (15 psi) and 121°C. The final pH of the medium will be 7.0.

Then the medium is to be prepared with skim milk; 10 mL of skim milk is added to 1 L of medium just before sterilization.

LABORATORY TECHNIQUES

Gel Strength Method[43]

1. Weigh out 3.00 g of agar analytically.
2. Add the agar quantitatively to a previously tared 400-mL beaker.
3. Add 230 mL of distilled water (hand stirring with a glass stirring rod) and uniformly disperse the agar. The amount dispersed will vary according to temperature and humidity.
4. Heat the beaker rapidly on an asbestos wire gauze with a gas flame and with constant hand stirring until the boiling point is reached. This heating should be regulated to a total time of 10 min (±1 min).
5. After boiling starts, continue to boil over a small flame (to avoid foaming) for exactly 1⅓ min (15 min for flakes). Remove the beaker from the flame and break the foam by swirling and stirring for a few minutes. This will give a smooth top gel surface for testing of sol characteristics. Make sure that the weight of the hot solution is at least 220 g immediately after boiling; add water if necessary.
6. Place the solution away from air currents, and allow it to stand at room temperature for 2 h.
7. Place the solution-containing beaker into a 25°C water bath and keep it there for approximately 22 h.
8. Weigh the gel in the tared beaker and record the weight of the gel. It should be between 195 and 205 g.
9. Separate the gel that adheres to the sides of the beaker by inserting a spatula.
10. Place the gel on the right-hand pan of the two-pan balance (see Fig. 7.1).
11. Balance the gel's weight with a 1-L beaker or Erlenmeyer flask and convenient weights.
12. Press down the right pan and lower the plunger (diameter, 1.13 cm; area, 1 cm²) against the top gel surface at a position approximately 10 to 14 mm from the outer edge of the gel, and, at the same time, keep the beaker centered on the balance pan.
13. Allow water to run into the beaker from the graduated cylindrical separatory funnel at a rate of approximately 200 mL/min. The capillary tube should be adjusted for this rate with stopcock fully open.
14. Close the stopcock when the gel breaks.
15. Record the amount of water used.
16. Pour the water back into the funnel.
17. Readjust the gel as in step (12), and repeat steps (13) to (16). Repeat the procedure for a total of four holes. These holes should be symmetrically placed about a circle of 0°, 180°, 90°, and 360°, in this order.
18. Average the readings and calculate the gel strength according to the following formula:

$$G = X\frac{W}{200}$$

where G = gel strength in grams per square centimeter
W = milliliters of water necessary to break the gel
X = weight of gel in grams (to ±0.1 g) after 24 h

For additional tests, see the Notes at the end of the chapter.

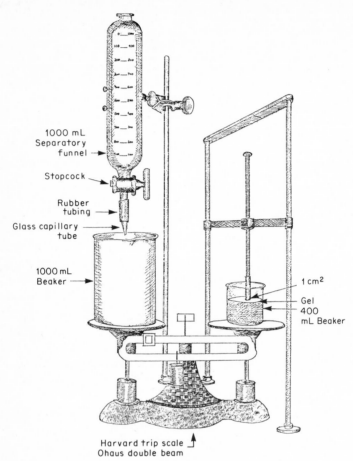

1000 mL
Separatory
funnel

Stopcock

Rubber
tubing

Glass capillary
tube

1000 mL
Beaker

1 cm²

Gel
400
mL Beaker

Harvard trip scale
Ohaus double beam

Figure 7.1

REFERENCES

1. American Society of Microbiology Specification (1958).
2. Araki, C., *Bull. Chem. Soc. Jpn.*, **29**, 543 (1956); *Chem. Abstr.*, **51**, 3462 (1957).
3. Araki, C., *Nippon Kagaku Zasshi*, **58**, 1351 (1937).
4. Araki, C., and K. Arai, *Bull. Chem. Soc. Jpn.*, **29**, 339 (1956); *Chem. Abstr.*, **51**, 3465 (1957).
5. Araki, C., and S. Hirase, *Bull. Chem. Soc. Jpn.*, **27**, 105, 109 (1954); *Chem. Abstr.*, **49**, 9517–9518 (1955).
6. Ariyama, H., and K. Takahasi, "The Relative Nutritional Value of Various Carbohydrates and Related Compounds," *Bull. Agric. Chem. Soc. Jpn.*, **6**, 1–5 (1931).
7. Azhitskii, G. Y., and B. V. Kobozev, *Lab. Delo*, 143 (1967); *Chem. Abstr.* **66**, 06120s (1967).
8. Barnett, H. W., and C. H. Perrin, U.S. Patent No. 2,905,561 (1959).
9. Barteling, S. J., *Clin. Chem.* **15**, 1002 (1969).
10. Bjerre-Beterson, E., Litex, Inc., personal communications, 1968.
11. Blethen, J., U.S. Patent 3,281,409 (1966); *Chem. Abstr.* **66**, 12116 (1967).
12. Booth, A. N., A. P. Hendrickson, and F. De, "Physiologic Effects of Three Microbial Polysaccharides on Rats," *Toxicol. Appl. Pharmacol.*, **5**, 5,478–484 (1963).
13. Chernenko, L. E., E. P. Gilinskaya, I. N. Patilova, and I. N. Smitnova, *Izv. Vyssh. Uchebn. Zaved., Pishch. Tekhnol.*, 49 (1968); *Chem. Abstr.* **69**, 53824c (1968).
14. DeLoach, W. S., O. C. Wilton, H. J. Humm, and F. A. Wolf, *Marine Sta. Bull.* Duke University (3), 31 (1946).

15. DeWaele, H., *Ann. Physiol. Physicochim. Biol.,* **5,** 877 (1929).
16. Difalco Laboratories, Inc., *Manual of Dehydrated Culture Media and Reagents for Microbiological and Clinical Laboratory Procedures,* 9th ed., 1974.
17. Duckworth, M., K. C. Hong, and W. Yaphe, *Carbohydr. Res.,* **18,** 1 (1971).
18. Duckworth, M., and W. Yaphe, *Carbohydr. Res.,* **16,** 189 (1971).
19. Duncan, W. F., and C. T. Bishop, *Can. J. Chem.,* **46,** 3079 (1968).
20. Elsner, H., German Patent 667,279 (1938); *Chem. Abstr.,* **33,** 22852 (1939).
21. Ershoff, B. H., and H. B. McWilliams, "Effects of Polysaccharides on the Appetite and Efficiency of Food Utilization in the Growing Rat," *Am. J. Dig. Dis.,* **13**(12), 385–386 (1946).
22. Fairbrother, F., and H. Mastin, *J. Chem. Soc.,* **123,** 1412 (1923).
23. Fischer, J. E., "Effects of Feeding Diets Containing Lactose, Agar, Cellulose, Raw Potato Starch or Arabinose on the Dry Weights of Cleaned Gastrointestinal Tract Organs in the Rat," *Am. J. Physiol.,* **188**(3), 550–554 (1957).
24. Friedman, A. I., and A. A. Alessi, *J. Am. Med. Assoc.,* **154,** 1273 (1954).
25. Frohberg, H., H. Oettel, and H. Zeller, "Comments on the Mechanism of the Fetotoxic Effect of Tragacanth," *Archiv. Toxikol.,* **25,** 268–295 (1969).
26. Gatti, F., U.S. Patent 2,485,043.
27. Gerendasy, J., *J. Med. Soc. N.J.,* **43,** 84 (1946).
28. Ghetie, V., D. Motet-Grigoras, and H. Schell, Romanian Patent 48,707 (1967); *Chem. Abstr.,* **69,** 19482e (1968).
29. Guiseley, K. B., *Carbohydr. Res.,* **13,** 247 (1970).
30. Hikaka, T., and M. Shameshima, *Kogoshima Daigaku Suisan Gakubakiyo,* 3, 158 (1953).
31. Hirase, S., and C. Araki, *Chem. Soc. Jpn. (Nippon Kagaku) Abstr. Papers,* **15,** 176 (1962).
32. Hjerten, S., *Biochim. Biophys. Acta,* **62,** 445 (1962).
33. Hjerten, S., *J. Chromatogr.,* **61,** 73 (1971).
34. Hoffman, W. F., and R. A. Gortner, *J. Biol. Chem.,* **65,** 376 (1925).
35. Hove, E. L., and J. F. Herndon, "Growth of Rabbits on Purified Diets," *J. Nutr.,* **63,** 193–199 (1957).
36. Hyland Laboratories, British Patent 1,070,770 (1967); *Chem. Abstr.,* **67,** 65704k (1967).
37. Ishimatsu, K., K. Minami, and I. Fujita, *Kagaku To Kogyo (Osaka),* **35,** 429 (1961); *Chem. Abstr.,* **57,** 14276 (1962).
38. Izumi, K., *Carbohydr. Res.,* **17,** 227 (1971).
39. Lantsevitskaya, S. L., and A. V. Vimberg, *Tr. Azerb. Nauchno-Issled. Inst. Dobyche Nefti* (10), 328 (1960); *Chem. Abstr.* **56,** 4374 (1962).
40. Lemer & Cie., Etablissements, French Patent 1,385,276 (1965); *Chem. Abstr.,* **62,** 14150 (1965).
41. Lewin, R. A., *Proc. of the 6th International Seaweed Symposium,* Santiago de Campostela, Spain, 1968 (1969).
42. Meer Corp., Technical Bulletin No. F-5.
43. Meer Corp., *Gel Strength Method,* Technical Bulletin No. G-29.
44. Meer Corp., Technical Bulletin No. P-252.
45. *Food Chemicals Codex,* National Academy of Sciences and National Research Council, 1966, p. 17.
46. Nilsen, H. W., and J. W. Schaller, "Nutritive Value of Agar and Irish Moss," *Food Res.,* **6**(5), 461–469 (1941).
47. Percival, E. G. V., and J. C. Somerville, *J. Chem. Soc.,* **137,** 1615–1619 (1937).
48. Percival, E. G. V., and W. S. Sim, *Nature,* **137,** 997 (1936).
49. Poller, A., U.S. Patent 1,672,776 (1929); *Chem. Abstr.,* **22,** 2644 (1928).
50. Rice, J. C., U.S. Patent 3,028,258 (1962); *Chem. Abstr.,* **57,** 1132 (1962).
51. Russell, B., T. H. Mead, and A. Polson, *Biochim. Biophys. Acta,* **86,** 169 (1964).
52. Salurai, and Y. Kato, *Nogaku,* **1,** 185 (1947); *Chem. Abstr.,* **44,** 2612 (1950).
53. Schultz, J. A., and B. H. Thomas, "Effect of Various Adjuvants to the Diet of Rats on the Changes in Body Fats Induced by Feeding Soybean Oil," *Iowa Agric. Exp. Stn. Bull. No. 336,* 629–647 (1945).
54. Schultz, J. A., and B. H. Thomas, "The Fecal Excretion of Lipides by Rats as Influenced by Diet," *J. Nutr.,* **42,** 175–187 (1950).
55. Seip, W. F., "Specifications and Experience in the Use of Bacteriological Grade Agar-Agar by a Leading Manufacturer of Dehydrated Media," Proceedings of the *8th International Seaweed Symposium,* Bangor, Wales, August 17–24, 1974.
56. Smith, F., and R. Montgomery, *The Chemistry of Plant Gums and Mucilage,* Reinhold Publishing Corporation, New York, 1959, p. 426.
57. Sordelli, A., and E. Mayer, "Agar as Antigen in Vitro," *Rev. Soc. Argent. Biol.,* **7**(5/6), 416–422 (1931).
58. *Standard Methods for the Examination of Dairy Products,* 9th ed., 1948, p. 132.
59. *Standard Methods for the Examination of Water and Sewage,* 9th ed., 1946, p. 186.
60. Steiner, A. B., and L. Rothe, U.S. Patent 2,471,019 (1949).
61. Stoloff, L. S., *U.S. Fish & Wildlife Service,* Fishery Leaflet No. 306, 1948.
62. Stoloff, L. S., *Fish. Mark. News,* **5,** 114 (1943).
63. Sugahara, Z., Japanese Patent 13,401 (1961); *Chem. Abstr.,* **57,** 14888 (1962).

64. Tarasova, L. S., *Biokhim., 26,* 736 (1961); *Chem. Abstr., 56,* 1741 (1962).
65. Tarver, M. L., and L. M. Marshall, *Anal. Chem., 36,* 1401 (1964).
66. *United States Pharmacopoeia,* 17th ed., Mack Publishing Co., Easton, Pa., 1965, pp. 17–18.
67. Veldkamp, H., *J. Gen. Microbiol., 26,* 331 (1961).
68. Vohra, P., and F. H. Kratzer, "Growth Inhibitory Effect of Certain Polysaccharides for Chickens," *Poult. Sci.,* 43(5), 1164–1170 (1964).
69. Whistler, R. L. (ed.), *Industrial Gums,* 1st ed., Academic Press, Inc., New York, 1959, pp. 28–31.
70. Whistler, R. L. (ed.), *Industrial Gums,* 2d ed., Academic Press, Inc., New York, 1973, p. 36.
71. Wierda, J. L., "A Comparison of the Weight of the Intestine with the Body and Kidney Weights in Rats which Were Fed Artificial Diets," *Anat. Rec.,* **107,** 221 (1950).
72. Wierda, J. L., "Measurements and Observations Upon the Intestine of Rats Fed Unbalanced and Supplemented Diets," *Am. J. Anat.,* **70,** 433–435 (1942).
73. Yenson, M., *Rev. Fac. Sci. Univ. Istanbul,* 13A, 97 (1948); *Chem. Abstr.,* **42,** 6560 (1948).
74. Yoshikawa, M., and K. Watanabe, *Hyogo Noka Daigaku Kenkyu Hokoku,* 3, 53 (1957); *Chem. Abstr.,* **52,** 19198 (1958).

NOTES

Dr. Tetsujiro Matsuhashi of the Nagano State Laboratory of Food Technology, Japan, has reviewed the material in this chapter and has provided added material to reflect the Japanese experience with agar.   This added information, with its associated figures and tables, is given in the following pages, with footnote numbers to identify where in the main text the information is directly related.

[1] Dependence of gel melting point and sol setting point on concentration of agars from *Gelidium amansii* and alkali-treated *Gracilaria verrucosa.* Reproduced by permission of S. Tagawa. [Y. Kojima, S. Tagawa, and Y. Yamada, *J. Shimonoseki College of Fisheries,* 10(1), 43–56 (1960).]

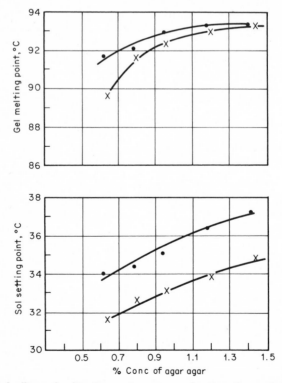

Typical gelling and melting temperature curves, as functions of concentration, for different types of agar are shown above.

[2] **Japanese Agricultural Standards for Agar (1953)***

			Classification of agars		
			Stringy agar		Powdered agar
	Grade	Bar style	Y-standard	K-standard	
Gel strength, g/cm²	Superior	≥ 300	—	—	≥ 600
for 20 s on 1.5%	1st	≥ 250	≥ 300	≥ 350	≥ 350
gel aged 15 h	2nd	≥ 200	≥ 200	≥ 250	≥ 250
at 20°C	3rd	≥ 100	≥ 100	≥ 150	≥ 150
Hot water–insoluble	Superior	≤ 2	—		≤ 0.5
substances, %	1st	≤ 2	≤ 2		≤ 2
	2nd	≤ 3	≤ 3		≤ 3
	3rd	≤ 4	≤ 4		≤ 4
Crude protein, %	Superior	≤ 1.5	—		≤ 0.5
	1st	≤ 1.5	≤ 1.5		≤ 1.5
	2nd	≤ 2.0	≤ 2.0		≤ 2.0
	3rd	≤ 3.0	≤ 3.0		≤ 3.0
Crude ash, %	All grades	No standard	No standard		≤ 4.0
Moisture, %	All grades	≤ 22	≤ 22		≤ 22
Weight, per	Superior	≥ 4875	No standard		No standard
600-bar	1st	≥ 4500	No standard		No standard
bale, g	2nd	≥ 4125	No standard		No standard
	3rd	≥ 4125	No standard		No standard

* In addition to the objective tests listed in the table, subjective tests are made, but are not included here. These include uniformity of shape and size, color, glitter, and other appearance factors.

NOTES: Similar standards are included in "The Governmental Standards for Exporting Commodities," 1958.

[3] Because of the variations within raw materials and processing conditions in different manufacturing plants, quality differences exist, as indicated in the accompanying table. Such differences are normally inconsequential since the agar is sold by grades, and differences within grades can be smoothed out through blending.

Range of Gel Strength Values of Powdered Agar from Seven Agar Factories in Nagano State, Japan, 1961–1968*

	A	B	C	D	E	F	G
Number of samples tested (606 total)	49	71	92	125	153	98	18
Gel strength, max.	1200	1250	950	1000	1070	1170	240
Gel strength, min.	550	300	150	220	<100	440	110
Mode of gel strength distribution	850–890	950–990	250–290	300–340	500–540	600–640	150–190

* T. Matsuhashi, B. Takahashi, and T. Kitazawa, *J. Jpn. Soc. Food Sci. Technol.*, 18(6), 284–287 (1971).

NOTES: A through D and F were powdered agars produced by the pressing-dehydration method. E was coarse agar powder produced by mechanical-freezing method. G was a powdered mixture of bar-style and stringy-style agars.

[4] Examples of the hysteresis of Japanese *Gelideum* agar and Argentine *Gracilaria* agar, without alkaline pretreatment, are given in the following table.

Distribution of Melting and Gelling Temperatures of Two Groups of Commercial Agars*

	Argentine Gracilaria agar	Japanese Gelidium agar
Avg. melting temp., t_m, °C	79.3 ± 1.7	87.3 ± 4.2
Avg. gelling temp., t_s, °C	38.7 ± 1.7	35.5 ± 1.7
Avg. hysteresis, $t_m - t_s$, °C	40.5 ± 2.5	51.8 ± 5.2
Number of samples tested	96	152

* T. Matsuhashi, B. Takahashi, and T. Kitazawa, *J. Jpn. Soc. Food Sci. Technol.*, 18(6), 291–295 (1971).

NOTE: No alkali treatment of any of the raw materials; prepared by natural freeze-thaw method in Nagano, Japan, December 1968 to March 1969.

[5] Mitsumame

This is a sweet dessert especially popular among women. Because of its colorful, attractive appearance, it is usually served in a glass bowl. It is prepared by soaking 237 mL (1 cup) of 1% agar gel cubes (approximately 1 cm to a side) in syrup, together with pieces of sweetened mandarin orange, pineapple, cherries, and peas (*Pisum sativum*, var. arvense; "mame" in mitsumame means peas). Agar is unique in its application here because of its ability to maintain the straight edges of the cubes, which other gels cannot do, and to permit sterilization of the canned product, provided that the agar selected has a sufficiently high gel melting point (85°C minimum).

Yokan

The largest single usage of agar in Japan is probably in yokan, a sweet, jellylike confection made from sugar, mashed Azuki beans *(Phaseolus angularis)*, and agar. In place of the Azuki beans, the following may be used: powdered green tea, brown sugar, honey, the juice of yuzu rind *(Citrus junos)*, grapes, red plums, or jams such as apricot, apple, or persimmon. Occasionally, chestnuts are used in place of the beans, making a very expensive yokan.

Following is a formulation that could be used to make yokan. The values were obtained by analysis of seven samples from major manufacturers [T. Matsuhasi and T. Shimada, *J. Jpn. Soc. Food Sci. Technol.*, 18, 370 (1971)]:

	Percent	
Moisture	26.5	
Total sugar	54.3	
(Reducing sugar 1.3%)		
Salt	0.2	
Water-insolubles	12.4	18% dry weight mashed beans
Unrecovered	5.6	
Agar	1.0	

Yokans are molded into ingot-shaped bars, and generally are packed and sealed in laminated-film containers.

Condensed Sweet Agar Jelly

This is a highly sweetened, mild-flavored candy of low acidity which, in a sense, is a modernized version of the more traditional Japanese confection, yokan. Because of its extreme stickiness, it is usually wrapped in oblate (or edible paper), which actually contains no paper at all but is also made from agar, as described in the next formulation.

In processing the jelly, agar is dissolved in boiling water to which sugar and invert sugar are then added. The thick solution is cooked to 106°C for condensation. After a brief cooling, fruit juice and/or synthetic organic acid, flavoring agents, and color additives are added, and the mixture is allowed to gel at room temperature overnight. At this stage, the solids content of the gel will be approximately 70 to 75%. Further drying is carried out by placing the gel in a 55°C oven for 15 to 30 h to reduce the moisture content to 16 to 18%, that is, to increase the solids level up to about 83%. The gel is then cut into small pieces which are manually wrapped, first in oblate, then cellophane.

The ingredients for condensed sweet agar jelly are:

	Weight
Agar	50 g
Water	2,000–3,000 g
Sugar	1,000–2,000 g
Invert sugar	1,500–3,000 g
Citric acid	2 g
Color and flavor as desired	

Oblate or Edible Paper

"Oblate" is a registered tradename for edible paper, or edible film, which is made from starch and agar [(M. Kobayashi, Japanese Patent No. 5543 (1902); 5543-Improved (1912)]. Its thickness is 10 to 15 μm. The product is generally sold as 93-mm-diameter circles at drug stores in Japan.

Oblate is prepared by adding 100 parts of a 5% aqueous starch solution to 200 parts of a 2.5% aqueous agar solution and brushing the hot mixed solution onto the glossy surface of a metal or glass body in order to produce a dry, rigid film which is then removed and cut into the desired shape.

It is a convenient and useful material for wrapping the condensed, sticky jelly (previously described), and for dosing powdered medicines.

[6] **Gel Melting Point Determination**

The general procedure was reported by T. Matsuhashi in *Bull. Jpn. Soc. Sci. Fish.*, 37(5), 441–448 (1971).

1. Determine the moisture content of the agar, using an infrared moisture balance or an oven at 105°C and an analytical balance.
2. Prepare 1.0% and 1.5% agar solutions by first soaking the agar at least 1 h in distilled water, then bringing it to a boil and boiling it for 10 min. The total weight may be kept constant by using a flask and reflux condenser, or water lost by evaporation may be replaced after dissolution is complete. In the reflux method, 200 g of sol are prepared at each concentration.
3. Pipette 5 mL of the hot solution into each of two 10 mm × 100 mm test tubes, filling them to approximately two-thirds full.
4. Allow the sol to set with the tubes in an upright position, loosely covered with gauze. Gelation is usually complete in 1 h, observable by the increase in cloudiness.
5. Wipe off condensate above the gel, and close the tubes with a rubber stopper. Let stand at 20°C overnight.
6. Mount the pairs of tubes upside down using burette clamps attached to a short, heavy stand. Four or five clamps may be placed on one stand.
7. Set the assembly in a 5-L water bath filled with 20°C water, and adjust the levels of the tubes to be certain that they are totally submerged.
8. To minimize evaporation, cover the surface of the bath with paraffin oil or plastic balls designed for that purpose, then heat to 80°C moderately rapidly. Reduce the heating rate to no more than 0.5°C/min until within 5°C of the expected melting temperature, then reduce the rate to 0.2°C/min, or less.
9. When the melting point is reached, the gel will suddenly drop down and melt away. Record the temperature at which the gel drops.
10. Calculate the average for the pairs of samples, expressing it to the nearest 0.1°C.
11. For more precise data, useful for melting temperature–concentration plots, the exact concentration of the gel can be determined from the unused portion of the agar solution.

Gelling Temperature Determination

This procedure was reported by Yanagawa and Nishida, National Ind. Expl. Station, Osaka, *Research Report 11*, 14, 22–32 (1930).

1. Dissolve 1.5 g of agar in 100 mL of water by heating under reflux.
2. Transfer 5 mL of the hot solution to a test tube having an inside diameter of 15 mm, and let stand at room temperature (about 20°C).
3. Carefully incline the tube at appropriate intervals to determine if the sol flows or has solidified.
4. Insert a thermometer into the agar at the time of solidification, and record the maximum temperature observed. The experimental error is ±0.5°C.

Chapter **8**

Gum Arabic

William Meer
Meer Corporation

General Information 8-2
  Chemical Nature 8-2
  Physical Properties 8-3
  Manufacture 8-3
  Biological/Toxicological Properties 8-4
  Rheological Properties 8-4
  Additives/Extenders 8-5
  Handling 8-6
  Applications 8-6
  Application Procedures 8-7
  Compatibility 8-7
Commercial Uses 8-8
  Food Applications 8-8
  Pharmaceuticals 8-11
  Medicines 8-12
  Cosmetics 8-12
  Adhesives 8-13
  Paints 8-13
  Inks 8-13
  Lithography 8-14
  Textiles 8-15
  Miscellaneous Uses 8-15
Industries Using Gum Arabic 8-16
  Food Industry 8-16
  Pharmaceutical Industry 8-16
  Cosmetics Industry 8-16
  Other Industries 8-16
Formulations 8-17
  Confectioneries 8-17
  Food Emulsions 8-17
  Beverages 8-18
  Beverage Stabilizers 8-19
  Nut Coating 8-19
  Inks 8-19

Laboratory Techniques 8-20
  30% Viscosity Method 8-20
  Insoluble Residue 8-20
  Sediment and Color 8-20
  Peroxidase Content 8-21
References 8-21

GENERAL INFORMATION

Gum arabic, sometimes known as acacia gum or acacia mucilage, is unique in that it forms solutions of greater than 50% concentration. It is the most widely used gum in industry. It has a comparatively low viscosity, but its function is based primarily on its properties as a protective colloid and emulsifier. The adhesive property of gum arabic is not related to its viscosity.

This gum is the amber, amorphous, dried exudate of the acacia tree. Although there are more than 500 species of acacia tree distributed throughout the tropical and subtropical areas of the world, most commercial gum arabic is derived from *Acacia senegal.*

Almost all of the world output of gum arabic is from the sub-Sahara or Sahel Zone of Africa. More than 90% of commercial gum arabic is collected from wild trees. A movement is underway in the Sudan to cultivate the tree, but as of 1975, plantation gum arabic represents less than 5% of the total output.

The acacia exudate occurs when the tree is in an unhealthy condition. The gum is collected by natives who scar the tree with a knife, then return days later to remove the "tear" of gum that has formed at the scar. There is a definite collection season, with variations in the physical properties of the exudate as each season progresses, and with a slight variation from year to year. The crude gum is carried to the local village until there is enough accumulated to be sent to market. El Obeid is the major market. Here, the gum is cleaned and sorted into two grades: *cleaned amber sorts* and *hand-picked selected.*

The crude but cleaned and sorted gum is transported by rail to Port Sudan. From there it is shipped to the major markets of the world in 100-kg-capacity jute bags. Importers in the United States and Europe then process the crude gum into the various sizes and grades which can be readily used by industry. The total world output of gum arabic averages 50,000 metric tons annually. The peak year was 1963 with an output of 75,000 metric tons. The 1973 crop year was the low in recent history with only 30,000 metric tons produced.

Chemical Nature

Gum arabic as found in nature exists as a neutral or slightly acidic salt of a complex polysaccharide containing calcium, magnesium, and potassium cations. Molecular weight determinations of gum arabic give variable results. Generally accepted values of average molecular weights for gum arabic are M_n of 250,000 and M_w of 580,000. Anderson and coworkers[5] reported weight-average molecular weights based on viscosity data ranging from 260,000 to 1,160,000. That gum arabic is a mixture of many components is indicated by its broad-skew molecular weight distribution.[5] The gum can be fractionated by careful precipitation with near-saturated sodium sulfate solutions.

Light-scattering data[126,129] suggests that gum arabic molecules are in the shape of short, stiff spirals with numerous side chains. The length of the coil is 1050 nm. Hydrolysis of gum arabic yields L-arabinose, L-rhamnose, D-galactose, and D-glucuronic acid. Gum arabic from different sources contains the same sugars, but the proportions vary greatly.

Arabic acid and its salts have been prepared by a number of workers. Arabic acid was prepared originally by the method described by Thomas and Murray.[123] Moorjani and Narwani[80] prepared arabic acid by electrodialysis. This arabic acid, when dried at 110°C, was insoluble. Schleif and coworkers[109] prepared arabic acid by ion exchange. This procedure was extended by Swintosky and coworkers[121] to the preparation of other polysaccharidic acids, and a similar ion-exchange process for this purpose was patented by Wood.[137]

Arabic acid is a moderately strong acid whose aqueous solutions have a pH range of 2.2 to 2.7. Its K_a is equal to 2.0×10^{-4} at 22°C. Arabic acid solutions show very little change in pH upon the addition of acid, indicating a buffering action. Arabic acid produces a higher solution viscosity than does its salts. Emulsions prepared with arabic acid cream rapidly, and are not as stable as those made with its salts. Pure arabic acid is nonreducing toward Fehling's solution and has a specific optical rotation of -27 to $-30°$ in water.[115,124]

Several procedures have been reported for preparing sodium arabate from gum arabic. Krantz and Gordon[66] precipitated the calcium ions in a gum arabic solution by the addition of sodium carbonate, and then filtered and evaporated the solution to yield sodium arabate. Briggs[21] directly neutralized an arabic acid solution with sodium hydroxide. Schleif, Higuchi, and Busse[109] prepared sodium arabate by titrating arabic acid with sodium hydroxide solution. This method was extended to prepare spray-dried arabic acid salts of potassium, calcium, magnesium, zinc, iron(III), and aluminum.[110]

A one-step, ion-exchange process for preparing gum arabic salts was patented by Adams,[2] who passed a gum arabic solution through a cation-exchange resin in the salt form to prepare potassium, sodium, lithium, and ammonium salts. Organic salts and esters of arabic acid were prepared by Schleif and coworkers[110] by reacting a solution of arabic acid with the appropriate amine or alcohol, and drying the neutral mixture.

Gum arabic as the calcium, magnesium, and potassium salt reacts with many reagents. Solutions of gum arabic will produce precipitates on addition of borax, ferric chloride, basic lead acetate, mercuric nitrate, gelatin, potassium silicate, sodium silicate, Millon's reagent, and Stoke's acid mercuric nitrate reagent.[71,72,133] Trivalent metal ion salts will cause precipitation of gum arabic.[31]

The flocculation or thickening of gum arabic can be prevented or retarded[49] by addition of soluble alkali polyphosphates. Gum arabic is hydrolyzed by dilute acids to yield a mixture of L-arabinose, L-rhamnose, D-galactose, and aldobiouronic acid composed of D-glucuronic acid and D-galactose.[102]

Gum arabic contains both oxidases and peroxidases that are inactivated by heating the gum solution to 80°C or higher for 1 h.[62] Diastase[102] and pectinases have also been reported to be present in gum arabic.

Physical Properties

Gum arabic is nontoxic, odorless, colorless, and tasteless. It is usually water-soluble, and it does not affect the odor, color, or taste of the system in which it is used. Gum arabic is insoluble in oils and in most organic solvents, but usually dissolves completely in hot or cold water, forming a clear, mucilaginous solution.

That gum arabic which is collected at the beginning of the season forms stringy, mucuslike fluids that separate into two phases upon standing. The solubility of gum arabic in organic solvents was examined by Taft and Malm.[102] Hot ethylene glycol and glycerol were effective solvents. Gum arabic was soluble also in aqueous ethanol up to about 60% of alcohol. Gum arabic mucilage can be mixed with glycerol to give a compatible solution with a viscosity equal to the resultant of the viscosities of both components.

The addition of gum arabic to water results in a lowering of the surface tension. The freezing point of gum arabic solutions decreases as the concentration of gum arabic increases.

Manufacture

Gum arabic is imported as tears that are approximately 2.5 to 5 cm (1 to 2 in) in size which have been sorted and cleaned at their origin and packed in 100-kg burlap (jute) bags. Of the two grades that are available (clean amber sorts and selected sorts), the selected sorts, as its name implies, is further selected by color for better appearance, and by the fact that it contains less adhering wood and other foreign matter.

The gum from both grades is processed into several forms: grains or crystals, milled powder, and spray-dried powder. The processes generally include grading, cleaning, aspiration, sizing, and blending. In the case of spray-dried gum, the solution of the gum is clarified by centrifugation, then filtered (and, in the case of enzyme-inactivated gum, pasteurized), and spray-dried. The powder is then screened to assure uniformity of particle size.

Biological/Toxicological Properties

Gum arabic is widely used in the food and drug fields. It is nontoxic, odorless, colorless, tasteless, usually water-soluble, and does not affect the color, odor, or flavor of the food or drug it is used in. It is in the GRAS (Generally Recognized As Safe) list under the Federal Food, Drug and Cosmetic Act. It is listed in the *United States Pharmacopeia* and *Food Chemicals Codex.*

Digestibility. With respect to human nutrition, it is generally agreed that gum arabic has a low level of digestibility, and in a food system it will not contribute to caloric

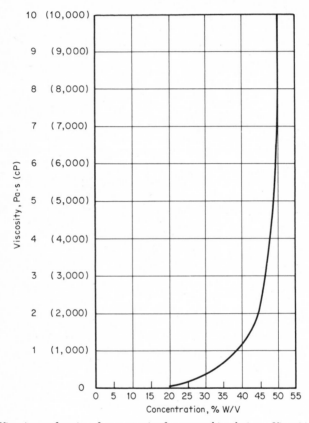

Figure 8.1 Viscosity as a function of concentration for gum arabic solutions. Viscosities determined by a Brookfield SynchroLectric viscometer at 25.5°C ± 0.5°C on solutions prepared with powdered gum arabic.

intake. Short-term studies[20,114] indicate that digestibility in rats ranges from 71 to 80%. Studies with guinea pigs[19] and rabbits[71] also indicate at least partial digestibility.

Acute Toxicity. No data available.

Sensitization. Gelfand[43] examined subjects suffering from allergic disorders and definitely implicated gum arabic as a cause of clinical manifestation in man.

Fetotoxicity. Studies[41] with pregnant mice produced no fetotoxic effects.

Rheological Properties

Gum arabic is extremely soluble and can form solutions over a wide range of concentrations[101] (see Fig. 8.1). It is easy to prepare solutions containing up to 50% of gum arabic at 25°C. The properties of the gum exudates are affected by the age of the parent tree, the amount of rainfall, the time of exudation, and the type of storage conditions.   Viscosities of solutions of gum arabic can vary as much as 50% and are affected by pH (see Figs. 8.2 and 8.3), salts or other electrolytes (see Fig. 8.4), and temperature.   At concentrations below 40%, gum arabic solutions exhibit Newtonian behavior.   At concentrations of 40% or higher, pseudoplastic characteristics are observed, denoted by a decrease in viscosity with increasing shearing stress.[11]

Taft and Malm[122] showed that, as the temperature increased, the relative viscosity and density of gum arabic solutions decreased.   The viscosity is inversely proportional to temperature.   When heated at 170°C and then put into water, gum arabic does not dissolve, but swells to form a nonsticky gel.

The viscosity of dilute sodium arabate solutions decreases with increasing pH, but falls rapidly above pH of 9 (see Fig. 8.5).   The viscosity of gum arabic solutions decreases with the addition of ethanol.[18]   In 60% ethanol, gum arabic precipitates.

The viscosity of gum arabic solutions varies with the pH, with a maximum viscosity developing between pH 6 and 7.   Electrolytes lower the viscosity of gum arabic solutions.

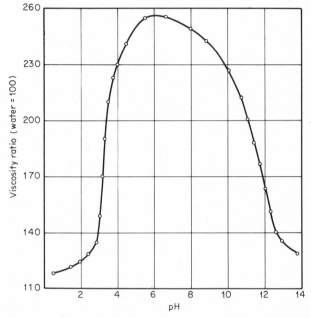

Figure 8.2 Effect of pH on the viscosity of gum arabic solutions.[123]

However, sodium citrate will increase the viscosity.   The viscosity of gum arabic solutions decreases with time, but this decrease can be minimized by the addition of preservatives.

Mechanical treatment of gum arabic solutions does not affect their viscosities (see Fig. 8.6).   Gum arabic solutions are depolymerized by ultrasonic waves,[72] and continued exposure to ultrasonic vibration releases monosaccharides.[111]   Ultraviolet radiation also reduces the viscosity of gum arabic solutions.

The comparative viscosities of several plant exudates can be seen in Fig. 8.7.

Additives/Extenders

Additives   The effects of additives on gum arabic solutions are as follows:
- Electrolytes lower solution viscosities.
- Polyphosphates prevent thickening and flocculation.
- Other water-soluble gums are compatible.
- Gelatin is incompatible and forms coacervates.

Materials that are incompatible with gum arabic solutions are listed in the section on Compatibility.

Extenders   The extenders with which gum arabic solutions can be combined include starch, other water-soluble gums (especially gum ghatti), mesquite gum, and larch gum. For emulsions, materials that can be used include other water-soluble gums, especially gum tragacanth and gum ghatti.   For encapsulation, modified starch is recommended; and for lithography, mesquite gum and larch gum can be used.

Handling

Gum arabic is normally packed in polyethylene-lined bags or fiber drums.   The product should be stored in a cool, dry area to avoid lumping and to prolong shelf life.

The material is nontoxic and free of odors.   It is available in lumps, in granular forms, in milled powder and spray-dried powder forms, the latter of which is free-flowing and easy to handle with conventional equipment.   Gum arabic is nonflammable.   It does not require any special safeguards or precautions.   Milled gum arabic and spray-dried gum arabic do dust, however, and normal antidust precautions should be used.

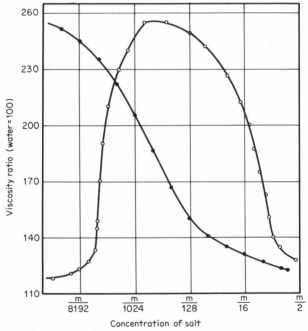

Figure 8.3   Viscosity of gum arabic solutions as a function of pH and salts.   *(Redrawn from Thomas & Murray, 1928.*[123]*)*

Applications

Emulsification.   Gum arabic is used as an emulsifier for drug preparations, for fat in caramels, for flavor oils in carbonated beverages, for citrus oils in bakery emulsions, for cod-liver oil, and for paraffin-water emulsions.

Colloid Stabilization.   Gum arabic is used by the brewing industry as a foam stabilizer for beer, as well as in other beverages.   It is also used to produce protective colloids in vinyl resin emulsions and liquid soaps.

Encapsulation.   Gum arabic is used as an encapsulation agent for flavors, vitamins, spray-dried fats for bakery mixes, and clouding agents.

Suspension.   Gum arabic is used as a suspending agent for soluble inks, water colors, quick-drying inks, typographic and hectographic inks, and lithographic inks (as a fountain solution).

Other Applications.   Other varied uses for gum arabic include protective films for the storage of printing plates, as an excipient in pharmaceutical tablets and pills, as a binding agent in cosmetics and rouges, and as a coacervate with gelatin for microencapsulation of inks, time-delay aspirins, and pressure-rupturable capsules.[46]

Application Procedures

There are no special procedures to be followed for dry-blending gum arabic, except that the mix should not be allowed to become moist or damp. And, although gum arabic is very soluble in water, finely divided particles tend to clump when dispersed in water and do not dissolve readily unless vigorous agitation is used.

In dissolving gum arabic, one or more of the following procedures should be used:
1. Apply vigorous agitation while adding the gum to water.
2. Allow two or more hours for hydration.
3. Disperse the gum into water as a mixture with another solid, such as sugar.
4. Premix the gum with one to five times its weight of a water-miscible liquid, such as ethanol or glycerol, prior to the addition to water.

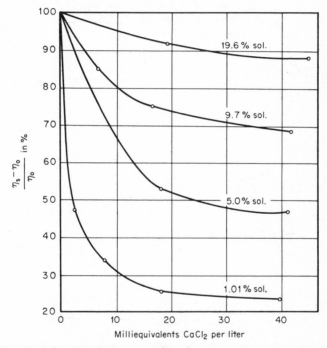

Figure 8.4 Effect of calcium chloride on gum arabic solutions as gum concentration increases. *(Redrawn from Bungenberg de Jong, Kruyt, and Lens.)*

Compatibility

Gum arabic is the most widely used of all the plant hydrocolloids. It is compatible with most other plant hydrocolloids and proteins, carbohydrates, and starches. Electrolytes tend to reduce the viscosity of gum arabic solutions. Some reagents give precipitates or heavy jellies, for example, borax, ferric chloride, and basic lead subacetate.

Solutions of gum arabic are incompatible with soap in making emulsions. The viscosity of a solution of a mixture of gum arabic and gum tragacanth tends to be lower than that of either constituent solution. A minimum viscosity is attained in a mixture consisting of 80% gum tragacanth and 20% gum arabic.

Gum arabic is compatible with alkaloids and is soluble in concentrated and dilute hydrochloric acid; concentrated and dilute acetic acid; concentrated and dilute ammonium hydroxide; dilute sodium hydroxide; and 10% solutions of sodium chloride, mercuric chloride, bismuth chloride, and silver nitrate; but it is insoluble in 10% ferric chloride and concentrated sodium hydroxide.   Gum arabic is completely compatible with 1:1000 dilutions of morphine sulfate, ephedrine hydrochloride, quinine sulfate, strychnine sulfate, and caffeine.

The *United States Dispensatory*[102] reports that gum arabic is incompatible with aminopyrine, pyrogallol, morphine, vanillin, phenol, thymol, carvol, α- and β-naphthol, pyrocatechol, guaiacol, cresols, eugenol, acetyleugenol, apomorphine, eserine, epinephrine, isobarbaloin, caffeotannic acid, gallic acid, and tannin.   But carboxymethylcellulose (CMC) is compatible with gum arabic in both solutions and films.

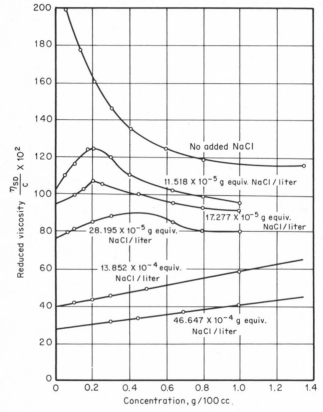

Figure 8.5   Reduced viscosity of sodium arabate as a function of sodium chloride concentration. *(Redrawn from Bungenberg de Jong, Kruyt, and Lens.)*

COMMERCIAL USES

Food Applications

The uses of gum arabic in foods depend upon its action as a protective colloid or stabilizer, the adhesiveness of its water solutions, its ability to thicken, and its low digestibility level in formulating low-calorie food products.   The main function of gum arabic in food is to impart desirable qualities through its influence over the viscosity, body, and texture.

Many citrus oils and other beverage flavor emulsions utilize gum arabic emulsification powers.   Gum arabic is also used as a foam stabilizer in beer, and when used as a fixative in spray-dried flavors, it forms a thin and impenetrable film around the flavor particle, protecting it from oxidation and evaporation, and prevents it from absorbing moisture from the air. While other colloids are also used for these purposes, gum arabic is the most important and most widely used because of its ease of solution, its low cost, and its superior performance.

Confectioneries Gum arabic is used extensively in the confectionery industry, primarily because of its ability to prevent crystallization of sugar, but also because of its thickening power.[23] It is used as a glaze in candy products and as a component of chewing gum, cough drops, and candy lozenges. Originally, gum drops were produced from gum arabic.

In most confectionery products, gum arabic has two important functions.[67] The most important function is to retard or prevent crystallization of sugar, and thus gum arabic finds its greatest application in confections where the sugar content is high and the moisture content comparatively low, such as in jujubes and pastilles. The second func-

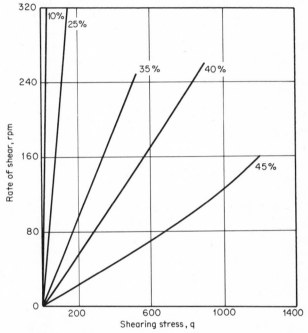

Figure 8.6 Rheological characteristics of various concentrations of gum arabic in solution. *(Redrawn from Aranjo.)*

tion is to act as an emulsifier to keep fat uniformly distributed throughout the product so as to prevent it from moving to the surface and forming an easily oxidizable, greasy film.[134]

For the preparation of jujubes, pastilles, or gum drops, gum arabic is dissolved in water and the filtered solution is mixed with sugar and boiled. Lozenges are prepared by mixing finely ground or powdered sugar with a thick mucilage of gum arabic. The resulting stiff dough is rolled into sheets, and lozenge shapes are stamped out.[58]

Dairy Products Gum arabic has been used as a stabilizer in frozen products, such as ice creams, ices, and sherbets, because of its water-absorbing properties.[105,128] The addition of gum arabic prevents the formation of ice crystals by combining with large quantities of water and holding it as water of hydration, thus producing a finer texture in the ice cream. The chief objection to this use of gum arabic is that the ice cream does not melt readily.[70]

Scholz[112] patented a process using gum arabic for the preparation of packageable milk or cream. The proper amount of milk or cream is mixed with gum arabic; the mixture is heated mildly, poured into molds, cooled, and packaged. The product dissolves easily in hot beverages and keeps better under refrigeration than plain milk or cream. Walder[128] recommends gum arabic as a protective colloid in the preparation of processed baby food.

Bakery Products   Gum arabic is widely used in the baking industry because of its viscosity and its adhesive property. It is used in glazes and toppings. It also bestows smoothness when used as an emulsion stabilizer.

When used in a bun glaze, gum arabic imparts stability in conjunction with free-flowing and adhesive characteristics. The glaze is applied to the bun while still warm, and adheres firmly to the bun upon cooling.[132]   A stable icing base utilizing gum arabic has been patented by Wagner.[127]

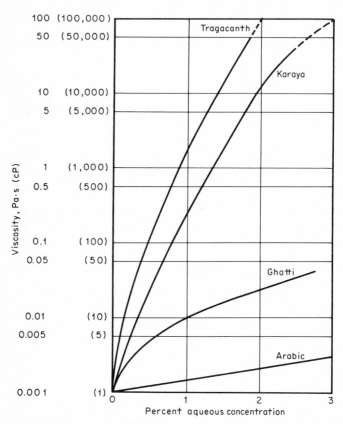

Figure 8.7   Comparative viscosities of plant exudates. *(Redrawn from Smith, 1965.)*

Flavor Fixation   With the introduction of spray-dried flavors into foods,[107] an extensive use was found for gum arabic as a fixative for flavors. Spray drying is a fairly simple production operation relying on proper selection of spray-dryer nozzles and inlet pressures, and different drying times. Rapid drying at comparatively low temperatures is particularly advantageous for flavoring materials that are volatile or easily oxidized.

When used as a flavor fixative, gum arabic forms a thin film around the flavor particle, protecting it from oxidation and evaporation and preventing it from absorbing moisture from the air. This is advantageous when hygroscopic materials are spray-dried. Flavors spray-dried with protective colloids last 10 to 20 times longer than the same flavors air-dried on a solid. Tests made on spray-dried emulsions of easily oxidized aldehydes showed that the film protected the products from oxidation for years, whereas unprotected materials oxidized in seconds.[131]

Citrus oils and imitation flavors are prepared by emulsifying them in gum arabic solution before spray-drying them. The particle size can range from 1 to 50 μm. Other colloids are also used for these purposes, but gum arabic is the most important and most widely used.

More recently, a new technique of microencapsulation has been developed for specialized flavor fixation.[13,52] Gum arabic is reacted with gelatin in a medium containing the flavor to form an insoluble, protective film surrounding or encapsulating the sphere of flavor.

Flavor Emulsification Many flavor emulsions for beverages are prepared with gum arabic as the emulsifier.[60] Some beverage flavor emulsions make use of the additive properties of a blend of gum arabic and gum tragacanth. Here, gum tragacanth is the emulsion stabilizer and gum arabic provides smoothness.[132]

In the preparation of baker's citrus oil emulsions consisting of citric acid, lemon oil, glycerol, water, and coloring matter, the best emulsions are prepared with gum arabic–gum tragacanth mixtures.

Beverages Gum arabic is an effective foam stabilizer for beverages and is largely responsible for the "lace curtain" effect on the sides of beer glasses. This is an important application.

Eye-appealing opacity in beverages and beverage dry mixes are produced by spray-dried combinations of vegetable oil and gum arabic sold commercially as a clouding agent to give the effect of fruit juices.

Pharmaceuticals

The inherent emulsifying and stabilizing properties of gum arabic plus its demulcent and emollient characteristics have led to a number of applications, ranging from the stabilization of emulsions to the preparation of tablets. Gum arabic applications are further extended because it retains its viscosity and stabilizing properties over a wide pH range.[7]

Suspending Agent Gum arabic is an effective suspending aid, and it has been employed to suspend insoluble drugs, as well as to prevent the precipitation of heavy metals from solution through the formation of colloidal suspensions.

Gum arabic is one of the best emulsifying and suspending agents for calamine suspensions, kaolin suspensions, liquid petrolatum emulsions, and cod-liver oil emulsions.[100] It is unsatisfactory, however, for suspension of paraldehyde (2,4,6-trimethyl-1,3,5-trioxane),[103] but it has been found to be an excellent agent for preparing a stable, nonsettling magnesia suspension.[8,45]

Intestinal absorption of poorly soluble medicinal substances suspended in gum arabic, such as steroids, fat-soluble vitamins, and barbiturates, can be facilitated by the incorporation of wetting agents or other emulsifiers in the preparation.[117]

Demulcent Agent Gum arabic's demulcent characteristics have led to its use in many pharmaceutical syrups.[10] A gum arabic syrup that contains sodium benzoate, vanillin tincture, and sucrose is recommended for use as a flavored vehicle because of its demulcent effect and its protective colloid action. The preparation of diabetic syrups has become increasingly important. Various nonsugar recipes have been developed, such as nonsugar cherry syrups, based on gum arabic.[9] A good syrup is made from saccharin, methyl *p*-hydroxybenzoate, water, and gum arabic.[110]

Emulsification Gum arabic is superior to gum tragacanth for preparing cottonseed oil emulsions because the average diameter of the oil globules in emulsions prepared with gum arabic is much smaller than that of the oil globules in emulsions prepared with gum tragacanth.[65] With both vegetable and mineral oils, gum arabic emulsions are stable over the pH range of 2 to 10.

Because gum arabic contains the potassium, calcium, and magnesium salts of weak acid groups, its buffer action is greater in the neutralization of acids than in the neutralization of alkalies.

Lotzkar and Maclay[68] compared pectin, gum tragacanth, gum karaya, and gum arabic as emulsifying agents for olive oil, cottonseed oil, and mineral oil in water while varying the pH, the oil/water ratio, and the gum concentration. They found that gum arabic stabilized emulsions over a wide pH range, and that there was no relationship between initial viscosity and stability. Higher viscosity may prevent creaming, but does not necessarily improve stability, and gum arabic emulsions are less viscous than those of other gum emulsions.

Paraffin oil-and-water emulsions for use as laxatives have been prepared with gum arabic, agar, and gum tragacanth.[22,97,98] Gum arabic has also been recommended for use in oral laxatives[64] and in laxative suppositories.[14]

Gum arabic has been used in the preparation of cod-liver oil emulsions.[7] Used in conjunction with gum tragacanth, stable emulsions with excellent shelf life can be made of cod-liver oil, linseed oil, and mineral oil. These emulsions are better than those prepared with either gum alone.[56]

The oxidase present in gum arabic may destroy 54% of the vitamin A content of cod-liver oil emulsified with it.[61] Therefore, the oxidases must be destroyed by heating gum solutions before such use. Gum arabic is effective in the preservation of vitamin A in vitamin-enriched margarine[94] and vitamin C in aqueous solutions.[4]

Antiseptic Preparations Antiseptic preparations have been made with a mixture of colloidal silver bromide and gum arabic. Silver arabate has antiseptic properties that make it suitable for use as a substitute for silver nitrate and organic silver compounds in the treatment of ophthalmic infections.[38] Silver compound preparations for the internal treatment of mucous membranes have been patented.[96]

It was reported that the use of gum tragacanth and other vegetable gums in pharmaceutical jellies can reduce the bactericidal action of the incorporated preservative.[36] It was found that gum tragacanth exerts a strong neutralizing effect upon the bactericidal activity of chlorbutanol (1,1,1-trichloro-2-methyl-2-propanol), the p-hydroxybenzoates, benzalkonium chloride, and, to a much lesser extent, a phenol, phenylmercuric acetate, and Merthiolate, [(o-carboxyphenyl)thio]ethyl mercury. The greater the concentration of gum, the less effective is the preservative agent. Although no work of this nature has been done with gum arabic, it seems highly probable that it would also reduce the antibacterial effectiveness of preservatives.

Miscellaneous Applications Gum arabic has been used as an adhesive or binder for pharmaceutical tablets, and also as an excipient in the manufacture of pills and plasters. In addition, many types of coatings for pills employ gum arabic in their manufacture.[7]

Because of its emollient and demulcent properties, gum arabic is used in the production of cough drops and cough syrups. It produces a smooth viscous syrup, and also prevents crystallization of sugar in both of these products.

Schwarzmann[113] has made a therapeutic gold–gum arabic–peptone complex, and another gum arabic–peptone complex without the addition of gold.[25] Mason[74] incorporated gum arabic with alkaloids for anesthetics. Iron arabate has been suggested for use as a medicinal hematinic preparation.[38]

Medicines

Gum arabic has been used for the treatment of low blood pressure caused by hemorrhage or surgical shock.[6,77,78] Intravenous saline injections alone were not successful because the salt escaped too rapidly from the blood vessels. The addition of a 7% gum arabic solution reduced the dissipation rate of the sodium chloride solution, and this treatment was successfully used in the 1920s. However, the subsequent development of blood plasma extenders such as dextran and polyvinylpyrrolidone eliminated the use of gum arabic.

In 1933, intravenous injections of gum arabic solutions were recommended for the treatment of nephritic edema.[6,78,102]

Cosmetics

In lotions and protective creams, gum arabic stabilizes emulsions, increases their viscosities, assists in imparting spreading properties, adds a smooth feel to the skin, and forms a protective coating.[125] It is also used as an adhesive in the preparation of facial masks[50] and as a binding agent in the formulation of compact cakes[51] and rouges. A typical compact cake is composed chiefly of a color vehicle, mineral oil, and an aqueous solution of gum arabic. Gum arabic is also used as a foam stabilizer in the production of liquid soap.[116]

One of the advantages of gum arabic in cosmetics is its nontoxicity and its comparative freedom from dermatological and allergic reactions.[38,47]

Nonoily hair fixatives based on vegetable gums have the advantage of vastly improving fixative properties without forming greasy stains on clothing. Gum arabic mucilages have been used in the preparation of these hair dressings.[69]

Adhesives

Powdered gum arabic is considered to be a safe, simple adhesive for miscellaneous paper products. It is commonly used by dissolving it in two or three times its weight of water.[30,136] A 40% solution makes an excellent mucilage for general office purposes.

Gum arabic glues are easy to prepare, light in color, odorless, and very stable. In addition, the adhesive strength of the gum can be improved by the addition of certain metal salts, such as calcium nitrate and aluminum sulfate. However, when these glues are heated too long or at too high a temperature, the gum arabic is degraded, resulting in a decrease in viscosity and a loss of adhesive strength. This degraded gum arabic will discolor bronze and colored papers.

The problem can be minimized by using dilute solutions that are then made more viscous by the addition of compatible thickening agents, such as gum tragacanth or methylcellulose. Gum arabic makes a brittle film, which is usually plasticized by the addition of 8 to 10% glycerol based on the weight of dry gum. Other suitable plasticizers are diethylene glycol, ethylene glycol, and sorbitol. Spoilage is prevented by glycerol or by the addition of a little calcium hydroxide.

Adhesives for envelopes, labels, and postage stamps are made by using gum arabic. A typical formula contains 100 parts of gum arabic, 2.5 parts of sodium chloride, 2.0 parts of glycerol, 2.0 parts of starch, and 130 parts of water.[16]

A good, transparent adhesive can be made with gum arabic, but to reduce costs, smaller amounts of gum arabic can be used and the viscosity built up with small amounts of gum tragacanth. Wallpaper paste can be based on a mixture of gum arabic, bentonite, and starch.[33] Gum arabic is compatible with flour, starch, and dextrins, and is used in conjunction with them in postage stamp adhesives,[15] mounting pastes,[16] and Lunel's paste for artificial flowers.[136] Gum arabic is sometimes used as a binder for water cements for castings. In the preparation of laminated papers, gum arabic has found specialized applications. The bonding of regenerated cellulose films can be accomplished ·by a special gum arabic paste.[17,53]

Some adhesives for bonding aluminum foil to paper have been prepared from gum arabic solutions modified with small amounts of tartaric or similar acids, but they are corrosive.[29] Gum arabic–based adhesives are frequently used for smooth, hard-surfaced glassine papers.[29]

Paints

The incorporation of protective colloids into a pigment-vehicle system is usually done to obtain improved particle size during precipitation of pigments, to control pigment aggregates and wetting qualities, and to control consistency and settling.[3,39,59] Acheson[1] prepared graphite dispersions in water by the use of tannic acid and gum arabic.

In paints and similar formulations, flocculation prevents hard settling of pigments. Water has been used for many years to prevent hard settling of the pigments in paints. Such controlled flocculation has been achieved by means of additives that presumably affect the coherence of particles in an immiscible liquid.[119] Gamble and Grady[42] treated pigments with water-soluble hydrophilic colloids such as gum arabic to give controllable thixotropic properties to paints.

Less than 1% gum arabic, based on the weight of the pigment, was recommended. By varying the quantity of gum arabic, the thixotropic properties of the paint could be adjusted to any desired degree. Pigments that are readily wetted by water, such as titanium dioxide, zinc oxide, clays, and silica, required very small quantities. The amount of water required to produce suitable thixotropic properties in a paint is usually less than 1%. In water-emulsion paints, the use of gum arabic is limited because of its unfavorable effect on the water resistance of the final paint film,[76] but it has found some use in the preparation of vinyl resin emulsions.[135]

Inks

Gum arabic is a constituent of many special-purpose inks[37] because of its excellent protective colloid properties. The effectiveness of this protective colloidal action of gum arabic or other gums is designated by its gold number. Early inks were simply dispersions of lampblack in water. Soon gum arabic was added routinely as a suspending agent or a protective colloid. Sometimes, the lampblack and gum were mixed into a thick paste and allowed to harden in molds or ink sticks. To use an ink stick, the operator moistened a small brush with water and then rubbed the brush on the end of the stick. To make liquid ink, the end of the stick was rubbed in a little water in a shallow dish until the ink solution developed the proper color.[95,130]

Record Ink. Government record ink had long used gum arabic in the formula as a protective colloid.[130]

Soluble Inks. Easily soluble inks are used by textile workers for temporary marking of cloth for cutting or sewing operations. When the garment is finished, a hot-water wash is used to remove the ink. A typical ink is composed of a mixture of dilute acetic acid, albumin, a basic dye, gum arabic, molasses, and triethanolamine.[104]

Watercolor Inks. Inks suitable for watercolor work are maintained in suspension by gum arabic. A typical ink base[54] is made by dissolving 1 part of gum arabic in 9 parts of glycerol and 1 part of water, then evaporating the mixture to a specific gravity of about 1.28. Pigment is subsequently added by grinding it into the ink base with a colloid mill.

Quick-Drying Inks. Inks designed to dry faster than ordinary writing inks are usually based on solvents other than water. Ethanol is commonly used in conjunction with gum arabic. However, some fast-drying formulas using water as the solvent are known, such as one consisting of water, Lysol (as a preservative), sodium nitrate, gum arabic, and a water-soluble dye.[38,108]

Fabric- and Laundry-Marking Inks. Many formulas for inks for marking cloth are known.[37] Gum arabic is frequently added to these formulas to bring the ink to the desired viscosity for smooth writing on cloth. Other indelible inks make use of metallic pigments, but also contain gum arabic as a thickener and protective colloid.

Pigmented Inks. The suspension of a pigment in a gum arabic solution makes a satisfactory ink that can be applied with a brush or pen. Pigments, such as titanium dioxide or bronze powder used for gold inks, are preferably moistened first with a small amount of ethanol or acetone before mixing with the gum solution.

A typical white ink is made by mixing a filtered solution of 20 parts of gum arabic in 160 parts of water with 30 parts of titanium dioxide paint pigment and 0.5 part of sodium salicylate preservative.[130]

Emulsion or Typographic Inks. Some inks consisting of emulsions use gum arabic.[37] Many emulsion inks are oil-in-water emulsions with the pigment dissolved in the oil phase. A typical ink of this class consists of lampblack, mineral oil, rosin, catechu black, Formalin, sodium silicate, sodium carbonate, gum arabic, aluminum resinate, coloring agent, and water.[105,106]

Hectographic Inks. An ink that has been used for many years in hectograph duplicating equipment is prepared by mixing methyl violet dye with water or ethanol. Some formulas using hydrochloric acid, oxalic acid, lactic acid, tannic acid, or procatechol employ gum arabic to obtain the proper viscosity.[37]

Electrically Conductive Inks. Inks that conduct electricity are becoming increasingly important because of their use in the manufacture or repair of printed circuits in electronic applications. Activation of electronic calculators by conductive areas on a card rather than by punched holls is also attracting increasing interest.

Most conductive inks are made of carbon black, powdered graphite, and finely divided silver or powdered copper. The suspending agent can be a lacquer, but water-based inks function well when gum arabic, sodium chloride, and citric (or tartaric) acid are used.[120]

Lithography

Gums are important in lithographic applications where they have many functions as sensitizers for lithographic plates, elements in the light-sensitive composition, ingredients of the fountain solution used to moisten plates during pointing, and protectors during storage of plates.

The desensitizing use of gum arabic takes advantage of the good wettability imparted to the solution by the gum and also of the viscosity control that allows the wash solution to cling to the plate without running off or forming isolated droplets or pools on the plate. On metal plates, the desensitizing effect might be caused by the formation of an insoluble film of, e.g., aluminum or zinc arabate. A more plausible explanation is that a film of gum is absorbed by the plate. Studies have shown that such films occur on plates of zinc, aluminum, copper, silver, iron, tin, lead, glass, and fused silica. These films are not monomolecular, but are composed of many molecular layers.[37]

Measurement of the wettability of desensitizing solutions is conveniently evaluated by measurement and study of the contact angles. In this process, a section of a plate is partially immersed in water or in a solution of the gum to be tested. The plate is then turned at an angle to the surface of the liquid until the meniscus appears to be eliminated. The resulting angle of the plate to the surface of the liquid is known as the *contact angle* and is a measure of the wettability of that particular plate with the solution being tested.[106]

Mixtures of gum arabic and dichromates are tanned by the action of light to form water-insoluble substances. Thus, a wettable, plastic-impregnated paper sheet, a metallic sheet, or even a stone surface can be made light-sensitive by application of a gum and a dichromate solution that is allowed to dry.[79] Exposure of a pattern through a negative will cause tanning or hardening of all areas that are struck by light, and the unexposed material can be removed by a water or dilute acid wash. The surface can be used either for planographic printing or for an etching process if the plate is used for direct printing.

A related use is in the preparation of etched circuits and in the manufacture of small metallic parts. In the latter process, the part is sketched or photographed onto a transparent sheet, exposed onto a metal sheet made light-sensitive with gum arabic and a dichromate, washed in water, and etched to remove unwanted parts of the sheet. In this manner, intricate and delicate parts are readily produced.

Gum arabic used in lithography should be of the best quality. Preferably, it should be converted to the sodium salt by ion exchange. Sodium arabate is superior to gum arabic for stability, viscosity characteristics, and resistance to bacterial and mold attack. The purified gum solution can be used directly for deep etch–sensitive coatings. To each 3.65 parts of gum solution, 1 part of ammonium dichromate is added. The solution is made slightly alkaline (pH 8.8 to 10.9) with ammonium hydroxide. The final viscosity is preferably 0.026 to 0.029 Pa·s (26 to 29 cP). Coating, drying, and exposure follow conventional procedures.

It is important to control the viscosity of coatings since a high viscosity will not flow uniformly over the plate surface and can produce streaks. The strength of the solutions can be controlled by specific gravity readings with a Baumé hydrometer, and the viscosity can be adjusted by thinning with water.

Textiles

Gums are widely used in the textile industry as sizing and finishing agents and in printing formulations for imparting designs or decorations to fabrics. Gum arabic is an efficient sizing agent for cloth. Gum arabic finds limited use in finishing silk and rayon by giving body to the fabric without interfering with the transparency.

Although gum arabic can be used to create desirable effects on cotton, the cost is prohibitively high.[73] However, sizings are sometimes added to cotton fabrics to give them a fuller finish and to make them feel more firm and compact. Most cloth sizings are lost in washing.[34]

Typical low-cost sizes are made by simply heating a slurry of rye starch. A considerable improvement can be made by adding gum arabic. Thus, a mixture of 45 kg of rye flour starch, 24 kg of gum arabic, and 700 L of water is heated to boiling. A separate solution of 16.5 kg of castor oil, 2.5 kg of oleic acid, 7.5 kg of 15% sodium hydroxide, and 23.5 kg of water is added to the starch–gum arabic solution. After mixing, 1 L of oleic acid is added. The mixture is stirred and boiled for 20 min, and acetic acid is added until the mixture is neutral.[81]

Gum arabic from *Acacia decurrens* has been found to be an effective thickening and binding agent for vat sizes.[48] The gum is light yellow to red in color and is free from tannins. It has excellent adhesive properties and will not interfere with the colors or dyes used in printing.

Miscellaneous Uses

A coating of gum arabic on typing paper allows erasures to be made easily.[28] A double coating is preferred, with a short drying period between applications. Gum arabic–based transfer inks can also be used for the preparation of good-grade carbon papers.[32]

Eichorn[35] described a novel photographic method for making templates utilizing gum arabic. A woven glass cloth is impregnated with a polyester or alkyl resin, and an opaque drawing is made upon the surface, which has excellent heat- and light-dimensional stability. Conventional blueprint salts containing gum arabic (or other gums) and a tanning agent are poured over the work, which, for example, might be a sheet of metal being formed into an airplane wing. Then, when the pattern has been printed photographically upon the metal sheet, the sheet can be cut, drilled, or otherwise mechanically manipulated, according to the pattern printed on its surface.

Gum arabic phthalate and gum arabic succinate are used as overcoating and backing layers for photographic material.[40] Gum arabic is used to inhibit metal corrosion in some applications. The addition of the gum, up to approximately 2% of the weight of the negative plate of a storage battery, increases the life of the battery by reducing the growth of surface projections. The gum arabic intimately mixed with the active negative plate material, for example, zinc oxide, produces an increase in electrical resistance and reduces the rate of corrosion.[118]

The corrosion of iron is considerably accelerated by light and the presence of salts, but it is inhibited by the presence of gum arabic.[24] The gum also effectively controls the corrosion of aluminum.[27]

The gum is useful in the immersion plating of copper on aluminum.[99] A bright, dense, strongly adhering copper coating is formed on aluminum by dipping the aluminum with an electrically connected piece of iron into a copper sulfate bath containing 0.75 to 1.25% by weight of gum arabic. Gum arabic has been utilized in the preparation of drilling fluids[12] and oil-well cement.[26] It has been used as a glaze binder[63] in ceramics, where it has had to meet rigid specifications pertaining to bonding power, absence of slip, and stability for long periods at elevated temperatures.

A Japanese patent[75] recommends gum arabic as a binder for insecticides. Horn and Samko[55] have developed a nonglare coating for automobile windshields based on a water-soluble dye dissolved in a gum arabic solution. A typical formulation contains 0.25% of brilliant green dye, 0.25% of the dioctyl ester of sodium sulfosuccinate, 5.00% of gum arabic, and 94.50% of water.

INDUSTRIES USING GUM ARABIC

A brief summary of the industries that use gum arabic is given in the following:

Food Industry

- *Confectioneries.* Used to prevent crystallization of sugar.
- *Dairy products.* Used as a stabilizer in frozen products.
- *Bakery products.* Used for its viscosity and adhesive properties.
- *Flavor fixative.* Used in spray-dried encapsulation.
- *Flavor emulsifier.* Used as an emulsifier and protective colloid.
- *Beverages.* Used as a foam stabilizer in beer, and as a clouding agent to give opacity.
- *Diabetic and dietetic products.* Used because of its low level of metabolization.

Pharmaceutical Industry

- *Emulsions.* Used as a stabilizer.
- *Tablets.* Used as an binder.
- *Tablet coatings.* Used as a mucilage.
- *Cough drops and syrups.* Used as an emollient and demulcent.
- *Medicine.* Used intravenously for nephritic edema.

Cosmetics Industry

- *Lotions and protective creams.* Used to give smooth feel.
- *Facial masks.* Used as an adhesive.
- *Face powders.* Used as an adhesive.

Other Industries

- *Adhesives.* Used as a mucilage; as a simple adhesive and glue.
- *Inks.* Used as a protective colloid and suspending agent.

■ *Lithography.* Used as a sensitizer for lithographic plates; as a protector and ingredient of fountain solutions.

■ *Paper.* Used as a coating for specialty papers; as a coacervate in carbonless paper.

■ *Paints.* Used as a protective colloid; as a flocculant and emulsifier in vinyl resin emulsions.

■ *Textiles.* Used as a sizing agent for cloth; to give body to certain fabrics; as a thickener and binding agent for vat sizes.

■ *Metals.* Used as a template to prevent metal corrosion.

FORMULATIONS

This section contains selections of formulations that are representative of those in which gum arabic is used for confectioneries, food products, beverages, nut coatings, and inks.

Confectioneries

Dietetic or Sugarless Candies[88]   Dietetic or sugarless candies utilize a wide range of concentrations of sorbitol, mannitol, and gum arabic. In the following formulas, formula I will give a hard candy drop; formula II will give a hard candy drop if the gum arabic content is low, and a soft drop or jujube if the gum arabic content is high; formula III gives a jujube product. The greater the percentage of gum arabic, the softer and more chewable the candy.

	I	II	III
Sorbitol	88%	64%	15–20%
Mannitol	10%	1–10%	0–5%
Gum arabic	1%	1–35%	70%
(Add flavor, saccharin, color, citric acid as desired.)			

To prepare, add sufficient water to dissolve all the ingredients, then pressure-cook to 193°C (380°F). Cast drops in mold. Cool. The final concentration of water will have to be regulated to a low concentration (5 to 10%) for rapid drying of the candy.

If pressure-cooking equipment is not available, the water content must be adjusted by boiling sorbitol (70% solution) and mannitol to drive off the water, then the gum arabic and other ingredients added before casting. If foam is a problem, a trace of food-grade antifoam material can be used.

Marshmallows[82]   Addition of 1 to 2.5% gum arabic to a marshmallow mix results in greater body and more chewiness. Also, the drying time is reduced considerably, and the production rate is increased. The following is a basic formulation:

	Percent
Gum arabic (powdered, USP)	1–2.5
Gelatin	0.35
Sugar	37.00
Dextrose	19.30
Albumen	1.80
Salt	0.35
Vanilla	0.35–1.00
Corn syrup, 42 D.E.	23.40
Water (hot) as needed to make 100.00%	

To prepare, dry-mix the gum arabic, gelatin, albumen, and salt into the sugar and dextrose. Then add the corn syrup and water mixture. Mix until thick. Add vanilla while mixing. Mold and dry.

Food Emulsions[84]

In many applications, the plant hydrocolloids act as emulsifiers. In other applications, they act as thickeners or stabilizers of emulsions which would otherwise separate. Thus the function of plant hydrocolloids in food emulsions is twofold: (1) they form protective coatings around the particles of dispersed oil, thereby preventing these particles from coalescing and separating from the water phase; (2) they increase the body or viscosity of the emulsion, thereby causing more resistance against rising oil particles.

Gum arabic, being a truly water-soluble and emulsifying gum, renders emulsion stability mainly by the formation of a protective film around each oil particle. Gum tragacanth, unlike gum arabic, is a swelling or suspending gum and depends primarily upon its thickening action to prevent the dispersed oil from coalescing and rising. Other gums used as protective emulsifiers or stabilizers are agar, Irish moss extract, and karaya. In addition to acting as stabilizers, the water-soluble gums must not discolor from age or oxidation, nor change in consistency upon standing, nor adversely affect the properties of the final product.

Flavor emulsions are the most common oil-in-water emulsions prepared for the food industry. Other food emulsions where the vegetable colloids act as protective emulsifiers are salad dressings, mayonnaise, dressings, gravies, mustard, cocktail sauces, spaghetti and other sauces, pickling oils, and whipped foods.

The following flavor formulations are suggested as basic, and are not meant as finished products. They can be varied to suit individual or regional taste requirements. All percentages are by weight.

Pickle Oil Emulsion

	Percent
Gum arabic (powdered)	15–20
Water	55–65
Oil with cassia, clove, pimento, and other spices	20–25

Hydrate the gum before adding the oil and spices. Then homogenize to effect a better emulsion.

Pickle Juice

	Percent
Sugar	35–45
Salt	2–4
Vinegar	2–4
Water to make total of 100%	

Use 0.1 to 0.5% of the pickle oil emulsion in the pickle juice, and allow the pickles to soak for several days before bottling.

Beverages

The following three formulations are suggested as starting points for fruit drinks. The first formulation is for a fruit drink flavored with natural ingredients; the second formulation is for a dry-mix imitation orange drink; and the third formulation is used to give a cloudy look to otherwise clear fruit drinks.

Stabilized Fruit Drink[83]   Following is a suggested basic formulation for a finished and stabilized fruit drink in which the flavoring is derived from citrus oil and citrus juice concentrate. Gum tragacanth is used as a thickening agent, and gum arabic is used as a stabilizer.

	Percent
Gum tragacanth (powdered)	0.04
Gum arabic (powdered)	0.20
Orange oil	0.10
Orange juice	7.00
Citric acid	0.15
Sodium citrate	0.10
Sugar	11.00
Water as needed to make 100.00%	

To prepare, dry-mix the sugar with the gums, citric acid, and sodium citrate, and then disperse the mixture in the water. Then add orange juice and orange oil and mix well.

Dry-Mix Imitation Orange Drink[86]

	Percent
Granular sugar (fine)	85.0
Spray-dried gum arabic	5.0
Citric acid	4.3
Cloud gum	1.9
Gum tragacanth (powdered)	0.8
Sodium citrate	0.7
Vitamin C	0.4
Tricalcium phosphate	0.2
Color:	
Yellow No. 5	0.016
Yellow No. 6	0.004
Flavor	q.s.

To prepare, make a 10% aqueous solution of the two colors. Mix with two parts of alcohol. Spray and intimately mix this solution with the granular sugar. Allow to dry. Then dry-mix the remaining ingredients. Use 123 g (4⅓ oz) of mixture to make 0.95 L (1 qt) of finished beverage.

Cloud Gum[85]   Cloud gum is a combination of gum arabic and vegetable fat, spray-dried to give a free-flowing, uniform product. This material can be dispersed almost instantly in cold water to give a uniform cloudy solution to make imitation fruit drinks more nearly resemble the real fruit juice.

The following is suggested as a basic cloud gum formula:

	Percent
Citric acid	4.00
Tricalcium phosphate	1.50
Orange or lemon flavor	0.10–0.30
Color	0.005–0.01
Cloud gum (gum arabic)	1.50
Sugar as needed to make 100.00%	

To use, put approximately 3 to 4 ml (2 heaping tsp), more or less to taste, of this formulation into 116 g (4 oz) of cold water. Stir and drink.

Beverage Stabilizers[87]

In the manufacture of beer, gum arabic is an ideal foam stabilizer. It gives a creamy, appetizing, and stable foam that resists breakdown in the presence of foam inhibitors. Gum arabic has no other effect on beer, except stabilization of the foam. It does not change in any way the palatability, brilliance, or shelf-life stability of the beer.

For beer, spray-dried gum arabic from which all insolubles have been removed is dissolved in a small amount of water or beer to make a 20% solution of the gum arabic. This solution may be prepared as a stock solution in advance, and then it can be metered into the beer after transfer from the fermenting tanks and prior to aging. The level of use is 8.8 to 13.2 kg/16 m³ of beer (4 to 6 lb/100 bbl).

Nut Coating[89]

For the coating of nuts, an adhesive base is needed. The coating itself may vary from a thin syrup to a heavy chocolate. The nut coating described here is a vegetable gum blend that forms a thin, clear, transparent adhesive film around the nuts and seals in the oil and flavor of the nuts. In addition, this film forms an excellent adhesive base for any further coating that may be applied to the nuts.

To prepare the nut coating, slowly sift 2.3 kg (5 lb) of gum arabic into 3.8 L (1 gal) of cold water while mixing rapidly. The mixing should be continued until the solids are completely in solution; then heat the solution to 71°C (160°F). Allow to cool. This quantity of coating is applied to about 90.7 kg (200 lb) of nuts.

Inks

Gloss-Finish Inks   Decorative inks with a glossy finish for show cards or other display purposes can be prepared by incorporating a dye into the following formula:

Gum arabic	11.3 kg (25 lb)
Liquid soap	3.8 L (1 gal)
Dextrin	45.4 kg (100 lb)
Starch	9.1 kg (20 lb)
D-glucose	18.2 kg (40 lb)
Lysol*	457 mL (0.5 pt)
Water	37.8 L (10 gal)

* Or other preserving agent.

Wood-Grain Inks Inks devised to imitate expensive grains by printing a pattern of wood on metal, paper, stone, or inexpensive woods use gum arabic as a suspension aid. Similarly, wallpaper or metallic objects are frequently decorated with wood-grain patterns.

A typical formula for a wood-grain ink is:

	Percent
Ethylene glycol	60
Water	7.5
Pigment	30
Gum arabic	2.5

LABORATORY TECHNIQUES

30% Viscosity Method[93]

1. If whole gum arabic is used, take a representative sample of at least 200 g. Crush it in a large mortar or in a meat grinder until the entire sample passes through a U.S. Standard Sieve No. 6. (If granular gum arabic is to be tested, use as is.)
2. Weigh a representative sample of 120.0 g of the gum arabic into a 500-mL (about 1 pt) jar.
3. Measure 230 mL of distilled water into a 500-mL graduate cylinder, and add to the gum.
4. Stir the mixture slowly until the gum is dissolved. Avoid the formation of air bubbles. If no stirrer is available, rotate the mixture on a roller for 3 to 4 h until the gum arabic is dissolved.
5. Place solution into a 250°C water bath overnight.
6. After allowing the solution to stand overnight, note the color. Also note the cleanliness of the solution.
7. Invert the container gently several times, and allow to settle for a short time.
8. Measure the viscosity of the unfiltered solution with a Brookfield Viscometer, Model LVF, spindle #3, at 60 rpm and 25°C. Factor is 20.

Insoluble Residue[92]

1. Weigh out a 5.0000-g sample of gum arabic.
2. Add 100 mL of distilled water to a 250-mL Erlenmeyer flask.
3. Add 10 mL of 10% hydrochloric acid, and boil the mixture gently for 15 min.
4. Filter the solution by suction while still hot, using a tared, medium-sintered glass crucible.
5. Wash the captured residue thoroughly with about 200 mL of hot distilled water.
6. Dry the crucible with filter cake at 105°C for 1 hr.
7. Cool in a desiccator; then weigh.
8. The weight of the residue thus obtained shall not exceed 50 mg (1%) if the gum is to pass U.S.P. specifications.

Sediment and Color[91]

Note: Always run a control sample of similar material to the test sample.
1. Weigh out a 20.0-g sample of gum arabic into a 250-mL beaker.
2. Measure 20.0 mL of alcohol into a 25-mL graduate cylinder, then add the alcohol to the gum.
3. Measure 75.0 mL of distilled water into a 100-mL graduate cylinder.
4. With a glass stirring rod, manually mix the alcohol and gum, then add the water

while stirring rapidly.   If lumps form, stir and mash the lumps until they are dissolved.

5. After the gum arabic is completely dissolved, allow the solution to stand for 2 to 5 min, then transfer the solution into a 100-mL centrifuge tube (oil tube ASTM long form, graduated, Fisher Catalog No. 5–605).   During the transfer, scrape the side of the beaker with the stirring rod to remove as much as possible of the mixture.   Wash the remainder out of the beaker and into the centrifuge tube with a few milliliters of distilled water from a wash bottle.

6. Place the centrifuge tube into a suitable centrifuge, and spin the tube for about 15 min at 2000 rpm.

7. The mixture will separate into three layers.   These should be observed as follows:

   a. Upper liquid layer
      (1) Water clear, transparent, almost transparent, slightly hazy, hazy, very hazy.
      (2) Color: colorless, very light tan, light tan, tan, tan-brown.
   b. Upper sediment layer
      (1) Read total amount of upper and lower sediment.
      (2) Color: white, off-white, light tan, tan, tan-brown.
      (3) Transparent to opaque.
   c. Lower sediment layer
      (1) Read amount of lower sediment.
      (2) Color: off-white, light tan, tan, tan-brown, black.

Peroxidase Content[90]

The presence of trace amounts of the enzyme peroxidase in gum arabic is determined by the blue color reaction of a gum arabic solution with benzidine [p,p'-diaminobiphenyl; $(C_6H_4NH_2)_2$].

The test solution used is hydrogen peroxide T.S. (2.5 to 3.5 g of H_2O_2 in 100 mL water).   Dilute 1 mL of concentrated H_2O_2 to 10 mL total benzidine Test Solution USP (1 g benzidine to 50 mL water and 5 mL concentrated hydrochloric acid).   Solution of ingredients should be complete or nearly complete.

Add five drops of the hydrogen peroxide solution and five drops of benzidine to 5 mL of a 1% gum arabic solution.   Observe for a few minutes, utilizing a white background.   The presence of peroxidase is indicated by the development of a blue color.

REFERENCES

1. Acheson, E. G., U.S. Patent 964,478 (1910); *Chem. Abstr.*, **4**, 2717 (1910).
2. Adams, D. N., U.S. Patent 2,694,057 (1954); *Chem. Abstr.*, **49**, 9310 (1955).
3. Alexander, J., U.S. Patent 1,259,708 (1918); *Chem. Abstr.*, **4**, 1502 (1918).
4. Ali, A., B. Kahn, and B. Ahmad, *Pak. J. Sci. Res.*, **6**, 58 (1954).
5. Anderson, D. M. W., I. C. M. Dea, and K. A. Karamalla, *Carbohydr. Res.*, **6**, 97 (1968).
6. Anderson, H. H., F. Murayama, and B. E. Abreu, *Pharmacology and Experimental Therapeutics*, Univ. of Calif. Press, Los Angeles, 1947, p. 1.
7. Andon, A. J., *Drug & Cosmetics Ind.*, **79**, 762 (1956).
8. Anonymous, French Patent 1,002,540 (1952); *Chem. Abstr.*, **51**, 4606 (1957).
9. Anonymous, *J. Am. Pharm. Assoc. Pract. Pharm. Ed.*, **8**, 83 (1947).
10. Anon., *United States Pharmacopeia*, 15th ed., Mack Publishing Co., Easton, Pa., 1955.
11. Arauji, O. E., *J. Pharm. Sci.*, **55**, 636 (1966).
12. Badar-Ud-Din, *Pak. J. Sci. Res.*, **2**, 28 (1950).
13. Balassa, L. L., and J. Brody, *Food Eng.*, **40** (11) (1968).
14. Baron, L., U.S. Patent 1,621,186 (1927); *Chem. Abstr.*, **21**, 1523 (1927).
15. Bennett, H. (ed.), *The Chemical Formulary*, vol. 1, Chemical Publishing Company, Inc., New York, 1933, pp. 7, 10.
16. Bennett, H. (ed.), *The Chemical Formulary*, vol. 4, Chemical Publishing Company, Inc., New York, 1939, pp. 19, 24.
17. Bennett, H. (ed.), *The Chemical Formulary*, vol. 5, Chemical Publishing Company, Inc., New York, 1941, p. 17.
18. Beutaric, A., and M. Roy, *Bull. Soc. Chim. Fr.*, **6**, 316 (1939); *Chem. Abstr.*, **33**, 3658 (1939).
19. Booth, A. N., C. A. Elvehyem, and E. B. Hart, "The Importance of Bulk in the Nutrition of the Guinea Pig," *J. Nutr.*, **32**(2), 263–274 (1949).
20. Booth, A. N., A. P. Hendrickson, and F. DeEds, "Physiologic Effects of Three Microbial Polysaccharides on Rats," *Toxicol. Appl. Pharmacol.*, **5**, 478–484 (1963).
21. Briggs, D. R., *J. Phys. Chem.*, **38**, 867 (1954).

22. Brown, C. L., and E. A. Lum, *Pharm. J.*, 131, 341 (1933).
23. Bungenberg de Jong, H. G., E. G. Hoskam, and B. H. van den Brandhof-Schlaegen, *Proc. K. Ned. Akad. Wet.*, 44, 1104 (1941); *Chem. Zentralbl.*, I, 2969 (1942).
24. Charmandarian, M., and N. M. Andronikova, *Bull. Soc. Chim. Fr.*, 99 (1952).
25. Chemische Fabrik Schweizerhall, Swiss Patent 307,867 (1954); *Chem. Abstr.*, 51, 3937 (1957).
26. Clark, C. T., U.S. Patent 2,620,279 (1952); *Chem. Abstr.*, 47, 2454 (1953).
27. Colegate, G. T., *Metallurgia*, 39, 316 (1948).
28. Constance, J. A., U.S. Patent 2,676,119 (1954); *Chem. Abstr.*, 48, 14208 (1954).
29. Debruyne, N. A., and R. Houwink, *Adhesion and Adhesives*, Elsevier Publishing Company, New York (1951).
30. Delmonte, J., *The Technology of Adhesives*, Reinhold Publishing Corporation, New York, 1947, p. 291.
31. Deuel, H., and J. Solms, *Kolloid-Z.*, 124, 65 (1951).
32. Dixon, H. W. A., and R. S. Moore, U.S. Patent 2,022,276 (1935); *Chem. Abstr.* 30, 792 (1936).
33. Dumas, J., *Ann. Chim. Anal.*, 25, 214 (1943).
34. Dyer, E., *Textile Fabrics*, Houghton Mifflin Company, Boston, 1923, p. 62.
35. Eichorn, A., U.S. Patent 2,801,919 (1957); *Chem. Abstr.*, 51, 15317 (1957).
36. Eisman, P. C., J. Cooper, and D. Jaconia, *J. Amer. Pharm. Assoc. Sci. Ed.*, 46, 144 (1957).
37. Ellis, C., *Printing Inks*, Reinhold Publishing Corporation, New York, 1940, pp. 230, 334, 346, 398, 399, 417.
38. Feinberg, M., and B. B. Schoenkerman, *Wis. Med. J.*, 39, 734 (1940).
39. Fisher, E. K., *Colloidal Dispersions*, John Wiley & Sons, Inc., New York, 1950, p. 264.
40. Fordyce, C. R., and J. Emerson, Brit. Patent 554,758 (1943); *Chem. Abstr.*, 39, 467 (1945).
41. Frohberg, H., H. Oettel, and H. Zeller, "On the Mechanism of the Fetotoxic Effect of Tragacanth," *Arch. Toxicol.*, 25, 268–295 (1969).
42. Gamble, L. D., and D. L. Grady, U.S. Patent 2,135,936 (1938); *Chem. Abstr.*, 33, 1524 (1939).
43. Gelfand, H. H., "The Allergenic Properties of the Vegetable Gums," *J. Allergy*, 14(3):203–219 (1943).
44. Gelfand, H. H., "The Vegetable Gums by Ingestion in the Etiology of Allergenic Disorders," *J. Allergy*, 20(5), 311–321 (1949).
45. Gerding, P. W., and G. D. Sperandio, "Drug Standards," Soc. pour l'ind. chim. à Bâle, 21, 215 (1953).
46. Green, B. K., and L. Schleicher, U.S. Patent 2,800,457 (1957); *Chem. Abstr.*, 51, 15842 (1957).
47. Greenberg, L. A., and D. Lester, *Handbook of Cosmetic Materials*, Interscience Publishers, Inc., New York, 1954, p. 19.
48. Gutheil, N. C., and M. Formoso, *An. Assoc. Quim. Bras.* 10, 335 (1951).
49. Haller, R., and B. Frankfurt, *Kolloid-Z.*, 80, 68 (1937).
50. Harry, R. G., *Modern Cosmeticology*, pp. 56, 287, 314, 316, Chemical Publishing Company, Inc., New York (1947).
51. Helfrich, J. H., U.S. Patent 1,655,369 (1929); *Chem. Abstr.*, 22, 1017 (1929).
52. Herbig, J. A., in Kirk and Othmer (eds.), *Encyclopedia of Chemical Technology*, 2d ed., Vol. 13, 1967, p. 436.
53. Hodgman, C. D. (ed.), *Handbook of Chemistry and Physics*, 35th ed., Chemical Rubber Co., Cleveland, 1953, p. 2984.
54. Hoeffler, J., U.S. Patent 1,660,196 (1928); *Chem. Abstr.*, 22, 1486 (1928).
55. Horn, E. N., and J. Sanko, U.S. Patent 2,651,583 (1953); *Chem. Abstr.*, 48, 339 (1954).
56. Husa, W. J., and C. H. Becker, *J. Am. Pharm. Assoc. Sci. Ed.*, 30, 171 (1941).
57. Ishida, K., Brit. Patent 194,156 (1922); *Chem. Abstr.*, 17, 3615 (1923).
58. Jacobs, M. B., *Amer. Perfum Estent. Oil Rev.*, 54, 54 (1949).
59. Johnson, H. M., U. S. Patent 99,907 (1870).
60. Johnstone, C., *Manuf. Confect.*, 19, 14 (1939).
61. Kedvessy, G., *Ber. Ungar. Pharm. Ges.*, 17, 607 (1941); *Chem. Abstr.*, 37, 4531 (1943).
62. Kieft, J. P., *Pharm. Weekbl.*, 76, 1133 (1939).
63. Knapp, T., *Am. Ceram. Soc. Bull.*, 33, 11 (1954).
64. Knight, W. A., *Pharm. J.*, 121, 297 (1928).
65. Krantz, J. C., and N. E. Gordon, *J. Am. Pharm. Assoc.*, 15, 93 (1926).
66. Krantz, J. C., and N. E. Gordon, *J. Am. Pharm. Assoc. Sci. Ed.*, 18, 463 (1929).
67. Langwill, K. E., *Manuf. Confect.*, 19, 37 (1939).
68. Lotzkar, H., and W. D. Maclay, *Ind. Eng. Chem.*, 35, 1294 (1943).
69. Maas, J. A., *Q. J. Exp. Physiol.*, 28, 315 (1938).
70. Mack, M. J., *Ice Cream Trade J.*, 32, 33 (1936).
71. Mantell, C. L., "Natural Plant Hydrocolloids," *Adv. Chem. Ser.*, 11, 20 (1954).
72. Mantell, C. L., *The Water-Soluble Gums*, Reinhold Publishing Corporation, New York, 1947.
73. Marsh, J. T., *An Introduction to Textile Finishing*, John Wiley & Sons, Inc., New York, 1943, p. 23.
74. Mason, C. F., *Chem. Ind.*, 53, 630 (1943).
75. Matsudaira, T., Japanese Patent 2997 (1953); *Chem. Abstr.* 48, 7840 (1954).

76. Mattiello, J. J., *Protective and Decorative Coatings,* vol. 4, John Wiley & Sons, Inc., New York, 1944, p. 311.
77. Maytum, C. K., and T. B. Magath, *J. Am. Med. Assoc.,* 99, 2251 (1932).
78. *Merck Index,* 8th ed., Merck & Co., Rahway, N.J., 1968, pp. 2, 3.
79. Mertle, J. S., *Graphic Arts Mon.,* 9, 32 (1937).
80. Moorjani, M. N., and C. S. Narwani, *J. Indian Chem. Soc.,* 25, 503 (1948).
81. Mukoseev, N. A., *Tekst. Prom.,* 23 (1943).
82. Meer Corp., Technical Bulletin No. F-2.
83. Meer Corp., Technical Bulletin No. F-19.
84. Meer Corp., Technical Bulletin No. F-21.
85. Meer Corp., Technical Bulletin No. F-22.
86. Meer Corp., Technical Bulletin No. F-23.
87. Meer Corp., Technical Bulletin No. F-25.
88. Meer Corp., Technical Bulletin No. F-27.
89. Meer Corp., Technical Bulletin No. F-31.
90. Meer Corp., Technical Bulletin No. G-60.
91. Meer Corp., Technical Bulletin No. G-61.
92. Meer Corp., Technical Bulletin No. G-63.
93. Meer Corp., Technical Bulletin No. G-64.
94. Naito, K., and T. Mori, *Nippon Nogei Kagaku Kaishi,* 27, 268 (1953); *Chem. Abstr.* 49, 8519 (1955).
95. Neal, R., C. F. Bailey, and R. S. Casy, *J. Chem. Educ.,* 24, 429 (1947).
96. Neergaard, K. van, Brit. Patent 218,323 (1923); *Chem. Abstr.,* 19, 560 (1925).
97. Nitardy, F. W., F. F. Berg, and P. Georgi, U.S. Patent 1,913,561 (1933); *Chem. Abstr.* 25, 3130 (1931).
98. Nitardy, F. W., F. F. Berg, and P. Georgi, U.S. Patent 1,913,561 (1933); *Chem. Abstr.* 27, 4348 (1933).
99. Norwitz, G., U.S. Patent 2,680,711 (1954): *Chem. Abstr.,* 48, 9240 (1954).
100. Osborne, G. E., and H. G. DeKay, *J. Am. Pharm. Assoc. Pract. Pharm. Ed.,* 2, 420 (1941).
101. Osborne, G. E., and C. O. Lee, *Drug Stand.,* 19, 13 (1951).
102. Osol, A., and G. E. Farrar, *The Dispensatory of the United States,* 25th ed., J. B. Lippincott Company, Philadelphia, 1955, p. 1.
103. Pfeiffer, C. C., and H. L. Williams, *J. Am. Pharm. Assoc. Pract. Pharm. Ed.,* 8, 572 (1947).
104. Poschel, A. B., British Patent 393,132 (1933); *Chem. Abstr.,* 27, 5553 (1933).
105. Pyenson, H., and C. D. Dahle, *J. Dairy Sci.,* 21, 169 (1938).
106. Reed, R. F., *Mod. Lithogr.,* 47, 62 (1951).
107. Revie, C. N., and B. R. J. Thomas, *Food Manuf.,* 37, 40 (1962).
108. Rochat, J., Swiss Patent 153,496 (1931); *Chem. Abstr.,* 27, 1216 (1933).
109. Schleif, R. H., T. Higuchi, and L. W. Busse, *J. Am. Pharm. Assoc. Sci. Ed.,* 40, 98 (1951).
110. Schleif, R. H., T. Higuchi, and L. W. Busse, *J. Am. Pharm. Assoc. Sci. Ed.,* 40, 221 (1951).
111. Schlemmer, J., *Chemie (Prague),* 3, 73 (1948); *Chem. Abstr.,* 46, 2888 (1952).
112. Scholz, L. A., U.S. Patent 2,568,369 (1951); *Chem. Abstr.,* 45, 10433 (1951).
113. Schwarzmann, A., Swiss Patent 241,167 (1946); *Chem. Abstr.* 43, 8104 (1949).
114. Shue, G. M., C. D. Douglass, and L. Friedman, "Nutritional Studies of Complex Carbohydrates," *Red. Proc.,* 21:91 (1962).
115. Smith, F., *J. Chem. Soc.,* 744 (1939).
116. Smith, P. I., *Am. Perfum. Aromat.,* 67, 67 (1956).
117. Societe de Recherches et d'Enterprises Industrielles et Chimiques, Swiss Patent 225,886 (1943); *Chem. Abstr.,* 43, 2375 (1949).
118. Strauss, H. J., U.S. Patent 2,692,904 (1954); *Chem. Abstr.,* 49, 2222 (1955).
119. Summer, C. G., *Clayton's Theory of Emulsion and Their Technical Treatment,* 5th ed., Churchill, London, 1954, p. 425.
120. Suzuki, K., Japanese Patent 109,595 (1935); *Chem. Abstr.,* 29, 4958 (1935).
121. Swintosky, J. V., L. Kennon, and J. Tingstad, *J. Am. Pharm. Assoc. Sci. Ed.,* 44, 109 (1955).
122. Taft, R., and L. E. Malm, *J. Phys. Chem.,* 35, 874 (1931).
123. Thomas, A. W., and H. A. Murray, *J. Phys. Chem.,* 32, 676 (1928).
124. Tieback, F. W., *Pharm. Weekbl.,* 59, 547 (1922); *Chem. Abstr.,* 16, 2433 (1922).
125. Thompson, H., Compagnie Francaise pour l'exploitation des Procedes, French Patent 692,757 (1930); *Chem. Abstr.,* 25, 1940 (1931).
126. Veis, A., and D. N. Eggenberger, *J. Am. Chem. Soc.,* 76, 1560 (1954).
127. Wagner, W. W., U.S. Patent 2,682,472 (1954); *Chem. Abstr.,* 48, 10256 (1954).
128. Walder, W. O., *Food,* 18, 86 (1949).
129. Warburton, B., *The Chemistry and Rheology of Water Soluble Gums and Colloids,* Soc. Chem. Ind. Monograph (1966), pp. 24, 118.
130. Waters, C. E., *Circular No. C426,* National Bureau of Standards, 1940, pp. 3, 33, 34, 45, 53.
131. Wenneis, J. M., *Proc. Flavor, Extract Manuf. Assoc.,* 47, 91 (1956).
132. Werbin, S. J., *Baker's Dig.,* 27, 21 (1953).

133. Whistler, R. L., and C. I. Smart, *Polysaccharide Chemistry,* Academic Press, Inc., New York, 1953.
134. Williams, C. T., *Confect. Manuf.,* **6,** 299 (1961).
135. Wilson, W. K., U.S. Patent 2,508,343 (1950); *Chem. Abstr.,* **44,** 7583 (1950).
136. Wolfe, H. F., U.S. Patent 1,983,650 (1934); *Chem. Abstr.,* **29,** 857 (1935).
137. Wood, W. H., U.S. Patent 2,666,759 (1954); *Chem. Abstr.,* **48,** 3716 (1954).

Chapter **9**

Gum Ghatti

George Meer
Meer Corporation

General Information 9-1
   Chemical Nature 9-2
   Physical Properties 9-2
   Manufacture 9-4
   Biological/Toxicological Properties 9-5
   Handling 9-5
   Application Procedures 9-5
Commercial Uses 9-5
Industries Using Gum Ghatti 9-6
Formulations 9-6
   Wax Emulsion 9-6
   Table Syrup Emulsion 9-6
Laboratory Techniques 9-7
   Bark and Foreign Organic Matter (BFOM) 9-7
   Viscosity (5% Solution) 9-7
   Viscosity (7½% Solution) 9-7
Product Glossary 9-8
References 9-8

GENERAL INFORMATION

Gum ghatti (also known as Indian gum) is a complex water-soluble polysaccharide. It is a plant exudate that has long been in use and whose name is derived from the word *ghats*, which means "passes." The name was given to the gum because of its ancient mountain transportation routes. Gum ghatti is approved for food use and is in the GRAS (Generally Recognized As Safe) list under the Federal Food, Drug, and Cosmetic Act. This gum has a unique commercial position because of its viscosity, intermediate between those of gum arabic and gum karaya, and because it is an excellent emulsifier for difficult-to-handle mixtures.

Chemical Nature

Gum ghatti occurs in nature as a calcium-magnesium salt. Asphinall and coworkers[7] have shown that it is composed of L-arabinose, D-galactose, D-mannose, D-xylose, and D-glucuronic acid in a molar ratio of 10:6:2:1:2, with traces (less than 1%) of 6-deoxyhexose. Partial hydrolysis gives two aldobiouronic acids, namely 6-O-(B-D-glucopyranosyluronic acid)-D-galactose, and 2-O-(β-D-glucopyranosyluronic acid)-D-mannose.[7] A previous investigation reported the presence of 50% pentose and 12% galactose or galacturonic acid.[1,10,17]

Further work[2] has shown that the gum contains a backbone of 1,6-linked β-D-galactopyranosyl units, and that partial acid hydrolysis[6] produces two homologous series of oligosaccharides together with small amounts of 3-O-β-D-galactopyransoyl-D-galactose and 2-O-(β-D-glucopyranosyluronic acid)-D-mannose.[4] Acid-labile side chains are attached to the backbone through L-arabinofuranose residues.[3,5]

Apparently, gum ghatti is similar in structure to *Anogeissus leiocarpus* gum which contains two distinct polysaccharides, one a major component and one a minor component, that can be separated from each other.[14] Partial structures for the interior[20] and exterior[19] chains of the major fraction have been proposed.

In solution, the molecules may have an overall rod shape.[34] The heterogeneity of the polysaccharide is revealed by electrophoresis on glass-fiber paper[8] as well as on DEAE-cellulose.[30] A 1% dispersion of the gum has a pH of 2.63 and no buffering activity.[26] The equivalent weight[26] is 1340 to 1735. Riboflavin and thiamine are present in traces.[32]

Table 9.1 indicates the interaction of certain reagents with solutions of gum ghatti.

TABLE 9.1
Character of Precipitates from Gum Ghatti Solutions

Reagent	Precipitate
Milion's reagent	Fine
Lead acetate, 20% solution	None
Basic lead acetate (A.O.A.C.)	Translucent, flocculent
Potassium hydroxide, 10% solution	None
Ferric chloride, 5% solution	None
Ethanol, 4 vol/vol water	Fine, flocculent, nonadherent
Sodium tetraborate, 4% solution	None
Stokes acidic mercuric nitrate reagent	None
Cationic soak, 1% solution	Fine
Saturated ammonium sulfate	None
Papain, 2% solution	None
Gelatin, 2% solution	Precipitate

Physical Properties

Gum ghatti forms viscous dispersions in water when in concentrations of 5% or greater. The dispersions are thixotropic and non-Newtonian in behavior, as is true with most of the water-soluble gums, and viscosity increases geometrically with concentration (see Table 9.2). The dispersions are less viscous than those of gum karaya, although they are more viscous than those of gum arabic, and they do not have shear stability. A higher apparent viscosity can be obtained by dispersing the gum in 25% ethanol or

TABLE 9.2
Viscosity of Gum Ghatti Solutions Measured at 25°C with a Type LVF Brookfield Viscometer at 30 rpm

Concentration	Viscosity	
	Pa·s	cP
1%	0.002	2
2%	0.035	35
5%	0.288	288
7.5%	1.012	1012
10%	2.444	2444

by increasing the pH to about 8.0 (see Fig. 9.1).   A dispersion in 50% ethanol does not hydrate completely.   Gum ghatti is a good emulsifying agent and can emulsify more difficult systems than can gum arabic.   Aging of gum ghatti dispersions increases their viscosity (see Table 9.3).

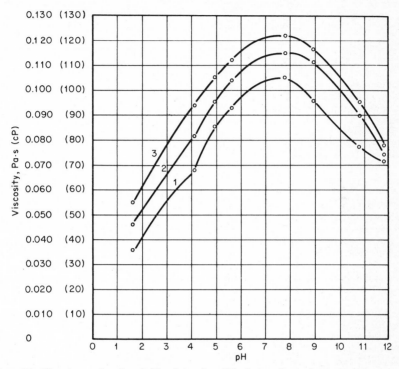

Figure 9.1   Viscosity as a function of pH and time for a 5% solution of gum ghatti; viscosity measured at 25°C and 30 rpm with a Type LVF Brookfield viscometer.

Gum ghatti solutions may be slightly colored because of traces of pigment remaining in the gum.   The normal pH of the dispersions is 4.8.   There may be more incompletely dissolved material than observed in dispersions of either karaya or arabic gums.   Insoluble matter can be removed from gum ghatti by filtering its aqueous solutions, and spray-dried gum ghatti in which no insolubles are present is currently available.   The

TABLE 9.3
Viscosity of 5% Gum Ghatti Solutions Measured at 25°C with a Type LVF Brookfield Viscometer

	Viscosity					
	1 day		1 week		2 weeks	
pH	Pa·s	cP	Pa·s	cP	Pa·s	cP
1.6	0.036	36	0.046	46	0.055	55
4.1	0.068	68	0.082	82	0.094	94
4.9*	0.086	86	0.096	96	0.105	105
5.6	0.093	93	0.104	104	0.112	112
7.8	0.105	105	0.115	115	0.122	122
8.9	0.095	95	0.112	112	0.116	116
10.8	0.077	77	0.090	90	0.095	95
11.8	0.072	72	0.074	74	0.078	78

* Control sample.

viscosity of the spray-dried gum is somewhat lower than that of natural dry-milled gum.

The adhesiveness of gum ghatti dispersions is similar to that of gum arabic. Because of its higher viscosity, it is not possible to prepare dispersions as concentrated as can be prepared with gum arabic. Gum ghatti does not form a true gel. Films prepared from gum ghatti dispersions are relatively soluble and brittle. Gum ghatti has good emulsifying properties, which serve as the basis for most of its applications.

Manufacture

Gum ghatti is an amorphous, translucent exudate of the *Anogeissus latifolia* tree of the Combretaceae family. The tree is quite large and is found abundantly in the dry, deciduous forests of India and, to a lesser extent, Ceylon. It has a gray bark and red leaves during the dry season. In addition to its use as a source of gum ghatti, the tree is widely used for timber, and tannin is extracted from its leaves. The gum has a glassy fracture and frequently occurs in rounded tears, which are normally less than 1 cm in diameter, but it more often occurs in larger, vermiform masses.

The color of the exudate varies from very light to dark brown; the lighter the color, the better the quality and grade. Factors that determine the color are the proximity of the tear to the bark and the length of time it has remained on the tree before being picked, because tannins from the bark darken the gum. Airborne impurities also adhere to the gum while it is still plastic. Artificial incisions can be made in the tree bark to increase the yield; however, these incisions must be well planned so as not to destroy the tree. The gum is thought to act as a sealant when the bark is damaged.

Harvesting and grading of gum ghatti are done by methods similar to those used with gum karaya, near which it grows. The tonnage of gum karaya imported into the United States, however, far exceeds that of gum ghatti.

After picking, the gum is dried in the sun for several days. It is moved to a classification and storage depot where it is hand-sorted according to color and amount of impurity. The grades are then placed in burlap bags and exported. Normally, three grades are exported to the United States: No. 1, No. 2, and unassorted. Number 1 grade is of the lightest color (off-white to buff); No. 2 grade is light amber to brown; and the unassorted grade is dark brown to nearly black.

The better the grade, the more effective is the gum and the less the amount of impurities. The best crop is obtained in the dry season, and normally the largest crop is picked in April. Table 9.4 shows average impurities (bark and sand), total ash, acid-insoluble ash, and moisture content of the three grades, determined according to methods described in Laboratory Techniques.[24]

TABLE 9.4
Commercial Grades of Gum Ghatti as Imported into the United States

Grade	Impurities, %	Total ash, %	Acid-insoluble ash, %	Viscosity (5%), Pa·s (cP)		Moisture, %
No. 1	0.9–1.6	1.4–1.9	0.02–0.2	0.30–0.400	(30–400)	12–15
No. 2	1.4–3.6	2.2–3.9	0.2 –1.0	0.30–0.350	(30–350)	12–15
Unassorted	11.0–15.0	6.0–10.0	3.7 –5.8	0.30–0.300	(30–300)	12–15

Ghatti tears are processed in the United States. This processing is mainly a grinding operation in which the gum is pulverized to a fine powder. However, various other mesh separations can also be made to satisfy the demands of the consumer. During the process of particle breakdown, impurities are removed from the gum by sifting, aspiration, and density-table separation. From 1967 through 1971, between 450 and 600 tons of ghatti gum were imported into the United States per year. The price has varied from $0.78 to $1.51 per kilogram ($0.35 to $0.68 a pound), depending on the grade and the market conditions at the time of purchase.[25] It is estimated that approximately 900 to 1150 tons are available on a worldwide basis.

The amount of gum harvested depends on the economic conditions in the region in which the trees grow. If other crops pay a higher price for harvesting, the natives naturally prefer to harvest these, and ghatti is neglected. There are large areas in which the trees are as yet uncultivated, but which could supply gum if the demand should justify it.

It is well known to dealers of gum ghatti that gums of a different botanical origin are sometimes collected as ghatti and sold as such by collectors.

Ghatti has a bland taste and practically no odor. Only about 90% of the gum disperses in water, and this portion forms a colloidal dispersion.

Biological/Toxicological Properties

Gum ghatti is GRAS, as mentioned earlier in this chapter, and is on the FDA list. It is not toxic, and it is not metabolized in humans.

Handling

Handling recommendations include warehousing in a cool, dry place for extended storage. If the gum becomes damp, it tends to agglomerate and form lumps.

Application Procedures

There are no special procedures to be followed for dry-blending gum ghatti, except that the mix should not be allowed to become moist or damp. In preparing sols, cold water should always be used for easier dispersion and to avoid lumping. Ghatti sols are generally compatible with other water-soluble gums and starches and related products. Ghatti is insoluble in organic solvents.

COMMERCIAL USES

Gum ghatti is used in applications also served by gum arabic. Ghatti is often used in pharmaceutical preparations as an emulsifying agent. In the United States, ghatti is used in the preparation of stable, powdered, oil-soluble vitamins.[13] Gum ghatti has been used in combination with proteins as a means of stimulating the formation of eosinophils.[11]

Another use for gum ghatti is to stabilize table syrup emulsions containing about 2% butter. In such an application, about 0.4% ghatti is used in combination with 0.08% lecithin.[33,36] The refractive index of table syrup containing emulsified butter may be adjusted by additional quantities of ghatti to produce clarity.

Used at low concentrations, gum ghatti prevents fluid loss in oil-well drilling muds in neutral mixtures with high salt concentrations.[15] The gum also prevents fluid loss at elevated temperatures, which leads to its use in the maintenance of thin wall-cakes in oil wells where high temperatures occur.[28,35] In dispersions at high pH, however, gum ghatti is not as effective as are special starches.

Oil-well acidizing is another use for gum ghatti. The gum is moistened with a water-insoluble nonaqueous liquid that is inert both to the gum and to the acid solution. Then acid is added with mixing to form a uniform dispersion, which is pumped under pressure to permeate the earth formation. This results in enlarged passageways, or worm holes, which increase the productivity of the well. The drilling mud and other fracture-clogging deposits are washed out, and the oil flows freely.[9]

Gum ghatti is used to emulsify petroleum and nonpetroleum waxes to form liquid and wax paste emulsions which find wide use in the paper industry as coatings and as moisture barriers. Powdered gum ghatti is used in ammonium nitrate–semigelatin mixtures and powdered explosives to improve their resistance to water damage.[21] Gum ghatti is used in old and new varnishes[22] and for emulsifying oils, such as 40% kerosene oil,[23] and other technical emulsions. It acts as a stabilizer in auto polishes and wallpaper gum sizings.

Gum ghatti has been used in combination with polyacrylamide to aid in the polymerization and formation of uniform and discrete prills of crosslinked polystyrene.[12] Salts of ghatti acid have reportedly been used for light-sensitive papers, pigments, and fungicides.[18] Because of its high L-arabinose content, gum ghatti is hydrolyzed to prepare pure L-arabinose on a commercial scale.[29] L-arabinose is used as a flavor adjunct in food products and in the preparation of nucleosides used as antitumor drugs.

Another application of gum ghatti is to stabilize the Prussian blue color in spectrophotometric determinations,[16,27] and it is used in the polarographic determination of copper, lead, and iron.[31] It shows some promise in forming protective hydration layers around clay particles, thus maintaining a dispersion for the particle-size analysis of soils.[37]

INDUSTRIES USING GUM GHATTI

As mentioned earlier in this chapter, gum ghatti is particularly effective as an emulsifier for difficult-to-handle materials. This property is the basis for many of its commercial applications. A summary of these applications follows:

- ▪ *Food Products:*
  Table syrup emulsions
- ▪ *Petroleum and Mining:*
  1. Drilling muds
  2. Oil-well acidizing
  3. Emulsification of waxes
- ▪ *Miscellaneous:*
  1. Polymerization aid in forming prills of polystyrene
  2. Aid in analytical procedures:
     a. As a stabilizer for Prussian blue in photoelectric determinations
     b. As protective hydration layers around clay particles for soil analysis
  3. As a source for production of arabinose

FORMULATIONS

Wax Emulsion[21]

	Percent by weight
Gum ghatti	1–5
Wax (petroleum)	40–50
Water	59–45

To avoid lumping, add the gum ghatti slowly with sifting into well-agitated water. Heat the resulting gum solution to about 20°C above the melting point of the wax; remove any sediment by settling, centrifugation, or filtration. Then add the wax to the clarified gum solution, and maintain the temperature above the melting point of the wax to keep the wax melted.

Homogenize the hot mixture to produce a thick but free-flowing emulsion having a viscosity of about 0.4 to 1.0 Pa·s (400 to 1000 cP) at 25°C and 30 rpm (Brookfield viscometer, type LVF).

Table Syrup Emulsion (Patented)[36]

	Percent by weight
Cane syrup (67°Brix)	60.0
Cane syrup–corn syrup blend (67°Brix)	33.0
Butter	2.0
Maple syrup	2.0
Gum ghatti	0.40
Lecithin	0.08
Trisodium citrate	0.06
Citric acid	0.02
Water	2.44

Add the gum ghatti to cane syrup at 93°C (200°F) with the aid of a Venturi, and hold for 60 min at 77°C (170°F), with constant agitation. The cane syrup has a sugar solids content of 67°Brix and is in a quantity sufficient to constitute 60% of the finished product. Add gum ghatti in an amount equal to 0.40% of the finished product. Form another syrup separately by mixing and blending for 5 min a cane syrup–corn syrup blend, maple syrup, citric acid, and sodium citrate. The cane–corn syrup blend is at 73°Brix and amounts to about 33% of the finished product. The amounts of maple syrup, citric acid, and sodium citrate constitutes 2.0, 0.02, and 0.06% by weight of the finished product. Blend the two syrups together and adjust the sugar solids to 67°Brix by the addition of water.

This resulting syrup constitutes the aqueous phase of the finished product; hold this syrup at a temperature of 77°C (170°F). Next, add 2% butter and 0.08% lecithin to the aqueous phase, and blend the oil and aqueous phases for 5 min at a temperature of 77°C (170°F). Pass the blended syrup into a Manton-Gaulin homogenizer at 77°C (170°F) with a first-stage pressure of 27.64 MPa (4000 psig) and a second-stage pressure of 3.46 MPa (500 psig). The diameter of each oil globule, by which is meant the oil droplet, its immediately surrounding and adhering aqueous phase, and the interface zone between the droplet and its aqueous phase, is less than 2 μm. The buttered syrup so produced has a pH of approximately 5.1. Bottle and cap it. It is ready for use.

LABORATORY TECHNIQUES

The laboratory tests of primary importance to the users of gum ghatti are: total ash (TA); acid insoluble ash (AIA); bark and foreign organic matter (BFOM); viscosity; color (dry, visual); and pH (5% solution). Refer to the *Food Chemicals Codex* for details on methods.

Bark and Foreign Organic Matter (BFOM)

1. Use 10-g samples on crude whole lots; use 5-g samples on finished powders.
2. Quarter and grind a representative portion of the whole gum to about −80 mesh.
3. Accurately weigh 10.0000 g of the laboratory-powdered crude gum.
4. Add 50 mL of 5% hydrochloric acid (prepared by diluting 117 mL of 37.7% hydrochloric acid, 1.191 sp gr, to 1000 mL) to a 250-mL Erlenmeyer flask.
5. Add the 10.0000 g of gum to the aqueous hydrochloric acid in the flask.
6. Connect the flask to a water-cooled reflux condenser, clamp the Erlenmeyer flask ½ in above the wire gauze, and heat with a low gas flame. After the gum has been dissolved, reflux the mixture for 30 min.
7. Filter the hot solution through a tared sintered glass crucible of medium porosity.
8. Wash the residue with water until the washings are free from acid (approximately 200 mL of hot water is necessary).
9. Dry the crucible and residue to constant weight at 105°C for 2 h.
10. Calculate percent dried residue in the sample of gum taken:

$$\% \text{ BFOM} = \left[\frac{\text{weight of dry residue}}{\text{sample weight}} / \right] \times 100$$

Viscosity (5% Solution)[22]

1. Weigh accurately 10.0000 g of gum on an analytical balance.
2. Measure 190 mL of distilled water into a 250-mL graduate.
3. Add half of the water to a 250-mL milk reactivity jar, then add the gum and the rest of the water.
4. Cap the jar and shake it vigorously. Continue to shake periodically throughout the day, or until the gum is in solution. If the sample is too viscous, mechanical agitation may be used to speed the solution of the gum into the water.
5. Place the gum solution into a 25°C water bath overnight.
6. Observe the gum solution for color, odor, bark, dirt or fiber, foam or crust, and sediment, if any.
7. Measure viscosity using a type LVF Brookfield viscometer at spindle #2 and 30 rpm. Also use NCV pipette. Check pH.

Viscosity (7½% Solution)

1. Weigh accurately 15.0000 g of gum on an analytical balance.
2. Measure 185 mL of distilled water.
3. Procedure from here on is exactly as for the 5% solution viscosity, except for the elimination of the NCV pipette and the replacement of spindle with #3 at 60 rpm, instead of #2 at 30 rpm.

PRODUCT GLOSSARY

The following list is for Meer Corporation grades of gum ghatti:

Powdered Ghatti, Nos. 1, 2, 1M
Stafome P (Spray-dried Gum Ghatti, Industrial Use)[23]
Stafome P-1 (Spray-dried Gum Ghatti, Pharmaceutical Use)
Stafome P-2 (Spray-dried Gum Ghatti, Food Use)
Powdered Gum Ghatti Technical

REFERENCES

1. Aspinall, G. O., B. J. Auret, and E. L. Hirst (Univ. Edinburg, Scot.), *J. Chem. Soc.*, 221–230 (1958).
2. Aspinall, G. O., B. J. Auret, and E. L. Hirst (Univ. Edinburg, Scot.), *J. Chem. Soc.*, 4408–4414 (1958).
3. Aspinall, G. O., V. P. Bhavanandan, and J. B. Christensen (Univ. of Edinburg, Scot.), *J. Chem. Soc.*, 2677–2684 (1965).
4. Aspinall, G. O., J. J. Carlyle, J. M. McNab, and A. Rudowski (Univ. of Edinburg, Scot.), *J. Chem. Soc.* **5**, 840–845 (C-1969); G. O. Aspinall and J. M. McNab (Univ. of Edinburg, Scot.), *J. Chem. Soc.*, 845–851 (C-1969); G. O. Aspinall and J. J. Carlyle (Univ. of Edinburg, Scot.), *J. Chem. Soc.*, 851–856 (C-1969).
5. Aspinall, G. O., and J. B. Christensen (Univ. of Edinburg, Scot.), *J. Chem. Soc.* (672), 3461–3467 (August 1961).
6. Aspinall, G. O., and T. E. Christensen (Univ. Edinburg, Scot.), *J. Chem. Soc.*, 2673–2676 (1965).
7. Aspinall, G. O., E. L. Hirst, and A. Wick-Strom (Univ. Edinburg, Scot.), *J. Chem. Soc.*, 1160–1165 (1955).
8. Broker, R., and J. V. Bhat (St. Xavier's Coll., Bombay), *Current Sci. (India)*, **22**, 343 (1953).
9. Cardwell, P. H., and L. H. Ellers, U.S. Patent No. 2,824,833 and No. 2,824,834.
10. Carhart, H. W., and E. H. Shaw, Jr., *Proc. S.D. Acad. Sci.*, **19**, 130 (1939); **16**, 40 (1936).
11. Chapman, J. S. (Univ. Texas Southwestern Med. School, Dallas), *Am. J. Clin. Pathol.*, **40**(14), 357–62 (1963).
12. Chomitz, N., U.S. Patent No. 3,172,878; *Chem. Abstracts*, **62**, 13345 (1965).
13. Dunn, H. J., U.S. Patent No. 2,897,119 (1959).
14. Elworthy, P. H., and T. M. George (Royal Col. of Sci. & Tech., Glasgow, Scot.), *J. Pharm. Pharmacol.*, **15**(12), 781–793 (1963).
15. Gray, G. R. (Natl. Lead Co., Los Angeles, Calif.), *Bull. Agric. Mech. Coll. Tex. Eng. Exp. Sta. Bull. No. 96*, 63–74 (1946).
16. Greig, M. E., *Science*, **105**, 665 (1947).
17. Hanna, D., L. McReynolds, and E. H. Shaw, Jr., *Proc. S.D. Acad. Sci.*, **19**, 130 (1939); D. Hanna and E. H. Shaw, Jr., *Proc. S.D. Acad. Sci.*, **21**, 78 (1941).
18. Harris-Seybold Co., British Patent No. 689,623 (1953); *Chem. Abstr.*, **47**, 10256 (1953).
19. Jermyn, M. A. (Div. Protein Chem., C.S.I. R.O., Parkville), *Aust. J. Biol. Sci.*, **15**, 787–791 (1962).
20. Lewis, B. A., and F. Smith (Univ. Minnesota, St. Paul), *J. Am. Chem. Soc.*, **79**, 3929–3931 (1957).
21. Meer Corp., *The Use of Gums in Wax Emulsions*, Technical Bulletin No. 1-11.
22. Meer Corp., *5% Ghatti Viscosity*, Technical Bulletin No. G-74.
23. Meer Corp., *Marketing Grades*.
24. Meer Corp., *BFOM Method*, Technical Bulletin No. G-127.
25. Meer Corp. purchase and sales records.
26. Newburger, S. H., J. H. Hones, and G. R. Clark (Food & Drug Admin., Washington, D.C.), *Proc. Sci. Sect. Toilet Goods Assoc.*, **19**, 25–29 (1953).
27. Nussenbaum, S., and W. Z. Hassid, *Anal. Chem.*, **24**, 501 (1952).
28. Owen, W. L., *Sugar*, **45**(11), 35–37 (1950).
29. Pfanstiehl Laboratories, Inc., personal communication.
30. Proszyorski, S. T., A. J. Mitchell, and C. M. Stewart (Commonwealth Sci. Ind. Res. Org., Melbourne), *Aust. C.S.I.R.O. For. Prod. Technol. Pap. No. 38*, 19 pp. (1965).
31. Reynolds, C. A., and L. B. Rogers, *Anal. Chem.*, **21**, 176 (1949).
32. Sadtler Research Laboratories, Inc., *Infrared Spectrogram #D-976.*
33. Smith, F., U.S. Patent No. 3,362,833 (1968).
34. Srivostava, V., and R. Shanker Rai (Chem. Labs., Univ. Gorakhpur, India), *Kolloid-Z.*, **190**(2), 140–143 (1963).
35. Tchillingarian, G., and C. M. Beeson (Univ. S. Calif., Los Angeles), *Pet. Eng.*, **24**(4), B45–52 (1952).
36. Topalian, H., and C. Elsesser, U.S. Patent No. 3,282,707 (1966); *Chem. Abstr.*, **66**, 1748 (1967).
37. Wentermyer, A. M., and E. B. Kinter, *Public Roads*, **28**(3), 55 (1954).

Chapter **10**

Gum Karaya

William Meer
Meer Corporation

General Information 10-2
   Chemical Nature 10-2
   Physical Properties 10-3
   Manufacture 10-6
   Biological/Toxicological Properties 10-8
   Rheological Properties 10-8
   Handling 10-9
   Applications 10-9
   Application Procedures 10-9
   Compatibility 10-9
   Future Developments 10-10
Commercial Uses 10-10
   Pharmaceuticals 10-10
   Pulp and Paper 10-10
   Food Products 10-10
   Textiles 10-11
   Petroleum and Gas Recovery 10-11
Industries Using Gum Karaya 10-11
   Pharmaceutical Industry 10-11
   Food Industry 10-11
   Textile Industry 10-11
   Pulp and Paper Industry 10-11
   Petroleum Industry 10-11
Formulations 10-12
   Pharmaceuticals 10-12
   Cosmetics 10-12
   Food Products 10-13
Tradename Glossary 10-13
References 10-13

GENERAL INFORMATION

Gum karaya, sometimes known as Sterculia gum, is unique in that it forms extremely strong adhesives with small amounts of water. It is widely used in pharmaceutical and dental adhesive preparations. It has a high viscosity when freshly milled. The viscosity drops appreciably when the gum is stored in the powdered form over a period of time. The adhesive property of gum karaya is not related to its viscosity.

Gum karaya is a complex water-soluble polysaccharide. It is a hydrophilic colloid prepared from the exudate of the *Sterculia urens* tree. Gum karaya is approved for food use and is in the GRAS (Generally Recognized As Safe) list under the Food and Drug Act. It has a monograph in the *Food Chemicals Codex* and was official in the 11th edition of the *National Formulary.*

Chemical Nature

Gum karaya is a partially acetylated complex polysaccharide with a high molecular weight. It contains approximately 8% acetyl groups and around 37% uronic acid residues.

Partial acid hydrolysis yields D-galactose, L-rhamnose (6-deoxy-L-mannose), and D-galacturonic acid, along with the aldobiouronic acids [2-0-(α-D-galactopyranosyluronic acid)-L-rhamnose, and 4-0-(α-D-galactopyranosyluronic acid)-D-galactose] and the acidic

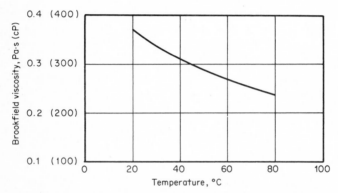

Figure 10.1 Effect of temperature on the viscosity of a fully hydrated 0.5% solution of high-grade gum karaya. Rate of temperature increase is 2°C/min.

trisaccharide [0-(α-D-glucopyranosyluronic acid)-(1→3)-(β-D-galactopyranosyluronic acid)-(1→2)-L-rhamnose].

Methylation and hydrolysis of gum karaya before and after reduction of carboxyl groups furnishes information leading to the postulation of several structures. The isolation of 3,4-di-0- and 3-0-methyl-L-rhamnose, as well as of 2,3-di-0-, 2-0-, and 3-0-methyl-D-galactose, following hydrolysis of the carboxyl-reduced, methylated gum karaya shows that D-galacturonic acid and L-rhamnose units are branch points in the polysaccharide structure.

Degradation of the gum yields a degraded polysaccharide which, upon further hydrolysis, gives 2-0-(α-D-galactopyranosyluronic acid)-L-rhamnose and a trisaccharide believed to be β-D-glucopyranosyluronic acid)-(1→3)-α-D-galactopyranosyluronic acid-(1→2)-L-rhamnose.[2]

Apparently, *Sterculia caudate* gum (a species related to gum karaya) is structurally similar to *Sterculia urens* gum.[10] Partial hydrolysis of *S. caudate* gum yields, in addition to the acidic oligosaccharides present in the hydrolyzate of *S. urens* gum, 2-0-acetyl-4-0-(β-D-galactopyranosyluronic acid)-D-galactose and 3-0-(β-D-glucopyransyluronic acid)-D-galacturonic acid. Thus, D-glucuronic acid residues are present as nonreducing end groups in both gums, whereas D-galacturonic acid residues are 4-0-substituted, with the majority being also 2-0- or 3-0-substituted.[1] It is not yet possible to propose a complete structure for the repeating unit of sterculia gum.

Physical Properties

Viscosity   Gum karaya absorbs water very rapidly to form viscous mucilages at low concentrations of the gum.   However, it is the least soluble of the commercial exudates. The rapidity with which powdered gum karaya swells can be seen in the high initial rate of viscosity increase.   The effect of particle size on viscosity development is shown by the fact that the rate of viscosity increase in a dispersion of 80- to 200-mesh gum karaya is slower than that in a dispersion of a more finely powdered gum of the same quality.

When the temperature of a fully hydrated gum karaya solution is gradually raised from 20 to 85°C, the viscosity decreases (see Fig. 10.1).   Boiling reduces the viscosity of gum karaya solutions, particularly when they are held at this temperature for more

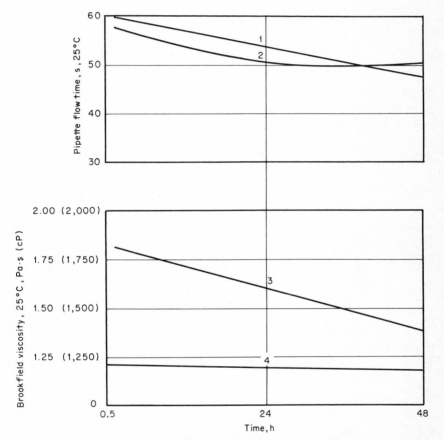

Figure 10.2   Viscosity development of 1% high-grade gum karaya dispersions versus time.   Curves 1 and 3, gum karaya hydrated at 25°C; curves 2 and 4, gum karaya hydrated initially at 80°C for 10 min.

than 2 min.   Higher maximum viscosity is obtained by cold hydration of gum karaya than is afforded by hot hydration (see Fig. 10.2).

The reduction in viscosity that is obtained by cooking gum karaya suspensions, especially under pressure, is accompanied by an increase in the solubility of the gum. Under these conditions, it forms a smooth, homogeneous, translucent, colloidal dispersion.   Concentrations as high as 15 to 18% can be prepared in this manner.

In dilute solutions of gum karaya, the viscosity increases linearly with concentration, up to about 0.5%. Thereafter, gum karaya dispersions behave as non-Newtonian solutions (see Fig. 10.3).

Gum karaya maintains its solubility with changes in pH. The viscosity decreases upon the addition of acid or alkali (see Fig. 10.4). Higher viscosities and pH stability over a wider range are obtainable when the gum is hydrated prior to pH adjustment. The solution color lightens in acidic media and darkens in alkaline solutions because of the presence of tannins. At a pH of 7 or higher, the characteristic short-bodied

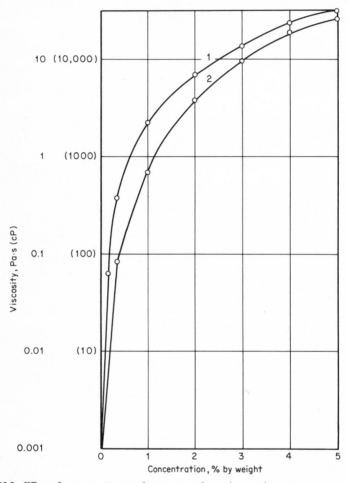

Figure 10.3 Effect of concentration on the viscosity of gum karaya dispersions as measured with a Brookfield viscometer at 20 rpm and 25°C. Curve 1 is for high-grade gum karaya; curve 2 for average grade.

gum karaya solutions become ropy mucilages. This irreversible transformation, which has been ascribed to deacetylation, is accompanied by an increase in viscosity.

The viscosity of gum karaya dispersions decreases when electrolytes, such as sodium, calcium, and aluminum chlorides and aluminum sulfate, are added. When gum karaya is hydrated in solutions that contain as much as 25% of a strong electrolyte, the viscosity is stable, although there is an initial depression that is accompanied by a separation of solids from solution. The normal viscosity of a 0.5% high-grade gum karaya solution [0.400 Pa·s (400 cP)] drops below 0.100 Pa·s (100 cP) when electrolytes are added (see Fig. 10.5). Sensitivity of gum karaya to strong electrolytes begins at low salt concentrations; the gum is not as sensitive to solutions of weak electrolytes.

Stability   The viscosity of gum karaya suspensions remains constant for several days. Increased stability can be provided by the addition of preservatives, such as chlorinated phenols, formaldehyde, mercuric salts, and benzoic or sorbic acid. Gum karaya loses viscosity-forming ability when stored in the dry state; the loss is greater for a powdered material than for the crude gum.[23] This decrease is most noticeable in the first few weeks after the gum has been ground, especially if it is stored under conditions of high humidity and temperature. Cold storage inhibits this degradation. It has been suggested that the decrease in viscosity is related to the loss of acetic acid.[9]

Gels   Gum karaya forms heavy, nonflowing pastes at concentrations higher than 2 to 3%. Its dispersions do not pass through a noticeable gel stage, although they exhibit

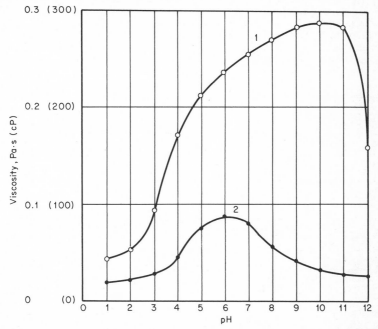

Figure 10.4   Effect of pH on viscosity of 0.5% gum karaya dispersions. Curve 1 is for gum karaya hydrated prior to pH adjustment; curve 2 is for unhydrated gum karaya added to acidic or alkaline solutions. Measurements were made with a Brookfield viscometer at 25°C and 20 rpm. The pH was adjusted with hydrochloric acid or sodium hydroxide, as needed.

the absorption characteristics of gels, especially at high concentrations. Its compatibility with a wide variety of other gums and proteins indicates that it could be useful in cosmetic, pharmaceutical, and food products.

Films   Gum karaya forms smooth films. When plasticized with compounds such as glycols, it finds use in hair-setting preparations. However, practically all other applications of the gum depend on its viscosity and water-absorbing characteristics.

Adhesiveness   At concentrations of 20 to 50% in water, gum karaya exhibits strong wet-adhesive properties. Its use as a binder in denture powders is due to its good gel strength and the strong bonds produced. Its ability to absorb water enables the gel to resist loss of strength when diluted. Blending the powdered gum with a mild alkali improves adhesiveness.

Hydrolysis   Gum karaya undergoes hydrolysis slowly, and resists hydrolysis in 10% hydrochloric acid solutions at room temperature for at least 8 h (see Fig. 10.6). Acid concentrations of 4% at 50°C for 10 to 24 h, or of 90% for 1 to 2 h, are required to hydrolyze a gum karaya solution to water-thinness.

Pastes   The short, gel-like body of a gum karaya mucilage can be transformed into a coherent, ropy paste by alkaline treatment. An empirical measurement of a solution's ropiness is obtained by determining the volume of a dilute gum karaya solution that pours into a graduate cylinder as an unbroken string. Viscosity is, to a considerable degree, correlated with the degree of ropiness.

Grades   Gum karaya is available in coarse grades, commonly 8 to 30 mesh. A coarse-mesh product has a slower viscosity build-up than a finer mesh grade and produces a discontinuous mucilage. A modified osmometric procedure has been used to determine the water-retention properties of a coarse gum karaya which is used as a bulking agent.[22]

Manufacture

Gum karaya is the name applied to the dried exudation of the tree *Sterculia urens* which is widespread in India. It is a large bushy tree that grows to a height of 9 m (30 ft). It is found on the dry, rocky hills and plateaus of central and northern India.

The best quality of gum is collected during April, May, and June, before the monsoon season. As the weather becomes warmer, the gum yield and quality improves. The collection cycle is repeated. Gum gathered in the fall has a grayish color and usually gives less viscous solutions than does gum collected earlier in the year.

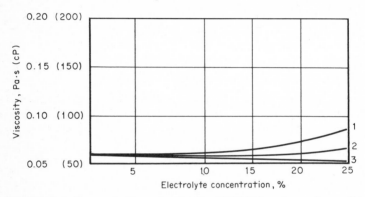

Figure 10.5   Effect of salt concentration on the viscosity of 0.5% high-grade gum karaya dispersions. Curve 1 is for aluminum chloride or sulfate; curve 2 is for calcium chloride; curve 3 is for sodium chloride. Measurements were made with a Brookfield viscometer at 25°C and 20 rpm.

Gum exudes from *Sterculia* trees after they are tapped or blazed. Usually, only the younger trees are tapped. The gum begins to exude immediately, and exudation continues for several days. The maximum amount of exudation occurs within the first 24 h. The gum is in the form of large, irregular tears, some of which weigh a kilogram or more (several pounds). The dried tears are picked by countryside natives and delivered to village collection points. Price is governed by both world market conditions and the gum quality. The yield of gum from mature trees is estimated at 1 to 4.5 kg (2 to 10 lb) per tree per season. The average tree can be tapped about five times during its lifetime.[12]

First the bark is removed from the large tears of field-grade gums. The tears are then broken up, and the fragments are sorted into grades on the basis of color and amount of adhering bark.

The processor evaluates incoming raw materials according to impurity content, solution viscosity, and color. Purification is accomplished by size reduction and air flotation of loose bark.

Gum karaya can be separated into two classes: (1) crude grades, the gum as it is imported (see Table 10.1), and (2) processed grades, the gum sold to the industrial user. There is no uniformity in the nomenclature of commercial grades of gum karaya. However, in spite of the differences in nomenclature, many of the grades offered are based on similar processing techniques and assays.

Crude gum karaya is imported into the United States in grades that are based entirely on bark and foreign organic matter content and color. A processor usually offers three to five grades of gum karaya, varying in bark and foreign organic matter content from 0.5 to 3.0% and in color from white to tan. In order to be classified as NF-grade material and be acceptable for food and drug use, the gum must not contain more than 3% of bark and foreign organic matter.[2] Grades that contain more than 3% are offered for industrial uses and are called technical grades. See Table 10.2 for imports of gum karaya.

Although designated as a water-soluble gum, gum karaya is one of the least soluble of the exudation gums. A gum particle placed in water does not dissolve; rather it

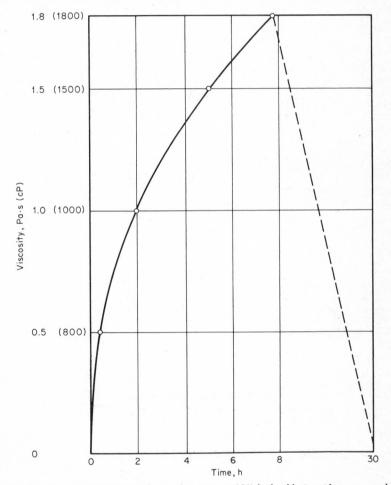

Figure 10.6 Viscosity of 1.5% gum karaya dispersion in 10% hydrochloric acid as measured with a Brookfield viscometer at 25°C and 12 rpm.

absorbs water and swells to many times its original size. Therefore, the particle size influences the type of dispersion obtainable. A coarse granulation yields a discontinuous, grainy dispersion; a finely powdered gum yields an apparently homogeneous dispersion.

Granulated or crystal gum karayas are usually processed so that the particle size is between 6 and 30 mesh. These granulated gum karayas are used principally as bulk laxatives.

Finely powdered grades are the most common types of processed gum karaya. The gum is prepared so that the powder completely passes through a 150-mesh screen and so that 80 to 90% also passes through a 200-mesh screen. The particle size of such a gum is so small that, in spite of individual particle swelling, an apparently homogeneous dispersion results.

Some grades of gum karaya have a particle size that lies between those of the granular and powdered types. These products are usually 80 to 200 mesh in size. They disperse in water with fewer lumps than produced by the finely powdered grades, and are more dust-free. They represent a compromise between grades that are easily dispersed and those that form homogeneous dispersions.

There is no distinct correlation between viscosity and grade. Cost is determined on the basis of color and purity.

Biological/Toxicological Properties

Gum karaya is GRAS and is on the FDA list. It is nontoxic and is generally considered not to be metabolized in humans.

Short-Term Studies Gum karaya has no acute toxicity. Biological data is available on the effect of gum karaya in short-term studies on several animals, including humans.

Tested with rats, gum karaya had only 30% of the available caloric value of cornstarch.[24] After 60 days, both karaya-fed and control female albino rats maintained practically the same growth rate.[13] Results for tests on dogs is given in reference 13.

In humans, karaya gum is a laxative agent that has no observable harmful effect.[13]

TABLE 10.1
Gum Karaya Grade Standards

Grade*	Bulk and foreign organic matter, %	Color	Percent material minus 10 U.S. std. mesh
WSGA No 1	0.5 max.	White with slight gray cast	2
WSGA No. 2	1.5 max.	Very light tan	3
WSGA No. 3	3.0	Tan	3
Technical siftings	6.0	Brown	

* Water-Soluble Gum Association grades.[2]

Long-Term Studies A group of nine male hooded rats were fed karaya gum as a 10 to 25% component of their diets for their entire life spans. Serious intestinal lesions were found in three of five test subjects fed the gum from 396 to 711 days. The experiment was designed to test the nutritive value of gum karaya; however, it did appear that there was a deleterious fraction in the gum. At inclusion of more than 25% of the diet, karaya gum showed marked depression of the growth rate.[11]

Six Wistar rats (three male, three female) were fed karaya gum as a 10 to 25% component of their diets for their entire life spans. They were observed for cecal ulceration and life span. Karaya-fed rats exhibited a decreased food intake and a depressed growth rate. However, contrary to previous findings, no cecal ulceration was observed.[3]

Special Studies *Sensitization.* Sixteen cases of allergy to gum karaya were reported. General symptomology consisted typically of an asthmaticlike condition, atopic dermatitis, hay fever syndrome, and/or gastrointestinal distress. The allergen was confirmed by scratch test and subcutaneous injection of karaya antigen. Ingestion of karaya-containing toothpaste, gum drops, and laxatives resulted in gastrointestinal distress and other symptoms.[5,23]

Other studies have confirmed these observations on the allergenic nature of gum karaya for certain individuals.[7,8]

Fetotoxicity. A 1% aqueous suspension of gum karaya injected intraperitoneally between the eleventh and fifteenth day of gestation produced no fetotoxic effects in CBI mice.[6]

Rheological Properties

Gum karaya sols are thixotropic, with viscosity increasing with concentration. These sols do not have sheer stability. See further discussion of viscosity under Physical Properties.

Handling

If the gum becomes partially wet or damp, it tends to agglomerate or form lumps. Therefore, handling recommendations include storage in sealed polyethylene-lined containers. For extended storage, material should be warehoused in a cool, dry place. Normal dust-handling equipment should be used when dumping the dry material.

Applications

Gum karaya has been produced commercially only since 1920. Since then, it has been imported into this country in increasing amounts. The original incentive for its use was its price, which was much lower than that of gum tragacanth. Recently, however, gum karaya has been found to be superior to other gums for some purposes:[1]

- As an adhesive when partially wet with water, particularly in denture adhesives and colostomy rings.
- As a gelling agent when used with alkali such as sodium borate to form soft gels.
- As a bulking agent used in coarse particle size; absorbs water and swells to 60 to 100 times it original volume.

TABLE 10.2
Gum Karaya U.S. Imports and Prices

	U.S. imports		Price of Gum Karaya, U.S. ¢/lb (U.S. ¢/kg)					
Year	Total pounds	(Kilograms)*	WSGA Grade no. 1		WSGA Grade no. 2		WSGA Grade no. 3	
1951	7,696,225	(3,490,351)	35–47	(77–104)	26–27	(57–60)	19–27	(42–60)
1956	7,014,000	(3,180,952)	39	(86)	11–24	(24–53)	21–29	(46–64)
1961	7,345,125	(3,331,122)	31–33	(68–73)	27–30	(60–66)	24–26	(53–57)
1962	7,461,805	(3,384,039)	33–41	(73–90)	30–34	(66–75)	25–33	(55–73)
1963	7,132,397	(3,234,647)	41–46	(90–101)	35–38	(77–84)	33–35	(73–77)
1964	7,682,267	(3,484,021)	45–50	(99–110)	36–38	(79–84)	31–33	(68–73)
1965	9,061,322	(4,109,443)	45–47	(99–104)	36–38	(79–84)	32–33	(71–73)
1966⎤	7,344,074	(3,330,646)	45†	(99)	36	(79)	34	(75)
1966⎦			33‡	(73)	27	(60)	24	(53)
1967	7,201,505	(3,265,989)	35–36	(77–79)	26	(57)	23–24	(51–53)
1968	7,344,074	(3,330,646)	36–50	(79–110)	26–38	(57–84)	23–34	(51–75)
1969	8,281,000	(3,755,556)	62–70	(137–154)	53–58	(117–128)	46–49	(101–108)
1970	7,717,000	(3,499,773)	59–64	(130–141)	52–56	(115–123)	47–51	(104–112)
1971	7,650,000	(3,469,388)	58–62	(128–137)	48–52	(106–115)	41–43	(90–95)
1972	7,353,000	(3,342,273)	57–60	(125–132)	45–47	(99–103)	35	(77)
1973	6,976,000	(3,170,909)	57–63	(125–139)	48–53	(106–117)	38	(84)
1974	7,846,000	(3,566,364)	68–74	(150–163)	59–62	(130–136)	50	(110)
1975	5,016,000	(2,280,000)	52–59	(114–130)	43–48	(95–106)	38–40	(84–88)
1976	6,314,000	(2,870,000)	71–75	(156–165)	58–60	(128–132)	50	(110)
1977	5,799,000	(2,635,900)	72–80	(166–175)	68–70	(150–155)	54	(120)

* These figures were compiled from U.S. Department of Commerce Bulletins entitled "U.S. Import of Merchandise for Consumption," from the Chemical and Rubber Industry Report of the U.S. Department of Commerce, As-OP, June 1955, and from the U.S. Department of Commerce Bulletin FT 135 (called FT 125 through 1966).
† Before the June 1966 devaluation.
‡ After the June 1966 devaluation.

- As a modified gum (deacetylated or partially hydrolyzed) using mild alkali; its properties are pH-irreversible.
- As a binder in the paper and textile industries.

Application Procedures

To achieve a uniform dispersion when starting with finely powdered gum karaya, one or more of the following procedures should be used:

1. Apply vigorous agitation while adding the gum to water, or
2. Allow two or more hours for hydration, or
3. Disperse the gum into water as a mixture with another solid, such as sugar, or
4. Premix the gum with one to five times its weight of water-miscible liquid, such as ethanol or glycerol, prior to the addition of water.

Boiling reduces the viscosity of gum karaya solutions.

Compatibility

Gum karaya is compatible with other plant hydrocolloids as well as with proteins and carbohydrates. Gum karaya gels are apparently incompatible with pyrilamine maleate, a strong hydrotrope and antihistaminic. Electrolytes, as well as excessive acid, cause a drop in the gel viscosities. Alkalies cause the gels to become stringy.

Future Developments

The market conditions for gum karaya are generally stable, although some specialty adhesives using gum karaya have been developed recently.

COMMERCIAL USES

Gum karaya is generally useful as an adhesive gel–former and bulking agent, as a modified gum, and as a diacetylated binder and deflocculent.

Pharmaceuticals

A large proportion of the gum karaya imported into this country is utilized by the pharmaceuticals industry. As a bulk laxative, gum karaya is second only to psyllium seeds in importance. When used in bulk laxatives, gum karaya is usually processed so that it is 8 to 30 mesh in size. These coarse gum particles absorb water and swell to 60 to 100 times their original volume, forming a discontinuous type of mucilage. This type of mucilage is very effective as a laxative. The gum is not digested, nor is it absorbed by the body.

Another important pharmaceutical application of gum karaya is as a denture adhesive. The powdered gum is usually dusted onto the dental plate, and it swells when it touches the moist surfaces of the mouth. This results in a more comfortable and tighter fit of the plate. The rapid swelling of the gum particles, their relative insolubility, and their unusual resistance to bacterial and enzymatic breakdown make the gum suitable for this use.

Pulp and Paper

In the paper industry, gum karaya is largely confined to use with long-fibered, lightweight papers, such as with condenser tissues and fruit-wrap tissues[15,16] in which the gum acts as a binder. The principle behind its use is the removal of the acetyl groups by treatment with alkali, thereby exposing more active carboxyl and hydroxyl groups. This increase in the polarity of the molecule makes the binding of the gum to hydrated cellulose fibers more efficient. Removal of acetyl groups also affords an internally cohesive or ropy solution.

A weak alkali, such as ammonia or triethanolamine, is then added to raise the pH of the solution from its normal pH of 4.5 up to a pH of 8.0 to 8.5. The properties imparted to the gum by deacetylation are pH-irreversible, and are maintained even when the solution is added to an acidic pulp suspension.

The deacetylated gum solution can be added to the pulp suspension at the fan pump. Usual rates of addition are 1 kg of gum/200 to 300 kg of pulp (10 to 15 lb of gum/ ton of pulp). The gum deflocculates the fibers, producing lightweight sheets having improved formation and strength. However, a deacetylated gum karaya solution has the disadvantage of decreasing the freeness and thereby slowing the production of the sheet. In lightweight papers, the slowing of the stock flow becomes sufficient to prohibit the use of deacetylated gum karaya. The use of alum or other cationic materials in the finish impairs the efficiency of the gum as a deflocculent.

Food Products

Powdered gum karaya is used in French dressing, ice pops, sherbets, cheese spreads, ground meat preparations, and meringue products. In French dressing, it is used as a stabilizer. It is used sometimes in conjunction with gum arabic, which acts as a protective colloid.

Gum karaya is used in concentrations of 0.2 to 0.4% in the manufacture of ice pops and sherbets. The gum prevents the bleeding of free water and the formation of large ice crystals. Its water-absorbing and water-holding capacities and its excellent acid compatibility make it suitable for this use.

Concentrations of 0.8% or less of the gum are used in cheese spreads. Its acidic nature is not objectionable in this type of dairy product. It is added to prevent water separation and to increase the ease of spreading.

Gum karaya is also used as a meringue-stabilizer because of its binding properties. In addition, its incorporation enables a greater volume of meringue to be prepared from a fixed amount of protein. Ground meat products, such as bologna, require an efficient water-holding substance that has a small amount of adhesiveness. Gum karaya in concentrations of 0.25% provides these characteristics and gives the meat product a smooth appearance.

Textiles

In the textile industry, gum karaya is modified so that it can be used for printing operations in which it was considered unsatisfactory until a method of increasing its solubility was developed. This is done by cooking a water suspension of gum karaya under pressure. The rate of dissolution varies with the pressure.[14] In commercial practice, solutions containing 15 to 18% of solids are marketed as textile gum solutions.

An alternative method of solubilization of gum karaya consists of treatment with sodium peroxide, persulfate, or persilicate. The textile gum solution is used principally as a thickening agent for the dye in direct color printing on cotton fabrics.

Petroleum and Gas Recovery

Gum karaya has been used as an ingredient in a temporary plugging agent for a natural gas–well bore.[4] An old gas well that was producing gas at a daily rate of 14,160 m³ (0.5 million ft³) from the Edwards formation in Texas was acidified using a diverting plug in water. The plug's composition was:

	Kilograms	Pounds
Whole Austrian peas	22.05	10
Milo maize	44.10	20
Flaked phthalic acid	66.15	30
Guar meal	55.13	25
Soybean meal	33.10	15
Gum karaya	6.60	3
Poly(styrene sulfanate) solution*	2.205	1

* 30% active poly(styrene sulfanate) in 636 L (168 gal) of water

Preceding the agent, 3785 L (1000 gal) of 28% hydrochloric acid were pumped down the tubing. The 636 L (4 bbl) of diverting agent followed, and the final 3785 L (1000 gal) of 28% inhibited hydrochloric acid were injected. The acid was flushed from the tubing and into the underground formation with water.

The initial 3785 L (1000 gal) of acid solution entered the formation at a rate of 11 kg/min (5 lb/min) at 19.3 MPa (2800 psi) pressure. This pressure increased to 27.0 MPa (3900 psi) as the diverting agent reached the formation and plugged the existing channels. The final 3785 L (1000 gal) of acid entered a new section of the formation at a rate of 397 L/min (2.5 bbl/min) at 26.3 MPa (3800 psi). After this treatment, gas production from the well increased to 70,800 m³/day (2.5 million ft³/day).

INDUSTRIES USING GUM KARAYA

A brief summary of the industries that use gum karaya is given as follows:
- *Pharmaceutical industry*
  a. As a binder in denture adhesives and for colostomy rings.
  b. As a bulking agent in bulk laxatives.
  c. As a gelling agent in cosmetics.
- *Food industry*
  a. As a stabilizer in sherbets, ices, meringues, toppings, and whipped cream products.
  b. As an emulsifier in processed meats and sausages and in cheese spreads.
- *Textile industry:* As a dye thickener for direct-color printing operations.
- *Pulp and paper industry:* As a deflocculent and binder for lightweight paper manufacture.
- *Petroleum industry:* As a thickening agent for drilling muds and as a plugging agent for secondary recovery operations.

FORMULATIONS

Pharmaceuticals

Denture Adhesive[21]

	Parts by weight
Powdered gum karaya, 95%	85
Borax and/or magnesium oxide, 10%	3
Silica, fine powder	q.s.
Magnesium stearate	q.s.
Flavor (peppermint or spearmint oil)	q.s.

Dry-mix this formulation, and incorporate the flavor into the mixture.

Colostomy Rings[17]   The following formulation is for sealing pads or a postsurgical drainage pouch. The approximate composition is:

	Parts by weight
Superfine karaya gum	45
Propylene glycol	55
Tween 60	0.1

The pad is formed out of a mixture of a nonevaporative, oleaginous liquid carrier and a healing powder which is miscible with the carrier and forms a gel when mixed in sufficient quantity with the carrier. In the preferred embodiment, karaya gum powder, a well-known healing powder, is mixed with glycerol. A suitable mixture of karaya gum with glycerol should be at least 33% karaya gum to 67% glycerol, or may be as much as 50% karaya gum with 50% glycerol. The more glycerol used, the softer and less solid is the ring. With the addition of heat of 93 to 149°C (200 to 300°F), the gel time is shortened.

The gel of karaya gum and glycerol is relatively viscous; therefore, in forming the sealing ring out of this material, an open-face dished mold and a centrally upright boss can be used. Preferably, the inside of the mold (that portion which will contact the mixture) is precoated with a suitable wax or oil, such as paraffin, in a hexane solution. After the application, the hexane will evaporate, leaving the paraffin base. The mixture is poured into the coated mold and is allowed to stand and solidify into a relatively stiff, semisolid, tacky ring which contains elastic and resilient properties.

After the mixture has gelled, a pressure adhesive, such as is made from a vinylethyl ether base, may be applied to one end of the ring to promote attachment of the ring to a drainage bag or similar appliance. When the other face of the ring is placed about the stoma and against the body of the patient, the body heat will increase its tackiness, and thus a very effective seal with the patient's skin is formed. Perspiration from the body of the patient, or water from the stoma discharge, which contacts the karaya ring will increase the tackiness of the ring and further promote the seal with the patient's skin.

For use with a postsurgical drainage pouch adapted to receive drainage from the stoma of a patient, the sealing ring would be made up of a mixture containing at least a minor part of karaya gum powder with an equal part of glycerol. This would have a central opening of a size suitable to receive the stoma, with the ring adapted to be telescoped on the stoma between the skin of the patient and the drainage pouch to form a seal between them.

Cosmetics

Typical formulations for wave sets are given in the following two examples:[20]

Alcohol Wave-Set Concentrate

	Weight percent
Gum karaya, superfine grade	19.5
Borax	6.5
Alcohol	74.0

Dry-mix the gum karaya and borax, then add the mixture to the stirred alcohol. Continue stirring for at least ½ h. There will be a separation of the ingredients later upon standing and it must be well mixed before use. Use 170 g (6 oz) of this concentrate to make up to 15.14 L (4 gal) of the finished wave set.

Typical Wave-Set Formula

	Weight percent
Tartrazine	0.05
Once-distilled water	88.90–89.40
Lanolin	2.00
Sodium hydroxide (or borax)	0.40
Formalin	0.20
Gum karaya	0.75–1.00
Alcohol	7.70–6.95
Total	100.00

Dissolve the tartrazine in the once-distilled water; then, in sequence, dissolve first the lanolin and then the sodium hydroxide (or borax) in the tartrazine-water solution; then add the formalin. Next, add the gum karaya to the alcohol and make a slurry. Transfer the karaya-alcohol slurry to the aqueous solution while stirring, and continue to stir for a minimum of ½h.

Food Products

Sherbet Stabilization[19]   Sherbets that contain not less than 1% by weight of milk fat and not less than 4% by weight of total milk solids are stabilized by a combination of stabilizers and emulsifiers at not more than 0.5% of the total mix.

The use of a combination of karaya gum and guar gum in a 50:50 mixture (or either gum alone) with sugar will eliminate crystallization that would otherwise be caused by heat shock. It also imparts firm body and smooth texture and provides a cool, refreshing, full-flavored sherbet. Further benefits afforded by the use of a stabilizer are to overcome bleeding, promote and stabilize the even distribution of color, control overrun, and minimize shrinkage.

The stabilizer-sugar mixture should be added slowly, with vigorous agitation, to all of the other ingredients. If the mix is pasteurized, heat should be applied when thoroughly mixed. If the mix is not pasteurized, homogenization is desirable.

TRADENAME GLOSSARY

Following is a list of the common grades of gum karaya available in powdered form, along with the designations of several manufacturers:

Grade	Hercules	Meer	S. B. Pennick	Stein Hall
Karaya #1	Alligator	Superfine XXXX	Super Initial	˜Superior
Karaya #2	Elephant	Superfine #1	Supreme	K-3
Karaya #3	Cobra	Select #1	Initial	K-4

REFERENCES

1. Aspinall, G. O., and R. N. Fraser, "Plant Gums of the Genus *Sterculia,* Part II: *Sterculia caudata* Gum," *J. Chem. Soc.,* 4318 (1965).
2. Aspinal, G. O., and Nasir-ud-din, "Plant Gums of the Genus *Sterculia,* Part I. The Main Structural Features of *Sterculia urens* Gum," *J. Chem. Soc.,* 2710 (1965).
3. Carlson, A. J., and F. Hoelzel, "Prolongation of the Life-Span of Rats by Bulk Formers in the Diet," *J. Nutr.,* **36,** 27–40 (1948).
4. Eilers, L. H. (assigned to Dav Cleem Co.), U.S. Patent 3,480,084 (November 25, 1969).
5. Figley, K. D., "Karaya Gum (Indian Gum) Hypersensitivity," *J. Am. Med. Assoc.,* **114** (9), 747–748 (1940).
6. Frohberg, M., H. Oettle, and H. Zeller, "On the Mechanism of the Fetotoxic Effect of Tragacanth," *Arch. Toxikol.,* **25,** 268–295 (1969).
7. Gelfand, H. H., "The Allergenic Properties of the Vegetable Gums," *J. Allergy,* **14,** 203–219 (1943).

8. Gelfand, H. H., "The Vegetable Gums by Ingestion in the Etiology of Allergic Disorders," *J. Allergy,* 20, 311–321 (1949).
9. Goldstein, A. M., "Natural Plant Hydrocolloids," *Adv. Chem. Ser.,* 11, 35 (1954).
10. Hirst, E. L., and Vanhandlugsber, *Kolloid-Ges.,* 18, 104 (1958).
11. Hoelzel, F., E. Da Costa, and A. J. Carlson, "Production of Intestinal Lesions by Feeding Karaya Gum and Other Materials to Rats," *Am. J. Dig. Dis.,* 8(7), 266–270 (1941).
12. Howes, F. N., *Vegetable Gums and Resins,* Chronical Botanica Co., Waltham, Mass., 1949, pp. 41–44.
13. Ivy, A. C., and B. L. Isaacs, "Karaya Gum as a Mechanical Laxative: An Experimental Study on Animals and Man," *Am. J. Dig. Dis.,* 5, 315–321 (1938).
14. Knecht, E., and J. B. Fothergill, *The Principles and Practice of Textile Printing,* 2d ed., Griffin and Co., London, 1924, pp. 123, 124.
15. LeCompte, T. R., U.S. Patent 2,069,766 (1937); *Chem. Abstr.,* 31, 2434 (1937).
16. LeCompte, T. R., U.S. Patent 2,088,479 (1937); *Chem. Abstr.,* 31, 6881 (1937).
17. Marsan, A. E. (to Hollister, Inc.), U.S. Patent 3,302,647 (1967).
18. Meer Corp. product grades.
19. Meer Corp., Technical Bulletin F-43.
20. Meer Corp., Technical Bulletin P-4.
21. Meer Corp., Technical Bulletin P-253.
22. Monaco, A. L., and E. J. Dehnen, "An In Vitro Evaluation of Some Hydrophilic Colloids as Bulking Agents," *J. Am. Pharm. Assoc., Sci. Ed.,* 44, 237 (1955).
23. Money, R. W., "Karaya Gum: Physico-Chemical Properties with Particular Reference to the Effect of Dry Grinding," *J. Sci. Food Agric.,* 2, 385 (1951).
24. Wisconsin Alumni Research Foundation Laboratory, unpublished report No. 3110860/1 to Stein, Hall & Co., 1964.

Chapter **11**

Gum Tragacanth

Kenneth R. Stauffer
Cook College, Food Sciences Department, Rutgers University

General Information 11-2
  Chemical Nature 11-3
  Physical Properties 11-7
  Preservatives 11-13
  Biological/Toxicological Properties 11-14
  Manufacturing and Quality Control 11-14
  Application Procedures 11-15
  Future Developments 11-15
Commercial Uses: Compounding and Formulating 11-16
  Pharmaceutical and Medical 11-16
  Food Products 11-16
Commercial Uses: Processing Aids 11-20
  Crayon Manufacture 11-20
  Ceramics Manufacture 11-20
  Leather Curing 11-20
  Textiles Processing 11-20
  Paper Processing 11-20
  Wooden Match Manufacture 11-20
Industries Using Gum Tragacanth 11-20
  Food Industry 11-21
  Pharmaceutical and Cosmetic Industries 11-21
  General Industrial Uses 11-21
Formulations 11-21
  Italian Dressing 11-21
  Russian Dressing 11-22
  Blue Cheese Dressing 11-22
  French Dressing 11-22
  Low-Calorie Italian-Type Dressing 11-23
  Sweet and Sour Sauce 11-23
  Low-Calorie French-Type Dressing 11-23
  Barbecue Sauce 11-24

Dietetic (Artificial) Fruit Jelly 11-24
Citrus-Flavor Beverage Emulsion 11-24
Low-Calorie Chocolate Syrup 11-25
Low-Calorie Chocolate Pudding 11-25
Marshmallow Topping 11-25
Nondairy Sour Cream 11-25
Toasted Onion-flavored Chip Dip 11-26
Mexican-flavored Chip Dip 11-26
Tuna, Chicken, and Ham Salad Spreads 11-27
Cole Slaw Dressing 11-27
Imitation Mayonnaise Dressing 11-27
Mustard Sauce 11-28
Spaghetti Sauce 11-28
Pickle Relish 11-28
Laboratory Techniques 11-29
Identification 11-29
Viscosity Testing 11-29
Product Tradename/Term Glossary 11-29
References/Further Useful Reading 11-30

GENERAL INFORMATION

Gum tragacanth is a natural vegetable gum exudated from various species of shrubs belonging to the genus *Astragalus* (of the order Leguminosea).[14] The single most important consideration of gum tragacanth is that its usefulness is the result of all its functional characteristics. Some gum systems show good initial viscosity but degrade in low pH systems (i.e., gum karaya). Gum tragacanth does not. Some gums will emulsify oil-in-water by thickening the aqueous phase, but require surface-acting agents to lower the resistance between the oil and water (i.e., xanthan); gum tragacanth does not. Gum tragacanth is a bifunctional emulsifier[23] in that it will increase the aqueous phase viscosity and also lower the interfacial tension between the oil-in-water emulsion, and thus eliminate the need to incorporate surface-active agents. Gum tragacanth (0.5%) shows pseudoplastic flow which has an effect on the pouring and textural properties of the finished products. Other considerations are: Gum tragacanth is shear thinning with respect to elevated temperature, has a wide range of viscosity measurements resulting from the grade from which it was derived, and will suspend finite particles in solution.

The suspending property can be attributed to the electrochemical nature of the gum (anionic) since the galacturonic acid units found in the gum will repel each other much like sulfate half esters do in carrageenans.[18] Also, suspension properties are enhanced by tragacanth's pseudoplastic flow properties. The solutions at rest will have greater apparent viscosity than when in motion. Therefore, when gum tragacanth is employed as a thickener, emulsifier, suspending agent, moisture retainer, film former, binder, bodying agent, sizing agent, or adhesive, several considerations of the gum characteristics may be important for the finished product.

Although the gum is grown and harvested in Asia Minor, Syria, Iran, and Turkey, the largest marketplaces are the United States and England, with the remainder going to Russia, West Germany, France, Italy, and Japan. The estimated importation for gum tragacanth to the United States for 1976 is 499,000 kg (1,100,000 lb).

Gum tragacanth is listed in the *United States Pharmacopoeia* and the *Food Chemicals Codex*,[8] is approved for use in food, and is on the GRAS (Generally Recognized As Safe) list, which was recently exhaustively revised (1972).[5] No new findings have reduced its stature. It is also listed under the Federal Food, Drug, and Cosmetic Act.

In bakery products, gum tragacanth is used in meringues, bakers' citrus oil emulsions, and frozen pie fillings. One of the largest food applications is in regular and low-calorie dressings and sauces. This includes French, Italian, Thousand Island, Roquefort, and blue cheese dressings, and pizza, spaghetti, mustard, and barbecue sauces.

Gum tragacanth is also used in the confectionery industry as a binder in cough drops and lozenges, gum drops, jujubes, and pastilles.

Gum tragacanth is used in the pharmaceutical and cosmetic field in gynecological jellies, ointments and salve bases, syrup, fish and mineral oil emulsions, tablets, pills, cosmetic creams and lotions, facial clays, toothpastes, and hair dressings.

Industrially, gum tragacanth finds applications for textile sizing, printing pastes, furniture and auto polishes, crayons, ceramics, insecticide emulsions, and cigars.

Chemical Nature

Gum tragacanth as found in nature exists as a slightly acidic salt, a complex mixture of polysaccharides containing calcium, magnesium, and potassium. After acid hydrolysis, the major sugars produced are D-galacturonic acid, D-galactose, L-fucose, D-xylose, L-arabinose, and L-rhamnose.

Structure Most investigators believe that gum tragacanth consists of at least two components:[15,20] a water-swellable major component, bassorin (60 to 70%), and a water-soluble component, a colloidal hydrosol, tragacanthin.

There has been some confusion on the nomenclature of these two constituents, and one author[6,7] refers to the soluble part as an araban and the swellable part as tragacanthin, which is in direct conflict with the majority opinion.

The water-swellable bassorin contains the tragacanthic acid polymer, while the water-soluble portion, tragacanthin, is a neutral polysaccharide which, at best, contains only small amounts (3%) of uronic acid.[3] This finding is in contrast to an earlier work which shows uronic acid as a major constituent in the water-soluble portion.[13]

Also, this neutral polymer has been referred to in early works as a galactaraban,[22] but more recent work[3] refers to it as an arabinogalactan, and rightfully so, since the arabinose is the sugar found in abundance and galactose is the core repeating unit.

Therefore, hereafter in this chapter, the water-soluble, neutral polymer will be referred to as an arabinogalactan, and the water-swellable polymer as tragacanthic acid.

The neutral polysaccharide, arabinogalactan, may be separated from the tragacanthic polymer in an aqueous solution with the addition of ethanol. The arabinogalactan is soluble in ethanol at concentrations as high as 70%, while the tragacanthic acid will precipitate. It is not certain whether the arabinogalactan is in a physical admixture with, or chemically bonded to, the polysaccharide acid, although its ease of separation favors the former view.

The partial structure of tragacanthic acid is shown in Fig. 11.1.[2] The arabinogalactan is highly branched (although strict evidence of homogeneity is not available), and this suggests that the structure is based on a core of D-galactose residues to which are attached highly ramified chains of L-arabinofuranose residues.[3] The majority interior chains of D-galactopyranose are joined by 1,6 linkages, and the smaller portions by 1,3 linkages. The L-arabinofuranose residues are mutually joined by 1,2, 1,3, and 1,5 linkages.

Also, the general rule of thumb that the more highly branched a polysaccharide is, the more water-soluble it becomes, supports this reasoning, since the arabinogalactan is very water-soluble.

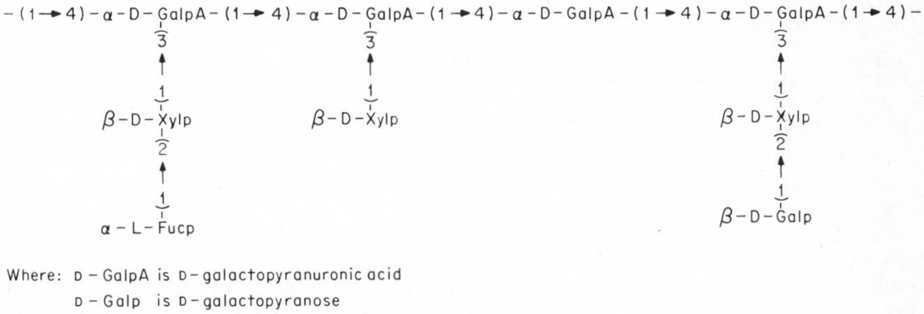

Where: D-GalpA is D-galactopyranuronic acid
D-Galp is D-galactopyranose
D-Xylp is D-xylopyranose
L-Fucp is L-fucopyranose

Figure 11.1 Structure of tragacanthic acid. *(Ref. 2.)*

The analysis also shows that the polysaccharide is notable in containing a high ratio of arabinose to galactose residues. The polysaccharide also contains small proportions of D-galacturonic acid and L-rhamnose residue, whose modes of linkage, if any, are not clear.

Molecular Weight   The molecular weight recorded for a high-grade flake-type tragacanth was obtained by the viscosity method using Svedberg's formula:[11]

$$M = \frac{R\,Ts}{D(1 - vd)}$$

where M = molecular weight of solute
   R = gas constant = 8.314 J/(mol · deg)
   T = absolute temperature
   d = density of solution
   D = diffusion constant
   s = sedimentation constant
   v = partial specific volume

The nondegraded tragacanth molecular weight was recorded as 840,000.

Molecular Shape   The molecular shape may be calculated by:[11]

$$\frac{ld^2}{6} = \frac{mv}{N}$$

where l = elongated shape, nm
   d = cross section, nm
   m = molecular weight of solute
   v = partial specific volume
   n = Avogadro's number

For nondegraded high-grade flake-type gum tragacanth the l = 4500 nm and the d = 19 nm.

Reactivities   Gum tragacanth as a calcium, magnesium, and potassium salt of D-galacturonic acid reacts with many reagents in a manner similar to that of other anionic polymer salts. Solutions of gum tragacanth will produce precipitates or heavy gels on addition of Millon's reagent, lead acetate, potassium hydroxide, ferric chloride, Schweitzer reagent and Cetavlon. In general, trivalent metal ion salts will cause precipitation of gum tragacanth.

Acid-Labile Sugars   Acid-labile sugars found in polysaccharides are those that show preferential cleavage of the glycoside bonds joining them in the polymer. However, it should be noted that some glycoside bonds are more acid-stable than others; that is, starch α-1,4 bonds are more acid-labile between their respective D-glucopyranose sugars than are cellulose β-1,4 bonds.

With this in mind, some generalizations can be made with respect to acid-labile sugars:

1. Pentose (C_5) is more acid-labile than is hexose (C_6) [that is, ribose (C_5) versus galactose (C_6)].

2. Five-membered rings (furanose) are more acid-labile than are six-membered rings, such as pyranose (that is, arabofuranose versus arabopyranose).

3. 6-Deoxyhexoses are more acid-labile than are their counterparts containing an OH at the 6 position [that is, fucose (6-deoxygalactose) versus galactose].

4. Uronic acids are much less acid-labile than are their corresponding sugars (that is, galacturonic versus galactose).

5. *l*-isomers are more acid labile than are *d*-isomers (that is, *l*-galactose versus *d*-galactose).

Examination of the partial structure of tragacanthic acid (see Fig. 11.1) shows that the acid-labile groups *d*-xylopyranose and *l*-fucopyranose are located in the branches of the polymer rather than in the main backbone chain.

In contrast, the main backbone chain contains the very acid-stable D-galactopyranouronic acid groups. Also the L-fucopyranose, which should be the most acid-labile sugar with respect to the above considerations, is found at the terminal position in one of the branches, and thus, if hydrolyzed, it would not contribute to a reduction in molecular weight as drastically as it would if it were located in the middle of a chain.

The arabinogalactan has acid-stable D-galactopyranose along with acid-labile L-arabinofuranose sugars. The exact positions of these sugars in the molecules are unknown, but studies[3] indicate that galactose makes up the backbone chain and the acid-labile L-arabinofuranose groups are in the branches.

Since gum tragacanth is an anionic gum and does contain acid-labile sugars, it will autohydrolyze. This autohydrolysis is more pronounced when heat is applied during the initial hydration of the gum (see Fig. 11.2), as opposed to heating fully hydrated gums (see Fig. 11.3). This could be explained in terms of the location of acid-stable and acid-labile sugars. When heat is applied to the hydrating gum as opposed to the hydrated gum, the reactive sites for hydrolysis are closer together.

Also, practical experience has shown that gum tragacanth will develop higher viscosities in the absence of vinegar when making salad dressings. This also relates to the hydrolysis of acid-labile sugars. But since gum tragacanth is one of the better acid-stable gums, it must be said that even though hydrolysis and autohydrolysis occur, there is a considerable amount of acid stability resulting from the acid-stable sugars (mainly

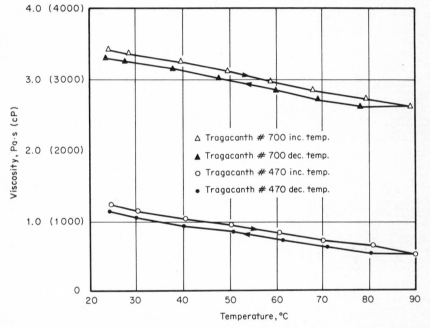

Figure 11.2 Effect of temperature on fully hydrated 1% solutions of gum tragacanth.

galacturonic acid), and of the way in which they are linked to the polysaccharides by glycosidic bonds.

Electrochemical Properties Tragacanthic acid, the major component of gum tragacanth, has a main backbone chain of α-1,4-galacturonic acid (see Fig. 11.1) which is analogous to pectic acid. One difference between pectic acid and tragacanthic acid is that the latter has neutral side chains bonded to the electroactive backbone. Both are polyelectrolytes which show electrochemical properties such as the Donnan effect, salt concentration dependency, and reactivity. Differences in functionality of the two may be equated to the location of galacturonic acid, the side chains, and to other components present (e.g., pectic acid will gel easily by crosslinking with calcium, whereas tragacanth will not).

Chain conformations of polysaccharides have been predicted by conformational analysis in a computer. Sodium polygalacturonate is considered to be a twisted, buckled ribbon.[19]

By similar computer prediction, tragacanthic acid is shown as a twisted, buckled ribbon with neutral side chains extending (or wrapping around) the ribbon itself. This viewpoint is also based upon the following analogy:

The primary structure of xanthan (see Chap. 24, Fig. 24.1) has a β-1,4 cellulosic backbone with a three-membered sugar side chain attached to alternate glucose units. Computer analysis suggests that β-1,4 glucose has extended and ribbonlike conformation, and side-chain location would be similar to that proposed for tragacanthic acid.

One way in which tragacanthic acid differs from xanthan is that the electroactive components of xanthan are located within the side chains instead of along the main backbone. This in turn will relate to different functional properties of the two (i.e., xanthan will gel when hydrated with locust bean, whereas tragacanth will not).

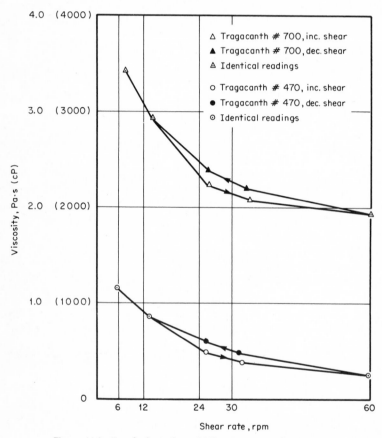

Figure 11.3 Pseudoplastic flow of 1% gum tragacanth solutions.

Hydration of gum tragacanth to a certain extent is the result of the D-galacturonic acid group being ionized, and the $R—COO^-$ groups ionically rearranging (ordering) the water molecules. The reaction is:

$$R—COOH + R—COOM \underset{}{\overset{H_2O}{\rightleftharpoons}} 2R—COO^- + H + M^n$$

This illustrates, in general, the ionization of the uronic acids and the utonate salts when dispersed in water, with a lowering of the pH from 7 to between 5 and 6 for a 1% tragacanth solution. The attraction of the $R—COO^-$ groups with the water causes long-distance as well as short-distance ordering, which, in turn, contributes to the overall thickening effect of the tragacanth.

When the above reaction goes to the acidic by titrating with HCl (see Fig. 11.4), the R—COOH groups produce less ordering of the water, and thus less viscosity.

The titration curve for around the pH of 3 is not ideal for galacturonic acid because of its geometry with respect to the linkage in the tragacanthic acid polymer and inherent salts present. This deviation from the norm causes the curve to become smoother. At higher pH (greater than 7), all the galacturonic acid groups are ionized. When the environment surrounding the hydrocolloid is alkaline with excessive OH ions, the ordering effect on the water by the tragacanth is less effective and a decrease in viscosity also occurs due to beta elimination.

Hydration Figure 11.5 shows the effect that heat has on gum tragacanth while being hydrated, as opposed to hydration at room temperature (25°C). When heat is incorporated into the system, there is a certain amount of degradation (up to one-third loss of viscosity). The introduction of heat into tragacanth solutions promotes rapid hydration. After reaching 90°C (15 min), both tragacanth #700 and #470 are fully hydrated.

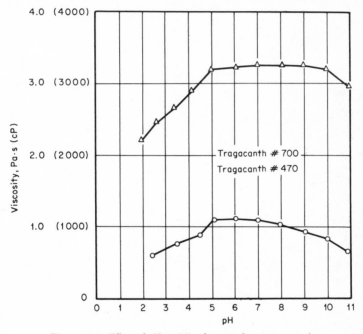

Figure 11.4 Effect of pH on 1% solutions of gum tragacanth.

The continuing increase in viscosity observed in Fig. 11.5 occurs in the cooling process, and is attributable to the shear-thinning phenomenon that heat has on tragacanth, and the shear thickening which occurs upon cooling.

Physical Properties

Rheological Properties *Viscosity of 1% Solutions.* Both the ribbon and flake grades of gum tragacanth are graded by hand. This may cause quality variance from lot to lot. Thus, tragacanth has traditionally been standardized by viscosity, thickness of exudate, and color rather than by ribbon or flake type.

Figure 11.6 shows the variation of typical tragacanth viscosities of 1% solutions with respect to flake and ribbon grades. There is an overlapping of viscosity between high-grade flake and low-grade ribbon. As shown, there is a broad viscosity range. The lowest grade (technical) non-U.S.P. is located to the extreme left. This low-viscosity gum is the result of poor growing and harvesting conditions, as discussed in the introduction. Gum tragacanth #470 and #700 are illustrated as narrow bands in the spectrum of the respective flake and ribbon grades, and will be used herein as representative of flake and ribbon qualities, respectively. The highest ribbon-grade quality is at the extreme right in Fig. 11.6 and has the highest viscosity known for any of the commercially sold gums.

Viscosity vs. Concentration. Figure 11.7 illustrates the logarithmic nature that tragacanth solutions exhibit when solution concentrations are increased to 3.0%. This in not unusual since most gums (except for gum arabic) exhibit this relationship. However, this demonstrates the relationship of flake and ribbon gum when building solution viscosity by using increased proportions of flake tragacanth, or lesser proportions of ribbon grade.

Pseudoplastic Flow. Figure 11.3 shows the effect that shear rate has on 1% tragacanth solutions.   As shear rate (rpm) increases, the apparent viscosity decreases.   This

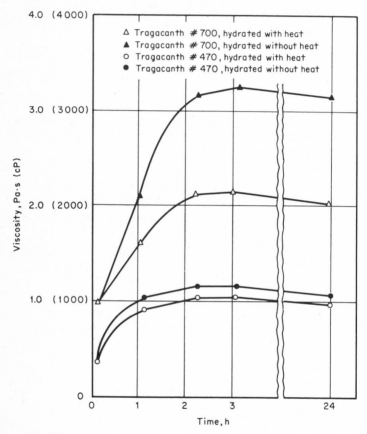

Figure 11.5 Effect of time and hydration on the viscosity of 1% gum tragacanth solutions.

phenomenon is *pseudoplasticity,* a reversible process, with the original viscosity obtained once again upon decreasing shear rate.   Aside from the differences in viscosity, tragacanth #700 and #470 exhibit similar pseudoplastic flow characteristics. Pseudoplasticity assists the suspending properties of tragacanth, in that the application of shear force will ease pourability yet allow the solution to revert back to its original viscosity upon the cessation of force.

The solution structure is sufficient to prevent fine solid particles from settling or oil droplets from rising, yet both solutions pour readily and appear to be thinner when poured, with tragacanth #470 being more pronounced.

The organoleptic properties of this pseudoplastic flow is said to be slightly slimy, as opposed to slimy or nonslimy.[10]

Viscosity versus Temperature.   Elevated temperature has a viscosity thinning effect on fully hydrated solutions of both tragacanth #470 and #700 (see Fig. 11.2).   Upon cooling, the solutions revert to nearly their original viscosities, indicating that the elevated temperature did not degrade the macromolecules structurally.

Viscosity Stability with Acetic Acid.   Gum tragacanth has historically been noted for its acid stability when used in such products as pourable dressing and low-pH sauces. Figure 11.8 illustrates the viscous stability of 1% tragacanth solutions with 1% acetic acid when held for 21 days.   Both tragacanth #470 and #700 demonstrate viscosity stability in this acid medium.

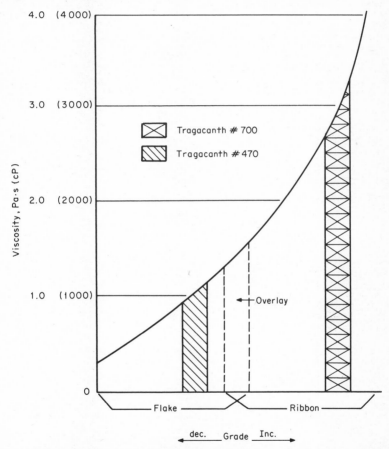

Figure 11.6 Viscosity differences in 1% gum tragacanth solutions going from low-grade flake to high-grade ribbon grade. Tragacanth #470 and #700 properties are represented by the narrow bands, and represent flake and ribbon grades, respectively.

Intrinsic Viscosity.   The intrinsic viscosity [η] recorded for a high-grade flake-type tragacanth[11] measured in an Ostwald viscometer at 20°C and 0.2 N NaCl was calculated to be 12.3 from the formula:

$$[\eta] = \lim_{c \to 0} \frac{\ln \eta/\eta_0}{c} = \lim_{c \to 0} \frac{\eta_{sp}}{c'}$$

where $\eta_{sp} = \eta/\eta_0 - 1$,
   η = viscosity of the solution
   η_0 = viscosity of the solvent
   c = concentration (g/100 g)

Surface Tension   Surface tension readings for tragacanth as determined by a DuNuoy tensiometer, in which a platinum ring of radius r is attached to a tension balance, show that both tragacanth #470 and #700* lower the surface tension of water (see Fig. 11.9).   The force required to remove the ring from the surface or interface is measured directly.   The surface tension acts over the circumference of the ring, so that the applied force $f = 4\pi r\gamma$, or surface tension $\gamma = f/4\pi r$.

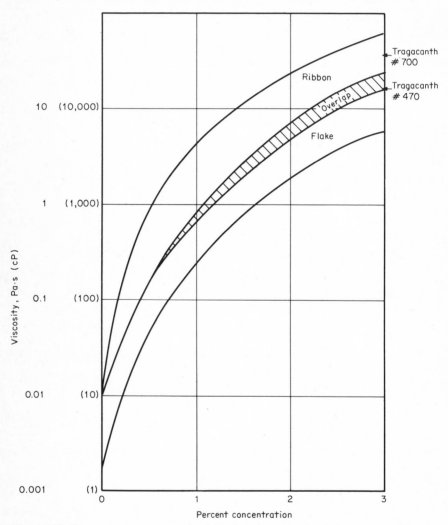

Figure 11.7   Viscosity vs. gum tragacanth solution concentration. The shaded area represents an overlap between the properties of flake and ribbon grades.

A number of correction factors are necessary for calculation.   Type III curves (typical of surface-active agents) are obtained for both tragacanth #470 and #700.   A type III curve defines a rapid lowering of surface tension at low concentrations, whereas a type I curve defines an increase in surface tension above that for water with increasing concentrations of surface active agents.   A type II curve is a decrease in surface tension with increasing concentration of hydrocolloid.

* Grades of gum tragacanth marketed by Tragacanth Importing Corp., New York.

Interfacial Tension Figure 11.10 shows the interfacial tension between vegetable oil and gum tragacanth #470 and #700 solutions. Both solutions reduce the interfacial tension between the oil surface and the aqueous surface, with #470 solution being the most effective. This property, along with the viscous nature of gum tragacanth solutions, promotes stable emulsions. Interfacial tension reading recorded with a DuNuoy tensiometer are determined by measuring the interface between the oil and the solution in a like manner as for surface tension.

Surface activity and interfacial tension reduction are associated with the hydrophilic-lipophilic balance (HLB) of a molecule. Gum tragacanth has a reported HLB of 11.9.[12]

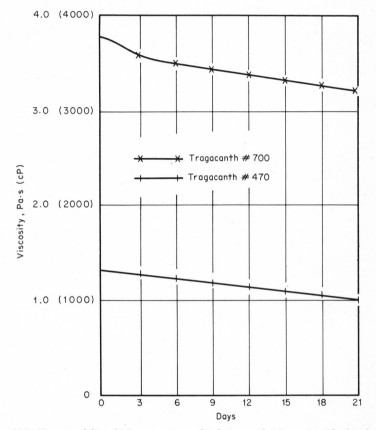

Figure 11.8 Viscous stability of 1% gum tragacanth solutions with 1% acetic acid when held for 21 days.

This HLB cannot be accounted for by looking at the molecular makeup of the gum. It is difficult to find any lipophilic constituents that would account for the lowering of interfacial tension or for an HLB number of 11.9.

However, if consideration is given to the amorphousness and crystallinity of crude ribbon and flake exudates, this phenomenon can be rationalized. Upon hydration, the amorphous regions will have a greater affinity for water than the crystalline regions will, and they will hydrate first; thus at least some of the crystalline regions will stay intact and act as lipophilic (hydrophobic) bodies when in solution.

Based upon interfacial tension reductions, tragacanth #470 should have more unhydrated crystalline regions than tragacanth #700. This is also indirectly supported by the fact that tragacanth #470 solutions are more opaque than tragacanth #700 solutions.

It should be pointed out, however, that other factors, such as degree of purity, would be more likely to affect the reflection of light than unhydrated crystalline bodies.

Compatibility Gum tragacanth is compatible with most gum systems. When solutions are combined with other gum solutions (except for gum arabic), a mathematical viscosity relationship exists. No apparent reactivity exists between gum tragacanth and guar gum, locust bean gum, starch, carboxymethylcellulose propylene glycol alginate, or xanthan.

However, gum tragacanth may be combined with xanthan or propylene glycol alginate to produce stable emulsions that give improved flow and cling characteristics to pourable dressings.

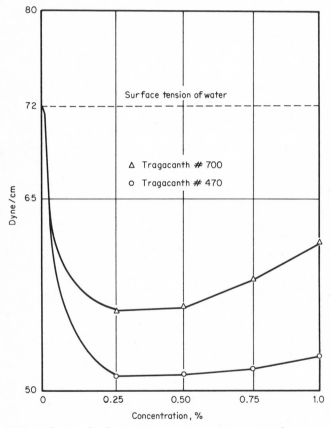

Figure 11.9 Reduction of surface tension in water vs. concentration of gum tragacanth.

Synergism with Gum Arabic. Gum tragacanth, when combined with gum arabic, is very interesting. Unlike some synergistic systems that promote increased viscosity or produce gels,[21] gum arabic lowers the viscosity of gum tragacanth and, in turn, produces emulsions with superior, smooth quality from citrus, cod-liver oil, linseed oil, and mineral oil.

One theory is that there is an ion exchange of salts between molecules, which reduces viscosity. Another idea is that gum arabic is similar to the arabinogalactan polymer in gum tragacanth, in that they are both very water-soluble due to their highly branched structure. This could also have an effect on the final viscosity, since the two highly branched polymers (arabic and arabinogalactan) have greater affinities for available water than does the swellable polymer tragacanthic acid and thus depress its ability to swell. In turn, when this system is used to promote emulsion stability, the water-soluble netlike polymers produce a protective coating film around the noncontinuous oil phase, whereas the tragacanthic acid would swell and thicken the aqueous phase. It is known that gum arabic reduces the surface tension of water,[4] as does gum tragacanth, and thus functionality would also enhance emulsions.

Modified Tragacanth Sulfuric acid monoesters of gum tragacanth can be prepared by treating the gum with a sulfating agent (e.g., sulfuric acid) under conditions wherein there are substantially no other undersized changes in the substrate; i.e., hydrolysis of polysaccharide bonds.[15]

The sulfuric acid monoesters of tragacanth possess a strong pepsin-inhibitory effect, and can be employed for treatment of stomach ulcers and gastritis.

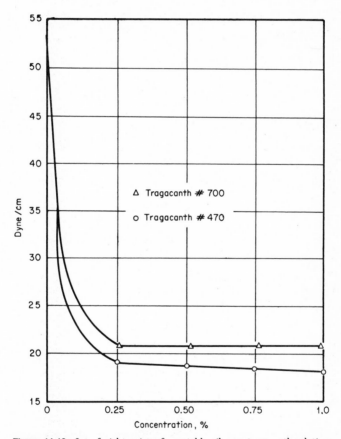

Figure 11.10 Interfacial tension of vegetable oil–gum tragacanth solutions.

Preservatives

Gum tragacanth solutions are unique in that they have extremely long shelf lives without viscosity loss or microbial growth. Possibly the L- sugars found in tragacanth gums (L-arabinose and L-fucose) are responsible for their extraordinary resistance to microbial attack, since most organisms would be unable to metabolize these foreign sugars (most sugars found in nature are D- sugars). A patent has been issued[16] using gum tragacanth for preserving milk, and as a stabilizer for vitamin C.[1] However, when preservatives are needed in a system, benzoic acid (and/or its salt, sodium benzoate), chlorobutanol, or methyl and propyl esters of *p*-hydroxybenzoic acid are effective preservatives when used under the proper conditions.

Biological/Toxicological Properties

The Bureau of Foods, Food and Drug Administration, completed an evaluation of the health aspects of gum tragacanth in December 1972.[5] A select group of qualified scientists were chosen as consultants to make a continuing review, analysis, and evaluation of the available information on each of the GRAS substances, including gum tragacanth. Their conclusions were: "The available information contains no evidence demonstrating that gum tragacanth constitutes a hazard to the public when used in the manner and quantity now practiced."

Consumer Exposure Data　The National Research Council (NRS) has provided information on the possible daily human intake of gum tragacanth.[5] These consumption levels, although considered to be slightly high when compared to the amount of gum imported into the United States, were compared to biological test levels.

Caloric Value　When a stock diet containing 2% gum tragacanth was compared with a control diet containing 2% cellulose, and fed to day-old chicks, the results showed that tragacanth exhibited a 34% depression of growth, whereas the control showed little or no depression.　Similar effects were found for guar gum, locust bean gum, gum karaya, carrageenan, dried okra, and psyllium husk.　The cause of the inhibitory effect was not reported.

Hypercholesterolemia　When 3% each of gum tragacanth and cholesterol were fed to cockerels, hypercholesterolemia was inhibited.

Tumors　Tragacanth powder inhibits ascites tumor growth in mice, acting indirectly as a mitotic block by becoming attached to the cellular membrane occurring in the interphase or early prophase cell.　No carcinogenic or mutagenic properties of gum tragacanth were reported by the Select Committee.

Allergenic Properties　Gum tragacanth, like gum acacia and gum karaya, is known to induce an allergic response by ingestion, contact, or inhalation by susceptible individuals.

However, when given orally at 3 g/day to patients suffering from dermatitis, the gum was reported to act as an antiallergenic therapeutic substance, although it should be stated that the data was incomplete.

Lethal Effects　Pregnant animals were fed by oral intubation with suspensions of gum tragacanth with no discernible effect on maternal or fetal survivors.

The following levels were used:

    Mice, 1200 mg/kg for 10 consecutive days

    Rats, 210 mg/kg for 10 consecutive days

    Hamsters, 900 mg/kg for 5 consecutive days

    Rabbits, 33 mg/kg for 13 consecutive days

However, at higher levels of feeding, significant toxic effects were produced in some groups of animals.

Manufacturing and Quality Control

Gum tragacanth as received from the marketplace after hand grading and sorting is packed in wooden cases or burlap bags.　The initial grading of the gum is based upon color, texture, opacity, and presence of foreign matter.　Representative samples of the specific lots of gum offered by the dealers are forwarded to United States laboratories for more precise evaluation.　Test results are used to blend specific lots to produce a finished powder of known quality.　The most important considerations in the selection of gum tragacanth are viscosity and emulsion stability characteristics.

Crude gum lots are not milled in anticipation of being sold, but are milled as the demand dictates.　This is done to prevent viscosity loss that would occur while the gum is stored as a powder.　After proper selections of crude gum lots are determined for milling, they are subjected to an initial cleaning process to remove bark and other foreign matter.

Attrition mills are used for grinding, in which the gum is ground into itself to reduce the build-up of heat and degradation of the gum.　After milling, the gum is sieved and screened into specific crystals, grains, or powders of various mesh sizes to meet particular customer requirements.

Next the gum is sampled, retested, packed in 45- to 136-kg (100- to 300-lb) polyethylene-lined drums, and stored in a cool, dry place.　If sterilization is needed, the gum is packed in 22.7-kg (50-lb) drums and introduced into a gas-filled chamber to reduce the standard plate count of the gum, but not destroy the viscosity. Ethylene oxide or propylene oxide are the preferred gases. The temperature and humidity must also be controlled throughout the sterilization.

Handling Gum tragacanth is usually packed in polyethylene-lined fiber drums and kept in cool, dry, dark storage prior to shipment. The crude gum is conveyed by express, rail, or air to the designated site of manufacture.

The gum should be used within 3 months to 1 year and should be handled similarly to the handling of spices in a food factory. It is not poisonous, toxic, or explosive, but care should be taken to prevent water or moisture from entering the product.

Additives and Extenders Gum tragacanth is sold as crude exudate, coarse crystals, or free-flowing powder. No additives or extenders should be in the crude gum. In the early days, some devious brokers would adulterate the gum with gum karaya. This is a cheaper gum and is not acid-stable, so provisions are made in the *United States Pharmacopeia* for the detection of the contaminant.

Application Procedures

Like other hydrocolloids, gum tragacanth is very hydrophilic, and thus care must be taken in the incorporation of the gum into solutions.

To avoid lumping and/or fish eyes, the gum should be dispersed by conventional methods such as:

1. Prepare a slurry with a portion of the insoluble nonaqueous material used in the product (i.e., vegetable oil, glycerol, glycerine, propylene glycol, citrus oil, or other alcohol or lipid). The ratio of immiscible liquids to the gum tragacanth is usually from 3:1 to 5:1.

   This slurry then may be introduced into the aqueous phase. The nonmiscible solvent will slow down the penetration of the water and thus prevent lumping, caking or fish eyes, allowing the hydrocolloid to swell and imbibe water uniformly.

2. Or the gum may be dry-blended with dry ingredients such as salt (NaCl or other monovalent salts) or sugar (sucrose or others) at a 1:5 to 1:15 ratio of the gum to the dry solids. This dry blend is usually dusted into the vortex of the aqueous phase in an agitation system. Heat may be employed to facilitate hydration, but caution should be taken to compensate for the viscosity loss due to the introduction of the heat.

When acid is used in the system, it is desirable, if possible, to hold it back until the gum tragacanth has had time to hydrate. If this is not possible, again an adjustment must be made to compensate for the viscosity loss due to acid.

Trivalent salts should be avoided in formulations of gum tragacanth, unless some protective additives are employed to react with the salts instead of the galacturonic acid found in the gum.

Specialties Gum tragacanth solutions may be dried to produce water-soluble films when cast on stainless steel or Teflon surfaces. This could be expanded into water-soluble packaging material if proper plasticizers are used. If the films are painted with a basic or neutral lead acetate, they become insoluble in water.

Future Developments

The world supply of food has become of major importance. The United Nations World Food Conference held in November 1974 in Rome clearly showed a new awareness of the balance of world supply and demand on foodstuffs.

Although gum tragacanth is not used for its nutritional value as such, it should play an active role in stabilizing new food products made from soy isolates, marine sources, or other new food products not yet on the drawing board.

Gum tragacanth should be used in such systems where acid stability, salt compatibility, and heat stability are factors in the stabilization. Such products might include nutritional acidulant drinks, acidulated emulsions, creamfills, compressed bars, or tablet-type products.

Gum tragacanth should retain its status as a prominent stabilizer in pourable dressings, the reasons being, aside from acid stability, long shelf life and salt compatibility. Gum tragacanth gives to pourable dressings the rheological flow and cling properties that are preferred over those resulting from other gum systems.

While the demand for foodstuffs increases, the supply of natural gums is relatively fixed. Coupled with the observation that the microbial biopolymers (e.g., xanthan) or synthetic gums (e.g., CMC, methylcellulose) have not been able to successfully stopgap the demand, and coupled with a greater demand still to come, gum tragacanth supplies should be ample to meet demand in the foreseeable future. However, we should be aware that the nations currently harvesting gum tragacanth are agriculturally oriented, and as industry moves into these areas, the local gum pickers may choose to seek more lucrative industrial employment.

With the continuous concern for lowering the caloric intake of diets in the Western Hemisphere, it appears that the low-calorie food market has only been scratched. As new and improved flavorants develop for pourable dressings with low oil content and increased amounts of hydrocolloids, the demand for gum tragacanth along with the other acid-stable gums will increase proportionately to the market growth.

Again, gum tragacanth's rheological flow and cling properties in low-calorie products are more closely related to the rheological properties of the high-oil dressings than are those of the other gum systems.

COMMERCIAL USES: Compounding and Formulating

Table 11.1 illustrates the broad range of applications for gum tragacanth: from food products to pharmaceuticals and cosmetics, to industrial uses.

Pharmaceutical and Medical

Gum tragacanth mixed with glycerin and water and thoroughly mixed until homogeneous will result in a thick paste. This paste, tragacanth glycerite, is a useful excipient to bind tablet masses. The water-swelling property of tragacanthic acid along with the surface tension reduction properties of gum tragacanth and glycerin are two functional considerations. The adhesiveness is derived from the synergism of the glycerin and the gum tragacanth.

Oral Ingestion Gum tragacanth is one of the oldest drugs in the *Materia Medica*. It has an official listing in every edition of the *United States Pharmacopeia* since 1820. Therefore many oral-intake suspensions, pills, and emulsions have been formulated with gum tragacanth. Oral suspension of procaine penicillin is one such product.

Toothpastes Gum tragacanth acts as a thickener, suspender, binder, and emulsifier in toothpaste formulations. Functional properties such as viscosity, pseudoplastic flow, interfacial tension between aqueous and oil phase, suspension properties, and room temperature hydration all play important roles for the stabilization of toothpastes.

Mineral and Fish Oil Emulsions Gum tragacanth used alone or in combination with gum arabic will function as an emulsifier in oil-water systems because of its ability to thicken the aqueous phase and because it lowers the interfacial tension between the oil and water. This, combined with its long shelf life and smooth texture, has resulted in the use of gum tragacanth for generations. The absorption of poorly soluble substances, such as fat-soluble vitamins and steroid glycosides, is facilitated by this emulsion.

External Lotions Gum tragacanth mucilage is used to stabilize hair and hand lotions by giving the products the proper body needed for application. In hair lotion in which there is 10% isopropanol, gum tragacanth is able to withstand a certain amount of alcohol without precipitating out. When an external lotion system contains insoluble oils or fats, then the emulsification properties of gum tragacanth are utilized. Spermicidal jellies and creams may also be formulated with gum tragacanth.

Food Products

Pourable Dressings Historically, gum tragacanth has been used as a thickener of the aqueous phase of oil-in-water systems, but more recently it has been shown that gum tragacanth also lowers the resistance between the oil and aqueous phase.[23]

Both properties keep the oil from coalescing with eventual oil-water separation. This, plus flow properties, acid resistance, and heat stability, has resulted in gum tragacanth's use in pourable dressings for generations. The resulting emulsions show heat-cool stability, good cling and pseudoplastic flow properties, and long shelf life.

Such dressings, regular as well as low-calorie, are French, Thousand Island, Roquefort, blue cheese, and other specialty oil-in-water emulsion dressings. The main difference between regular and low-calorie dressings is that the oil is substantially reduced in the latter (i.e., from 35% or more, down to 5 to 15%), and increased amounts of the gum are used to provide the body that the higher oil would give.

Mayonnaise-Type Products Although stabilizers such as gum tragacanth are not allowed by the government in their standard of identity for mayonnaise, they can and have been successfully incorporated into imitation mayonnaise and/or low-calorie mayonnaise. Usually starch is also included in such products. The gum tragacanth is used to give the product body and emulsification, whereas the starch is used to give body and textural properties similar to mayonnaise (i.e., setting and peeking).

Mustard Sauce Ground mustard seed, vinegar, spices, water, and salts have a tendency to crack and bleed (serum off) when milled into a sauce. This shortcoming can be corrected by the inclusion of gum tragacanth. The acid stability, pseudoplastic flow, emulsion (mustard oil), and thickening properties all contribute to the end product's long shelf life, texture properties, and control of syneresis.

Barbecue, Spaghetti, and Pizza Sauces Gum tragacanth has similar functional properties for these products as for mustard sauce, except that here the purée is the vegetable component to be stabilized in a low-pH system. Also heat of hydration must be considered.

Low-Calorie Milkshakes Gum tragacanth is used in low-calorie milkshake-type products to build viscosity, which is needed because of the reduced level of ice cream used. The gum will suspend solids and help emulsify the immiscible fats in the system. The pseudoplastic flow gives the end product a mouth-feel similar to milkshakes made with increased solids.

Pulpy Beverages The suspending and pseudoplastic properties of gum tragacanth are utilized to stabilize insoluble solids in fruit juices, nectar, papaya juice, and apple and fruit punches. Also, when gum arabic is incorporated with gum tragacanth, the suspension property is enhanced in much the same manner as when combined in flavor emulsions. The highly branched polymers, the arabinogalactan of gum tragacanth and gum arabic, along with the electrochemical properties of the tragacanthic acid, promote Brownian motion of the suspended particles. Also, the gum arabic lowers the viscosity of gum tragacanth and eliminates the undesirable slimy, high-viscosity mouth-feel.

Sandwich Spreads Most sandwich spreads are mayonnaise-based products to which flavor or solid masses (meat or texturized protein) are added. The extra load that the mayonnaise must carry is helped by the introduction of gum tragacanth. Preferably, gum tragacanth should be added directly during the manufacture of the mayonnaise, but this would exclude it as a true mayonnaise by the standard of identity. However, this will produce a stronger mayonnaise by reason of the previous discussion. Thus when the flavor components are added to this base, the spread will have a longer shelf life.

Meringues Meringues are air-in-liquid emulsions which are stabilized by the egg white protein. Gum tragacanth may be incorporated as a cold-process stabilizer to reduce or slow down the leaky meringue and increase the shelf life by several hours, as well as to enhance the protein-stabilizing effect.

Gum tragacanth, as well as egg white protein, reduces the surface tension of the liquid phase, promotes the incorporation of air sacks into the liquid, and slows down the movement of the air sacks to the outer perimeter of the meringues. If this movement were not slowed, it would cause serum separation (leaks).

The acid stability is not a consideration in this type of product, which has a nearly neutral pH.

Pickle Relish Gum tragacanth is used in pickle relish to increase the drained weight of the relish. Its more subtle characteristics include retention of natural color, impartation of "cling" to the product, and promotion of flavor release. The gum also gives the product a clean breakaway mouth-feel because of its shear-thinning nature.

The functional characteristics of gum tragacanth utilized in pickle relishes are resistance to acid (i.e., vinegar), compatibility with salts (monovalent, that is, NaCl), pseudoplastic flow (shear thinning), and long shelf life.

Bottle Emulsions See Synergism with Gum Arabic for a theoretical discussion on citrus emulsions stabilized with arabic and tragacanth gums. These emulsions must have the ability to be diluted 50- to 200-fold, yet still maintain the flavor oils in an emulsion suspension throughout the beverage.

Ice Cream Stabilizers In a review of ice cream stabilizers,[17] it was found that gum tragacanth is one of the best stabilizers among several evaluated, and the suggested

TABLE 11.1
Functions of Gum Tragacanth in Various Products and Processes

Functional characteristics of gum tragacanth*
(See key at end of table)

Product	1	1a	1b	2	2a	2b	3	3a	3b	4	4a	4b	5	5a	5b	5c	5d	6	6a	7	8	9	10	11	12
Food																									
Dairy products:																									
Ice cream stabilizer	4	x	x	5	x	x	3	x	x	7	x		6			x	x				2				1
Ice milk	6	x	x	4	x	x	3	x	x	7	x		5	x		x	x				2				1
Milkshake	5	x	x	3	x	x	2	x	x				1	x		x	x								4
Sherbert	6	x	x	4	x	x	3	x	x	7	x		5	x		x	x	5	x		2				1
Ice pops and water ices				5	x	x	6	x	x				3				x	4	x		2				1
Chocolate milk drinks				1	x								2		x		x				4				3
Flavored milk drinks				2	x	x	3	x	x				1	x	x	x	x								
Puddings							1	x	x				3	x	x	x	x								2
Neufchatel-type process cheese	1	x	x							6	x		4				x	3	x		2				5
Cheese spread	1	x	x							4	x		5	x							2				3
Beverages:																									
Soft drink with fruit pulp				1	x								4		x		x	3	x		2				
Fruit juices and nectars				1	x								4	x	x	x	x	3	x		2				
Dry beverage mixes				3	x		1	x	x												2				
Bakery products:																									
Meringues	1	x	x																		2	3	4		
Bakers' citrus oil emulsions	1	x	x				4	x	x				3	x	x	x	x				2				
Frozen pie fillings				1	x	x	2	x	x				3			x	x	4	x			5			6
Miscellaneous:																									
Whipped toppings	1	x	x																					2	
Quick cooking cereal				3	x	x	1	x	x				2	x		x	x								
Dressings and sauces:																									
French dressing	2	x	x	4	x	x	3	x	x	6	x		5	x		x	x	1	x						
Salad dressing	2	x	x	4	x	x	3	x	x	6	x		5	x		x	x	1	x						
Syrups and toppings	5	x	x				3	x	x				1		x	x	x				2				4
Relish										1	x		3	x	x	x	x	2	x						
White sauces and gravies	3	x	x				1	x	x				2			x	x								
Cocktail, barbecue, and spaghetti sauces				3	x	x	1	x	x	5	x		4	x	x	x	x	2	x			3			
Catsup				1	x	x	3	x	x				5	x		x	x	2	x			4			
Confections:																									
Candy gels and jellies																		6	x		2	3	4	5	1
Caramels, nougats, taffy																					2	3	4	5	1
Candy glaze																					2	3		4	1
Cough drops and lozenges																					1		2		
Gum drops, jujubes, pastilles																					1		2		
Dietetic Foods:																									
French and salad dressings	2	x	x	4	x	x	3	x	x	5	x		6	x		x	x	1	x						
Syrups	4	x	x				1	x	x				3		x	x	x								2
Puddings	5	x	x	4	x	x							3	x	x	x	x								2
Sauces	4	x	x				1	x	x	3	x	x	2	x	x	x	x								
Beverages							1	x	x				2		x		x								
Spreads (high polyunsaturated)	1	x	x				2	x	x				3				x								
Pharmaceuticals/cosmetics																									
Salves and jellies																									
Gynecological jellies	1	x	x				2	x	x				3	x			x								
Water-soluble ointment	3	x	x	2	x	x	1	x	x																
Water-soluble salve	3	x	x	2	x	x	1	x	x																
Emulsions:																									
Mineral oil	1	x	x				2	x	x				3				x								
Fish oil	1	x	x				2	x	x				3				x								
Cosmetic creams	2	x	x	3	x	x	1	x	x																
Cosmetic lotions	2	x	x	2	x	x	1	x	x																

Functional characteristics of gum tragacanth*
(See key at end of table)

	1	2	3	4	5	6	7	8	9	10	11	12
	a b	a b	a b	a b	a b c d	a						
Pharmaceuticals/cosmetics												
Suspensions:												
Facial clays		1 x x	2 x x						3	4	5	
Makeup	3 x x	1 x x	2 x x									
Antacids (liquid)	2 x x	1 x x	4 x x		3 x x x x							
Miscellaneous:												
Pills							1		3	2		
Tablets							1		3	2		
Toothpaste	2 x x	1 x x	3 x x		4 x							
Hair dressing	2 x x	1 x x	3 x x		4 x x x							
Textiles												
Sizing agents		1 x x	2 x x									
Printer's paste	2 x x	1 x x				3						
Printer's ink	2 x x	1 x x				3						
Paper												
Printer's ink	2 x x	1 x x				3						
Sizing agents		1 x x	2 x x						5	4	3	
Miscellaneous products												
Auto polishes	1 x x	3 x x	2 x x	4 x x								
Furniture polishes	1 x x	3 x x	2 x x	4 x x								
Insecticide sprays	1 x x	2 x x			3 x							
Insecticide emulsions	1 x x	2 x x			4 x x x							
Cigar wrapper leaf									3	1	2	
Industrial processes												
Crayon manufacture		1 x x	3 x x		2 x x x x							
Wooden matchstick manufacture		1 x x	3 x x		2 x x x x							
Ceramics manufacture		1 x x	3 x x		2 x x x x							

* Arranged in order of approximate importance by number.

Key to Column Numbers for Functional Characteristic of Gum Tragacanth
  1. *Emulsification agent*
     a. By increasing aqueous phase viscosity
     b. By lowering interfacial tension between oil and water
  2. *Suspending agent*
     a. From repelling action of galacturonic acid salt found in tragacanthic acid
     b. From pseudoplastic flow (shear thinning)
  3. *Thickening agent*
     a. By the ordering of water molecules
     b. By the hydration of the water-swellable fraction
  4. *Compatibility with salts*
     a. Mono- and divalent
     b. Precipitation by trivalent
  5. *Rheological properties*
     a. Pseudoplastic flow (shear thinning) at 0.5% and greater concentrations
     b. Newtonian flow of dilute solutions
     c. Shear thinning of elevated temperatures
     d. Viscosity characteristics
  6. Acid *stability properties*
     a. Presence of acid-stable saccharides
  7. *Synergistic effect with gum arabic*
  8. *Water-retention properties*
  9. *Binding properties*
  10. *Adhesive properties*
  11. *Film-forming properties*
  12. *Alter crystallization of other components (e.g., sugar)*

Ice Cream Stabilizers   In a review of ice cream stabilizers,[17] it was found that gum tragacanth is one of the best stabilizers among several evaluated, and the suggested concentration range was 0.2 to 0.5%. The surface tension of ice cream mixes ranges from 48 to 53 dyn. Gum tragacanth #470 and #700 are in or near this surface tension range at 0.25 to 0.5% concentration (see Fig. 11.8 relative to the recommended range for stabilizers in ice cream formulations). It is likely that when incorporating stabilizers into ice cream systems, it is more beneficial to use gums and/or surface-active agents that closely compare in this respect to the system in question. Gum tragacanth does this without the addition of surface-active reagents (i.e., mono- and diglycerides). Other stabilizers (i.e., guar gum, locust bean gum, and carrageenan) usually require the presence of such surface-active reagents. This leads to the belief that the surface tension characteristics of the stabilizing systems is an important consideration when selecting ice cream stabilizers.

Other functional considerations of gum tragacanth in ice cream are cold-mix hydration, hydration with heat [high-temperature, short-time (HTST)], and reduction of crystal size (lactose or ice). Such considerations result in smooth texture, heat-shock resistance, slow and creamy melt-down, and no masking of the flavor of the finished product.

Water Ices   Flavored ices, when made without the incorporation of a stabilizer, have a tendency to be "sucked out." That is, the flavors solubilize first and leave an ice core. By the incorporation of gum tragacanth into the system, the ice crystals are reduced in size and the flavors are simultaneously released with the melting of the ice block.

COMMERCIAL USES: Processing Aids

Crayon Manufacture

Gum tragacanth is used in the manufacturing process of crayons to promote initial adhesiveness between the color pigment, Chinese clay, and water. However, its main function is its ability to be burned out of the product, after extrusion, but while in the furnace. This burning process helps keep the clay pores open. Thus, when hot, molten paraffin is introduced into the dried clay-pigment core, it is imbibed more readily and uniformly.

Ceramics Manufacture

Gum tragacanth is used in ceramics manufacture for the same functionality as for crayons in that it acts as an initial adhesive or suspending agent of components, then is burned out in the furnace.

Leather Curing

Gum tragacanth is used in the tanning industry as a processing aid. Gum solutions (0.5 to 1.5%) are brushed onto the side of the hide. The hides are stuck onto heating plates to be cured, thereby becoming leather. After the hides have been properly cured on the plates, they are peeled off. Thus gum tragacanth must have good adhesive properties to hold the hide onto the vertical plates during processing and have a good release characteristic (workability), resulting in a smooth, clear, porous product to which leather conditioners may then be added.

Textiles Processing

Gum tragacanth is used to a limited extent in the textile industry. However, many other hydrocolloid systems are more economical. Thus, less expensive gum systems are preferred, since such functionality as acid stability and prolonged shelf life are not needed.

Paper Processing

Gum tragacanth plays a similar role in paper manufacturing as it does in textiles.

Wooden Match Manufacture

Historically, gum tragacanth has been used in the wooden match industry to suspend solids, prior to coating the wooden sticks. The gum also improves the flow properties of the liquid and enhances the adhesion of animal glue used in the system. Therefore the tragacanth is used as a processing additive as well as a final-product conditioner.

INDUSTRIES USING GUM TRAGACANTH

The following is a summary of the various commercial and industrial uses of gum tragacanth as discussed in the preceding sections of this chapter and as shown in Table
11.1.

Food Industry

Gum tragacanth is used in the food industry to:
- Stabilize air-in-liquid emulsions; meringues and ice cream mixes.
- Thicken and emulsify oil-in-water emulsions; citrus oil emulsions, Italian, French,
Thousand Island, Roquefort, blue cheese, and Russian dressings, pizza, spaghetti, mustard
and barbecue sauces, processed cheese, and milkshakes.
- Suspend insoluble solids; soft drinks (with fruit pulp) fruit juices, nectar, papaya
juice, apple drink, and fruit drink punch.
- Bind and act as an adhesive in products such as cough drops, lozenges, gum drops,
jujubes, pastilles.
- Control moisture in ice cream, ice milk, sherbert, ice pops, water ices, and frozen
pie fillings.

Pharmaceutical and Cosmetic Industries

Gum tragacanth is used in pharmaceutical and cosmetic manufacture to:
- Thicken and lubricate in gynecological jellies.
- Thicken for water-soluble ointment and salve bases.
- Suspend solids and liquids in water-based syrups and medicants for oral ingestion.
- Stabilize mineral oil and fish oil emulsions.
- Bind tablets and pills.
- Thicken cosmetic creams and lotions.
- Bind makeup and facial clays.
- Thicken toothpastes.
- Thicken and act as film-former in hair dressings.

General Industrial Uses

Industrially gum tragacanth finds many applications such as:
- Sizing of textiles.
- Thickening of textile printing pastes.
- Thickening and binding in furniture and auto polishes.
- Temporary binding for ceramics manufacture.
- Film forming and stabilizing in insecticide emulsions.
- Adhesion in cigar-wrapper leaves.

FORMULATIONS

Twenty-two typical food-product formulations that use gum tragacanth are given in
the following pages, including dressings, sauces, syrups, puddings, dips, relishes, toppings,
spreads, and low-calorie foods.

Italian Dressing

	Percent
Vegetable oil	66.0
Vinegar (10% white)	12.0
Water	16.0
Salt (fine flake)	3.5
Sugar (granulated)	1.0
Onion (minced)	0.2
Garlic (chopped)	0.3
Black pepper (coarse)	0.3
Red pepper (crushed)	0.2
Gum tragacanth,	
440 USP powder	0.5

Pretested gum tragacanth provides product stability during mixing, and then settles down.

Mix all dry ingredients into water.   Add vinegar and oil, with continued agitation. Continue agitation during filling.   No milling or homogenizing is necessary.

Russian Dressing

	Percent
Vegetable oil	50.0
Tomato paste (28–30% solids)	5.0
Sugar (granulated)	3.2
Water	14.8
Vinegar (10% white)	12.5
Carotenal solution #2 or	
oleoresin paprika (40,000 C.V.)	0.2
Salt (fine flake)	1.2
Onion powder	0.4
Mustard powder	0.5
Garlic powder	0.1
Celery (soluble)	0.1
Black pepper (ground)	0.1
Egg yolk (dry)	2.0
Dill pickle relish (drained)	8.5
Gum tragacanth, "C"	
USP powder	1.4

Mix egg yolk with water.   Add dry ingredients and vegetable oil.   Add tomato paste and rest of liquid ingredients with agitation.   Incorporate drained relish before pumping through colloid mill.

Blue Cheese Dressing

	Percent
Vinegar (10% white)	16.8
Vegetable oil	45.0
Salt (fine flake)	3.4
Egg yolk (dry)	3.5
Sugar (granulated)	3.3
Water	16.6
Onion powder	0.3
Garlic powder	0.1
Celery (soluble)	0.4
Black pepper (ground)	0.2
Gum tragacanth, 440	
USP powder	0.7
Blue cheese (crumbled)	9.5
Potassium sorbate	0.1
Sodium benzoate	0.1

Make emulsion of egg yolk with water, dry ingredients, and vegetable oil.   Add vinegar and blue cheese.   Pump through colloid mill.

French Dressing

	Percent
Water	28.8
Vinegar (10% white)	13.2
Sugar (granulated)	9.5
Salt (fine flake)	3.0
Paprika	1.3
Mustard powder	0.6
Onion powder	0.4
Garlic powder	0.2
Vegetable oil	39.5
White pepper (ground)	0.2
Tomato paste (28–30% solids)	1.3
Tomato catsup (30–33% solids)	1.3
Gum tragacanth, "C"	
USP powder	0.7

Procedures in mixing and emulsification depend on available equipment, ranging from Hobart mixer to colloid mill. Spicing may be varied. If no paprika specks are desired, oleoresin paprika may be substituted; soy oil is generally used as the vegetable oil.

Low-Calorie Italian-Type Dressing

	Percent
Vinegar (10% white)	20.00
Vegetable oil	4.00
Water	69.50
Salt	3.40
Gum tragacanth, 440 USP powder	1.75
Celery (soluble)	0.10
Garlic (chopped)	0.25
Onion (minced)	0.65
Black pepper (coarse)	0.13
Calcium saccharin	0.02
Potassium sorbate	0.10
Sodium benzoate	0.10

This is a good basic formula with low oil component. Viscosity can be varied by modifying amount of gum tragacanth and type of homogenizer.

Mix dry ingredients. Add to water with agitation. Next, add vinegar and oil, continuing agitation. Pump through colloid mill or homogenizer.

Sweet and Sour Sauce

	Percent
Water	28.8
Vinegar (10% white)	13.2
Sugar (granulated)	9.5
Salt (fine flake)	3.0
Paprika	1.3
Mustard powder	0.6
Onion powder	0.4
Garlic powder	0.2
Vegetable oil	39.5
White pepper (ground)	0.2
Tomato paste (28–30% solids)	1.3
Tomato catsup (30–33% solids)	1.3
Gum tragacanth, "C" USP powder	0.7

Procedures in mixing and emulsification depend on available equipment, ranging from Hobart mixer to colloid mill. Spicing may be varied. If no paprika specks are desired, oleoresin paprika may be substituted; soy oil is generally used as the vegetable oil.

Low-Calorie French-Type Dressing

	Percent
Water	49.70
Vinegar (10% white)	25.00
Gum tragacanth, "C" USP powder	2.80
Mustard powder	0.85
Celery (soluble)	0.35
Onion powder	0.55
Garlic powder	0.45
White pepper (ground)	0.28
Calcium saccharin	0.02
Vegetable oil	7.50
Tomato paste (28–30% solids)	8.80
Potassium sorbate	0.10
Sodium benzoate	0.10
Carotenal solution #2 or oleoresin paprika (40,000 C.V.)	0.10
Salt	3.40

This is a good basic formula with low oil content. Procedure depends on equipment. Note use of preservatives due to high moisture/low solids. Increasing mustard within limits will aid emulsion stability.

Barbecue Sauce

	Percent
Tomato paste (28–30% solids)	5.0
Tomato catsup (30–33% solids)	6.0
Prepared mustard	3.0
Vinegar (10% white)	8.0
Water	15.0
Sugar (granulated)	40.0
Corn syrup	10.0
Garlic powder	0.4
Onion powder	1.5
Celery (soluble)	0.5
Black pepper (ground)	0.3
Cayenne pepper (ground)	0.3
BBQ seasoning	3.0
Smoke flavoring	0.2
Gum tragacanth, "C" USP powder	1.0
Salt (fine flake)	5.8

Mix dry ingredients with water. Add tomato products, mustard, and vinegar. Pump through colloid mill.

Dietetic (Artificial) Fruit Jelly

	Percent
Gum tragacanth, 500 USP powder	0.15
TIC colloid 600	0.05
Low methoxy pectin	0.20
Sodium saccharin	0.35
Sorbitol	4.00
Sodium benzoate	0.10
Potassium sorbate	0.05
Dicalcium phosphate	0.15
Citric acid	0.30
Water	40.00
Fruit juice (14–15°Brix)	54.65

Mix all dry ingredients. Heat water and fruit juice to 82 to 88°C (180 to 185°F). Add dry ingredients with agitation, and mix for 5 min. Fill at temperature of at least 79°C (175°F).

Citrus-Flavor Beverage Emulsions

	Percent
Orange or lemon oil	15.00
Spray-dried gum arabic, USP powder	10.00
Gum tragacanth, 400 USP powder	2.50
Citric acid	0.40
Sodium benzoate	0.10
Water q.s. to equal	100.00

Add gum arabic and gum tragacanth to oil with agitation. Gradually add this mixture to the water into which the citric acid and benzoate have been dissolved. Homogenize this combination at 34.5 MPa (5000 psi) using piston-type homogenizer.

Low-Calorie Chocolate Syrup

	Percent
Water	90.25
Calcium saccharin	0.15
Cocoa (18% fat)	9.00
Gum tragacanth, 500 USP powder	0.50
Sodium benzoate	0.10

Homogenize. This is a basic formula for a low-fat chocolate pudding without added sugar. It can be used with added whole milk or low-fat milk, or as a beverage base.

Low-Calorie Chocolate Pudding

	Percent
Gum tragacanth, 440 USP powder	0.50
TIC colloid 700	2.50
Calcium gluconate	4.00
Diabasic potassium phosphate	4.50
Calcium saccharin	0.02
Vanilla flavor (dry)	0.23
Nonfat milk (dry)	66.25
Cocoa (18% fat)	22.00

This dry mixture is used for a packaged pudding. Mix 28.5 g (1¼ oz) into 237 mL (1 cup) of water for 1 to 2 min. Gel forms in 30 min under refrigeration.

Marshmallow Topping

	Percent
Cane sugar	25.50
Corn sugar	12.75
Corn syrup	38.25
Gum tragacanth, "C" USP powder	0.75
Water	12.25
Egg white solution (1 part dried egg whites + 3 parts water)	10.00
Vanilla flavor	0.50

Premix gum tragacanth and part of cane sugar. Mix this into the water with agitation. Add remainder of dry ingredients, followed by the syrup and egg white solution. Boil to 111°C (232°F). Cool to 88°C (190°F). Whip stiff. Slowly add syrup at 88°C (190° F). Beat to desired stiffness. Fold flavor in.

Nondairy Sour Cream

	Percent
Part A:	
Water	76.15
Avicel RC-510*	5.50
Sorbitol solids	2.00
Gum tragacanth #470 USP powder	0.30
Part B:	
Coconut fat, M.P. 37°C (98°F)	15.00
Monoglyceride (from hydrogenated cottonseed oil)	0.75
Polysorbate 60†	0.30

* Microcrystalline cellulose, American Viscose Div., FMC Corp.
† Manufactured by ICI United States, Inc.

To make Part A, disperse the Avicel into the water with agitation. Mix until the dispersion thickens. Add gum tragacanth and Sorbitol. Begin heating with stirring. Combine the melted fat and emulsifiers of Part B with the heated Part A mix. Cool to 43°C (110°F) under agitation, and then homogenize at 10.3 MPa (1500 psi). Cool to 10°C (50°F), and add flavoring ingredients. Careful attention to these steps is important.

Suggested flavoring for a 1000-g batch is as follows:
2 g of table salt.
6 mL of Stabilac No. 2 (water, citric acid, and imitation flavor), made by Meyer-Blanke Company, St. Louis, Mo. 63110.
10 mL of a 10% solution of Imitation Butter Flavor V-14, 431 (butter flavor added to oil phase), made by International Flavors and Fragrances, New York, N.Y. 10017.

Toasted Onion-flavored Chip Dip

	Percent
Part A:	
Water	77.15
Avicel RC-591	4.50
Sorbitol solids	2.00
Gum tragacanth #470	
U.S.P. powder	0.30
Part B:	
Coconut fat, M.P. 37°C (98°F)	15.00
Monoglyceride (from hydrogenated	
cottonseed oil)	0.75
Polysorbate 60	0.20

To make Part A, disperse the Avicel into the water with agitation. Mix until the dispersion thickens. Add tragacanth and Sorbitol. Begin heating with stirring. Combine the melted fat and emulsifiers of Part B with the heated Part A mix. Cool to 43°C (110°F) under agitation, then homogenize at 10.3 MPa (1500 psi). Cool to 10°C (50°F), and add flavoring ingredients.

Suggested flavoring for a 100-g batch is as follows:
2 g of table salt.
6 mL Stabilac No. 2 (see Nondairy Sour Cream).
67 g of toasted onion dip mix #63730, made by McCormick & Co., Inc., Cockeysville, Md. 21030.

Mexican-flavored Chip Dip

	Percent
Water	58.03
Vegetable oil	15.00
Avicel RC-591	4.50
Gum tragacanth #470 USP powder	0.30
Polysorbate 60	0.20
Vinegar (50 grain)	9.69
Spice mix	8.08
Tomato paste (26% solids)	2.74
Dehydrated red bell peppers	0.90
Dehydrated onion	0.54
Hickory smoke flavor	0.02

Disperse the Avicel into vigorously agitated water, and mix for 5 min. Add gum tragacanth combined with oil and emulsifer. Continue mixing until the mixture is smooth. Pass the mixture through a homogenizer or colloid mill to form an emulsion. Hydrate all spices and flavorings in vinegar; then slowly blend them into the emulsion. Heat the emulsion under agitation to 88°C (190°F) in a steam-jacketed kettle. Hot-fill the product into a sterile container, and seal.

Tuna, Chicken, and Ham Salad Spreads

| | Percent | | |
	Tuna	Chicken	Ham
Water	35.55	33.74	34.70
Corn oil	30.00	30.00	30.00
Vinegar (50 grain)	14.00	14.00	14.00
Sorbitol solution	10.00	10.00	10.00
Avicel RC-501	6.00	6.00	6.00
Chicken broth	—	2.00	—
Ham broth	—	—	2.00
Salt	1.50	1.50	1.40
Monosodium glutamate	0.60	0.60	0.30
Onion powder	0.60	—	—
Lemon juice	0.50	0.50	0.30
Mustard powder	0.40	0.50	0.50
Gum tragacanth #470			
USP powder	0.30	0.30	0.30
Polysorbate 60	0.30	0.30	0.30
Celery (soluble)	0.20	0.50	—
Clove	—	—	0.20
Garlic powder	0.04	—	—
White pepper	0.01	0.06	—

Disperse the Avicel into water with a planetary mixer or other suitable high-shear device. Add the gum tragacanth to the Avicel gel. Combine the corn oil and the Polysorbate 60. Add this combination slowly to the gel with mixing. Add the Sorbitol solution. Dry-blend the MSG and spices, then add to the emulsion. Add broth, salt, and vinegar. Homogenize at 13.8 MPa (2000 psi).

Cole Slaw Dressing

	Proportions
Mayonnaise	75.7 L (20 gal)
White vinegar (25 grain)	11.4 L (3 gal)
Sugar (granulated)	1.81 kg (4 lb)
White pepper	113.4 g (4 oz)
Salt	340.2 g (12 oz)
Celery (soluble)	56.7 g (2 oz)
Onion powder	226.8 g (8 oz)
Garlic powder	56.7 g (2 oz)
Gum tragacanth #440	340.2 g (12 oz)

To manufacture, dry-blend the gum, sugar, and salt. Next, mix the vinegar with the remaining dry ingredients, using a mixing tank. Add the gum, sugar, and salt combination to the mixer with the vinegar mixture, and mix for 10 to 15 min. Add the mayonnaise last, and mix until homogenized.

Imitation Mayonnaise Dressing

	Percent
Water	57.47
Vinegar (50 grain cider)	5.50
Vinegar (100 grain white)	10.50
Salt	2.25
Mustard (dry)	0.80
Onion powder	0.04
Garlic powder	0.02
White pepper	0.05
Gum tragacanth #470	0.75
Saladizer #11*	7.00
Saccharin	0.02
Sodium benzoate	0.10
Dried egg yolk (F-flo)	3.50
Vegetable oil	12.00

* Tragacanth Import Corp., New York.

To manufacture, hydrate the Saladizer #11 and salt in the available water and vinegar under vigorous agitation for 6 to 8 min. Next, add the remaining dry ingredients during mixing until the mixture is smooth. Add the vegetable oil slowly, and mix thoroughly; then mill.

Mustard Sauce

	Percent
Water	55.00
Vinegar C (120 grain)	25.00
Mustard seed	14.00
Salt	3.00
Proplyene glycol	0.30
Oil of dill	0.30
Granulated turmeric	0.60
Spice mix	1.60
Gum tragacanth "C" powder	0.25

To manufacture, dry-blend the gum tragacanth with the salt; then meter the water and vinegar into the mixing tank. Add ingredients to the agitating system; dust in the gum-salt premix, and mix until uniform. Mill the resulting mixture through a J-hawk, wet-stone mill.

Spaghetti Sauce

	Percent
Tomato paste (26%)	75.00
Chopped onion	5.00
Garlic	0.35
Pepper	0.45
Salt	7.00
Thyme	0.15
Hydrolyzed vegetable protein	6.00
Monosodium glutamate	0.75
Spice mix	4.10
TIC gum tragacanth 470	1.20

To manufacture, add 2.5 oz (70 g) of the spaghetti sauce mix to 473 mL (2 cups) of water in a small saucepan. Add two tablespoons (10 mL) of olive oil. Heat to a simmer while stirring continuously. Simmer until the sauce has reached the desired consistency (will require approximately 8 to 12 min).

Pickle Relish

	Percent
Chopped pickles	43.00
Vinegar (120 grain)	16.60
Chopped onions	4.00
Sugar	32.00
Mustard seed	.75
Mustard flour	.15
Celery seed	.05
Tomato paste (26%)	2.00
Spice mix	1.25
Gum tragacanth #470	.20

To manufacture, dry-blend the gum tragacanth with the sugar in a 1:5 ratio. Add the remainder of the ingredients into the mixer, and mix. Dust in the dry blend slowly while the mixer is running, and continue to mix for 10 to 15 min.

LABORATORY TECHNIQUES

Identification

There are several approaches for the detection or confirmation of gum tragacanth:

Classical Chemistry Gum tragacanth may be identified by boiling a 0.5% solution with a 10% concentrate of potassium hydroxide, which yields a yellow stringy precipitate and a solution tinged with yellow.

There are other schemes of identification that will qualitatively differentiate tragacanth from other gums,[9] such as pectin, alginate, carrageenan, agar, methylcellulose, starch, carboxymethylcelluose, locust bean gum, arabic gum, ghatti gum, and gelatin.

Microscope When examined microscopically in water mounts, gum tragacanth shows numerous angular fragments with irregular lamellae. There should be very few or no fragments of lignified vegetable tissue.

Instrument Gum tragacanth films cased for infrared analysis exhibit a strong carbonyl absorption at 5.75 μm. The spectra of gum tragacanth, pectin, and gum karaya show similar absorption, but vary at 7.5 to 8 μm, 9 to 10 μm, and 12 μm. This difference is enough to serve as a means of identification and classification.

Viscosity Testing

Equipment:

Waring commercial blender: two speed, high-torque motor; blades revolve at 20,500 rpm on high speed, 12,600 rpm on low speed

Viscometer: Brookfield Synchro-lectric RVF

Balance: Any approved balance that is accurate to the one-hundredth place ($\pm$ 0.01)

Disperse 4.0 g of gum with 10 mL of alcohol (ethanol, 1-propanol, 2-propanol, etc.) in the Waring blender bowl. Add 390 mL of water, and blend at low speed for 1 min. Blend an additional 6 min at high speed, then transfer solution into suitable closed container (i.e., 1-L or 1-qt jar). Let solution age 24 h in a 25°C water bath. After aging, mix solution with a spoon to promote uniformity. Test solution with an RVF viscometer, spindle # 2 to 6, at 20 rpm.

PRODUCT TRADENAME/TERM GLOSSARY

General Information About Gum Tragacanth

Grades: There are five ribbon and five flake grades, ranging from high viscosity, and translucent to medium-low viscosity and cloudy when mixed with water (1%). The crude ribbons are long (5 to 10 cm), flat, flexible, and curled. The crude flakes are oval, dark, thick, and brittle and measure 1.25 to 5 cm in diameter. Both grades have a short fracture, are odorless, and have an insipid, mucilaginous taste.

Properties: It is strongly hydrophilic, insoluble in alcohol, and precipitates with Millon's reagent, lead acetate, potassium hydroxide, alcohol. It forms a gelatins with ferric chloride and exhibits a strong carbonyl absorption at 5.75 U (I.R.). It is combustible and nontoxic.

Constituents: It is made up of polysaccharides. There is a soluble fraction, arabinogalactan, and a swelling fraction, tragacanthic acid, made of D-galacturonic acid, D-galactose, L-fucose, D-xylose, and arabinose monomers. Its molecular weight is 840,000 and it has an elongated shape (4500 nm by 19 nm). It forms salts with calcium, magnesium, and potassium uronate.

Commercial grades: Its grades are U.S.P., F.C.C., and technical.

Containers: It is packaged in cases, bags, and barrels.

Uses: It is used for thickening, emulsification, suspension, moisture regulation, binding, film forming, sizing, and plasticizing.

Tradenames	Suppliers
Bassora Gum	
Goats Thorn	
Gog's Gum	
Leaf Gum	
Syrian Gum	
Tragacanth Gum p-17 USP	S. B. Penick & Co.

Tradenames	Suppliers
Tragacanth Gum Type C USP	S. B. Penick & Co.
Tragacanth Gum—T—300 USP Powder	Stein & Hall
Tragacanth Gum—T—400 USP Powder	Stein & Hall
Tragacanth Gum Type G-3 USP Powder	Meer Corp.
Tragacanth Gum—Type W USP Powder	Meer Corp.
Tragacanth Gum—Type L USP Powder	Meer Corp.
Tragacanth Gum—Type A.10 USP Powder	Meer Corp.
Tragacanth Gum—Technical Powder N	Meer Corp.
TIC Pretested—Gum Tragacanth "C" U.S.P. Powder	Tragacanth Importing Corp.
TIC Pretested—Gum Tragacanth "D" Powder Industrial	Tragacanth Importing Corp.
TIC Pretested—Gum Tragacanth 440 U.S.P. Powder	Tragacanth Importing Corp.
TIC Pretested—Gum Tragacanth 470 U.S.P. Powder	Tragacanth Importing Corp.
TIC Pretested—Gum Tragacanth 500 U.S.P. Powder	Tragacanth Importing Corp.
TIC Pretested—Gum Tragacanth 600 U.S.P. Powder	Tragacanth Importing Corp.
TIC Pretested—Gum Tragacanth 700 U.S.P. Powder	Tragacanth Importing Corp.
TIC Pretested—Gum Tragacanth 800 U.S.P. Powder	Tragacanth Importing Corp.
TIC Pretested—Gum Tragacanth #1 Powder U.S.P.	Tragacanth Importing Corp.
"Tragtex" R, Number 1	I.C.I. Organic Inc.
Tragacanth Gum FT-20	Hercules Inc.
Tragacanth Gum FT-40	Hercules Inc.
Tragacanth Gum FT-60	Hercules Inc.
Tragacanth Gum FT-100	Hercules Inc.
Tragacanth Gum FT-140	Hercules Inc.
Tragacanth Gum FT-180	Hercules Inc.
Tragacanth Gum FT-250	Hercules Inc.

REFERENCES/FURTHER USEFUL READING

References

1. Ali, B., B. Khan, and B. Ashmad, "Study of Substances Inhibiting the Aerobic Oxidation of Ascorbic Acid Solutions," *Pak. J. Sci. Res.*, **6**, 58 (1954).
2. Aspinall, G. O., and J. Baillie, "Gum Tragacanth. Part I. Fractionation of the Gum, and the Structure of Tragacanthic Acid," *J. Chem. Soc.*, 1702–1714 (1963).
3. Aspinall, G. O., and J. Baillie, "Gum Tragacanth. Part II. The Arabinogalactan," *J. Chem. Soc.*, 1714–1721 (1963).
4. Banerji, S. N., "Gum Arabic," in R. L. Whistler (ed.), *Industrial Gums*, 2d ed. Whistler, Academic Press, Inc., New York, 1973, p. 229.
5. Federation of American Societies for Experimental Biology, *Evaluation of the Health Aspects of Gum Tragacanth as a Food Ingredient*, PB-223-835, National Technical Information Service, U.S. Department of Commerce, p. 10.
6. Ferri, C. M., "Factors in Selecting Water-Soluble Gums," *Manuf. Confect.*, 39, 37 (1959).
7. Ferri, C. M., "Gum Tragacanth," in R. L. Whistler (ed.), *Industrial Gums*, Academic Press, Inc., New York, 1959, pp. 511–515.
8. *Food Chemicals Codex*, 1st ed., National Academy of Sciences and National Research Council, Washington, D.C., 1966, pp. 7, 8, 359, 360, 698, 699.
9. Glicksman, M., "Analysis and Identification of Hydrocolloids," in *Gum Technology in the Food Industry*, Academic Press, Inc., New York, 1969, p. 530.
10. Glicksman, M., and E. H. Farkas, "Hydrocolloids in Artificially Sweetened Goods," *Food Technol.*, 20(2), 58–61 (1965).
11. Gralen, N., and M. Karrholm, "The Physicochemical Properties of Solutions of Gum Tragacanth," *J. Colloid Sci.*, 5(1) 21–36 (1950).
12. Griffin, W. C., and M. J. Lynch, in T. E. Furia (ed.), *Handbook of Food Additives*, 2d ed., The Chemical Rubber Co., Cleveland, 1972, p. 397.
13. Meer, G., W. A. Meer, and T. Gerard, "Gum Tragacanth," in R. L. Whistler (ed.), *Industrial Gums*, 2d ed., Academic Press, Inc., New York, 1973, p. 291.
14. Mentell, C. L., *The Water-Soluble Gums*, Reinhold Publishing Corp., New York (1947).
15. O'Sullivan, C. J., in *The Chemistry of Plant Gums and Mucilages*, ed. by F. Smith and R. Montgomery, p. 281, Reinhold Publishing Corporation, New York, 1959.
16. Perrin, P. H., French Patent 860,210 (1941).
17. Potter, F. E., and D. H. Williams, "Stabilizers and Emulsifiers in Ice Cream," *Milk Plant Mon.*, 39(4), 76–78 (1950).
18. Rees, D. A., "Polysaccharide Gels: A Molecular View," paper given at joint meeting of the Food Group and the Colloid & Surface Chemistry Group of the Society of the Chemical Industry, Oct. 11, 1971.

19. Rees, D. A., and A. W. Wright, "Carbohydrates" in *Organic Chemistry*, ser. 1, vol. 7, Butterworth University Park Press, London, p. 267.
20. Rowson, J. M., in F. Smith and R. Montgomery (eds.), *The Chemistry of Plant Gums and Mucilages*, Reinhold Publishing Corporation, New York, 1959, p. 281.
21. Schuppner, H. R., Canadian Patent 797,202 (1968).
22. Smith, F., and R. Montgomery, *The Chemistry of Plant Gums and Mucilages*, Reinhold Publishing Corporation, New York, 1959, pp. 102, 281, 284.
23. Stauffer, K. R., and S. A. Andon, "Comparison of the Functional Characteristics of Two Grades of Tragacanth," *Food Technol.*, 4(46) (1975).
24. Unger, R., G. Seitz, M. Klockow, and W. Mehrhof, U.S. Patent 3,686,164 (1972).

Further Useful Reading

Glicksman, M., *Gum Technology in the Food Industry*, Academic Press, Inc., New York, (1969). This book is considered to be the best food-hydrocolloid material available.

Klose, R. E., and M. Glicksman, "Gums," in T. E. Furia (ed.), *Handbook of Food Additives*, The Chemical Rubber Co., Cleveland, pp. 313–375. A good cross-reference.

Lawrence, A. A., *Edible Gums and Related Substances*, Noyes Data Corp., London, England, 1973, pp. 46–52. A patent-oriented discussion of gum tragacanth with respect to sulfate monoester and reducing microbial contamination of tragacanth.

Smith, F., and R. Montgomery, *The Chemistry of Plant Gums and Mucilages*, Reinhold Publishing Corp., 1959. Insight into the chemistry of gums, including gum tragacanth.

Whistler, R. L., *Industrial Gums*, Academic Press, Inc., New York, 1974. Good gum reference book.

Whistler, R. L., *Methods in Carbohydrate Chemistry*, vols. I–VI. A must for the chemist interested in experimental laboratory work in carbohydrate chemistry.

Chapter **12**

Hydroxyethylcellulose

George M. Powell, III*

General Information 12-2
   Chemical Nature 12-2
   Physical Properties 12-3
   Manufacture 12-6
   Biological/Toxicological Properties 12-6
   Rheological Properties of Solutions 12-7
   Additives/Extenders 12-10
   Handling 12-10
   Applications 12-10
   Application Procedures 12-11
   Specialties 12-12
   Future Developments 12-12
Commercial Uses: Compounding and Formulating 12-13
   Protective Colloid in Latex 12-13
   Thickener for Latex Compositions 12-13
   Cosmetics and Pharmaceuticals 12-14
   Paper Sizes and Coatings 12-14
   Carpet and Textile Dye Pastes 12-14
   Special Applications 12-14
Commercial Uses: Processing Aids 12-15
   Crude-Oil Drilling and Recovery 12-15
   Electroplating and Electrowinning 12-15
   Miscellaneous Binders 12-15
   Other Specialty Uses 12-16
Industries Using Hydroxyethylcellulose 12-16
   Adhesives 12-16
   Agricultural Products 12-16
   Building Products 12-16
   Cosmetics 12-16
   Oil and Gas Extraction 12-17
   Paints and Coatings 12-17
   Paper and Allied Products 12-17

* Prepared under a consulting agreement with Union Carbide Corporation.

Synthetic Resins 12-17
Textile Mill Products 12-17
Formulations 12-18
Copolymer Latex 12-18
Latex Interior Flat Wall Paint 12-19
Textile Printing 12-20
Oil-Well Workover Fluid 12-20
Roll-on Antiperspirant 12-21
Liquid Shampoo 12-21
Laboratory Techniques 12-21
Product/Tradename/Term Glossary 12-21
Further Useful Reading 12-21

GENERAL INFORMATION

Hydroxyethylcellulose (HEC) belongs in the general category of water-soluble cellulose ethers. By chemical modification, enough hydroxyethyl ether side groups are formed at reactive hydroxyl groups along the cellulose molecular chains to convert this insoluble and relatively intractable material into a series of soluble, easily handled industrial products. The major commercial types are completely soluble in either hot or cold water, and are produced in a wide range of solution viscosity (molecular weight) grades.

These HEC products are versatile, non-ionic hydrocolloids, which find numerous uses as thickeners, flow-modifiers, protective colloids, stabilizers, water-loss control agents, binders, and the like. Their largest single industrial use at present is in latex paints as thickeners and protective colloids. In total sales of water-soluble cellulose derivatives HEC ranks second only to sodium carboxymethylcellulose. In 1978 total domestic production of HEC was of the order of 18,000 metric tons (40 million pounds) (per industry estimates), with Union Carbide Corp. and Hercules Inc. being the major producers.

Chemical Nature

In hydroxyethylcellulose the substituent groups which attach to the cellulose molecular chains and impart valuable new properties to these derivatives are generated by reaction of alkali-swollen cellulose with ethylene oxide. Thus:

$$\text{Cellulose—OH} + H_2C\underset{\diagdown \diagup}{\overset{}{\text{—}CH_2}} \rightarrow \text{cellulose—O—CH}_2\text{—CH}_2\text{—OH}$$
$$O$$

Because the hydroxyethyl groups attached in this manner also can undergo reaction with ethylene oxide and these compete as reaction sites with the remaining cellulose hydroxyls, side chains of two or more poly(oxyethylene) units can (and usually do) form also.

$$\text{Cellulose—O—CH}_2\text{—CH}_2OH + H_2C\underset{\diagdown \diagup}{\overset{}{\text{—}CH_2}}\rightarrow$$
$$O$$
$$\text{cellulose—O—CH}_2\text{—CH}_2\text{—O—CH}_2\text{—OH}_2OH$$

The ratio of molar substitution (MS) to degree of substitution (DS) for any particular HEC product expresses the average number of oxyethylene groups per side chain.

A relatively small amount of substitution serves to render the products soluble in dilute aqueous alkali (MS about 0.3 to 0.5, DS about 0.2 to 0.3). Alkali-soluble grades of HEC have been used in the past as textile sizes and treatments. For the much more important water-soluble products the level of substitution must be higher. The minimum MS for water solubility is about 1.0 (DS about 0.65), but in actual practice MS values fall in the 1.5 to 3.0 range (DS values about 0.85 to 1.35), with MS levels of 1.8 to 2.5 being most popular. These MS and DS values are averages per anhydroglucose (β-O-glucopyranosyl) unit, meaning that both highly substituted and completely unsubstituted units can occur in any one molecular chain. Producers make every effort to minimize the proportion of unsubstituted units in their products.

Because the hydroxyethyl groups are non-ionic in water solution, HEC does not interact with positively or negatively charged ions. Its solutions are relatively insensitive to the presence of dissolved salts, which is an advantage in a number of commercial applications. But HEC molecules can enter into chemical reactions typical of other polymers containing hydroxyl groups, such as esterification, etherification, acetal formation, etc. Thus, crosslinking reactions can be carried out to render them insoluble in water or to modify their properties in other ways, as desired.

Physical Properties

Hydroxyethylcellulose is available on the market as an off-white, powdered solid in a number of specific grades. These differ principally in molecular weight as indicated by viscosity determinations on dilute water solutions. For example, the viscosity of a 2% solution may be as low as 0.002 to 0.003 Pa·s (2 to 3 cP) or as high as 100 Pa·s (100,000 cP), depending upon the specific grade selected. Weight-average molecular weights are estimated to range from about 6.8×10^4 for the low-viscosity grades to about 8.0×10^5 and above for the highest-viscosity grades. Some are available in several degrees of fineness and at several levels of molar substitution (MS). In addition, many of the viscosity grades can be obtained with a special surface treatment which makes them easier to disperse and dissolve in water. The various commercial grades of the two principal United States suppliers and some of their general characteristics are shown in Tables 12.1 to 12.4.

TABLE 12.1
Viscosity Grade Ranges for Cellosize* Brand HEC Solutions

Viscosity grade	Concentration, wt % (dry)	Brookfield viscosities at 25°C., cP†	
		L Range	H Range
09	5	75–112	113–150
3	5	215–282	283–350
40	2	80–112	113–145
300	2	250–324	325–400
4,400	2	—	4801–5600
10,000	2	—	6000–7000
15,000	1	—	1100–1450
30,000	1	—	1500–1900
52,000	1	—	2400–3000
100M	1	—	4400–5600

* Registered Trademark, Union Carbide Corp.
† Multiply by 0.001 to convert from centipoises (cP) to pascal seconds (Pa·s).
NOTES: Prefix WP used for untreated type, QP used for easy-dissolving type. Grades 09, 3L, and 300 available in either WP or QP types. Grade 40 and Grades 4400 and higher are available only in QP type.

TABLE 12.2
Typical Properties of Cellosize* Brand HEC as Shipped

Ash content (calculated as Na_2CO_3)	3.0% by wt.
Volatile matter	Less than 5.0% by wt.
Particle size	95% thru U.S. 20 mesh
	85% thru U.S. 40 mesh
Apparent density	0.35–0.61 g/mL
Bulking value	1.32 g/mL
Color	Cream to white
Softening point	Above 140°C
Decomposition temperature	About 205°C
Approximate degree of substitution (DS)	0.9–1.0
Molar substitution (MS)	1.8–2.0

* T.M. Union Carbide Corporation

Solubility in Water Aside from the minor specialty products of low substitution intended for textile sizing and soluble only in dilute alkali, all the various grades offered will dissolve completely in hot or cold water to form clear, colorless solutions substantially free of residual insoluble matter. The solutions can be frozen and thawed, or heated to boiling, without any gelation or precipitation taking place. Some of the typical proper-

TABLE 12.3
Specifications for Viscosity Types of Natrosol* Brand HEC

Viscosity type	Brookfield viscosities† at 25°C at varying concentrations in water, cP		
	1% (dry)	2% (dry)	5% (dry)
HH	3400–5000		
H4	2600–3300		
H	1500–2500		
MH	1000–1500		
M		4500–6500	
K		1500–2500	
G		150–400	
E		25–105	
J			150–400
L			75–150

* Registered Trademark, Hercules Inc.
† Multiply by 0.001 to convert from cP to Pa·s
NOTES: Above types all available in Grade 180 (MS = 1.80) and Grade 250 (MS = 2.50). Grade 150 (MS = 1.50) and Grade 300 (MS = 3.0) are available in some but not all viscosity types. All grades and types are available with fast-dispersing treatment (suffix R), or without.

TABLE 12.4
Typical Properties of Natrosol* Brand HEC as Shipped

Ash content (calculated as Na_2SO_4)	4.0%, max.
Moisture content	5.0%, max.
Particle sizes available:	
Regular grind, retained on 40 U.S. mesh,	10%, max.
X Grind, retained on 60 U.S. mesh,	0.5%, max.
W Grind, retained on 80 U.S. mesh,	0.5%, max.
Bulk density	0.55–0.75 g/mL
Color	White to light tan
Softening range	135–140°C
Browning range	205–210°C

* T.M. Hercules Incorporated

TABLE 12.5
Typical Properties of HEC Aqueous Solutions

Specific gravity, 2% solution 20°C	1.0033
Refractive index, 2% solution	1.336
pH of 2% solution	6.0–7.0
Bulking factor in solution, mL/g	0.835
Color, platinum-cobalt scale, 2% solution	30
Surface tension, dyn/cm, 25°C, 0.01 to 1.0% solutions	64–69

ties of HEC aqueous solutions are given in Table 12.5. Note that HEC aqueous solutions are only mildly surface-active, as indicated by their relatively high surface tension.

Solubility in Organic Solvents Solubility in organic solvents is rather limited. Dimethyl sulfoxide, dimethyl formamide, N-acetylethanolamine, phenol, ethylenediamine, diethylenetriamine, formic acid, and ethylene chlorohydrin are among the few which will dissolve HEC. Many of the commonly used organic solvents are without effect, but water solutions of HEC will tolerate dilution with moderate proportions of water-soluble solvents such as acetone, alcohols, formaldehyde, glycols, etc.

Dissolving Methods   The commercial value of hydroxyethylcellulose depends largely upon its behavior when dissolved in water.   While it does dissolve completely, some care is needed in preparing solutions on a commercial scale.   Since HEC is a true high polymer, time is needed for water to penetrate the particles, then swell and dissolve the molecules.   Dissolving methods recommended by the producers are described later under Application Procedures.

Viscosity Properties   It is also important to understand the viscosity characteristics of solutions of the various grades of HEC at practical use concentrations.   Many of these are non-Newtonian in behavior; that is, the viscosities observed vary with the rates of shear at which they are measured.   These characteristics are discussed in more detail in the section on rheological properties.   It is of interest that even when exposed to extremely high rates of shear, HEC solutions show little or no permanent loss of viscosity or molecular degradation.

Compatibilities   Solutions of HEC are compatible or partially compatible with most water-soluble gums and resins of importance, as shown in Table 12.6.   In addition, they can be blended successfully with many of the commercial polymer latexes; emulsions; and water-reducible, water-dispersible, and water-soluble resins, modifiers, and additives.   Because of the non-ionic nature of HEC, tolerance for other materials is

TABLE 12.6
**Aqueous Solution Compatibilities
HEC with Other Gums and Resins**

Compatible
  Gum arabic
  Gum karaya
  Gum tragacanth
  Guar gum
  Borated dextrin
  Hydroxypropylmethylcellulose
  Methylcellulose
  Sodium alginate
  Sodium carboxymethylcellulose
Partly compatible
  Casein
  Starch
  Poly(vinyl alcohol)
  Gelatin
Incompatible
  Zein

quite broad.   This is also true of dissolved inorganic salts.   When the proportions of such salts in solution are less than the proportion of HEC, few problems are encountered. Even when the salt proportions are much greater than the proportions of HEC or when concentrated or saturated salt solutions are used, as when thickening brines, many salts cause no problems (among them most chlorides and nitrates).   Those which do cause precipitation or "salting out" under these conditions are generally the multivalent salts such as the sulfates, phosphates, thiosulfates, ferrocyanides, and some sulfites.

Interactions   Interactions leading to unusual viscosity increases are encountered with a few gums, resins, and modifiers.   Some of these can be used to form stable gels when desired, such as blends of HEC with sodium alginate, sodium rosinate residues, or sodium pentachlorophenol and related chlorinated phenol salts.   Sodium carboxymethylcellulose, some grades of poly(vinyl alcohol), and soy protein may interact also. Gelatin and sodium caseinate are examples of additives which interact more slowly. On rare occasions other ingredients in complex formulations containing HEC can cause viscosity increases, phase separation, or precipitation.   Should this occur, varying concentrations and order of mixing, or using substitute materials, will often avoid the trouble.

Film Formation   Clear, colorless, light-stable, and moderately strong films can be deposited from HEC solutions.   Such films are odorless and tasteless and have very good resistance to oils, grease, fats, and many organic solvents.   Some of the properties of HEC films are given in Table 12.7.

Physical properties of films vary considerably with the relative humidity to which they are exposed, for the adsorbed water they contain acts to plasticize them. Because of this hygroscopicity, static charges do not tend to build up, even at low humidities. When softer, more flexible films are desired, relatively nonvolatile plasticizers can be incorporated at 10 to 30% by weight. Suitable additives include glycerine, sorbitol, the lower glycols and polyglycols, ethanolamines, sulfonated oils, etc. While dry films of HEC are not thermoplastic and cannot be heat-sealed, some of the plasticized films can be heat-sealed to form good bonds.

HEC films and coatings remain permanently water-soluble unless crosslinking additives are deliberately added before application. Reactants such as the dialdehydes or the water-soluble urea- or melamine-formaldehyde resins are effective. Photosensitive compositions can be formulated if desired. Other additives can be used to modify other properties. Thus, adhesion to metal, glass, fiber, or other surfaces can be improved by the inclusion of small proportions of protein glues or other water-soluble gums or resins. Raising pH can also improve adhesion.

TABLE 12.7
Some Properties of Hydroxyethylcellulose Films
(Cast from aqueous solutions)

General appearance	Transparent, colorless
Specific gravity (50% R.H., 25°C)	1.34–1.38
Refractive index (sodium "C" line, 20°C)	1.50–1.51
Tensile strength (50% RH, 25°C)	260–281 kg/cm²
	(3700–4000 psi)
Elongation at break (50% RH, 25°C)	14–41%
Tensile impact, low-viscosity grade, (50% RH, 5°C)	
Unplasticized	28–30 kg·cm/cm²
	(13–14 ft·lb/in²)
Plasticized with 20% diethylene glycol	99 kg·cm/cm²
	(46 ft·lb/in²)
Equilibrium moisture content*	
10% RH, 25°C	1.5–3.0%
45% RH, 25°C	7.5–10.0%
65% RH, 25°C	18–19.5%
90% RH, 25°C	25–28%

* Low values are for dried films gaining weight; high values are for damp films losing weight.

Manufacture

Commercial processes for making HEC generally start by treating shredded, purified wood pulp or cotton linters with the proper ratio and strength of aqueous caustic to promote optimum swelling, then adding ethylene oxide to make the desired product. Volatile organic liquids which are nonsolvents for the final product are usually present also to promote heat transfer and ease of mixing in the stirred pressure reactors used. Upon completion of the reaction, the caustic must be neutralized to salts which can be largely removed, along with by-products of other ethylene oxide reactions, during washing of the product with water and carrier liquids. After being washed, the products are dried, ground, and packaged for shipment. Special treatments of the particles to promote more rapid dissolving action are applied to some grades in these product-recovery steps.

Biological/Toxicological Properties

Hydroxyethylcellulose is generally regarded as nontoxic when ingested and is substantially nonirritating in contact with human skin. The manufacturers of HEC cite acute and subacute oral toxicity tests with rats, showing no detectable harm even after 2-year feeding tests. Eye contact tests of powder and dilute solutions in both human and albino rabbits showed no more than moderate inflammation, which cleared up within a day or two. Patch tests on human skin showed HEC to be no more irritating than wheat starch, with no evidence of sensitization. Successful commercial uses in cosmetics and pharmaceuticals support these conclusions. HEC solutions should not be used in parenteral injections, however. HEC is cleared by the Federal and Drug Administration (FDA) for use in adhesives and coatings in contact with food, but has not been approved as a direct food additive. It is exempt under Environmental Protection Agency (EPA) regulations when used in pesticide formulations applied to growing crops. The pertinent references to FDA and EPA regulations can be obtained from the suppliers.

The biochemical oxygen demand (BOD) of HEC solutions is relatively low. Under simulated-river test conditions with a low concentration of "nonacclimated" microorganisms, less than 10% biooxidation occurs within 5 days. Its theoretical oxygen demand for complete biodegradation is in the range of 1.37 to 1.47 mg oxygen/mg HEC, depending upon the MS (1.5 to 3.0). Practical BOD values for the powdered product as shipped are in the 7000- to 18,000-ppm range after 5 days incubation, for high- to low-molecular-weight grades (versus over 800,000 ppm for cornstarch).

Like cellulose itself and many of its derivatives, hydroxyethylcellulose can be attacked and degraded by enzymes generated by many fungi and other microorganisms. Presumably, such attack occurs most readily in portions of the molecular chains where two or more unsubstituted anhydroglucose units occur in sequence, causing chain scission and lowering of average molecular weight. In most applications the resulting drop in solution viscosity is highly undesirable, so that precautions such as the use of good housekeeping practices to avoid accidental contamination of HEC, its solutions, or products containing it, and the addition of appropriate preservatives are recommended. It should be realized that the cellulolytic enzymes responsible for degradation can retain their activity after the organisms generating them have been destroyed, so that avoiding contamination in the first place is most important. Phenylmercuric compounds reportedly can deactivate enzymes, as can prolonged heating, but such methods are undesirable or impractical for a number of end-use products containing HEC. Manufacturers or suppliers of preservatives should be consulted for recommendations in specific instances.

Rheological Properties of Solutions

It is important to realize that the viscosity of aqueous solutions of HEC is not just a simple function of the usual factors such as molecular-weight grade, concentration, temperature, and the like, but is also highly dependent upon the shear rate at which viscosity is measured. The apparent viscosity drops off as the rate of shear at which it is measured increases. This mayonnaise-like behavior, called *pseudoplasticity,* is often most desirable in a number of commerical products. And, sometimes, when solutions are subjected to high shear rate conditions, there may be a time lag in the tendency of the apparent viscosity to increase again to its earlier "at rest" level. This time-delay behavior is called *thixotropy.* Because of both of these flow anomalies, much care is needed when determining the viscosity of HEC solutions to follow standardized preparation and measurement methods so as to obtain reproducible results.

Figure 12.1 illustrates a typical viscosity–shear rate rectangular coordinate plot for

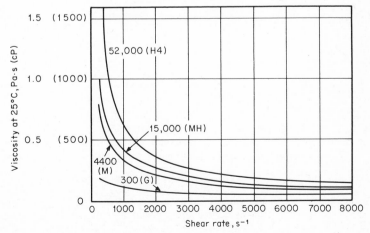

Figure 12.1 Typical viscosity-shear rate curves for 2% hydroxyethylcellulose in water. Numbers are Union Carbide "Cellosize" grades; letters are Hercules "Natrosol" grades.

four different molecular weight grades of HEC at 2% concentrations in water. From this it is readily apparent that pseudoplasticity is much more pronounced in solutions of the higher-molecular-weight, higher-viscosity grades than in solutions of the lower grades. The other flow anomaly, thixotropy, is likewise more pronounced in solutions at higher concentrations of the higher-viscosity grades, and is generally absent from solutions of the lower grades. Temperature, too, has an important effect on both anomalies: lower temperatures make them more pronounced, and higher temperatures just the reverse.

The shear rates shown in Fig. 12.1, ranging up to 8000 s⁻¹, put more emphasis on the high shear rate range. Those above 1000 s⁻¹, especially toward 8000 s⁻¹ and above, are representative of the shear rates encountered during application of compositions

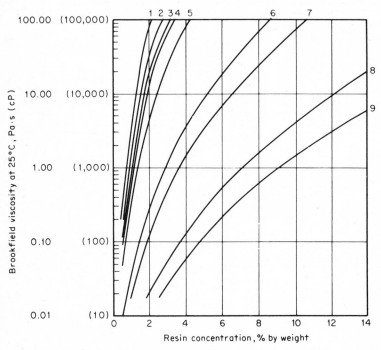

Figure 12.2 Typical viscosity-concentration curves for various hydroxyethylcellulose grades dissolved in water. Midpoints of specification ranges. *(Union Carbide Corporation laboratory data.)*

Curve	Grade	Curve	Grade
1	100-M	6	300
2	52,000	7	40
3	30,000	8	3
4	15,000	9	09
5	4,400		

thickened by HEC, using methods such as wiping, spreading, rolling, brushing, or spraying. Typical pouring and stirring shear rates are usually in the range of 10 to 1000 s⁻¹. Shear rates which are important in keeping particles in suspension, or which are active in flow and leveling of coatings, are much lower, from essentially zero to about 0.10 s⁻¹.

Figures 12.2 and 12.3 show average viscosity-concentration curves for typical grades of HEC, from the lowest to the highest molecular weights. Notice that this time the viscosity is plotted on a logarithmic scale for convenience. The viscosity values are all determined with the Brookfield viscometer, which generally operates at quite low shear rates (below 10 s⁻¹). Usually, the solutions are subjected to strong agitation just before measurement to minimize thixotropic effects. When viscosity levels intermediate between established grades are desired, they can be obtained by blending the next higher and lower grades in suitable proportions. (Manufacturers can supply appropriate blending charts.)

Not unexpectedly, temperature can exert a strong influence on the viscosity of HEC solutions, as shown on a semi-log plot in Fig. 12.4 for a range of grades. But on the

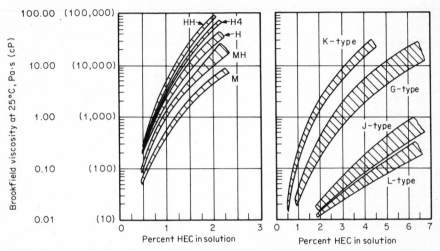

Figure 12.3 Typical viscosity-concentration curves for various hydroxyethylcellulose grades dissolved in water. Specification ranges shown. *(Hercules data.)*

other hand, pH in the range of pH 2 to pH 12 has very little effect. At very low pH some acid hydrolysis is possible, especially at higher temperatures; at very high pH levels some oxidative degradation may occur on storage. The role of pH in promoting rapid dissolving of the surface-treated grades is important, however, and this is discussed later under Application Procedures.

The viscosity of aqueous HEC solutions can be influenced dramatically when other dissolved material is present. For example, if the viscosity of a sucrose-water solution

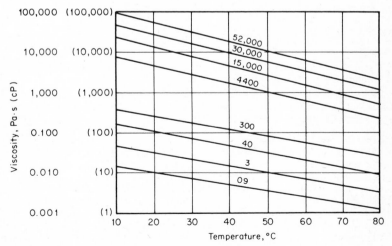

Figure 12.4 Viscosities vs. temperature for 2% aqueous solutions of various viscosity grades of hydroxyethylcellulose. *(Union Carbide Corporation laboratory data.)*

is some 50 times that of water, an HEC solution in this mixture will have about 50 times the viscosity it would have had in water. Of course, the presence of dispersed solids can also influence observed viscosities of HEC solutions, depending largely upon particle size and volume concentration of solids. In those rare situations where interactions of other ingredients with HEC can occur, viscosity changes can be dramatic, as discussed under Physical Properties.

Additives/Extenders

Hydroxyethylcellulose is frequently the only water-soluble resin used in specific formulations, but it can be modified or extended with various other materials, such as adhesion-promoters, flow-modifiers, and plasticizers. Formulations containing HEC should normally contain a minor amount of a fungicide or preservative to avoid inadvertent spoilage, since maintaining completely sterile conditions is often inconvenient.

Handling

Hydroxyethylcellulose is normally packaged in 22.7-kg (50-lb) multiwall bags by both major suppliers, although it can also be obtained in fiber drums if desired. Because of its rather innocuous nature, no specific warning labels are required. Storage presents few problems, but the material is flammable (it burns in a manner similar to cellulose) and hygroscopic. As shipped, it contains less than 5% volatiles; but when exposed to the air, humidity changes can cause it to take up or lose moisture. For example, equilibrium moisture contents may vary from about 5% at 40% relative humidity to over 20% at 80% relative humidity. The original shipping bags usually provide adequate protection from humidity changes during storage. Partially emptied bags should be resealed or their contents transferred to airtight containers. Spills should be cleaned up promptly, for the material can absorb water and become quite slippery.

Dusts of HEC, like those of cellulose itself and of many powdered organic solids, are explosive when mixed with air in critical proportions and exposed to an ignition source. Minimum explosive concentration in air for a dust cloud is 0.7 mg/mL; autoignition temperature for a dust cloud is 390°C, for a dust layer 340°C. For this reason, care in handling the dry, powdered HEC to minimize dust formation is important. Workers in confined areas handling dry HEC should wear dust respirators and safety glasses.

HEC is sterile and free of all microorganisms as packaged, and care should be taken during storage and subsequent handling to avoid accidental contamination. Partly emptied packages should be resealed promptly. When solutions are being prepared, all equipment should be clean and the water and other ingredients should be uncontaminated. Containers should be covered to minimize airborne contamination, and they should be rinsed and drained thoroughly when emptied. Since even with careful housekeeping practices accidental contamination is possible, it is usually desirable to add suitable preservatives.

Applications

Hydroxyethylcellulose has achieved commercial acceptance in the manufacture of a wide range of products by a number of different industries. Most often, it is used as a modifier or additive, making up only a small proportion of the total ingredients in these products, yet contributing significantly to their final properties. Among the desirable results obtained by the inclusion of HEC are the following:
- Effective thickening action at low concentrations.
- Pseudoplasticity and flow control.
- Stabilization of dispersed systems (suspending aid, protective colloid, etc.).
- Ready formation of moderately strong films of good grease resistance and good binding, sizing, sealing, and penetration-control properties.
- Water-retaining capabilities (water-loss control, hydrophilicity).

Other water-soluble gums and resins can contribute at least some of these properties, of course. However, to these HEC can add the following additional characteristics:
- Ready and complete solubility in hot or cold water (no "cloud" points).
- Retention of properties in the presence of many dissolved salts because of its non-ionic character.
- Wide compatibility with many other gums, modifiers, surfactants, etc.

- Availability in a broad range of solution viscosity grades.
- Dependable, reproducible quality because of careful manufacturing control.
- Good stability against degradation under high-shear conditions.
- Moderate surface activity with low foaming tendencies.
- Almost complete freedom from toxicity and skin irritation problems.
- Low biochemical oxygen demand, yet subject to digestion in waste treatments.

With the foregoing combination of properties, it is hardly surprising that HEC is selected from among competing materials for inclusion in many industrial and consumer products. For example, HEC is widely used in latex paints, both as a protective colloid in latex synthesis and as a thickener and flow-control agent in the paints themselves. Similarly, it serves as a thickening, flow-control, and water-binding additive in other latex compositions such as paper-coating colors; textile adhesives, sizes, and coatings; tape-joint cements; and sealants. In cosmetics and pharmaceutical gels, lotions, creams, hair dressings, and shampoos, etc., it contributes thickening, body, stability, and smoothness. Other kinds of products often using HEC are plywood adhesives, aqueous printing inks, lithographic dampening fluids, laundry aids (detergents, starches), agricultural sprays, reconstituted tobacco, textile and carpet dyeing and printing compositions, water-soluble packaging films, to mention but a few.

As a processing aid an important use of HEC is as a thickener and water-loss control agent in crude-oil drilling and recovery fluids. It finds a variety of small uses in electroplating baths, ceramic glazes, cement and concrete mixes, suspension process polyvinyl chloride, graphite extrusion (for lead pencils), foundry cores, welding-rod coatings, etc. The manufacturers of HEC can supply more detailed information on many of the various uses selected for mention above as well as on many others omitted because of space.

Application Procedures

Special care is needed in preparing aqueous solutions of hydroxyethylcellulose. Because of its high molecular weight, time is needed for water to penetrate the resin particles and solvate the individual molecules. If HEC is added to water too rapidly, clumps of resin particles can be sealed over by gelatinous layers of solvated resin formed on the surface, delaying penetration of the water and tending to form lumps which then dissolve very slowly. These problems are more troublesome with the higher-molecular-weight thickener grades, but are encountered to some extent even in the lowest-viscosity grades.

The obvious remedy is to sprinkle or sift the HEC slowly into vigorously stirred water. This isn't always practical or convenient, however, so that both major producers offer specially treated grades of HEC which disperse more readily. In these, the slight degree of modification actually delays the initial dissolving action of the water on the resin, allowing time for the water to wet and disperse all the resin particles before gelatinous lumps can be formed. As a result, the time needed for complete solution can be greatly reduced.

This is illustrated in Fig. 12.5, which is based on laboratory test conditions. Note that the untreated grade starts to dissolve in water almost immediately, but because of lumping often requires stirring for a considerable period before solution is complete. The treated product in this example required 40 min at a pH of 6.8 before any dissolving action started. Obviously, this allows ample time for thorough dispersion and wetting to occur. Even at a pH of 8 there is time for wetting and dispersion, but the so-called hydration time is much shorter. This hydration time, then, depends not only upon the extent of the treatment imparted by the producer, but also upon pH and temperature. This is illustrated in Fig. 12.6, again for laboratory conditions. Hydration time is longest at pH 4 to 5, and shortens dramatically at pH 7, 8, or 9. Temperature also has a pronounced effect, with hydration times becoming much shorter as this is raised.

When multicomponent compositions are being prepared, it is often possible to add an ingredient which raises the pH as soon as the HEC has been dispersed but not dissolved. When the pH is raised, then, the hydration time is greatly reduced and the solution preparation is accelerated. A good example of this is adding an alkaline preservative such as sodium pentachlorophenol or phenylmercuric acetate to water-dispersed HEC. Heating the mixture as soon as the HEC has been dispersed is another useful method.

Other methods can also be used to speed up the dissolving process, whether or not the surface-treated grades are being used. The HEC can be blended first with other dry ingredients before it is added to water, thus reducing the lump-forming tendencies. Similarly, a slurry of HEC in a water-miscible organic liquid can be prepared first, then added to the water to accomplish the same result. (Some organic liquids can swell HEC on standing, so it is often best to prepare such slurries just before use.) Prior addition of a small amount of a surfactant to the water can also speed up wetting and dissolving. Some prefer to use special mixing devices or eductors to entrain the dry powder in a rapidly moving stream of water entering the mixing tank. Specific suggestions are offered by both of the major suppliers for various end uses.

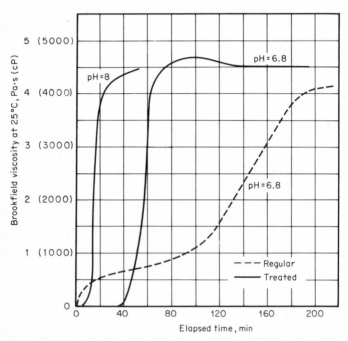

Figure 12.5 Typical dissolving behavior of regular and treated hydroxyethylcellulose (4,400 or M viscosity grade), 2% in distilled water at 25°C and pH indicated. *(Union Carbide Corporation laboratory data.)*

Specialties

While both major manufacturers of HEC offer a range of viscosity and surface-treated grades and one (Hercules) offers a choice of molar substitution levels, both will also prepare special grades if the need warrants.

Late in 1975, Hercules Inc. announced the availability of a new series of Natrosol B hydroxyethylcellulose thickeners in a range of viscosity grades. These show increased resistance to enzyme attack, as demonstrated by viscosity stability tests of typical latex paints inoculated with a cellulolytic enzyme preparation.

A modified hydroxyethylcellulose containing cationic groups has been introduced (Polymer JR, Union Carbide). This shows good substantivity to protein substrates such as human hair, good solubility in water-alcohol mixtures, and enhanced hygroscopicity.

Future Developments

The history of ongoing product improvement and modification in the general field of water-soluble cellulose ethers suggests that further product improvements in hydroxyethylcellulose can be anticipated in the future. Two performance areas receiving attention are in the latex paint market, where methods for obtaining improved flow and leveling and better inherent bioresistance are under development.   Other end-use markets are under similar investigation as well.

The entire field of water-soluble and water-dispersible polymers is undergoing rapid development and change, and a wide range of competing products can be anticipated with confidence.   Natural resins and gums will continue to find many uses, and new, totally synthetic products can be expected to emerge also.   The chemically modified natural products, and especially the water-soluble cellulose ethers, with the important advantages of being based upon a renewable basic raw material yet having carefully designed and reproducible characteristics, can be expected to maintain a significant share of these expanding markets in the foreseeable future.

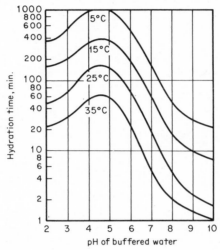

Figure 12.6   Effects of pH and temperature on hydration time of R-grade hydroxyethylcellulose. *(Hercules data.)*

COMMERCIAL USES: Compounding and Formulating

Protective Colloid in Latex

An early and still important and growing use of HEC is as a protective colloid in the synthesis of polymer latexes.   It can serve as the sole protective-colloid surfactant, or can be used in combination with conventional non-ionic and anionic surfactants.   It remains in the latex afterwards to contribute good mechanical and freeze-thaw stability.

Current theories of the role of HEC in emulsion polymerization involve its ability to provide sites for graft copolymerization with the polymer chains being formed and to provide protective sheaths around the polymer particles as they grow.   The lower-viscosity grades are usually preferred and are used at concentrations of about 1 to 2 parts per 100 parts of monomers, depending upon other surfactants included, particle size and latex viscosity desired, etc.   While a typical latex recipe is given in a later part of this section, both HEC suppliers can provide more detailed information and suggestions.

Thickener for Latex Compositions

The largest present volume of use for HEC products is as thickeners and flow regulators for a variety of end-use compositions based on synthetic polymer latexes, especially those containing vinyl acetate and acrylic ester polymers and copolymers.   HEC is generally only a minor ingredient, but it provides the viscosity and stabilizing action needed to keep pigments and fillers in suspension, the pseudoplasticity to permit easy application under shear, and the water-retention properties to control penetration into porous substrates and regulate drying.

Latex Paints Almost from their first commercial acceptance in interior flat wall paints the vinyl acetate polymer and copolymer latexes have relied upon the water-soluble cellulosics, and more specifically upon HEC, for thickening and flow control. Now most types of latex paints and enamels made with almost any of the available latexes may well contain HEC.

The viscosity grade(s) and proportions of HEC used in any particular paint product depend upon a number of different factors. For maximum thickening action at lowest cost, the highest-viscosity HEC grades are generally used. When higher "brush drag" during application (i.e., higher high–shear rate viscosity) is desired, larger proportions of medium- or lower-viscosity grades can be used. Flow and leveling after application are also influenced by this choice. For latexes such as the vinyl acetate copolymers containing no ionizable functional groups, the incorporation of a larger proportion of a lower-viscosity grade generally gives better leveling. With acrylic latex copolymers containing anionic functional groups (such as carboxyls), interactions with HEC may occur, and lower proportions of higher-viscosity grades may well give better leveling. In general, the proportions of thickener used tend to fall in the range of 36 to 72 kg/ 100 L (3 to 6 lb/100 gal) of paint for most formulations.

Color Coats for Paper Latex-based clay coatings to improve the smoothness, printing properties, and opacity of paper and paperboard often contain HEC as a thickener and water-retention aid. This latter property is often quite important because large proportions of clay and pigment are used in proportion to binder and because application is over porous surfaces. Ratios are generally in the 0.1 to 1.0% of total pigment weight range, depending upon solids content of the formula, method of application, other ingredients, etc.; 0.5% of a moderately low viscosity grade of HEC (grade QP-300 or 250 GR) would be a starting point.

Textile Binders and Adhesives Acrylic latexes for many textile industry uses are modified with HEC for viscosity regulation and water-loss and penetration control. These serve as flock adhesives, nonwoven fabric binders, laminating adhesives, and fabric-back coatings principally. These products are usually formulated to do highly specific jobs, often on specific kinds of equipment, so that the HEC suppliers should be consulted for recommended methods of use in specific instances.

Building Specialties Synthetic resin latexes are finding increasing use in products such as tape-joint compounds, crack-fillers, texturized coatings, surface treatments in wallboard manufacture, plaster compositions, caulks, and sealants. In all of these some proportion of thickener and water-loss control additive is needed, and HEC is quite often the product selected.

Cosmetics and Pharmaceuticals

The properties of HEC make it useful in a range of hair dressings and shampoos, creams, lotions, ointments, toothpastes, soaps, bubble baths, gels, makeup preparations, and the like. Its non-ionic nature and tolerance for water-miscible solvents, plus its highly efficient thickening and stabilizing action, often dictate its use in preference to other water-soluble gums and resins as minor modifiers in such products.

Paper Sizes and Coatings

Thin surface sizes and coatings are used on paper products to control penetration, provide grease resistance, improve printing characteristics, and upgrade gloss and smoothness. Combined with glyoxal or urea-formaldehyde, water-soluble polymers such as HEC can form the base for treatments to improve the wet-strength of towels and paper products in general.

Carpet and Textile Dye Pastes

HEC can serve as an effective thickener and penetration-control agent in a variety of dye compositions, especially for those used on continuous machines. Good stability with acid or basic dyes and other ingredients helps HEC promote good dye acceptance, uniformity, and print definition, and good fastness in washing, steaming, and drying operations.

Special Applications

Among the many products and product classes in which HEC finds small commercial uses are the following.

- Size coating for glass fiber manufacture.
- Water-based duplicating inks.
- Detergents and cleaners (thickener and stabilizer).
- Agricultural sprays (to suspend toxicants, stick them to surfaces, coat seeds, etc.).
- Binder for reconstituted tobacco.
- Binder for fluorescent tube coatings.
- Temporary protective coating for items in transit or storage.
- Base for water-soluble packaging films.
- Fabric finishes and spray starches.
- Asphalt emulsions.
- Wind-loss prevention treatment for fine ores, minerals, etc.
- Modifier in specialty adhesives such as wallpaper pastes and plywood adhesives (with wheat flour).
- Binder for graphite used in pencils.
- Binder for welding-rod flux coatings.

COMMERCIAL USES: Processing Aids

The distinction between commercial products in which HEC is used as a compounding ingredient and those in which it serves only as a processing aid, and not as an integral material in the final product, is not easily drawn. Since HEC is usually only a minor ingredient in most of its large-volume uses, perhaps some of the applications mentioned earlier could be listed here with equal justification.

Crude-Oil Drilling and Recovery

With the greatly increased activities in well drilling and in petroleum (and gas) recovery, the demand for water-soluble resins in drilling muds and in various kinds of treatment and recovery fluids has also increased considerably. The cellulosics, and especially HEC, represent one of the frequently selected additives to serve as thickeners and viscosity-control agents, water-loss control additives, and suspending and stabilizing modifiers. The technology in this business is highly complex, and must be adapted to meeting a wide range of individual field conditions. Hence, it is difficult indeed to provide general suggestions for selecting and using specific ingredients in the many kinds of fluids needed.

HEC has demonstrated considerable merit in actual drilling muds, especially those using the so-called low-solids, nondispersed concept. Increased penetration rates with good bore-hole stability are obtained, even when drilling hydratable shales. The insensitivity of HEC to sodium or calcium brines up to complete saturation contributes to its success. Drilling speeds are increased because bit viscosities are quite low, but transport viscosities remain high for efficient cuttings removal. HEC also serves to encapsulate hydratable shales and to maintain bore-hole stability in troublesome sloughing or heaving shales. The need for viscosity-building clays is eliminated in polymer muds, thus greatly reducing transportation costs to remote drilling sites.

In workover, completion, and packer fluids HEC not only provides the desired rheology but, because of its freedom from gels and suspended matter, causes very little formation plugging. It can also be used in acid-soluble systems with graded calcium carbonate. Likewise, HEC has proven most useful in fracturing fluids for transporting sand or glass-bead propping agents into producing formations. Because HEC solutions can be degraded by the addition of enzyme concentrates, viscosity decline can be timed to permit easy removal of the fluids. Furthermore, no insoluble residues remain to cause formation damage. In cementing operations a little HEC controls fluid loss and aids in setting rate regulation to yield improved bond strengths.

Electroplating and Electrowinning

In these processes colloids are often included in the metal-plating baths to promote the deposition of smoother, more uniform metal films or layers. Because of its tolerance for electrolytes and dissolved salts and solubility over a wide range of temperature and pH, HEC is frequently the colloid selected.

Miscellaneous Binders

In a number of specialty applications HEC finds use as a binder and behavior modifier. For example, in ceramic glazes and refractory shapes it is used as a workability and green-strength promoter, and is subsequently burned out.   Metal foundry core binding is a related but small-volume use.

Other Specialty Uses

■ Some HEC is used as a suspending aid in polyvinyl chloride manufacture by the aqueous suspension process.   HEC generally promotes the formation of denser particles than some of the other colloids often used.

■ Ingredient in mold-release coatings.

■ Additive to dampening solutions in lithographic printing.

■ Cure-retarder and water-retaining agent in cements, mortars, plasters, concrete, etc.   Also used to improve pumpability of concrete mixes.

■ Stabilizer in fire-fighting foams and thickener in "thickened water" for fire fighting.

INDUSTRIES USING HYDROXYETHYLCELLULOSE

Adhesives

HEC finds use as a protective colloid in synthetic resin latex manufacture, principally because of the very good mechanical and freeze-thaw stability it confers.   It also increases viscosity to retard settling.   It is often postadded to latex adhesive compositions to increase viscosity still further, to impart a desirable pseudoplastic-type of flow, and to reduce water-loss tendencies, especially over porous surfaces.   It finds some uses in wallpaper pastes (flow control) and in wheat flour and other plywood adhesives (penetration and water-loss control).

Agricultural Products

HEC has been ruled exempt from the requirement of a residue tolerance by the Environmental Protection Agency when it is used as an inert ingredient in pesticide formulations applied to growing crops or to raw agricultural commodities after harvest.   As a result, HEC finds some commercial use as a suspending agent and binder-sticker in water-based slurries and emulsions for spray application of toxicants.   It is also useful as a binder in seed coatings.

Building Products

HEC finds a variety of uses in a number of products going into the building industry. For the most part, these are auxiliary or specialty products in which HEC is used in small proportions for its thickening, flow-control, and water-retention properties.   For example, it is used in latexes or emulsions for surface treatment of gypsum wallboard to improve press-release and paint-acceptance properties.   It is used in the latex-based tape-joint compounds for covering joints between wallboard panels in place, in crack-fillers, and in compositions for producing texturized, three-dimensional surface effects. Latex caulks and sealants often include HEC as the thickener.   Similarly, it is used in wallpaper pastes.

In cement and concrete mixes, a little HEC acts as a retarder and moisture-retaining agent.   It also improves the "pumpability" of concrete mixes.   In fabricating ceramic tiles, HEC improves green-strength and forming properties of the mixes; it contributes similar properties to glaze coats.

Cosmetics

This industry makes a large number of specialized products for many kinds of specific uses.   Frequently, these compositions require the inclusion of a thickener, suspending agent, emulsion stabilizer, emulsification aid, film-former, binder, or water-retention agent.   Many of the commercially available water-soluble gums and resins find at least some use here, provided they are innocuous in contact with human skin or when inadvertently ingested.   HEC is often selected when a non-ionic polymer unaffected by dissolved salts or changes in pH is needed.   Its high thickening efficiency, complete solubility, and tolerance for water-miscible organic solvents are often advantageous also. A modified HEC having cationic functionality is of interest in hair-care products because of its improved substantivity.

Oil and Gas Extraction

HEC is finding increasing use, along with a variety of other water-soluble gums and resins, as a thickener, stabilizer, flow-modifier, friction-reducer, and water-loss control agent in a range of fluids used in crude-oil and gas production. It is used in drilling muds, in workover, completion, packer, and fracturing fluids, and in compositions for cementing casings in place. These formulations necessarily vary with the particular formations and recovery problems encountered, but HEC, with its efficient thickener action, freedom from insoluble matter, tolerance for acidification and for brines of all types, and controlled viscosity-reduction capability via enzyme addition, clearly warrants careful consideration.

Paints and Coatings

The paints, varnishes, lacquers, enamels, and allied products subdivision of the major chemicals and allied products industry consumes a larger proportion of the total HEC production than any other industry, section, or group of manufactures. HEC finds two important areas of use: as a protective colloid in latex synthesis, and as a thickener, flow-control agent, and stabilizer in latex trade sale paints.

Latexes containing HEC as protective colloid generally show superior mechanical, freeze-thaw, and storage-stability properties which carry over into compounded products containing such latexes. And in latex paints HEC provides efficient, reproducible thickening action and imparts a desirable pseudoplastic character. Furthermore, it promotes good color acceptance and development in tints, tolerates a wide range of other ingredients and additives because of its non-ionic nature, imparts good storage stability, and minimizes any separation tendencies. All in all, it is frequently the thickener of choice in latex paints.

Paper and Allied Products

Many of the latex and modified latex clay-coating compositions applied to improve the smoothness, opacity, and printing properties of paper and paperboard rely upon HEC as a primary thickener and flow-control agent. It serves also as a protective colloid and water-retention agent in these coatings. It minimizes pigment-binder separations, lessening streaking during blade application, and retards migration of the aqueous phase into porous base stock.

HEC is also of interest in a number of clear coatings and sizes normally applied in very low coating weights. These impart grease and solvent resistance, improved gloss, better ink and coating holdout, pick resistance, curl resistance, etc. In these, HEC competes with starch and other natural products and modified natural products, which are generally lower in price. However, the good film strength, the high solution viscosity, and the water-retention properties of HEC often will dictate its use, either alone or blended with these other products, on a cost-performance basis.

Synthetic Resins

Much of the large volume of poly(vinyl chloride) resin in commercial use is manufactured by the so-called suspension process. In this, monomer droplets are suspended in a water phase; then the resulting polymer particles are recovered, washed, and dried. Small amounts of water-soluble resins are incorporated in the water phase to maintain the monomer droplets in suspension during the reaction period. The selection of these suspending agents depends upon a variety of factors, usually related to the characteristics of the recovered PVC-resin particles. HEC is used alone or, more frequently, in combination with others when it is desired to produce relatively denser, more impervious particles. HEC also finds some use in the production of polystyrene by the aqueous suspension process.

When synthetic resins are produced by the emulsion-polymerization process and are marketed as latexes, HEC is often the protective colloid of choice. However, since these seldom go into fabricated plastics or rubber articles, this use is described under the industries which process these latexes further into paint adhesives and the like.

Textile Mill Products

The uses for HEC in the textile industry fall mainly into two major categories: (1) thickeners and modifiers for a range of latex coatings, adhesives, and binder compositions, and (2) thickeners and penetration control agents for various fabric and carpet dyeing pastes and printing inks, particularly continuous dyeing and screen printing systems.

Here, again, the industry has a wide range of natural and synthetic water-soluble gums and resins from which to choose for its specific jobs. But HEC offers the advantages of efficient thickening at low concentrations, reproducible quality, ready solubility in hot or cold water and freedom from insoluble portions, non-ionic character (meaning tolerance for dissolved salts, acidic or basic dyestuffs, and other materials), and good water-loss and penetration control properties. It can also be used as a softening and finishing agent for fabrics and cloth items.

FORMULATIONS

Copolymer Latex

HEC can function as a protective colloid when used in a vinyl acetate-2-ethylhexyl acrylate copolymer latex of the following composition:

	Percent by weight
Water (deionized)	46.23
HEC (grade WP-09 or 25OL)	1.20
Non-ionic surfactant (nonyl phenol plus 10 to 11 moles ethylene oxide)	1.20
Anionic surfactant (sodium di-2-ethylhexyl sulfosuccinate)	0.90
Borax	0.35
Potassium persulfate	0.32
Vinyl acetate (3–5 ppm hydroquinone)	44.80
2-Ethylhexyl acrylate (60 ppm monomethyl ether of hydroquinone)	5.00
	100.00

Latex properties:	
Viscosity, Brookfield, 25°C	1.5–2.0 Pa·s (1500–2000 cP)
Total solids content, % by wt.	55+
pH	4.7
Average particle size, μm	No. 5
Freeze-thaw stability, 5 cycles	OK
Mechanical stability	Excellent

Equipment needed includes stainless steel or glass-lined reactor equipped with agitator, reflux condenser, jacket or other means of heating and cooling, thermometer or thermocouple, inlets for adding ingredients, discharge valve, provision for inert gas (N_2) purge, monomer reservoir, etc.

The procedure is to charge all of the water to the autoclave (except that needed to prepare a 4% solution of the potassium persulfate catalyst). Add the HEC protective colloid, the surfactants, and the buffer with agitation. When the HEC is completely dissolved, purge the reactor with nitrogen to displace all oxygen, and heat to 68 to 70°C. Add about 25% of the catalyst as a 4% solution, then about 10% of the premixed monomers while continuing agitation. The temperature should rise to about 70°C as the polymerization begins. When the reflux rate slows, begin feeding the balance of the monomer mix uniformly over a 2- to 4-h period at a temperature of 75 to 80°C. The balance of the catalyst solution can be added continuously also, or in several small increments. In either case, add about 10% of the catalyst after all the monomers are in. Then, raise the temperature to 90 to 95°C for about 30 min to complete the reaction and reduce the residual monomer content below 1%. Cool and check pH, adjusting to 4 to 5 with additional buffer if needed.

The foregoing is intended as a starting-point recipe only, and can be varied to obtain specific latex characteristics and film properties as desired. Variations in amount, grade, and MS level of the HEC can influence latex viscosity; but since these can change colloidal stability, this factor should be checked also. Variations in proportions and types of surfactants can regulate particle size. While in some systems it is possible to omit all surfactants and use only HEC as emulsifier-stabilizer, this often leads to unusually high viscosities, and so a three-component system of the type illustrated above is preferred.

Variations in polymerization procedures are often used also. The example cited above is classed as a *semibatch* process. It could also be operated as a continuous process, using several reactors in series and feeding from one to the other as the reaction proceeds. Other commercial techniques sometimes used (although not recommended for the above recipe) are the *staged adiabatic* and the *full-batch* processes. In the former the monomers and catalysts are added in portions at about room temperature and the heat of polymerization is absorbed by an increase in temperature in the bulk of the charge. When the reaction subsides, the temperature is lowered again and the monomer-catalyst addition is repeated. In the full-batch process, essentially all the monomers and other ingredients are charged initially, and the reaction is regulated to stay within the available cooling capacity by the rate of feeding one of the components of a so-called redox-type of catalyst mixture.

Emulsion polymerization processes can have many ramifications, and the uninitiated are advised to seek assistance from textbooks and from the suppliers of protective colloids, monomers, equipment, etc.

Latex Interior Flat Wall Paint

HEC functions as a thickener in latex interior flat wall paints. Shown below is the formulation for a typical medium-quality product:

	Percent by weight
Water	33.17
Hydroxyethylcellulose (QP-15000 or 250 MHR)	0.41
Pigment dispersant	0.50
Surfactant (nonyl phenol + 10 to 11 mol ethylene oxide)	0.25
Ethylene glycol	1.24
Filming aid (such as butyl ether of diethylene glycol acetate)	1.00
Defoamer	0.17
Preservative	0.08
Rutile titanium dioxide	20.73
Calcium carbonate	10.36
Calcined clay	6.22
Talc	4.15
Defoamer	0.16
Vinyl acetate/acrylic ester latex (55% solids)	21.56
	100.00

Paint properties:	
Total solids content	54%
Pigment volume concentration in dried film	55%
Viscosity of paint (modified Stormer viscometer)	90–95 Krebs units
Weight per unit volume	1.44 kg/L (12 lb/gal)

Add ingredients in order listed, first dissolving the HEC, then adding the other minor ingredients. Add the pigments, and mix under high shear in a high-speed disperser or other intensive disperser, holding back a portion of the defoamer and the latex until all pigments and extenders are dispersed. Stir in latex, and make any viscosity adjustments as needed.

While this formulation represents a typical medium-quality latex paint, it is obvious that many modifications and substitutions can be made, if desired. Consult with raw material suppliers for specific suggestions.

In making latex paints, a good reference is G. N. Danziger, in N. I. Gaynes (ed.), *Formulation of Organic Coatings,* D. Van Nostrand Co., Inc., Princeton, N.J., 1967.

The moderately high viscosity grade of HEC shown can be replaced with other grades, such as lower grades or a blend of high and low grades, to obtain somewhat better leveling, if desired. Quantities should be modified, however, to maintain paint viscosity. And the HEC can be incorporated in various ways. Sometimes, a portion is held back and added at the very end to adjust final viscosity.

Textile Printing

Another application of HEC is as a thickener in acrylic latex print flocking adhesives used for rotary-screen printing of textiles, a formulation for which is shown below.

	Percent by weight
Premix:	
Water	2.80
All-acrylic latex,* 60% solids by wt. (T.S.)	2.20
Polypropylene glycol (MW 2000)	0.09
Hydroxyethylcellulose (250 H4R or QP-52000H)	0.74
Master Mix:	
All-acrylic latex,* 60% T.S.	84.26
Polypropylene glycol (MW = 2000)	0.09
Premix	as above
Oxalic acid (10% solution)	5.30
Tri (2-ethylhexyl)phosphate	.96
Ammonium hydroxide (14% NH_3)	2.51
Melamine resin (water-soluble)	1.05
	100.00
Properties:	
Total solids content, % by weight	55
pH at 25°C	6.5
Viscosity (Brookfield RVF#6, 20 rpm)	approx. 41 Pa·s (41,000 cP)

* Latex is a high-solids content, carboxylated, self-cross linking, soft all-acrylic, such as UCAR Latex 874, Union Carbide Corp.

Prepare the premix separately, then add this slowly to the bulk of the preweighed latex in an agitator-equipped mixing tank. Add the oxalic acid solution and the phosphate plasticizer; then allow the mixture to thicken while stirring. When the blend has a smooth, shiny appearance (after an hour or more), blend in the ammonium hydroxide and continue agitation while the HEC goes completely into solution. Add the melamine resin, and mix well.

This formulation has been designed specifically for application to fabrics in a rotary-screen printer. In this composition the HEC gives superior performance because it prevents plugging of the screens and affords excellent print definition. After application of the adhesive, the fabric goes immediately to the flock-application stations, and thence to a zoned curing oven. Here again, it is found that the high water-holding capacity of the HEC and the long "open time" it confers combine to produce better flock acceptance, greater flock density, and reduced blistering tendencies during curing. Of course, this formulation would have to be modified if printing equipment other than rotary screens were to be used instead.

Oil-Well Workover Fluid

The following oil-well workover fluid composition has been designed to be completely acid-soluble, meaning that it can be removed from the hole by a simple acid treatment. Thus, it has the important advantage of causing a minimum of formation damage in the process.

	Quantity
Water	As needed
High-viscosity grade HEC (surface treated)	5.7 g/L
Ground calcium carbonate (approx. 10 μm)	5.7 g/L
Biocide (glutaraldehyde)	0.1%
Shale inhibitor (potassium chloride)	3%
Sodium hydroxide	To adjust pH to 9.0–9.5

Dissolve the potassium chloride in the water. Disperse the HEC thoroughly, then add sodium hydroxide to raise pH and accelerate the dissolving of the HEC. When solution is complete, add the ground calcium carbonate and biocide. Sometimes a corrosion inhibitor is also included. For deep wells the workover fluid can be weighted by dissolving calcium chloride as needed.

Roll-on Antiperspirant

	Percent by weight
Water	56.2
HEC (medium viscosity, surface treated)	0.5
Glyceryl monostearate (pure)	2.0
Emulsifier (nonyl phenol + 30 mol ethylene oxide)	1.3
Aluminum chlorohydroxide	40.0
	100.0

Add the water to the mixing tank, and agitate as the HEC is sifted in slowly and the tank contents are being heated. After the HEC has dissolved completely, add the glyceryl monostearate and the emulsifier. Continue heating to 80°C with constant stirring until a fine dispersion is formed; then allow mix to cool to 35°C with agitation. Then add the aluminum chlorohydroxide, and stir well.

Liquid Shampoo

	Percent by weight
Water	72.3
HEC (high viscosity, surface treated)	1.0
Conditioner (substituted fatty-acid amide)	1.6
Detergent (lauryl sulfate)	25.1
Perfume	As desired
	100.0

Add the HEC slowly to the water while stirring. Heat to 70 to 80°C and continue stirring until HEC is completely dissolved. Melt the conditioner, and add to the hot solution. Add detergent and cool. Add perfume at 40°C, then allow to stand for several days. Filter through glass wool, and package.

LABORATORY TECHNIQUES

For methods of testing for moisture, ash, viscosity, density, and molar substitution (MS), see "Standard Methods of Testing Hydroxyethylcellulose," *1978 Annual Book of ASTM Standards,* Part 21, ASTM Designation D2364-75 American Society for Testing and Materials, Philadelphia, pp. 277–284.

There are no generally recognized standard methods suitable for the determination of degree of substitution (DS) or of proportion of unsubstituted anhydroglucose units in HEC. A recent research publication of interest is the following: E. D. Klug, D. P. Winquist, and C. A. Lewis, "Studies on the Distribution of Substituents in Hydroxyethylcellulose," in N. M. Bikales (ed.), *Water-Soluble Polymers,* Plenum Press, New York, 1973.

If information about methods of test for other properties and characteristics of HEC is needed, it can be obtained from either of the United States producers, Hercules Inc. or Union Carbide Corp.

PRODUCT/TRADENAME/TERM GLOSSARY

Cellosize: Trademarked name for hydroxyethylcellulose sold by Union Carbide Corp.
Hydroxyethylcellulose: Common industrial name for (2-hydroxyethyl)cellulose.
(2-Hydroxyethyl)cellulose: Chemical name for hydroxyethylcellulose.
Natrosol: Trademarked name for hydroxyethylcellulose sold by Hercules Inc.
Ucare Polymer JR: Trademarked name for cationic hydroxyethylcellulose sold by Union Carbide Corp.

FURTHER USEFUL READING

Technical Bulletins

CELLOSIZE Hydroxyethyl Cellulose. Obtain from Union Carbide Corp., Chemicals and Plastics, 270 Park Avenue, New York, NY 10017
NATROSOL Hydroxyethyl Cellulose, a Non-ionic Water-Soluble Polymer: Physical and Chemical

Properties. Obtain from Hercules Inc., Coatings and Specialty Products Department, 910 Market Street, Wilmington, Del. 19899

These bulletins, published by the two domestic producers of HEC, provide much useful and detailed information about the properties and behavior of their lines of HEC products. In addition, both producers have available other technical bulletins giving more specific information and suggestions for various end uses.

Books

Wiley-Interscience Series of Monographs on "High Polymers," vol. V, "Cellulose and Cellulose Derivatives" Parts IV and V, Edited by N. M. Bikales and L. Segal, 1971

These two parts represent an updated and revised version of one of the recognized standard publications on cellulose chemistry, and can be consulted for a more basic understanding of this subject.

Chapter **13**

Hydroxypropylcellulose

R. W. Butler and E. D. Klug
Hercules, Inc.

General Information 13-1
  Chemical Nature 13-2
  Physical Properties 13-2
  Manufacture 13-7
  Toxicological Properties 13-8
  Additives 13-8
  Handling 13-10
  Applications 13-10
  Application Procedures 13-11
  Specialties 13-11
Commercial Uses: Compounding and Formulating 13-11
Commercial Uses: Processing Aids 13-11
Industries Using Hydroxypropylcellulose 13-11
Formulations 13-13
  Cosmetics 13-13
  Paint Removers 13-14
  Pharmaceuticals 13-15
  Thermoplastics 13-15
Laboratory Techniques 13-16
Tradename Glossary 13-16
References 13-16

GENERAL INFORMATION

The hydroxypropylcellulose (HPC) of commerce is a non-ionic cellulose ether with an unusual combination of properties among cellulose derivatives. These include solubility in both water (below 40°C) and polar organic solvents, high surface activity, and plastic-flow properties that permit its use for molded and extruded articles.

The product is sold in powder form, is available in a wide range of viscosity types, and has Food and Drug Administration (FDA) clearance for use as a direct additive to foods. It is manufactured in the United States under the name of Klucel*[2,9] by Hercules, Inc., and in Japan under the name of Nisso HPC by Nippon Soda Kabushikigaisha.

The organic-solvent solubility of HPC enables it to be used in alcohol-based hair dressings, grooming aids, colognes, perfumes, inks, medicinals, and in methylene chloride–methanol-based paint removers[3] and solvent-based adhesives. Films that are flexible, glossy, and nontacky can be laid down from water or alcohol solvents. HPC is used to coat pharmaceutical tablets,[14] as well as candies and other confections.

Its surface activity enables it to be used as a stabilizer and whipping aid in whipped toppings (i.e., whipped cream substitutes)[1,6] and cosmetic lotions and creams. It provides protective colloid action as a suspending agent in poly(vinyl chloride) polymerization.

The excellent burn-out properties and low ash content of HPC make it a useful burn-out binder for glazes, ceramics, and electrical insulators.

The plastic properties of HPC[10,12,13] enable it to be used as the base material in extrusion, blow, or injection molding, and film-making operations. Thermoformed products retain their original solubilities, are biodegradable, and can be made to be edible.[12]

Chemical Nature

The empirical formula for hydroxypropylcellulose may be written as

$$[-C_6H_{10-n}O_5(CH_2CHOHCH_3)_n-]_x$$

where n has a range of 3 to 4.5, and x has a range of 150 to 3000. Thus the molecular weight of HPC ranges between 60,000 and 1,200,000.

The hydroxyl groups ($-OH$) present can be reacted with chemicals to give polymer modifications, as well as produce insoluble derivatives.

Stability   HPC in water solution, like other water-soluble cellulose ethers, is susceptible to both chemical and biological degradation. This degradation generally results in a reduction of the viscosity of the solution. The high level of hydroxypropyl substitution (MS = 4.0) generally associated with HPC does, however, enhance its resistance to both acid hydrolysis and biological degradation.

Chemical Stability.   HPC in solution is susceptible to acid hydrolysis as well as to alkali-catalyzed oxidative degradation, both of which can result in a decrease in solution viscosity. To minimize acid hydrolysis, solutions should be buffered to a pH of 6 to 8 and held at low temperatures. Oxidative degradation is also minimized by buffering to a pH of 6 to 8 and by the use of antioxidants.

Ultraviolet (UV) radiation will degrade HPC in solution. Thus, for best viscosity stability, solutions of HPC should be protected from light, or stabilized with UV absorbers.

Biological Stability.   HPC in water solution is susceptible to degradation by certain microorganisms and by enzymes. If prolonged storage of HPC in an aqueous system is contemplated, a preservative is required. On the other hand, solutions made in organic solvents do not generally require preservatives.

Insolubilization   The HPC polymer may be reacted through its hydroxyl groups with certain resins to give insoluble compositions. The rate of reaction and degree of water resistance of the final product is dependent upon the type and amount of crosslinking resin used, the pH of the system, and the temperature of cure.[2]   Resins such as dimethylolurea, Cymel 301, and Aerotex M-3 may be used.

Physical Properties

HPC is a white, odorless, tasteless powder, and is available in a wide range of viscosity types. Heating at 450 to 500°C in nitrogen or oxygen leaves only very small amounts of inorganic ash as a residue. The bulk density of commercial-grade HPC is approximately 0.5 g/mL with a purity (dry basis) of better than 99.5% and a moisture content generally less than 5%.

* Registered trademark of Hercules, Inc.

Moisture Content   The equilibrium moisture content of powdered HPC at any given humidity is very low compared to most cellulose ethers. Typical values for the effect of relative humidity on the moisture content of HPC at 23°C are:

Percent relative humidity	Percent moisture content
20	1.5
40	3.5
50	4.5
70	8.5

Solutions   HPC is soluble in water and in a wide range of organic liquids. It is infinitely soluble in water at temperatures below 40°C, but it is precipitated from solution at temperatures above 45°C. Typical values for the properties of HPC solutions in water are given in Table 13.1.

TABLE 13.1
Typical Values of Water Solutions of Hydroxypropylcellulose

Specific gravity, 2% solution at 30°C	1.010
Refractive index, 2% solution	1.337
Surface tension, 0.1% solution	43.6 dyn/cm
Interfacial tension, 0.1% solution	
(HPC vs. Fractol)	12.5 dyn/cm
Bulking value of HPC in water	0.333 cm³/g (0.04 gal/lb)
pH of solution, 1% solution	5.0 to 8.8

In most polar organic liquids, HPC generally has complete solubility at ambient or elevated temperatures.

Rheology.   Solutions of HPC are smooth-flowing, and they exhibit little or no structure or thixotropy. Solutions are pseudoplastic at high shear rates, but have good resistance to shear degradation. A wide range of viscosity types is available to provide solution viscosities of from 0.004 to 2.500 Pa · s (4 to 2500 cP) at a 1% concentration in water or ethanol. For data on viscosity versus concentration for HPC solutions, see Table 13.2. For data on viscosity versus temperature, see Fig. 13.1.

TABLE 13.2
Commercial Grades of Hydroxypropylcellulose

Grade designation*	Concentration, %	Viscosity, cP†	Approximate molecular weight, M_w
H	1	1500–2500	1,000,000
M	2	4000–6500	800,000
G	2	150–400	300,000
J	5	150–400	150,000
L	5	75–150	100,000
E	10	300–700	60,000

* Grade designations used for Hercules HPC and tradenamed Klucel. This product has a hydroxypropyl substitution of approximately 4. Grades having F in the designation (for example, HF) are suitable for use in foods.
† Measured with a Brookfield viscometer; multiply by 0.001 to convert to Pa · s.

Organic Solutions.   The molecular weight (viscosity type) of HPC influences its solubility, and this becomes an important consideration when borderline solvents are involved. The lower the molecular weight, the better is the solubility in borderline solvents.

For typical organic solvents for HPC, see Table 13.3. The viscosity of ethanol solutions of HPC versus concentration is similar to that given for water solutions of HPC in Table 13.4.

Hot Melts and Waxes. HPC is compatible with a number of high-molecular-weight, high-boiling waxes and oils, and can, therefore, be used to modify the properties of these materials. The addition of HPC to wax or oil systems will increase the system viscosity and will improve the hardness and crack resistance of coatings made from them.

At elevated temperatures, certain of these waxes are sufficiently good solvents that the HPC is readily dissolved by the addition of the dry powder to the stirred, molten wax. Materials that are good solvents for HPC at elevated temperatures include acetylated monoglycerides (e.g., Myvacet 5–00 and 7–00 series); glycerides (e.g., Myverol

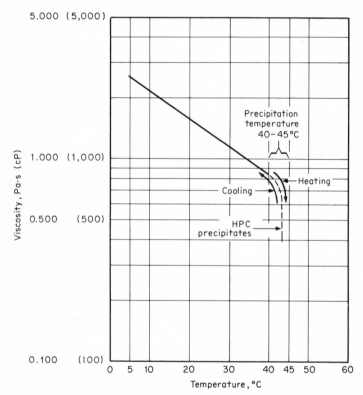

Figure 13.1 Effect of temperature upon the viscosity of aqueous solution of 1% high-viscosity hydroxypropylcellulose.

18–07); polyethylene glycols (e.g., Carbowaxes); polypropylene glycol, pine oil, and tall-oil fatty acids (e.g., Pamak 985).

Compatibility HPC is compatible with most water-soluble resins and gums. Homogeneous solutions and dried compositions are obtained with carboxymethylcellulose, hydroxyethylcellulose, methylcellulose, gelatin, sodium caseinate, polyethylene oxide, Carbowax 1000, poly(vinyl alcohol), guar gum, sodium alginate, and locust bean gum.

Some unexpected viscosity effects may be obtained when HPC is combined with certain polymers in water solution. For example, in combination with anionic polymers such as carboxymethylcellulose and sodium alginate, a synergistic viscosity increase is obtained. With non-ionic polymers such as hydroxyethylcellulose, methylcellulose, and guar gum, a lower than expected viscosity is obtained.

By use of common solvents, HPC can be incorporated with water-insoluble resins,

TABLE 13.3
Typical Organic Solvents for Hydroxypropylcellulose*

Good solvents†	
Acetic acid (glacial)	Ethylene chlorohydrin
Acetone:water (9:1)	Formic acid (88%)
Benzene:methanol (1:1)	Glycerin:water (3:7)
t-Butanol:water (9:1)	Isopropyl alcohol (95%)
Cellosolve	Methanol
Chloroform	Methyl Cellosolve
Cyclohexanone	Methylene chloride:methanol (9:1)
Dimethylformamide	Morpholine
Dimethyl sulfoxide	Propylene glycol
Dioxane	Pyridine
Ethane	Tetrahydrofuran
	Toluene:ethanol (3:2)

Fair solvents‡	
Acetone	Isopropyl alcohol:xylene (3:1)
t-Butanol	Lactic acid
Butyl acetate	Methyl acetate
Butyl Cellosolve	Methylethyl ketone
Cyclohexanol	Methylene chloride
Isopropyl alcohol (99%)	Naphtha:ethanol (1:1)

* Solvents were tested using a medium-viscosity grade of HPC at a 2% solids concentration. Ratios in the table for solvent composition are on a weight basis.
† Solutions are clear and smooth.
‡ Solutions may be somewhat granular and/or hazy.

TABLE 13.4
Effect of Hydroxypropylcellulose Concentration upon Solution Viscosity
(Values are in centipoise* measured on a Brookfield viscometer at 25°C)

HPC conc., %	Viscosity grade of HPC		
	High	Medium	Low
0.5	300	15	3
1.0	2,000	40	4
2.0	25,000	300	6
3.0		1,000	10
5.0		10,000	25
10.0			250
HPC grade:	Klucel H†	Klucel G†	Klucel E†
M_w of HPC:	Approx. 1,000,000	Approx. 300,000	Approx. 60,000

* Multiply by 0.001 to convert to Pa·s.
† Tradename of Hercules, Inc.

such as zein, shellac ethylcellulose, and cellulose acetate phthalate, to prepare homogeneous coatings.

The compatibility of HPC with inorganic salts in water solutions varies with the type of concentration of salt involved. Relatively high concentrations of salts tend to *salt out* the HPC as a highly swollen precipitate. The compatibility of a medium-viscosity HPC with various salts is given in Table 13.6.

Film Properties   Clear, flexible films of HPC may be prepared from aqueous or organic solvent solutions of the polymer. Films may also be prepared by thermoforming procedures. Typical properties of thin film cast from water solutions are given in Table 13.7.

Thermoplasticity   The good plastic-flow properties of HPC[10,12,13] enable it to be thermoformed into flexible film or rigid articles without the aid of plasticizers or other additives. Thermoformed products retain their water solubility. They are nontacky at high humidity, are biodegradable, and can be made to be edible.[12]

Normally, small amounts of plasticizers, internal lubricants, and antioxidants are used to improve the melt-flow, mold-release, and stability properties of the polymer during

TABLE 13.5
Typical Viscosity Values for HPC Solutions in Organic Solvents

Solvent	Solution viscosity	
	cP†	Pa·s
A. Solutions of high-viscosity HPC* at 1% concentration		
Acetic acid (glacial)	3,600	3.600
Acetone	1,120	1.120
Acetone:water (9:1)	560	0.560
t-Butanol	2,800	2.800
t-Butanol:water (9:1)	2,320	2.320
Butyl acetate	950	0.950
Butyl Cellosolve	2,800	2.800
Cellosolve	2,160	2.160
Chloroform	8,000	8.000
Cyclohexanol	6,800	6.800
Cyclohexanone	2,720	2.720
Dimethyl formamide	1,160	1.160
Dimethyl sulfoxide	2,360	2.360
Dioxane	2,080	2.080
Ethanol	1,280	1.280
Formic acid	140	0.140
Glycerin:water (3:7)	2,880	2.880
Isopropyl alcohol (95%)	1,480	1.480
Isopropyl alcohol (99%)	2,000	2.000
Isopropyl alcohol:xylene (3:1)	1,520	1.520
Lactic acid	11,800	11.800
Methanol	600	0.600
Methanol:benzene (1:1)	920	0.920
Methyl Cellosolve	1,880	1.880
Methylene chloride	9,000	9.000
Methylene chloride:methanol (9:1)	3,800	3.800
Methylethyl ketone	1,280	1.280
Propylene glycol	9,400	9.400
Pyridine	1,560	1.560
Tetrahydrofuran	1,600	1.600
Toluene:ethanol (3:2)	960	0.960
B. Solutions of low-viscosity HPC* at 6% concentration		
Acetone	75	0.075
n-Butanol	250	0.250
t-Butanol	600	0.600
Cellosolve	220	0.220
Ethanol	110	0.110
Ethyl acetate	270	0.270
Isopropyl alcohol	250	0.250
Methyl acetate	350	0.350
Methylene chloride	1,200	1.200
Methylethyl ketone	175	0.175
Tetrahydrofuran	105	0.105

* Values obtained using Klucel H and Klucel L as representatives of high- and low-viscosity grades of HPC, respectively.
† Measured on a Brookfield viscometer.

processing. Fillers such as clay, talc, and starch may be used to reduce material costs and to increase rigidity, hardness, and rate of solution. Functional fillers, such as dry foodstuffs, detergents, fragrances, and water-treatment chemicals, may also be used.

The rate at which a molded HPC item will dissolve is to a great extent dependent upon the molecular weight of the HPC used. The lower the molecular weight, the faster the rate of solution. Inert fillers and plasticizers generally increase the rate of solution, whereas materials such as glycerol monostearate and high-molecular-weight polyethylene glycol, decrease the rate of solution.

HPC molding formulations exhibit a melt flow very much like that of low-density

TABLE 13.6
Compatibility of Hydropropylcellulose with Some Salts*

	Salt concentration, wt %			
	2	5	10	50
Disodium phosphate	I	—	—	—
Sodium carbonate	C	I	—	—
Aluminum sulfate	C	I	—	—
Ammonium sulfate	C	I	—	—
Sodium sulfate	C	I	—	—
Sodium sulfite	C	I	—	—
Sodium thiosulfite	C	I	—	—
Sodium acetate	C	C	I	—
Sodium chloride	C	C	I	—
Potassium ferrocyanide	C	C	I	—
Calcium chloride	C	C	C	I
Sodium nitrate	C	C	C	I
Ferric chloride	C	C	C	I
Ammonium nitrate	C	C	C	I
Silver nitrate	C	C	C	I
Sucrose	C	C	C	I

* Tests were conducted by adding a 2% solution of HPC to salt solutions of various concentrations. The salt concentration in the system is indicated in the table, and the final HPC concentration was approximately 0.1% by weight in all cases.
NOTES:
  I = incompatible
  C = compatible

polyethylene. The effect of the molecular weight of HPC upon properties of molded tensile bars is given in Table 13.8.

Preparation of molding-powder pellets of HPC is described in a later section of this chapter. A typical formulation for a filled and a nonfilled HPC composition for injection molding is given under Formulations. Temperature and pressure conditions for injection and compression molding are given in Table 13.9.

Manufacture

The chemical cellulose used to prepare HPC is derived from wood pulp or cotton linters. The cellulose is treated with aqueous sodium hydroxide, and the resulting alkali cellulose is reacted with propylene oxide:

$$R_{cell}OH + CH_3\overset{O}{\overset{\wedge}{CHCH_2}} \rightarrow R_{cell}OCH_2CHOHCH_3$$

The secondary hydroxyl group in the hydroxypropyl group is capable of hydroxypropylation to give a side chain:

$$R_{cell}OCH_2CHOHCH_3 + CH_3\overset{O}{\overset{\wedge}{CHCH_2}} \rightarrow R_{cell}\overset{\overset{\displaystyle CH_2CHOHCH_3}{|}}{\underset{|}{\overset{O}{}}}CHCHCH_3$$

The sodium hydroxide functions as a swelling agent and a catalyst for the etherification. A side reaction in which propylene oxide reacts with water to form a mixture of propylene glycols also occurs. To minimize this side reaction, the water input is kept as low as possible. However, a certain amount of water is necessary as a swelling agent and carrier for the sodium hydroxide.

TABLE 13.7
Typical Properties of Hydroxypropylcellulose Film*

Appearance		Clear
Specific gravity		Approximately 1.1
Tensile strength		13.82 MPa (2000 psi)
Modulus of elasticity		414.6 MPa (60,000 psi)
Elongation at break		Approximately 50%
Flexibility [MIT double folds, 5.08×10^{-5}-μm (2-mil) film]		Greater than 10,000
Refractive index		1.559
Heat sealing on a bar sealer		0.5 s at 210–249°C (410–480°F)
Blocking resistance		No blocking at 90% RH
Oil and grease resistance		Good
Transmission rates: O_2	25 mL	Per 645 cm²/24 h/atm (100 in²/24 h/atm)
N_2	3 mL	for 5.08×10^{-5}-μm (2-mil) film at 24°C
CO_2	170 mL	(75°F)
Water vapor	20 g	As above (ASTM Method E96B)
Moisture content (equilibrium):		
at 23°C (73°F) and 50% relative humidity		4%
at 23°C (73°C) and 84% relative humidity		12%

Electrical properties at 24°C (75°F):	*760 mm, 38% RH*	*1 mm, 0% RH*
Dielectric constant (1000 cycles)	9.07	6.71
Dissipation factor (1000 cycles)	0.0706	0.0408
Volume resistivity ($\Omega \cdot$cm)	5×10^9	9×10^{11}

Solubility:	
Water below 40°C	Soluble
Water above 45°C	Insoluble
Polar solvents (hot or cold)	Soluble in many
Nonpolar organic solvents	Insoluble in most

* Films were nonplasticized and cast from water solution to 2.54 to 5.08×10^{-5} μm (1 to 2 mil) thickness.

The etherification is conducted at 70 to 100°C, and the reaction time may be 5 to 20 h, depending on the temperature. Since the reaction is exothermic, sufficient cooling capability must be incorporated into the reactor. One method of accomplishing this is to carry out the reaction in the presence of an inert diluent, such as an aliphatic hydrocarbon.

At the end of the hydroxypropylation, the alkali is neutralized and the product is purified by washing with hot water (70 to 90°C) to remove sodium salts and glycols. It is then dried and ground to a fine white powder.

Different viscosity grades of HPC are available, as shown in Table 13.2 Highest-viscosity products are obtained from high-molecular-weight cotton linters. Lower-viscosity types are obtained from wood pulp with an oxidative degradation treatment.

Toxicological Properties

HPC is classified as a food additive. It is permitted in foods for human consumption as set forth by the Food and Drug Administration in Section 172.870 of the Food Additives Regulations (21 CFR 172.870). HPC is listed in the *Food Chemicals Codex.*

There is no evidence that HPC is either a primary skin irritant or a skin-sensitizing agent,[4] and it is physiologically inert following oral administration.[4]

Additives

Solutions of this non-ionic polymer have wide ranges of compatibilities with relatively low concentrations of soluble salts, resins, and other polymers. At higher concentrations, certain salts may salt-out the HPC polymer from solution. The following are examples of preservatives, defoamers, and plasticizers that are recommended for use with HPC:

TABLE 13.8
Typical Thermoplastic Properties of Hydroxypropylcellulose

Properties	ASTM Test Method	Molecular weight of hydroxypropylcellulose			
		High 1,250,000	Medium 300,000	Low 75,000	Extra low 50,000
Nominal melt flow:					
I_2 at 160°C	D-1238	0.8			
I_2 at 180°C			13.6		
I_2 at 200°C				18	40
Density, g/mL	D-792	1.15	1.15	1.15	1.15
Tensile modulus, MPa (psi)	D-638	635.7 (92,000)	1,845.0 (267,000)	2,349.6 (340,000)	2,487.6 (360,000)
Tensile yield strength, MPa (psi)	D-638	22.11 (3,200)	44.22 (6,400)	45.61 (6,600)	46.30 (6,700)
Tensile yield elongation, %	D-638	4	3.5	4.3	4.0
Flexural modulus, MPa (psi)	D-790	5.53 (800)	14.03 (2,030)	24.88 (3,600)	26.95 (3,900)
Rockwell hardness, R scale	D-785	*	*	7.3	9.0
Softening temperature, °C		150	130	110	100

* Values below Rockwell R scale.

TABLE 13.9
Molding Conditions for Hydroxypropylcellulose

	High	Medium	Low	Extra low
Compress-molding temperature,* °C (°F)	166–171 (330–400)	160–193 (320–380)	143–177 (290–350)	132–171 (270–340)
Compression-molding pressure, MPa (psi)	3.95–13.82 (500–2,000)	2.07–13.82 (300–2,000)	1.38–13.82 (200–2,000)	0.69–6.91 (100–1,000)
Injection-molding temperature,* °C (°F)	160–182 (320–360)	166–177 (330–350)	138–177 (280–350)	127–171 (260–340)
Mold shrinkage, cm/cm (in/in)	0.005–0.010	0.003–0.009	0.001–0.005	0.001–0.004

* The maximum temperatures given should not be exceeded; excessive heat results in degradation and discoloration of the polymer.

Preservatives   Where prolonged storage of aqueous solutions of HPC is contemplated, a preservative is recommended. The following are among the many that have been found effective:[2] formaldehyde, phenylmercuric acetate, phenol, and dioxin. Those sanctioned by the FDA for use in food include sodium benzoate, sodium propionate, and salts of sorbic acid.

Defoamers   Water-dispersible defoamers are effective when used to reduce foaming or air entrainment during processing. Recommended are: Hercules Defoamer 1512; Colloids 581-B; Nopco NDW; Antifoam A, AF, or C.

Plasticizers   Many plasticizers are effective in modifying the properties of HPC films, coatings, or molded articles.[2,12] Examples are: propylene glycol, glycerin, Carbowax 600 to 1500, and glycerol monostearate.

Handling

For safety in handling, HPC should be treated as an inert organic particulate. It should be stored under dry conditions, away from heat.

Applications

The applications of HPC are derived from its ability to bind, emulsify, foam, stabilize, suspend, and thicken. Products that can benefit from HPC include foods, cosmetics, medicinals, cleaners, paper and textiles, tablets, and many others. A more detailed listing of HPC applications is given in Table 13.10.

TABLE 13.10
Applications for Hydroxypropylcellulose by Product Function

Function	Application
Binder	Alcohol core-wash compounds
	Burn-out binder for ceramics, glazes, and insulators
	Extruded and fabricated foods
	Matrix-board manufacture
	Tablets
Coatings	For paper, film, textiles
	For foods, nuts, confections
	For tablets
Emulsification aid	Cosmetic emulsions
Film forming	Roll-on antiperspirants
	Cosmetic creams and lotions
	Hair dressings and grooming aids
	Perfumes and colognes
Foaming aid	Whipped toppings
	Cosmetic emulsions
Stabilizer	Cosmetic emulsions
	Whipped toppings
	Medicinal emulsions and suspensions
	Suspending aid for PVC manufacture
Suspending agent	Medicinal and cosmetic suspensions (aqueous- to alcohol-based)
	Alcohol- and glycol-based inks
	Cleaners and polishes
	Alcohol core-wash compounds
Thermoplastic	Extruded and fabricated foods
	Molded and extruded items
	Film and sheet
Thickener	For aqueous, alcohol, glycol and other solvent systems
	Cosmetic creams, lotions, antiperspirants
	Perfumes and colognes
	Hair dressings and grooming aids
	Medicinal elixirs, emulsions, suspensions
	Paint removers (acid- and solvent-based)
	Inks (alcohol- and glycol-based)
	Cleaners and polishes

Application Procedures

Water Temperature   HPC is insoluble in water at temperatures in excess of 45°C; therefore solutions cannot be prepared above this temperature. Also, if an aqueous solution is heated above 45°C, the HPC will be precipitated out of solution in a highly swollen form. The insolubility of HPC in hot water is utilized as an aid to solution preparations. The HPC powder can be dispersed in a small volume of hot water (50 to 60°C) without lumping of the powder. Water temperature in excess of 60°C will melt the powder and cause lumping. This hot slurry is then added to a larger volume of cold water with stirring, and rapid dissolution of the particles results.

Compatibility with Salts   The tendency for HPC to be salted-out by relatively high concentrations of salts should be kept in mind when formulating. See the earlier discussion of compatibility under Physical Properties.

Molding-Powder Preparation   Dry ingredients are charged into a high-intensity mixer (e.g., Henschel or Welex) and mixed for 20 s. Liquids are then added through the top port of the mixer, and mixing is continued for 2 min at high speed. Low speed can be used if liquids are omitted.

For optimum extrusion characteristics, a vented twin-screw extruder should be used. Extrusion temperatures and properties of the plastic are given in Table 13.9. Strands can be cooled in air before being chopped into pellets of the required size.

Specialties

A modification of HPC that gives a product which is soluble in water at temperatures up to 90°C if the solution pH is held below 6 is available as Klucel 6, in various viscosity grades, from Hercules.

COMMERCIAL USES: Compounding and Formulating

Table 13.11 gives details on 13 general and 38 specific end uses of HPC for the compounding and formulating of products. These applications are quite varied, derived from the versatility of this material.

COMMERCIAL USES: Processing Aids

The primary applications of HPC as a processing or manufacturing aid fall into two main categories: (1) as a green-strength burn-out binder in the manufacture of ceramics, electrical insulators, or ceramic glazes, or (2) as a suspending aid for the suspension polymerization of poly(vinyl chloride). Details of these applications are given in Table 13.12.

INDUSTRIES USING HYDROXYPROPYLCELLULOSE

For quick reference purposes, the following tabulation gives in capsule form the various industries using HPC and the reasons for such uses:

Adhesives	Thermoplastic
	Solvent-soluble binder and thickener
Ceramics	Burn-out binder with negligible ash residue
Cosmetics	Solvent-soluble thickener
	Nontacky, flexible film-former and binder
	Non-ionic stabilizer and suspending agent
	Surface-active protective colloid
Encapsulation	Solvent-soluble, oil and fat barrier
	Insoluble in hot water
Food	Non-ionic, surface-active emulsion stabilizer
	Edible, thermoplastic binder and film-former
	Solvent-soluble, nontacky coatings
Household items	Thickener, suspending aid, film-former for cleaners/polishes
Inks	Thickener, binder, suspending agent
	Soluble in water, alcohol, and glycols
Matrix board	Solvent-soluble thickener, binder, and suspending agent for coatings
Paint	Thickener for solvent- and acid-based paint removers

TABLE 13.11
Compounding and Formulating Uses for Hydroxypropylcellulose

Product category	Specific product	HPC properties utilized	Properties from HPC
Aerosols	Cosmetic emulsions	Surface-active stabilizer	Emulsion stability, foaming aid
	Solvent-based	Solvent-soluble binder	Film-former, binder
	Whipped toppings[1,6]	Surface-active stabilizer	Emulsion stability, foaming aid
Adhesives	Hot-melt	Thermoplastic binder	Binder
	Solvent-based	Solvent-soluble polymer	Thickener, binder
Binders	Alcohol core-wash compounds	Solvent-soluble polymer	Thickener, binder, suspending agent
	Burn-out binders		
	Fabricated foods		
	Matrix boards	Solvent-soluble, thermoplastic binder	Thickener, binder, suspending agent for board coatings
	Tablet binders	(see Pharmaceuticals)	
Cleaners	Solvent- and aqueous-based	Solvent-soluble thickener	Thickener, suspending agent
Coatings	Film	Solvent-soluble, thermoplastic film-former	Solvent-soluble, oil and fat barrier, heat-sealable coating
	Foods	(see Foods)	
	Tablets	(see Pharmaceuticals)	
	Paper	Solvent-soluble, thermoplastic film-former	Oil and fat barrier, thermoplastic coating
	Textiles	Solvent-soluble, thermoplastic film-former	Solvent-soluble, thermoplastic coating
Cosmetics[5]	Alcohol-based	Alcohol solubility	Thickener, binder
	Antiperspirants	Non-ionic polymer	Thickener, film-former
	Emulsion creams, lotions	Surface-active stabilizer	Thickener, stabilizer, binder
	Hair dressings and grooming aids	Alcohol solubility	Thickener, binder, film-former
	Perfumes and colognes	Alcohol solubility	Thickener, binder, film-former
	Shampoos	Non-ionic polymer	Thickener, stabilizer
Encapsulation	All types	Non-ionic polymer with unique solubility	Edible, soluble barrier
Foods	Coatings for nuts and confections	Solvent-soluble film-former	Nontacky, oil and fat barrier
	Fabricated foods	Thermoplastic binder	Binder
	Glazes	Solvent-soluble film-former	Nontacky, oil and fat barrier
	Whipped toppings[1,6]	Surface-active stabilizer	Emulsion stability and foaming aid
Inks	Alcohol- and glycol-based	Solubility in alcohol and glycols	Thickener, binder, suspending agent
Paint removers[3]	Acid- and solvent-based	Solvent solubility, acid resistance	Thickener
Pharmaceuticals	Medicinals		
	Elixirs	Non-ionic, solvent-soluble polymer	Thickener
	Emulsions	Surface-active stabilizer	Emulsion stability, film-former
	Suspensions	Surface-active stabilizer	Thickener, suspending agent
	Tablets		
	Film coating	Solvent-soluble polymer	Nontacky coating
	Binder, granulation aid	Solvent-soluble binder	Applied from nonaqueous or water solution
Polishes	Aqueous- and solvent-based	Solvent-soluble, surface-active polymer	Thickener, stabilizer, suspending agent
Thermoplastics[10,12,13]	Fabricated foods	Soluble, flexible thermoplastic	Edible binder
	Film and sheet	Soluble, flexible thermoplastic	Nontacky, flexible, soluble plastic
	Molded and extruded items	Soluble, flexible thermoplastic	Soluble, nontacky plastic

Paper	Solvent-soluble, thermoplastic coating
	Oil and fat barrier
Pharmaceuticals	Edible, solvent-soluble thickener
	Surface-active protective colloid
	Non-ionic stabilizer and suspending agent
	Nontacky, flexible film-former and binder
PVC manufacture	Processing aid for increased rate of plasticizer absorption
	and narrower particle-size distribution for PVC resin
	Suspending aid for less build-up on reactor walls and easier
	dewatering of PVC resin before drying
Textiles	Solvent-soluble, thermoplastic coating
Thermoplastics	Edible, water-soluble binder and film-former
	Nontacky, water- and solvent-soluble plastic
	Clear, flexible film and sheet
	Oil and fat barrier

TABLE 13.12
Hydroxypropylcellulose Used as a Processing and Manufacturing Aid

Application	End-product	HPC property used	Property from HPC	Use procedure
Burn-out binder	Ceramics Electrical insulators Glazes (ceramics)	Strong binder, ready burnout in N_2 or air, negligible ash residue, water and solvent-soluble	Green-strength binder of high efficiency with negligible ash residue	HPC is dissolved in water or organic solvent
Suspending aid	Poly(vinyl chloride) by suspension polymerization	Surface-active protective and suspending aid	Increased rate of plasticizer absorption and narrower particle-size distribution for PVC resin. Less build-up on reactor walls and easier dewatering during manufacturing	Used in conjunction with other suspending aids[8]

FORMULATIONS

The sample formulations of this section are presented to illustrate some of the functions served by HPC in the preparation of products for cosmetics, paint removers, pharmaceuticals, and thermoplastics.

Cosmetics

Antiperspirant (Roll-On)   HPC is used in a lotion antiperspirant as a thickener as well as a stabilizer in the presence of acid salts. The HPC also functions as a film-former and reduces the tackiness of the aluminum salt residues left on the skin. The formulation is:

Ingredient	Percent by weight
Polyoxyethylene (75) lanolin ether	2.0
Klucel M (HPC)	0.7
Ethanol	20.0
Aluminum chlorhydrol (50%)	40.0
Water	37.3
Perfume	q.s.

Hair-Grooming Aid   HPC serves as a thickener and film-former in a hair-grooming aid that contains alcohol and propylene glycol. The formulation is:

Ingredient	Percent by weight
Water	39.0
Alcohol	12.0
Klucel H (HPC)	1.0
Propylene glycol	48.0

Dissolve the HPC in the water-alcohol mix, then add the glycol and stir until the solution is uniform. This formulation has a viscosity of approximately 10 Pa · s (10,000 cP), which may be varied by changing the concentration of HPC.

Lotion (Body and Hand) HPC serves to thicken and stabilize cosmetic emulsions, such as the hand and body lotion given as follows:

Ingredients	Percent by weight
Klucel M (HPC)	0.25
Water	86.75
Glyceryl stearate	2.0
CTFA quaternium 7 (e.g., Emcol E 607S)	1.0
Mineral oil	4.0
Glycerin (U.S.P.)	5.0
Lanolin (U.S.P.)	1.0
Perfume	q.s.

Dissolve the HPC in room-temperature water. Prepare the oil phase at 85°C, cool to below 40°C, and add with perfume to the HPC solution under moderate agitation. The consistency may be varied by changing the concentration of HPC.

Shampoo (Gel) HPC can be used to thicken shampoos that contain high concentrations of anionic and amphoteric surfactants. For example:

Ingredients	Percent by weight
Water	45.0
Klucel M (HPC)	1.0
Triethanolamine lauryl sulfate (46% active)	40.0
Miranol C2MSF	8.0
Lauric diethanolamide	5.0
Polyoxyethylene (75) lanolin ether	1.0

Dissolve the HPC in water. Add the oil phase with mild agitation.

Paint Removers

HPC is used as the thickener for many types of solvent- and acid-based paint and varnish removers,[3] as illustrated in the following two examples.

Nonflammable Solvent–Type Remover

Ingredients	Percent by weight
Methylene chloride	73.0
Toluene	2.1
Paraffin wax	1.7
Klucel H (HPC)	1.0
Methanol	8.0
Potassium oleate	5.0
Triethylammonium phosphate	0.1
Di-tri-isopropanolamine	7.8
Water	1.3

Dissolve the melted paraffin in the methylene chloride–toluene solvent.   Add the HPC slowly while stirring.   Add methanol and stir until the HPC is dissolved.   Add the remaining ingredients.   This formulation has a viscosity of approximately 5 Pa · s (5000 cP), which may be varied by changing the concentration of the HPC.

Acid-Type Remover

Ingredient	Percent by weight
Methylene chloride	65
Paraffin wax	1
Klucel H (HPC)	1
Potassium oleate	3
Phenol	20
Formic acid (90%)	10

Dissolve the melted paraffin in the methylene chloride solvent.   Add the HPC slowly while stirring.   Add the remaining ingredients, and stir until the solution is uniform. This formulation has a viscosity of approximately 3 Pa · s (3000 cP), which may be varied by changing the concentration of the HPC.

Pharmaceuticals

For the application of tablet coatings from organic solvents, HPC is used as the film-former and binder for the coating.[4,13]   As an example:

Ingredient	Percent by weight
Klucel LF (food grade of HPC)	3.0
Ethylcellulose (pharmaceutical grade)	1.0
Opaspray Lake Pigment (100% solids basis)	93.8
Total solids	6.2

The viscosity of this solution is 0.040 Pa · s (40 cP Brookfield) and 10 s (Zahn cup No. 3).   To apply the coating to the tablets as they tumble in a revolving coating pan, use an air-spray gun (e.g., Binks 46 Model).

Thermoplastics

HPC can be handled easily in thermoforming operations.   It can be extruded and molded with or without high levels of fillers to form articles that retain their water solubility, are biodegradable, and can be made edible.   Typical formulations for unfilled and filled compositions of HPC suitable for injection molding at 166 to 177°C (330 to 350°F) are:

Injection-Molding Formulation (Unfilled)

Ingredient	Percent by weight
Klucel J or JF* (HPC)	95.6–98.1
Plasticizer:	
Propylene glycol† ⎤	
Polyethylene glycol 4000‡ ⎦	0.5–3.0§
Mold-release agent:	
Glycerol monostearate† ⎤	
Polyethylene glycol 4000‡ ⎦	1.0
Stabilizers:	
Butylated hydroxytoluene	0.2
Lauryl thiodipropionate	0.2

* For food or pharmaceutical use.
† For food use.
‡ For pharmaceutical or general use.
§ Level depends upon the flexibility required.

Injection-Molding Formulation (Filled)

Ingredient	Percent by weight
Klucel G, medium molecular weight (HPC)	44.7–46.3
Filler*	44.7–46.3
Plasticizer:	
Polyethylene glycol 400	2.0– 5.0†
Mold-release agent:	
Polyethylene glycol 4000	3.0
Whitener:	
Titanium dioxide	2.0
Stabilizers:	
Butylated hydroxytoluene	0.2
Lauryl thiodipropionate	0.2

* Fillers such as starch, clays, talc, asbestos, calcium carbonate, and microcrystalline cellulose may be used.
† Level depends upon the flexibility required.
NOTE: It may be necessary to predry the composition in a hopper drier before injection molding.

LABORATORY TECHNIQUES

The hydroxypropyl substitution (MS) of HPC can be determined by the method of Lemieux and Purves[4] in which terminal methyl groups are measured by recovery of acetic acid formed by treatment with hot chromic acid.

Moisture content and solution viscosity are determined as for carboxymethylcellulose under ASTM Method C1439–65.

TRADENAME GLOSSARY

Tradenames	Chemical Identity	Manufacturer
Aerotex M-3	Melamine-formaldehyde resin	American Cyanamid Co.
Antifoam A, AF, C	——	Dow Corning Corp.
Carbowax 1000	Polyethylene glycol	Union Carbide Corp.
CMC (cellulose gum)	Sodium carboxymethylcellulose	Hercules, Inc.
Colloids 581-B	——	Colloids, Inc.
Cymel 301	Hexamethoxymethylmelamine	American Cyanamid Co.
Dioxin	6-acetoxy-2,4-dimethyl-*m*-dioxane	Givaudan Corp.
DMU	Dimethylolurea	Glycol Chemicals, Inc.
Dowicides A, B, G	Substituted phenols	Dow Chemical Co.
Dowicil 100	*N*-(3-chloroallyl)-hexaminium chloride	Dow Chemical Co.
Emcol E607S	——	Witco Chemicals
HEC	Hydroxyethylcellulose	Hercules, Inc.
HPC	Hydroxypropylcellulose	Hercules, Inc.
Klucel	Hydroxypropylcellulose	Hercules, Inc.
Kroniflex	Tris(tetrahydrofurfuryl) phosphate	FMC Corp.
Miranol C2MSF	Amphoteric surfactant	Miranol Chemical Co., Inc.
Myvacet	Acetylated monoglyceride	Eastman Chemical Products
Myverol	Monoglyceride	Eastman Chemical Products
Natrosol	Hydroxyethylcellulose	Hercules, Inc.
Nisso HPC	Hydroxypropylcellulose	Nippon Soda
Nopco NDW, KFS	——	Nopco Chemical Co.
Onyxide 172	——	Onyx Chemical Co.
Opaspray Lake Pigment	——	Colorcon, Inc.
Pamak 985	Tall-oil fatty acids	Hercules, Inc.
Quilon S	Stearatochromyl chloride complex	E. I. du Pont de Nemours
Sodium Omadine	Sodium 1-hydroxypyridine-2-thione	Olin Mathieson Chemical Corp.

REFERENCES

1. Ganz, A. J. (to Hercules, Inc.), U.S. Patent 3,479,190 (1969); *Chem. Abstr.*, **72**, 5391w (1970).
2. Hercules, Inc., *Chemical and Physical Properties of Klucel, Hydroxypropyl Cellulose*, 800–804.

3. Hercules, Inc., *Klucel, Hydroxypropyl Cellulose in Paint and Varnish Removers*, VC 477.
4. Hercules, Inc., *Klucel, Hydroxypropyl Cellulose: Summary of Toxicological Investigations*, T-122.
5. Hercules, Inc., *Klucel for Use in Cosmetic Products*, VC 439B.
6. Hercules, Inc., *Klucel as a Stabilizer in Whipped Toppings*, VC 469B.
7. Hercules, Inc., *Klucel in Tablet Coatings*, VC 476.
8. Hercules, Inc., unpublished data.
9. Klug, E. D. (to Hercules, Inc.), U.S. Patent 3,278,521 (1966).
10. Klug, E. D. (to Hercules, Inc.), U.S. Patent 3,314,809 (1967).
11. Lemieux, R. U., and C. B. Purves, *Can. J. Res.*, **25B**, 485 (1947).
12. Rossman, J. M., "Now: A Plastic to Eat," *Package Eng.*, **16**(7), 54 (1971).
13. Rossman, J. M., and A. J. Desmarais, *Hercules Chem.*, (61), 9 (1970).
14. Signorino, C. A., *Food Sci. Technol. Abstr.*, **2**, 667 (1970).

Chapter **14**

Locust Bean Gum

James K. Seaman
Celanese Plastics and Specialties Company

General Information 14-1
   Manufacture 14-2
Properties 14-3
   Biological Properties 14-6
   Handling 14-8
Commercial Uses: Compounding and Formulating 14-8
   Food Products 14-8
Commercial Uses: Processing Aids 14-8
   Textiles Processing 14-8
   Paper Products 14-10
   Mining Industry 14-13
Industries Using Locust Bean Gum 14-13
   Food Industry 14-13
   Mining Industry 14-14
   Paper Industry 14-14
   Textiles Industry 14-14
Formulations 14-14
   Ice Cream 14-14
   Ice Milk 14-15
   Sherbet 14-15
   Sour Cream 14-15
   Buttermilk 14-15
   Yogurt 14-15
   Instant Imitation Bakery Jelly 14-15
   Whipping Cream Composition (for Frozen Desserts) 14-15
   Product Term Glossary 14-16

GENERAL INFORMATION

Locust bean gum is the common name applied to the gum found in the seeds of the carob tree *(Ceratonia siliqua)*. The tree grows extensively throughout the Mediterranean region.   The fruit of the tree, the locust bean or carob, is harvested annually. The locust bean is a dark brown pod 10 to 20 cm (4 to 8 in) long.   The several seeds or kernels contained within are the source of an edible carbohydrate polymer which is useful as a thickening agent for water and as a chemical reagent for reaction with mineral and cellulosic surfaces.

Manufacture

The production of locust bean gum is basically a series of crushing, sifting, and grinding steps designed to separate the seeds from the pod and then to separate the valuable gum from the seeds.

Seed Structure   The gum is contained within a portion of the seed known as the *endosperm*.   This is the reserve food supply for the developing embryo during germination.   Since the seed is dicotyledonous, there are two endosperm halves obtained from each seed.   These endosperm halves surround the embryo, and they are in turn surrounded by a dark, brown hull.   The endosperm is from 42 to 46% of the weight of the seed.

Purification   Hull is removed from the seed by various processes which may use acid to loosen it, heat to embrittle it, mechanical means to abrade or crack it, and sifting to effect separation of the hull parts.   The endosperm halves separate readily to reveal the embryo, which is also cracked and separated by sifting.

Grades   The endosperm is ground to powder form, and depending on the efficiency of the purification, high-grade, industrial, or technical gum is produced in different mesh sizes.   See Table 14.1 for typical analysis.   Determination of moisture, protein, and acid-insoluble residue content is generally considered to be the minimum analysis to monitor the efficiency of the purification process.   The presence of embryo is reflected in high protein content.   Hull fragments show up as a high acid-insoluble residue. Moisture contents above 14% or below 8% exceed normal equilibrium moisture and indicate abuse of the seed or powder.

A crude-fiber determination is sometimes used in place of the acid-insoluble residue. Crude fiber is the acid- and alkali-insoluble residue and typically amounts to about one-half of the acid-insoluble residue.   It is recognized that other natural impurities are present in commercial locust bean gum.   The natural mineral content of the seeds

TABLE 14.1
Locust Bean Gum

	Typical Analyses of Locust Bean Gum Powder			
	High grade, 175 Mesh	High grade, 100 mesh	Industrial	Technical
Percent moisture	10–12	10–12	10–14	10–14
Percent protein	4–6	4–6	5–7	6–13
Percent acid-insoluble residue	1.5–2.0	1.5–2	2–4.5	5–8
1% viscosity, Pa·s (cP)*	3.2–3.5 (3200–3500)	2.3–3.7 (2700–3700)	1.5–1.8 (1500–1800)	0.5–1.0 (500–1000)
Mesh distribution:				
Minimum through 80	99%	98%	99%	99%
Maximum through 200	20%	10%	25%	uncontrolled

Food Chemicals Codex Specifications	
Galactomannans	73.0%
Limits of impurities:	
Acid-insoluble matter	5%
Protein	8%
Loss on drying	15% max.
Ash (total)	1.2% max.
Heavy metals (as Pb)	20 ppm max.
Lead	10 ppm max.
Arsenic (as is)	3 ppm max.
Starch	Negative

* At 25°C as measured on a Brookfield Synchro-Lectric viscometer at 20 rpm.

gives rise to an ash content. Low-molecular-weight soluble sugars are present, but not usually assayed, and a small ether-extractable fraction reflects the natural presence of fats and oils.

Properties

Locust bean gum is a carbohydrate polymer containing galactose and mannose as the structural building blocks. The ratio of the two components may vary slightly depending on the origin of the seed, but the gum is generally considered to contain one galactose unit for every four mannose units.

Structure It has long been accepted that the structure of locust bean gum is a linear chain of β-D-mannopyranosyl units linked 1,4 with single-membered α-D-galactopyrano-syl units occurring as side branches. The α-D-galactopyranosyl units are linked 1,6 with the main chain. Recent work points to the conclusion that the side branches are not spaced uniformly, as was previously thought. It now seems more likely that there are segments of the chain composed of unsubstituted β-D-mannopyranosyl units alternated with chain segments where the α-D-galactopyranosyl side branches substitute on each of the main-chain units. Rather than being an alternating copolymer, it is more of a natural block polymer.

Solubility in Water Locust bean gum has limited solubility when added to water at ambient temperature. The gum will hydrate and develop its properties when its solutions are heated. It will reach full hydration if heated for 10 min at 80°C. Upon cooling, it remains in solution in dissociated, extended molecular form, where it can serve as a thickener or chemical reagent. This behavior is probably explained as a result of the side-branch distribution. Where the mannose units of the backbone chain come in contact with mannose units of an adjacent chain, they bond strongly to one another, and sufficient heat energy must be applied before these secondary bonds are broken. Water is the only common solvent for locust bean gum, although it will tolerate limited concentrations of water-miscible solvents such as alcohol. Dimethylformamide and dimethylsulfoxide are solvents for locust bean gum, as they are for most other polymers.

Commercial locust bean gum solutions are typically turbid. The turbidity is caused largely by the presence of the endosperm. Polymer solutions intensively purified in the laboratory by noncommercial methods approach water clarity.

Rheology Locust bean gum is one of the more efficient water-thickening agents among water-soluble gums. Its solutions are non-Newtonian classed as pseudoplastic. They thin out reversibly when heat is applied and degrade irreversibly with time when elevated temperature is maintained. The solutions resist shear degradation as well as many other water-soluble polymers do, but they will degrade progressively under high shear.

Viscosity. Rarely is this gum, in its unmodified form, used at concentrations above 1%. At this concentration it is a thick solution; at 3% it looks more like a gel than a solution. (See Figs 14.1 and 14.2.)

Shear Response. Solutions of the gum have zero yield value. That is, they begin to flow as soon as the slightest shear is applied. The apparent viscosity of the solution decreases sharply as the rate of shear is increased, then levels off and approaches a minimum limiting value that is dependent on the concentration of the solution. (See Figs. 14.3 to 14.9.) The apparent viscosity of the locust bean gum solution at any given rate of shear is independent of time and prior shear testing. It makes no difference whether the final rate of shear is approached from low shear rate to high shear rate or vice versa, provided shear rates are not high enough to degrade the molecular struc-ture. Its molecular structure can be irreversibly degraded by high rates of shear that would be found, for example, in pumps used to transfer a solution from one tank to another. Such breakdown is related to the degree of shear and the time over which it is applied. Degraded solutions progressively lose viscosity at a given rate of shear. They become less pseudoplastic, and they tend toward the linear response of a Newtonian fluid. Thus the shape of the viscosity versus shear rate curve of degraded solutions changes shape and flattens out.

Reactivity The cyclic, neutral sugar structures which make up the polymer contain numerous hydroxyl groups (an average of three per sugar unit).

Derivatives. Etherification and esterification reactions are readily carried out through

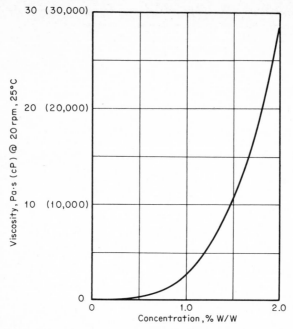

Figure 14.1   Effect of concentration on viscosity of locust bean gum dispersions at 25°C as measured on a Brookfield Synchro-Lectric viscometer at 20 rpm.   Solutions prehydrated at 85°C for 15 min, then cooled.

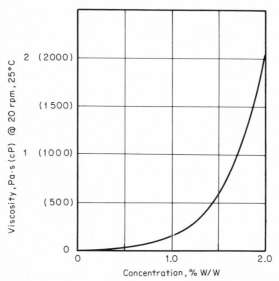

Figure 14.2   Effect of concentration on viscosity of locust bean gum dispersions at 25°C as measured on a Brookfield Synchro-Lectric viscometer at 20 rpm.   Solutions hydrated at 25°C.

these hydroxyl functionalities. The commercially important derivatives to date have been confined, however, to the etherification reactions: carboxymethylation with sodium monochloracetate, hydroxyethylation with ethylene oxide, and hydroxypropylation with propylene oxide.

Borax Reaction. Each of the sugar residues in the polymer have two hydroxyl groups positioned in the cis form. This leads to an interesting and valuable reaction with dissociated borate ion that is characteristic of such polymers. The reaction is fully reversible with changes in pH. An aqueous solution of the gum will gel in the presence of borate when the solution is made alkaline and will liquefy again when the pH is dropped below 7. If the dry, powdered gum is added to alkaline borate solution, it will not hydrate and thicken until the pH is dropped below 7. The critical pH at which gelation occurs is modified by the concentration of dissolved salts. The effect of dissolved salts is to change the pH at which a sufficient quantity of dissociated borate ion exists in solution to cause gelation.

Reaction with Xanthan Gum. It has been known for some time that locust bean

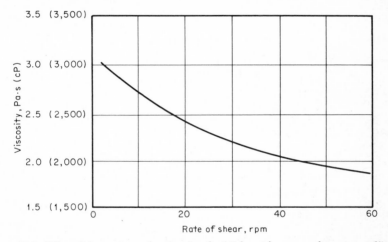

Figure 14.3 Effect of shear rate on the viscosity of a 1% locust bean gum dispersion at 25°C as measured on a Brookfield Synchro-Lectric viscometer. Solutions prehydrated at 85°C for 15 min, then cooled.

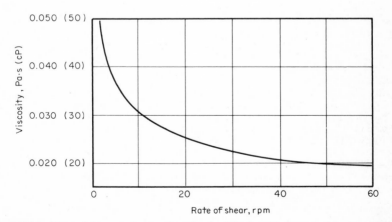

Figure 14.4 Effect of shear rate on the viscosity of a 0.3% locust bean gum disperion at 25°C as measured on a Brookfield Synchro-Lectric viscometer. Solutions prehydrated at 85°C for 15 min, then cooled.

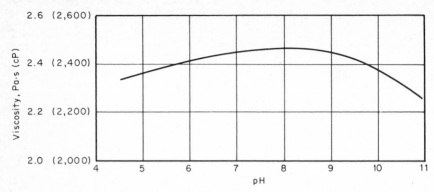

Figure 14.5 Effect of pH on viscosity of a 1% locust bean gum dispersion at 25°C as measured on a Brookfield Synchro-Lectric viscometer at 20 rpm. Solutions prehydrated at 85°C for 15 min, then cooled.

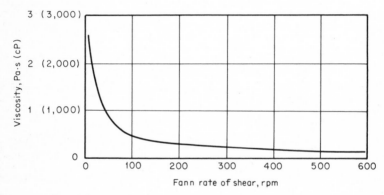

Figure 14.6 Effect of shear rate on the viscosity of a 1% locust bean gum dispersion at 25°C as measured on a Fann V.G. Model 35 viscometer. Solutions prehydrated for 15 min at 85°C, then cooled.

gum will react strongly in solution with the xanthan polysaccharide developed by the Peoria laboratories of the U.S. Department of Agriculture. Gum with similar structure to locust bean gum (e.g., guar gum and tara gum) also show some reactivity, but the intensity of the reaction with locust bean gum is such that thick solutions and gels can be obtained at very low total concentrations of the two components. The intensity of the reaction is thought to be a reflection of the structure of the two polymers.

Salt Reactions. Strong reactions are also obtained with solutions of certain inorganic cations. The addition of a high concentration of calcium salt, for example, will cause a gel to form under alkaline conditions. If dry powder is added to the salt solution, the gum will not hydrate and thicken. In general, the gum will react with polyvalent cations much as it does with the borate anion. Significant differences, however, are that many of the reactions lead to insolubilization, precipitation, or unstable, nonuseful gels.

The polymer has a strong affinity for hydrated mineral and cellulosic surfaces. This is attributed to its numerous hydroxyl functions capable of strong hydrogen bonding and to its linear molecular structure which allows it to make contact at many points.

Biological Properties

Locust bean gum has the status of a direct food additive. Gum which meets *Food Chemicals Codex* specifications is affirmed as generally recognized as safe (GRAS) as a direct food additive under FDA Regulation 184.1343. Technical locust bean gum has

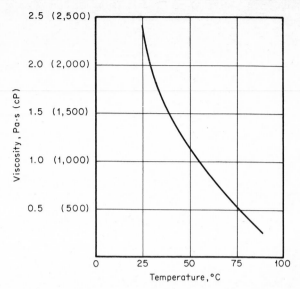

Figure 14.7 Effect of temperature on the viscosity of a 1% locust bean gum dispersion as measured on a Brookfield Synchro-Lectric viscometer at 20 rpm. Solutions prehydrated at 85°C for 15 min, then cooled.

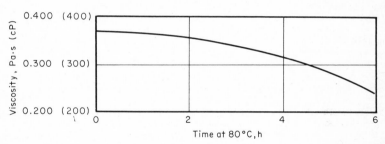

Figure 14.8 Stability versus time of a 1% locust bean gum dispersion at 80°C as measured on a Brookfield Synchro-Lectric viscometer at 20 rpm. Solutions prehydrated at 85°C for 15 min, then cooled.

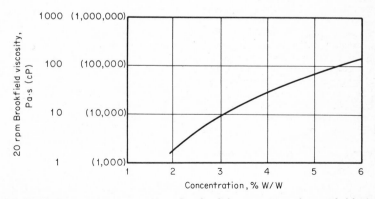

Figure 14.9 Viscosity versus concentration of oxidized, low-viscosity, carboxymethyl locust bean gum. Solutions hydrated at 25°C.

been affirmed as GRAS in food contact surfaces under FDA Regulation 186.1343. The gum must meet the specifications listed in Section 186.1343. Locust bean gum is also accepted as a food additive by the WHO and the *Codex Alimentarius.*

Solutions of locust bean gum serve as a food source for the growth of common microorganisms. Where it is necessary to hold solutions for a length of time, common food preservatives such as benzoic or sorbic acid may be added to inhibit bacterial growth. In nonfood applications quaternary ammonium preservatives, chlorinated phenolics, or formaldehyde are effective bacterial inhibitors.

In those cases where bacterial contamination has been allowed to get out of hand and manufacturing equipment contains the residue of preceding batches, it is much easier to thoroughly clean the equipment and start again than to try to control fermentation with preservative.

Handling

The powdered commercial grades of locust bean gum are all stable in the dry form. In common with other water-soluble polymers, the one condition that adversely affects the gum is to allow it to get wet. As long as it is stored in a dry place there should be no problem from fermentation or lumping of the powder.

COMMERCIAL USES: Compounding and Formulating

Food Products

The use of locust bean gum in formulated or compounded products is primarily limited to use in food products. This industry has made wide use of the ability of locust bean gum to bind large amounts of water.

Ice Cream   Small amounts of locust bean gum do not greatly affect the viscosity of the mix during manufacture but impart a smooth, chewy texture to the finished product. A slow meltdown of the product and increased heat-shock resistance are other benefits. Ice cream stabilized with locust bean gum is free of the graininess caused by ice crystal formation.

Cheese   In soft-cheese processing, locust bean gum controls the consistency and spreading properties of the product. Smoother, more homogeneous cheese spreads are possible with high water content because of the gum's water-binding properties.

Sauces and Salad Dressings   These products make use of the high viscosity at low concentration that is a basic property of locust bean gum.

Canned Pet Food   Locust bean gum normally functions as a water-binding component of a formulated product, and as such it aids in control of product texture or consistency. There are occasions, however, such as in the preparation of canned pet foods, where locust bean gum does double duty and also serves as a processing aid.

High free-water content is characteristic of canned pet food. It is desirable to immobilize water in the final product form of most canning operations, and locust bean gum is well suited to this type of processing. However, during the can-filling operation it is often desirable to limit the viscosity of the product to control its consistency so that it is fluid enough to fill the cans fully without air pockets, but not so fluid as to splash from the cans when fed by high-speed automatic can fillers.

Locust bean gum has limited thickening power in cold water. It can be added to a pet food at a level sufficient to impart a desirable can-filling viscosity, and then when the product is retorted and cooled, the gum develops its full water-holding power, thus bringing the final product to a higher viscosity.

An example of locust bean gum being used in this fashion would be the addition of 1% by weight of gum to a meat and vegetable mix containing up to 74% water. The cans are filled and retorted at 121°C (250°F). The finished product has the appearance of a thick stew.

COMMERCIAL USES: Processing Aids

Textiles Processing

Use of Derivatives   Derivatives of locust bean gum are used more widely in this industry than is the gum itself. These derivatives are for the most part either carboxymethyl or hydroxyalkyl ethers of the gum. They are often oxidized as well to deliberately reduce their molecular weight. When required, the oxidation is carried out under controlled conditions so that the thickening power of the product versus concentration is brought to a predetermined level.

There are multiple reasons for the derivatization and oxidation of the gum. Locust bean gum will not fully hydrate in cold water. Most of the industry needs are for dye-bath thickeners that will hydrate in cold water. Carboxymethylation or hydroxyalkylation renders the gum soluble in cold water.

In many printing and dyeing operations the penetration of the dye into the fiber or fabric is controlled primarily by the way the thickened dye solution responds to varying rates of shear. Oxidation of the derivatives of locust bean gum changes the rheology of the product. For instance, solutions of an oxidized and unoxidized derivative of the gum may be prepared so that their viscosities are equal when measured by a Brookfield viscometer at a constant rpm. However, if the rate of shear is varied so that plots are obtained for viscosity versus rate of shear, it will be seen that the oxidized derivative is less pseudoplastic. Its viscosity will vary much less with rate of shear than will the unoxidized derivative. This in turn will affect the manner in which the dye solution responds to successive stages of shear as it is applied to rollers or squeezed through screens in the course of being applied to carpets or flat goods of varying thickness, construction, and fiber content. There is no one product right for all uses. The answer lies in choosing the most economical thickener consistent with rheology for the mechanical problem at hand.

Yet another reason for the derivatization and oxidation of locust bean gum is that these processes lead to greater solubility of the end product. This increased solubility is reflected in the greater ease with which dye solutions can be pumped through fine filters and the lower amounts of insoluble particles that accumulate on the printing screens.

Locust bean gum derivatives are widely used in flat and rotary-screen printing and roller printing applications where they impart sharpness of print, uniformity of shade, and brilliance of color. They can be easily removed by washing when desired.

Carpets Processing Locust bean gum derivatives are widely used in printing and dyeing applications as thickeners to control the mobility of dyestuffs. In some cases the viscosity they develop controls the volume of dye feed. The derivatives are largely hydroxyalkyl and carboxymethyl products oxidized to varying degrees to control their thickening power. Locust bean gum products compete directly with guar gum products in this area, and the choice is usually governed by economics since such things as flow, viscosity, and solubility can be altered at will.

The following discussion of applications in the carpet industry illustrates the wide range of techniques in current practice. Similar techniques would apply to flat goods, after taking into account the differences in construction.

Dyeing Carpets. Some carpet yarns (primarily the spun yarns) are stock-dyed or package-dyed. In the package dyeing, gums are used at very low concentrations (0.1 to 0.2%) as migration-control agents to produce a more even distribution of dye in the package. Most of the carpet yarns are dyed after the carpet is formed, either by the beck method or, where long runs of the same shade are involved, by the newer continuous-dyeing machines. Beck dyeing is done by the exhaustion of dyes onto the fiber and requires no gum. With but one exception all of the newer continuous-dyeing machines use gum as the viscosity control to regulate the pickup of dye on a polished stainless steel roller immersed in the dye feed pan. This film of dye is continuously knifed off and cascaded in a sheet across the width of the carpet just prior to the fixation process, which takes place in a steaming chamber. During the steaming operation the gum content of the dye solution serves to control the migration of the dye molecules so that a level shade is obtained from side to side of the carpet and also from the surface to the backing of the carpet. Very high viscosity gums are used in this application so that an extremely low percentage of gum can be employed to maintain the critical viscosity level. The actual viscosity of the dye solution is low, in the range of 0.020 to 0.030 Pa·s (20 to 30 cP), and the concentration of gum used is in the area of 0.20 to 0.25%. This low level of gum is an advantage since there is less gum to wash off after the dyeing operation.

Printing Carpets. Carpet printing is done by four methods, and the gum requirements vary with the method. Space printing is a technique originated some years ago.

Essentially, space printing is the production of a completely random distribution of a number of colors to produce a "pepper and salt" type pattern, and the original patented process consisted of printing a geometrical distribution pattern on knotted fabric, either flat or tubular, and after fixation of the color, deknitting and winding on cones. This printed yarn was then supplied to tufters to produce the patterned carpet. Other methods have been developed to produce similar effects by warp sheet printing, splatter printing of droplets of color, and computer programmed distribution of color through nozzles or by pressure bars. All these methods require a gum thickener, and the thickener must have very special characteristics for optimum performance. The thickener should be essentially protein-free, low in insolubles content, high in viscosity, easily dispersible to a lump-free paste, and extremely soluble for easy washout; produce good apparent color value; and be compatible with the auxiliary chemicals and dyes being used.

Gums for space printing are generally used at 0.35 to 0.5%, and the viscosity will be 0.250 to 0.500 Pa·s (250 to 500 cP), depending on the method, fiber, and pattern.

Pattern designs on carpet are produced by two different screen printing techniques. The first and oldest method is by use of flat-screen stencils, and the second is by means of rotary-screen stencils which are tubular in shape and made of metal such as nickel or stainless steel.

The flat-screen method is somewhat slower than rotary printing but has a wider range of pattern choices. It also takes up more floor space. However, if a screen damage or blockage should occur, it is easier and quicker to rectify, and so this method produces less damaged carpet. Flat-bed machines run at speeds in the range of 4.5 to 6 linear m/min (15 to 20 linear ft/min) for most patterns and types of carpet, while rotary machines are capable of speeds of 18 or more linear m/min (60 or more linear ft/min). Flat-bed machines use medium-viscosity gums for most patterns and may use high-viscosity gums at lower solids under certain special conditions or on certain pattern or carpet types. Rotary machines can run most patterns on the more economical high-viscosity gums, but do require medium-viscosity gums for finer line patterns.

Flat-bed machines generally operate with dye pastes in the range of 0.800 to 7.500 Pa · s (800 to 7500 cP) viscosity depending on the pile type and pattern. The gum percentages may be anywhere between 0.75 and 3% depending on the type gum and viscosity requirements.

Rotary machines operate in a narrower range of viscosities 6.500 to 10.000 Pa·s (6500 to 10,000 cP) using gum concentrations of from 1.25 to 2.5%, depending on the gum type and viscosity requirement.

There are several printing machine types which do not fit the description of either the flat-bed or rotary type but which are also capable of producing pattern effects. These are: the modified roller type which uses raised patterns on a series of rolls arranged radially around a central cylinder; the jet nozzle type which feeds color through jet tubes arranged 1.55 per square centimeter (10 per square inch) and which may be programmed to produce the pattern required; the reverse-dip-method machine which feeds carpet with the pile face down and applies color by compartmented reservoirs which rise and fall as the carpet transverses; and the pressure-bar segmented roller machine which prints a warp sheet of yarn run between a number of pairs of such rolls using one pair of rolls for each color. These machines all use relatively low viscosity dye pastes made up with high-viscosity gum at low solids. The gum viscosities will vary from 0.050 to 0.300 Pa·s (50 to 300 cP), and the gum concentration will be in the range of 0.3 to 0.5%.

This area of carpet processing is growing rapidly, and innovations are the rule rather than the exception in this particular segment of dye application.

The viscosities mentioned are all by Brookfield viscometer RVF model using the appropriate spindle at 25°C and 20 rpm. Viscosities measured by other means will be considerably different.

Paper Products

The paper industry has become the largest user of galactomannans. It is estimated that about 7,000 metric tons (15 million pounds) of galactomannans (locust bean gum and guar gum) are used annually by this industry. Locust bean gum is used less frequently than guar gum but represents a substantial percent of the total consumption.

Wet-End Addition The major use of locust bean gum is as a wet-end additive. This means that the gum is added to the pulp suspension just before the sheet is formed on either a Fourdrinier machine or a cylinder machine. The pulping process, which is designed to remove lignin and thereby produce a fibrous cellulosic pulp, also removes a large part of the hemicelluloses normally present in the wood. These hemicelluloses, which are mostly mannans and xylans, could contribute greatly to the hydration properties of the pulp and the strength of the paper formed from the pulp. Galactomannans replace or supplement the natural hemicelluloses in paper bonding.

It is generally agreed that the hydrogen-bonding effect is one of the major factors affecting fiber-fiber bonding. An examination of the molecular structure of galactomannans reveals a rigid long-chain polymer molecule with primary and secondary hydroxyl groups. Such molecules are capable of bridging across and bonding to adjacent fibers.

It is difficult to compare the efficiency of locust bean gum and guar gum in paper products because too many variables exist, such as type of gum formulation, pulp type, beating time, and machine conditions. However, many in the paper industry believe that guar gum is the more efficient additive.

Commercial locust bean gum and guar gum contain a small amount of hull as a contaminant. Locust bean hulls are red-brown; in white sheets, these particles appear as small but visible specks. However, guar hull is almost white and, when present in the paper sheet, is indistinguishable from the bleached pulp.

There are several important advantages in the addition of galactomannans to pulp. In sheet formation, there is a more regular distribution of pulp fibers, and consequently fewer fiber bundles. The Mullen bursting strength is increased, as well as the fold and tensile strengths. And pick, the measure of the force required to pull a fiber from the surface of a sheet (important for printing-grade paper), is increased.

Also important is the effect on pulp that is normally passed through jordans or wet mills which increase the surface of the fiber, thereby allowing it to bind more water. The galactomannans added to the pulp aid in the binding of water and thus decrease the amount of refining necessary, which in turn lowers the power consumption. Machine speed is increased without losing desirable end-product test properties, and the retention of fines during processing is increased.

Two other end-product properties benefit: finish is improved, as measured by the smoothness and a count of protruding fibers, and porosity is decreased.

Gum Preparation In general, there are two ways to prepare guar or locust bean gum for use: by cooking or by dry addition.

For mills that do not have nor wish to install cooking equipment, specially processed products are available which can be added dry to the beaters or pulpers. These products are formulated for ease of dispersibility in the stock and designed to suit the specific conditions of stock preparation, i.e., temperature, pH, and retention time. They develop full hydration while in the stock at specific rates determined by individual product design.

However, most mills find it easier and more practical to prepare solutions of guar or locust bean gum before addition to the pulp slurry. This enables them to maintain a finer control on the amount of gum being added to their pulp system and also allows them to quickly increase or decrease the amount of additive as needed.

Guar or locust bean gum solutions are prepared by one of two methods: the batch system or a continuous cooking system. Solutions can be prepared in the batch system by dispersing 0.5 to 1.0% (10 to 20 lb/ton) of the dry powder in cold water under mechanical agitation, heating the dispersion to 90 to 95°C (195 to 205°F), and holding the solution at that temperature for 10 to 20 min. The benefit of a continuous cooker is the preparation of a gum solution on an automatic, continuous basis with lower manpower requirements.

After the guar or locust bean gum solution is prepared, it can be added at any convenient point in the stock system that lies after the point of refining and provides good distribution. Since these materials have such a high affinity for absorption on cellulose, it is recommended that the solution be diluted as much as possible before addition so that a more uniform distribution rate of gum to cellulose can be accomplished. In most mills, addition in the vicinity of the fan pump offers best control of the additive process. The rate of delivery of the prepared guar or locust bean gum to the paper machine can best be regulated by use of a rotometer. The normal requirement for attaining measurable improvements is in the range of 0.1 to 0.6 wt% (2 to 12 lb/ton) dry basis, but conditions of the pulp furnish, basis weight, machine speed, stock freeness, etc., will ultimately govern the proper level of addition.

General Benefits   The general benefits gained through the use of guar or locust bean gum vary with the particular paper product being manufactured.

Kraft Papers. A growing use of gums by makers of multiwall bag and gumming grades of kraft paper has shown specific improvements. The rate of addition varies considerably from mill to mill, but is usually 0.25 to 0.45 wt%. Benefits include:

- Improved tensile strength
- Improved tear resistance, brought about by reduction of refining made possible by tensile improvement
- No adverse effect on porosity
- Higher production speeds

Kraft Linerboard. This industry was among the earliest users of guar and locust bean gum and remains so because of the introduction of newer high-speed, high-production paper machines. Guar and locust bean gum products have proved to be the most reliable and effective materials to improve Mullen test on this grade of board. Addition rates of 0.1 to 0.3 wt% are normal in order to accomplish:

- Higher speed at equivalent Mullen due to less need for refining
- Higher freeness and, thus, better drainage
- Higher Mullen at reduced basis weights
- More uniform Mullen values between top and wire sides
- Optimum balance of refining, drainage, drying rates, and Mullen tests in 31.3-kg (69-lb) and heavier weight boards

Recycled Linerboard. Galactomannans are almost universally used by producers of this grade of paper, largely because of the use of reclaimed paper and board used in the pulp furnish of this grade. Gums provide the necessary upgrading of quality needed in the finished product. The need for guar and locust bean gum is particularly great in paper mills employing high-temperature, high-pressure dispersion units, but in all cases, gums provide these needed benefits at addition rates of 0.3 to 0.6 wt%. Benefits include:

- Higher Mullen
- Lower basis weight
- Higher production speeds
- More economical pulp furnish through use of more recycled paper
- Improved finish

Corrugating Medium. Almost all medium made from recycled stock requires an additive to meet their strength tests. Galactomannans are the commonly used additives. The addition rate normally varies from 0.3 to 0.6 wt%. The benefits include:

- Higher concorra
- Higher speed
- Better crush resistance in the finished container

Boxboard. Depending on the individual mill's conditions for stock preparation, specific products are recommended for addition to either beaters or hydropulpers. Usually, mills making these grades prefer a product designed for dry addition to the stock at addition rates of 0.3 to 0.6 wt% based on liner stock. Ordinarily no gum is used in the filler portion. Resultant production or product benefits include:

- Improved ply bonding
- Increased board stiffness
- Better formation of top liner
- Better bonding strengths on coated grades

Offset News Stock. Through the use of galactomannans paper mills have been able to realize marked benefits in the production of stock for continuous web printing of newsprint at a use level of 0.10 to 0.15 wt%. Benefits include:

- Reduction of linting and dusting during paper manufacture
- Higher pick values
- Less build-up in printing on offset press blankets, permitting more impressions between cleanups

White Papers. Guar and locust bean gum are used in a wide variety of white papers at addition rates of 0.15 to 0.40 wt%. By increasing sheet strength, it is possible to reduce the size concentration at the size press, with resultant improvement in tear, opacity, and calendering.

Gums used in high hardwood furnishes reduce surface picking, improve bonding of vessel segments, and increase fold.   Guar or locust bean gum is used in neutral or alkaline furnishes to improve the retention of calcium carbonate.

Galactomannans are added dry to the beater to reduce the refining time required to develop strength characteristics of rag-content paper, particularly banknote.

These products are also very effective in decreasing two-sidedness and thus reducing the tendency of papers to curl.

Where formation is uppermost, gums are used to deflocculate the fiber bundles, closing up the sheet and markedly improving the appearance.

Mining Industry

In this industry locust bean gum is used as a chemical flotation agent rather than a thickener or water-binding agent.   The property used is the ability of the gum to adsorb to hydrated mineral surfaces.

The principal application for locust bean gum is in the recovery of base metals by flotation.   Locust bean gum acts as a depressant for the talc or gangue that is mined along with the valuable minerals.

In flotation operations, the ore is ground to a desirable particle size and then slurried in water prior to treatment.   When locust bean gum is added as a dilute solution to the slurry, the gum adsorbs to the hydrated surfaces of the unwanted talc or gangue. When reactive chemicals called collectors are then added to the slurry, they react with the desirable minerals rather than the talc or gangue.   The collectors condition the mineral surfaces so that the minerals will rise to the surface when a frother is added and air is bubbled through the slurry.   The froth is skimmed off, separating the values from the talc or gangue that has been depressed by the locust bean gum.

As an illustration of a typical application, consider the flotation of pentlandite (Ni-Fe-sulfide) from a high-talc ore containing chaleopyrite pyrrhotite, pyrite, serpentine, and biotite.

After crushing and grinding to 90% minus 325 mesh, a nickel-pyrrhotite concentrate is removed magnetically, followed by treatment of the nonmagnetic fraction by flotation.

Flotation feed is diluted to 25% solids at a natural pH of 8.5 to 9.0.   Locust bean gum is added at 0.02 to 0.03 wt% (0.4 to 0.6 lb/ton) to depress talc.   The gum is added as a 0.5% (10 lb/ton) solution after cooking to 71 to 88°C (160 to 190°F) for 15 min.   This is followed by addition of 0.065 to 0.075 wt% (1.3 to 1.5 lb/ton) sodium ethyl xanthate as a collector and triethoxybutane as a frother.   Rough concentrate is removed with a total float time of 5 min.

Rougher tails are scavenged for 9 min after the addition of 0.005 to 0.0075 wt% (0.1 to 0.15 lb/ton) locust bean gum and 0.03 to 0.035 wt% (0.6 to 0.7 lb/ton) sodium ethyl xanthate.   Subsequently, the rougher and scavenger concentrates are combined and floated in the cleaners to produce a final concentrate and a middling recycle. No additional locust bean gum is used in the cleaners, but 0.015 to 0.02 wt% (0.3 to 0.4 lb/ton) of sodium ethyl xanthate and triethylbutane are added for a float time of 6 min.

The final concentrate assays 12 to 15% (240 to 300 lb/ton) Ni, and the final tails 0.02 to 0.4% (0.4 to 8 lb/ton) Ni.

INDUSTRIES USING LOCUST BEAN GUM

Food Industry

Locust bean gum is used in the food industry for its ability to bind and immobilize large amounts of water, thereby contributing to viscosity, inhibiting ice crystal formation in frozen products, modifying product texture, or stabilizing product consistency to changes in temperature.

An illustration of the use of locust bean gum can be found in the manufacture of ice cream, where small amounts of gum promote smooth chewy texture, inhibit ice crystal formation caused by heat shock, and impart slow meltdown qualities to the product.

In canned pet food use is made of the limited viscosity of locust bean gum in cold water. A level of gum concentration can be chosen to make a homogeneous mix with a desirable consistency for can filling by automatic equipment. When the sealed cans are retorted and cooled, the gum develops additional thickening power, which gives the finished product a pleasing appearance with respect to sheen and consistency.

Other food products which use locust bean gum are baked goods and baking mixes, nonalcoholic beverages and beverage bases, cheeses, gelatins, puddings and fillings, jams and jellies, and dairy products such as sour cream, yogurt, and buttermilk.

Mining Industry

Locust bean gum is used at very low concentrations, below the level at which it thickens water. It is used as a chemical reagent which bonds to hydrated mineral surfaces. In its major application as a depressant in base-metal flotation it functions as an agent to block the adsorption of other reagents on talc and gangue surfaces.

Paper Industry

In the paper-making process locust bean gum adsorbs on hydrated cellulosic surfaces. It is a wet-end additive that promotes dry strength, improved sheet formation, and easier and faster manufacturing.

Textiles Industry

Derivatives of locust bean gum thicken the dye baths that are used in printing and dyeing of fibers, fabrics, and carpets. The derivatives control the rheological characteristics of the dye formulations and permit complex, multiple dye patterns that are sharp, bright, and controlled as to penetration.

FORMULATIONS

The use of locust bean gum in formulated or compounded products is primarily limited to use in food products. Table 14.2 lists permissible use levels under FDA Regulation 184.1343. In all cases the function is as stabilizer and thickener.

TABLE 14.2
Maximum Usage Levels Permitted

Food (as served)	Percent
Baked goods and baking mixes	0.15
Beverages and beverage bases, nonalcoholic	0.25
Cheeses	0.8
Gelatins, puddings, and fillings	0.75
Jams and jellies, commercial	0.75
All other food categories	0.5

It is worth noting that dairy products are otherwise regulated at a maximum 0.5% total stabilizer. No stabilizer is composed entirely of locust bean gum. Other typical food-type formulations are given in the following pages.

Ice Cream

10.0%	Butterfat
12.0%	Milk solids, nonfat
11.0%	Cane sugar
5.0%	Corn syrup solids (42DE)
0.3%	Stabilizer/emulsifier
38.3%	Total solids

Ice Milk

3.5%	Butterfat
12.0%	Milk solids, nonfat
11.0%	Cane sugar
8.0%	Corn syrup solids (42DE)
0.45%	Stabilizer/emulsifier
34.95%	Total solids

Sherbet

1.0%	Butterfat
3.0%	Milk solids, nonfat
20.0%	Cane sugar
8.0%	Corn syrup solids (42DE)
0.25%	Stabilizer
32.25%	Total solids

Sour Cream

18.0%	Butterfat
9.0%	Milk solids, nonfat
0.4%	Stabilizer
27.4%	Total solids

Buttermilk

1.0%	Butterfat
10.0%	Milk solids, nonfat
0.4%	Stabilizer
11.4%	Total solids

Yogurt

1.0%	Butterfat
12.0%	Milk solids, nonfat
0.8%	Stabilizer
13.8%	

Instant Imitation Bakery Jelly

30.00%	Modified food starch
0.50%	Locust bean gum
1.30%	Sodium hexametaphosphate
1.60%	Dicalcium phosphate
8.20%	Adipic acid
0.30%	Raspberry flavor
0.25%	Raspberry shade
0.01%	Purple grade shade
0.70%	Potassium sorbate
57.14%	Sodium benzoate
100.00%	

Whipping Cream Composition (for Frozen Desserts)

Parts by weight

50–70	Sucrose
3–11	Modified food starch
3–6	Gelatin (200 Bloom)
0.5–1.5	Soy protein (degraded)
1.0–1.5	Organic food acid
0.1–0.5	Sodium hexametaphosphate
0.2–0.6	Locust bean gum
65.0	Fructose
10.5	Modified food starch
4.3	Gelatin (200 Bloom)
1.0	Soy protein
0.6	Flavor
1.2	Citric acid
0.25	Salt
0.2	Sodium hexametaphosphate
1.5	Color
0.2	Locust bean gum

PRODUCT TERM GLOSSARY

Locust bean gum is a glactomannan polymer found in the endosperm of the seeds of the carob tree *(Ceratonia siliqua)*. The straight-chain mannan backbone of the polymer has single-membered galactose side branches. The branches are thought to be arranged in blocks on consecutive D-mannopyranosyl units of the main chain. The blocks are separated by unsubstituted lengths of the main chain. The ratio of mannose to galactose in the polymer is approximately 4:1.

Chapter **15**

Pectins

Jens K. Pedersen
The Copenhagen Pectin Factory Ltd.

General Information 15-1
  Chemical Nature 15-2
  Physical Properties 15-3
  Manufacture 15-5
  Biological Properties 15-6
  Electrochemical Properties 15-7
  Additives/Extenders 15-7
  Handling and Storage 15-7
  Applications 15-7
  Application Procedures 15-8
  Future Developments 15-9
Commercial Uses: Compounding and Formulating 15-10
Formulations 15-10
  Jams and Jellies 15-10
  Pectin Fruit Jellies 15-15
  Aerated Confectionery Products 15-16
  Cold-Setting Flan Jelly 15-17
  Orange-Drink Concentrate 15-18
  Sour-Milk Drink Stabilization 15-18
  Cold-Setting Milk Desserts 15-19
Laboratory Techniques 15-20
Product/Tradename Glossary 15-20
Recommended Reading 15-21

GENERAL INFORMATION

Pectin is the designation for a group of water-soluble, industrially isolated, partially methoxylated polygalacturonans which are able to form gels under suitable conditions. Pectins are extracted from plant materials which have pectic substances as important structural polysaccharides that occur mainly in the mid lamella between the cells.

Pectins are acidic polysaccharides which show maximum stability in acidic conditions. The main use of pectin is as a gelling agent in acidic foods. The most important group of pectins commercially, the so-called high-methoxyl pectins, requires in addition to a low pH the presence of a minimum of 55% soluble solids in order to gel. Jams, jellies, and marmalades still absorb the major part of the world's pectin production, which in 1974 was estimated at approximately 10,000 metric tons. The percentage of pectin production which finds use in jams is, however, decreasing as new uses are being developed for pectin as a gelling agent, a thickener, and a protective colloid in foods.

Pectin has always been a substantial part of man's diet, and the amount of pectin consumed as a part of fruit and vegetables is many-fold the quantity of commercially isolated pectin. The fact that pectin is a natural product with a long record of safety favors the use of pectin in applications, particularly in food, where other hydrocolloids might be used at a lower cost.

Chemical Nature

Structure Pectin is essentially a linear polysaccharide containing from a few hundred to about one thousand building blocks per molecule, corresponding to an average molecular weight from about 50,000 to about 180,000.

D-Galacturonic acid is the principal constituent of the pectin molecule, but some neutral sugars are also present in pectin. L-Rhamnose is present in the galacturonan chain. Small amounts of D-xylose, L-arabinose, and D-galactose may be part of the side chains or of accompanying neutral polysaccharides.

Degree of Methylation The polygalacturonic acid is partly esterified with methoxyl groups, and the free acid groups may be partly or fully neutralized with sodium, potassium, or ammonium ions. The ratio of methoxylated galacturonic acid groups to total galacturonic acid groups is termed the *degree of methoxylation* (DM; see Fig. 15.1), and is of vital influence on the properties of pectin, especially the solubility and gelation requirements which are directly derived from solubility.

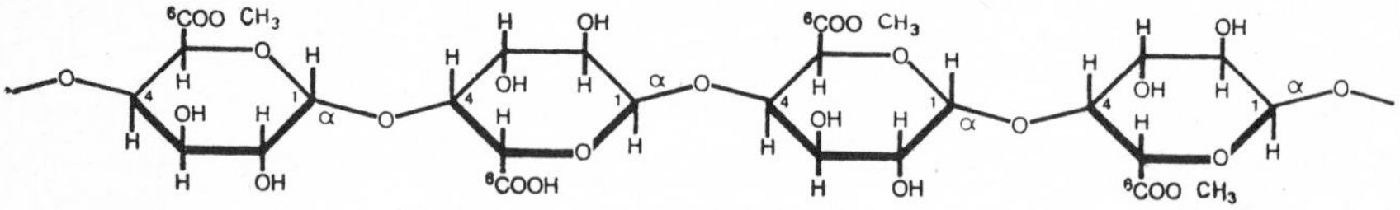

Figure 15.1 Segment of high-methoxyl pectin, degree of methoxylation (DM) of 75%.

The DM of 50% divides commercial pectins into high-methoxyl (HM) pectins and low-methoxyl (LM) pectins. These two groups of pectins gel by different mechanisms (see Fig. 15.2). HM pectins require a minimum amount of soluble solids and a pH within a pretty narrow range around 3.0 in order to form gels. LM pectins require the presence of a controlled amount of calcium ions for gelation, and do not require sugar and/or acid.

The DM of HM pectins control their relative speed of gelation as reflected by the designations *slow-set* and *rapid-set* high-methoxyl pectin (see Fig. 15.2). The degree of methoxylation of low-methoxyl pectins controls their calcium sensitivity.

The highest DM that can be achieved by extraction of natural raw material is about 75%, yielding a so-called rapid-set, high-methoxyl pectin. HM pectin with a DM from 75 to 85% is produced by the methoxylation of HM pectin with methanol. Pectins with a DM of from 0 to 70% are produced by controlled demethoxylation in the manufacturing process.

Reactions The reactions which pectins may undergo in use will tend to degrade them. Pectins have maximum stability at around pH of 3.5. At low pH values, pectins are deesterified and glycosidic links are hydrolyzed. Which of these reactions will dominate depends on the temperature. At room temperatures, deesterification dominates. The effect of the deesterification depends on the DM of the starting material. A rapid-set HM pectin becomes slower-gelling by deesterification, whereas a slow-set pectin gradually adopts LM pectin characteristics. Hydrolysis of glycosidic links invariably leads to gradual loss of physical properties, such as viscosity and gel strength.

At near-to-neutral pH (5 to 6), pectins are stable only at room temperature. As the temperature (and pH) is increased, a so-called beta elimination starts (see Fig. 15.3). This beta elimination requires that the galacturonic acid unit is esterified. It involves formation of a double bond in the molecule and results in chain cleavage and very rapid loss of physical properties.

At basic pH values, pectins are rapidly deesterified, even at room temperature. Simultaneous beta elimination may be minimized by conducting the deesterification at very low temperatures. If ammonia is used to deesterify pectin, some of the methoxyl groups are converted into amide groups. So-called amidated, low-methoxyl pectins

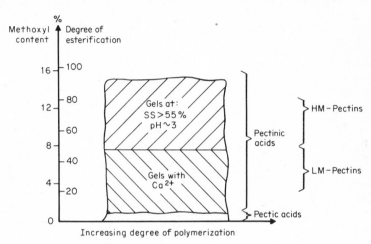

Figure 15.2 Nomenclature of pectic substances as related to degree of polymerization and degree of esterification (methoxyl content).

show better physical properties than other low-methoxyl pectins and are used extensively as gelling agents. The degree of amidation of low-methoxyl pectins varies from about 15 to 25%, which has been set as the maximum content in the specifications laid down for the amidated pectins. High-methoxyl pectins with a degree of amidation of about 5% are extremely slow-gelling.

Physical Properties

Powdered pectins are isolated from solution by alcohol or metal-salt precipitation, which results in a somewhat fibrous product with flow properties inferior to crystalline material. A typical specific gravity of a powdered pectin is about 0.7. Color of powdered pectins

Figure 15.3 The β-elimination reaction chain cleavage of the pectin molecule from increases of pH or temperature.

varies from off white for some citrus pectins to light brown for apple pectins. Pectins have a slightly acidulous taste, but may taste slightly sweet if standardized with sucrose or dextrose. Tables 15.1 and 15.2 give typical properties of HM and LM pectins, respectively.

Pectins are soluble in water and insoluble in organic solvents such as alcohols, ethers, and hydrocarbons. The solubility of HM pectins in sugar solutions decreases as the sugar concentration increases. Pectin manufacturers recommend that pectins be dissolved at soluble-solids concentrations of below 25%. Similarly, LM pectins should be dissolved in water without any calcium salts added. Most LM pectins are soluble in

TABLE 15.1
Typical Properties of HM (High-Methoxyl) Pectin

Type designation	Typical specification			Typical gelling range, °C†	Typical DM range‡	Typical use
	Grade, USA-SAG	Gelling time, s*	pH of 4% solution			
HM pectin, ultra-rapid-set					80–85	Household pectin powders, liquid household pectin
HM pectin, rapid-set	150 ± 5	<70	3.4–4.2	85–95	72–75	Jams filled into jars at high temperatures (above 80°C)
HM pectin, medium-rapid-set	150 ± 5	100–135	3.4–4.2	75–80	67–70	Jams filled at intermediate temperatures (60–80°C)
HM pectin, slow-set	150 ± 5	150–200	3.4–4.2	60–75	64–66	Jellies, jams filled into large containers at temperatures below 60°C
HM pectin, extra-slow-set	150 ± 5	>200	3.4–4.2	60 or lower	60–64	Fruit jellies with 76–80% soluble solids

*Joseph and Baier method: Soluble solids of test jelly, 65.0%; pH of test jelly, 2.2–2.4; cooling rate in standard test jelly glass in 30°C water bath.
† Test conditions: Soluble solids, 65.0%; pH, 3.05; cooling rate in standard test tube in water bath allowed to cool from boiling point in air.
‡ DM = degree of methoxylation.

TABLE 15.2
Typical Properties of LM (Low-Methoxyl) Pectin

Type designation	DM°	DA°	DFA°	Typical specification		Typical use
				Grade*	pH of 1% solution	
LM pectin, acid-demethylated	35–40		65–60		3.5	Low-solids fruit jams, calcium salt added
LM pectin, amidated	30–35	18–22	45–50	100 ± 5	4.1	Low-solids fruit jams and gels
LM pectin, amidated	30–35	10–15	50–60	100 ± 5	4.5	Gelled fruit/milk desserts

* Food Chemicals Codex; conditions of 31% soluble solids, pH of 3.1.

hard water. The presence of calcium salts increases the viscosity of pectin solutions, thereby reducing the concentration of pectin that can be dissolved and handled. When pectin is dissolved, sugar may be added without gelation. The exact concentration of soluble solids which can be reached without gelation will depend on the pH in the system, the particular type of pectin, and the temperature.

Pectin is normally predissolved in water before being added to other components of a food. If efficient dissolving equipment is used, pectin may be dissolved to a concentration of 6 to 12%, depending on the pectin type. Pectin solutions are opaque.

Rheological Properties   Pectin solutions have low viscosities when compared to other plant gums, and pectin consequently has little use as a thickener. Thin pectin solutions are close to Newtonian in their flow properties, but this changes to pseudoplastic when the pectin concentration increases or when calcium ions are added.

HM-Pectin Gels.   HM pectin forms gels at concentrations starting from as low as 0.3%. Soft pectin gels as used in jams and preserves are rather brittle and nonelastic. Pectin gels with greater than 1% pectin are more coherent, but pectin gels never acquire the stiffness that is typical of agar gels, nor the elasticity that is typical for gelatin gels. An advantage of the specific HM-gel character is an unsurpassed flavor release.

HM-pectin gels are in practice not heat-reversible. On the other hand, only undisturbed, strong pectin gels may be characterized as heat-resistant, a property which is desired in bakery jams.

LM-Pectin Gels.   LM-pectin gels vary in rheological and other properties with the calcium content. At low calcium concentrations, LM-pectin gels are soft, coherent, and almost transparent. As the calcium concentration increases, the LM-pectin gels become harder, more brittle, and less transparent. Also, LM-pectin gels with relatively high calcium concentrations have a strong tendency to syneresis, whereas gels with low calcium content are syneresis-free.

LM-pectin gels are heat-reversible. The gelling and remelting temperatures depend on many factors, such as calcium concentration, pH, pectin type, and soluble solids.

Manufacture

Raw Materials   Many plant materials contain sufficient content of pectic substances to appear attractive as pectin sources. At present, however, only apple pomace, containing 10 to 15% pectin, and citrus peel (lime, lemon, orange, and grapefruit), containing 20 to 30% pectin, both on a dry weight basis, are being used as raw materials for pectin manufacture. These materials qualify as pectin raw materials because of their high contents of pectic substances which can be converted to pectins with desirable gelling properties. Also, apple pomace and citrus peel are available as by-products from concentrated geographical areas, justifying installation of drying plants to preserve their quality. Further, the pectins in apple pomace and citrus peel are relatively easy to isolate by industrial processes from associated cellulosic and starchy materials and from the natural colors of the raw materials.

The Process   Pectins may be extracted from fresh apple pomace or citrus peel as such material appears from juice pressing. If the raw material is left untreated for some time, pectic enzymes which are liberated during the juicing operations will destroy the pectins. All pectin manufacturers therefore base their processes at least partly on dried pomace or peel; some pectin manufacturing plants are situated at locations remote from the raw material sources, and their processes are consequently based exclusively on dried raw materials.

Pectins are extracted with acidified soft water, a typical extraction pH being 2.5. Extraction temperatures vary from 60 to 95°C, with extraction time varying correspondingly from several hours to 30 min. The quantity and quality of pectin from a specific raw material will depend heavily on the proper selection and control of extraction conditions.

Separation of pectin liquor and exhausted plant residue is probably the most crucial step in the production of pectin. Centrifugation and/or various combinations of filtrations are used, the last step being a polishing filtration with diatomaceous earth to secure complete transparency in jellies.

Pectins may be isolated from solution by precipitation with alcohol from a concentrated sol (approximately 2%) or with aluminum salt from a dilute solution (0.3 to 0.5%).

When a pectin is isolated as aluminum pectinate, isolation must be followed by washing with acidified alcohol to convert the insoluble aluminum pectinate into pectinic acid, and then a subsequent washing with slightly alkaline alcohol to adjust the pH to 3 to 4.

Various production paths for HM and LM pectins are shown in Fig. 15.4. It should be noted that HM pectin is an intermediate product in the LM-pectin process. Two methods for demethylation are in use, giving names to the LM pectins produced by them: acid-demethylated LM pectins, and ammonia-demethylated (or amidated) LM pectins.

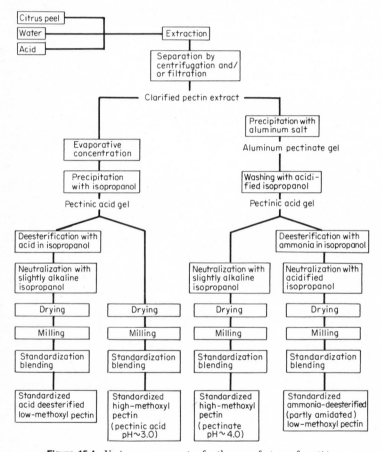

Figure 15.4   Various process routes for the manufacture of pectins.

Biological Properties

Human metabolism produces no pectolytic enzymes, and pectins therefore pass unchanged to the large intestine, where bacteria are capable of using pectin as their carbohydrate source. Although pectins are degraded in the intestinal tract, they have no caloric value.

A number of valuable biologic effects of pectins have been reported, including an antidiarrhea effect, a cholesterol lowering effect, and an effect on the nitrogen metabolism, where some of the nitrogen that would otherwise have been excreted via the urine is excreted instead via the feces.

FAO/WHO Expert Committee reports found no reason to limit the use of pectins as food additives. Amidated pectins are regarded as separate products, which in 1974 were assigned a preliminary ADI (acceptable daily intake) value of 25 mg/kg of body-weight per day.

Electrochemical Properties

A pectin is a polyuronic acid, and the chain molecule is negatively charged at neutral pH. The pK value of pectins is approximately 3.5.

Pectins react with positively charged macromolecules, e.g., proteins, at pH values below their isoelectric pH. Pectin precipitates gelatin at low pH values, but this reaction can be prevented by addition of salt. When pectins are added to milk at the pH of milk (6.6), a milk protein–pectin reaction product precipitates out. If the pH is subsequently reduced, the precipitate is redissolved to produce a stable acidified milk which may even be heat-treated to extend the shelf life of the product. Without the pectin, the milk protein would behave in exactly the opposite way.

Additives/Extenders

Pectins used in foods normally are standardized to a certain jelly grade by blending the pure pectin with fine dextrose or sucrose. Where these standardizing agents are undesired in the food product, e.g., when a sugar-free jelly must be made, pure pectin may be obtained from the pectin manufacturer.

Pectins used in synthetic jellies are often blended with buffer salts to control the pH in the application. Buffer salts may be citrates, tartrates, or phosphates as required by the food legislation and the specific requirements in the application. Bicarbonate may be added in a small amount to facilitate dispersion of the pectin in water. LM pectins may have a pyro- or polyphosphate added to allow full dissolution under adverse conditions, i.e., high calcium content or high soluble solids.

Handling and Storage

Pectins carry the handling recommendation "Store Cool and Dry" on the label. Most pectins are manufactured with only 6 to 8% moisture and are liable to pick up moisture if not protected by a suitable vaportight packaging. Equilibrium moisture content of a typical 150 jelly-grade HM pectin will be approximately 12% at 70% relative humidity. Most pectin manufacturers pack their products in a package that is a combination of a polyethylene bag to protect the pectin from moisture and a fiber drum for external protection. This package, if it is opened and the contents are not completely used up on the same day, should be closed again.

HM pectin (150 grade) loses approximately five jelly grades per year when stored at room temperature. Also, HM pectin is slowly deesterified in storage; a rapid-set pectin, e.g., turns into a medium rapid-set pectin over a period of a year. Degradation velocity is more than doubled when the storage temperature is increased from 20 to 30°C.

LM pectins are more stable in storage than are HM pectins, and degradation is normally not detectable over a period of a year at room temperature.

Applications

Gelation Pectins found use as gelling agents in fruit preserves before they were ever isolated in pure form. Probably more than 75% of the world's production of HM pectin is presently used as a gelling agent for jams and jellies to make up for the deficiency of natural pectins in the fruits themselves.

Many national and proposed *Codex* jam and jelly standards allow only pectins as gelling agents, although there is a tendency to allow other toxicologically safe gelling and thickening agents. Nevertheless, HM pectins are almost unrivaled as gelling agents in high-sugar, low-pH products for several reasons: the texture produced by pectins is organoleptically preferred; pectins have their maximum stabilities at the pH of fruit preserves (contrary, for instance, to agar, carrageenan, and gelatin, which are all less stable at low pH values); and pectins are less expensive to use than are other gelling agents. Other gelling agents have been able to find use in jams and jellies only where specific product or process requirements cannot be met by pectins.

HM-pectin gels may be produced by a cold process by adding a small amount of citric acid solution to a sugar syrup containing HM pectin and having a pH above the gelling range.

When the soluble solids in fruit preserves fall below approximately 55%, gelling agents other than HM pectins must be resorted to. LM pectins are the best choice in the upper part of the soluble-solids range of 55 to 10%, as they still have the advantages of superior organoleptic and stability properties, compared with other gelling agents. At the low end of the soluble-solids range, insufficient water retention of LM-pectin gels demands that they be supplemented by water binders such as carboxymethylcellulose or locust bean gum.

LM pectins might seem the ideal gelling agents for milk products, as milk supplies a high and constant calcium level. LM pectins cannot, however, compete in price with carrageenan and furcellaran, which are effective at much lower use concentrations. Also, the high calcium level of milk makes it difficult to dissolve LM pectins without the use of calcium-complexing agents.

The specific properties of LM pectins are unrivaled in a canned liquid fruit dessert preparation which, by blending with an equal amount of cold milk, is made into a fruit-flavored light pudding. This base is intended for blending with two or three parts of milk to make a chocolate pudding with all the advantages of cooked pudding, i.e., the texture, the full chocolate flavor, and the stability of the base.

Thickening The viscosity or mouth-feel building properties of HM pectins find some use in recombined juice products to restore the mouth-feel of the product to that of the freshly pressed juice.

Stabilization Gelation of HM pectins may be used as a means of stabilizing a multiphase system, if gelling conditions can be achieved at some step in the process. Gelation supplies the yield value that is necessary to obtain permanent stabilization of emulsions, suspensions, and foams. HM pectins are used in this way as stabilizers in fruit drink concentrates and in aerated confectionery products. LM pectins are used in light (i.e., specific gravity of approximately 0.20) confectionery foams with soluble solids on the order of 70%.

HM pectins may be used to stabilize sour-milk products which are produced by direct acidification, or are cultured. The HM pectin reacts with the casein and prevents coagulation when the pH is reduced to below the isoelectric pH of casein (approximately 4.6). This colloid protective effect of the HM pectin allows pasteurization of sour-milk products to extend the shelf life.

Application Procedures

As is true for any gelling agent, pectins cannot dissolve in a medium where gelling conditions exist. Pectins are normally dissolved in water before being combined with other ingredients of the end products.

Standard procedures for dispersion of gums are used, i.e., blending with inert material to prevent lumping, dispersion in media where the pectin is insoluble (such as alcohol or sugar syrup), or dispersion by special mixers. Dispersion of pectin is facilitated by incorporation of approximately 2% of sodium bicarbonate with the pectin. When such pectin mixtures are dispersed in water, they will lump initially; then when carbon dioxide is developed by reaction between the acidic pectin and the bicarbonate, the lumps will gradually disappear completely.

The viscosity of pectin solutions follows a common logarithmic temperature relationship; thus higher concentrations of pectin solutions can be made at elevated temperatures. The temperatures that may be used depend on the pH and the holding time. If the water or other medium into which pectin must be dissolved has any significant buffer capacity, and a pH outside of the 3 to 4.5 range, it may be necessary to adjust the pH to the above range before the pectin is dissolved. Pectin solutions are completely stable at room temperature for weeks if no pectolytic enzymes are present. At 60°C, pectin solutions may be safely stored for an 8-h period. At higher temperatures, longer holding times should be avoided, as illustrated by Table 15.3.

LM-pectin solutions tend to gel when made with higher pectin concentration in hard water. Gelation may be avoided by addition of calcium-complexing salts.

Care must be exercised to avoid local pregelation when pectin stock solutions are mixed with other ingredients. As an illustration, pectin pregelation is liable to occur in jam manufacture if a 5% pectin solution with a pH of approximately 3 is mixed with a fruit syrup containing about 70% solids at 70°C. In such a case, pregelation is avoided by using a pectin with a higher pH (for example, 3.6 to 4.0).

HM pectins may be gelled at room temperature, as gelation is a function of time as well as of pH and temperature.   A flan jelly may be set at room temperature by blending a sugar syrup containing HM pectin and having a pH of 4.0 with a small measured quantity of citric acid solution.   Crystalline acid should never be added to set a pectin-sugar syrup; pregelation will invariably occur around the acid crystals, and the resulting gel will be grainy.

Another way of utilizing the cold-setting ability of HM pectins is by blending 30 parts of fruit syrup containing approximately 2% of HM pectin and having a pH of 2.9 with 70 parts of liquid sugar (66.5% soluble solids) to obtain a jelly with soluble solids of 53 to 54% and a pH of 3.0.   Such a formulation will set within a few hours after blending.

Gelation of LM pectins with calcium ions is temperature-dependent.   LM pectins must therefore be combined with the soluble calcium source at a temperature above the gelling temperature of the system.   If the calcium source is a food ingredient, such as milk or fruit juice, LM pectins are normally added in water solution.   If calcium is added as a gelling salt to a medium containing LM pectin, it must be predissolved in water to a rather dilute solution (for example, 4% $CaCl_2$ in water).   Crystalline calcium salts and concentrated calcium salt solutions cause pregelation when added to LM pectin solutions.

Gelation of LM pectin without application of heat may be obtained by controlled release of the calcium ion or by diffusion of calcium ions into the LM pectin–containing medium.   The practical methods are the same as for alginate gelling systems.

TABLE 15.3
Effect of pH, Time, and Temperature on Stability of HM Pectin in Solution

Temperature, °C	Holding at: pH	Time, min	Resulting loss in jelly grade of HM Pectin, %
80	3.4	120	7–9
80	3.4	240	11–15
100	4.0	10	3–6
100	4.0	20	7–11
100	4.6	10	7–13
100	4.6	20	17–20
100	5.1	5	More than 17

Future Developments

As jams and jellies account for the major part of the pectin consumption, any trend in the consumption or the composition of these products will be immediately reflected in the pectin market.

Recent jumps in sugar prices are not expected to influence the consumption of HM pectins if sugar is replaced by other lower-priced carbohydrate sweeteners.   If the effect of higher prices on the consumer product is an overall lower consumption, then HM pectins will suffer from lower consumption in this application.

Where jam standards have allowed, there has been a tendency to reduce the soluble solids from the traditional 65% to approximately 50%.   In Sweden, two-thirds of the jam is produced with soluble solids below the 55% concentration; in Denmark, one-third of the jam is so produced.   This tendency is likely to spread to other countries, particularly if international jam standards (*Codex* and EC) provide for lower soluble-solids contents.

If, for economic or nutritional reasons, the carbohydrate sweetener content is reduced, more pectins will automatically be required to set the products, and a different type will be required (see Table 15.4).

The second largest application of HM pectins is in sugar confectionery products. The pectin consumption for this purpose is therefore found to be affected by trends for products that compete with sugar confectioneries, such as chocolate and various snack products.   If chocolate prices continue to increase at a higher rate than the prices of sugars, a change to higher consumption of sugar sweets is likely to occur.   This view is supported by the consumption pattern in the U.S.S.R., where the chocolate prices have been set very high. The Soviet people consume 11 kg of sugar sweets per capita per year.

COMMERCIAL USES: Compounding and Formulating

Pectins are used almost exclusively in foods. A very minor quantity is used in pharmaceutical and cosmetic products.

Pectin grades for food applications are normally standardized for specific functions by blending with an inert material, such as sucrose or dextrose. Standardization ensures a constant effect in the end product and at the same point in the production process. Pectins for jams and jellies are standardized to constant *jelly grade* which expresses the amount of sugar that can be made into a 65% soluble-solids jelly with a defined strength. Pectins for jams and jellies are made with varying relative speed of gelation properties, expressed by type designations of, for instance, *rapid-set* or *slow-set* pectin. For specific uses, such as for "confectionery jellies," HM pectin may be blended with a suitable buffer salt to control the pH in use.

Application conditions for LM pectins show a wide variation; thus, to make standardization methods relevant, they must differ just as widely. The most widely used standardiza-

TABLE 15.4
Pectin Type and Relative Concentration at Various Percentages of Soluble Solids

Percent soluble solids in jam	Recommended pectin type*	Relative pectin† concentration to produce jam of equal firmness
75	D or DD, slow-set	77
72.5	Same	83
70	Same	88
67.5	⎧ B or BB, rapid-set	⎧ 94
65	⎨ A or AA, medium-rapid-set	100 (basis)
62.5	⎩ D or DD, slow-set	108
60		⎩ 117
57.5	B or BB, rapid-set	126
55	⎧ B or BB, rapid set	135
	⎩ or LM 15–20% amidated	
35	LM 20–25% amidated	175
10	Carrageenan	225

* Letters designate Genu brand grades.
† Standardized pectin: HM, 150 USA-SAG at 65% SS; LM, 100 jelly grade at 31% SS.

tion method of some general value is derived from the jelly-grade method for HM pectins. Jelly composition is different, as soluble solids are at 31%; gelation is achieved by adding a controlled amount of calcium salt.

Many LM-pectin grades are sold on the basis of specifications and performance tests developed in cooperation between the user and the pectin manufacturer. So-called N.F. *(National Formulary)* pectins, or pharmaceutical-grade pectins, are pure HM pectins that conform with the purity requirements of the latest edition of the *National Formulary*. Such pectins are available in various molecular weights.

Table 15.5 gives typical applications of HM pectins; Table 15.6 gives similar information for LM pectins.

FORMULATIONS

Jams and Jellies

The function of a pectin in jam is to provide a gelled texture that allows the jam to be spread easily without running. Further, the mouth-feel must be pleasant, the gel must not break during transport and handling, and syneresis must not occur. During manufacturing, the pectin must set at the desired point in the process. In jam manufacture, the pectin must set soon after filling to prevent separation of fruit particles and syrup; in jelly manufacture, to produce a completely homogeneous and transparent product, the pectin must not set before all the air bubbles have escaped. All these requirements for jam and jelly manufacture obviously cannot be met by one pectin, especially when the wide ranges of soluble solids, pH, and processing variations are considered.

The technological requirements are met by proper selection of HM or LM pectin (see Tables 15.1 and 15.2). Typical jam formulations are as follows:

Fruit Jam The following recipe is for a 650-kg batch with 65.2% soluble solids and a pH of 3.10.

Order of addition	Ingredients	Amount, kg
A	Frozen fruit	227.50
	Sugar	294.00
	Glucose syrup, 81% SS	130.00
B	Pectin, rapid-set, 150° USA-SAG	1.95
	Water, 60–80°C	58.50
C	Citric acid, 50% w/v sol.	2.60
	Total ingredients	715
	Evaporation	65
	Yield	650

To prepare, follow the sequence of instructions:

1. Measure hot water (B) to the pectin mixer; start the mixer and add the pectin (B). Mix for 5 min.
2. Suck the glucose syrup (A) into the vacuum pan, and start heating by opening the steam valve.
3. Add the fruit (A) to the vacuum pan.
4. Suck the sugar (A) into the vacuum pan, break the vacuum and start the stirrer, and dissolve the sugar at 70 to 80°C.
5. Boil to the desired soluble-solids concentration (see below) under vacuum; use a temperature of 75 to 60°C.
6. Suck the pectin solution into the cooling pan.
7. Break vacuum, and start heating to 90°C; when reached, close steam valve.
8. Add the citric acid solution (C), and transfer to holding tank/filler.
9. Fill into 0.5-kg jars, and cap.
10. Cool in cooling tunnel with cold water spray to an internal temperature of 40 to 45°C.
11. Label and pack in shipping container.

Note for step 5: Desirable soluble solids at this stage may be calculated as follows:

$$\frac{65.2 - 0.3}{100 - (9.00 + 0.30)} \times 100 = 71.6\%$$

This calculation assumes that the addition of the pectin solution, 9.00 kg water + 0.30 kg pectin/100 kg, brings the yield to 100 kg and the soluble solids to 65.2 kg without any further evaporation. The factor $65.2 - 0.3$ is the amount (kg) of solids per 100-kg yield before the pectin solution is added. The $[100 - (9.00 + 0.30)]$ value is the weight (kg) of the jam batch per 100-kg yield before the pectin solution is added.

Low-Sugar-Content Jam The following jam recipe contains 55% soluble solids and incorporates LM pectin. The pH is 3.2 to 3.5.

Order of addition	Ingredients	Amount
A	Strawberries	40 kg
	Sugar	51 kg
B	LM pectin	0.55 kg
	Water	20 L
C	Citric acid, 50% w/v sol.	0.250 L
	Total ingredients	111.80 kg
	Evaporation	11.80 kg
	Yield	100 kg

TABLE 15.5
Typical Applications of HM (High-Methoxyl) Pectins

Industry	Product	Function of pectin	Preferred pectin type	Reason for specific pectin type	Typical use concentration, %
Fruit-preserving industry	Jams and preserves in jars	Gelling agent	Rapid-set	Allows filling and capping at high (85–95°C) temperature (no need to sterilize filled jar) without risk of fruit floating	0.2–0.5
	Jams and preserves in jars and 5- to 10-kg containers	Gelling agent	Medium-rapid-set	Allows cooling to 70–75°C before filling if heat-sensitive packaging used, or jam filled in 5- to 10-kg containers which cool slowly	0.2–0.6
	Jams and preserves in large containers	Gelling agent	Slow-set	Allows cooling to 60°C before filling; required for filling in large containers	0.2–0.7
	Jellies	Gelling agent	Slow-set	Allows all air bubbles to escape from jelly before gelation starts	0.3–0.5
	Oven-resistant jams and bakery jellies	Gelling agent	Special rapid-set	Allows filling in large containers at 70°C; jam is stable at oven temperatures when gel structure is undisturbed	0.5–1.0
Confectionery industry	Pectin jellies	Gelling agent	Extra-slow-set	High solids (76–78%) requires slowest pectin type to allow sufficient depositing time	1.0–2.5
	Aerated confectionery products	Gelling agent, texturizer, and foam stabilizer	Special slow-set	High solids (70–78%) and low processing temperatures in certain equipment requires slow-setting pectin	0.7–2.0
Bakery supply industry	Cold-setting flan jellies	Gelling agent	Rapid-set	Allows controlled setting of cold syrup with SS of 61%, pH ca 4.0 by adding measured amount 50% w/v citric acid	0.6–1.0

Industry	Application	Function	Type	Comments	Dosage
				Allows controlled setting of cold flan jelly base with SS ca 30%, pH ca 2.9 by addition of two to four times liquid sugar	0.6–0.8 of end product (2–4% of flan jelly base)
Dairy industry	Long-life, pasteurized, cultured-milk drinks and milk-fruit juice drinks	Stabilizer for milk proteins	Special rapid-set	Allows pasteurization of acidic milk products without curdling. This pectin must be controlled for protective colloid effect	0.4–0.7
Beverage industry	Citrus-drink concentrate	Stabilizer of pulp suspension and oil emulsion	Rapid-set	Stabilization is obtained by gel formation in manufacturing process. This gel formation must be achieved rapidly	0.1–0.2
	Fruit juice	Bodying agent to restore mouth-feel if natural pectin insufficient	Special HM-type	Standardized to constant viscosity	0.05–0.10
	Fruit beverage	Bodying agent to give mouth-feel	Special HM-type	Standardized to constant viscosity	0.05–0.10
Pharmaceutical industry	Pharmaceutical suspensions and emulsions	Stabilizer	N.F.-type	N.F. types are guaranteed to meet the purity requirements of National Formulary XIII	0.1–3.0
Cosmetic industry	Hair shampoo	Stabilizer	N.F.-type	(Same as above)	0.1–1.0

TABLE 15.6
Typical Applications of LM (Low-Methoxyl) Pectins

Industry	Product	Function of pectin	Preferred pectin type	Reasons for specific pectin type	Typical use concentration, %
Fruit-preserving industry	Low-solids jams (SS below 55%)	Gelling agent	Amidated	Gelation without addition of calcium salt	0.4–1.0
Confectionery industry	Soft confectionery jellies with neutral taste	Gelling agent	Acid-hydrolyzed (in combination with starch)	Tolerates high (76–78%) soluble solids	1.0–2.0
Bakery supply industry	Heat-reversible bakery jelly for glazing	Gelling agent	Amidated	Gelation within rather broad limits of soluble solids (40–65%)	1.0–2.0
Dairy industry	Gelled sour-milk products	Gelling agent	Amidated		0.5–1.0
Convenience food industry	Canned fruit preparation to make fruit/milk desserts	Gelling agent	Amidated or acid-hydrolyzed	Must be controlled to gel by addition of milk to fruit preparation	0.8–1.2
	Instant pudding powder	Gelling agent	Amidated	Must dissolve rapidly in water and gel rapidly by subsequent addition of milk	0.8–1.2
	Syrup base for instant chocolate or vanilla pudding	Gelling agent	Amidated	Must not thicken syrup excessively. Must gel rapidly at neutral pH by addition of milk to syrup	0.8–1.2

To prepare, follow the sequence of instructions:
1. Weigh fruit and transfer to cooking pan, either open or vacuum; add sugar and heat to dissolve the sugar. Allow enough time for the sugar to penetrate into the fruit. Adjust the soluble solids to 60 to 61% by boiling.
2. Dissolve the pectin in water at 55 to 70°C in a high-speed mixer. Add the pectin solution in open pan or in vacuum boiler, and boil to mix the pectin solution intimately with the fruit and sugar. Adjust the soluble solids to 55 to 56%.
3. Break the vacuum if cooking is done in vacuum boiler; heat to sterilizing temperature (about 90°C).
4. Add citric acid solution. Transfer to holding tank and filling station.

Pectin Fruit Jellies

Pectin jellies are delicate confectionery products with water content of approximately 20%. These jellies may contain fruit pulp or juice, but are more often pure sugar jellies flavored with synthetic flavors and acidified with citric acid. Pectin is used in concentrations of from 1 to 3%. Pectin competes with agar and to some extent with gelatin and starches in this application. The advantage of pectin is superior texture and flavor release. The problem in using pectin is mainly in control of the process, as pectin gelation is rather critically influenced by variations in pH and soluble-solids content.

The recipe following is for a 100-kg batch with 78% soluble solids and giving an end product with a pH of 3.5.

Order of Addition	Ingredients	Amount
A	Water	30.000 L
	Sodium citrate, $2H_2O$	0.400 kg
	Citric acid, H_2O	0.370 kg
B	HM pectin, slow-set, confectionery grade, 150° USA-SAG	1.500 kg
	Sugar	5.000 kg
C	Sugar	46.400 kg
	Glucose syrup, DE approx. 40%, SS approx. 80%	30.000 kg
D	Citric acid, H_2O (must be predissolved)[2]	0.370 kg
	Total ingredients	114.0 kg
	Evaporation	14.0 kg
	Yield	100.0 kg

To prepare, follow the sequence of instructions:
1. Measure water (A) into boiling kettle. Add citric acid (A) and sodium citrate, and dissolve. Water temperature may be 10 to 70°C.
2. Dry-blend pectin with sugar (B) in order to avoid building of pectin lumps, which are difficult to dissolve. Add to the water in the kettle while stirring.
3. Heat slowly while stirring until the solution starts to boil. Boil for 2 min to secure complete dissolution of the pectin.
4. Add sugar (C) gradually to avoid too heavy cooling of the batch; dissolve by boiling. Add the glucose syrup.
5. Boil to 78% refractometer soluble solids.
6. Add citric acid solution (D), and deposit immediately in molds (starch, ceramic, metal, or plastic).

It is important that the acid is added as a solution at the step 6 stage. Any addition of crystalline acid to the already cooked batch will invariably cause pregelation to occur around each acid crystal and will make the jelly grainy.

The addition of acid reduces the pH of the jelly batch from 4.1 to 3.5. At this pH, a maximum depositing time of 20 min is allowed on the condition that the temperature of the batch exceeds 90°C.

Aerated Confectionery Products

Sugar confectionery products may be aerated by incorporating a whipping agent and beating at atmospheric or higher pressure. Pectin may be used to stabilize the aerated product, and at the same time it provides texture to the end product.

If aeration is obtained with egg albumen, high temperatures must be avoided to prevent coagulation. If whipping takes place in open equipment, the product will be cooled quite noticeably at the product-air interface. In both cases, pectin gelation is likely to occur if correct soluble solids and pH conditions prevail.

Special slow-setting, high-methoxyl pectins may be used in an open-bowl beating process if the soluble-solids content is reduced to about 70% and if gelling acid is added immediately before depositing. This type of formulation is illustrated in the recipe for open-bowl confectionery foam shown below.

If closed pressure-beating equipment is available, coupled with automatic acid dosing and efficient mixer equipment, HM pectin may be used at soluble-solids contents of up to about 80%, as shown in the recipe for medium-density confectionery foam.

Confectionery Foam, Open-Bowl Process   The following recipe is for a fruit-flavored, light confectionery foam produced by the previously described open-bowl process. The batch size is 10 kg, soluble solids are 70%, pH is 3.6, and density is 0.45 g/mL.

Order of Addition	Ingredients	Amount
	Water	3.000 L
	Potassium citrate, H_2O	0.043 kg
A	Citric acid, 50% w/v sol.	0.075 L
	Pectin, confectionery grade	0.100 kg
	Sugar	0.430 kg
	Sugar	5.100 kg
B	Glucose syrup, DE = 42,	
	SS = 80%	1.500 kg
C	Egg white, liquid	0.500 L
	Color	Optional
D	Flavor	Optional
	Citric acid, 50% w/v sol.	0.050 L
	Yield	10.0 kg

To prepare, follow the sequence of instructions:

1. Measure water (A) into boiling kettle. Add citric acid and potassium citrate, and dissolve. Water temperature may be 10 to 70°C.
2. Dry-blend pectin with sugar (A) in order to avoid build-up of pectin lumps, which are difficult to dissolve. Add dry blend to water in the kettle while stirring.
3. Heat slowly while stirring until the solution starts to boil. Continue boiling for 2 min to secure complete dissolution of the pectin.
4. Add sugar (B) gradually, and dissolve by boiling. Add glucose syrup, then boil until the solution evaporates to 73% soluble solids as measured with a refractometer.
5. Whip the egg white for 5 to 7 min in an open bowl to obtain a stiff foam. Cool the boiling batch to 95 to 100°C, and gradually add to the egg white. Whip for another 4 min.
6. Add ingredients (D), mix carefully, and deposit immediately. The addition of acid (D) reduces the pH of the batch from 4.05 to 3.60. At this pH, a maximum depositing time of 15 min is allowed on the condition that the temperature of the batch exceeds 50°C.

Confectionery Foam, Medium Density   The following recipe is for a fruit-flavored, medium-density confectionery foam made with pressure beating. Batch size is 100 kg, soluble solids are 75%, pH is 3.5, and density is 0.5 to 0.6 g/mL.

Order of Addition	Ingredients	Amount
	Water	30.00 L
A	Sodium citrate, $2H_2O$	0.40 kg
	Citric acid, H_2O	0.20 kg
	Pectin, extra-slow-set,	
B	confectionery grade	1.50 kg
	Sugar	4.00 kg

C	⎰ Sugar	56.00 kg
	⎨ Glucose syrup, DE = 42,	
	⎱ SS = 80%	15.00 kg
D	Egg white, liquid	5.00 L
E	⎰ Color	Optional
	⎨ Flavor	Optional
	⎱ Citric acid, 50% w/v sol.	0.50 L
	Total ingredients 112.7 kg	
	Evaporation 12.7 kg	
	Yield 100.0 kg	

To prepare, follow the sequence of instructions:
1. Measure water (A) into boiling kettle. Add citric acid (A) and sodium citrate, and dissolve.
2. Dry-blend pectin with sugar (B) in order to avoid formation of pectin lumps, which are difficult to dissolve. Add dry blend to the water in the kettle while stirring.
3. Heat slowly, while stirring, until the solution starts to boil. Boil for 2 min to dissolve the pectin completely.
4. Add sugar (C) gradually, and dissolve by boiling. Add the glucose syrup, and boil until the solution evaporates to 79% soluble solids as measured by a refractometer.
5. Measure the egg white into the beater bowl. Cool the pectin-sugar-syrup mixture to 80°C, and add slowly to the egg white while stirring slowly. (Cooling is not required when the whipping agents used are not heat-coagulating.)
6. Whip for 4 min at 1.5 atm pressure.
7. Add color, flavor, and citric acid (E). Mix carefully, and deposit immediately. (Longer depositing time is possible if the contents of the pressure beater can be maintained at a constant high temperature.)

Further flexibility is achieved if gelling acid is added "in-line" between the beater and the depositor. Proper mixing of the whipped material and the gelling acid (plus color and flavor) is obtained in a "static" mixer.

Cold-Setting Flan Jelly

The fact that HM-pectin gelation is time-dependent is utilized in a cold-setting pectin jelly. An HM-pectin–sugar syrup is made with 61% soluble solids and a pH of about 4.0 (which is above the gelling range). When a concentrated fruit acid solution is added to the pectin-sugar syrup, gelation is induced within one to several minutes, depending upon the exact amount of acid.

The following recipe is for a 100-kg batch with a final concentration of 61% soluble solids and a pH of approximately 4.0.

Order of Addition	Ingredients	Amount
A	⎰ Pectin, rapid-set, pH 2.7–3.2,	
	⎨ 150° USA-SAG	0.700 kg
	⎱ Sucrose	3.500 kg
B	⎰ Water, calcium-free for	
	⎪ max. jelly clarity	42.00 L
	⎨ Sodium citrate, 2H$_2$O	0.080 kg
	⎱ Citric acid, 50% w/v sol.	0.200 L
C	⎰ Sucrose	47.700 kg
	⎱ Glucose syrup	11.000 kg
D	Sodium benzoate, 20% w/v sol.	0.250 L
	Total ingredients 105 kg	
	Evaporation 5 kg	
	Yield 100 kg	

The procedure for preparation is as follows:
1. Blend pectin carefully with five parts of sugar (A).
2. Dissolve sodium citrate (B) and citric acid in water, and add the pectin-sugar blend to the solution. Heat to boiling, and continue to boil for 2 min to dissolve the pectin completely.
3. Add sugar (C) and dissolve. Add glucose syrup. Adjust the soluble-solids concentration to 61.0% as sucrose on refractometer reading.

4. Add sodium benzoate solution (D).
5. Cool the jelly base and fill.

Directions for application are:

(a) Weigh (or measure) as much jelly base as is needed for the flan work or as can be used within the time available for depositing (shown in following).

(b) Measure the required amount of citric acid solution, and add to the jelly base. Stir carefully to secure an even distribution.

Gelling time required	Citric acid sol. (50% w/v) to be added per kg (0.78 L) jelly base	pH in the flan jelly
5–6 min	10 mL	ca 3.10
1–2 min	15 mL	ca 3.00
45 s	20 mL	ca 2.85

(Gelling time is time available for depositing without serious gelation. After this time, the jelly will no longer flow.)

Orange-Drink Concentrate

It is possible to obtain permanent suspension of fruit particles and an oil emulsion in orange-drink concentrates with 46 to 65% soluble-solids concentration if the product is gelled with HM pectin and the gel texture is subsequently destroyed mechanically to obtain a free-flowing liquid.   Such a recipe follows:

Order of Addition	Ingredients	Amount
A	Water	4.65 L
	Pectin, rapid-set, pH 2.7–3.2, 150° USA-SAG	0.15 kg
B	Sugar syrup, 66.5% SS	61.00 kg
	Juice concentrate	7.00 kg
C	Citric acid, 50% w/v sol.	2.00 L
D	Sodium benzoate, 20% w/v sol.	0.50 L
	Water	24.30 L
	Yield of stabilized fruit-drink concentrate	100.00 kg

The procedure for preparation is as follows:

1. Dissolve pectin (A) in 30 parts of 50 to 60°C water using an efficient high-speed mixer.   Essential oils may be added at this stage and emulsified into the pectin solution.
2. Blend the sugar syrup (B) and the fruit concentrate.
3. Add the pectin solution to the fruit syrup, and blend.
4. Add the citric acid (C) solution, and blend.   Allow the gelation to proceed for 30 min.
5. Stir to break up the gel, and add sodium benzoate (D) and water to bring the solution to 46.0% soluble-solids concentration.
6. Deaerate, if necessary.
7. Fill into bottles, and cap.

Sour-Milk Drink Stabilization

Drinks that combine milk and juice as pasteurized cultured-milk drinks may be made with HM pectin as a stabilizer to prevent coagulation, as in the following recipe for a milk–orange juice drink.

Order of Addition	Ingredients	Percent of product
A	Pectin	0.45
	Sodium citrate, $2H_2O$	0.20
B	Water	9.00
C	Milk	45.00
D	Orange juice	40.35
	Sugar	5.00
		100.00

The procedure for preparation is as follows:
1. Mix the pectin (A) with the sodium citrate, and dissolve in 50 to 60°C water (B) with a high-speed mixer. Cool to 20°C, or lower.
2. Add the pectin solution to the milk (0 to 10°C) while stirring.
3. Add the milk to the sweetened orange juice (0 to 10°C).
4. Pasteurize by heating to 70°C in a plate heater. Homogenize at 150 atm pressure.
5. Cool and fill.

Cold-Setting Milk Desserts

Milk desserts with fruit, chocolate, vanilla, or caramel flavor and with a gelled texture can be made with LM pectin as the gelling agent. The pectin must be dissolved in a syrup containing the sugar sweetener and the flavor, e.g., cocoa or fruit juice. The pectin-containing syrup is mixed with cold milk which provides the calcium ions necessary for gelation. The next two recipes illustrate this type of dessert.

Milk-Fruit Dessert Preparation   The following recipe is for a 100-kg batch containing 30.28% soluble solids and having a pH of 3.9 to 4.1.

Order of addition	Ingredients	Amount
A	LM Pectin	2 kg
	Sugar	6 kg
	Modified starch	1 kg
	Soft water	52.6 L
B	Fruit pulp or juice	10 kg
	Whole fruit	6 kg
	Sugar	19 kg
C	Sodium citrate, 20% w/v sol.	1.8 L
	Citric acid, 20% w/v sol.	1.6 L
	Yield	100 kg

The procedure for preparation is as follows:
1. Mix pectin (A), sugar, and starch. Disperse in water at 20°C, and stir for about 10 min.
2. Add fruit pulp or juice, whole fruit, and sugar (B). Mix thoroughly.
3. Adjust pH to 4 by addition of sodium citrate solution (C).
4. Mix the remainder of sodium citrate with citric acid solution. Add to the batch under stirring. Transfer to holding tank and filling station.

Filling is usually done in tins that are adequately sterilized and cooled down to 20°C (room temperature). The dessert is prepared by mixing, while stirring, the contents of one tin with the same quantity of cold milk taken from the refrigerator.

Intermediate-Moisture Chocolate Pudding Base

Order of addition	Ingredients	Amount
A	Water	36.00 L
	LM pectin	3.30 kg
	Sodium citrate, 2H$_2$O	0.50 kg
B	Dextrose, H$_2$O	31.60 kg
	Cocoa	6.60 kg
	Invert sugar syrup, 75% SS	22.00 kg
C	Citric acid	(as needed to reach pH approx. 5.0)
	Yield	100.0 kg

The procedure for preparation is as follows:
1. Blend sodium citrate (A) and pectin, and dissolve in hot water using a high-speed mixer.
2. Blend dextrose (B) and cocoa, then add to the pectin solution.
3. Heat to 80 to 85°C, then add the invert sugar syrup. Maintain at 80 to 85°C for 5 to 10 min. Add citric acid solution (C) needed to bring the mixture to a pH of approximately 5.0.
4. Fill while hot, or add a preservative and cool and fill.

To use the pudding base:
   (a) Stir one part by weight of the chocolate syrup into three parts by weight of cold milk, and mix until homogeneous.

(b) Pour immediately into serving dishes, and place into the refrigerator until served.

LABORATORY TECHNIQUES

Official methods as published in, for example, *Food Chemicals Codex,* are used for chemical analysis of pectins. Physical properties are evaluated by official jelly-grade and gelling-time methods as published in the *Codex* and/or specific tests developed by pectin manufacturers or users, and very often in cooperation between the two parties.

Specific tests may be for performance, for instance if an HM pectin is to be used as a protective colloid in a sour-milk drink. A high-methoxyl pectin may be standardized to 150 jelly grade, and may be controlled to maximum viscosity in solution if viscosity is critical during the manufacturing process of a fruit jelly.

PRODUCT/TRADENAME GLOSSARY

Manufacturer	*Pectin Type(s)*	*Tradenames*
Sunkist Growers, Inc. 720 East Sunkist St. Ontario, Calif. 91761	High-methoxyl Amidated, low-methoxyl	Sunkist Exchange Jet-Sol Exchange Jet-Sol Hi-Poly
General Foods Corp. Atlantic Gelatin Pectin Plt. 1127 N. Mansfield Ave. Hollywood, Calif. 90038	High-methoxyl	Atlantic Citroflash
The Copenhagen Pectin Factory Ltd. DK 4623 Lille Skensved Denmark	High-methoxyl Amidated, low-methoxyl	Genu Genu
POMOSIN-Verkaufsbüro D-6 Frankfurt am Main 60 Postfach 600 640 West Germany	High-methoxyl	
Cesalpinia S.p.A. Via Pinamonte da Brembate 3 I-24100 Bergamo Italy	High-methoxyl Acid-demethylated, low-methoxyl Amidated, low-methoxyl Polygalacturonic acid	
Obipektin AG CH-9220 Bischofszell Switzerland	High-methoxyl Acid-demethylated, low-methoxyl Amidated, low-methoxyl Polygalacturonic acid Polygalacturonates	Obipektin Obipektin Obipektin Obipektin Obipektin
Unipectine, S.A. 26 Avenue de l'Opera 75001 Paris France	High-methoxyl Amidated, low-methoxyl Acid-demethylated, low-methoxyl	Unipectine Unipectine Unipectine
HP Bulmer Limited Ryelands Street Hereford HR4 OLE England	High-methoxyl Acid-demethylated, low-methoxyl	Bulmer Bulmer
Pektin-Fabrik Hermann Herbstreith KG D-754 Neuenbürg (Württ.) West Germany	High-methoxyl Acid-demethylated, low-methoxyl	
KUK Nährmittelindustrie und Handelsgesellschaft mbH A-4910 Ried/Innkreis Austria	High-methoxyl	KUK

Yakhin Canning Company Ltd.   High-methoxyl
P.O. Box 332
Tel-Aviv
Israel

RECOMMENDED READING

The following books and articles on pectin do not provide detailed information about formulations for foods or pharmaceutical and cosmetic preparations. For such information, as well as for detailed information on application procedures, commercial literature as published by the pectin manufacturers must be referred to, e.g., The Copenhagen Pectin Factory Ltd., *Genu Pectins,* 3d ed., February 1976.

Christensen, O., and G. A. Towle, "Pectin," in R. L. Whistler (ed.), *Industrial Gums,* 2d ed., Academic Press, Inc., New York, 1973, pp. 429–461. Review, covering all aspects of pectin as an industrially isolated polysaccharide, with emphasis on its application in foods. An extensive list (199) of references.

Doesburg, J. J., *Pectic Substances in Fresh and Preserved Fruits and Vegetables,* I.B.V.T. Communication No. 25, Institute for Research on Storage and Processing of Horticultural Produce, Wageningen, Netherlands, 1965. A 150-page book (in English) on the role of pectin in plants and the properties of pectin relevant for use in foods.

Kertesz, Z. I., *The Pectic Substances,* Interscience Publishers, Inc., New York, 1951. The standard book (628 pages) on pectin; covers anything published about pectin up to 1950.

Pilnik, W., and P. Zwiker, "Pektine," *Gordian,* **70,** 202–204, 252–257, 302–305, 343–346 (1970). Condensed review (in German); covers all the aspects of pectin which are essential for the user of pectin in foods, its structure and properties, and methods of evaluation. Extensive (157) reference list.

Chapter 16

Polyacrylamide

H. Volk and R. E. Friedrich
Dow Chemical U.S.A., Midland, Michigan

General Information 16-2
    Chemical Nature 16-2
    Physical Properties 16-3
    Manufacture 16-5
    Biological/Toxicological Properties 16-7
    Rheological Properties 16-9
    Additives/Extenders 16-10
    Handling 16-10
    Applications 16-10
    Application Procedures 16-11
    Future Developments 16-11
Commercial Uses: Compounding and Formulating 16-11
    Pulp and Paper 16-11
    Tape-Joint Cement 16-11
Commercial Uses: Processing Aids 16-12
    Processing Industries 16-12
    Industrial Reclamation 16-12
    Water Treatment 16-13
    Sludge Treatment 16-13
    Hot and Cold Lime Softening 16-13
    Pulp and Paper Manufacture 16-13
    Mining and Ore Processing 16-13
    Oil Production 16-15
    Miscellaneous Applications 16-16
Industries Using Polyacrylamide 16-16
Formulations 16-16
    Methylolated Polyacrylamide 16-16
    Gelled Explosives 16-16
    Gelled Hydrazine 16-18
    Explosives Gelled via Ionic Bonds 16-18
Laboratory Techniques 16-19
    Analytical Methods 16-19
    Laboratory Evaluation 16-19

Product/Tradename Glossary 16-25
References 16-25
Further Useful Reading 16-25

GENERAL INFORMATION

The polyacrylamides are versatile synthetic polymers, readily water-soluble over a broad range of conditions. They can be engineered to fit a large number of uses ranging from adhesion to drag reduction. The polymers of acrylamide are unique in their strong hydrogen bonding, linearity, and very high molecular weights. These properties have enabled polyacrylamides to find utility as dry-strength resins, retention and freeness aids in the paper industry, as flocculents in water clarification and mining application, as flooding aids in secondary oil recovery, and as binders for foundry sand. These are but some of the major applications.

Ionic polyacrylamides are readily obtained by copolymerization or by chemical reactions on the amide functionality. Polymers which are predominantly acrylamide will be classified as polyacrylamide for the purposes of this chapter.

The polymers are usually prepared by the free-radical polymerization of acrylamide in aqueous solution and sold as such or dried and sold as a powder. The very-high-molecular-weight polymers cannot be handled conveniently in solutions over several percent concentration. The polymers are also available in nonaqueous dispersions.

Polyacrylamides became commercially significant in the early 1950s, although a polyacrylamide product was prepared by Moureu in 1893. World production, excluding U.S.S.R., is estimated to be approximately 4500 metric tons (100 million pounds) annually, from about 15 companies.

For in-depth reading, see Further Useful Reading, page 16-25. References 1 and 2 are recommended; references 3 and 4 are reviews of somewhat narrower scope.

Chemical Nature

Most of the reactions of polyacrylamide are typical amide chemistry. The most important is the hydrolysis to the carboxylate ion, which converts the essentially neutral polyacrylamide to a strongly anionic molecule.

$$
\begin{array}{c}
\text{H} \quad \text{H} \\
-\text{C}-\text{C}- \ + \ \text{H}_2\text{O} \\
\text{H} \quad | \\
\quad\quad \text{C}=\text{O} \\
\quad\quad | \\
\quad\quad \text{NH} \\
\quad\quad \text{H}
\end{array}
\xrightarrow[\text{H}^+]{\text{OH}^- \text{ or}}
\begin{array}{c}
\text{H} \quad \text{H} \\
-\text{C}-\text{C}- \ + \ \text{NH}_4^+ \\
\text{H} \quad | \\
\quad\quad \text{C}=\text{O} \\
\quad\quad | \\
\quad\quad \text{O}_-
\end{array}
$$

Most alleged pure polyacrylamides are hydrolyzed to some extent, but many to less than 1%. The hydrolysis rate is lowest at neutral pH. Under alkaline conditions, the hydrolysis progresses readily to about 35%, after which the rate slows appreciably. To get hydrolysis to proceed beyond 70%, stringent conditions are required. Acid hydrolysis is autocatalytic but develops imides at partial conversions, leading to insolubilization. The amide functionality is generally more resistant to hydrolysis than the esters in acrylamide–acrylic ester copolymers.

Polyacrylamides react with aldehydes to form the methylol at neutral or slightly alkaline pH, but at acid pH (<4) the methylols will react with other amides to form methylene crosslinks, a common route to water-swellable gels. Formaldehyde and bisulfite ion will react with polyacrylamide to form the anionic sulfomethyl polyacrylamide. Formaldehyde in combination with dimethylamine will form the dimethylamino-N-methylacrylamide polymer. This is commonly referred to as the *Mannich reaction*.

The Mannich reaction is a route to the preparation of cationic polyacrylamides. The tertiary amine can be quaternized by reaction with dimethyl sulfate. The Mannich reaction exhibits some reversibility to the amide methylol, which tends to crosslink the polymer.

A Hofmann degradation can be performed on the polyacrylamide with sodium hypobromite or hypochlorite followed by neutralization to yield a reaction mixture which contains polyamines. This product has found utility in flocculation applications. Exposure to ionizing radiation will cause gelation of concentrated polyacrylamide solutions. Dilute solutions will be degraded.

A growing number of commercial polyacrylamides contain ionic comonomers such as dimethylaminoethylmethacrylate, diallyldimethylammonium chloride, and vinylbenzyltrimethylammonium chloride. These impart a positive charge to the molecule. The cationic polymers are particularly effective flocculents. Anionic character can be imparted by comonomers such as acrylic acid, methacrylic acid, 2-acrylamido-2-methylpropanesulfonic acid and sodium styrene sulfonate. Polymers containing diacetoneacrylamide form water-insoluble films upon heating.

The anionic polyacrylamides such as the hydrolyzed polymer will interact or "coacervate" with cationic polymers such as the Mannich product or polyvinylbenzyltrimethylammonium chloride. The hydrolyzed polyacrylamide also interacts with the hydrous oxides of polyvalent metals such as iron and aluminum to form gel structures (via ionic crosslinking) which are pH-sensitive.

Physical Properties

Polyacrylamide is supplied commercially in a number of forms. The most common is a powder which usually dissolves within several hours depending on the particle size, method of mixing, and the temperature of the water. Raising the temperature increases the rate of dissolution. Some of the low-molecular-weight polymers are supplied as viscous aqueous solutions, usually less than 30% concentration, which facilitate dilution at time of use. Some polymers are supplied in a dry-bead form, having been prepared in a water-in-oil suspension. This form has the advantage of low dusting and higher bulk densities. Polyacrylamide emulsions in hydrocarbons are also available. These contain approximately 30 to 50% solids and are a very convenient way to get fast solubility.

Polyacrylamide is soluble in water in all proportions, but at concentrations above 70% it would be more appropriate to consider the water to be soluble in the polymer. Molecular weight does not appear to affect the water solubility, but high-molecular-weight polymer tends to form gel-like structures at concentrations over 10%. This is believed to be due to intermolecular hydrogen bonding. Polyacrylamide is insoluble in most organic liquids except for a few polar ones such as acetic acid, acrylic acid, chloroacetic acid, ethylene glycol, glycerin, molten urea, and formamide. Such solubilities are limited, frequently requiring heating, and have not found much application.

Polyacrylamide can be purified by precipitation in methanol or acetone. Impurities can also be removed by washing the dry polymer with methanol or acetone containing 20 to 30% water.

The rate of solution of pure polyacrylamide is relatively unaffected by pH, but as many of the commercial products are partially hydrolyzed, the rate is frequently enhanced by slight alkalinity. At pH above 10.5, hydrolysis occurs on standing. Dilute solutions of unhydrolyzed polyacrylamide are unaffected by most inorganic salts.

The dried forms are generally not thermoplastic. Dry heating to high temperatures usually results in crosslinking, presumably by imide formation.

Effect of Concentration   The viscosity of polyacrylamide increases nearly logarithmically with concentration, as shown in Fig. 16.1. The high-molecular-weight polymers are quite unmanageable at concentrations of over a few percent. Increasing the temperature decreases the viscosity, but not dramatically.

Effect of Shear   The high-molecular-weight polyacrylamides exhibit non-Newtonian character, with increasing shear rates causing large decreases in measured viscosities, as shown in Fig. 16.2. As the shear rate is reduced, the higher viscosities recur. Prolonged shear can result in chain scission, however. Many polymers are evaluated by single viscosity determinations. However, if care is not taken to standardize shear conditions, significant error will occur.

Effect of Hydrolysis   The sensitivity of viscosity to pH of hydrolyzed polyacrylamide is demonstrated in Fig. 16.3 and Table 16.1. As the pH is shifted from the acid to the alkaline region, the viscosity increases dramatically. The non-ionized carboxylic acid groups are converted to anionic carboxylate ions which exert repulsive forces along the chain, causing uncoiling and stiffening of the macromolecule. The addition of salts to the aqueous solution increases the ion concentration, partially shielding the anion repulsion with a resulting decrease in viscosity. Viscosity measurements are usually made with a standard level of NaCl (4%) or NaNO$_3$ (1 N).

Spectra. For rapid identification of polyacrylamide and for determination of the degree of hydrolysis, the infrared spectra are very useful. Copolymers of acrylamide

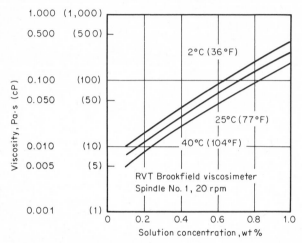

Figure 16.1   Viscosity-concentration relationship of high-molecular-weight polyacrylamide. (*From The Dow Chemical Co.,* Separan® Polymers Settle Process Problems, *Midland, Mich., 1971.*)

are also readily analyzed by infrared. The spectrum of a film of polyacrylamide on AgCl plate is shown in Fig. 16.4.

Flocculation   The great size of the polyacrylamide molecule allows it to bridge between particles. In Fig. 16.5, curves B and D show the sharp increase in viscosity or gelation which occurs when hydrous aluminum oxide is formed *in situ* in the hydrolyzed

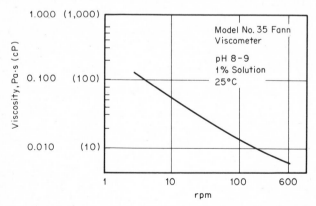

Figure 16.2   Viscosity and shear rate properties of high-molecular-weight polyacrylamide. (*From The Dow Chemical Co.,* Separan® Polymers Settle Process Problems, *Midland, Mich., 1971.*)

polyacrylamide solutions. The anionic sites on the polymer adsorb on the cationic sites on the aluminum hydroxide. This is similar to the mechanism for flocculation, where one molecule can simultaneously adsorb on several particles, causing them to draw together and settle rapidly. In Fig. 16.6 is seen the flocculation of uranium ore. The rate of settling depends both on the flocculant concentration and on the suspended-solids concentration.   Each affects the bridging kinetics.   By overdosing, the rate of flocculation can be reduced because of rapid occupation of the adsorption sites on the particles and consequent decrease in the probability for bridging.

Viscous Drag Reduction   Polyacrylamides, as well as other high-molecular-weight, water-soluble polymers, have the ability to greatly reduce the energy required to pump

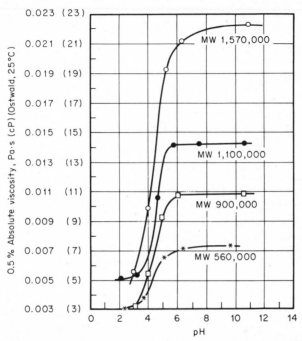

Figure 16.3 Effect of pH on the viscosity of 0.5% solutions of polyacrylamide (25% hydrolyzed, 4% NaCl). *(From The Dow Chemical Co., unpublished data.)*

TABLE 16.1
Viscosity of Two Commercial Polyacrylamides*

Separan®	Concentration, %	Approx. hydrolysis, %	Viscosity, cP at 25°C†		
			pH 3	pH 5	pH 7
NP10	0.5	6	13	33	56
	1.0		62	127	216
	2.0		950	1,400	1,480
AP30	0.5	30	17	440	1,460
	1.0		58	2,100	4,400
	2.0		440	8,700	16,000

* The Dow Chemical Co., *Let Separan® Settle Things In Your Solid-Liquid System*, Midland, Mich., 1968.
† Brookfield viscometer #4 spindle, 6 rpm, no salt added.

water through pipes.   In Fig. 16.7 the reduction of friction is shown to be as much as 80%.   The drag reduction is dependent on the polymer concentration and on linear velocity.   It does not appear to be very dependent on pipe diameter.

Manufacture

Acrylamide monomer is prepared by the hydrolysis of acrylonitrile and is sold commercially both as dry crystalline powders and as aqueous solutions.

The polymers are prepared by free-radical polymerization of the monomer in aqueous solution, usually between 10 to 50% concentration. The resulting gels are dried by drum drying, precipitation in methanol, or by azeotropic distillation. The latter method is used for water-in-oil suspensions and emulsions, or with bulk gels dispersed in hydrocarbons.

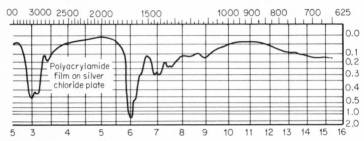

Figure 16.4 Infrared spectrum of polyacrylamide. (*From D. C. MacWilliams, "Acrylamide and Other Amides," in R. H. Yocum and E. B. Nyquist (eds.), Functional Monomers, Vol. 1, Marcel Dekker, Inc., New York, 1973, p. 1. Published with permission.*)

The impurities which are carried along with the polymers will be trace amounts of monomer; catalyst residues; inhibitors; by-products of the monomer, such as the ammonium ion from hydrolysis; Michael addition products of the monomer; drying aids; and surface-active compounds. Some patents read on the use of large amounts of sodium sulfate (about 20%) as an adjunct to polymerization. Inert compounds such as sodium sulfate have been found helpful in drying and in dissolving the polymer. While such

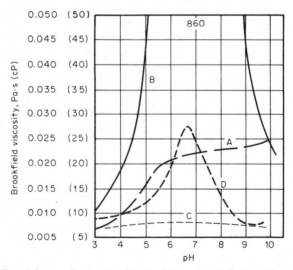

Figure 16.5 Effect of alum on the viscosity of 0.5% solutions of partially hydrolyzed polyacrylamide. A = 3% hydrolysis; B = 3% hydrolysis, 300 ppm sodium aluminate; C = 0.8% hydrolysis; D = 0.8% hydrolysis, 300 ppm sodium aluminate. *(From The Dow Chemical Co., unpublished data.)*

additives can be expected in a given polymer, they would probably not be deleterious, as the products are generally formulated for the appropriate end use. Detailed specifications are usually not published by the manufacturers, but the molecular weight, functionality, and residual monomer are described in a general way. Customer requirements are coordinated with more refined quality control.

Biological/Toxicological Properties

Polyacrylamide and hydrolyzed polyacrylamide exhibit low toxicity.   Most commercial polyacrylamides are nonirritating to the skin.   However, some of the hydrolyzed polyacrylamides may have residual alkalinity which could cause skin irritation on repeated

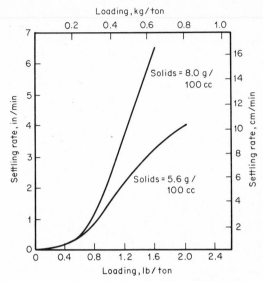

Figure 16.6   Effect of solids concentration on rate of settling of Plateau uranium ore at 1.5 pH by polyacrylamide, molecular weight of approximately 2×10^6.   *(From The Dow Chemical Co., unpublished data.)*

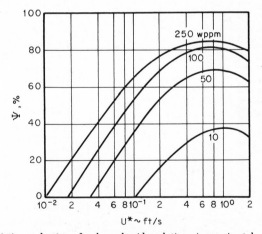

Figure 16.7   Friction reduction of polyacrylamide solutions (approximately 25% hydrolyzed) vs. velocity and concentration.   (*From N. F. Whitsitt, L. J. Harrington, and H. R. Crawford, "Effect of Wall Shear Stress on Drag Reduction of Viscoelastic Fluids," in C. S. Wells (ed.),* Viscous Drag Reduction, *Plenum Press, New York, 1969, p. 265. Published with permission.*)

or prolonged contact.   Good industrial handling practice dictates that such contact be avoided and that suitable eye protection be worn.   Adequate basis exists for establishing the safety of these products, from the public health aspect, for many applications where small amounts may possibly occur in the food or drink of animals or humans.   A number of commercial polymers have been cleared by the U.S. Environmental Protection Agency

TABLE 16.2
Constants for the Mark-Houwink Equation for Polyacrylamide and Derivatives[1]

Polymer	Method	Solvent and temperature	Molecular average	Weight range	K	a	Reference
Acrylamide (polymerized in solution)	(1) Sedimentation	Mixed, 25°C	$M_z \simeq M_w$	10^4 to 10^6	6.31×10^{-5}	0.80	2
	(2) Light scattering	Mixed, 30°C	M_w	4×10^4 to 3×10^6	3.73×10^{-4}	0.66	3
	(3) Kinetic	H_2O, 25°C	M_n	10^4 to 3×10^5	6.8×10^{-4}	0.66	4
	(4) Light scattering	10% NaCl, 20°C	M_n	1.3×10^5 to 5×10^6	2.5×10^{-4}	0.70	5
	(5) Light scattering	1% NaCl	M_w	—	3.02×10^{-3}	0.82	6

References: [1] D. C. MacWilliams, "Acrylamide and Other Amides," in R. H. Yocum and E. B. Nyquist (eds.), *Functional Monomers*, vol. 1, Marcel Dekker, Inc., New York, 1973, p. 1. Used by permission.; [2] W. Scholtan, *Makromo. Chem.*, 14, 169 (1954).; [3] M. V. Norris, "Acrylamide Polymers," in F. D. Snell (ed.), *Encyclopedia of Industrial Chemical Analysis*, vol. 4, Interscience, New York, 1967; p. 168; [4] A. M. North and A. M. Scallan, *Polymer*, 5, 447 (1964); [5] O. Z. Korothkina, V. V. Magdinets, M. N. Savitskaya, and V. E. Eskin, *Zh. Prikl. Khim.*, 38, 2533 (1965); [6] V. A. Myagchenkov, E. V. Kuznetsov, and V. F. Kurenkov, *Tr. Kazdh Khim.-Tekhnol. Inst.*, 1967, no. 36,336, from *Ref. Zh. Khim.*, 1968, Abstr. No. 9520.

or the Food and Drug Administration for such sensitive uses as potable water treatment, sugar juice clarification, and fruit and vegetable washing. The maximum dosage level for these applications is defined, and the residual amount of acrylamide is held to less than 0.05% in the polymer. Essential data on the polyacrylamides are supplied by the manufacturer.

Cationic copolymers have shown some evidence for fish kill. This is not toxicity in the usual sense but is believed to be due to the coating of the anionic gill surfaces with the cationic polmer, causing the fish to lose oxygen-absorption capabilities.

The toxicity of each copolymer of acrylamide must necessarily be considered individually. Most manufacturers alert their customers to any potential hazards.

Rheological Properties

The physical property of polyacrylamide of greatest interest is the viscosity of its aqueous solutions. Intrinsic viscosity is used to determine molecular weight, which is the major factor in flocculation effectiveness and thickening. Table 16.2 shows the Mark-Houwink constants determined by various methods. The corresponding intrinsic viscosity plots are given in Fig. 16.8. Lines 1 and 2 are in good agreement for the more accepted

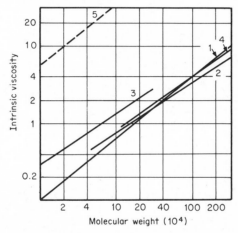

Figure 16.8 Correlations of intrinsic viscosity and molecular weight for solution-polymerized polyacrylamide. Line 1: $M_z \simeq M_w$; line 2: M_w; line 3: M_n; line 4: M_n; line 5: M_w. (*From D. C. MacWilliams, "Acrylamide and Other Amides," in R. H. Yocum and E. B. Nyquist (eds.), Functional Monomers, Vol. 1, New York, 1973, p. 1. Published with permission.*)

weight-average molecular weights. Line 5 was determined at the relatively low NaCl concentration of 1%, and is an indication that the solution may have been partially hydrolyzed.

The viscosities of unhydrolyzed polyacrylamide solutions are pH-insensitive, but as little as 3% hydrolysis can cause an increase in viscosity from 0.007 Pa·s (7 cP) at pH 3 to 0.021 Pa·s (21 cP) at pH 7. The incorporation of ionic sites by copolymerization causes marked increases in viscosity as a result of chain expansion by charge repulsion (Fuoss effect). These high viscosities are sensitive to dissolved salts, but adding as much as 4% NaCl will not completely restore the non-ionized viscosity.

The viscosities are usually studied between 0.1 and 2.0% concentration. At the lower concentrations, the viscosity bears an approximate log relationship to the concentration. The dilute solution viscosities are non-Newtonian for the polymers in the molecular weight range of one million, and attention needs to be given to shear corrections or strict standardization of measurement.

There does not appear to be good correlation between Ostwald- and Brookfield-type viscosities. A consistent method for measurement is essential. Correlations for molecular weight by intrinsic viscosity measurements usually require the use of NaNO$_3$ or measurement at pH 3 in order to suppress the Fuoss effect. Polymers with molecular weights in the million range are very shear-sensitive, degrading at significant rates when fast shakers and stirrers are used. One can observe the loss in viscosity on sequential passes through a capillary viscosimeter. Use of high-speed blenders for rapid dissolving will destroy a major part of the viscosity within minutes. This shear sensitivity puts stringent requirements on polymer solution feed systems. Centrifugal pumps should not be used.

Ionic crosslinking can thicken and gel polymer solutions. Generation of hydrous oxides of the polyvalent metals such as iron and aluminum in the presence of hydrolyzed polyacrylamide will cause gelation. Addition of cationic polymers will also cause gelation or coacervation.

Additives Extenders

Most polyacrylamides are used without the incorporation of additives. Some products, however, may contain sodium sulfate and polyglycols to assist in dispersing and dissolving. The anionic polymers will not tolerate cationic soaps, polymers, or dispersions. The reverse is also true for the cationic copolymers, which will precipitate and gel with anionic soaps, etc. The uncharged polyacrylamide is compatible with most water-soluble or colloidal systems. Mineral dispersions may be flocculated, however.

Handling

Most solid polyacrylamides are shipped in moisture-resistant bags lined with polyethylene or in steel drums with polyethylene liners. The liquid products are usually shipped in bulk via tank truck or tank car, but are also available in plastic-lined drums. Storage of unopened bags and drums presents no special problems. Bags should be stored in an indoor location. Drums of the liquid materials should be stored in a heated building to minimize viscosity changes due to fluctuations in outdoor temperature. Storage of liquid polymers at temperatures greater than 32°C (90°F) and less than 0°C (32°F) should be avoided. Caking in opened bags or drums may occur if precautions to exclude moisture are not taken.

Because of low corrosivity of the anionic polymers, mild steel can sometimes be used for the dry powders, but not for the cationic polymers. Noncorrosive alloys and plastic materials are preferred. Zinc, aluminum, or magnesium should never be used in contact with the polymers.

Possibly the greatest hazard in handling polyacrylamides is that of slipping. Polyacrylamide solution or even wet polyacrylamide powders have excellent lubricity and have caused numerous falls. All spilled solutions should be washed down, and powders should be swept up immediately. Adsorbents and antiskid compounds are helpful.

Applications

The polyacrylamides have several properties which lead to a multitude of uses in diverse industries. Table 16.3 lists the function and uses in the various industries.

TABLE 16.3
Applications of Polyacrylamide

Function	Application	Industry
Flocculation	Solids recovery	Mining
	Waste removal	Sewage
	Water clarification	General
	Retention aid	Paper
	Drainage aid	Paper
Rheology control	Waterflooding	Petroleum
	Viscous drag reduction	Petroleum
		Fire fighting
		Irrigation pumping
Adhesion	Dry strength	Paper
	Wallboard cementing	Construction

Application Procedures

The most troublesome step in getting the high-molecular-weight polymer to the point of use is the dispersing of the powder to avoid agglomeration. At least one manufacturer (Dow) has developed a continuous disperser that aspirates the powder with a water jet to make up the stock solution. Failure to get good dispersion results in formation of agglomerates which take an inordinately long time to dissolve. After dispersion is attained, only sufficient agitation is required to keep the particles suspended in gentle motion. Elevated temperatures speed dissolution, but temperatures over 66°C should be avoided. Excessive shear should be avoided to prevent polymer breakdown; for this reason, positive-displacement pumps are preferred over centrifugal pumps.

The highest-molecular-weight polymers are made up to 0.5% solutions, and the lower-molecular-weight polymers are made up to 1% or more. High concentrations can result in unmanageably high viscosities. Once the stock solution is prepared, it must be further diluted to 0.02% or less before being used in flocculation processes. In flocculation applications, improved efficiency is frequently obtained by multipoint addition of the stock solution to maximize dilution.[1]

The polyacrylamides are good culture media for microorganisms. For this reason, it may be necessary to use an appropriate slimicide, providing its use does not result in toxicity to humans.

Future Developments

Because of the difficulties in dissolving the solid polymers, it is to be expected that polymers of unique dispersing and dissolving qualities will be forthcoming. Recent appearance in the marketplace of polymer emulsions in hydrocarbons is a case in point. It is anticipated that polyacrylamide will find utility in medical and personal-care applications. It will be necessary to exercise great caution as regards toxicity, however. To facilitate such applications, improved techniques for residual-monomer analysis and removal should be forthcoming.

COMMERCIAL USES: Compounding and Formulating*

Pulp and Paper

Polyacrylamides are useful in the paper industry as processing aids and in compounding and formulating.

Filler-Retention Aid   Polyacrylamides are widely accepted in the paper industry as filler-retention aids. The polymers are normally added at the headbox or fan pump. Their use provides savings to the mill through improved filler and fiber retention. Other advantages include improved sheet formation, lower white-water solids, increased save-all efficiencies, less two-sidedness, and longer wire life. The polymers have utility on all types of paper-making machines.

Dry-Strength Improvement   Polyacrylamide is used as a dry-strength additive for paper and paperboard. When the polymer is incorporated in paper, such dry-strength properties as tensile burst, folding endurance, surface, pick, and ply bonding (board) are improved. In addition, usage may provide increased production, tearing strength, and porosity, as well as improved optical and printing properties.

Tape-Joint Cement

Polyacrylamide has utility as an adhesive when incorporated in drywall tape-joint cement formulations. Formulations containing polyacrylamide require less than 5 min soak time and are stable for extended periods of time. Formulations are insensitive to makeup water hardness, pH, or temperature. Joint cements based on polyacrylamide exhibit excellent binding and adhesion over a wide range of temperature and relative humidity.

---

* Based on data in *Separan Polymers*, The Dow Chemical Company, 1975. Reprinted with permission of the copyright owner.

COMMERCIAL USES: Processing Aids*

Processing Industries

Sugar Clarification Anionic polyacrylamides are utilized in the sugar industry for clarification of raw sugar juice. The addition of polymer to the filter feed promotes drier cakes and increases the clarity of the filtrate.

Phosphoric Acid Purification Anionic polyacrylamides aid in the separation of gypsum from wet-process phosphoric acid. When added to dilute acid, the polymer not only flocculates the gypsum solids but also reduces evaporator scaling and inhibits heavy-metal postprecipitation. Recirculation of flocculated dilute and concentrated acid sludges to the gypsum filter often improves clarity and filtration rates.

Aluminum Sulfate Manufacture Polyacrylamides can reduce settling times for slimes from hours to minutes in aluminum sulfate production plants. Overhead clarity is greatly improved, maintenance is reduced, and plant throughput and efficiency are increased.[2]

Coal Washeries Settling and filtration of coal fines and slimes in coal-washer operations are major economic and pollution control problems with many coal operations. Polyacrylamide polymers help provide effective solutions. These flocculents help reduce clay staining, plugged lines, pump maintenance, and stream pollution. The polymers also make possible cleaner overheads, closed circuit operations (where water is scarce), larger throughput, economical recovery of fines, and better filter rates.

Cement Manufacture Polyacrylamide has made possible greatly increased thickener capacity in wet-process cement manufacture. Other important results of treatment with polyacrylamide include increased filterability and improved homogeneity of kiln feed.

Borax Manufacture Clay impurities are efficiently settled from hot borax streams by the addition of polyacrylamide. Improved filtration efficiency and much greater thickener capacity are the benefits gained.

Inorganic-Fiber Slurries *Asbestos-Cement Products.* Polyacrylamide is in wide use in the asbestos-cement industry. The most important advantage gained is improved drainage, which in turn results in increased wet web strength, increased plant capacity, and savings in drier costs. Polymer use also results in increased retention of both the cement and asbestos.

Insulation Board. Polyacrylamide is added to both inorganic and wood fiber–based insulation boards, providing improved retention of both fiber and additives.

Electrorefining Processes In the electrolytic refining or electrowinning of copper and zinc, it is common practice to add chemicals to improve the quality of metal deposition on the cathode and increase current efficiency. Superior results and increased current efficiencies can be obtained by employing polyacrylamide alone or in combination with naturally occurring gums.

Lime Sulfur Filtration Treatment with polyacrylamide has improved filtration rates five- to sixfold in the production of lime sulfur for agricultural purposes.

Magnesia Production Magnesia production entails large washing systems with resultant dilute (5 to 10% solids) magnesia slurries. Treatment of thickener feed streams with anionic polyacrylamide improves settling and overhead clarity.

Industrial Reclamation

Flue-Dust Recovery Smelters frequently use some form of wet scrubbing to remove solids from their flue gases. Recovery of these solids by settling is important to prevent metal loss and to clarify the scrub water. Polyacrylamides may be used to eliminate such metal loss.

Petroleum Refinery Oil Reclamation When used in conjunction with alum, slightly anionic polyacrylamide is effective in flotation units for the reclamation of oil from combined refinery oily effluents. When added to effluent treated in a flotation-type clarifier, the polymer will provide improved floc stability, greatly reducing the oil content in the underflow and the water content of the oil. The increased stability of the floc blanket allows for more efficient removal by the scraper blades.

---

* Based on data in *Separan Flocculating Agents,* copyrighted by The Dow Chemical Company. Reprinted with permission of the copyright owner.

Water Treatment

Polyacrylamides are used as primary flocculents or coagulant aids for the clarification of water for industrial use. In the treatment of raw river water and the recovery of waste water where the solids level is high, the polymers are useful as primary flocculents.

Potable Water Treatment   Certain anionic polyacrylamides are approved by the U.S. Environmental Protection Agency for clarification of potable water. The polymers are used as the sole flocculent or with inorganic primary flocculents. Polymer treatment also allows filters to operate at higher hydraulic rates.

Raw Water Treatment   Polyacrylamides are effective for clarification of raw river water. The capacity of water clarifiers can be increased when the polymer is used as a secondary coagulant in conjunction with lime and ferric chloride.[3]

Filler Retention and Drainage Improvement   Slightly anionic and also cationic polyacrylamides are useful in the paper industry as filler-retention aids. Addition of these products at the headbox or fan pump can result in improved filler and fiber retention.[4]

Cationic polyacrylamide can function as a drainage aid. Improved drainage results in faster machine speeds and increased production. Decreased drying requirements and more uniform moisture profile may also be realized.[5]

Flotation-Type Save-Alls   Polyacrylamides are in commercial use for solids recovery in flotation systems, yielding improved effluent clarity and increased equipment capacity.

White-Liquor Clarification   Settling rates in white-liquor clarification have been doubled by treating with slightly anionic polyacrylamides.

Kraft mills report that the addition of polymer to causticized liquor improves clarity. Increases can also be obtained in lime mud filtration rates and wash rates by application of polymer to the filler feed line.

Sludge Treatment

Polyacrylamides, and especially cationic polyacrylamides, are used for conditioning municipal and industrial sludges for dewatering. This includes processes that employ vacuum filters, sand beds, centrifuges, and other mechanical devices. The polymer is also used for sludge thickening in processes using gravity settlers, air-flotation units, and centrifuges. Polymer treatment results in increased efficiency of the elutriation process.

Hot and Cold Lime Softening

Increased throughput, improved effluent clarity, and less postprecipitation in the clear wells may be obtained by treating well water for cooling-tower makeup in a cold lime softener. The use of polyacrylamides in hot processing water softeners results in reduced chemical costs and improved settling of calcium carbonates.

Pulp and Paper Manufacture

Polyacrylamide polymers have shown excellent utility in a number of operations in the pulp and paper industry. Their versatility is demonstrated by the results obtained in the following uses.

Green-Liquor Clarification   Polyacrylamide has shown excellent effectiveness in green-liquor clarification.

White-Water Clarification   Combined de-inking and white-water effluent wastes from bond mills and similar operations are efficiently clarified with polyacrylamides.

Mining and Ore Processing

Polyacrylamides having a range of hydrolysis have application in the mining and metallurgical industries.[6]

Underground Mine Water   Many underground mines have problems with high solids contents in mine water. Some domestic and foreign operators have applied polyacrylamides for slime flocculation in both underground and surface ponds. Treatment has resulted in improved clarity of water for mill use, reduction of stream pollution, and reduced pump maintenance.

Settling of Slimes   Fine grinding is usually required for the concentration of low-grade iron ores. Such grinding often results in excessive slime formation, thus causing water-recovery and stream-pollution problems. Polyacrylamide settles these slimes and produces a clear overflow in the clarification of saturated KCl-NaCl brines. Plant capacity is increased because of better separation of clay slimes.

Thickening Flotation Concentrates   Mill expansion frequently results in overloaded concentrate thickeners and filters.   Treatment with polyacrylamide makes it possible to thicken several times the design volume of flotation concentrates without reducing overflow capacity.   The possibility of application exists in all flotation plants, cement copper precipitators, and some gravity concentrators.

Treatment with polyacrylamide also aids in potash ore flotation circuits, permitting increased settling along with improved flotation.

Leaching   *Copper.*   Copper oxide ores are leached with sulfuric acid, and the resultant solution is treated with iron scrap to precipitate cement copper.   Polyacrylamide has proved effective for clarification steps following the leach and the precipitation stages.

Gold.   In the cyanide leaching of gold ores, treatment with polyacrylamide results in increased settling rate during the pulp-thickening and the filtration and decantation steps.   Usage increases the capacity of existing equipment.

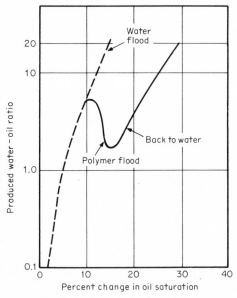

Figure 16.9   Effect of injecting a 0.05% solution of partially hydrolyzed polyacrylamide into a water flood in an oil field.   (*From D. C. MacWilliams, J. H. Rogers, and T. J. West, "Water-Soluble Polymers in Petroleum Recovery," in N. M. Bikales (ed.),* Water-Soluble Polymers, *Plenum Press, New York, 1973, p. 105.   Published with permission.*)

Uranium.   Uranium is almost universally recovered from its ores by leaching with an acid or carbonate solution.   The separation of the leached residues from the liquor is accomplished by multistage filtration or settling and constitutes a major process problem.   Other liquids-solids separation problems occur in thickening preleach ore slurries and in precipitating uranium salts and other materials from plant solutions.

Treating with polyacrylamide has proven highly effective in thickening preleach concentrates, yielding up to fourfold increases in filtration rates.

In acid-leach filtration, treatment with polyacrylamide gives greatly increased filter capacity, decreased cost of flocculation, and increased uranium recovery.

Polyacrylamides are used as settling aids in acid-leach plants that are designed with countercurrent decantation instead of filtration.

Following the leaching operation, most plants precipitate uranium from pregnant solutions with a base.   Treating with polyacrylamide results in increased thickening and filtration rates.

Zinc Calcines In zinc electrolytic plants, acid leaching of flotation concentrate calcines is carried out in a countercurrent decantation operation. Increased settling and more efficient washing are obtained by use of polyacrylamide.

Following this leaching stage, a number of electrolytic purification steps which involve slime removal are required. Polyacrylamide flocculents are used in this part of the process.

Tailings Disposal The problem of tailings disposal has been complicated by the severe water shortages that have developed in many areas, as well as by increasing antipollution legislation. As a result, most mill operators now find it necessary to settle mill tailings in thickeners or tailing ponds in order to reclaim water or prevent stream contamination. The use of polyacrylamide makes it possible to reclaim larger amounts of process water, alleviate water shortages, and reduce the quantity of flotation agent required. Polyacrylamide also aids in the transport of tailings slurry and the build-up of tailings dams.

Hydraulic Backfill Polyacrylamide flocculents have proved effective for improving drainage rates on flotation tailings used for mine backfill.

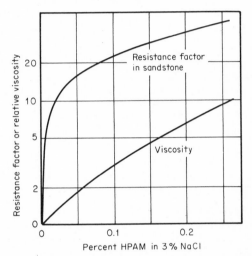

Figure 16.10 Comparison of resistance factor and viscosity of hydrolyzed polyacrylamide (HPAM) solutions. (*From D. C. MacWilliams, J. H. Rogers and T. J. West, "Water-Soluble Polymers in Petroleum Recovery," in N. M. Bikales (ed.),* Water-Soluble Polymers, *Plenum Press, New York, 1973, p 105. Published with permission.*)

Oil Production

Secondary Oil Recovery It is common practice in petroleum recovery to apply water pressure to the oil-bearing rock some distance from the producing well in order to force the oil out of the rock into the well. Water, being of low viscosity, will tend to finger ahead of the more viscous oil. Addition of as little as 0.05% polyacrylamide to the water flood reduces the by-passing of the oil and gives significantly higher oil/water ratios at the producing wellhead, as shown in Fig. 16.9. The effectiveness of hydrolyzed polyacrylamide as measured by its flooding resistance far exceeds its ability to increase the flood viscosity, as seen in Fig. 16.10. Greatly increased yields of oil are reported from use of polymer in waterflooding as compared with use of water alone.[7]

Drilling-Water Clarification In some areas it is possible to use water as a drilling fluid for the first several thousand feet of a well.

Solids are removed from the recycled drilling water in settling basins. The use of polyacrylamide improves the settling operation and provides longer drill-bit life and less plugging in the hold during bit changes.

Miscellaneous Applications

Friction Reduction   Solutions containing polyacrylamide are very slippery and can be used where water-based lubricants are required. Small amounts of the polymer, when added to an aqueous solution, can significantly reduce the friction generated in moving this solution through piping, thereby either increasing the throughput or reducing the power demand.

Viscosity Agent   Anionic polyacrylamide is an efficient viscosity improver or thickener of aqueous solutions.

These properties are useful for the suspension of solids in high-viscosity systems.

Film Forming   Polyacrylamide can be used to form clear, strong, water-soluble films.

Adhesives   Aqueous solutions containing polyacrylamide may be used as water-soluble adhesives.

Other   Other proven applications for polyacrylamide are found in the production of ammonia, barium, brines, clay, drugs, fine chemicals, inorganic salts, lithium, soda ash, and pigments.

INDUSTRIES USING POLYACRYLAMIDE

Polyacrylamide and its modifications are used in many commercial applications. Table 16.4 provides general guidelines for selecting the best product for a specific application. The multiplicity of processes and raw material or substrate variables will sometimes necessitate testing of several products to ensure selection of the proper one. Of course any use of polyacrylamides should be in strict compliance with the manufacturer's recommendations as regards toxicity, safe handling and/or Federal Drug Administration and U.S. Environmental Protection Agency requirements, proper method of solution makeup and application, spills and disposal, etc.

FORMULATIONS

The use of polyacrylamides in process improvements far exceeds its use in formulations or finished products. A possible exception is its use as a dry-strength resin in paper; but even there it is added in the conventional paper-making process.

Methylolated Polyacrylamide

The ability of polyacrylamide to gel or thicken has been most useful in formulations. The crosslinking of polyacrylamide to effect a gel, cured coating, or adhesive can be accomplished by methylolation or by ionic bridging.

To methylolate polyacrylamide, proceed as follows:

Four parts of potassium hydroxide are introduced into 20.3 parts of 37% formalin in a suitable reaction vessel. The resulting solution is added to 177.5 parts of 10% polyacrylamide having an average molecular weight of about 40,000. Two hundred parts of deionized water are added. The resulting mixture is held at 30°C for 30 min at pH 11 and then cooled. The pH is then adjusted to 8.0, and the product is dried on a double-drum drier at 168°C with a contact time of 15 s. The product is water-soluble and can be used as a curable adhesive, in sizing and creaseproofing textiles, and as backing for pile fabrics. The methylolated polymer is dried for convenience of storage and shipping. The methylolated polymer solution may be used directly, without the drying step.[8]

Gelled Explosives

Acrylamide copolymers with methylolated acrylamide and methylenebisacrylamide produce three-dimensional structures which can be used to gel explosive compositions. For example:

    9.5% nitroglycerin

  71.5% NH_4NO_3

   1.2% $NaNO_3$

    4.0% of a mixture of a dried copolymer of methylolacrylamide (10%), acrylamide (75%), and methylmethacrylate (15%) with oxalic acid (3% of the mixture) and the usual combustibles.[9]

TABLE 16.4
Applications for Polyacrylamide*

Application	Type polyacrylamide†	Molecular weight	Suggested dosage
Acid leach, high slimes	Sl. A	V. high	0.0025–0.1 kg/ton (0.005–0.2 lb/ton)
Acid leach, CCD, uranium	Non	High	0.005–0.1 kg/ton (0.01–0.20 lb/ton)
Acid mine. waters	Sl. A	V. high	1–5 ppm
Alum slimes washing	Sl. A	Med.	0.05–0.25 kg/ton (0.1–0.5 lb/ton)
Asbestos-cement fines retention	Sl. A	Med.	0.05–0.25 kg/ton (0.1–0.5 lb/ton)
Base-metal sulfides or oxides	Sl. A	V. high	0.0005–0.005 kg/ton (0.001–0.01 lb/ton)
Base-metal sulfide concentrate thickening and filtration	Non	High	0.0005–0.005 kg/ton (0.001–0.01 lb/ton)
Brine clarification, high slimes	Sl. A	V. high	0.05–0.4 kg/ton (0.1–0.8 lb/ton)
Carbonate leach, filtration, uranium	Non	High	0.05–0.25 kg/ton (0.1–0.5 lb/ton)
Carbonate leach, CCD, uranium	Non	High	0.05–0.10 kg/ton (0.1–0.2 lb/ton)
Cement fines thickening	A	Med. to V. high	0.0025–0.005 kg/ton (0.005–0.01 lb/ton)
Coal-washery effluents	A	Med. to V. high	0.0025–0.005 kg/ton (0.005–0.01 lb/ton)
Fine coal concentrate filtration	A	Med. to V. high	0.025–0.005 kg/ton (0.05–0.01 lb/ton)
Fruit and vegetable washing	Sl. A	Med. to V. high	Up to 5 ppm
Friction reduction	Non	V. high	Small amounts
	A	V. high	Small amounts
Gypsum filtration	A	V. high	0.05–0.25 kg/ton (0.1–0.5 lb/ton)
Hot and cold lime softening	A	Med to V. high	0.01–1 ppm
Hydroxide precipitates thickening and filtration	A	Med. to V. high	0.0025–0.005 kg/ton (0.005–0.01 lb/ton)
Iron ore tailings water recovery	A	Med. to V. high	0.0025–0.005 kg/ton (0.005–0.01 lb/ton)
Lime sulfur filtration	Sl. A	Med.	0.025–0.05 kg/ton (0.05–0.1 lb/ton)
Mill water supply (clarification)	Non	High	0.5–2.0 ppm
Mill tailings backfill drainage aid	Non	High	0.01–0.05 kg/ton (0.02–0.1 lb/ton)
Mine and waste water clarification	Non	High	1–6 ppm
Oil flotation	A	V. high	0.5–2.0 ppm
Paper manufacture Drainage aid	C		0.15–0.75 kg/ton (0.3–1.5 lb/ton)
Retention aid	C	Med.	0.1–0.375 kg/ton of paper (0.2–0.75 lb/ton of paper)
	C	Med.	0.25–0.5 kg/ton of solids (0.5–1.0 lb/ton of solids)
Dry-strength improvement	Sl. A	Med.	0.1–0.5% ton of furnish
Pigment thickening	Sl. A	Med.	0.005–0.05 kg/ton (0.01–0.1 lb/ton)
Phosphate plant tailings (water recovery)	A	Med. to V. High	0.0025–0.05 kg/ton (0.005–0.1 lb/ton)
Phosphoric acid clarification	A	V. high	1–5 ppm
Preleach slurry thickening	Non	High	0.005–0.01 kg/ton (0.01–0.02 lb/ton)
Saturated KCl–NaCl brine clarification	Non	High	0.05–0.75 kg/ton (0.1–1.5 lb/ton)

TABLE 16.4
Applications for Polyacrylamide* *(Continued)*

Application	Type polyacrylamide†	Molecular weight	Suggested dosage
Sugar beet washing	A	Med. to V. high	1–3 ppm
Sugar juice clarification	A	Med. to V. High	1–4 ppm
Sugar muds filtration	A	Med. to V. High	1–10 ppm
Secondary oil recovery, mobility control	A	Med. to V. High	0.05%
Sewage sludge dewatering	C	Med.	0.25–5.0 kg/ton (0.5–10.0 lb/ton)
Raw sewage flocculation	A	V. high	0.05–2.0 ppm
Trona plant operations	A	Med. to V. high	0.00025–0.005 kg/ton (0.0005–0.01 lb/ton)
Tailings water recovery	Non	High	0.0005–0.005 kg/ton (0.001–0.01 lb/ton)
Water pollution problems	A	Med. to V. high	0.0025–0.005 kg/ton (0.005–0.01 lb/ton)
Industrial water clarification and filtration:			
Primary coagulant	Sl. A	Med.	0.1–1.0 ppm
Secondary coagulant	Sl. A	Med.	0.01–0.1 ppm
Filtration aid	Sl. A	Med.	0.001–0.1 ppm
Potable water clarification Direct human consumption and industrial processing (brewings, soft drinks, foods):			
Primary coagulant	Sl. A	Med.	0.1–1.0 ppm
Secondary coagulant	Sl. A	Med.	0.01–0.1 ppm
Filtration aid	Sl. A	Med.	0.001–0.1 ppm
Raw-water classification:			
Primary coagulant	A	V. high	0.1–1.0 ppm
Secondary coagulant	A	V. hign	0.01–0.1 ppm
Filtration aid	A	V. high	1–3 ppm
Wet-cement thickening	A	V. high	0.0025–0.005 kg/ton (0.005–0.01 lb/ton)

* Based on data in *Separan Flocculating Agents* copyrighted by The Dow Chemical Co. Reprinted with permission of the copyright owner.
  † Type polyacrylamide:
    A = anionic
    Sl. A = slightly anionic
    Non = nonionic
    C = cationic

Gelled Hydrazine

Hydrazine may also be gelled with methylenebisacrylamide-crosslinked polyacrylamide.

Three 100-g samples of 95% hydrazine (Eastman Chemicals, anhydrous) were mixed with varying amounts of a finely ground copolymer of 70% acrylamide and 30% sodium acrylate crosslinked with 1000 ppm N,N'-methylenebisacrylamide and having a gel capacity of about 1100 in water. These samples contained 0.1, 0.5, and 1.0% by weight of the copolymer, respectively. Each of these samples gelled within a minute of the time the copolymer was added.

Gelled hydrazine has utility in rocket fuel applications.[10]

Explosives Gelled via Ionic Bonds

Gelation of polyacrylamide can also be obtained by generation of hydrous oxides of multivalent cations such as Al^{3+} and Fe^{3+}. The cationic particles form ionic bridges between anionic sites on the partially hydrolyzed polymer molecules. This is illustrated in the next example.

Place the dry ingredients (ammonium nitrate, urea, and 2.8% of sodium chloride) in a gel mix bowl, add the trinitrotoluene and polyacrylamide, and mix for 2 min.

Add sodium chloride solution containing 5.2% of sodium chloride in 14.2% of hot water, and mix for 5 min. Add aluminum sulfate dissolved in 0.8% of water, and mix for 2 min to provide the final gel product. The final ratio of ingredients is as follows:[11]

	Percent
NH_4NO_3, uncoated whole prill	18.50
NaCl	2.80
NaCl ⎤ as saturated solution	5.20
H_2O ⎦	14.20
$NaNO_3$	17.78
Trinitrotoluene	35.00
Polyacrylamide*	0.62
Aluminum sulfate	0.10
Water	0.80
Urea	5.00

   * Contains 1.5% carboxy groups and has a molecular weight of about 5 million.

LABORATORY TECHNIQUES

Analytical Methods

Two methods for the determination of trace amounts of polyacrylamide in water are available from The Dow Chemical Company.

The first method involves the use of Nessler's reagent and a Beckman model DU spectrophotometer or its equivalent.[12]   The water sample is concentrated by boiling and distillation; an aliquot is prepared with Nessler's reagent; the absorbance is measured in the spectrophotometer; and the amount of polyacrylamide, as organic nitrogen, is determined through reference to a standard curve.

The second method also measures the absorbance of a sample in a spectro-photometer.[13]   In this method the polymer is converted to N-bromoamide, which is then used to convert iodide to triiodide.   The triiodide can be measured using an ultra-violet (UV) spectrophotometer, and the resulting absorbance can be related to polymer concentration through reference to a standard curve.

Laboratory Evaluation*

Systems in which polyacrylamide polymers are used vary greatly in the nature of the solids present, particle size, concentration, and amenability to treatment.   The liquid can also contain chemicals that might influence treatment.   In addition, other variables such as pH, temperature, and ionic strength may greatly affect the degree of activity obtained with the polymer.

Because of these many variables, laboratory evaluation is extensively used to determine the degree of activity obtained with different polymers in each specific system.   While activity can often be predicted with reasonable accuracy, laboratory tests are needed to determine the most suitable polymer, the proper use level, and the effect that treatment will have on the process being examined.

Four major types of laboratory tests are employed: (1) settling, (2) filtration, (3) drainage improvement, and (4) retention improvement.   These are described in the sections that follow Laboratory Solution Preparation.

Laboratory Solution Preparation   Like many other high-molecular-weight, water-soluble powders, polyacrylamides must be well dispersed when added to water to prevent the formation of lumps.   If lumps are formed, they usually take a very long time to dissolve.   There are two methods commonly used to effectively disperse the powder in the preparation of laboratory solutions.

   1. The amount of polyacrylamide required for the solution is accurately weighed and placed in a beaker or bottle.   Then, 3 or 4 mL of methanol is added to the polymer to coat or prewet the surface of the particles without causing them to

* Based on data in *Separan Polymers*, copyrighted by The Dow Chemical Co., 1975.   Reprinted with permission of the copyright owner.

dissolve. The required amount of water is then quickly added to bring the solution to the desired volume. The solution is then shaken or stirred until solution is complete. As an example, to prepare 200 mL of a 0.5% solution, 1 g of polyacrylamide is placed in a bottle followed by 4 mL of methanol. Then 196 mL of premeasured water is quickly added to the bottle, followed by shaking or stirring until the polymer particles are completely dissolved.

2. The polymer and the water required for the solution are first measured. Before adding the polymer to the bottle or beaker, the person preparing the solution should breathe or blow into it so some of the moisture in the breath will condense on the inside surface. The polymer is then added, and the container is quickly shaken or rolled around so that the particles will adhere to the moistened walls. This results in the polymer particles being separated or dispersed inside the container. The water is quickly added, and the solution is shaken or stirred until the polymer is completely in solution.

Settling Tests Settling tests are used when the solids will settle rapidly and produce a clear supernatant liquid of sufficient volume to permit rapid decantation. These tests measure clarity, sedimentation rate, and/or the degree to which the solids can be concentrated. There are two different settling tests: the clarification test and the thickening test. The first of these is designed to determine the degree of clarification that can be achieved in a low-solids system. The second is designed to show the rate of settling and concentration achieved in a medium-to-high–solids system.

Clarification Test. This test should be used with systems containing less than 2% solids by weight. Primary flocculents such as alum or lime are sometimes used in such systems to prepare the solids for subsequent treatment with polyacrylamide flocculent. The effectiveness of the treatment is measured by the clarity, in Jackson turbidity units (JTU), obtained after settling.

Equipment

    1500-mL beaker
    Phipps and Bird variable-speed gang stirrer
    Primary flocculent chemicals
        Examples: alum, ferric chloride, lime
    Pipettes (graduated)
    0.02% solution of polymer
        Solution should be freshly made before running the evaluation, since the polymer activity decreases if the solution is held more than 1 day.
    Turbidimeter (Jackson Candle, Hach, or Hellige brands)

Procedure

1. Add 1 L of sample to a 1500-mL beaker
2. Agitate at 100 rpm.
3. Add primary flocculent in standard amounts (optional).
4. Agitate 2 min.
5. Add 1 ppm of polymer (5.0 mL/0.02% solution) drop by drop.
6. Agitate 1 min.
7. Reduce the agitator speed to 50 rpm.
8. Agitate 15 min.
9. Stop the agitator, and note the floc size.
10. Wait 15 mins, and then check the overhead for turbidity.
11. Successively change the flocculent loadings in 10% steps to determine the optimum economic conditions.

Thickening Test. A thickening test is recommended for slurries having a solids content between 2 and 40% by weight. Solids content can be determined by a pulp density, moisture content, refractive index, or other test. Two thickening procedures are available. Normally, procedure A in this section will provide adequate flocculent distribution. However, when the solids content is on the high side or the particle size is very small (and if the process being investigated permits), procedure B is recommended. Procedure B provides the increased dilution of flocculent that is needed with large surface areas to ensure adequate distribution of flocculent.

Equipment

Stoppered graduate (100 to 1000 mL)
Stopwatch
0.02% solution of polymer
  Solution should be freshly made before running the evaluation, since the polymer
  activity decreases if the solution is held more than 1 day.
Pipettes or syringes

Procedure A

1. Add the slurry to the graduate.
2. Invert the graduate several times to distribute the solids evenly.
3. Add one-third of the total polymer flocculent (0.02% solution).
4. Rotate or rock the graduate end over end three times in approximately 0.5 min.
5. Repeat step (3) and step (4) twice to complete the addition of polymer.
6. Set the graduate upright, and record the time it takes for the solids-liquid interface
   to settle to various levels.
7. Determine the clarity of the supernatant liquid by observation or use of an instrument such as a turbidimeter.
8. Evaluate a number of flocculent loadings to determine the optimum, minimum,
   and maximum dosages.

Procedure B

1. Add the concentrated solids portion of the slurry to the graduate.
2. Add the flocculent solution to the liquor-water portion of the slurry contained in
   a separate graduate.
3. Follow steps (3) through (8) of Procedure A.

Filtration Tests. These tests are for slurries containing more than 20% solids by
weight and where separation is by pressure filter, rotary-drum filter, vacuum filter, or
centrifuge. In addition, filtration tests are sometimes used in conjunction with drainage-
improvement tests in the evaluation of paper-stock furnishes. Two test procedures
are in common use. The first employs a 93 cm^2 (0.1 ft^2) filter leaf which is immersed
in 1 L or more of slurry. Vacuum is then applied for a predetermined period of time,
and the cake weight, cake moisture content, entrained values, and filtrate volume are
measured. The second procedure measures the time necessary to filter a predetermined
volume of slurry, using a vacuum technique and a Buchner funnel or other suitable
filter surface. It is this procedure that is used with the paper-stock furnishes.

Equipment

Two 1500-mL beakers or graduated cylinders
One laboratory filter leaf or Buchner funnel
One aspirator or laboratory vacuum pump
0.02% solution of polymer (0.1% solution of cationic polyacrylamide for checking
  paper-stock furnishes. Solution should be freshly made before running the evaluation, since the polymer activity decreases if the solution is held more than 1
  day).
Pipette

Procedure

1. Add a measured volume of slurry to beaker A.
2. Add one-third of the solution containing polymer to beaker B.
3. Pour beaker A rapidly into beaker B to permit turbulent mixing of flocculent and
   solids.
4. Agitate and pour B into A.
5. Repeat steps (2), (3), and (4) twice.
6. Insert the filter leaf into the beaker or pan containing the slurry. (If using Buchner
   funnel, pour contents of beaker A onto the funnel.)
7. Apply a standard vacuum for a predetermined time, maintaining a slight agitation
   with the filter leaf. The cake can then be washed and air-dried to simulate the
   filter operation in the process being examined.

8. Check the cake for total weight and moisture content, values retained, and check filtrate volume as desired.
9. Repeat at a number of different flocculent loadings to determine the optimum loading.

Procedure For Paper-Stock Furnishes

1. Take a convenient volume of headbox stock in a graduated cylinder. This may be 300, 400, or 500 mL, depending on consistency.
2. Determine the amount of solution of cationic polyacrylamide needed as described in the drainage-improvement test procedure (next).
3. Add the necessary amount of dilute polymer to an equal volume of furnish in a graduated cylinder.
4. Invert the cylinder several times to mix the contents thoroughly.
5. Start the aspirator or vacuum pump, and filter the contents of the cylinder through the Buchner funnel. Record the time necessary to collect the filtrate.
6. Repeat the filtration procedure with an equal volume of furnish that has not been treated with polymer.

Drainage-Improvement Test. This test is carried out on paper furnish to determine the utility of cationic polyacrylamide as a drainage aid and the optimum level required for satisfactory results. It is carried out with the use of a Canadian standard freeness tester or similar type apparatus. The test does not require diluting the furnish with water to the standard 0.3% consistency. However, if the consistency of the headbox or vat furnish is too high or the white water contains too many fines, the furnish should be diluted with water to a consistency range of 0.3 to 0.6%. The consistency must be known so that the kilogram per ton loading of the polymer can be determined. To find the consistency, filter a standard volume of furnish through a Buchner funnel, then dry and weigh the pad obtained. The dry weight will give the consistency.

Equipment

Buchner funnel
Canadian standard freeness tester or similiar device
Graduated cylinder, 1-L capacity.
Pipettes
0.1% solution of polymer. (Solution should be freshly made before running the evaluation, since the polymer activity decreases if the solution is held more than 1 day).
Sample bottles

Procedure

1. Add 1 L of headbox or vat stock to a graduated cylinder, and measure the freeness on any standard freeness tester. If it is impossible to get headbox or vat furnish, obtain concentrated stock from the stuffbox and dilute with white water from the machine.
2. Save a sample bottle of the overflow water for comparison with overflow water from a treated sample.
3. Dilute concentrated Separan® CP7 (Dow Chemical Co. cationic polyacrylamide) with clear water to 0.1%.
4. Determine the amount of dilute polymer to be added to 1 L of furnish to give an addition rate of 500 g/metric ton (1 lb/ton). If, for example, the consistency is 0.5%, then 2.5 mL of the 0.1% polymer in 1 L of furnish will be equivalent to an addition rate of 500 g/metric ton (1 lb/ton).
5. Add the necessary amount of dilute polymer to 1 L of furnish in a graduated cylinder, and mix well by inverting the cylinder several times.
6. Measure the freeness of the treated sample, and compare the results with those obtained for the untreated one. Compare the clarity of the overflow water from the two samples. An increase in freeness of 50 to 100 mL or more indicates that the polymer will perform well on the machine as a drainage aid. An improvement in the clarity of the overflow water indicates good performance as a retention aid. To check the performance of polymer in removing water from the furnish

under a vacuum, the filtration test outlined in the preceding section should be followed.

Retention-Improvement Test. This test is conducted to determine the utility of cationic polyacrylamide in increasing the amount of filler retained in paper. Handsheets are made on a laboratory paper machine such as a British or Noble and Wood handsheet mold, the sheets are ashed, and the percentage of ash determined. Improvement is calculated quantitatively by comparing the percentage of ash in paper made from furnish that has been treated with polymer with that in paper made from furnish that has not been treated with polymer.

The test does not require diluting the furnish to the standard 0.3% consistency. However, if the consistency is too high, the furnish should be diluted with water to a consistency range of 0.3 to 0.6%. The consistency must be known so that the kg/ton loading of the polymer can be determined. To find the consistency, filter a standard volume of furnish through a Buchner funnel, then dry and weigh the pad obtained. The dry weight will give the consistency.

Equipment

Buchner funnel
British or Noble and Wood handsheet mold
Two graduated cylinders
Two 1-L beakers
Pipettes
0.5% solution of polymer
0.05% solution of polymer (Solutions should be freshly made before running the evaluation, since the polymer activity decreases if the solution is held more than 1 day.)

Procedure

1. Take several liters of the actual machine furnish from the headbox. If the headbox is inaccessible, take the stock from the stuffbox and dilute with white water to the desired consistency. The furnish used for evaluation should contain all the constituents of the actual machine furnish.
2. Dilute concentrated cationic polymer retention aid to 0.05% or lower. This should be done each day the evaluation is run.
3. Fill the mold almost to the top with the water that feeds the mold.
4. Adjust the pH of this water to the pH of the furnish by adding the required amount of base or acid. Once the pH is determined, the same amount of acid or base can be added at each trial without rechecking the pH. pH is an important factor in filler retention.
5. To make blanks, add 500 mL of furnish, the standard volume, directly to the mold. Agitate the stock in the mold with the plunger. Always use the same number of strokes for each sheet made during the evaluation.
6. For the treated samples, pipette the desired amount of cationic polymer into the bottom of a 1-L beaker. Add 500 mL of furnish to the beaker; then using another beaker, transfer the treated furnish back and forth three times. This gives proper mixing of the polymer with the furnish.
7. Repeat steps (3), (4), and (5) for the treated furnish.
8. Run a series of polymer loadings from 120 to 500 g/metric ton (0.25 to 1.0 lb/ton). If the consistency of the furnish is 0.5%, then 500 mL will contain 2.5 g of solids, and 2.5 mL of the 0.05% polymer solution will be equivalent to a loading of 500 g/metric ton (1.0 lb/ton).
9. Make three handsheets at each polymer-loading level so errors can be averaged.
10. Ash the handsheets, and weigh the ashes.

Ash-retention improvement is calculated quantitatively by comparing the percentage of ash in paper made from furnish that has been treated with polymer with that in paper made from furnish that has not been treated with polymer. See first paragraph of this section.

One-pass filler-retention values are calculated both for untreated and polymer-treated handsheets by dividing the percent ash in the handsheets made on a paper-forming machine (Noble-Wood, British, etc.) by the percent ash found by Buchner funnel filtration where all the fillers are known to be retained.

Example: A Buchner funnel handsheet contains 20% ash. An untreated handsheet formed on sheet mold contains 5% ash. Therefore,

$$5/20 = 25\% \text{ one-pass retention.}$$

A handsheet formed using 0.3 lb/ton of polymer contains 11% ash. Thus,

$$11/20 = 55\% \text{ one-pass retention.}$$

These results will not necessarily be the same as those obtained on the paper machine. They will, however, indicate the relative increase in filler retention that can be expected on the machine.

Joint Cement (Taping)   Procedure for evaluation of joint cement (taping) based on Dow Resin 164 polyacrylamide:

Formulation A:

Material	Parts by weight, %
Calcium carbonate	84.17
Mica	5.81
Asbestos	5.81
Talc	0.96
Methocel* 228 cellulose ether	0.50
Dow Resin 165 polyacrylamide	3.00
	100.00

Formulation A-3:

Material	Parts by weight, %
Calcium carbonate	87.10
Mica	5.80
Asbestos	2.90
Talc	0.95
Methocel* 228 cellulose ether	0.50
Dow Resin 164 polyacrylamide	3.00
	100.00

* Trademark of The Dow Chemical Co.

Test Procedure

1. Prepare sample by adding the powder to the desired amount of water and mixing thoroughly.
2. Allow paste to stand for 30 min.
3. Apply paste to wallboard with a broad knife, and adjust thickness of paste to 635 μm (25 mils) with shim.
4. Place tape over compound (scarfed side down), and push tape into compound with several strokes of the broad knife. Make sure tape is firmly embedded into compound.
5. Place finished test panel into appropriate temperature and humidity chamber.
6. Remove test panel from testing chamber after desired time interval.
7. Use the following procedure to determine the initial adhesion of the test panel immediately upon its removal from the testing chamber.
    a. Mark three 10-cm (4-in) segments of tape on the test panel. This should result in three 10 cm × 5 cm (4 in × 2 in) rectangles.
    b. Make an x cut through one of the rectangles and the tape with a sharp wallboard knife.
    c. Pull up tape sharply from the points of the intersecting knife cuts.
    d. Observe and record the approximate percentage area on tape adhering to the compound. A bond of 100% is defined as the condition in which the tape delaminates within itself over the entire area over which it was originally placed.
8. Allow test panel to dry at room conditions until tape surface shows a moisture reading of less than 27 units according to a moisture meter (ring type). Evaluate remaining rectangular segments of test as in step (7a) to (7d).

PRODUCT/TRADENAME/GLOSSARY
Commercial Sources of Polymers of Acrylamide*

Manufacturer	Tradenames for polyacrylamide-containing products†
Allied Colloids Mfg. Co. Ltd., England	Magnafloc, Percol, Tiofloc, Gen Floc‡
American Cyanamid Co., U.S.A., Netherlands	Accostrength, Accurac, Aerofloc, Cyanamer, Magnifloc, Superfloc, Cyanagum (cross-linked), AM-9 Chemical Grout
BASF A.-G., Germany	Sedipur
Betz Laboratoratories, Inc., U.S.A.	Betz, BTI§, Polyfloc
The Calgon Corp., U.S.A.	Calgon, Calgon Coagulant Aid, Hydraid
Cassella Farbwerke Mainkur A.-G., Germany	Euflotal
Chemische Fabrik Stockhausen and Cie., Germany	Praestol
The Dow Chemical Company, U.S.A., Netherlands, Mexico	Separan, Purifloc, Gelgard, Norbak, Bonaril
Drew Chemical Corp., U.S.A.	Drewfloc
Farbenfabriken Bayer, Germany	Levamid
General Mills, Inc., U.S.A.	Gen Floc
Halliburton Co., U.S.A.	FR-14
Hercules, Inc., U.S.A.	Reten, Hercofloc
Hodag Chemicals Corp., U.S.A.	Hodag Flocs
Ingenieria Industrial Internacional, S.A. (Mexico)	Complex
Konan Chemical Co. Ltd., Japan	Konafloc
Kyoritsu Yuki, Japan	Hi-moloc
The Mogul Corp., U.S.A.	Clarocel
Nalco Chemical Co., U.S.A., Mexico	Nalco, Nalcolyte
Precision Scientific, U.S.A.	Speed Floc
Sankyo Chemical Industries Ltd., Japan	Accofloc, Sepalon
Standard Brands Inc., U.S.A.	Tychem, Tylec
Stein, Hall and Co., U.S.A.	Stein-Hall, Polyhall (MRL), Purgol

* D. C. MacWilliams, "Acrylamide and Other Amides," in R. H. Yocum and E. B. Nyquist (eds.), *Functional Monomers,* vol. 1, Marcel Dekker, Inc., New York, 1973, p. 1.
† Tradenames may also be used for products which do not contain acrylamides.
‡ Sold through General Mills Inc., in the United States.
§ From BTI Chemicals Ltd., England.

REFERENCES
1. D. J. Pye, U.S. Patent 3,087,890 (1963).
2. R. B. Booth, U.S. Patent 3,425,802 (1969).
3. R. M. Hedrich and D. T. Mowry, U.S. Patent 3,516,932 (1970).
4. E. K. Stilbert, M. W. Zembol and L. H. Silvernail, U.S. Patent 2,972,560 (1961).
5. W. A. Foster and J. E. Stout, U.S. Patent 3,323,979 (1967).
6. R. B. Booth and J. M. Dobson, U.S. Patent 3,418,237 (1968).
7. K. R. McKennon, U.S. Patent 3,039,529 (1962).
8. R. H. Hunt and D. E. Nagy, U.S. Patent 3,214,420 (1965).
9. M. Scalera and M. Bender, U.S. Patent 2,826,485 (1958).
10. B. L. Atkins and B. G. Harper, U.S. Patent 3,359,144 (1967).
11. J. A. Arbie, U.S. Patent 3,321,344 (1967).
12. W. B. Crummett and R. A. Hummel, "The Determination of Traces of Polyacrylamide in Water," *J. Am. Water Works Assoc.,* 55(2), 209–219 (February 1963).
13. The Dow Chemical Co., *Dow Method ML-AM 73–25.*

FURTHER USEFUL READING
MacWilliams, D. C., "Acrylamide and Other Amides," in R. H. Yocum and E. B. Nyquist (eds.), *Functional Monomers,* vol. 1, Marcel Dekker, Inc., New York, 1973, p. 1.

Thomas, W. M., "Acrylamide Polymers," in N. Bikales (ed.), *Encyclopedia of Polymer Science and Technology*, vol. 1, Interscience Publishers, a division of John Wiley & Sons, Inc., New York, 1964, p. 177.

Norris, M. V., "Acrylamide Polymers," in F. D. Snell (ed.), *Encyclopedia of Industrial Chemical Analysis*, vol. 4, Interscience Publishers, a division of John Wiley & Sons, Inc., New York, 1967, p. 168.

Montgomery, W. H., "Polyacrylamide," in R. L. Davidson and M. Sittig (eds.), *Water-Soluble Resins*, Reinhold Publishing Corp., New York, 1962, pp. 153–168.

Chapter **17**

Poly(Acrylic Acid) and Its Homologs

Harold L. Greenwald and Leo S. Luskin
Rohm and Haas Co.

General Information 17-1
   Physical and Chemical Nature 17-2
   Methods of Preparation 17-5
   Polymer Reactions 17-7
Commercial Applications 17-9
   Thickening 17-10
   Suspending and Dispersing 17-11
   Flocculation 17-13
   Binders 17-14
   Coatings 17-14
   Leather Paste 17-16
   Ion-Exchange Processes 17-16
   Pharmaceuticals 17-17
   Adhesives 17-17
   Miscellaneous 17-17
   FDA Clearances 17-18
   Tradename Glossary 17-18
   References 17-19

GENERAL INFORMATION

Poly(acrylic acid) and other water-soluble acrylic polymers possess a wide range of physical and chemical properties which make them candidates for many applications. Extreme variations in hydrophilicity, hardness, toughness, adhesion, and complex formation constitute the basis for such wide use. Included within the scope of this chapter are poly(acrylic acid) and poly(methacrylic acid), their salts, and copolymers of acrylic acid and methacrylic acid with hydrophobic or hydrophilic comonomers.

These copolymers, either upon neutralization or other solubilizing reaction, are the base materials for much interesting chemistry and applications. Outside the scope of this discussion are the many water-insoluble copolymers whose properties have been purposefully modified by the incorporation of water-soluble acrylic monomers. In these cases, necessary and useful property modifications are obtained, but water solubility is not achieved, although variations in water sensitivity are possible.

Physical and Chemical Nature

The following structural representation shows a typical all-acrylic polymer:

$$-CH_2-CR-CH_2-CR-$$
$$\qquad\quad |\qquad\qquad |$$
$$\qquad\quad COOR'\quad\ COOR'$$

where R is H or CH_3 and R' is H, alkyl, or metal cation. Such polymers exhibit various degrees of hygroscopicity, and many are hydroplastic in accordance with the proportion of hydrophilic groups present in the molecule.

Solid polymers of acrylic or methacrylic acids and their salts are hard, clear, brittle materials. Films of such materials often have Tukon hardness ratings of 40 to 50, as compared with 22 for poly(methyl methacrylate). Incorporation of comonomers can reduce the hardness rating to values even below 5. Comonomers can also influence other properties, such as the glass transition temperature (T_g). Table 17.1 lists values of T_g for various copolymers of styrene and methacrylic acid in the acid and sodium salt forms.

TABLE 17.1
Glass Transition Temperatures for Methacrylic Acid:
Styrene Copolymers

Methacrylic acid: styrene, mol %	Glass transition temperature, °C	
	Acid form	Sodium salt form
100:0	185	...
40:60	175	185
20:80	150	...
10:90	132	143
6:94	122	120
2:98	117	120

Eisenberg obtained higher values of T_g by dehydrating the polymeric acids to the corresponding anhydrides or by salt formation. Table 17.2 lists values of T_g for poly-(acrylic acid) and some derivatives, including salts at various degrees of neutralization. For the series of monovalent alkali salts, T_g falls with increasing charge size of the cations.

Nielsen compared the dynamic mechanical properties of the polymeric acids and their salts with those of polystyrene, a typical covalent polymer. The increased intramolecular forces in poly(acrylic acid) resulting from hydrogen bonding raise the shear

TABLE 17.2
Values of T_g for Poly(Acrylic Acid) and Derivatives

Polymer	T_g, °C
Poly(acrylic acid)	102
Poly(acrylic anhydride)	140
20% Sodium polyacrylate	108
40% Sodium polyacrylate	132
60% Sodium polyacrylate	162
80% Sodium polyacrylate	198
Sodium polyacrylate	251
1:1 Sodium:potassium polyacrylate	219
Potassium polyacrylate	194
Cesium polyacrylate	174

modulus by a factor of 2.3 times that of polystyrene. Neutralization to the zinc salt forms even stronger ionic intramolecular forces and gives a shear modulus 5.2 times that of polystyrene.

Salts of polyvalent cations show no glass transition or softening below the decomposition point (about 300°C). The rigidity of these salts also appears in a sixfold drop in the thermal coefficient of expansion and a quadrupling of the compressive strength as compared with polystyrene. Unlike the highly water-soluble salts of monovalent metals, the zinc salts are also completely unaffected in mechanical properties by soaking 90 days in water or after 180 h of immersion in boiling water.

Solutions   As Table 17.3 indicates, poly(acrylic acid) is a stronger acid than poly (methacrylic acid), and each polymer is a weaker acid than the corresponding monomer. The properties of copolymers containing acid components may be modified by the ionic nature of the polyelectrolyte chains. The incorporation of hydrophobic moieties in the copolymer molecules often further complicates their solution behavior.

TABLE 17.3
Strengths of Acrylics as Acids

	pK_A*	
Acid	Monomer	Polymer
Acrylic acid	4.26	4.75
Methacrylic acid	4.66	5.65

*By definition, the general form of pK reads like the pH definition ($pK = -\log k$, where k is the ionization constant). The subscript A indicates that it is the acid ionization constant. For weak acids (such as acrylic and methacrylic acids), pK equals pH at the half-neutralization point, since at this point the salt-over-acid term vanishes from the expression:

$$pK = pH + \log \left(\frac{\text{conc. acid ion}}{\text{conc. un-ionized acid}}\right)$$

Conventional polymers of acrylic acid ($CH_2{=}CHCOOH$) and methacrylic acid [$CH_2{=}C(CH_3)COOH$] are soluble in such polar solvents as methanol, ethanol, dioxane, ethylene glycol, β-methoxyethanol, and dimethylformamide, but are insoluble in solvents such as saturated hydrocarbons (hexane, decane), aromatic hydrocarbons (benzene, toluene), and other nonpolar solvents. Monovalent metal and ammonium salts of such polymers are generally soluble in water.

The syndiotactic forms of the polyacids and their salts show similar solubility to the conventional polymers. The isotactic polyacids, by contrast, are almost insoluble in water (solubility 0.1%), but partial neutralization with sodium hydroxide dissolves them.

Experimental data are satisfactorily explained if it is postulated that aqueous solutions of the polyacids contain the polymer in a very slightly ionized, tightly coiled shape, with a resultant low solution viscosity. This polymer coiling can be increased and the ionization reduced if the polymer is present in dilute hydrochloric acid solution. Increasing the degree of ionization of the polymer by neutralization with a base such as sodium hydroxide results in the gradual increase of viscosity.

Data in Table 17.4 show the variation of η_{sp}/C with pH for an aqueous solution of poly(methacrylic acid) with a degree of polymerization of 2510 (according to data of Katchalsky and Eisenberg). It can be seen that as sodium hydroxide is added, the value of the specific viscosity over concentration is increased, but that as the sodium hydroxide is added in excess, this value is reduced. These effects are accentuated at lower polymer concentrations.

In the presence of hydrochloric acid solution, ionization of the polyacid is apparently reduced from its initial low value, and the polymer formed presumably possesses a nonpermeable, coiled configuration. With the first increment of neutralizing base, some ionization occurs, providing sodium counterions and the polymeric ion. As more and more of the carboxyl groups on the polymer chain become ionized, mutual repulsion of the charges forces the polymer chain to uncoil and assume a more nearly rodlike configuration with a resultant increased resistance to flow and increased solution viscosity. After complete neutralization has been achieved, the addition of more sodium hydroxide results in a build-up of sodium counterions and hence a repression of the effective ionization of the polymer chain.   The effect is some chain coiling and a consequent reduction in solution viscosity.

Neutralization   A conformation change in the poly(methacrylic acid) molecule takes place in the neighborhood of 20 to 30% neutralization.   This change is associated with intrachain attractive forces and is influenced by the environment and the tactic configuration of the chains.   Evidence for the change is given by discontinuities in the neutralization curves; far-ultraviolet, infrared, and Raman spectra; plots of specific viscosity with concentration; binding with acridine orange dye; and spectral perturbations with inserted hydrophobic chromophores.[1]   Poly(acrylic acid) shows very little, if any, change in configuration.[2]   Poly(methacrylic acid), unlike poly(acrylic acid), has a negative heat and entropy of dilution ascribed to attractive forces and a shorter statistical chain element.[3]

The conformational change for poly(methacrylic acid) which occurs in water disappears with increasing alcohol content in mixtures of methanol and water.[4]   The break in the titration curve is absent at methanol contents of 40 to 50%.   The presence of methanol breaks up the compact structure of the chains, and the molecules then unfold regularly during neutralization.

Shifts in the Raman spectra caused by neutralization differ between syndiotactic and isotactic poly(methacrylic acid).   In the syndiotactic polymer, a sharp line appears at

TABLE 17.4
Viscosity Versus pH for Aqueous Solutions of Poly(Methacrylic Acid)

pH	η_{sp}/C	
	0.172 g/100 mL	0.011 g/100 mL
4.0	5	5
5.8	10	10
6.0	30	30
6.2	50	50
6.3	59	—
7.0	64	240
7.5	70	—
8.3	73	255
9.0	68	200

780 cm^{-1} at 10% neutralization, decreases strongly in intensity at 20%, broadens and shifts toward a higher frequency up to 50%, and finally sharpens at 70% neutralization to give the line at 832 cm^{-1} characteristic of the neutral salt.   The change of configuration of the syndiotactic molecule therefore occurs gradually over a wide range of increasing ionization.   The isotactic polyacid has two lines already evident at 30% neutralization, the minimum degree of salt formation to give solubility in water and a measurable spectrum.   The lines at 733 cm^{-1} and 765 cm^{-1} show changes in relative intensity but remain unchanged in their position with increasing degree of neutralization.

The solution properties of these polyacids are therefore affected by the following factors: strength of the acid, degree of ionization, nature and amount of counterion, concentration, presence of cosolvents, temperature, and tactic configuration.

Acid Strength   A comparison of the acid strength of poly(acrylic acid) with various copolymers of acrylic acid and acrylamide has shown that the acid strength of the block copolymer is the same as that of a linear homopolymer of acrylic acid.   In contrast, a random copolymer exhibits greater acid strength and graft and branched copolymers show weaker strengths than the linear homopolymer.

A graphic demonstration of the postulated coiling and uncoiling of polyacids has been developed by Katchalsky.   He obtained mechanical work from strips of poly(methacrylic acid) by changing the ionic atmosphere in the solution in which the polymer strip is suspended while attached at each end to a device for measuring mechanical work.   Ion atmosphere changes result in alternate contraction and stretching of the polymer strip.   Similar tieno or pH muscle effects have been reported for specific physical mixtures of poly(acrylic acid) and polyvinyl alcohol[5] and similar fibers with a central platinum counterelectrode.[6] Interactions of the polymeric acids with alkylene oxide polymers have also been reported.[7]

The nature of the counterion has been investigated extensively. In addition to the repression of coiling by mass effects of excess counterions, certain ions can form complexes with polyions. For example, copper ions form an insoluble complex with the polyacrylate ion. The effect of polyvalent counterions thus may be one of mass-law repression or of insoluble-complex formation. Transport of an electric current by a variety of counterions has shown that with each kind of counterion, certain proportions are bound to the polyion, while the remainder are present as freely moving single-ion species.

Molecular Weight Clearly, the determination of such properties as electric anisotrophy, streaming birefringence, and molecular weight are greatly affected by the ambient ionic atmosphere in which such determinations are made. Molecular weights have been determined by light-scattering techniques in hydrochloric acid solution, and reasonable agreement of values has been obtained viscometrically. The molecular weight of poly(sodium acrylate) has been determined by osmometric methods using an external medium of sodium chloride solution to compensate for the Donnan term in the dialysis.

What appears to be the most satisfactory method for determining the molecular weight, however, consists of converting the polyacid to the corresponding methyl ester by reaction with diazomethane, and then determining the molecular weight of the formed ester by more reliable methods. Katchalsky and Eisenberg have demonstrated that no change in molecular weight takes place in this esterification step. In the course of this and of subsequent Japanese work, it was further brought out that no molecular weight degradation results from hydrolysis of the polymeric methyl ester to the acid with either aqueous sodium hydroxide or p-toluene sulfonic acid in aqueous acetic acid.

Other workers have derived expressions for the molecular weight of polyacids from viscometric analysis in various solvent media designed to eliminate the effect of ionization of the polyelectrolyte. In Table 17.5 are listed the polymer, solvent, and k and a values

TABLE 17.5
Data for Molecular Weight Determinations

Polymer	Solvent	k	a	Temp., °C
Poly(acrylic acid)	Dioxane	85×10^{-5}	0.5	30
Poly(methacrylic acid)	$2M$ NaNO$_3$ solution	49.4×10^{-5}	0.65	25
Poly(methacrylic acid)	$0.002M$ HCl solution	66×10^{-4}	0.5	30
Poly(methacrylic acid)	Methanol	2.42×10^{-3}	0.51	26

to be used in the calculation of the molecular weight from the formula: intrinsic viscosity $= kM^a$, in which M is the molecular weight, and intrinsic viscosity is in units of deciliters per gram.

Methods of Preparation

Polymerization Poly(acrylic acid) and poly(methacrylic acid) may be prepared by direct polymerization of the corresponding monomer. A conventional polymerization recipe involves water, monomer, potassium persulfate as the polymerization initiator, and such an activator as sodium thiosulfate. The polymerization may be carried out at a temperature of 50 to 100°C. To regulate the polymer chain length, a number of chain transfer systems have been used. These include secondary alcohols, mercaptosuccinic acid, and combinations of sodium hypophosphite and copper acetate. The solution viscosity of linear polymers is a direct function of the molecular weight of the polymer. Similar polymerizations in aqueous solutions have been carried out by irradiation of the aqueous solutions with ultraviolet light; very high molecular weights are produced by this technique.

Both acrylic acid and methacrylic acid have been polymerized in a solvent such as benzene, heating the solution in the presence of an initiator, such as benzoyl peroxide, and recovering the precipitated polyacid by filtration.

The acids have been polymerized by exposing the monomers to ^{60}Co radiation. The polymer is presumably formed by a free-radical mechanism, and the polymer continues to form for long time intervals following the cessation of irradiation. This subsequent polymerization is probably initiated by the trapped, free radicals. Polymer has been formed in this way from frozen methacrylic acid, and nuclear magnetic resonance studies suggest the presence of an atactic configuration.

Salts of acrylic and methacrylic acids may be polymerized by treatment with initiators in aqueous media. Pinner has reported data which show that the methacrylate ion polymerizes at a rate which is lower than that at which the un-ionized acid polymerizes, but there are numerous techniques which employ the polymerization of the monomeric salt as the means of obtaining the polymer. The polymerization of a concentrated paste of an acrylic salt in the presence of a carbonate has been described for the preparation of a readily pulverizable polymeric cake. Acrylic salts have been polymerized successfully by gamma-ray irradiation.

Because of the nonvolatility of the acrylic salts, certain special techniques have been developed for their polymerization. Simultaneous polymerization and spray drying produce relatively high-molecular-weight polymers from concentrated solutions of monomer. A similar technique has been employed for the simultaneous drum drying and polymerization of acrylic salts. The use of such materials as polyethylene glycol monolaurate has been described for making a drum-dried polymer more readily pulverizable.

Hydrolysis   In addition to the polymerization of the appropriate monomer, the hydrolysis of a suitable polymer constitutes a major method of preparation of both the polymeric acid and salt. To produce poly(sodium acrylate), saponification may be carried out by heating a suspension or an emulsion of a polymeric acrylic ester, such as methyl acrylate, with aqueous sodium hydroxide at 100°C for a few hours. The formed alkanol can be removed by stripping operations.

The hydrolysis of conventional and syndiotactic poly(isopropyl acrylate) in aqueous acetic acid medium with p-toluenesulfonic acid results in the formation of a water-soluble poly(acrylic acid). On the other hand, a similar hydrolysis of isotactic polymer produces a poly(acrylic acid) which is insoluble in both water and aqueous acetic acid. Hydrolysis of the isotactic polymer has been accomplished without destroying polymer tacticity and gives isotactic poly(acrylic acid).

Crystalline poly(acrylic acid), identified by x-ray and electron microscopy, has been obtained by the hydrolysis of $tert$-butyl acrylate polymer of definite tactic form.[8]

The relative ease with which acrylate ester polymers hydrolyze does not hold for the corresponding methacrylate polymers.[9] The quantitative studies corroborate earlier qualitative findings that conventional methods give complete hydrolysis of poly(methyl acrylate), but no appreciable hydrolysis of poly(methyl methacrylate) is obtained under similar conditions. Modified hydrolyses, however, can produce poly(sodium methacrylate) from polymethacrylates. Suitable techniques involve the use of molten caustic or a heterogeneous system of sodium hydroxide and aqueous isopropanol. In addition to the different hydrolytic rates for polyacrylates and polymethacrylates, marked differences have been observed for conventional, syndiotactic, and isotactic poly(methyl methacrylate). The isotactic polymer is hydrolyzed much more rapidly than the others.[10]

Hydrolyses have also been carried out on copolymers with hydrolyzable component monomers or with one or more of the monomers not susceptible to hydrolysis. Hydrolysis of a homopolymer has been conducted with mixtures of bases (sodium hydroxide and calcium hydroxide) to produce double salts.

The preparation of acidic polymers by hydrolysis can be conducted by saponification, as described above, followed by removal of the counterions. This removal has been carried out by means of dialysis, although complete removal of cations by this procedure is difficult. An alternative method is the use of a mixed-bed ion-exchange resin which removes the counterions and any contaminant monomeric anions while leaving the polymeric acid in solution. The treatment can be carried out by column techniques or by agitation of the solid resin with the polymer solution with subsequent removal of the resin by filtration.

Acid hydrolysis may be carried out directly by dissolving the polymeric ester in a solution of acetic acid and water in a ratio of about 80 to 20. After solution has been completed, *p*-toluenesulfonic acid is added and the batch heated to 110 to 120°C while passing a stream of nitrogen through the solution and removing formed alkyl acetate. The polymer may be isolated by precipitation.

Using increased amounts of *p*-toluenesulfonic acid and increased hydrolysis time can achieve the hydrolysis of poly(methyl methacrylate), especially isotactic poly(methyl methacrylate), which can be hydrolyzed completely. Hydrolysis of the polymeric esters has also been accomplished by dissolving the polymer in concentrated sulfuric acid, warming, and isolating the product by pouring the reaction mixture into water.

Copolymerization The preparation of copolymers can be carried out to produce soluble or solubilizable types. Soluble copolymers can be produced by copolymerization with hydrophilic monomers. For example, acrylic acid and/or methacrylic acid have been copolymerized with maleic anhydride, itaconic acid, acrylamide, sodium salts of acrylic or methacrylic acids, and methyl vinyl ether to produce water-soluble polymers. Water-soluble polymers have been produced by partial hydrolysis of polymeric esters, amides, and nitriles, or by cohydrolysis with mixed bases. Depending upon the monomers employed, it is also possible to obtain soluble copolymers containing small amounts of hydrophobic monomers.

Solubilizable copolymers have been prepared by solution, emulsion, and suspension techniques from mixtures of acrylic or methacrylic acid with larger amounts of hydrophobic monomers. For instance, copolymers prepared from 65% methyl methacrylate and 35% methacrylic acid are water-insoluble, but aqueous suspensions or emulsions of such copolymers are readily solubilized by the addition of ammonium hydroxide or sodium hydroxide.

The preparation of polyampholytes from acrylic monomers has been reported. Such polyampholytes have been produced from dimethylaminoethyl methacrylate and methacrylic acid in various ratios.[11] Another class of polyampholytes has been prepared from acrylic acid and vinylpyridine. These polymers can be quaternized with alkyl bromides, and when mixtures of bromides such as ethyl and dodecyl are employed, interesting polysoaps are formed. They are effective surfactants by virtue of possessing hydrophilic and hydrophobic moieties in the same molecule.

Polymer Reactions

Neutralization Neutralization of polymeric acids has been accomplished with a variety of bases. Many inorganic monovalent and polyvalent bases have been used. In the case of polymeric salts with both monovalent and polyvalent counterions, several effects are produced. When counterions are present in excess of stoichiometry, both repression of effective ionization, resulting in viscosity decrease, and salting out of the polymer are encountered.

In both instances, isolation of the polymer and separation from the excess counterions permit resolubilization of the formed salt in fresh water. In other cases, as with copper ions, complex formation occurs; the precipitated polymer salt then cannot be redissolved. This complex formation is so rapid that fibers of the copper salt of poly(acrylic acid) can be formed by introducing a thin filament of an aqueous solution of the sodium salt of the polymer into a solution of copper chloride. As suggested earlier, the binding of counterions takes place in different degrees with the different ions utilized for the neutralization.

An extension of this idea is the formation of the zinc-ammonium or zirconium-ammonium salt of an acrylic polyacid. Such a salt is water-soluble. Upon drying down a film formed from a aqueous solution of such a polymer, the ammonia is lost with the water during evaporation and the residual polyvalent metal salt of the polyacid is extremely resistant to attack by water. Werner complexes of chromium with poly(acrylic acid) may also be used to form insoluble residues.

The formation of double salts by neutralization with mixed bases has been mentioned previously. Similarly, *in situ* formation of insoluble salts may be used by first introducing a soluble form of the polymer for maximum penetration, followed by insolubilization *in situ* to achieve the required impenetrability. A case in point is the addition of calcium chloride solution to a previously impregnated material in which the impregnant is poly(acrylic acid).

Organic bases may be used for the neutralization step, especially when modifications in film properties or polymer solubility are needed. Representative neutralizing organic bases include triethylamine, tri(hydroxyethyl)amine, tetramethylammonium hydroxide, morpholine, and choline. Polyfunctional amines are also suitable.

In this latter category, polymeric bases have been used to produce insoluble films. Representative polymers are those derived from acrylic acid and from dimethyl-aminoethyl methacrylate.

Reversal of neutralization is often desired. For this purpose, dialysis has been employed, but as suggested earlier, the complete removal of cations is difficult by this means. Alternatively, ion-exchange resins are useful, especially in small-scale operations. With an appropriate cation-exchange resin, column techniques can provide poly(acrylic acid) from its soluble salts.

A wide variation has been noted in the base used for the neutralization operation, and as already indicated, the same latitude exists in the choice of the acidic polymer. By varying the hydrophobic component of the polymer molecule, water-soluble polymers can be made which form films or other final shapes having a wide range of physical properties involving resolubility of the film, hardness, toughness, gloss, clarity, and other features.

Esterification   Esterification of the polymeric acids has been carried out by the preparation of the silver salt and subsequent reaction with the appropriate alkyl halide. A more convenient laboratory method involves the use of diazomethane or diazoethane. Suspension of the polyacid in a benzene solution of the diazo compound results in the evolution of nitrogen and the gradual solution of the polymeric ester in the solvent from which it can be removed by conventional techniques.

Esterification is also conducted by treatment of the polyacid with an alcohol at elevated temperatures. This has been used with glycols and glycerin to induce crosslinking and insolubilization of films of poly(acrylic acid), especially in the formation of a durable finish for textile-sizing application.

Still another important ester-forming reaction employs an alkylene oxide. The polymeric acids and their salts are treated with alkylene oxides to produce polymeric esters by the following reaction:

$$-CH_2-CH(COOH)-CH_2-CH(COOH)- + CH_2\!\!-\!\!CH_2 \rightarrow$$
$$\diagdown\!O\!\diagup$$

$$-CH_2-CH(COOCH_2CH_2OH)-CH_2-CH(COOCH_2CH_2OH)-$$

With such catalysts as piperidine or sodium hydroxide, the reaction proceeds smoothly. The β-hydroxyethyl ester has sites for the further reaction of alkylene oxide groups, resulting in the formation of grafted polyoxyethylene side chains on a backbone of poly(acrylic acid). The likelihood of crosslinking and insolubilization by interaction be-tween terminal hydroxyl groups and acidic portions of other polymer chains offers an opportunity for controlled film insolubilization.

Conditions have been reported for control of the preparation of the alkylene oxide adducts to avoid premature insolubilization by interchain reactions. This involves the continuous adjustment of the pH of the reaction medium and a careful regulation of the ratio of the alkylene oxide to the poly(acrylic acid). The reaction should be run at a temperature of about 100°C.

Complex Formation   An association complex of poly(acrylic acid) with polyethers, such as polyoxyethylene glycols, has been described. This complex forms readily at ordinary temperatures and appears to involve formation of strong hydrogen bonds. The resultant products are tough, hard, and water-insoluble. Possibly similar complexes of poly(acrylic acid) with poly-N-vinylpyrrolidone have also been made.

Another complex has been reported between poly(acrylic acid) and clays of various kinds. There is no doubt that stable aggregates can be formed by mixing such a polyacid with soil. The mechanism of such an aggregate stabilization may involve the formation of stable complexes at various sites of soil particles at different points on an extended polymer chain, thus effectively holding together groups of the individual particles.

A reaction complex has also been found to exist between nylon and poly(acrylic acid). When the latter is used as a sizing agent for nylon, and the bulk of the nylon is dissolved away from the sized fiber, a network of nylon to which the poly(acrylic acid) is spot-welded remains.

Dehydration and Degradation   Heating poly(acrylic acid) at temperatures of 150°C or higher drives off water and other volatiles. The principal product, obtained by a first-order process, is poly(acrylic anhydride) obtained by the intramolecular formation of a six-membered ring of the glutaric anhydride type:

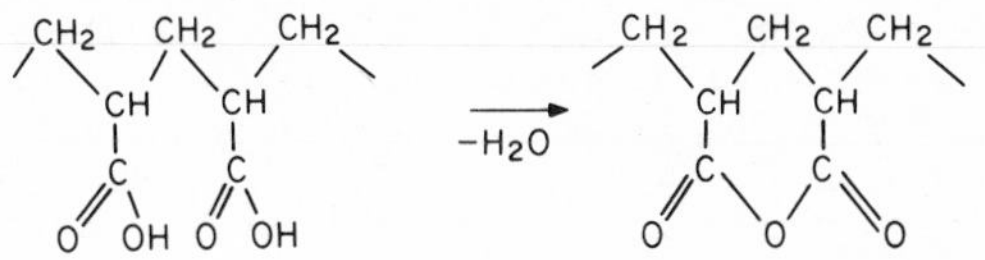

Intermolecular condensation to a network of isobutyric anhydride structures proceeds much more slowly:

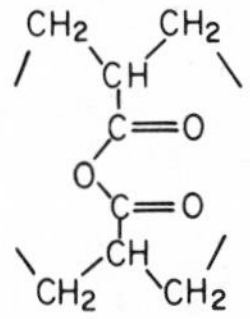

Infrared spectra and comparison with model compounds confirm the presence of both types of anhydrides in the dehydration product. The glutaric anhydride can react further at about 300°C with loss of carbon dioxide to form cyclic ketone structures and at 350°C or higher to form degraded structures in which unsaturation is increasingly apparent.

At temperatures up to 200°C, poly(methacrylic acid) also loses water and carbon dioxide to produce an insoluble polyanhydride, largely consisting of glutaric anhydride groups but insolubilized by intermolecular crosslinking. The isotactic polyacid loses water at a rate about four times faster than the conventional or syndiotactic acids. Since the measured energies of activation for the dehydration are about the same for the various tactic acids, the neighboring carboxyl groups of the isotactic polymer must be situated much more favorably with respect to anhydride formation.[12]

Other Reactions   Heating a mixture of poly(acrylic acid) with a polyamine such as tetraethylenepentamine results in the formation of substituted amides. The formation of a hydrazide, hydrazone, and azide from poly(acrylic acid) has also been observed.

What may be a fairly complicated series of reactions involves the polymerization of a basic monomer such as dimethylaminoethyl methacrylate in the presence of preformed poly(acrylic acid). This reaction could involve polymer grafting or a neutralization of the polyacid by the monomeric base with polymerization of the pendant monomer function. The possibility of homopolymer formation with subsequent mutual neutralization by the acidic and basic polymers should not be excluded.

COMMERCIAL APPLICATIONS

The breadth of the applications is reminiscent of the fable of the man who blew on his hands to warm them and blew on his porridge to cool it, for these polymeric acids are used as adhesives and release coatings, as flocculants and dispersants, as thickeners and fluidizers, as reaction inhibitors and promoters, as permanent coatings and as removable coatings, etc. The multiplicity of applications of poly(acrylic acid) and other water-soluble acrylic polymers is the direct result of the varied physical properties and many polymer reactions noted previously. Many of the applications depend on the ability of these polymers to form complexes, and the bonding action as such to substrates opens up additional fields of use.

The clarity, toughness, and durability of the polymers lead to still other end uses. While many of the applications described here are not clearly understood, often more than one of the properties or reactions of the polymers are involved in the successful applications.

Thickening

Latexes   The thickening of polymer latex systems is required for a wide variety of applications.   The mechanism of this thickening action is not clearly understood, although several mechanisms might be expected as a result of the wide range of water-soluble polymers and polymer latexes involved.   Brown and Garret have suggested that this complex phenomenon involved the ability of a water-soluble polymer to thicken the aqueous phase of the polymer latex, to interact with the polymer to produce a pseudoaggregate of polymer particles, and to interact with emulsifiers and other ingredients of the latex.[13]

Water-soluble acrylic polymers are effective latex thickeners.   With any given water-soluble polymer, thickening of the latex is increased with increased molecular weight of the thickener, which indicates some support for the requirement of thickening the aqueous phase.   However, very-low-molecular-weight polymers, which have a relatively small thickening effect in water, may be extremely effective in thickening certain latexes.   One common situation seems to be that the thickening ability increases and the stabilizing ability passes through a maximum with increasing concentration of the polymer.

This variability of thickening action is typical and may be attributed to the various latex compositions, particularly the nature of the emulsifier.   The phenomenon is made even more complex with compounded latex formulations in which the thickener may also react with clay, pigments, and fillers introduced into the system.

In handling certain rubber latexes, especially natural rubber, a convenient practice is to cream the latex in order to concentrate the rubber.   The addition of such a creaming agent as a low-molecule-weight sodium salt of poly(acrylic acid) to a 40% natural rubber latex will result in the separation into a clear serum layer and a more concentrated (approximately 60%) latex layer.

In textile applications, the thickening of latexes is required in the backing of rugs for the preparation of resilient, rubberized, nonslip floor coverings.   It is also used in upholstery fabric backing.   Thickened latexes are also required for dipping operations in which fabric forms may be dipped in thickened rubber latex and the covered form cured to prepare such articles as gloves.

The latex paint field is a rapidly expanding one.   The adjustment of such paints to a desired viscosity level is required for brushing and roller and spray applications.   Viscosity adjustment can be made with water-soluble acrylics.   In addition, the final formulation must be stable.   Both formulation stability and reduction of viscosity drift of the thickened paint are achieved with selected acrylic thickeners.

Nonflammable Hydraulic Fluids   There is a real interest in the development of a nonflammable fluid for use in power transmission and actuating devices.   Such hydraulic fluids find use in such industrial establishments as foundries.   In these cases, since the fluids are under considerable pressure, a leak or rupture in the lines results in the formation of a fine mist of a potentially flammable material, if the base for the fluid is a petroleum lubricant or other flammable stock.

One method of reducing the danger of such fires has been the use of blends of water with ethylene glycol.   In order to improve the lubricity of such systems and to adjust their viscosity to the desired level, water-soluble acrylic polymers have been employed.   Polymer modification or the addition of antitack agents is required to prevent the formation of hard films at valve fittings, such as would be formed by poly(sodium acrylate).

Viscous Flooding   In the secondary recovery of underground petroleum deposits, it is customary to employ flooding techniques.   In this operation, an injection well is drilled in a central location in which it is surrounded by producing wells.   Water is then injected underground, and this waterflood forces the petroleum deposits out toward the producing wells with consequent increase in oil recovery.

Success has been achieved in the addition of surface-active agents to such flooding waters.   In certain cases, however, where the petroleum crudes are more viscous, the flooding water is incapable of forcing the heavier material ahead of it, and the deposit is merely channeled.   Some evidence is suggested for the use of water-soluble polymers for the thickening of flooding waters to overcome this difficulty.   Such viscous-flooding studies have included water-soluble acrylic polymers.

The requirements for these polymers are that efficient thickening be achieved at very low polymer concentrations and that this thickening be affected relatively little by dissolved salts and brackish water. Modifications of poly(acrylic acid) are necessary to overcome its insolubilization by polyvalent metal ions in some waters.

Miscellaneous   The extent of thickening applications may be suggested by the following representative uses. In textile operations thickening of printing pastes has been carried out with the water-soluble acrylic polymers. Ceramic glazes have been thickened for use as patching compositions, and the thickener subsequently burned off. Jelly toothpaste and shaving cream have been bodied with such polymers. Stabilization of ice cream with alkali-metal (sodium) salts of poly(acrylic acid) has been claimed, as has the thickening of orange juice. Hydrazine fuel gelled with poly(acrylic acid) and crosslinked lightly with divinylbenzene does not adhere to the container walls. It shows an increase in low shear viscosity without a proportionate increase in high shear viscosity; thus it can support a high concentration of metal particles. The result is a high-energy rocket fuel. Microcellular closed-cell foams of urea resins were produced when sodium polyacrylate was incorporated into the resin precondensate along with a foaming agent. In an unusual application, a freeze-resistant cooling liquid for internal combustion engines was made by gelling a glycol by means of the polyacid, then peptizing the gel in water.

Suspending and Dispersing

Petroleum Production   Drilling for petroleum deposits involves the use of drilling muds. The well is drilled with a bit, and the shaft of the bit is increased in length, as the hole deepens, by adding sections. As each section is added, a section of pipe is also put in place around the shaft to lubricate and clean the drilling bit. In returning to the surface via the annulus, the mud also carries cuttings to the surface.

In the preparation of the muds, clays in various formulations are customarily used. The upgrading of such muds can be accomplished by the use of poly(acrylic acid) or its homologs. Such upgrading results in the formation of more barrels of mud from a given quantity of ingredients. In ordinary mud the acids are often used as extenders for bentonite to give high-yield muds. For deep-well use these acids serve as dispersants stable under the extreme conditions.

One of the exciting recent developments in drilling-mud technology is the use of low-viscosity polyacrylic acid–containing fluids circulated at very high speeds to remove the chips and cool the bit. This technique has resulted in remarkably increased drilling speeds.

Two other drilling applications will be mentioned here, although the action of such an additive as poly(acrylic acid) or other water-soluble acrylic polymers is more complex than that of a mere suspending agent. If the drilling mud passes through a porous structure in the underground formation as it returns to the surface, appreciable water loss results. This leads to deterioration of the formation and to the loss of raw materials. Fluid-loss additives are employed to seal porous formations and thus prevent this loss.

Large quantities of starch and carboxymethylcellulose are used for this operation. As drilling has proceeded to deeper and deeper levels, bacterial and thermal degradation have suggested the use of less susceptible fluid-loss additives. Hydrolyzed polyacrylonitrile has been used and has been able to withstand both bacterial and thermal conditions. However, the susceptibility to insolubilization by polyvalent metal ions has limited its usefulness in this application.

A related problem involved the prevention of water intake when the core passes through an underground deposit of water-bearing sands. Such water shut-off has been effected by either polymer insolubilization or *in situ* polymerization.

Metal Production   In the preparation of copper pellets, poly(acrylic acid) serves as both a dispersant and an activator for the reduction of the copper. In addition, the particle size of the copper pellets is controlled by the same additive.

Poly(acrylic acid) is also said to be advantageous in nickel-plating processes, presumably by virtue of dispersant activity.

Dispersing   Poly(acrylic acid) and certain of its derivatives serve as effective dispersants, primarily for inorganic pigments. For this purpose polymers of low molecular weight are preferred. Their use includes paint applications, in which improved homogeneity and stability of formulations is achieved, presumably by interaction of the polymer with sites on the suspended pigments or fillers. Use in adhesive formulations, automotive polishes, and cleaners follows the same pattern. Organic pigments are not dispersed effectively by the homopolymers. However, the lack of foaming in dispersed formulations in many cases offsets this restricted activity, which can also be overcome by the use of suitable copolymers. These polymers and copolymers serve as grinding media for the preparation of paint pastes and printing inks.

In other dispersant applications poly(acrylic acid), poly(methacrylic acid), and their salts were found to improve the dispersion of cement in water and, thereby, to decrease the amount of water needed for the mix and increase the strength of the concrete produced. When incorporated into a paper-size formulation, sodium polyacrylate led to optimal viscosity of the starch-based binder system with a minimum amount of dispersant. For reinforcing polyethylene sheet, gypsum or calcium sulfite has been used with 0.1% of poly(acrylic acid) added to the molding powder, presumably to improve the dispersion and the incorporation of the solids. A high-molecular-weight poly(methacrylic acid) dissolved in ethanol is used as a dispersion medium for the electrophoretic deposition of cathodoluminescent coatings. In another somewhat unusual development, pigments were dispersed in caprolactam by utilizing sodium polyacrylate as the dispersant and heating the mixture to produce a polyamide molding.

A fluidity titration procedure has been reported for determining dispersion activity. This procedure consists of titrating a nonfluid slurry of the pigment and extenders to be dispersed with small increments of a 10% dispersant solution and recording an arbitrary fluidity end point at which it is assumed the pigment is dispersed. In Table 17.6 are listed the amounts of low-molecular-weight poly(sodium acrylate) which are required to disperse the designated pigments.

TABLE 17.6
Fluidity Titrations for Determining Dispersing Activity

Pigment	Weight percent dispersant on pigment	Weight percent pigment concentration
$BaSO_4$	0.086–0.11	77
ZnO	0.11–0.14	69
Fe_2O_3	0.14–0.17	58
Kaolin	0.086–0.11	70
$CaCO_3$	0.04–0.06	71
Carbon black	8+	21 max.

Polymerization Aid   In many polymerization processes, a dispersing agent is required to maintain the initial monomer charge and the ultimate polymer particles in a relatively stable latex form. For this purpose, such derivatives as the ammonium salt of poly(methacrylic acid) have been reported effective. Some rather sophisticated processes have been developed based on these dispersants. In one, a stable, bimodal, particle-size polystyrene is produced. In another system, the polymer acid is used to stabilize a latex containing a polymer onto which styrene or methyl methacrylate is grafted.

The polyacrylic acids are often used in polymerization recipes wherein they play mixed roles and sometimes not very well-defined roles. Poly(vinyl chloride) prepared in suspension in the presence of poly(methacrylic acid), neutralized, and treated with a reactive tin stabilizer, is reported to have shown improved stress-whitening resistance and discoloration resistance. In the presence of poly(acrylic acid), the gamma-ray initiated polymerization of vinyl monomers, such as styrene and methyl methacrylate, in alcohols gave stable polymer organosols. Predominantly syndiotactic poly-2-vinylpyridine is produced by polymerization of the monomer in the presence of poly(methacrylic acid). An impact-resistant thermoplastic polymer is made by grafting vinyl monomers onto an ethylene-propylene-diene rubber in the presence of poly(acrylic acid) and a mixture of water and hydrocarbon solvents.

Miscellaneous   Poly(acrylic acid) and certain derivatives have been used as additives for boiler-feed water. In this case, they are reported to act as suspending agents for the developed boiler scale, allowing longer operating time between shutdowns for boiler cleanup and greater efficiency of heat transfer. For this use, the polyacids are often formulated with other materials such as amino phosphates.

A number of formulations have been offered in which the partial salts of these polyacids act as sequestering agents for the hard-water ions. The partial sodium salts have been found to be free-flowing powders when produced at the proper degree of neutralization and spray-dried. Heavy-duty cleaners have been made from anionic surfactants and perlite, with the poly(acrylic acid) serving the additional role of stabilizing the perlite dispersion. Fabrics washed in laundry detergents formulated with poly(acrylic acid) had less loss of reflection after repeated launderings than those washed in the formulation without the acid. Improved dishwashing detergent compositions which are moderate in foaming and nonirritating to the skin have also been reported. A blend of anionic surfactants and poly(acrylic acid) has been found which is stable at high and low temperatures and has good detergency.

Flocculation

A great variety of water-soluble polymers function as flocculants for many types of suspended materials. Polymers which have been used in this type of application include cationic, anionic, and neutral molecules derived from such monomers as dimethylaminoethyl methacrylate, acrylic acid, and acrylamide, respectively. Because of the divergence of base materials and effective polymer types, the action is not explained by a simple mechanism. It has been suggested that best results may come with the use of a fairly-long-chain linear polymer. This functions to encompass a number of the individual fine particles of the dispersed material, attaching itself to the particles at various sites by means of chemical bonds, electrostatic attraction, or other attractive forces. Relatively stable aggregates are thus produced.

Clarification   Flocculants have been used to clarify a number of liquids which have suspended fines contaminating them. Such aggregated materials may be removed by filtration, settling, or other convenient means. Acrylic polymers, such as poly(sodium acrylate), have been reported of interest in water clarification and in the treatment of sugar solutions. Fields of application include white-water treatment in paper manufacture and sewage-disposal problems.

The flocculation of synthetic and natural silica and alumina hydrosols by poly(acrylic acid) has been under active investigation in Russia. For purifying waste water, coagulants have been prepared by coating magnetic powder with a quaternary ammonium salt of poly(acrylic acid) to produce what is claimed to be a particularly efficient coagulant. The ferric salt of poly(acrylic acid) was used to flocculate the impurities in a waste water containing oils and anionic surfactants, after which the flocs were removed by filtration. In the flocculation of kaolin suspension by poly(acrylic acid), efficiency increased with certain levels of divalent cations; in the case of quartz, a continuous increase was found with the level of both monovalent and divalent salts. Both poly(acrylic acid) and poly(methacrylic acid) have been found to be efficient in flocculating small soil particles in dry-cleaning solvents, whereby the clogging of paper filters in dry-cleaning machines was reduced. The reaction product of poly(acrylic acid) with amines in less than stoichiometric amounts has produced products effective as flocculants and retention aids in papermaking and in water purification in an electroplating plant.

Metallurgy   The water-soluble acrylic polymers have found application in the recovery of ores from suspension in water. In addition to the clarification of the water, a further object in metallurgical processes is to recover the aggregated ore fines. The increased speed of settling and the enhanced rate of filtration are responsible for the successful application of these polymers to this type of operation.

This type of aggregation to improve settling and/or filtration rates has been expanded to include other than metallurgical fines. For instance, in the filtration of a precipitate such as calcium sulfate, improved filtration rates are obtained by prior flocculation with such polymers as the acrylics.

Soil Conditioning   At one time, appreciable publicity was given to synthetic soil conditioners, that is, materials designed to make stable aggregates in soil to improve the tilth of the soil and plant germination and to reduce water loss from the soil and run-off of top soil in heavy rains. The inability of these polymers to compete economically with other methods of soil treatment for large-scale agricultural uses does not alter the fact that they do indeed function as very efficient stabilization agents for soil aggregates. Table 17.7 shows data which support this statement.

TABLE 17.7
Aggregate Stability with Hydrolyzed Polyacrylonitrile*

| Soil type | Weight percent of soil particles finer than 0.006 mm | | |
	0.00 g polymer/100 g soil	0.05 g polymer/100 g soil	0.10 g polymer/100 g soil
Virginia sandy clay	15.5	0.0	0.0
New Hampshire salt	65.0	0.0	0.0
Texas alkali sand	37.4	6.1	9.4

* Data reported from Michaels and Lambe, "Soil Flocculants and Stabilizers," *Agric. Food Chem.*, 1, 835 (September 16, 1953).

Although many successful field trials were run suggesting that improved soil performance was obtained for certain treated plots as compared with untreated controls, the stabilization effect is readily noted in the laboratory by first forming soil aggregates, treating them with an aqueous solution of the polymer in question, placing the aggregates on the top screen of a set of screens of progressively decreasing opening size, and then jigging the assembly while it is immersed in water. After a given time, the loss in particle stability is observed by noting the amount of soil retained on each of the screens in the assembly. This demonstrated aggregate-stabilizing effect has also been suggested for application in erosion control on roadside banks, railroad rights of way, and temporary construction operations.

In contrast to this type of effect, actual soil stabilization, i.e., the more durable hardening and impermeabilization of soil structures such as walls and dikes, has been achieved by the *in situ* polymerization of soluble acrylic derivatives such as blended calcium acrylate and methylenebisacrylamide. In a recent development, repeatedly spraying soil with alternate sprays of polyoxyethylenes in water solution and polyacrylic acid in water solution yielded a stable soil; upon drying, the soil surface was hard and water-insoluble although water-permeable. Apparently, use is made here of the well-known secondary bonding between polyethers and polycarboxylic acids.

Binders

In certain applications, it is necessary to impart green strength to molded articles so that they may retain reasonable dimensional stability after formation in a wet condition until they have been fired or cured to their final form. For convenience, the temporary binder employed for this purpose should be capable of being burned out during the firing operation. Operations requiring such temporary binders include ceramic manufacture, asbestos-board preparation, and the formation of vitreous grinding wheels. A specific use is the formation of nickel briquettes, in which poly(acrylic acid) is especially effective as a binder. Poly(acrylic acid) and starch have been used with polyacrylate esters to form print paste binders for printing synthetic fiber and cellulose fiber fabrics. A binder for wood flour has been made from poly(acrylic acid) and basic aluminum acetate. A silver catalyst for the conversion of ethylene to ethylene oxide was prepared by the heat treatment of silver ion–treated poly(acrylic acid) supported on alumina.

Dental cements have been prepared based on poly(acrylic acid) solution and various powders, for example: (1) zinc oxide and tin fluoride, and (2) aluminum fluorosilicate glass with or without an additional chelating agent (also glasses containing additional phosphate ions).

Coatings

Textile Sizes In the textile operations of weaving and knitting, it is often desirable to size the yarn or thread. This is done to reduce machine shutdowns resulting from fouling of the shuttles and other working parts with fuzzballs formed by loose ends and fibrillation of the fiber. Sizing is accomplished by passing the thread through an aqueous bath containing the agent and thus placing a coating on the outside of the fiber.

After drying, the fiber can be woven or knitted; the size can then be removed by washing, or it may be left on the fabric as a temporary finish or feel-improver. Starch is widely used in the sizing of cotton. Blends of salts of poly(acrylic acid) with starch find application in this and finishing operations by virtue of a plasticizing action imparted to the starch.

With nylon and other synthetic fibers, such as Dacron, sizing has been most effectively carried out with poly(acrylic acid) or other suitable water-soluble acrylic polymers. Poly(acrylic acid), as suggested earlier, is apparently spot-welded to the nylon and thus serves as an effective sizing agent, which may be readily removed by subsequent washing techniques. Alternatively, a durable finish can be achieved by heat-curing the polyacid, with or without added glycerin or polyhydroxy compound, before it has been removed. The use of poly(methacrylic acid) as a size for rayon fibers in a continuous spinning process resulted in a decrease in fiber defects.

Various modified fibers have been made by the use of the polyacids. A cellulose acetate fiber with lower electrical resistance was produced by the introduction of poly(acrylic acid) into the cellulose acetate solution. Nylon 6 and nylon 66, prepared by polymerization in the presence of a poly(acrylic acid) salt, gave fibers showing improved antistatic properties and dyeability. The addition of poly(acrylic acid) to rayon solution before spinning resulted in a hydrophilic fiber made from regenerated cellulose. Modified wool fibers have been made during high-temperature dyeing by adding acrylic or methacrylic acid monomers to the dye bath to produce a resin-treated, crosslinked, dyed wool or wool-blend yarn.

Textile Soil-Release Finishes Durable press fabrics which have the soil-releasing properties of plain cottons have appeared on the market. One way of producing this result is to make the fabric hydrophilic by attaching to the fiber poly(acrylic acid) or poly(methacrylic acid) by using either a grafting process or crosslinking agents to bond a preformed polyacid to the fiber. In one process, the poly(acrylic acid) is bonded onto polyester-cotton fabric by irradiation and mild heating. Cotton and a very wide variety of synthetic fabrics have been made antistatic and soil-releasing by coating with polyacids fixed by polyglycols, including poly(vinyl alcohol)-type materials and simple glycols, and then heat-set. Bonding of the acids has also been accomplished by the use of siloxanes, alkaline silicates, and alkali fluorosilicate materials, followed by mild heating. Classical amino resins have also been used to bond the polyacid onto fabrics, again using a mild-heat treatment in the process. Polyesters have been modified with polyglycols to make a fiber which is receptive to treatment by poly(methacrylic acid) to give improved hydrophilicity and washfast dyeings.

Leather Finishing In the finishing of leather, base and top coats are applied in order to improve the appearance, feel, durability, and general performance of the leather article. Water-soluble acrylic polymers have been suggested for application as pigment binders and clear top-coat vehicles in the formulation of such finishing systems.

Waxes and Paints In addition to the previously mentioned use as thickeners and dispersants in paint formulations, certain acrylic polymers may offer possibilities as wax or paint vehicles. These are the zinc-ammonium and zirconium-ammonium salts of acrylic acid polymers which operate by virtue of the loss of ammonia during the volatilization of the water and drying of the film. Other volatile neutralizing bases may also be used, e.g., morpholine. Iron salts of poly(acrylic acid) have been found to decrease the corrosion of steel in electrolyte solutions when used as a component of a primer coating.

Paper The use of poly(acrylic acid) and some of its derivatives has been suggested for slush-stock or beater-additive application for the improvement of suspending properties in the stock or performance properties, i.e., dry strength, of the formed paper. The addition of poly(acrylic acid) to a system for the alum-induced coagulation of cellulose fiber–polymer latex mixtures produced improved polymer adhesion to the fiber, water separation, and properties of the paper, especially water resistance. Paper coatings with good hiding power and printing properties were prepared using poly(acrylic acid) or poly(methacrylic acid) and polyvalent metal salts, such as aluminum sulfate, along with lactic acid, as the pigment binder system.

Lithographic printing mats have been prepared with the ammonium salt of poly(acrylic acid) applied as a coating on the paper. After suitable treatment, the mat retains a proper combination of ink holdout and rewet properties to function satisfactorily in this application. In another development, thermochromic paper has been prepared by bonding a triphenylmethane lactone-type dye to the paper by means of poly(acrylic acid). Electrically conductive coatings for paper useful for electroreproduction papers were prepared by adding a limited amount of poly(acrylic acid) to a coating based on hydroxyethyl starch and a polymeric quaternary ammonium compound.

Cosmetics The water-soluble acrylic polymers are of interest in hair-finishing formulations in which they serve to coat the hair and permit setting and styling operations. More specifically, these acids have been used in fixer compositions for cold-wave preparations, in hair shampoos and grooming aids, and in hair dyes.

Film Formation The use of poly(acrylic acid) as a film former has been directed toward many applications. Representative of this type is the suggested use of the sodium salt of poly(acrylic acid) as the major component of a nonglare coating for headlights. Antifogging coatings of glass and transparent plastics for optical use have been made by crosslinking poly(acrylic acid) with aminoplast resins to produce scratch- and water-resistant coatings. These polymer acids have been used as release coatings for polymerization molds, as parting agents, and as soil-release coatings for hard surfaces subject to being spattered by printing ink. Polyion complexes (polylipo salts) with interesting properties have been made from poly(methacrylic acid) and polymeric quaternary ammonium salts in the form of free films which are water-insoluble but highly water-swollen. A poly(acrylic acid) polymer has been grown on glass microballoons to serve as a protective coating on the microballoons. In other microencapsulation applications, poly(acrylic acid) has been used to induce coacervation of other hydrophilic colloids to form the capsule wall.

The acrylic polyacids in the form of membranes have been used in a number of ways, particularly as semipermeable membranes useful in water desalination. In this kind of use, the polyacid is supported by another film, such as a cellulosic film. Multivalent metal salts have been used to crosslink poly(methacrylic acid). Other membranes have been formed by grafting polyacrylic acid onto polypropylene. In a related use, high-molecular-weight poly(acrylic acid) was employed in the insulating separator between the electrodes of electrolytic cells. Another exchange membrane consists of poly(acrylic acid) grafted onto polyvinyl alcohol.

Leather Paste

When animal hides have been unhaired, tanned, and washed, they are customarily dried by placing them on vertical panels which then move slowly through a drying tunnel. This operation requires several hours. The hides may be tacked peripherally to the panels; but drying shrinks the hides, which thus may pull away from the tacks and rip the edges in so doing.

It has become more desirable to paste the hides to the drying panels. This procedure requires a paste possessing good wet adhesion in order to prevent drop-off of hides in the drying tunnel. The paste must also have relatively poor dry adhesion so that the hides can be readily removed from the panels at the end of the operation without damaging the grain of the hide; and it must be readily removed from the dried hide by brushing or swabbing. Water-soluble acrylic polymers have found use in this paste application, either alone or in conjuction with starch or carboxymethylcellulose.

Ion-Exchange Processes

Water solubility of an acrylic polyacid can be reduced either by reducing the acid content of a copolymer with a hydrophobic comonomer or by crosslinking the acid polymer. In the latter case, complete water insolubility can be achieved. Such insoluble polyacids constitute a major group among the ion-exchange resins. The polyacrylic acids have also been used in ion-removal processes in a number of other ways. Poly-(acrylic acid) has been used to form soluble chelates with iron and manganese in processing paper pulp and in cleaning rusted iron, as well as to form insoluble salts of copper, cadmium, and lead to remove these ions from water. Cation-exchange membranes made from poly(acrylic acid) and polyaziridines impregnated onto a polypropylene fabric have been used in softening water for textile laundering. In a chemical process, polyalkylene ethers have been purified by poly(acrylic acid) used to remove the potassium hydroxide polymerization catalyst. A sodium polyacrylate flocculant has been used to remove hardness ions from salt water used for the production of chlorine by electrolysis.

Pharmaceuticals

Many of the uses of the polyacids in pharmaceutical technology are quite old; some of them have been referred to above and are typified by the formulation of ointments, eyewashes, hair preparations, tablet binders, etc. In recent years, other uses in research and manufacture have arisen for these versatile compounds.

A process has been developed which isolates a purified lactase by precipitation of the lactase by poly(acrylic acid) of high molecular weight. Mixtures of proteins and enzymes have been fractionated by a process utilizing precipitation with polyacrylic acid with high recovery of enzymatic activity; however, the fractionations were not complete. A high-molecular-weight sucrose polyallylether crosslinked with poly(acrylic acid) was reacted with urea to give an inclusion compound useful in treating uremic animals.

Interesting effects of injection of the acids have been found in cancer research studies on small animals. Examples of this work are the use of intraperitoneal or intravenous injection of poly(methacrylic acid). In one study this resulted in decreased metastasis, which may be related to a mobilization of malignant cells by poly(methacrylic acid) reported in other studies. Injection of poly(acrylic acid) before polyoma virus inoculation has served to protect against the tumor. Another rather suggestive experiment reports that in some cases poly(acrylic acid) of selected molecular weights is capable of releasing the DNA template restrictions. The primary immune response to an antigen was found to be stimulated by poly(acrylic acid). Poly(acrylic acid) has been found, in some cases, to induce virus resistance in both plants and animals.

Adhesives

The acrylate acid polymers are used as adhesives and as components of adhesives, in which they may play various roles. One role of particular importance is the control of the nature of the flow of the adhesive, i.e., its rheology, during application and use. Rheology during application is particularly important in many high-speed operations, such as those involving application of the adhesive by blades, rolls, spray, etc. The amount and nature of the thickening is particularly influenced by the molecular weight of the polymer and the pH of the system, as indicated above. Poly(acrylic acid) and acrylic acid–rich copolymers are used in preparing plywood adhesives and adhesives for laminating wood and surface films for wood. Low-molecular-weight poly(acrylic acid) has been sprayed onto polypropylene, to which aluminum foil is applied to form a strong laminate. In the fluidized bed process for applying protective coatings to glass bottles, poly(acrylic acid) has been employed both as a tackifier applied to the bottle and as a tackifier applied to the polyethylene-based particles. Addition of poly-(methacrylic acid) to a polyisoprene rubber improved the adhesion to steel more than twofold and improved the properties of the rubber.

Adhesives with controlled water sensitivity have led to a number of interesting applications for poly(acrylic acid). One example is a bandage adhesive for a painless-peeling bandage. Other examples are label adhesives, adhesives for photographic film purposely made water-and alkali-sensitive, and adhesives for baling scrap paper subsequently converted to pulp.

Miscellaneous

There are a number of applications for these polymers which do not fit in the categories outlined above. As demonstrated in some of the previous uses, the acids are commonly employed as chemical raw materials. They are often further reacted, as with compounds containing epoxide groups or aziridine groups, to form new and useful polymers. A protective coat for electrophotographic photoconductors is made by crosslinking poly(methacrylic acid) with an imine-isocyanate-polyglycol. Some of these polymerization systems are readily photoinitiated.

A transition metal complex of poly(acrylic acid) has been found useful as a catalyst for the condensation of urea and formaldehyde at pH's above 7. Catalysts for olefin hydrogenation have been prepared from multivalent salts of poly(acrylic acid) by drying under mild conditions to give a porous powder. A similar material pyrolyzed to decarboxylate the acid also gives a catalyst useful for the same purpose. A rhodium complex of poly(acrylic acid) has functioned as a homogeneous catalyst for the hydrogenation of terminal alkyne, and its triethylamine salt was effective for the hydrogenation of both internal and terminal alkynes.

The acids or their sodium salts have been found useful in stabilizer systems for sodium percarbonate and to control the rate of decomposition of dicyclohexyl-peroxydicarbonate. In photographic film practice, poly(acrylic acid) has been used in the production of an antistatic layer, as a coupler which produces better color definition, and as an agent giving pH control in the image layer of the film. A synthetic pulp for wet-papermaking was made from a melt blend of polypropylene and poly(acrylic acid).

FDA Clearances

Polymers	Regulation	Title
Poly(acrylic acid),	175.300	Resinous and Polymeric Coatings
Poly(methacrylic acid)	175.390	Zinc–Silicon Dioxide Matrix Coatings
	177.1210	Closures with Sealing Gaskets for Food Containers
	175.380	Xylene-Formaldehyde Resins Condensed with Epoxy Resins
	175.105	Adhesives
	175.320	Resinous and Polymeric Coatings for Polyolefin Films [Poly(acrylic Acid) Only]
	176.180	Paper and Paperboard for Dry-food Contact
Sodium Polymethacrylate,	176.170	Coating Adjuvant for Viscosity Control: Components of Paper and Paperboard in Contact with Aqueous and Fatty Foods
Sodium Polyacrylate		
	175.105	Adhesives
	176.180	Paper and Paperboard for Dry-Food Contact
Sodium polyacrylate (only)	175.300	Resinous and Polymeric Coatings
	177.1210	Closures with Sealing Gaskets for Food Containers
	176.200	Defoaming Agents Used in Coatings

TRADENAME GLOSSARY

The properties of commercially available grades of water-soluble poly(acrylic acid) and its salts are shown in the following for materials manufactured by Rohm and Haas Co..

Polymer	Properties and identity		
Acrysol® A-1 Acrysol® A-3 Acrysol® A-5	Aqueous solutions of poly(acrylic acid), 25% solids content, pH 2, increasing molecular weight series:		

		Brookfield viscosity*		
		25% solids	5% solids	Molecular weight
Clear colorless liquids	A-1	320 cP	15 cP	50,000
density 8.5 lb/gal	A-3	3000 cP	18 cP	150,000
	A-5	18000 cP	20 cP	

Acrysol GS	Aqueous solution of sodium polyacrylate, 12.5% solids, pH 8.5–9.8, viscosity 10,000–20,000 cP (as supplied), 1150 cP (5% solids), density 1.07 kg/L (8.9 lb/gal)
Acrysol G-110	Aqueous solution of ammonium polyacrylate, 25% solids, pH 8.5–9.5, viscosity 90–170 cP (5% solids), density 1.06 kg/L (8.8 lb/gal)
Acrysol HV-1	Aqueous solution of sodium polyacrylate, 10% solids, pH 8.5–9.5, viscosity 15,000–20,000 cP (5% solids), density 1.10 kg/L (9.2 lb/gal)
Acrysol A-41	Aqueous solution of poly(acrylic acid), 30% solids, pH 2.0, viscosity 1700–4500 cP (as supplied), density 1.02 kg/L (8.5 lb/gal)

* To convert cP to Pa·s, multiply by 0.001.

REFERENCES

1. Liquori et al., *J. Macromol. Chem.*, **1**, 291 (1966).
2. Mathieson and McLaren, *J. Polym. Sci.*, **A3**, 2555 (1965).
3. Eliasef, *J. Polym. Sci.*, **B3**, 767 (1965).
4. Braude et al., *J. Polym. Sci.*, **12**, 2767 (1974).
5. Walters, Kuhn, and Kuhn, *Nature*, **189**, 381 (1962).
6. Hamlen, Kent, and Shafer, *Nature*, **206**, 1149 (1965).
7. Smith, Winslow, and Petersen, *Ind. Eng. Chem.*, **51**, 1361 (1959).
8. Miller, Botty, and Rauhet, *J. Colloid Sci.*, **15**, 83 (1960).
9. Bevington et al., *J. Polym. Sci.*, **32**, 317 (1958).
10. F. J. Glavis, U.S. Patent 3,029,228 (April 10, 1962).
11. Doty, *J. Am. Chem. Soc.*, **76**, 3764 (1954).
12. Geuskens et al., *Eur. Polym. J.*, **7**, 561 (1961).
13. Garret and Brown, *J. Appl. Polym. Sci.*, **1**, 287–295 (1959).

Chapter **18**

Polyethylene Glycol

George M. Powell, III*

General Information 18-2
    Chemical Nature 18-3
    Physical Properties 18-6
    Biological/Toxicological Properties 18-14
    Manufacture 18-14
    Handling 18-15
    Applications 18-15
    Application Procedures 18-16
    Additives/Extenders 18-17
    Specialties 18-17
    Future Developments 18-17
Commercial Uses: Compounding and Formulating 18-19
    Chemical Intermediates 18-19
    Adhesives 18-19
    Agricultural Formulations 18-19
    Cellophane-Film Humectants 18-19
    Cosmetics and Toiletries 18-20
    Detergents and Cleaners 18-20
    Inks 18-20
    Paints and Coatings 18-20
    Pharmaceutical Products 18-20
    Rubber Compounds 18-21
    Miscellaneous Products 18-21
Commercial Uses: Processing Aids 18-21
    Ceramics 18-21
    Dialysis Operations 18-21
    Electroplating 18-21
    Heat-Transfer Baths 18-21
    Leather Treatment 18-22
    Metal-Working Operations 18-22
    Paper Products 18-22
    Petroleum Recovery and Processing 18-22

* Prepared under a consulting agreement with Union Carbide Corporation.

Plastics Compounding 18-22
Rubber Products 18-22
Textile Products 18-22
Wood Products 18-23
Industries Using Polyethylene Glycol 18-23
Adhesives 18-23
Agricultural Products 18-23
Ceramics Products 18-23
Chemical Specialties 18-23
Cosmetics and Toiletries 18-23
Electroplating and Electrowinning 18-23
Food Products 18-23
Inks and Printing 18-24
Leather Processing 18-24
Lubricants and Hydraulic Fluids 18-24
Medical Sundries 18-24
Metal Fabricating 18-24
Packaging Materials 18-24
Paints and Coatings 18-24
Paper Products 18-24
Petroleum Recovery and Processing 18-25
Pharmaceuticals 18-25
Photographic Products 18-25
Plastics Products 18-25
Rubber and Elastomers 18-25
Textile Products 18-25
Wood Processing 18-25
Formulations 18-25
Fatty Acid Esters. 18-25
Water-Dispersible Alkyd Resin for Paints 18-26
Suppository Bases 18-26
Ointment Bases 18-27
Cosmetic Cream 18-27
Hand Lotion 18-27
Brushless Shaving Cream 18-27
Cream Rouge (Vanishing) 18-28
Perfume Stick 18-28
Clay-Starch Paper Coating 18-28
Metal-Working Lubricant 18-28
Ball-point Pen Ink 18-28
Laboratory Techniques 18-29
Identification of PEGs 18-29
Determination of PEGs in Other Materials 18-29
Tradename Glossary 18-30
Addresses of Suppliers of Polyethylene Glycol 18-30
References 18-30
Further Useful Reading 18-30

GENERAL INFORMATION

The polyethylene glycols (PEGs) comprise a series of low- to medium-molecular-weight (M.W.), wholly synthetic water-soluble polymers. They can be regarded chemically as polyether diols, made by the stepwise addition of ethylene oxide to water or ethylene glycol starters to form long, linear chains of oxyethylene ($-CH_2CH_2O-$) units having primary hydroxyl groups at each end. Depending upon chain length, they range in physical form at room temperature from water-white viscous liquids (M.W. 200 to 700), through waxy semisolids (M.W. 1000 to 2000), to hard, waxlike solids (M.W. 3000 to

20,000 and above). All are completely soluble in water, very low in toxicity, quite bland and nonirritating, and possess wide compatibility, good solvent action, good stability, and good lubricity. This unusual combination of properties has enabled them to find a very wide range of commercial applications in fields such as pharmaceutical salves, ointments, and suppositories; cosmetic creams, lotions, and dressings; paper-coating lubricants; textile sizes; softeners and modifiers; metal-working lubricants; detergent modifiers; and wood impregnants. In addition, their chemical derivatives, the mono- and diesters of fatty acids, are widely used as emulsifiers and lubricants.

The PEGs should be distinguished from the high- and very-high-molecular-weight polymers of ethylene oxide, which are also water-soluble. These have molecular weights of about 100,000 to over 5 million and form strong, thermoplastic films. These resins are described in a separate section of this Handbook.

The PEGs should also be distinguished from the more general class of polyalkylene glycols in which propylene oxide is used in place of a portion or all of the ethylene oxide. The random copolymers of ethylene and propylene oxides are water-soluble when the proportion of ethylene oxide is at least 40 to 50% by weight. These are generally relatively low in molecular weight, and are quite valuable as functional fluids of various sorts. The block copolymers consist of sequences or "blocks" of all-oxypropylene or all-oxyethylene groups. Properties vary considerably, depending upon the lengths and arrangements of these blocks and on the functionality of the starting compounds used. Together, they comprise unique and valuable surface-active agents, for they can act as breakers for water-in-oil emulsions, as defoamers, as low-sudsing detergents, or as wetting and dispersing agents. Neither of these random or block types of copolymers is included in this discussion of the polyethylene glycols, however.

Polyethylene glycol was first made in 1859 by two different chemists working independently (Lourenco and Wurtz) and was studied by many investigators after that. Union Carbide introduced commercial quantities of PEGs to the United States market in 1939. Uses expanded steadily, and by 1948 other manufacturers of ethylene oxide began to offer PEGs also. There are now at least a dozen domestic producers, with about as many more outside the United States. Among these are: Union Carbide Corporation, The Dow Chemical Company, Jefferson Chemical Company, Olin Corporation, BASF Wyandotte Corporation, and overseas producers include Hoechst, Huels, Imperial Chemical Industries, Kuhlmann, Montedison, and Sanyo. The U.S. Tariff Commission reported domestic sales for 1978 as about 36,700 metric tons (81,000,000 pounds) worth $30.5 million, and sales in the rest of the world were at least another 20,000 metric tons (44,000,000 pounds).

Chemical Nature

The chemical composition of the PEGs can be viewed as an extension of the chemistry of the simple ethylene glycols. Thus, ethylene oxide reacts with water to form ethylene glycol (molecular weight 62, boiling point 198°C):

$$H_2C-CH_2 + HOH \rightarrow HOCH_2-CH_2OH$$
$$\diagdown_O\diagup$$

A second molecule of ethylene oxide can add to this to form diethylene glycol (M.W. 106, boiling point 246°C):

$$HOCH_2-CH_2OH + H_2C-CH_2 \rightarrow HOCH_2CH_2-O-CH_2CH_2OH$$
$$\diagdown_O\diagup$$

By the continuous addition of ethylene oxide the reaction can proceed in a stepwise fashion to make polyethylene glycols of higher and higher molecular weights. Thus, the general reaction can be written as:

$$HOH + nC_2H_4O \rightarrow HO(C_2H_4O)_nH$$

where the subscript n represents the degree of polymerization (DP).

The highest polymer made commercially as a pure compound is tetraethylene glycol, with $n = 4$. As the DP is increased further, it is no longer possible to separate out pure compounds (because volatility becomes much too low for practical distillation), and the resulting polymers can be described as mixtures of molecules of varying DPs, categorized by an average DP, n value, or molecular weight (M.W. = 18 + n times 44). The commercial grades are usually identified by numbers indicating their approximate average molecular weights. These are listed in Tables 18.1 and 18.2. Thus, for PEG 400, $n = 8$ to 9; for PEG 1000, $n = 22$ to 23; for PEG 2000, $n = 45$; for PEG 6000, $n = 136$; etc.

By the nature of polymer-forming reactions the products obtained under any one set of conditions contain a range of molecular weights, usually clustered about an average value. Some 35 years ago P. J. Flory[1] showed that for a stepwise polymerization reaction such as this, with a fixed number of "starter" molecules, the number of molecular species of various sizes in any particular product follows Poisson's distribution law. From this he calculated the weight percents of various species in some of the standard PEG grades. These are shown in Fig. 18.1. It is of interest that careful fractionation experiments usually fall quite close to these theoretical curves. It should be pointed out that these distribution curves for the PEGs are unusually narrow for polymeric materials; polymers

TABLE 18.1
Properties of Liquid Polyethylene Glycols

Grade number	200	300	400	600
Avg. molecular weight range	190–210	285–315	380–420	570–630
Suppliers (see code below)	1, 2, 3, 4, 5	1, 2, 3, 4, 5	1, 2, 3, 4, 5	1, 2, 3, 4, 5
Specific gravity, 20°/20°C	1.127	1.127	1.128	1.128
Freezing range, °C	Supercools	−15 to −8	4 to 8	20 to 25
Viscosity at 25°C (77°F), cSt*	40	69	90	131
Viscosity at 98.9°C (210°F), cSt*	4.3	5.8	7.3	10.5
Water solubility	Complete	Complete	Complete	Complete
Flash point, open cup, °C	171–182	196–224	224–243	246–252
Flash point, closed cup, °C	Over 150	Over 175	Over 175	Over 175
Heat of combustion, cal/g, 25°C	5656	6017	6145	6184
Hygroscopicity (glycerol = 100)	~70	~60	~55	~40
Refractive index, n_D^{20}	1.459	1.463	1.465	1.467

* Multiply by 10^{-6} to convert to m^2/s.
NOTE: The above numbers have been composited from suppliers' literature, and do not necessarily conform to specifications for any specific products, nor to data from any one supplier.

Supplier identification (incomplete listing):
1	CARBOWAX	PEG	Union Carbide Corporation
2	JEFFOX	PEG	Jefferson Chemical Company, Inc.
3	Polyglycol E	PEG	Dow Chemical U.S.A.
4	POLY-G	PEG	Olin Corporation
5	PLURACOL E	PEG	BASF Wyandotte Corporation

made by other reactions, such as condensation or free-radical polymerization, show a very much wider range of molecular weight species present.

All the PEG molecular chains, regardless of length, are in fact diols, with primary hydroxyl groups at each end. Thus, they can enter into all the chemical reactions characteristic of aliphatic hydroxyl groups, e.g., esterification, urethane formation, cyanoethylation, and the like. Commercially, the most useful chemical derivatives of the PEGs are the fatty acid mono- and diesters, which are valuable as emulsifiers, detergents, softeners, and lubricants.

As the polyether chains in the PEGs become longer and longer, the ether linkages become more numerous and exert more influence on overall properties. For example, the hygroscopicity of the PEGs drops off, since the hydroxyl groups are largely responsible for this. These and other property changes can be seen from Tables 18.1 and 18.2, and are discussed later. Another change with increasing numbers of ether groups comes from the electron-rich nature of these oxygen atoms, permitting them to be electron donors and to form relatively weak hydrogen bonds with electron-acceptor molecules.

The water solubility of these higher-molecular-weight PEGs results from association

TABLE 18.2
Properties of Solid Polyethylene Glycols

Grade number	Blend*	1000	1500†	2000	4000	6000	9000	14,000	20,000
Avg. molecular weight range	500–600	950–1050	1300–1600	1900–2200	‡	6000–8500	9700	12500–15000	18500
Suppliers (see code below)	1, 2, 3, 4	1, 2, 3, 4	1–5 incl.	3, 4§	1, 2, 3, 5	1, 2, 3	3	1	3
Density, 25°C., g/cm³	1.200	1.17	1.210	1.211	1.212	1.212	1.212	1.202	1.215
Freezing range, °C	37–41	37–41	43–47	50–54	53–60	57–63	59–62	61–67	56–64
Viscosity at 98.9°C (210°F), cSt‡	15	17–19	25–32	47	75–110	580–900	1120	2700–4800	6900
Water solubility, % by wt, 20°C	~73	~74	~70	~65	~62	~53	~52	~50	~50
Flash point, open cup, °C	221–232	254–266	254–266	266	268	271	271		288
Flash point, closed cup, °C	Over 175	Over 175	Over 175	Over 175	Over 175	Over 175	Over 174	Over 175	Over 175
Heat of combustion, cal/g, 25°C	6162	6251	6289		6306	6328		6333	
Hygroscopicity (glycerol = 100)	~35	~35	~30	Low	Low	V. low	V. low	V. low	V. low

* Blend identifications: 1500 (1), 550B (2), E500M (3), B1530 (4).
† Product identifications: 1540 (1), 1450 (2), 1500 (4) (5).
‡ Molecule weight ranges: 3000–3700 (1), 4500 reg, 3400 U.S.P. grade (3), ~3900 (5).
§ Available as 60 wt. % solution in water.
Multiply by 10^{-6} to convert to m²/s.
NOTE: The above numbers have been composited from suppliers' literature, and do not necessarily conform to specifications for any specific products, nor to data from any one supplier.

Supplier identification (incomplete listing):
1 CARBOWAX PEG Union Carbide Corporation
2 JEFFOX PEG Jefferson Chemical Company, Inc.
3 Polyglycol E PEG Dow Chemical U.S.A.
4 POLY-G PEG Olin Corporation
5 PLURACOL E PEG BASF Wyandotte Corporation

complexes or hydrates formed with water molecules. Phenols and various phenolic resins also complex readily with PEGs, as do a number of other simple and complex compounds. With some drugs and medicaments, complex formation can lead to incompatibility or inactivation, making these combinations impractical. With others, these loosely held complexes can reverse under body conditions to allow slow release. All these effects become more noticeable in higher-molecular-weight PEGs, and are even

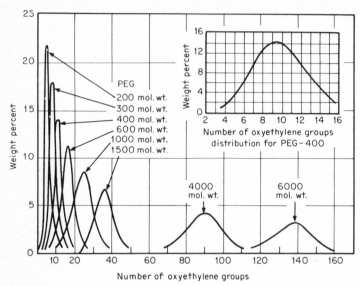

Figure 18.1 Calculated molecular weight distributions for liquid and solid polyethylene glycols *(after Flory)*, and number of oxyethylene groups distribution for PEG-400.

more significant in the high-molecular-weight poly(ethylene oxide) resins. While association complexes are seldom troublesome in most applications of the PEGs, users should be aware of their possible formation in specific types of formulations.

Physical Properties

The various commercially available grades of PEGs and related products sold by the larger United States producers, and a number of their more important physical properties, are summarized in Tables 18.1 to 18.3. Note that all are identified by numbers which indicate their approximate average molecular weights, with the single exception

TABLE 18.3
Some Additional Properties of PEGs

Appearance: Liquids or melts are clear and substantially free of foreign matter.
Weight: Liquids: about 1.13 kg/L (9.4 lb/gal) at 20°C
             Solids: lower-molecular-weight grades about 1.10 kg/L (9.17 lb/gal) at 55°C
                      higher-molecular-weight grades vary with form (flakes or powder)
Coefficient of volumetric expansion, 55°C.: Liquids, ~ 0.00074
Specific heat, room temp.: Liquids, ~ 0.5 cal/g/°C
Heat of fusion: Liquids and soft waxes, 35–37 cal/g
                 Solids, 40–45 cal/g
Volume contraction on solidification: Solids, about 7%
Surface tension: Pure liquids, about 44 dyn/cm
                  10% in water, liquids or solids, about 55 dyn/cm
pH, 5% in water: 4.0–7.0

NOTE: Special grades of PEGs meeting the specifications of the *United States Pharmacopeia, National Formulary, Food Chemicals Codex,* or other end-use requirements are available from most suppliers.

of a 50:50 blend of grades 300 and 1500 (which has an especially desirable petrolatum-like consistency).

There is some inconsistency among producers in the numbers they use for identifying several of their grades. This can be seen by reference to Table 18.2 and the footnote explanations, and applies especially to the 1:1 blend of 300 and 1500 molecular weight grades and to the 1500 molecular weight grade. For this reason, in this text the identifying numbers are all shown as molecular weight grade numbers. Note, too, that in some of the higher-molecular-weight grades the actual average molecular weights vary somewhat from producer to producer. These points should all be kept in mind when formulations calling for specific PEG grades by number alone are being prepared.

Since the PEGs form an homologous series of polymers, it is hardly surprising that many of their properties vary in a continuous manner with molecular weight. This is illustrated by Fig. 18.2, which shows how freezing ranges first rise rapidly through the lower-molecular-weight grades, then increase more gradually through the solids while approaching 66°C, the true crystallite melting point for very-high-molecular-weight poly(ethylene oxide) resins. Figure 18.3 shows how viscosity at 98.9°C (210°F) increases as molecular weights become higher. Other examples can be taken from the tables, such as the increase in flash points and fire points as molecular weights increase

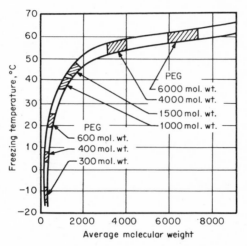

Figure 18.2 Approximate freezing ranges vs. molecular weight for polyethylene glycols.

and the slower increases in specific gravity and in heat of combustion. In a reverse relationship, hygroscopicity decreases as molecular weight increases, as does solubility in water at 20°C. Other properties not shown in these tables change also with molecular weight, as will be brought out later. By blending the individual grades in different ways, desirable combinations of properties often can be obtained.

When blends of different grades of PEG are being considered, it should be realized that the average molecular weights of the blends must be calculated on the basis of the number of moles of each grade, not the actual proportions by weight. For example, an equimolar blend of grades PEG 1000 and PEG 4000 would have an average molecular weight of 2500 and would correspond to a 20:80 PEG 1000:PEG 4000 mixture on a weight basis. The 50:50 weight basis mixture of these two would correspond to a 4:1 mixture on a molar basis and would have an average molecular weight of 1600.

Viscosity The viscosities of the various commercial grades of liquid and solid PEGs are shown in Figs. 18.3 and 18.4 as a function of temperature. The values for the normally solid grades relate to the molten state, of course. Corresponding curves for 50% aqueous solutions of the various grades as a function of temperature are shown in Fig. 18.5. In each set, note that viscosities are highly dependent upon both temperature and molecular weight grade. These curves are approximations only, and more accurate and much more detailed viscosity-temperature-concentration information is available from all the major producers on their own products.

Neither the pure materials nor their aqueous solutions show any unusual viscosity behavior such as the shear-dependent or time-dependent variations (pseudoplasticity, thixotropy, etc.) encountered often in much higher-molecular-weight polymers. Viscosity is a strictly linear coefficient expressing the relationship of shear stress to shear rate; hence simple capillary viscometers are quite acceptable for its measurement.

When it is desirable to blend to a specific viscosity intermediate between the viscosities of two established grades, a simple semilogarithmic plot can be helpful. The viscosities of the two established grades can be marked on the logarithmic scale at either end,

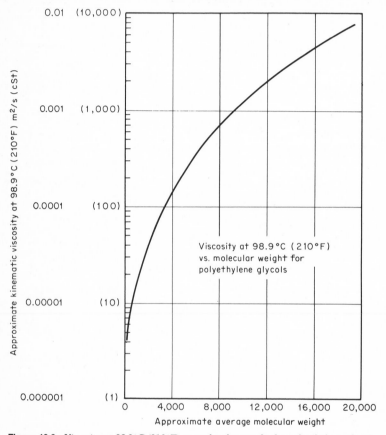

Figure 18.3 Viscosity at 98.9°C (210°F) vs. molecular weight for polyethylene glycols.

with the volume percent of the higher-viscosity component plotted from 0 to 100% on the rectangular coordinate scale. The viscosity of blends can be estimated from a straight line between the viscosities at 0 and 100% of the higher component.

Solubility in Water All the PEGs dissolve readily in water. Figure 18.6 shows, in a phase-diagram format, how the freezing points in water solution vary with concentration and molecular weight grade. The compositions above and to the left of the freezing lines represent true solutions. Note that the freezing points of 50 to 60% concentration solutions are very low and that the solubilities of the solid grades in water increase rapidly as temperatures are raised.

Viewed as maximum weight percent PEG soluble in water at a fixed 20°C temperature, the values drop from complete miscibility in all proportions for the liquids to more limited solubilities for the solids, but even for the highest-molecular-weight grades this remains above about 50 weight percent. As the temperature is raised, however, the solids rapidly become more soluble, and all grades become miscible with water in all proportions if the temperature is high enough (above about 60°C for PEG 6000, for example).

When liquid PEGs are mixed with water, there is a slight volume shrinkage (about 2% for 50:50 mixtures) and a noticeable amount of heat is released (12 to 14 cal/g for PEGs 300 to 600 mixed with equal weights of water). This is more than the heat of mixing for glycerol or ethylene glycol and comes largely from the heat of hydration of the ether oxygen bridges. For the solid PEGs, however, this heat of hydration is overbalanced by the larger negative heat of fusion (40 to 45 cal/g), so that the net result is a lowering of the temperature as the solid PEGs dissolve. Obviously, supplying heat in preparing such solutions will speed up the dissolving process.

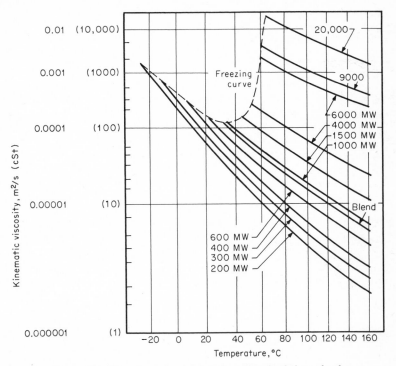

Figure 18.4   Viscosity vs. temperature for pure polyethylene glycols.

The PEGs are all non-ionic, hence do not acquire electrical charges in water solution. This means that they are relatively insensitive to the presence of dissolved salts or ionizable materials. The PEGs are soluble in pure water from their minimum solution temperature (see Fig. 18.6) up to the boiling point, but in the presence of appreciable amounts of dissolved salts they will often show "cloud points," or temperatures above which they tend to precipitate out of solution. For example, for a 0.5% PEG-6000 solution, 5% NaCl also dissolved shows no cloud point up to 100°C; when the NaCl concentration is 10%, there is a cloud point at about 86°C, and at 20% NaCl, there is a cloud point at 62°C. Interestingly, CaCl₂ shows a much more modest effect. Presumably, this is because the hydrated calcium ions can themselves complex with the polyether groups and not rob them of their solubilizing associated-water molecules.

Solubility in Organic Solvents   The PEGs are soluble in most of the more polar organic solvents, such as the alcohols, the ether-alcohols, the glycols, the esters, the ketones, the aromatic hydrocarbons, the nitroparaffins, and the like. They are not soluble in the aliphatic or cycloaliphatic hydrocarbons, nor in other solvents of low polarity.

Solubility data for three different grades of PEGs at two different temperatures in just a few of the available solvents are presented in Table 18.4.   Much more extensive data of this type can be obtained from the manufacturers of PEGs, if desired.

In addition, the PEGs are also soluble in a number of liquids not usually regarded as solvents, such as the lower aldehydes, amines, organic acids and anhydrides, and polymerizable monomers.   They are not soluble in some other types of liquids such

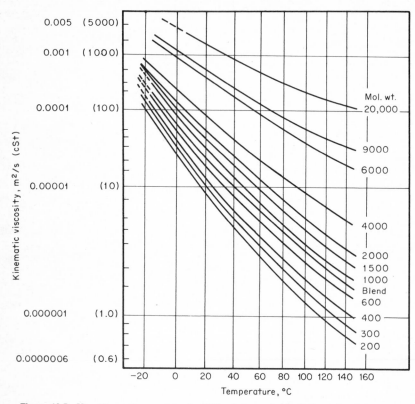

Figure 18.5   Viscosity vs. temperature for 50% solutions of polyethylene glycols in water.

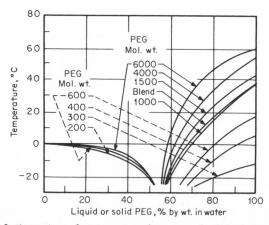

Figure 18.6   Approximate freezing points of aqueous polyethylene glycol solutions.

as the vegetable or mineral oils or similar liquid compounds in which the polar groups, if any, are heavily overbalanced by long hydrocarbon chains.

The individual PEGs differ considerably in their solubilities, with the lower-molecular-weight grades being generally much more soluble than the higher grades. Often, raising the temperature high enough to melt the solid PEGs greatly improves their solubility.

Solvency and Compatibility Especially useful characteristics of the PEGs are their own dissolving power for many substances and their compatibilities with others. In general, they show most solvent action and compatibility for the more polar types of materials, and least for those of low polarity. Tables 18.5 and 18.6 list the solubilities of some substances and pharmaceuticals in the PEGs.

TABLE 18.4
Approximate Solubilities of PEGs in Common Organic Solvents

Solvent	PEG 400		300/1500 Blend		PEG 4000	
	20°C	50°C	20°C	50°C	20°C	50°C
Water	M	M	69	97	60	84
Methyl alcohol	M	M	48	96	35	M
Ethyl alcohol (absolute)	M	M	<1	M	<1	M
Monoethyl ether of ethylene glycol	M	M	<1	M	<1	88
Mono(n-butyl)ether of ethylene glycol	M	M	<1	M	<1	52
Monoethyl ether of diethylene glycol	M	M	2	M	<1	63
Mono(n-butyl)ether of diethylene glycol	M	M	<1	M	<1	64
Acetone	M	M	20	M	<1	99
Dichloroethyl ether	M	M	44	M	25	85
Trichloroethylene	M	M	50	90	30	80
Ethyl acetate	M	M	15	M	<1	93
Dimethyl phthalate	M	M	30	90	13	74
Dibutyl phthalate	M	M	<1	M	<1	55
Ethyl ether	I	IBP	I	IBP	I	IBP
Isopropyl ether	I	I	I	I	I	I
Toluene	M	M	13	M	<1	M
n-Heptane	I	I	0.50	0.01	<0.01	<0.01

M = completely miscible; numbers = approximate g PEG/100 g solution; I = insoluble; IBP = insoluble at boiling point.

TABLE 18.5
Solubilities of Some Substances in PEGs at Room Temperature

Substance	PEG 400	300/1500 Blend	PEG 4000
Beeswax	I	I	I
Carnauba wax	I	I	I
Casein	S	S	PS
Castor oil (raw)	I	I	I
Chlorinated starch	S	S	S
Ester gum	I	I	I
Gelatin	I	I	I
Gum arabic	I	I	I
Methylcellulose	PS	I	I
Mineral oil	I	I	I
Nitrocellulose	S	S	PS
Olive oil	I	I	I
Paraffin wax	I	I	I
Pine oil	S	PS	I
Poly(vinyl acetate)	S	S	PS
Rosin	S	PS	PS
Shellac	PS	PS	I
Tung oil (raw)	I	I	I
Zein	S	S	PS

S = soluble (or compatible); PS = partly soluble; I = insoluble.

In addition, some metal salts can be dissolved in PEGs at about 100°C, and will remain stable at room temperature. These include calcium, cobaltous, cupric, ferric, magnesium, manganous, stannous, and zinc chlorides, plus mercuric acetate and potassium iodide.

Because of their solvent action and miscibility, the PEGs often will attack conventional paints and lacquers if spilled onto them. The high-bake, crosslinked coatings used as linings in shipping containers, collapsible tubes, etc., are generally immune to such attack, however. In contact with plasticized poly(vinyl chloride) the PEGs will extract plasticizer. At higher temperatures the polyamide and phenolic plastics may also be attacked. The nonpolar plastics such as polyethylene, polystyrene, and polypropylene are generally completely resistant.

TABLE 18.6
Solubilities of Some Pharmaceutical Materials in PEG 400 at Room Temperature

Organics			
Acetanilide	B	Menthol	B
Acetophenetidin	B	Paralydehyde	A
Acetylsalicylic acid	B	Phenobarbital	B
Aloin	A	Phenol	A
Antipyrine	B	Piperazine	B
Barbital	C	Quinine	B
Benzocaine	A	Resorcinol	A
Benzoic acid	B	Salol	A
Benzyl alcohol	A	Sulfadiazine	C
Caffeine	D	Sulfamerazine	C
Camphor	B	Sulfanilamide	B
Cetyl alcohol	D	Sulfathiazole	B
Chloral hydrate	A	Sucrose	F
Chlorobutanol	B	Tannic acid	A
Chlorothymol	A	Terpin hydrate	C
Citric acid	A	Thymol	A
Diethoxin	E	Urea	C
Ethyl carbamate	A	Vanillin	B
Hexamethylenetetramine	E	Zinc sulfocarbolate	D

Flavorings			
Anise oil	A	Lemon oil	D
Benzaldehyde	A	Methyl salicylate	A
Cinnamon oil	A	Peppermint oil	D
Clove oil	A	Sweet orange oil	D

KEY:

Code	Description	Parts PEG 400 to dissolve 1 part solute
A	Very soluble	<1
B	Freely soluble	1–10
C	Soluble	10–30
D	Sparingly soluble	30–100
E	Slightly soluble	100–1000
F	Very slightly soluble	>1000

Hygroscopicity The ability of the lower PEGs to pick up and retain moisture from the atmosphere, coupled with their good permanence and good plasticizing action for other substances, makes them industrially useful as humectants for a variety of hydrophilic products (such as glues or cellophane). The relative hygroscopicities, or water-holding capacities, for the various grades of PEG are shown in Tables 18.1 and 18.2. These are all expressed in relation to glycerin as 100% and are only approximate values, since much depends on test methods and conditions. Figure 18.7 shows approximate equilibrium moisture pickup as a function of percent relative humidity at normal temperatures for the liquid and softer-solid grades.

The tables and chart make it evident that hygroscopicity drops off rapidly as average molecular weight increases, and the influence of the hydroxyl end groups on overall properties becomes less. For all practical purposes, the solid-PEG grades do not act as humectants, but they can be blended with liquids to regulate the level of humectancy desired for specific uses. But even the highest-molecular-weight PEGs will take up moisture if stored for any length of time exposed to highly humid conditions.

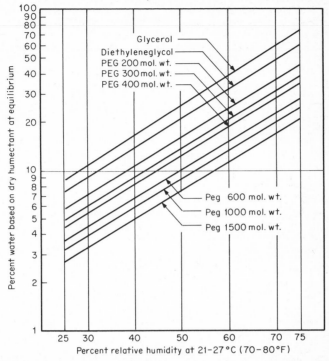

Figure 18.7   Hygroscopicities of polyethylene glycols at various relative humidities.

Surface Tension   The variation in surface tension as a function of concentration of PEGs in water solution at 25°C is shown in Fig. 18.8. The solid curve represents approximate values for all the liquid grades, while the dashed curve represents the solids. The solution-freezing concentrations for the separate solid grades are shown by the points as marked.

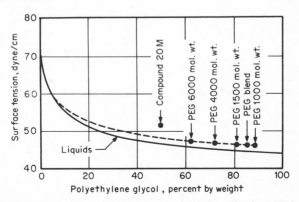

Figure 18.8   Surface tensions of aqueous solutions of polyethylene glycols at 25°C. Dashed curve for solid grades. Solid dots show freezing compositions at 25°C.

Volatility Only the two very lowest molecular weight grades of PEG show any measurable volatility at all, and this results from the small proportion of the lower glycol fractions they contain. By average molecular weight grade 400 such fractions are almost entirely missing. Any weight losses observed when heating the higher PEGs at elevated temperatures comes from volatile products of decomposition.

Thermal Stability While the PEGs are all quite stable under normal conditions, they will discolor when held at temperatures of 120°C or above in contact with air. If oxygen is displaced by an inert gas, such as nitrogen or carbon dioxide, temperatures of 200 to 240°C can be used without difficulty. The inclusion of an antioxidant such as 0.25 to 0.5% of phenothiazine is helpful; some producers include a little antioxidant (monomethyl ether of hydroquinone) in their PEGs 4000 and 6000 as a precaution.

Any thermal decomposition products of PEGs are volatile and do not form hard incrustations or sludge deposits. Furthermore, any equipment, tanks, heating coils, etc., can be cleaned easily with water. For this reason, the PEGs are often used as heat-transfer fluids, dip baths, tempering baths, etc. When exposed to air, some fumes and odors may be released, so that adequate ventilation is essential.

In the absence of air the PEG chains break down and pyrolyze only at temperatures above about 300°C. Their complete and clean decomposition without any residue makes them useful in foundry cores, molded ceramics, fluxes, and the like. In a muffle furnace they burn off with a blue, soot-free flame.

Biological/Toxicological Properties

The biochemical oxygen demand (BOD) of PEG solutions is rather low, especially in nonacclimated biological systems. For example, under simulated-river test conditions PEGs 400 and 600 showed less than 5% biooxidation in 5 days with either acclimated or nonacclimated organisms. In 10 days the values were 7 to 8% for nonacclimated and 20 to 26% for acclimated test organisms. In general, the higher-molecular-weight grades are attacked much more slowly than the lower ones. The ether groups are evidently much more resistant to such attack than the hydroxyl end groups.

The PEGs as a class are bland products possessing a very low order of toxicity, as demonstrated by any route of administration to test animals and by years of widespread successful use in products applied on or ingested by humans. Detailed reports of animal tests supporting this general statement can be obtained from any of the manufacturers, if desired.

It should be noted that special grades meeting the specifications established for various pharmaceutical, cosmetic, and food uses are available from the suppliers.

Under the Food Additives Amendment to the Federal Food Drug and Cosmetic Act, the *Food Chemicals Codex* grades of the PEGs are cleared for various direct and indirect food additive uses, provided that the amounts used do not exceed those required to produce the intended effect. For example, they can be used in boiler-water additives to make steam which contacts food, in defoaming agents in the processing of food, in coatings applied to fresh citrus fruit, as a coating or binder in tablets used for food, as an additive in nonnutritive sweeteners, as an additive in dispersing vitamins and minerals, and as a coating on sodium nitrite to reduce hygroscopicity. They may not be used in milk or in preparations added to milk. Generally, they are approved as components of articles intended for contact with food, as components of adhesives, and as surface lubricants on metal foil or sheet used for food packages. PEG suppliers can provide detailed specific references to the current regulations. Under EPA regulations they are exempt from any requirement of residue tolerance when used as inert ingredients in pesticide formulations applied to growing crops or harvested agricultural products.

It should be emphasized that in pharmaceutical applications each new use should be evaluated carefully first, for the PEGs may increase the absorption or activity of a therapeutic ingredient over that in other carriers and cause the composition to be classed as a "new drug" under applicable federal or state laws.

Manufacture

The PEGs are manufactured commercially by a batch process, using pressure reactors equipped with suitable agitation and heat-transfer means. Ethylene or diethylene glycol is often the "starter" for making the liquid PEGs, while the lower-molecular-weight

PEGs are often themselves the starters for the higher-molecular-weight grades. A little alkali, usually in concentrated water solution, is added as a catalyst, and after thorough nitrogen purging, the ethylene oxide feed is started. Temperatures and pressures vary, depending on the products to be made, but may be in the 150 to 180°C and 3 to 4 atm ranges.

It is important that proper precautions be taken to assure safe handling of the ethylene oxide, as ethylene oxide and air form explosive mixtures over a rather wide composition range. The vapor phase of the reactor should always be blanketed with enough inert gas to keep the ethylene oxide concentration below the explosibility range. The reaction proceeds by a stepwise addition of oxyethylene groups onto the active ends of the growing chains. When the desired molecular weight level has been reached, the oxide feed has been stopped, and the pressure allowed to drop, an acid is added to neutralize the catalyst. Alternatively, the catalyst can be neutralized by treatment with magnesium silicate, or the batch freed of inorganic materials by ion exchange. Afterwards, the batch is filtered, cooled, and packaged as desired.

The higher-molecular-weight grades are generally flaked and bagged, but may also be ground to powders for special uses.

Handling

The PEGs are relatively easy and safe to handle and store. Their flash points are high (see Tables 18.1 and 18.2), and their autoignition temperatures are even higher [about 400°C (750°F)]. Dusts of the powdered solids are relatively insensitive to ignition (compared to powdered coal), but can be exploded at 410°C or above in concentrations over 0.122 mg/cm³ in air. They are essentially free of any vapor under ordinary conditions, but if heated to high temperatures objectionable fumes may be emitted; adequate ventilation is needed in such instances.

Despite their bland nature, if any of the PEGs spill on the skin or splash into the eyes, prompt washing with water is recommended. Spills on the floor can be slippery and should be washed away. If PEGs are accidentally swallowed, vomiting should be induced. The use of safety glasses and rubber gloves when handling the PEGs is recommended.

From the standpoint of storage and transfer through pipes the problems encountered mainly concern the freezing points of all but the lowest-molecular-weight grades and their viscous nature. Mild-steel bulk-storage tanks are acceptable, but if iron pickup is to be avoided completely, stainless steel, aluminum, polyester-fiberglass, or resin-coated steel should be used. Maintaining a dry nitrogen atmosphere in storage tanks will minimize color development, even on long aging. The aqueous solutions of the PEGs have more tendency to corrode bare steel than do the pure products, so that corrosion-resistant materials should be used for these.

When the PEGs are to be transferred by pumping, it is usually desirable to heat them to 10 to 15°C above their freezing points. Drums can be moved into warmed rooms or heated with electric drum warmers. Storage tanks are usually insulated and have some means for heating them, such as by tempered water. In all cases, it is advisable that surface temperatures in contact with the PEGs be kept relatively low, preferably not over about 50°C (125°F), unless there is good agitation. Transfer lines should be insulated and heated for handling the higher-molecular-weight grades. Centrifugal or reciprocating pumps can be used. PEG grades 4000 and higher are normally handled in bags as flaked solids, and no special precautions are needed (other than to keep the bags tightly sealed and below about 45 to 50°C). Bulk deliveries of molten PEGs 1000 to 6000 molecular weight grades are available if desired. For these, the products are loaded at about 80 to 90°C into insulated (and coil-equipped) stainless steel containers, with a nitrogen atmosphere in the vapor space, and usually arrive for unloading with temperatures of 70 to 75°C, well above their melting points. Detailed handling recommendations can be provided by the principal suppliers.

Applications

Because of their unusual combination of properties—water solubility, lubricity, blandness, low toxicity, stability, solvent action, nonvolatility, etc.—and availability in a range of molecular weights, varying from viscous liquids to hard, waxy solids at room temperature, these PEG products find a very wide range of applications in which they serve many functions in many products in many industries. For this reason, it would be difficult indeed to list every one of their uses, let alone to describe and explain them.

Functions   One of the chief functions of PEGs is to impart water solubility or water sensitivity to a variety of products. As chemical intermediates, they supply hydrophilic properties to fatty acid esters, to alkyd and polyester coating resins, to polyurethane foams, etc. As ingredients, they combine water solubility with solvent ability, lubricity, blandness, and desirable consistency to serve as carriers for active components in pharmaceuticals, cosmetics, agricultural sprays, etc. They act as humectants to attract and hold moisture in various compositions, and serve as plasticizers in their own right. Allied to this is their value as antistatic agents.

Another important function is to provide lubricity to many products, as for example, in forming lubricants for metals, in extrusion of tile and ceramic shapes, to impart a pleasant slippery feel to cosmetics, as lubricants for textile yarns and fibers, in bag lubricants for making tires, etc. Related to this is their use as mold-release agents in making products from latex foam, and in molding or forming other rubber products. They often serve as dispersants, binders, bodying agents, spreader-stickers, foam stabilizers, ink vehicles, coatings, and the like.

End Products   Many of the end products containing the PEGs were alluded to in the foregoing discussion. PEGs are often found in pharmaceutical ointments, salves, lotions, suppositories, tablets, parenteral solutions, and in various cosmetics such as toothpaste, hair dressings, stick deodorants, cleansing creams, and the like. They can also be found in calendered paper, nylon fibers, regular cellophane film, natural and synthetic adhesives, silica-filled rubber compounds, detergents, fluxes, polishes, photographic developers, etc.

Industries   The industries which use the PEGs are also quite numerous: textile; paper; paint; metal casting, stamping, and forming; adhesive; food; pharmaceutical; cosmetic; rubber; leather; ceramics, electroplating; wood processing; packaging; and the manufacture of surfactants, dispersants, resins, and plastics, to mention only some of the more important.

Processes   The PEGs also serve as processing aids in making other products without serving as integral parts of such products. Some of these have been listed above, such as lubricants for tile and ceramic extrusions and metal shaping, binders for foundry cores and ceramic glazes, mold-release agents, and electroplating-bath additives. Others include pattern waxes, modifiers for viscose rayon spinning, dialysis solution modifiers for dewatering protein solutions and other liquids, heat-transfer fluids (dip baths) for forming thermoplastics or other purposes requiring hot-liquid immersion, etc.

Application Procedures

There are few problems in actually working with the PEGs in the preparation of industrial products. As discussed under Handling, it is often much easier to have them in the liquid form, even if this means heating them for the purpose. Other materials generally dissolve or blend in better under these conditions. For dry-compounding uses, the solid grades can be obtained as either relatively coarse or finely ground powders to facilitate mixing.

Blends of various molecular weight grades are often desired to obtain combinations of properties. Often these are combinations of liquid and solid PEGs. These, too, can be prepared readily by being melted first. To obtain the unctuous, petrolatum-like consistency often desired in pharmaceuticals and cosmetics, it is necessary to stir the co-melted PEG grades while they are cooling. PEG producers offer one such blend of grades 300 and 1500 molecular weight (under several different designations) already prepared.

As discussed in the section on chemical nature, the PEGs can form association complexes with certain compounds capable of acting as electron acceptors, and the possibility of forming such complexes should be kept in mind in preparing blends of the PEGs with other substances, notably some drugs and medicaments. In the case of phenolic compounds, mutual solubility is much enhanced by complex formation; more often, incompatible mixtures are the result. While such complexes are often relatively loose and can be reversed by changes in concentration, temperature, pH, etc., they can alter release properties. For example, phenol itself can be completely inactivated by complex formation in certain proportions (thus, PEG 400 is recommended for application to skin areas on which phenol has been accidentally spilled), but dilution with water releases it. In other cases, precipitation occurs, so that each case where such interactions are suspected should be investigated individually.

Additives/Extenders

For applications where the higher-molecular-weight resins are to be heated to any appreciable extent, the inclusion of a little antioxidant is often advisable to minimize any tendencies toward oxidation and discoloration. Some manufacturers actually include a little antioxidant, such as the monomethyl ether of hydroquinone, in their solid PEG products. Sometimes, too, the inclusion of a little buffer will inhibit the development of acidity from oxidation, and at other times specific corrosion inhibitors can be added if this should become a problem. But for most grades and most end uses, such additives are rarely needed.

Specialties

Several types of modified polyethylene glycols having somewhat different combinations of properties than the regular PEGs are available from Union Carbide Corporation. These include a higher-molecular-weight compound, a group of methoxypolyethylene glycols of various molecular weights, and a trifunctional polyethylene glycol having three hydroxyl groups. Some of the properties of these are shown in Table 18.7.

Polyethylene glycol Compound 20-M is made by joining several molecules of PEG 6000 together by reaction with a diepoxide, and has an average molecular weight in the 15,000 to 20,000 range. This differs from the regular PEG-20,000 grade, supplied in the United States by Dow Chemical U.S.A., in having a somewhat different molecular weight distribution, a slightly lower melting range, and a higher melt viscosity.

The methoxypolyethylene glycols comprise an important group of specialty products. These are essentially polyethylene glycols with methyl ether groups on one end of each molecule, so that the hydroxyl functionality is reduced to about one, instead of two for regular PEGs. They are available is several different molecular weight grades, ranging from about 350 to 5000, and they exhibit improved tolerance for hydrocarbons because of their modified structure. Furthermore, they are of special interest in the preparation of monoesters of fatty acids of high surface activity for use as detergents, emulsifiers, or dispersants that are not affected by hard water, dissolved salts, or mild acids. In preparing water-dispersible resins, the lower functionality of these methoxy PEGs is often an important consideration.

A third type of modified PEG is a triol based on glycerol, a trifunctional compound, instead of the diols used in making regular PEGs. This is identified as Triol TPEG 900 and has an average molecular weight of about 1000. With three hydroxyl groups it can serve as a crosslinking component in condensation polymers or as a base for preparing simple fatty acid esters. It is completely soluble in water, low in color and odor, and probably a little more hygroscopic than PEG 1000.

Future Developments

Since their initial introduction over 35 years ago, the industrial use of the polyethylene glycols has grown steadily, both in total volume of sales and in the multiplicity of end uses for these versatile, water-soluble polymers. With the emphasis now being placed on reducing pollution of all types, improving worker and consumer safety standards, and meeting increasingly stricter regulations of other sorts, it is probable that interest in water-based compositions of all kinds will increase significantly. This interest is quite likely to spark increased interest in the PEGs in many instances, so that continuing development and expansion of their markets is highly probable over the next few years.

Two general avenues of new product development by the producers of PEG will probably be followed to capitalize on market growth opportunities. These will be extensions of trends already in evidence and will involve either products of higher and higher molecular weights or products incorporating chemical modifications of the basic PEG structure.

For some of the newer end uses somewhat higher molecular weight grades of conventional PEGs are preferred. These are more difficult and expensive to manufacture in the normal manner, because viscosities and time in the reactor increase disproportionately and undesirable side reactions increase also. Novel polymerization processes will

TABLE 18.7
Some Modified Polyethylene Glycols*

Identification	Molecular weight range	Specific gravity, 20/20°C	Freezing range, °C	Viscosity at 98.9°C, cSt†	Water solubility at 20°C, % by wt	Flash point open cup, °C
Compound 20 M‡	15–20,000	1.207	50–55§	~14500	~50	>240
Methoxypolyethylene glycols:						
350	335–365	1.094	~5–10	4.1	Complete	227
550	525–575	1.089¶	15–25	7.5	Complete	238
750	715–785	1.094¶	27–32	10.5	Complete	246
2000	~1900	1.20	~52	54.6	~68	>260
5000	~5000	1.20	~59	613	~58	>260
Triol TPEG-990	~1000	1.138	~−5.4	16.8	Complete	277

* Available as CARBOWAX brand from Union Carbide Corp.
† Multiply by 10^{-6} to convert to m^2/s.
‡ Reaction product of PEG 6000 and a diepoxide.
§ Softening range.
¶ Measured at 40/20°C.

probably be developed to avoid these limitations, possibly involving suspension processes or chain extensions by condensation reactions.   The basic PEG molecular structure may be modified by substitutions on the hydroxyl end groups to change solubility, reduce transesterification tendencies, or to provide polymerizable or other functional groups, to cite but a few general possibilities.   It is understood that some new products of these types are already in the market development stage.

COMMERCIAL USES: Compounding and Formulating

The polyethylene glycols find application in a wide range of compounded products for consumer or industrial use, either as straight-chain polyglycols as produced and sold or as chemical modifications made by reactions at their terminal hydroxyl groups. For this reason a brief description of these chemical intermediate uses is presented first, before the various classifications of compounded products are discussed.

Chemical Intermediates

One of the largest categories of commercial uses for the PEGs is as intermediates for the synthesis of derivatives having specialized characteristics.   The PEGs themselves show little surface activity, but when converted to mono- or diesters by reaction with fatty acids, they form a series of widely useful non-ionic surfactants.   Hydrophil-hydrophobe balance can be adjusted as desired by the molecular weight of the PEG and the nature of the fatty acid.   Such products made from the PEGs obviously must compete with related products made by the direct addition of ethylene oxide to the fatty acids. For large-volume items this second route is often preferred, but it involves taking all the precautions needed for the safe handling of bulk ethylene oxide.   For this reason, esterification of the PEGs is still preferred by many, especially for synthesizing a wider range of more specialized surfactant products.

End uses for the fatty acid esters are largely as textile lubricants, softeners, and emulsifiers, and as emulsifiers in food products, pharmaceuticals, and cosmetics.   They find similar but smaller-volume uses in a range of other types of products.

Of course, the PEGs can be used in other types of syntheses.   Thus, water-soluble or water-dispersible alkyd and polyester coatings resins often include a polyglycol (or sometimes a methoxypolyglycol).   Polyurethanes containing the PEGs are softer and more hydrophilic.   Unsaturated polyesters from PEGs and maleic anhydride can be crosslinked via vinyl group polymerization to tough, resinous products.   Likewise, PEGs can undergo such reactions as the formation of ethers, acetals, halides, amides, urethanes, carbonates, etc., to make special products.

Adhesives

In a variety of natural and synthetic polymer adhesives the PEGs often serve as softeners, humectants, anticurl additives, or modifiers.   Thus starch, dextrin, or casein adhesives may contain liquid or solid PEGs to soften them or add lubricity.   Animal glues tend to dry out, crack, and curl, but the liquid PEGs serve as softener-humectants, especially when blended with urea, which complexes with the PEGs and overcomes any incompatibilities which might occur.   PEGs are used in some hot-melt adhesives to add ready remoistening capability.   They flexibilize latex adhesives based on poly(vinyl acetate), serve as ingredients in heat-activated label adhesives, and can be incorporated in pressure-sensitive adhesive compositions.

Agricultural Formulations

The PEGs and their fatty acid esters find some use as emulsifiers, spreader-stickers, or carriers in a variety of insecticide and herbicide sprays.   Chemical derivatives of 2,4-D or 2,4,5-T show much decreased volatility and permit better application control.

Cellophane-Film Humectants

The liquid PEGs find use as plasticizer-humectants in some grades of cellophane cellulose film, where they impart improved permanence and dimensional stability in packaging applications.   These films can be overcoated readily with nitrocellulose lacquers for moisture-vapor resistance for uses such as wrapping cigarette packages.

Cosmetics and Toiletries

For many of the same reasons which account for their uses in pharmaceutical products, the PEGs themselves and their fatty acid esters find many applications in cosmetic compositions. Their skin-smoothing, moisturizing, and softening characteristics are especially desirable. Alone or, more frequently, in combination with other bases, they go into creams, lotions, jellies, sticks, cakes, and powders of all sorts. Typical examples are vanishing creams and powder bases, brushless shaving creams, toothpastes, suntan lotions, shampoos, hair rinses, pomades and dressings, bar detergents, skin cleaners, deodorants, stick perfumes, powder cakes, rouge, mascara, and the like. Several typical formulations are shown later in this section.

Detergents and Cleaners

Although the PEGs are not surface-active themselves, they are sometimes used in detergent and cleaner compositions for their soil-suspending and dispersing action in combination with regular detergents. They impart a pleasant feel to the hands and to cleaned materials such as textiles. The non-ionic surface-active PEG fatty acid esters find uses also in a variety of cleaning compositions.

Inks

Thixotropic inks for ball-point pens can be made with liquid or solid PEGs as the internal phase. Water-based stencil inks also often use solutions of solid PEGs for viscosity and flow control. Steam-set printing inks benefit from the dye solvency and humectancy control afforded by the PEGs. Stamp-pad inks containing liquid PEGs retain their color and flow characteristics over a long period of time.

Paints and Coatings

As intermediates for preparing coatings resins, the PEGs provide a convenient method for modifying alkyd and polyester resins during synthesis to obtain water dispersibility. Usually about 10 to 30 weight percent of an intermediate-molecular-weight PEG or methoxypolyethylene glycol is required for this purpose. Obviously, too, the PEGs can be used as flexibilizing components in synthesizing other types of coatings resins. The straight PEGs are sometimes used as modifiers in poly(vinyl acetate) latex paints, in shellacs, and as binders in artists' watercolors. Very thin films of PEGs applied to glass greatly improve scratch resistance. Some nonflammable water-removable paint removers are formulated with a solid PEG.

Pharmaceutical Products

Ever since their initial commercial introduction, the PEGs have found a variety of uses in pharmaceutical products. An especially important example is as bases for suppositories, where the various molecular weight grades can be blended to provide almost any desired melting point, degree of stability, or rate of medication release. Their water solubility, blandness, and good solvent action for many medicaments are other reasons why they are frequently the products of choice. A few typical starting-point suppository formulations are shown later in this section.

Ointments, creams, salves, and lotions are also frequently based upon the PEGs. In addition to their water solubility, blandness, and solvent action, they spread easily, produce a pleasant, nongreasy feel on the skin, and are tolerant of body fluids. Blends of liquid and solid grades are often selected because of their desirable petrolatum-like consistency. Prepared blends of molecular weight 300 and 1500 grades are available from most suppliers for this use. While a few typical compositions are illustrated later, individual suppliers can supply many other suggested formulations.

It is important for potential users to realize that when the PEGs are substituted for other bases in pharmaceutical products it may be possible to reduce the proportion of active ingredients. This may even be necessary if the activity, absorption, or toxicity of the active ingredients is enhanced in these bases. The final products thus might be "new drugs" within the meaning of the Federal Food Drug and Cosmetic Act and similar laws. The compounder is responsible for complying with such regulations and for determining that the final products are safe for the intended uses.

The fatty acid esters of the PEGs are often used in pharmaceuticals as emulsifiers and suspending agents because of their blandness, non-ionic nature, and desirable surface activity. The solid PEGs find use as binders and lubricants in the manufacture of medicinal tablets of many sorts. Blended in powdered form with other solid ingredients and fillers, they permit more rapid manufacture of stronger, more chip-resistant tablets. Alcohol solutions are sometimes used to coat tablets, either as replacements or as supplements for sugar coatings. Solutions of the PEGs can serve as carriers for drugs and active ingredients for intramuscular injections. Many preparations intended for veterinarian use are also based on the PEGs.

Rubber Compounds

The solid PEGs are used in rubber mixes, where they generally act as dispersing aids and cure promoters. In carbon-filled compounds they are used principally as a dispersant, while in compounds containing silica-reinforcing fillers they promote cure rate by inhibiting adsorption of accelerators on the silica surface. In addition, they reduce the interference of absorbed water, found on all silicas, with ionic sulfur-curing mechanisms.

Miscellaneous Products

In addition to the product categories described in individual sections, some other kinds of compounded or formulated products in which the PEGs or their chemical modifications are often used are the following:

Cork products. The liquid PEGs serve as humectants and plasticizers in gaskets, floor tiles, etc.
Food products. PEGs and their fatty-acid esters are approved for miscellaneous uses as emulsifiers, carriers for vitamin preparations and nonnutritive sweeteners, binders for food tablets, and as components of food-packaging materials.
Lubricants and hydraulic fluids. Special-purpose lubricants, as for use on food-handling equipment, low-temperature service, etc., can be made with the PEGs or their esters.
Paper products. PEG treatments provide humectancy and softness to paper products, slip properties to cards, etc.
Photographic developers. PEGs in developing baths serve as sensitizers to improve speed and contrast.
Sponges. The lower-molecular-weight PEGs soften and plasticize synthetic cellulose and poly(vinyl alcohol) sponges.
Wood-swelling agent. The liquid PEGs are used in proprietary products for swelling wood, tightening joints, etc.

COMMERCIAL USES: Processing Aids

The industrial uses for the PEGs in which they serve primarily as processing aids in making other products, and do not remain as integral components afterwards, are quite numerous and varied. Some typical applications are described briefly below. These are arranged principally (but not entirely) by consuming industries.

Ceramics

The PEGs add green strength and good formability to various ceramic compositions to be stamped, extruded, or molded. They may be used alone or in combination with other resins or additives. Similarly, they serve as binders in color glazes and enamels. In all these uses, the PEGs can be burned out cleanly during subsequent firing operations.

Dialysis Operations

Because of their water-holding capacity, the PEGs are useful as dialyzing agents for removing water from protein, blood, and other biological solutions in a very gentle manner.

Electroplating

In electrowinning and electroplating baths small amounts of PEGs improve smoothness and grain uniformity of the deposited coatings. They improve the performance of electropolishing baths as well.

Heat-Transfer Baths

Molten PEG baths are useful for a variety of heat-transfer uses, either by direct immersion (as in rubber vulcanization, plastics forming, and annealing) or in circulating baths at temperatures in the 150 to 300°C range. If contact with air cannot be avoided, antioxidants are commonly added to reduce decomposition.

Leather Treatment

The PEGs serve as humectants and softeners for leather, reducing cracking and facilitating shaping and forming operations. The fatty acid esters are even more widely used as leather lubricants and softening agents, often in combination with various natural oils and fats.

Metal-Working Operations

The PEGs find a variety of uses in metal-working operations, ranging from preparing sand molds for casting various shapes to lubricants for metal stamping, rolling, and forming, to cutting and grinding fluids, to soldering fluxes, and to buffing and polishing compounds. Their water solubility, nonvolatility, blandness, and good lubricating abilities are just some of the reasons for their selection.

The "lost wax" investment casting process, in which the PEG is simply melted out once the foundry sand composition is packed in place and dried, is related to the pattern wax use of the solid PEGs to stiffen honeycomb structures for machining, after which they are just melted out. Soldering fluxes based on PEGs show good spreading, easy removal, and freedom from residues or charring.

Metal-working lubricants for all but the most severe forming operations are made with the PEGs or with their esters or derivatives. With increasing concern about pollution and employee health, there is growing interest in using water-based lubricants, cutting and grinding fluids, and the like wherever possible, so that the usefulness of the PEGs in such compositons can be expected to grow accordingly.

Paper Products

The solid PEGs are effective lubricants in paper-coating compositions, promoting better gloss and smoothness in calendering operations. Here, they can either supplement or displace the more conventional calcium stearate emulsions added to clay coatings based on starch, protein, or poly(vinyl alcohol) binders used on paper or paperboard. The liquids are useful as antistick treatments for waxed paper to facilitate processing.

Petroleum Recovery and Processing

The PEGs and their esters are components of metal-corrosion inhibitors in oil wells where corrosive brines are present. Their esters are useful demulsifiers for crude oil–water separation. Their selective solvency is of value in separating butadiene from 4-carbon hydrocarbon mixtures, and for color-stabilization treatment of distillate fuel oils. Other processes rely on the PEGs for the removal of halogens from hydrocarbons or of sulfur dioxide from stack gases.

Plastics Compounding

The PEGs are useful lubricants in a number of plastics compositions, providing improved handling behavior during compounding, calendering, extrusion, or molding.

Rubber Products

In addition to their uses as modifiers in rubber compounds, the PEGs are effective release agents for molding foam rubber articles, rubber mats, battery cases, shoe heels, etc., as lubricants for air bags in tire manufacture, as mandrel lubricants for forming hoses, and as temporary protection for white sidewalls during tire vulcanization. Their blandness and ease of complete removal with water often lead to their selection over oil-based materials.

Textile Products

The textile industries find many uses for the PEGs, and more particularly for their fatty acid esters, as sizes, as lubricants and softeners, and as emulsifiers for other additives. They serve as sizes and lubricants for carding, spinning, weaving, and knitting fibers and yarns, and can be easily removed afterwards. Small proportions (1 to 3%) are often added to viscose-spinning solutions, along with ethoxylated fatty amines, to modify fiber reorientation, permit faster spinning, and obtain higher-tenacity fibers or cellulose films. In sizes for rayon tire cord they promote improved rubber adhesion during tire manufacture. Hydrophobic fibers such as nylon and the polyesters are given a desirable combination of drag and slip along with antistatic properties. As finishing treatments for fabrics, they confer softness and a pleasant feel or "hand."

Wood Products

Green wood soaked in a solution of PEG of about 1000 M.W. will not shrink, crack, or warp under accelerated drying conditions. Subsequent workability of the seasoned, treated wood is improved also, and serviceability of the resulting wood products is enhanced.

INDUSTRIES USING POLYETHYLENE GLYCOLS

Adhesives

The PEGs provide water sensitivity and humectancy to both natural and synthetic resin adhesives. Their low-melt viscosity is of value in hot-melt compositions.

Agricultural Products

Both the PEGs and their fatty acid esters are useful components of agricultural sprays of various types. They serve as carriers, emulsifiers, dispersants, and spreader-stickers, coupling good solvency and low volatility with water solubility. PEG esters 2,4-D provide much closer application control by practically eliminating drift.

Ceramics Products

Their ready water solubility, good lubricity, and complete burn-off or volatilization during fusing make the PEGs useful in ceramics. They add green strength and formability to ceramic mixes and are useful binders for ceramic glazes and decorations. They reduce die wear during extrusion operations.

Chemical Specialties

The PEGs are valuable intermediates for the synthesis of a variety of chemical derivatives, especially the fatty acid esters and diesters, which are non-ionic emulsifiers or functional fluids. They compete with products made by direct ethylene oxide additions onto suitable starters, and offer simplicity, greater safety, and closer control.

Compounded detergents, cleaners, polishes, and similar products often contain the PEGs or their esters because of their soil-suspending action, blandness, viscosity control, and ready water solubility.

Cosmetics and Toiletries

The PEG liquids and solids, plus their fatty acid esters, find a wide range of uses in cosmetics of all types, primarily because of their combination of desirable properties: water solubility, nonvolatility, inertness, blandness, lubricity, stability, and pleasant feeling on the skin. Their texture, viscosity, and degree of hygroscopicity can be varied according to the molecular weight grade selected. They show good solvent action and good compatibility with many of the ingredients normally used in such products. Hence, they go into creams and lotions, makeup of various types, hair dressings and shampoos, shaving creams, toothpastes, skin conditioners, deodorant sticks, and the like.

Electroplating and Electrowinning

Small amounts of the PEGs in various metal-plating baths promote the formation of smooth, uniform, fine-grain deposits. They are also helpful in electropolishing baths, giving cleaner, brighter surfaces.

Food Products

The PEGs, providing they contain not over 0.2% ethylene and diethylene glycols, have been cleared under food additive regulations for use in a variety of food products with the exception of milk. They are substantially nontoxic and have little taste, odor, or color. They may also be used in packaging materials which come in contact with foods. The PEG fatty acid esters are especially useful emulsifying agents in food products.

Inks and Printing

Steam-set inks, widely used in high-speed printing, often contain PEGs to minimize hygroscopicity and premature ink setup under high humidities. Their solvency for dyes and low volatility account for their use in stencil and stamp-pad inks. As the internal phase in inks for ball-point pens, the PEGs provide convenient consistency control.

Leather Processing

The lower-molecular-weight PEGs serve as softeners and humectants in leather processing to avoid drying and cracking during handling. The PEG fatty acid esters are useful lubricants and softening agents for "fat-liquoring" baths.

Lubricants and Hydraulic Fluids

Special-purpose lubricants and hydraulic fluids are made from PEG esters of dibasic acids and related derivatives. These are useful where accidental contamination of food or other products can occur and would be harmful, where special temperature ranges are encountered, where easy wash-off is important, or where freedom from residues or sludge is desired. PEGs are particularly desirable as rubber lubricants because of their complete lack of solvent or swelling attack.

Medical Sundries

In addition to their uses in pharmaceuticals of many types, the PEGs have several minor uses in the medical area. They serve as lubricants for surgical sutures and as embedding materials for preparing and mounting histological specimens for microscope examination. Pathological specimens can be preserved by being soaked in liquid PEG, then immersed in melted solid PEG. After withdrawal, solidification, and removal of the solid PEG layer, the specimens can be stored almost indefinitely.

Metal Fabricating

Their good lubricity, nonstaining quality, and easy removal by water washing are among the properties of the PEGs accounting for their use in metal drawing, rolling, stamping, and shaping lubricants. They are noncorrosive, easily removed vehicles for soldering fluxes which spread readily and yield strong joints. The solid PEGs are useful in foundry cores and as machining stiffeners in honeycomb structures, in both cases because they can be removed easily by melting. The PEGs, and more particularly the PEG fatty acid esters, are valuable components in water-based cutting and grinding fluids, as well as in buffing and polishing compounds.

Packaging Materials

The PEGs serve the packaging industry indirectly by being incorporated in a number of the materials used in this industry. These materials include cellophane film (humectants), coated paper and paperboard (calender lubricants to improve smoothness and printability, humectants, antistick treatments, etc.), and modifiers for various adhesives and printing inks.

Paints and Coatings

The PEGs find some use as ingredients in coatings resins synthesis for imparting water dispersibility or improving flexibility. Other small uses include plasticizers for some latex vehicles and natural gums and resins and carriers for water-based pigment or colorant dispersions, artists' water colors, etc.

Paper Products

The largest use for the PEGs in this industry is probably as a lubricant in clay coatings or calender sizes based on starch, casein, protein, or poly(vinyl alcohol), promoting good flow-out and improved response to calendering or supercalendering operations. For this the higher-molecular-weight solid grades are preferred, and may be used in combination with calcium stearate to regulate water repellency, adhesive acceptance, and slip properties. Other uses are as humectants and softeners in various kinds of papers, as slip-promoters in computer cards, as dimensional stabilizers in some kraft papers, and as color stabilizers in electrolytic recording papers.

Petroleum Recovery and Processing

The PEGs serve the petroleum industry in a variety of minor ways, such as corrosion inhibitors in oil wells producing brines containing H_2S or CO_2, as decolorizers for distillate fuel oils, as selective solvents for removing butadiene or halogens from product streams, as modifiers for silica-filled petroleum asphalts, etc. The PEGs and especially their fatty acid esters or other derivatives serve as effective "breakers" for crude oil–water emulsions.

Pharmaceuticals

Pharmaceuticals represent an important outlet for the PEG products, where their combinations of properties such as water solubility, lack of toxicity or irritation potential, unctuousness, lubricity, solvent action, storage stability, and the like are major advantages. Their wide melting point range permits their formulation into suppositories, ointments, salves, and lotions as desired. They are useful binders and lubricants for making a variety of medicinal tablets. The PEG fatty acid esters are often used as superior emulsifying agents.

Photographic Products

The PEGs can be used as ingredients in developing baths to improve speed and contrast. They provide moisture control in photosensitive plates and coatings; add flexibility, slip, and anticurl to photographic papers; and serve as antisticks in film-stabilizing baths.

Plastics Products

The PEGs are used to some extent both as ingredients in the synthesis of polymer structures and as lubricants and plasticizers in plastics compounds. In polymer synthesis they add improved softness and flexibility to polyesters, add softness and water sensitivity to urethane foams, and can also be used in a variety of molecular structures in making casting resins. Their use in synthesizing water-dispersible coatings resins has already been mentioned.

Rubber and Elastomers

In both carbon black and finely divided silica-filled rubber compounds PEGs can serve as dispersants and cure promoters. They serve as mold-release coatings in making foamed rubber and other rubber articles, as tire air-bag lubricants, as mandrel lubricants for making hoses, and as heat-transfer fluids in vulcanizing rubber articles. The PEGs are compatible with rubber and do not separate like mineral oils, yet they do not swell or soften rubber. Ultra-soft open-cell polyurethane foams for textile coatings are made using a PEG in place of part of the usual polyester polyol component.

Textile Products

The PEGs are important in textile processing in several different ways. As additives in viscose-spinning baths, they improve fiber formation and properties. As sizes and lubricants, they aid in fiber processing and in weaving and knitting operations. They can also serve as softeners, conditioners, and antistatic treatments on fabrics. The PEG esters and other chemically modified PEGs are quite widely used as emulsifiers, lubricants, and softeners in textile processing, often being preferred to the unmodified grades.

Wood Processing

The PEGs are valuable for the dimensional stabilization of wood and wood products, preventing shrinkage, cracking, and drying out. In green woods, especially those of the highest quality, preliminary soaking in PEG solution can permit kiln drying at higher temperatures without danger of warping or cracking. This also facilitates subsequent cutting and wood-working operations and provides permanent dimensional stabilization. The preservation of wooden archaeological finds by immersion in PEG solutions has attracted much favorable publicity in recent years.

FORMULATIONS

Fatty Acid Esters

An important use for PEGs is as chemical intermediates for preparing useful derivatives by reaction at their terminal hydroxyl groups. The fatty acid esters constitute by far the largest class of derivatives, because of their valuable surface-active properties. These can be varied in hydrophil-hydrophobe balance, in hydrocarbon tolerance, and in other properties by variations in the molecular weight of the PEG starters, by the nature of the fatty acid selected, and by the molar ratio of PEG to fatty acid used, which influences the ratio of mono- to diesters formed. When little or no diester is desired, some prefer to use the methoxypolyethylene glycols, which have only one reactive hydroxyl group available. Unsaturated fatty acids tend to yield more fluid and more oil-tolerant esters than the saturated acids.

The actual preparation of these fatty acid esters follows conventional esterification practice. A closed, stirred, autoclave with provisions for temperature control, inert gas blanketing, partial vacuum operation, and a condenser for water removal is often selected. The desired ratios of PEG and high-purity fatty acid are charged, often with a little xylene for easier water removal, the reactor is purged of oxygen, and about 0.1 to 0.2% of an acid catalyst (such as p-toluenesulfonic acid) is added. The reaction temperature is usually in the range of 130 to 150°C, but can be higher if catalyst is omitted. Time may be 4 to 6 h, depending on the product desired. Since there are so many possible recipes which might be used, depending upon the properties desired, instead of listing a "typical" recipe here those interested are urged to consult with one of the major PEG suppliers for specific recommendations.

Water-Dispersible Alkyd Resin for Paints

While there are many variations possible in PEG-modified alkyd resins, normally some 10 to 30 weight percent of PEG is needed for water dispersibility. An early recipe[2] is as follows:

Linseed acids	1605 parts by weight
Pentaerythritol	555
PEG 1500 M.W.	336
Phthalic anhydride	711
Styrene	336
	3543

The mixed ingredients were reacted at 200°C, using a little xylene to aid in water removal, to an acid value of 20 and a viscosity of 0.000350 m^2/s (350 cSt) at 70% in xylene. The styrene could have been omitted, but was included here to improve the surface hardness of the dried films.

Suppository Bases

	Parts by weight			
Base	A	B	C	D
PEG 1000 M.W.	96		75	
PEG 1500 M.W.		94		88
PEG 4000 M.W.	4		25	
1,2,6-Hexanetriol		6		12
	100	100	100	100

In preparing suppositories, the above ingredients are melted together, the desired amount of specific medicinal agent is mixed in, and the preparations are cooled almost to their solidification temperatures and then poured into molds to solidify. To avoid shrinkage depressions, the molds should be overfilled. Base A is low-melting and disintegrates rapidly in use; the others are higher-melting and release medicament more slowly in use.

A great variety of medicaments can be used in these bases. Just a few are atropine, belladonna, bismuth subgallate, zinc oxide, ichthammol, iodoform, Peru balsam, benzocaine, and aminophylline. Agents such as sulfanilamide, tannic acid, and chloral hydrate tend to lower the melting points of PEGs, so higher-molecular-weight grades of PEG are needed. Still others, such as sodium phenobarbital and mercurochrome, are not soluble in PEGs and require the presence of water to carry them. Again, specific recommendations should be sought from suppliers.

Ointment Bases

Base	Parts by weight			
	I	II	III	IV
PEG 400 M.W.	50–60	47.5	40	37.5
PEG 4000 M.W.	50–40	47.5	50	42.5
Cetyl alcohol	–	5.0	–	–
"Span" 40 emulsifier	–	–	1	–
Water	–	–	9	–
1,2,6-Hexanetriol	–	–	–	20.0
	100	100	100	100

To prepare these ointment bases, the ingredients should be melted together and then stirred while cooling until they are congealed. This produces a smooth, creamlike consistency.

A range of medicinal and therapeutic agents can be used with these various bases, but PEG suppliers should be consulted for specific recommendations.

Cosmetic Cream

PEG 4000 M.W.	45 parts by weight
Propylene glycol	10
Zinc stearate	22
Water	23
	100

Melt the PEG, add the propylene glycol, and then add the zinc stearate. Add the water slowly with stirring to form a smooth cream.

Hand Lotion

PEG blend	30 parts by weight
Water	480
Propylene glycol	20
Monoethyl ether of diethylene glycol	30
Triethanolamine	1
Stearic acid	18
Quince seed mucilage	20
85% potassium hydroxide	0.5
Perfume	As desired
	599.5

Brushless Shaving Cream

PEG blend (300 M.W. : 1500 M.W. 1:1)	13.5 parts by weight
Stearic acid	11.5
Lanolin	4.0
Terpineol	0.1
Triethanolamine	1.0
Potassium hydroxide	0.5
Propylene glycol	10.0
Hydroxyethylcellulose (low viscosity)	0.7
Zinc stearate	5.0
Perfume	0.2
Water	53.5
	100.0

Melt together and heat the first four ingredients to 60°C. Add the triethanolamine, and stir until homogeneous. Dissolve the hydroxyethylcellulose separately in about 15 parts of water. Dissolve the potassium hydroxide in the rest of the water, heat to 60°C, add to the PEG mixture, and stir continuously. Add the hydroxyethylcellulose solution, and stir until cooled to about 45°C. Add the perfume in the propylene glycol, and then slowly add the zinc stearate. Continue stirring until the batch has cooled to room temperature.

This cream lubricates and smooths the skin well, and can be rinsed easily from the razor.

Cream Rouge (Vanishing)

Stearic acid	18.00 parts by weight
Cetyl alcohol	2.00
Propylene glycol	4.00
Isopropyl palmitate	8.00
PEG 4000 M.W. (U.S.P.)	4.00
Potassium hydroxide	0.70
Color lakes	8.00
Preservative	0.15
Perfume	0.50
Water	54.65
	100.00

Melt the stearic acid, cetyl alcohol, and PEG together to about 80°C. Heat separately the water, propylene glycol, and potassium hydroxide to the same temperature; then pour this into the molten waxes with agitation. Continue stirring until cooled to 60°C. Disperse the color pigments in the isopropyl palmitate, and run through a mill; then stir this into the cream, and continue stirring until the temperature drops to ambient.

Perfume Stick

Sodium stearate	10.00 parts by weight
Propylene glycol	0.25
PEG 300 M.W.	50.00
Perfume oil	15.00
	75.25

Heat the mixture without the perfume oil until a clear solution is formed at about 85 to 90°C. About 10 parts of water can be added to facilitate solution, but the mixture should be held at 90°C to evaporate off at least a portion of the water. Mix the perfume oil with the batch after partial cooling, just prior to molding.

Clay-Starch Paper Coating

No. 2 clay	100 parts by weight
Oxidized starch (low viscosity)	18
PEG 4000 M.W.	1
Water (approx.)	79
	198

The PEG acts as a lubricant. It can be added to the mixture directly or cooked with the starch solution; it should not be added directly to the concentrated starch solution without being diluted with water. This formulation is intended for trailing-blade application at about 60% solids. For air-knife application, solids content should be reduced to lower viscosity.

Compared with calcium stearate lubricant, the PEG gives improved gloss and smoothness, with other important properties being approximately equivalent, when applied at 4.5 to 7 kg per ream.

Metal-Working Lubricant

PEG 4000 M.W.	4.0 parts by weight
Sulfonated mineral oil	0.5
Water	9.5
	14.0

Ball-point Pen Ink

Oleic acid salt of basic dyestuff	56 parts by weight
PEG 1000 M.W.	15
Pigment	27
Non-ionic surfactant	2
	100

The oleic salt forms the external phase, the PEG the internal phase, and the pigment the solid phase. Viscosity behavior of the ink can be modified by modifying the molecular weight of the PEG internal phase.

LABORATORY TECHNIQUES

The specification tests generally applied to the PEGs by their manufacturers can be classified into four groups as follows:

1. Those which characterize the base polymer itself. These include average molecular weight, freezing (or melting) point, and viscosity [usually at 98.9°C (210°F)].

2. Those relating to appearance or general properties. These include appearance, form, particle size, color, specific gravity, and water solubility.

3. Those covering purity. These include pH of aqueous solution, solution or melt clarity, freedom from contamination, ash on ignition (usually sulfate ash), etc.

4. Optional additional tests to meet special requirements. These include such things as content of ethylene or diethylene glycols, arsenic, heavy metals (as lead): acidity or alkalinity; water content; saponifiable content; unsaturation; and odor and taste.

For the most part, the test methods applied are conventional laboratory procedures, and details can be obtained from the individual manufacturers.

Information on suitable test methods can also be obtained in the *ASTM 1973 Book of Standards*, Part 26, published by the American Society for Testing and Materials. ASTM Designation D 2849–69 entitled "Standard Methods of Testing Urethane Foam Polyol Raw Materials" lists a number of methods applicable to the PEGs. Since average molecular weights are usually calculated from hydroxyl numbers, attention is called to Sections 40 to 45 describing the determination of hydroxyl number by the pressure bottle technique using phthalic anhydride–pyridine reagent. The conversion to average molecular weight is made by multiplying 56,110 by the functionality (2 for PEGs) and dividing this product by the hydroxyl number.

Three other publications specify requirements for PEGs going into pharmaceutical, cosmetic, and food uses, and these also give approved test methods:

United States Pharmacopeia, 19th rev., official from July 1, 1975. (See especially pages 565, 566.)
Food Chemicals Codex, 2d ed., 1972. (See especially pages 626–629.)
National Formulary, 14th ed., 1975. (See especially pages 820–822, 1025.)

A wet chemical procedure for determining the ethylene oxide content of PEGs or their derivatives is described as a tentative method of test in ASTM Designation D 2959–71T (*ASTM 1973 Book of Standards*, Part 22).

A gas chromatographic method for determining the molecular weight distributions of the lower-molecular-weight grades (PEGs 200, 300, 400) is described by Fletcher and Persinger.[3]

Small amounts of ethylene and diethylene glycol in the PEGs can be determined by liquid chromatography using a method described by Carey and Persinger.[4]

Identification of PEGs

When a PEG is heated with sulfuric acid, dioxane is formed. If this is collected and passed into a solution of mercuric chloride through a curved tube, a characteristic white precipitate appears. Thus, this can serve as a qualitative test for identifying PEG.

Determination of PEGs in Other Materials

The quantitative determination of PEGs in biological materials by complex formation is described in a paper by Shaffer and Critchfield.[5]

Small amounts of the lower PEGs in water-bacteria media can be detected colorimetrically by the method of Crabb and Persinger.[6]

Brown and Hayes published a method for determining oxyethylene polymers and derivatives up to about 1000 M.W. based on the formation of a blue complex with ammonium cobaltothiocyanate, extracted into chloroform.[7]

Another test for oxyethylene chain structures is based on the formation of a white precipitate when a dilute solution of the unknown is added to a larger quantity of a proprietary reagent developed by Antara division of GAF Corporation (available from Fisher Scientific as Nonionic Reagent N-127).

TRADENAME GLOSSARY

CARBOWAX: Registered trademark of Union Carbide Corp. for its lines of liquid and solid polyethylene glycols and methoxypolyethylene glycols.

JEFFOX: Registered trademark of Jefferson Chemical Co. for its line of liquid and solid polyethylene glycols.

Methoxypolyethylene glycols: PEGs with a methyl ether group on one end of each molecular chain and a single hydroxyl group on the other end (available from Union Carbide Corp.).

Polyethylene glycol: Water-soluble oily to waxy solid polymers having the structure $HO-(CH_2CH_2O)_n H$, where n, the degree of polymerization, can vary from an average of about 5 to 500 or more.

PEG Compound 20M: Union Carbide Corporation's identification for a solid PEG of about 20,000 average molecular weight made by reacting PEG 6000 average molecular weight with a diepoxide to extend its chain length.

Poly-G: Registered trademark of Olin Corp. for its line of liquid and solid polyethylene glycols.

Polyglycol E: Product name used by Dow Chemical U.S.A. to identify its line of liquid and solid polyethylene glycols.

Polyethylene oxide: Term generally applied to high polymers of ethylene oxide (H_2C-CH_2) having molecular weights of 100,000 or more ($n =$ more than 2200).

Polyoxyethylene: Term synonymous with polyethylene glycol or polyethylene oxide, referring to the basic structure of long chains of oxyethylene ($-O-CH_2-CH_2-$) groups.

PLURACOL: Registered trademark of BASF Wyandotte Corp. for its lines of polyoxyalkylene glycols. Letter E preceding average-molecular-weight code number identifies products as liquid or solid polyethylene glycols.

Addresses of Suppliers of Polyethylene Glycol

Code, numbers used in Tables 18.1 and 18.2	Company and Address
1	Union Carbide Corp. Chemicals and Plastics 270 Park Avenue New York, N.Y. 10017
2	Jefferson Chemical Co., Inc. Box 53300 Houston, Tex. 77052
3	Dow Chemical U.S.A. Ag-Organics Department Midland, Mich. 48640
4	Olin Corp. 120 Long Ridge Road Stamford, Conn. 06904
5	BASF Wyandotte Corp. Industrial Chemicals Group Wyandotte, Mich. 48192

REFERENCES

1. Flory, P. J., *J. Am. Chem. Soc.,* **62,** 1561 (1940).
2. Armitage and Trace, *J. Oil Color Chem. Assoc.,* **40,** 849 (1957).
3. Fletcher and Persinger, *J. Polym. Sci., Part A-1,* **6,** 1025–1032 (1968).
4. Carey and Persinger, *J. Chromatogr. Sci.,* **10,** 537–543 (1972).
5. Shaffer and Critchfield, *Anal. Chem.,* **19,** 32–34 (January 1947).
6. Crabb and Persinger, *J. Am. Oil Chem. Soc.,* **41,** 752–755 (1964).
7. Brown and Hayes, *Analyst,* **80,** 755–767 (1955).

FURTHER USEFUL READING

Any person with an interest in the PEGs for particular uses would do well to obtain current technical bulletins on the characteristics and uses of their products from the major manufacturers (such as those identified in Tables 18.1 and 18.2). Often, more detailed suggestions about specific uses can be obtained from these sources also.

Books

G. O. Curme, Jr., and F. Johnston (eds.), *Glycols*, Reinhold Publishing Corporation, New York, 389 pages, 1953.
> This is a comprehensive book on all the glycols (as of its date of publication) and contains a useful discussion of the polyethylene glycols on pp. 176–193.

N. G. Gaylord (ed.), *Polyethers. Part I: Polyalkylene Oxides and Other Polyethers*, vol. 13, part 1, *High Polymer Series*, Interscience Publishers, New York, 491 pages, 1963.
> There are discussions of polymerization on pages 102–130, of the properties of the PEGs on pages 169–189, and uses on pages 239–274 (interspersed with other products). In addition, there is an extensive bibliography.

Technical Articles

C. P. McClelland and R. L. Bateman, "Technology of the Polyethylene Glycols and Carbowax Compounds," *Chem. Eng. News*, **23**, 247–251 (February 10, 1945).
> This is the first major technical publication on the properties of the PEGs. Much of the information is also covered in more recent supplier technical bulletins.

J. D. Malkemus, "Production of Alkylene Oxide Derivatives," *J. Am. Oil Chem. Soc.*, **33**, 571–574 (1956).
> This is a good, general description of current manufacturing processes.

C. E. Colwell and S. M. Livengood, "Association Reactions of Polyethylene Glycols and Derivatives," *J. Soc. Cosmet. Chem.*, **13**, 201–213 (June 1962).
> This describes some of the association complexes and salt interaction effects observed with the PEGs and related compounds.

Chapter **19**

Poly(Ethylene Oxide)

David B. Braun
Union Carbide Corp.

General Information 19-2
   Chemical Nature 19-2
   Physical Properties 19-4
   Manufacture 19-6
   Biological/Toxicological Properties 19-7
   Rheological Properties 19-8
   Additives/Extenders 19-9
   Handling 19-11
   Applications 19-11
   Application Procedures 19-12
Commercial Uses: Compounding and Formulating 19-13
   Adhesives 19-13
   Industrial Supplies 19-15
   Construction Products 19-17
   Paints and Paint Removers 19-17
   Pharmaceuticals 19-18
   Printing Products 19-18
   Soap, Detergents, and Personal-Care Products 19-18
   Water-Soluble Films 19-19
Commercial Uses: Processing Aids 19-20
   Binder 19-20
   Coatings and Sizes 19-21
   Dispersant 19-22
   Flocculation 19-22
   Hydrodynamic Drag Reduction 19-23
   Thermoplastics Manufacture 19-24
   Thickening/Rheology Control 19-24
   Water Retention 19-25
Industries Using Poly(Ethylene Oxide) 19-25
Formulations 19-27
   Aluminum and Metal Cleaner 19-27
   Calamine Lotion 19-28

Denture Fixative, Powder 19-28
Detergent Bars 19-28
Detergent Liquid 19-28
Lithographic Press Dampening Fluid 19-29
Microencapsulation 19-29
Paint and Varnish Remover 19-29
Thickened Acetic Acid 19-30
Thickened Hydrochloric Acid (Muriatic Acid) 19-30
Thickened Sulfuric Acid 19-30
Rubber Lubricant (for Mounting of Tires) 19-31
Toothpastes 19-31
Product Glossary 19-31
Further Useful Reading/References 19-31

GENERAL INFORMATION

Very-high-molecular-weight poly(ethylene oxide) resins were first investigated by I. G. Farbenindustrie in 1929. In 1958, Union Carbide began the first commercial production of these resins. Since then, two Japanese companies, Meisei Chemical Works, Ltd., and Seitetsu Kagaku Company, Ltd. have begun producing poly(ethylene oxide). These are the only three known worldwide producers.

Poly(ethylene oxide) resins are made commercially by the catalytic polymerization of ethylene oxide in the presence of any one of several different metallic catalyst systems. They are available with average molecular weights from as low as 200 up to 5×10^6. However, those products with a molecular weight below about 25,000 are viscous liquids or waxy solids commonly referred to as *poly(ethylene glycols)* and are discussed in Chap. 18. The poly(ethylene oxide) resins described here have a molecular weight range from about 1×10^5 to 5×10^6. They are dry, free-flowing white powders, completely soluble in water at temperatures up to 98°C and completely soluble in certain organic solvents. They have crystalline melting points from 63 to 67°C. Above the crystalline melting point the resins become thermoplastic materials which can be formed by molding, extrusion, or calendering. They form association complexes, through hydrogen bonding, with a number of monomeric and polymeric electron acceptors as well as certain inorganic electrolytes. Electrochemically, they are classified as non-ionic polymers. Aqueous solutions display increasing pseudoplasticity and pituitousness (stringiness) as the molecular weight of the resin increases. These solutions are susceptible to viscosity decay caused by oxidative and/or shear degradation of the polymer.

The major commercial uses for poly(ethylene oxide) resins include adhesives, water-soluble films, textile sizes, rheology-control agents and thickeners, water-retention aids, lubricants and emollients, hydrodynamic drag-reducing agents, flocculents, dispersants, and additives in medical and pharmaceutical products. The unique combination of properties of these resins has led to developing applications in detergents, solids transport, control of sewer surcharges, dredging, and metal-forming lubricants.

In 1978 worldwide capacity for production of these poly(ethylene oxide) resins was in excess of 15 million pounds, not fully utilized. Although growth of the market for these resins has been steady since their introduction in 1958, it has been slow. Even today, there is no single existing application which consumes more than 1 million lb/yr of poly(ethylene oxide) resin.

Chemical Nature

The chemical structure of poly(ethylene oxide) resins is:

$$+O—CH_2—CH_2+_x$$

Because of the very high molecular weight of these resins, the concentration of reactive end groups is extremely small, and therefore these resins show no end-group reactivity. However, they do possess the limited reactivity normally associated with polyethers (see Association Complexes, page 19-4).

The resins are available in several molecular weight grades. Table 19.1 shows the different commercial grades available from the largest of the three worldwide suppliers of poly(ethylene oxide) resins.

Poly(ethylene oxide) resins produced by Meisei Chemical Works Ltd. are sold under the trademark Alkox, and Seitetsu Kagaku Co., Ltd., markets their resins under the trademark PEO. The resins manufactured by these two Japanese suppliers generally have molecular weights nearer the high end of the molecular weight range in Table 19.1.

TABLE 19.1
Commercial Poly(Ethylene Oxide) Resins[43]

Approximate average molecular weight	Union Carbide Corp. Name
1×10^5	Polyox® WSR-N-10
2×10^5	Polyox WSR-N-80
3×10^5	Polyox WSR-N-750
4×10^5	Polyox WSR-N-3000
6×10^5	Polyox WSR-205
9×10^5	Polyox WSR-1105
4×10^6	Polyox WSR-301
5×10^6	Polyox Coagulant

Narrow-Molecular-Weight-Distribution Grades Union Carbide provides a group of poly(ethylene oxide) resins with a narrower distribution of molecular weights than normally obtained by conventional polymerization techniques. This is accomplished by a proprietary postpolymerization technique. These resins are designated WSR-N. The different average molecular weight grades of the WSR-N type are listed in Table 19.1. The molecular weight distribution, i.e., the ratio of weight-average to number-average molecular weight, of these special grade resins is typically less than one-half that of the unmodified grades. Also, the properties and performance of the resins are altered in several respects as the molecular weight distribution becomes narrower.

Hydrogels Poly(ethylene oxide) resins may be converted from linear molecules to crosslinked hydrogels by exposure of the resin to gamma irradiation or by chemically crosslinking them with a vinyl monomer and a peroxide initiator.

This modification causes a dramatic change in properties of the resin, making it essentially insoluble in water and other solvents commonly used with the unmodified resin. However, these hydrogels absorb large quantities of water, frequently as much as 50 times their own weight. In addition, they retain absorbed water for extended periods of time, and therefore are finding increasing utility in applications where these unique properties are extremely important, such as diapers, catamenial devices, and water-retention additives for agriculture.

Thermoplastic Compound A specially formulated compound, based on medium-molecular-weight poly(ethylene oxide) resin, is produced by Union Carbide for use in thermoplastic applications including extrusion, molding, and calendering. The exact composition of this compound, WRPA-3154, is proprietary, but it is known to contain plasticizers and stabilizers needed for thermoplastic forming operations.

Hydrodynamic Drag Reduction Slurry Union Carbide also produces a slurry of high-molecular-weight poly(ethylene oxide) resin for use as a hydrodynamic drag reduction agent in fire fighting. This product is designated UCAR® Rapid Water Additive. The exact composition of this slurry is proprietary. It is a dispersion of the resin in a water-miscible vehicle.

Oxidative Degradation *In Aqueous Solution.* High-molecular-weight polymers of ethylene oxide are susceptible to severe autooxidative degradation and loss of viscosity in aqueous solution. The mechanism involves the formation of hydroperoxides that decompose and cause cleavage of the polymer chain. The rate of degradation is increased by heat, ultraviolet light, strong acids, or certain transition metal ions, particularly Fe^{3+}, Cr^{3+}, and Ni^{2+}. Ethyl, isopropyl, or allyl alcohols, ethylene glycol, or the Mn^{2+} ion are known to be effective stabilizers of aqueous poly(ethylene oxide) resin solutions. The alcohols and glycols are used at a concentration of 5 to 10 wt% of the solution, while the Mn^{2+} ion is generally added as manganous chloride at a concentration between 1×10^{-5} and 1×10^{-3} moles Mn^{2+} per liter of solution. Properly stabilized aqueous solutions of these resins are generally quite stable in the pH range of 2 to 12.

Aqueous solutions of these resins are also very susceptible to degradation by shear forces. (See Rheological Properties, page 19-8.)

In Bulk. Above the crystalline melting point of 63 to 67°C, poly(ethylene oxide) resins become thermoplastic. Thus, they can be used in bulk and formed by typical thermoplastic processing techniques including molding, extrusion, or calendering. Thermoplastic processing of these resins is generally carried out at temperatures between 100 and 150°C. At these temperatures oxidative degradation of the resin can occur very rapidly, unless an oxidation inhibitor is added to the resin. Phenothiazine, thiourea, and 1-acetyl-2-thiourea are examples of effective stabilizers for poly(ethylene oxide) resins. They are generally used at a level of 0.05 to 0.5 wt% of the resin.

Association Complexes Because the paired, ether-oxygen electrons in these polyethers have a strong affinity for hydrogen bonding, poly(ethylene oxide) resins form association complexes with a wide variety of monomeric and polymeric organic compounds as well as certain inorganic electrolytes. Table 19.2 lists compounds which have been reported to form complexes with these resins.

TABLE 19.2
Compounds which Form Association Complexes

Organic	Inorganic
Carboxymethyl dextran	Ammonium fluoride
Catechol tannins	Ammonium thiocyanate
Dichlorophene	Bromine
Gelatin	Cadmium chloride
Hexachlorophene	Iodine
Lignin sulfonates	Mecuric halides
2-Naphthol	Potassium halides
Phenolic resins	Potassium thiocyanate
Poly(acrylic acid)	Rubidium halides
Polyureas	Sodium fluoride
Poly(vinyl methyl ether–maleic anhydride)	
Sodium carboxymethylcellulose	
Thiolignin	
Thiourea	
Trimethylol phenol	
Urea	

The formation of an association complex with one of these compounds generally produces a product with properties and utility quite different from those of either component. Table 19.3 summarizes the property modifications and uses of some of these complexes.

Physical Properties

Bulk Properties Table 19.4 summarizes the typical physical properties of poly(ethylene oxide) resins in the molecular weight range of 1×10^5 to 5×10^6.

Morphology. These poly(ethylene oxide) resins have a sharp crystalline melting point at about 65°C, and they retain a high degree of crystallinity far above the melting point. The glass transition temperature, T_g, decreases slightly with increasing molecular weight from −45 to −53°C.

Thermoplastic Properties. Above the crystalline melting point, these polymers become thermoplastic solids which can be formed by the common thermoplastic processing techniques, i.e., molding, extrusion, or calendering. However, the melt viscosity of the medium-molecular-weight, thermoplastic-processing grades at temperatures 100 to 150°C above the melting point is still very high, in the range of

10^4 to 10^5 poise (see Fig. 19.1). As a result, it is frequently necessary to incorporate a plasticizing agent in the resin in order to achieve practical thermoplastic processing conditions. Polymer blends of poly(ethylene oxide) resins with other thermoplastic polymers usually provide a combination of properties not possible with either component

TABLE 19.3
Association Complexes of Poly(Ethylene Oxide)[40]

Property modification	Electron acceptor
Modified crystallinity	Cadmium chloride Mercuric halides Poly (acrylic acid) Thiourea Urea
Improved tensile strengths and plastic processability/films	Gelatin Novolak resins Poly(acrylic acid) Polyureas Resole resins Trimethylol phenol
Adhesives: water soluble; solvent soluble; glass, metal, and wood	Catechol tannins Lignin sulfonates 2-Naphthol Novolak resins Poly(acrylic acid) Resole resins
Controlled release	Bromine Dichlorophene Hexachlorophene Iodine
Microencapsulation	Gelatin Poly(acrylic acid) Poly(vinyl methyl ether–maleic anhydride)
Electroviscous effect	Ammonium fluoride Ammonium thiocyanate Potassium halides Potassium iodide Potassium thiocyanate Rubidium halides Sodium fluoride

alone. Poly(ethylene oxide) resins are compatible with a number of plasticizers, additives, and other polymers (see Table 19.5).

The melt viscosity of these resins is also highly dependent on rate of shear. Figure 19.2 shows that melt viscosity decreases dramatically with increasing shear rate, indicating the pseudoplastic character of these resins above their melting point.

TABLE 19.4
Typical Physical Properties of Poly(Ethylene Oxide) Resins

Appearance	White powder
Average particle size	150 μm
Melting point	65°C
Density at 20°C	1.21 g/cm^3
Bulk density	0.384 g/cm^3 (24 lb/ft^3)
Volatility at 105°C	<1.0% by wt

Solution Properties In addition to complete solubility in water the resins are also very soluble in a number of other solvents, particularly chlorinated hydrocarbons. Limited solubility exists in a wide variety of other organic solvents, but solubility of the resin in these solvents can only be achieved at elevated temperatures; and the resin may precipitate from solution at or above room temperature (see Table 19.6).

Aqueous Solution Properties. Although poly(ethylene oxide) resins are completely soluble in cold or warm water, they do undergo an inverse solubility relationship near the boiling point of water.[35] Thus, at about 98°C the resin becomes insoluble in water and precipitates (see Fig. 19.3).

Effect of pH. The upper temperature limit of solubility is also a function of pH. The effect is shown in Fig. 19.4. Between pH 2 and 10, the upper temperature limit of solubility remains at 98°C; below pH 2 it rises to above 100°C, and above pH 10 it drops sharply.

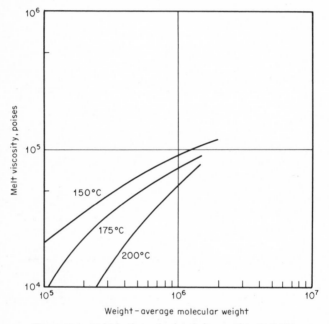

Figure 19.1   Melt rheology of poly(ethylene oxide). *(Ref. 35.)*

Effect of Salts. The presence of certain inorganic salts in the solution can have a dramatic effect on the solubility of poly(ethylene oxide) resins in water. The effect of salts on the upper temperature limit of solubility is essentially a linear function of the salt concentration, although different salts have quite different effects on the solubility, as shown in Fig. 19.5.

The rheological properties of aqueous solutions of these resins are discussed in detail in Rheological Properties, page 19-8.

Manufacture

Although no details of the methods of commercial manufacture of high-molecular-weight poly(ethylene oxide) have been made public, a study of the patent literature provides information about a great variety of successful methods for the polymerization of ethylene oxide. Polymers with molecular weights greater than 1×10^5 are prepared by heterogeneous catalytic polymerization in a low-boiling aliphatic hydrocarbon such as hexane or isooctane, using any one of several metal catalyst systems.

Early work involved the use of alkaline earth oxides as catalysts for the polymerization.

However, rates were extremely slow, and it was not until the late 1950s that more active and commercially feasible catalyst systems were developed. Since then, different catalyst systems have proliferated and include alkaline earth carbonates; hydrates of ferric chloride, acetate, and bromide; divalent metal amides; and hexammoniates and alkyls and alkoxides of aluminum, zinc, magnesium, and calcium, to name but a few.

In the heterogeneous catalytic polymerization of ethylene oxide using the above-mentioned catalysts, the reaction is believed to occur at the catalyst surface. It is suggested that the polymerization occurs in four steps, including physical absorption of the monomer on the catalyst, ring opening at the carbon-oxygen bond, formation of chemisorbed polymer, and growth of the polymer chains.[23]

TABLE 19.5
Compatibility of Poly(Ethylene Oxide) with Various Compounds[35]

Compound, plastic or rubber	Ratio of poly(ethylene oxide) to compound	Compatibility
Acetyltriethyl citrate	1:3	Incompatible
Acrylonitrile-butadiene-styrene terpolymer	1:9	Partially compatible
Acrylonitrite-butadiene rubber (NBR)	1:9	Partially compatible
Acrylonitrile-styrene copolymer	1:9	Partially compatible
Butadiene-styrene rubber (SBR)	1:9	Incompatible
Butyl phthalylbutyl glycolate	1:3	Partially compatible
Cellulose acetate butyrate	1:9	Partially compatible
Dextran	1:9	Partially compatible
Diisopropyl sebacate	1:3	Partially compatible
Dipropylene glycol	1:3	Incompatible
Glycerol monoacetate	9:1	Compatible
Glycerol diacetate	9:1	Compatible
Glycerol triacetate	9:1	Compatible
Methylphthalylethyl glycolate	1:3	Compatible
Poly(chloroprene)	1:9	Partially compatible
Poly(ethylene) (low density)	1:9	Compatible
Poly(ethylene) (high density)	1:9	Partially compatible
Poly(ethylene glycol) 400	1:3	Incompatible
Poly(ethylene glycol) 1500	9:1	Compatible
Poly(methyl methacrylate)	1:9	Partially compatible
Poly(propylene)	1:9	Partially compatible
Poly(propylene glycol) 2025	1:3	Incompatible
Poly(styrene)	1:9	Partially compatible
Poly(vinylbutyral)	1:9	Partially compatible
Propylene glycol linoleate	1:3	Incompatible
Sorbitol	1:3	Incompatible
Starch	1:3	Partially compatible
Vinyl chloride–vinyl acetate copolymer	1:9	Compatible

Biological/Toxicological Properties

Toxicological Studies Extensive testing with dogs, rats, and rabbits by Union Carbide and of a variety of marine life by the U.S. Naval Undersea Center indicates that poly(ethylene oxide) resins with molecular weights from 1×10^5 to 1×10^7 have a very low level of oral toxicity. They are not readily absorbed from the gastrointestinal tract. The resins are relatively nonirritating to the skin and have a low sensitizing potential. They caused only trace eye irritation when rabbits' eyes were flooded with a 5% aqueous solution of the resin. There was little or no effect on the death rate of fish, crabs, sea anemone, lerine shrimp, or algae when exposed to resin concentrations ranging from 250 to 2000 ppm for periods of 3 to 50 days. Hence both dry resin as well as aqueous resin solutions have a low level of toxicity and can be handled quite safely.

The U.S. Food and Drug Administration recognizes the use of Union Carbide's resins in certain food and related applications. Table 19.7 summarizes the status of these resins under the Food Additives Amendment to the Federal Food, Drug, and Cosmetic Act.

Biodegradability   High-molecular-weight poly(ethylene oxide) resins have a very low level of biodegradability. Data on the biochemical oxygen demand (BOD) as a function of concentration in aqueous solution confirm this. After 5 days the BOD is only 2% of the total oxygen demand (TOD) of the resin, and after 20 days the BOD is only 4% of the TOD.

Union Carbide's Polyox® resins are also approved by the U.S. Environmental Protection Agency for unrestricted use as an inert ingredient in pesticide formulations applied to crops up to the time of harvest. [See CFR Section 180.1001 (d).]

Rheological Properties

Aqueous solutions of poly(ethylene oxide) resins display pseudoplastic character at concentrations as low as a few hundredths of a weight percent polymer. High-molecular-

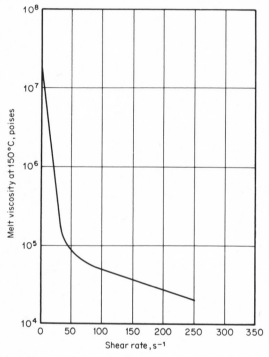

Figure 19.2   Melt rheology of poly(ethylene oxide) as a function of shear rate. *(Ref. 35.)*

weight grades ($> 1 \times 10^6$) form aqueous gels at concentrations greater than 10 wt% polymer. In addition, aqueous solutions tend to be pituitous and form long strings or filaments when immersed objects are withdrawn from the solution. The degree of pituity increases as the molecular weight increases and the molecular weight distribution becomes broader. Therefore, the lower-molecular-weight, narrow-distribution resins, such as the Polyox WSR-N grades, display very little pituity, and high-molecular-weight grades ($> 1 \times 10^6$) are quite pituitous.

Viscosity   The viscosity of aqueous solutions of these resins depends on molecular weight, concentration, temperature, and, because of their pseudoplastic character, the rate of shear applied during the viscosity measurement. Figure 19.6 shows that viscosity increases directly as either the molecular weight of the resin or its concentration in aqueous solution increases.

The relationship of the aqueous solution viscosity to temperature, presented in Fig. 19.7, shows that viscosity decreases significantly as the solution temperature is increased from 10°C up to 90°C.

As noted previously, aqueous solutions display pseudoplastic characteristics. Thus, the viscosity of these solutions decreases as the rate of shear increases (see Fig. 19.8). The degree of pseudoplasticity decreases as the molecular weight decreases, so that solutions of resin with a molecular weight of 1×10^5 are nearly Newtonian. However, irreversible shear degradation of the polymer can occur at very high shear rates. This irreversible degradation results from mechanical cleavage of the polymer chain. Aqueous solutions of highest-molecular-weight species are most susceptible to this type of shear degradation, with sensitivity decreasing as molecular weight decreases.

TABLE 19.6
Solubility of Poly(Ethylene Oxide)[43]
(Solution concentration approximately 1% by weight)

Solvents	Temperature at which poly(ethylene oxide) precipitates on cooling, °C	Temperature at which poly(ethylene oxide) dissolves on heating above 25°C, °C
Stay in solution at 0°C:		
Water	Below 0	
Acetonitrile	Below 0	
Ethylene dichloride	Below 0	
Trichlorethylene	Below 0	
Methylene dichloride	Below 0	
Stay in solution at room temperature:		
Benzene	2	
Isopropanol, 91%	2	
Dimethylformamide	12	
Tetrahydrofuran	18	
Methanol	20	
Methyl ethyl ketone	20	
Must be heated to dissolve:		
Toluene	20	30
Xylene	20	30
Acetone	20	35
Cellosolve acetate	25	35
Anisole	0	40
1,4-Dioxane	4	40
Ethyl acetate	25	40
Ethylenediamine	26	40
Dimethyl Cellosolve	27	40
Cellosolve solvent	28	45
Ethanol, 200 proof	31	45
Carbitol solvent	32	50
Butanol	33	50
Butyl Cellosolve	33	50
Butyl acetate	34	50
Isopropanol	37	50
Methyl isobutyl ketone	40	50
Diethyl Cellosolve	46	50

Additives/Extenders

In most of the applications for these resins the major additive used is an oxidative stabilizer to inhibit loss of molecular weight. The types of stabilizers used in solutions and the types used to stabilize the bulk resins were discussed earlier under Oxidative Degradation, page 19–3. Other additives used in combination with poly(ethylene oxide) resins are plasticizers for use in the thermoplastic processing of films, extrusions, and molded forms of the resins. These plasticizers perform two important functions: First, they lower the melt viscosity of the resin, making it less susceptible to degradation by shearing action during thermoplastic processing, and second, they improve its resistance to "stress cracking." Films of these resins tend to crack when only minor stress is applied. This is accelerated by exposure to ultraviolet (UV) light. However, this undesirable characteristic can be inhibited by the addition of plasticizers in combination with UV radiation and oxidation inhibitors. The most commonly used plasticizers are the nonylphenol polyglycol ether–type non-ionic surface-active agents. Preferably, these plasticizers should contain between 13 and 20 moles of ethylene oxide in the polyglycol chain. They are generally used in concentrations up to 20 wt% of the composition.[17]

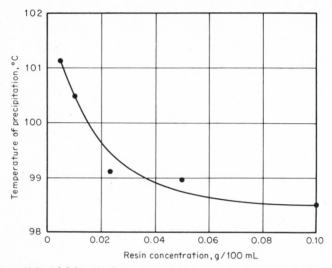

Figure 19.3 Solubility of poly(ethylene oxide) in water vs. temperature. *(Ref. 43.)*

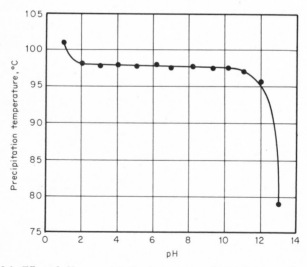

Figure. 19.4 Effect of pH on precipitation temperature of poly(ethylene oxide). *(Ref. 35.)*

Water may also be used as a fugitive plasticizer for poly(ethylene oxide) resins during thermoplastic processing. Concentrations of 10 to 20 wt% of the composition are typically used. However, care must be exercised when using water, since it volatilizes readily and causes bubbles or pinholes if the temperature of the melt exceeds about 100°C.

Handling

Poly(ethylene oxide) resins are dry, free-flowing powders which contain a small amount of very fine particle-size resin which causes dusting during handling. Dust respirators should be worn when handling these resins.

The resins are flammable, having a TAG open cup flash point of approximately 240°C (460°F) and an autoignition temperature, measured according to ASTM D-1929, of about 330°C (626°F). They present no unusual fire or explosion hazards and require no special firefighting procedures. Small fires should be extinguished with carbon dioxide or dry chemical extinguishers, larger fires with alcohol-type foam extinguishers. Addition of water will reduce the burning rate.

When resins are handled in a closed system, the possibility exists that the concentration

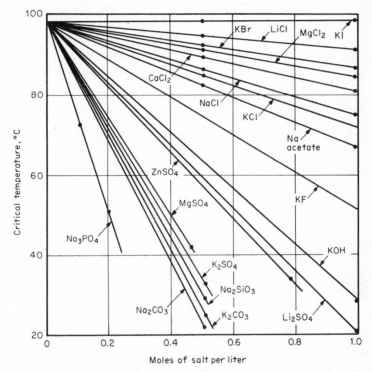

Figure 19.5 Solubility of poly(ethylene oxide) in aqueous salt solutions. *(Ref. 43.)*

of dust could build up to a level where a dust explosion could occur. This fact should be considered in designing handling systems such as bulk-storage bins and pneumatic conveyors. In open handling systems the hazard is negligible.

Spills of the dry resin must be very carefully handled. The spilled resin should first be thoroughly swept up and placed in containers. The residue should be treated with undiluted household bleach for a period of 30 min. Finally, the spill area should be thoroughly flushed with water until traces of slipperiness are gone. Care must also be taken to avoid stepping on the dry resin before or during cleanup since the resin may adhere to the bottoms of shoes and can result in later slipperiness if the bottoms of the shoes become wet.

Applications

Since their commercialization in 1958, the reported and established applications for poly(ethylene oxide) resins have been numerous and diversified. In this section we have attempted to classify these applications: first by effect or result obtained, then by differentiating between formulated end products and process additives, and finally by the industry which uses resins. Table 19.8 summarizes the applications of these resins using this format.

A detailed discussion of these applications for the resins appears later in this chapter.

Application Procedures

Because poly(ethylene oxide) resins are so very soluble in water, special techniques are required in the preparation of aqueous solutions of these resins.[41] The key to dissolution is good initial dispersion of the resin in the solvent. In water, the resin is instantly wetted. If the resin powder is not properly dispersed, the partially dissolved, solvated particles will agglomerate and form gels which will require prolonged stirring to dissolve. Organic solvents do not generally wet the resin particles as rapidly as does water, and therefore do not present as serious a problem with particle agglomeration and gel formation.

TABLE 19.7
Status of Polyox® Resins Under the Food Additives Amendment to the Federal Food, Drug and Cosmetic Act

Regulation	Use
Sec. 121.1161	To be used as a foam stabilizer in fermented malt beverages. This coverage applies to Polyox® WSR-301 or higher species. Polyox resins content shall not exceed 300 ppm of the total beverage
Sec. 121.2514	As a component of resinous and polymeric coatings applied as a continuous film or enamel over a metal substrate or applied as a continuous film or enamel to any suitable substrate provided that the coating serves as a functional barrier between the food and the substrate and is intended for repeated food-contact use (all grades)
Sec. 121.2526 Sec. 121.2571	As a component of the coated or uncoated food-contact surface of paper and paperboard in contact with fatty, aqueous, and dry foods (all grades)
Sec. 121.2548	Preparation of zinc–silicon dioxide matrix coatings used as food-contact surfaces for reusable bulk containers intended for storing, handling, and transporting food (all grades)
Sec. 121.2550	Manufacture of closures with sealing gaskets for food containers (all grades)
Sec. 121.2559	Adjuvant substance required in the production of or added to impart desired physical and technical properties to xylene-formaldehyde resins condensed with 4,4'-isopropylidene-diphenol-epichlorohydrin epoxy resins for use in coatings in contact with aqueous and dry foods, and beverages (all grades)
Sec. 121.2570	Adjuvant in the manufacture of ethylene–vinyl acetate copolymers used in food-contact applications (all grades)

Two techniques which are recommended for dissolving these resins in water are simple direct addition to boiling water, and predispersion in a water-miscible nonsolvent.

Boiling-Water Dispersion   Poly(ethylene oxide) resins are completely soluble in water at all temperatures up to about 98°C. At or near the boiling point of water, they precipitate from solution. This phenomenon of inverse solubility provides a convenient technique for dispersing and dissolving these resins.

For example, to make 1000 g of a 1 wt% solution, heat 700 g of water to the boiling point and slowly sift 10 g of the resin into the boiling water which is stirred at 700 rpm. After the resin is dispersed, remove the heat and add 290 g of room-temperature water. Reduce the speed of stirring stepwise from 700 to 300 to 60 rpm. Continue stirring at 60 rpm until solution is complete. This solution should be used within 24 h since it contains no stabilizer. If greater storage stability is required, an oxidative stabilizer should be added (see Oxidative Degradation, pages 19-3).

Nonsolvent Dispersion   The most convenient method of preparing large quantities of aqueous solutions of these resins is to first disperse the resins in a water-miscible nonsolvent such as anhydrous isopropyl alcohol, ethylene glycol, propylene glycol, anhydrous ethanol, glycerin, acetone, Cellosolve and Carbitol solvents, or inorganic salt solutions (see Fig. 19.5). The use of alcohols or glycols has the added advantage of stabilizing the solution viscosity against oxidative degradation.

For example, to make 1000 g of 1 wt% solution of the resin, disperse 10 g of resin in 50 g of anhydrous isopropyl alcohol. Add this to 940 g of water which is being stirred at 700 rpm. Then reduce the speed stepwise from 700 to 300 to 60 rpm. Continue stirring at 60 rpm until solution is complete. This technique can also be reversed so that the water is added to the stirred nonsolvent dispersion of resin.

COMMERCIAL USES: Compounding and Formulating

Adhesives

Water-Soluble Paper Adhesives   In the manufacture of paper products wound on a core, such as paper toweling or toilet tissues, it is necessary to adhere the start of

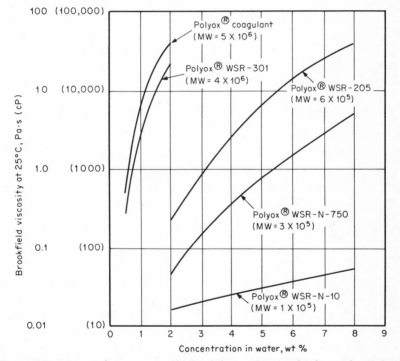

Figure 19.6   Aqueous solution viscosity of poly(ethylene oxide) vs. molecular weight. *(Ref. 43.)*

the paper sheet to the core with a suitable adhesive called the "pick-up" adhesive. Similarly, the last sheet on the roll must be bonded to the ply underneath to prevent unwinding during handling, packing, and other operations. This requires the use of a "tail-tie" adhesive. The compositions of the pick-up and tail-tie adhesives are usually very similar.

Traditionally, hot-melt–type adhesives have been used in this application. However, they have the disadvantage that they form a permanent bond between the paper and the core or between layers of paper at the end of the roll. This causes undesirable tearing of the sheets as paper is later removed from the roll.

Water-soluble pick-up and tail-tie adhesives are now manufactured using aqueous solutions of poly(ethylene oxide) resins in combination with other ingredients such as colorants and fillers. Exact compositions are proprietary, although it is known that medium-molecular-weight grades of resin (3×10^5 to 6×10^5) are preferred, and that a relatively broad molecular weight distribution is also desirable. These particular resins are used because they provide the required combination of thickening efficiency and pituitousness (stringiness). Pituitousness is important to the application because it enhances the wet tack of the composition.

Perhaps the most desirable feature of pick-up and tail-tie adhesives made with these resins is that when the adhesive dries it loses its adhesive properties. Thus, the problem of tearing of paper as it is removed from the roll is eliminated.

Adhesives from Association Complexes A variety of different types of adhesives can be produced by forming association complexes of poly(ethylene oxide) with catechol

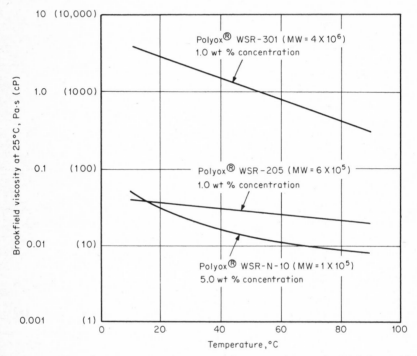

Figure 19.7 Effect of temperature on the viscosity of aqueous solutions of poly(ethylene oxide). *(Ref. 43.)*

tannin or phenolic resins.[46] A few examples of the types of adhesives which can be prepared from these association complexes include the following:

Wood Glue. A 5 wt% aqueous solution of 6×10^5 molecular weight poly(ethylene oxide) resin mixed with mimosa extract powder (a catechol tannin) in a 2:1 weight ratio produces an excellent quality wood glue. Douglas fir plywood panels were coated with this glue and allowed to air-dry several hours, forming a tacky coating which adhered readily when two pieces of plywood were pressed together. The plywood panels were then clamped together, and the glue was allowed to set overnight. When separated, failure occurred in the wood section rather than the glue joint.

Water-Soluble Quick-Set Adhesive. A 6 wt% aqueous solution of poly(ethylene oxide) resin (6×10^5 molecular weight) mixed with an equal amount of novolak-type phenolic resin produces an association complex that is a soft, gumlike material which makes a very good water-soluble quick-set adhesive.

Pressure-Sensitive Adhesives. Good pressure-sensitive adhesives are prepared by mixing a 5 wt% solution of 6×10^5 molecular weight poly(ethylene oxide) resin in

trichloroethylene with a 10 wt% solution of novolak-type phenolic resin to produce a
1:1 ratio of the two resins.   The clear yellow solution thus formed is cast into a film
which is baked at 150°C for 30 min.   The film is then redissolved in a mixture of
trichloroethylene and ethylene glycol monoethyl ether to produce a pressure-sensitive
adhesive solution.

Industrial Supplies

Thickened Cleaning Solutions   Poly(ethylene oxide) resins can be used to increase
the utility of many cleaning formulations, including organic acids, mineral acids,
ammonium hydroxide, and other liquids.   Small amounts of this resin dramatically
increase the viscosity and control the flow properties of such formulations, making
possible:

1. Compositions ranging from viscous liquids having a syrupy consistency to rubber-
like texture, giving the user versatility in handling.

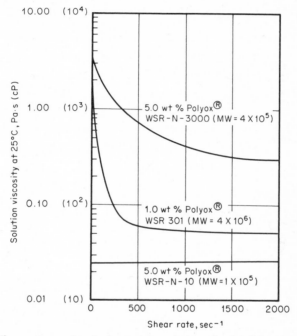

Figure 19.8   Shear rate vs. viscosity for aqueous solutions of poly(ethylene oxide).   *(Ref. 43.)*

2. "Fixing" the reactivity at the desired site.   This is especially useful on vertical
surfaces, where drip or run-off can be troublesome.

3. Controlled release of the thickened liquid to aqueous systems.

Examples of liquids that can be thickened or gelled are organic acids (citric acid, oxalic
acid), mineral acids (phosphoric acid, hydrochloric acid), bases (ammonium hydroxide),
and surfactants.

The control of viscosity of various liquids for specialized purposes is often very important, especially when the liquid is reactive.   For example, when cleaning vertical surfaces
with muriatic acid, the acid often runs and affects unintended areas.   The minimization
of spatter and the gain of controlled flow for convenient dispensability are important
properties for both industrial and consumer applications.   For example, sanitary cleaners
for vitreous surfaces can be dispensed from plastic containers safely and conveniently.
Solid cakes of other liquids, such as nonaqueous detergents or special-purpose alcohols,
can be prepared for slow release of the liquid.

These resins will not thicken the caustics, sodium or potassium hydroxides, because

TABLE 19.8
Applications of Poly(Ethylene Oxide) Resins

| Function/effect | Type of application | | Industry |
	Formulated end product	Process additive	
Adhesive	Denture fixative		Personal-care products
	Paper adhesives		Paper
	Association complexes		Construction
Binder		Ceramics	Ceramics
		Battery electrodes	Electrical
		Fluorescent lamps	Electrical
		Soil stabilization	Agriculture
Coating/sizing	Microencapsulated ink		Printing
		Fiber formation aid	Paper, construction
		Fiberglass size	Glass
		Tablet coatings	Pharmaceutical
Dispersant	Calamine lotion		Pharmaceutical
	Vinyl polymerization		Chemical
		Glass-fiber-reinforced concrete	Construction
Flocculation		Clay removal	Mining
		Coal washing	Mining
		Filler-retention aid	Paper
		Hydrometallurgy	Mining
		Silica removal	Mining
		Water clarification	Paper
Hydrodynamic-drag reduction		Fire-fighting additive	Municipal services
		Fluid-jet cutting	Industrial supplies
		Additive to prevent sewage surcharge	Municipal services
Lubricant/emollient	Detergents, hand soaps		Soap, detergent
	Creams, lotions,		Personal-care products
	Ophthalmic solution		Personal-care products
	Rubbing alcohol		Pharmaceutical
	Shaving preparation		Personal-care products
	Tire-mounting lubricant		Industrial supplies
	Toothpaste		Personal-care products
Thermoplastic	Purge dam		Construction
	Seed tape		Agriculture
	Water-soluble packaging		Agriculture
		Dyeability aid	Textile
		Fugitive textile weft	Textile
		Paper antistat	Paper
		Textile antistat	Textile
Thickener	Cleaning solutions		Industrial supplies
	Latex paint		Paint
	Press dampening fluid		Printing
	Paint and varnish remover		Paint Construction
	Paving composition		Paint
	Spatter finish		Food
		Beer foam stabilizer	Petroleum
		Drilling muds	Agriculture
		Drift-control agent	Industrial supplies
		Metal-working lubricant	Petroleum
		Secondary recovery fluids	
		Slurry for trenching	Construction
		Well-completion fluids	Petroleum
Water retention	Catamenial device		Personal-care products
	Diapers		Personal-care products
	Soil amendment		Agriculture
		Asbestos-cement extrusion	Construction

the resin will not dissolve in such solutions. However, it will thicken ammonium hydroxide. By selecting different grades of resin, a wide range of viscosities of acid cleaning solutions can be obtained.

The selection of the appropriate molecular weight resin sometimes involves a compromise. Although the high-molecular-weight resins have the greatest thickening efficiency, they also produce the greatest pituity in solution. So it is frequently desirable to use a low-molecular-weight, narrow-molecular-weight-distribution grade resin to avoid this problem. (See Formulations, page 19-27.)

Tire-Mounting Lubricant   Dilute solutions of these resins also produce very effective lubricants for rubber and have been used to formulate compositions which are effective lubricants in the mounting of tires on metal rims. The high-molecular-weight resins are preferred and are used in concentrations from 0.02 to 0.1 wt% (see Formulations, page 19-27).

Construction Products

Paving Composition   One problem frequently encountered with paving compositions, consisting of mixtures of aggregate and bituminous emulsion, is control of the viscosity of the bituminous emulsion. If the viscosity is too low, the emulsion will tend to run off the aggregate, failing to properly bind it. If it is too high, the emulsion will not flow properly around the aggregate so that the mixture will not be uniform in composition and will not bond properly to the subsurface. Furthermore, since the moisture content of the aggregate may vary broadly, it too can change the viscosity of the emulsion when the aggregate is added to it.

In order to achieve better control over the viscosity of the mixture, poly(ethylene oxide) resin is added to the bituminous emulsion at a concentration of 0.24 to 7.20 kg resin/1000 L emulsion (2 to 60 lb/10,000 gal), depending on the original viscosity of the emulsion and the water content of the aggregate. In addition, it may be possible to reduce the bitumen content of the emulsion without loss of viscosity by addition of these resins. The result is an emulsion which is less susceptible to viscosity loss due to changes in moisture content of the aggregate, giving a paving composition of more uniform performance.[25]

Water-Soluble Purge Dam   It is usually necessary to provide an inert atmosphere inside the piping or tubing and around the tungsten electrode during the tungsten-arc welding of piping or tubing. When the ends of the pipe or tube are not easily accessible, it is normal practice to install a dam inside the pipe to exclude air or other undesirable gases from the area and to contain an inert atmosphere around the weld site. Paper or cardboard dams have been used in the past, but they must be removed eventually by burning them away by applying an external flame to the pipe. This is a time-consuming practice which can also leave undesirable residues in the line.

These problems have been eliminated by using water-soluble film made from poly(ethylene oxide) resin to dam the pipe. After welding is complete, the pipe is flushed with water to dissolve the dam, leaving no residue in the pipe. The properties of water-soluble poly(ethylene oxide) films suitable for this use are presented in Table 19.9.

Paints and Paint Removers

Latex Paints   Although the pituitous nature of solutions of the high-molecular-weight poly(ehtylene oxide) resins eliminates them from consideration in thickening most latex paints, it is possible to use the low-molecular-weight, narrow-molecular-weight-distribution resins in conjunction with more efficient thickeners such as hydroxyethylcellulose to raise the true high-shear viscosity of latex paints, thus improving their behavior during application and permitting thicker and more uniform films to be applied. At the same time, these poly(ethylene oxide) resins will contribute significantly to improved flow, leveling, and gloss after application.

Spatter Finish   The pituitousness of the aqueous solutions of high-molecular-weight species can be used to advantage in the preparation of spatter finishes used to achieve decorative pattern effects.

In general, the processing of these paints follows standard pigmentation and let-down procedures, except that cellulosic thickeners should be omitted from the recipe. The finished but unthickened paint is modified with a 3 wt% aqueous solution of 4×10^6 molecular weight poly(ethylene oxide) resin and sufficient water to produce a sprayable consistency. The amount of resin added will depend on the specific effect to be produced.

Standard spray guns and typical air pressure, 0.21 to 0.28 MPa (30 to 40 psi), can be used for application. Considerable variation in the type of pattern and the amount deposited can be produced by changing the viscosity, air pressure, and distance between the gun and the work. Roller or brush application does not produce desirable results.[20]

Thickener for Paint and Varnish Remover   Because these resins are very soluble in chlorinated hydrocarbons, they can be used as effective thickeners for paint and varnish remover formulations which are primarily based on methylene chloride. Normally, the lower-molecular-weight, narrow-molecular-weight-distribution grades are desired to avoid problems with pituity. Resin concentrations in the range of 3 wt% produce formulations with good flow properties which do not readily drain off sloping or vertical surfaces (see Formulations, page 19-27).

Pharmaceuticals

Dispersant for Calamine Lotion   The tendency of calamine lotion to settle during storage is well known. Poly(ethylene oxide) resin has been successfully used as a suspending agent in calamine lotion. In addition to inhibiting the rapid settling of the solid components of the formula, it also provides good film-forming properties which help to spread the lotion evenly and leave a smooth, supple feeling on the skin (see Formulations, page 19-27).

Rubbing Alcohol   Rubbing alcohols have the disadvantages of being extremely volatile at room temperature, and they lack lubricity. Thus, they can irritate the skin and they do not permit adequate time for proper massage. The addition of small concentrations of high-molecular-weight poly(ethylene oxide) resins provides desirable lubricity, imparts a smooth, supple feel to the skin, and appears to retard volatilization of the alcohol. A concentration of about 0.1 wt% of 4×10^6 molecular weight resin in the final alcohol solution provides the desired effect.[4]

Printing Products

Microencapsulated Inks   Nonaqueous printing inks can be microencapsulated with an association complex of poly(ethylene oxide) and poly(acrylic acid). The products of this process can then be used as dry, free-flowing powders to produce "carbonless" carbon papers, for example. When pressure is applied to the paper coated with the dry, microencapsulated ink, the capsule wall is fractured and the ink is released.

Various water-immiscible liquids and solids can be encapsulated in a poly(ethylene oxide)–polycarboxylic acid association complex by taking advantage of the pH-controlled inhibition of the complex formation. The material to be encapsulated is emulsified using a suitable non-ionic surfactant in an aqueous solution of the polycarboxylic acid and poly(ethylene oxide) resin. The emulsion is then acidified to form the association complex at the interface between the phases of the emulsion, producing a coating around the water-immiscible phase and precipitation of the microcapsules, which are then dried by conventional techniques (see Formulations, page 19-27).

Lithographic Press Dampening Fluid   The volatility and misting of lithographic press dampening solutions is substantially reduced by incorporating low-molecular-weight (1×10^5) poly(ethylene oxide) resin in the solution. This improved dampening solution reportedly reduces the loss due to misting to less than 1% of the loss experienced with conventional solutions (see Formulations, page 19-27).

Soap, Detergents, and Personal-Care Products

Detergents   *Synthetic Detergent Bars.*   Poly(ethylene oxide) resins show promise as lubricants and emollients for synthetic detergent bars. Low concentrations of the resin completely overcome the tackiness of the alkyl aryl sulfonates and can give the detergent lather a lubricating effect that exceeds that of soap. Resin concentrations of 2 wt% of the bar contribute good foaming, foam stability, and improved dry and wet feel[4] (see Formulations, page 19-27).

Liquid Detergents.   Liquid detergents containing these resins exhibit thick lubricous lather and leave the skin with a smooth, supple feeling which is not characteristic of liquid detergents without poly(ethylene oxide) resin. Concentrations required are quite low, in the order of 0.05 to 0.1 wt% of the formulation. Low-molecular-weight resins with narrow-molecular-weight distribution are generally preferred because they do not impart any significant pituitousness to the product (see Formulations, page 19-27).

Toothpastes   Poly(ethylene oxide) resins are generally more compatible with a wide range of toothpaste components than are many other synthetic gums and resins. The addition of these resins to the formula contributes lubricating qualities to the toothpaste and provides a more pleasant feeling in the mouth than is experienced when other gums are used. These resins can be used in combination with many types of abrasives. Humectants, such as glycerol, propylene glycol, and sorbitol, generally added to prevent hardening of the paste, are compatible with the resins provided only small quantities of these components are used. Various anticaries agents may also be included if desired[4] (see Formulations, page 19-27).

Denture Fixative   One of the largest adhesive applications for these resins is in the personal-care industry as an additive in denture fixatives. Here, high-molecular-weight resin is combined with natural gums, fillers, and flavorings to provide a fixative to hold dentures firmly in place while providing a comfortable cushioning effect between the denture and the mouth. They are produced in both powdered and paste forms. Paste forms of these adhesives have liquid petrolatum added to the formulation. Generally, about 5% of the composition is poly(ethylene oxide). When placed in the mouth, saliva wets the natural gum and poly(ethylene oxide) resin, producing a solution with excellent wet-tack properties. The high-molecular-weight resins used are recognized by the U.S. Food and Drug Administration as an additive to foods and similar products.

Shaving Stick   High concentrations of poly(ethylene oxide) resin in water can form firm gels. This characteristic is the basis for the development of a preshaving preparation which is a substitute for conventional shaving creams.

By preparing a solution containing 20 wt% of 4×10^6 molecular weight resin, one obtains a stiff gel which can be stroked on the face to provide a lubricating coating on the skin prior to shaving. The gel may also contain about 20 wt% isopropyl or ethyl alcohol as a viscosity stabilizer and bactericide. Perfumes and colorants may also be added to the formula.[16]

Ophthalmic Solution   In a recent development[30] an aqueous solution of poly(ethylene oxide), in combination with other water-soluble additives, having a viscosity of 0.150 Pa·s (150 cP) is used as an ophthalmic solution for wetting, cleaning, cushioning, and lubricating "hard" contact lenses. A lower viscosity (0.010 Pa·s, or 10 cP) solution containing the same ingredients was found beneficial with "soft" contact lenses.

Absorbent Pads   Chemical or irradiation crosslinking of poly(ethylene oxide) resins produces products called *hydrogels* which are no longer water-soluble polymers, but rather are water-absorptive. Frequently these hydrogels can absorb 25 to 100 times their own weight of water.

Although details of the chemical crosslinking technique have not been made public, it is known that the irradiation crosslinked species can be prepared by exposing a 4 wt% aqueous solution of high-molecular-weight poly(ethylene oxide) to an electron beam of about 1 MeV intensity to provide a total irradiation dose from about 0.5 to 2 Mrad. This treatment converts the aqueous solution to a firm, crosslinked hydrogel which can be used directly or granulated. These hydrogels are reportedly useful in the manufacture of absorptive pads for catamenial devices and disposable diapers.[19]

Water-Soluble Films

Above their melting point these resins exhibit rheological properties typical of thermoplastics and can therefore be formed into various shapes using conventional thermoplastic processing techniques. Generally, the medium-molecular-weight resins (4×10^5 to 6×10^5) possess melt rheology best suited to thermoplastic processing. Plasticizers, stabilizers, and fillers may be added to the resin (see Additives/Extenders, page 19-9). Union Carbide produces a product called WRPA-3154 specially formulated for thermoplastic processing (see Thermoplastic Compound, page 19-3).

Commercially, thermoplastic processing of these resins has been limited almost exclusively to the manufacture of film and sheeting. These films are produced by calendering or blown-film extrusion techniques. The films are usually produced in thicknesses from 1 to 3 mils, and have very good mechanical properties combined with complete water solubility. (See Table 19.9 for film properties.)

Seed Tape   One of the earliest uses for these films was in the agricultural area, where they were used to produce seed tapes. These consist of seeds sandwiched between two narrow strips of poly(ethylene oxide) film sealed at the edges. The seed tape is planted, and water from the soil dissolves the water-soluble film within a day or two, releasing the seed for germination. The advantage of the seed tape is that it virtually eliminates the need for thinning of crops since the seeds are properly spaced along the tape at the time of manufacture.

Water-Soluble Packaging   Films of poly(ethylene oxide) are also used to manufacture water-soluble packages for preweighed quantities of insecticides, pesticides, fertilizers, detergents, dyestuffs, and the like. These packages dissolve quickly in water, releasing the contents. Water-soluble packages eliminate the need for weighing the active ingredients to control the concentration in solution and offer protection to the user from toxic or hazardous substances, i.e., dyestuffs.

In comparison with other commonly used water-soluble films, poly(ethylene oxide) films offer the advantage of good compatibility with most of the materials commonly

TABLE 19.9
Typical Properties of Poly(Ethylene Oxide) Film[42]

Thickness	38.1, 50.8, or 76.2 μm (1.5, 2.0, or 3.0 mils)
Specific gravity	1.20
Yield	12,656 cm^2/kg/μm (22,600 in^2/lb/mil)
Tensile strength:	
Machine direction (MD)	15.2–16.5 MPa (2200–400 psi)
Transverse direction (TD)	12.4–13.8 MPa (1800–2000 psi)
Secant modulus:	
MD	290 MPa (42,000 psi)
TD	483 MPa (70,000 psi)
Elongation:	
MD	500–600%
TD	600–700%
Tear strength:	
MD	9.84 g/μm (250 g/mil)
TD	23.62 g/μm (600 g/mil)
Dart impact at 50% failure	7.87 g/μm (200 g/mil)
Release time in water	15 s, approximate
O$_2$ transmission	430 cm^3/mil/100 in^2/24 h (263 cm^3/μm/100 m^2/24 h) at 760 mm
Melting point	66.6°C (151°F)
Heat-sealing capability	Excellent; equal to LDPE film
Heat-sealing temperature range	71–107°C (160–225°F)
Cold crack resistance	−45.6°C (−50°F)
Appearance	Translucent
Storage stability	Good; protect from excess light, heat, and moisture
Max. recommended storage temp.	51.7°C (125°F)

supplied in water-soluble packages, and they also have excellent heat sealability and low-temperature flexibility (see Table 19.9).

COMMERCIAL USES: Processing Aids

Although poly(ethylene oxide) resins are frequently used as components of formulated end products, more often these resins are used as additives to perform some essential function in a variety of industrial processes. This section discusses the application of poly(ethylene oxide) resins as process additives, and is categorized according to the function the resin performs in the process.

Binder

Poly(ethylene oxide) resins have been used in a wide variety of binding applications where the combination of water solubility, low level of inorganic impurities, and clean burn-out characteristics are critical to the process.

Ceramics   In ceramics, the resins are used to improve the green strength of clay and refractory compositions. They also provide lubrication of the ceramic body during forming processes. Upon firing of the ceramic, they burn out cleanly. The high-molecular-weight species, above 1×10^6, are generally preferred for this use, and the concentrations used may range from as low as 0.1 wt% to as much as 2 wt% of the ceramic formulation.

Battery Electrodes   By intimately mixing the poly(ethylene oxide) resin on a heated roll mill with another thermoplastic resin with which it is essentially incompatible, it is possible to produce a binder system suitable for battery and fuel cell electrodes. An electrochemically active material is then added to the mix, and it is formed into sheets, rods, etc.   The final step involves leaching-out the poly(ethylene oxide) resin with water, leaving voids and producing electrodes with a microporous structure and more efficient utilization of the electrode-active material.

In practice, medium-molecular-weight poly(ethylene oxide) resin is blended on a roll mill in about 1:1 weight ratio with another thermoplastic resin such as poly(ethylene), poly(styrene), or poly(propylene) at a temperature high enough to plasticize the two resins.   An electrode-active material is added to the plasticized mix at a level about 10 times the weight of the combined thermoplastic resins.   Compounds of cadmium, mercury, silver, nickel, and graphite have been successfully used as the active materials.[8,9]

Fluorescent Lamps   In the manufacture of fluorescent lights it has been found that the bonding of the phosphor to the glass tube is enhanced by using poly(ethylene oxide) resin as a temporary binder in combination with barium nitrate as a bonding agent. The coating is usually applied by preparing a suspension of the phosphor, binder, and dispersant.   The suspension is used to flush the glass tube, leaving a thin coating of phosphor on the inside.   Subsequent drying and lehring removes water, solvents, and organic additives and leaves a phosphor coating bonded to the inside of the tube.

The addition of about 2 wt% poly(ethylene oxide) resin based on the dry phosphor significantly improves the adherence of the phosphor suspension to the glass tube during drying and prior to lehring.   Resin with a molecular weight of 4×10^5 and narrow-molecular-weight distribution is preferred for this application.[3]

Soil Stabilization   The formation of a water-insoluble poly(ethylene oxide)–poly-(acrylic acid) association complex is the basis for a soil-stabilization process to prevent erosion of soil on hillsides and river banks.

An aqueous solution of poly(ethylene oxide) resin containing about 1 wt% medium-molecular-weight resin is sprayed on the soil.   This resin is deposited at a loading of about 0.054 kg/m² of soil (0.1 lb/yd²).   Then a second spray coating of 10 to 20 wt% aqueous poly(acrylic acid) solution is applied over the first coat, at a loading of about 0.272 kg/m² (0.5 lb/yd²).   Although the association complex which forms is insoluble in water, it is nevertheless water-permeable, and the thickness of the coating on the soil can be increased by repeated spraying.[34]

Other Binder Applications   These resins have also been utilized as water-soluble binders in such diverse products as explosives, pharmaceutical tablets, and tobacco filters. High-molecular-weight poly(ethylene oxide) resins have also been suggested as a binder for wood chips to prevent loss during transport in open railway cars.

Coatings and Sizes

Tablet Coatings   Coating of tablets is commercially important to mask tastes, improve appearance, protect tablets from atmospheric moisture, give tablets smooth, rounded edges for ease of swallowing, prevent irritation of mucous membranes, protect tablets from physical damage during handling and shipping, etc.

Among the many advantages of coating tablets with poly(ethylene oxide) resin is that it takes much of the art out of coating.   In addition, the film-coating process is fast; a special coloring operation is not required; neither a subcoating nor an additional moisture-resistant protective coating is usually needed; the size, shape, and weight of the original tablets are retained; embossing on the tablets is easily seen; and the coating techniques are simple, easy to learn, and reproducible.[5]

Glass-Fiber Size   In the production of yarn from staple glass fibers, it is necessary to use a size on the fibers which provides a desirable balance between (1) the lubricity needed to enable the individual fibers to shift lengthwise relative to one another without destruction of the fibers by abrasion, and (2) the drag characteristics required to permit controlled, lengthwise drafting of the yarn by as much as 300%.   Aqueous solutions of low-molecular-weight poly(ethylene oxide) resins in combination with other functional additives such as silane coupling agents and emulsifiers can be formulated to provide the desired combination of lubricity and drag in this sizing operation. The combination of lubricity and pituitousness of aqueous solutions of these resins is believed to hold the key to their successful use in this application.[24]

Dispersant

Vinyl Polymerization   In the polymerization of vinyl monomers using aqueous suspension polymerization processes, it is necessary to add a water-soluble polymer to keep the monomer droplets suspended and uniformly dispersed in the water phase. Poly(ethylene oxide) resins with a molecular weight of 3×10^5 have been successfully used as suspending agents in the suspension polymerization of vinyl chloride, styrene, and methyl methacrylate. Concentrations of resin from 0.1 to 0.2 wt% of the monomer are normally used.[14,32]

Glass-Fiber–Reinforced   Concrete   High-molecular-weight poly(ethylene oxide) resins are also used as dispersants in the manufacture of glass-fiber–reinforced concrete. Typically, 5 to 6 wt% chopped glass fiber is added to a 0.1 wt% solution of the resin. A relatively uniform dispersion of the fibers is obtained, and the usual problems of agglomeration and balling-up of the fibers are avoided. Cement and sand may then be added to this aqueous dispersion to produce the glass-fiber–reinforced concrete mix. Alternatively, the resin solution and glass fiber may be added sequentially to a cement slurry with the same desirable result.

This premixed material can be cast, sprayed, injection-molded, or extruded. The mixing operation produces a random three-dimensional distribution of fibers, but subsequent hand spraying or extrusion produces the preferred two-dimensional orientation.[11,18]

Flocculation

Clays   Hydrometallurgical operations, waste-water clarification, and similar processes present difficult solid-liquid separation problems. Clays are some of the most common constituents of the suspensions formed in these processes. High-molecular-weight poly-(ethylene oxide) resins are very effective flocculents for these clays. Suspensions of clay minerals such as kaolin, montmorillonite, illite, halloysite, sericite, attapulgite, and hectorite can be flocculated with these resins over a wide pH range using resin concentrations as low as 5 ppm, up to about 0.2 wt% of total clay suspension. Some clays react more favorably to flocculation with poly(ethylene oxide) resins than others. For example, the efficiency of the resins with calcium bentonite clay is significantly greater than with sodium bentonite clay.[7]

Coal   Poly(ethylene oxide) resins effectively flocculate many types of coal suspensions. An example of this is the treatment of effluent streams from coal-washing operations. Once again, concentrations as low as 5 parts of resin per million parts of suspension significantly increase the settling rate. The cleaner effluent stream thus obtained has greater washing capacity during recycling, and the sludge obtained is more compact and easily recovered. The resins are particularly effective in flocculating suspensions of oxidized coal without the need for pH adjustments.[39]

Silica   Suspensions of silica often occur naturally in surface waters, causing undesirable turbidity. Clarification of water containing silica is effectively accomplished by addition of high-molecular-weight poly(ethylene oxide) resin. Effective concentrations range from about 5 up to 100 parts resin per million parts of suspension.

In acid leaching of metal ores to solubilize desired metal values, the impure concentrates are sometimes contaminated with colloidal silica as well as solubilized silicas, which can interfere with subsequent processes designed to separate the pure metal from its ore. High-molecular-weight poly(ethylene oxide) resins are very effective in separating the undesirable silicas from these acid-leaching solutions. This technique has been found particularly effective in the acid leaching of uranium-, beryllium-, and molybdenum-containing ores.[12]

Filler-Retention Drainage Aid (Paper Making)   In the process of producing paper from wood pulp, fillers such as clays, titanium dioxide, and calcium carbonate are used to provide opacity and whiteness to the paper. However, a significant percentage of the filler added, as well as fiber fines, may be lost during draining of the wet paper web.

The loss of valuable fillers and fines during the drainage step is significantly reduced by adding 0.025 to 0.05 kg (0.05 to 1.0 lb) of high-molecular-weight poly(ethylene oxide) resin per 1000 kg (ton) of air-dried fiber. The resin is generally added just before the headbox of the paper-making machine in the form of a dilute aqueous solution. These resins also have been found to work synergistically with other common flocculents to improve filler and fines retention and to increase drainage rates.

An additional benefit of using these resins in paper making is that residual amounts of the polymer are very effective in clarifying the "white water" which is formed in the process. They improve the operation of the save-all systems which are frequently used as recovery and disposal mechanisms for paper-making white waters by allowing purer-quality water to be produced as well as effecting substantial increases in the amount of fiber and fillers recovered for reuse in the process.[22]

Hydrodynamic Drag Reduction

The turbulent flow of water through pipes and hoses or over surfaces causes the effect known as *hydrodynamic drag.* This hydrodynamic drag translates into pressure losses in hoses and pipes and frictional drag of water against surfaces. High-molecular-weight poly(ethylene oxide) resin is the most effective polymer currently available to reduce hydrodynamic drag.

Fire-fighting Additive This ability to reduce hydrodynamic drag has been applied in fire fighting where small concentrations of these resins (50 to 100 ppm) reduce the pressure loss in fire hoses and make it possible to transport as much as 60% more water through a standard 6.35-cm (2.5-in) diameter fire hose. Alternate methods of utilizing the hydrodynamic-drag-reduction effect in fire fighting include reducing the size of fire hoses from 6.35-cm (2.5-in) diameter to 4.45-cm (1.75-in) or 3.81-cm (1.5-in) diameter to achieve lighter weight and greater flexibility, decreasing pumping pressures required to obtain a given water flow rate or pumping water over much greater distances without the need for increased pumping pressures or relay pumping.

The addition of the resin to a turbulent stream of water reduces turbulence and provides a more laminar and cohesive stream. As a result, the stream also has a greater range and impact.

Union Carbide is the only producer currently supplying a hydrodynamic-drag-reducing additive for this application. It is called UCAR Rapid Water Additive (see Hydrodynamic-Drag-Reduction Slurry, page 19-3). This additive, which contains high-molecular-weight poly(ethylene oxide) resin as the active ingredient, is injected at a controlled rate into the pump on the fire truck and then into the hose lines, where the resin dissolves in the water, reducing frictional drag and increasing flow capability. Many municipal and volunteer fire companies have already adopted this system.[38]

Fluid-Jet Cutting As noted above, the presence of small concentrations of poly(ethylene oxide) in a turbulent stream of water produces a more cohesive stream with greater impact. This phenomenon has been applied to a new high-pressure jet system for cutting soft goods such as textiles, rubber, foam, cardboard, and the like. In these systems, specially designed nozzles produce a very small diameter water jet at a pressure of 200 to 400 MPa (30,000 to 60,000 psi). With plain water, the jet disperses significantly as it leaves the nozzle. When poly(ethylene oxide) resin is added, the stream becomes more cohesive and maintains its very small diameter up to 10 cm (4 in) from the nozzle.[10]

Additive to Prevent Sewer Surcharges The ability of poly(ethylene oxide) resin to reduce hydrodynamic drag has also been applied to the problem of handling urban waste-water overflows and surcharges. Adding small concentrations of these resins (about 200 ppm) to the sewer line can increase the flow rate by 50 to 100%. Thus, it is possible to substantially increase the capacity of existing sewage lines or to transport high-solids-content sewage. Similarly, the capacity of existing storm sewers can be increased to reduce flooding problems. In successful municipal operations currently in existence, the high-molecular-weight resin is metered as a dry powder into the sewer system whenever a serious surcharge problem is imminent. The reduction in hydrodynamic drag and consequent increase in flow rate have, in some cases, made possible the continuing use of already existing sewer lines and obviated the necessity of building new, larger systems to handle increased demand resulting from urban expansion.[45]

Other Drag Reduction Applications The hydrodynamic drag reduction effect has also led to the reported use of these resins in coatings for the hulls of ships and naval weapons, as additives in irrigation pipelines to increase water delivery capacity, and in dredging where the resin can be used to increase the concentration of solids in the stream as well as the flow rate of dredged material.

However, poly(ethylene oxide) resins are not suitable for all applications involving the flow of water through pipes and hoses. The susceptibility of these resins to molecular weight degradation by shearing forces makes them unsuitable for use in circulating water systems where the resin molecules are repeatedly exposed to the shearing effects of circulating pumps.

Because the resins reduce turbulence, they also decrease the heat-transfer coefficient of water, which may rule out their applicability in cooling-water systems and heat exchangers. However, this can only be determined by a careful analysis of the benefits of increased flow rate versus the decrease in heat-transfer coefficient for each particular system.

Thermoplastics Manufacture

Textile Antistat It has been reported that the addition of minor amounts of these resins to polyolefin, polyamide, and polyester compositions which are melt-spun into textile fibers produces significant improvements in the dyeability and/or antistatic properties of these fibers.

Fugitive Textile Weft Another application in the textile industry involves the use of a water-soluble monofilament of these resins produced by thermoplastic extrusion. These monofilaments are used as a periodic weft thread during the weaving of the fabric. This weft thread is located at a point where the fabric is to be subsequently cut or otherwise severed. Prior to severing, the fabric passes through an aqueous solution, for example a dyeing bath, which dissolves the weft thread, permitting easier cutting or tearing of the fabric at a predesignated location.

Thickening/Rheology Control

Antimist Additive Poly(ethylene oxide) resin lends antimisting quality to any aqueous solution to which it is added, including cutting and grinding fluids. When added in sufficient quantities to central coolant systems, extended machining runs have been made with no detectable misting of particulate matter. Although cutting fluids containing these resins may splatter slightly under high-speed machine conditions, they very effectively resist forming the small, discrete droplets that are required for stable airborne mists. Most work has been done with 3×10^5 molecular weight resin.

Poly(ethylene oxide) may be incorporated into the coolant or added directly to the central system as needed. Excellent results have been obtained with these resins when concentrations of 0.3% are maintained in the central system. As a rule of thumb, when the viscosity of diluted coolant is between 1.5 and 2.0 cSt (0.0000015 and 0.0000020 m^2/s) at 25°C, sufficient resin is present.[39]

Drift-Control Additive The addition of these resins to dispersions of solids in liquids also significantly reduces the tendency to form very fine mists during spraying operations. This effect has been applied to crop-dusting operations to control the drifting of insecticide and pesticide sprays during aerial spraying.

Oil-Recovery Fluids In the petroleum industry these resins have proved useful in secondary oil-recovery operations utilizing the waterflooding process. Small concentrations of the high-molecular-weight-grade resins are used alone or in combination with natural gums or other synthetic resins to produce fluids which usually match or exceed the viscosity of the residual oil in the underground formation. The use of a high-viscosity flooding fluid to push residual oil from the formation minimizes "fingering" of the fluid into the oil deposit and increases the efficiency of the recovery operation. The high-molecular-weight grades of resin are recommended for this application because of their thickening efficiency.[33]

The resins have also been used to increase the yield of well-drilling muds made with bentonite clays. When these resins are used in combination with poly(acrylic acid) or acrylate polymers and bentonite, a synergistic effect is obtained which increases the yield of 0.0125-Pa·S (12.5-cP) viscosity drilling mud from 110 barrels of mud/ton of bentonite to as much as 400 barrels/ton of bentonite.

Aqueous suspensions of bentonite are also used in other industries. For example, clay suspensions are used in so-called slurry trenching. A bentonite slurry is pumped into a trench as it is dug to prevent the side walls of the trench from caving in or collapsing, eliminating the need for timbering, cribbing, etc., and to seal the side walls of the trench. Concrete may then be pumped into the trench to displace the slurry, which flows out at the trench surface.[21]

A third application in the petroleum industry involves use of these resins in an aqueous dispersion of calcium carbonate or powdered marble to produce a well-completion and workover fluid to temporarily plug portions of subterranean formations or fractures with an impermeable filter cake. This plug can be easily removed by back-flushing, acidification, or waterflooding without disturbing or affecting the permeability of the underground formation.[2]

Water Retention

Asbestos-Cement Extrusion Aid Asbestos-cement compositions are extruded into various shapes and used by the construction industry. In this process considerable plastic flow of the composition is necessary to move the composition along the auger and through the extruder die. However, there is a very pronounced tendency for the water in these mixes to separate from the solids, causing compaction of the solids and loss of the desired plastic flow property. The addition of small quantities of poly(ethylene oxide) resins eliminates this dewatering problem by retaining the water in the interstices between the solid particles and preventing it from being squeezed out of the mix when pressure is applied. Another advantage of these resins in the extrusion of asbestos-cement compositions lies in the fact that the elimination of the dewatering problem makes it possible to substantially change the composition of mix to effect changes in the properties of the final product without loss of extrudability.

Although a concentration of 0.05 to 10.0 wt% of high-molecular-weight resin can be used, in practice a concentration of about 0.4 wt% of the total dry ingredients provides the desired result. Usually all the dry ingredients—cement, asbestos fiber, silica, and resin—are thoroughly blended in the dry state in a sigma blade mixer. Then water is added to yield a mix containing about 67 wt% solids which is ready to be extruded.[31]

Soil Amendment As noted elsewhere in this Handbook, a hydrogel prepared by the crosslinking of poly(ethylene oxide) resins can absorb as much as 100 times its own weight of water. This absorbed water is also readily desorbed by drying the hydrogel. This characteristic is the basis for the use of these hydrogels as so-called soil amendments. When mixed with ordinary soil, such as those used in farming and nurseries, in a concentration of about 0.001 to 5.0 wt% of the soil, these hydrogels will significantly reduce the rate of moisture loss due to evaporation, but still release water to the plants as needed. Thus, plants growing in soil containing these hydrogels do not need to be watered as frequently as do plants in ordinary soil.[15]

INDUSTRIES USING POLY(ETHYLENE OXIDE)

In the following table, the uses of poly(ethylene oxide) resins have been categorized by, first, the industry utilizing the resin and, second, the specific application. Each application is cross-referenced to more detailed discussions in other sections of this chapter.

Industry	Application	Cross-references
Agriculture	1. Water-soluble seed tapes	See Seed Tape, page 19-20 See Table 19.9
	2. Water-soluble packaging film for agricultural chemicals	See Water-Soluble Packaging, page 19-20 See Table 19.9
	3. Drift-control agent for crop sprays	See Thickening/Rheology Control, page 19-24
	4. Soil stabilization using association complexes	See Soil Stabilization, page 19-21
	5. Soil-amendment hydrogel to increase water retention	See Soil Amendment, page 19-25
Ceramics and glass	1. Binder for ceramics	See Ceramics, page 19-20
	2. Size for staple glass-fiber yarns	See Glass-Fiber Size, page 19-21

Industry	Application	Cross-references
Chemical	1. Dispersant and suspension aid in aqueous suspension poly-merization of vinyl monomers	See Vinyl Polymerization, page 19-22
Construction products	1. Extrusion aid for asbestos-cement compositions	See Asbestos-Cement Extrusion Aid, page 19-25
	2. Rheology-control agent for bituminous paving compositions	See Paving Composition, page 19-17
	3. Water-soluble purge dams for tungston-arc welding	See Water-Soluble Purge Dam, page 19-17; also Table 19.9
	4. Pressure-sensitive, quick-set and wood adhesives from association complexes	See Wood Glue and see Water-Solu-ble Quick-Set Adhesive, page 19-14
	5. Dispersant in glass-fiber-reinforced concrete	See Glass-Fiber-Reinforced Cement, page 19-22
Electrical	1. Water-soluble, fugitive binder for microporous battery and fuel cell electrodes	See Battery Electrodes, page 19-21
	2. Temporary binder for phosphor coating on fluorescent lamps	See Fluorescent Lamps, page 19-21
Industrial supply	1. Rheology-control agent for indus-trial cleaning formulations	See Thickened Cleaning Solutions, page 19-15; also Thickened Acetic Acid, page 19-30, and Thickened Hydrochloric Acid, page 19-30
	2. Antimist additive for metal-working fluids	See Antimist Additive, page 19-24
	3. Additive for fluid-jet cutting system	See Fluid-Jet Cutting, page 19-23
	4. Rubber lubricant for tire mounting	See Tire-Mounting Lubricant, page 19-17; also Rubber Lubricant, page 19-31
Metals and mining	1. Flocculent for removal of silicas and clays in hydro-metallurgical processes	See Flocculation, page 19-22
	2. Flocculent for clarification of effluent streams from coal-washing plants	See Coal, page 19-22
Municipal services	1. Drag-reducing additive for fire fighting	See Fire-Fighting Additive, page 19-23
	2. Drag-reducing additive to prevent sewer surcharges and overflows	See Additive to Prevent Sewer Surcharges, page 19-23
Paper	1. Pick-up and tail-tie adhesive for rolled-paper products	See Water-Soluble Paper Adhesives, page 19-13
	2. Filler-retention and drainage aid in manufacture of paper	See Filler-Retention Drainage Aid (Paper Making), page 19-22
	3. Flocculent for clarification of effluent white water from paper manufacture	See Filler-Retention Drainage Aid (Paper Making), page 19-22
Paint	1. Thickener for latex paint	See Latex Paints, page 19-17
	2. Thickener for decorative spatter finishes	See Spatter Finish, page 19-17
	3. Thickener for paint and varnish remover	See Thickener for Paint and Varnish Remover, page 19-18; also Paint and Varnish Remover, page 19-29
Personal-care products	1. Thickener for preshave preparation	See Shaving Stick, page 19-19
	2. Lubricant in toothpaste	See Toothpastes, page 19-19; also Toothpastes, page 19-31

Industry	Application	Cross-references
	3. Adhesion and cushioning ingredient in denture fixatives	See Denture Fixative, page 19-19 also Denture Fixative page 19-28
	4. Absorbent pads for catamenial devices and diapers	See Absorbent Pads, page 19-19
	5. Lubricant for ophthalmic solutions	See Ophthalmic Solution, page 19-19
Petroleum	1. Thickener for bentonite drilling muds	See Oil-Recovery Fluids, page 19-24
	2. Thickener for secondary oil-recovery fluids	See Oil-Recovery Fluids, page 19-24
	3. Thickener for well-completion and workover fluid	See Oil-Recovery Fluids, page 19-24
Pharmaceutical	1. Dispersant to inhibit settling of calamine lotion	See Dispersant for Calamine Lotion, page 19-18; also Calamine Lotion, page 19-28
	2. Lubricant and evaporation inhibitor for rubbing alcohol	See Rubbing Alcohol, page 19-18
	3. Water-soluble coating for tablets	See Tablet Coatings, page 19-21
Printing	1. Association complex for microencapsulated ink	See Microencapsulated Inks, page 19-18; also Microencapsulation, page 19-29
	2. Evaporation inhibitor for lithographic press dampening fluid	See Lithographic Press Dampening Fluid, page 19-18; also on page 19-29
Soap and detergent	1. Emollient and thickener for detergent bars and liquids	See Detergents, page 19-18; also Detergent Bars and Detergent Liquid, page 19-28
Textile	1. Fugitive weft thread for woven textiles	See Fugitive Textile Weft, page 19-24
	2. Additive to improve dyeability and antistatic properties of polyolefin, polyester, and polyamide fillers	See Textile Antistat, page 19-24

FORMULATIONS
Aluminum and Metal Cleaner

	Wt%
Isopropanol (anhydrous)	10.0
Poly(ethylene oxide) resin*	4.0
Water	67.5
Citric acid	2.0
Thiourea	1.0
Phosphoric acid (ortho, 85 wt%)	15.0
Tergitol® Nonionic 15-S-9	0.5

* 4×10^5 molecular weight.

Prepare a slurry by first adding the resin to the anhydrous isopropanol. The slurry should be agitated for about 5 min to ensure that the resin particles are thoroughly wetted. While agitating, add the water, citric acid, and thiourea, in that order. If desired, the citric acid and thiourea can be predissolved in water and then added to the resin slurry. The resin will require from 30 to about 90 min to dissolve completely, depending on how effectively the slurry and water are initially mixed. After the polymer is dissolved, add the phosphoric acid carefully, and mix for about 5 min. The Tergitol Nonionic 15-S-9 is added last to minimize foaming. It need be mixed for only a few minutes.[44]

In this formulation, the combination of citric acid and thiourea acts as the stabilizer for the composition. Another very effective stabilizer system for these acid solutions is a combination of thiourea and ethylenediaminetetraacetic acid, tetrasodium salt (EDTA Na₄). About 1 wt% of each of these should be used.

Calamine Lotion

	Wt
Calamine	40
Zinc oxide	40
Poly(ethylene oxide) solution (5 wt%)*	420

* 6×10^5 molecular weight.

Prepare poly(ethylene oxide) solution using recommended mixing procedure. See Application Procedures, page 19-12. Make a smooth paste of calamine, zinc oxide, and a small amount of resin solution. Then dilute with remaining resin solution.[36]

Denture Fixative, Powder

	Wt%
Karaya gum	94.60
Poly(ethylene oxide)*	4.76
Calcium silicate	0.24
Flavoring	0.40

* 4×10^6 molecular weight.

The four ingredients are mixed together in any suitable dry-blending equipment.[1]

Detergent Bars

	Parts by weight	
	A	B
Dodecylbenzene sodium sulfonate	41	—
Sodium lauryl sulfate	15	—
Coconut acid ester of sodium isothionate	—	47
Sodium tallow-cocoate soap (80:20)	30	15
Stearic acid	—	30
Calcium stearate	6	—
Titanium dioxide	1.5	1.5
Poly(ethylene oxide) resin*	2.0	2.0
Water	4	4
Perfume	0.2	0.2

* 6×10^5 molecular weight.

In preparing the above formulations, the poly(ethylene oxide) resin should be dispersed in a manner which avoids lumping or agglomeration. The resin may be melted in combination with soap, detergent base, or other ingredients, and then other materials incorporated into the melt; or the resin may be blended with dry solids to which water is added, followed by thorough homogenization and blending. Homogeneity of the mix, and therefore infinite distribution of the poly(ethylene oxide), is important prior to any pressing or molding of the bars.[28]

Detergent Liquid

	Wt%
Potassium coconut glyceryl ether sulfonate	3.0
Sodium coconut glyceryl ether sulfonate	4.0
Sodium tallow glyceryl ether sulfonate	3.0
Sodium salt of sulfated condensation product of 1 mol of nonylphenol with 4 mol of ethylene oxide	2.0
Coconut dimethylamine oxide	2.0
Potassium pyrophosphate	1.0
Sodium toluene sulfonate	1.0
Sodium carboxymethyl cellulose	0.3

Poly(ethylene oxide) resin*	0.05–0.1
Color	0.0001
Salts (sodium and potassium chlorides and sulfates from detergents)	1–2
Perfume	0.5
Water (approximate)	81.6

* The poly(ethylene oxide) resin preferably should have a molecular weight of approximately 4×10^5 and a narrow molecular weight distribution.[29]

Lithographic Press Dampening Fluid

Formulation A:

	Wt%
Poly(ethylene oxide) resin*	5.26
Silicone surfactant	0.45
Isopropyl alcohol	11.19
Diethylene glycol	0.36
Glycerine (99% pure)	0.70
Water	80.11
	98.07 (for A)

* 1×10^5 molecular weight.

Formulation B:

	Wt%
Silicone emulsion antifoam agent	0.17
Water	1.76
	1.93 (for B)
	100.00 (for A + B)

Slurry the poly(ethylene oxide) into the isopropyl alcohol. Add the slurry to the water in formulation A following mixing procedure in Application Procedures, page 19-12. Then gradually add silicone surfactant, diethylene glycol, and glycerin with gentle stirring. Mix ingredients in formulation B and gradually add to formulation A mixture with gentle stirring.[27]

Microencapsulation

Various water-immiscible oils and solids can be microencapsulated in a poly(ethylene oxide) and polycarboxylic acid complex by taking advantage of pH-controlled inhibition of the complex formation. A typical procedure for the preparation of microcapsules with poly(ethylene oxide) is as follows: 40 g of a mixture of tributylphosphate and undecane (1:2 by weight) are mixed with a suitable non-ionic surfactant and emulsified in 100 g of a 1% solution of poly(vinyl methyl ether/maleic anhydride) copolymer, half amide, ammonium salt. The resulting emulsion is mixed with 200 g of a warm (50°C) 5 wt% solution of 4×10^6 molecular weight poly(ethylene oxide). The mixture is diluted with 200 mL of water and stirred and cooled in an ice bath for 30 min. It is then treated with 100 g of a 10% solution of ethylene–maleic anhydride copolymer half methyl ester, which contains 1 mL of concentrated ammonia. The resulting emulsion is further diluted with 200 mL of cold water and slowly stirred in an ice bath for 1 h longer. At this time, 39 to 40 mL of 0.3 N hydrochloric acid is added dropwise. The microcapsules thus formed are stirred in an ice bath for 15 min longer. The capsules can be dried and isolated by a variety of techniques. These include spray drying, centrifugation, and filtration.

Paint and Varnish Remover

	Wt%
Methylene chloride	95.0
Poly(ethylene oxide) resin*	3.0
Paraffin wax	2.0

* 3×10^5 molecular weight, narrow molecular weight distribution.

Melt the paraffin wax at about 70°C and slowly add the methylene chloride to it with gentle stirring. Allow the mix to cool to room temperature, and add the resin with gentle stirring. Continue stirring until the resin is dissolved. *Note:* The vessel should be covered during the dissolving process to minimize loss of the very volatile methylene chloride.[26]

Thickened Acetic Acid

	Wt%
Glacial acetic acid	50.0
Water	39.2
Poly(ethylene oxide) resin*	3.0
Isopropanol (anhydrous)	7.8

* 4 × 10^5 molecular weight, narrow molecular weight distribution.

Prepare a slurry by first adding the resin to the anhydrous isopropanol. The slurry should be agitated for about 5 min to ensure that the resin particles are thoroughly wetted. While agitating, add the water. The resin will require from 30 to about 90 min to dissolve completely, depending on how effectively the slurry and water are initially mixed. After the polymer is dissolved, add the acetic acid carefully, and mix for about 5 min. *Note:* The viscosity of this formulation was 1150 cSt (0.001150 m²/s) at 25°C using a Cannon-Fenske capillary viscometer.[44]

Thickened Hydrochloric Acid (Muriatic Acid)

	Wt%
Hydrochloric acid	10.00
Water	80.20
Poly(ethylene oxide) resin*	2.00
Isopropanol	7.80

* 4 × 10^6 molecular weight.

Slurry the resin in the isopropanol using mechanical agitation to promote wetting. To this agitated slurry, add the water, and allow sufficient time (about 1 h) for the polymer to completely dissolve. As the last step, carefully add the hydrochloric acid to the aqueous polymer solution. Cooling may be required, depending upon the rate of acid addition. *Note:* Viscosity of this formulation is 21 Pa·s (21,000 cP) at 25°C using Brookfield Model LVT viscometer, spindle #3 at 1.5 rpm.[42]

Thickened Sulfuric Acid

	Wt%
Sulfuric acid (98.0% by wt)	10.0
Poly(ethylene oxide) resin*	4.0
Water (containing 0.126 g of manganous chloride stabilizer per liter of water)	86.0

* 4 × 10^5 molecular weight, narrow molecular weight distribution.

This formulation illustrates the use of a manganous ion stabilizer rather than an alcohol stabilizer; consequently, a somewhat different dissolving technique is needed. Initially, the manganous chloride is dissolved in water to yield an aqueous solution containing 0.0010 mol of Mn^{2+} (or 0.126 g of $MnCl_2$) per liter of water. The resin is carefully dispersed and dissolved in this very dilute manganous chloride solution.

After the resin is completely dissolved, the concentrated sulfuric acid is slowly and carefully added. The formulation is kept cool and agitation is continued long enough to uniformly dilute the acid, i.e., about 5 min. *Note:* Viscosity of this formulation was 475 cSt (0.000475 m²/s) at 25°C using a Cannon-Fenske (Series 500) capillary viscometer.[44]

Rubber Lubricant (for Mounting of Tires)

	Amount
Poly(ethylene oxide) resin*	0.46 kg
Isopropanol (99.5%)	4.71 L
N,N-diethanol coconut fatty acid amide	8.13 L
N,N-diethanolamine oleate	3.70 L
Water (approximately)	59.14 L

* 4×10^6 molecular weight.

Disperse the resin in the isopropanol and pour the dispersion into 59 L of rapidly stirred water. After a few minutes of stirring add the remaining ingredients, and continue stirring until homogeneous. For use as a tire-mounting lubricant, dilute the above concentrate in the ratio of 1 part concentrate to 15 parts water.[13]

Toothpastes

	Wt%				
	A	B	C	D	E
Poly(ethylene oxide) resin*	0.5–1.0	0.3–1.0	1.0	0.3–1.0	0.3–1.0
Propylene glycol	17.0	19.0	17.0	19.0	17.0
Glycerin	4.5	9.3	8.5	9.6	8.4
p-Hydroxymethyl benzoate	0.2	0.27	0.27	0.3	0.27
10% saccharin solution	0.5	0.63	0.63	0.7	0.63
Mineral oil	1.0	1.2	1.1	1.2	1.1
Sodium lauryl sulfate	3.5	—	2.9	—	3.0
Sodium lauroyl sarcosinate	—	2.0	—	—	—
Tricalcium phosphate	43.0	42.2	—	—	—
Dicalcium phosphate	—	—	51.0	—	—
Calcium carbonate	—	—	—	42.0	40.0
Sodium soap of tallow fatty acids	—	—	—	2.8	—
Sodium soap of coconut oil fatty acids	—	—	—	0.7	—
Flavor	1.0	1.2	1.1	1.2	1.1
Water (approximate) %	28	23	16	21	27

* 6×10^5 molecular weight.

The resin should be blended with all dry ingredients first, and then the dry ingredients should be mixed into the nonsolvents for the resin. With rapid stirring, water is slowly added to the mix. As the resin dissolves, the blend will become uniform. After complete solubilization of the resin, a stable, smooth paste of light density results.[6]

PRODUCT GLOSSARY

Polyox: Trademark for homopolymer poly(ethylene oxide) resins produced by Union Carbide Corp. Includes the WSR grades with molecular weights of 1×10^5 to 5×10^5, plus Coagulant, WRPA 3154, and UCAR Rapid Water Additive.

Alkox: Trademark for homopolymer poly(ethylene oxide) resins produced by Meisei Chemical Works Ltd., Kyoto, Japan.

PEO: Trademark for homopolymer poly(ethylene oxide) resins produced by Seitetsu Kagaku Co. Ltd., Osaka, Japan.

Poly(ethylene oxide), $+CH_2-CH_2-O+_x$: A linear, high-molecular-weight, water-soluble homopolymer made from ethylene oxide. White, free-flowing powder. Commercially available in several different molecular weight grades from about 100,000 to over 5 million. Density, 1.21 $\pm$ 0.03 at 25°C. Melting point, 63–67°C. Completely soluble in water at temperatures up to 98°C. Very soluble in chlorinated hydrocarbons and partially soluble in aromatic hydrocarbons, alcohols, cellosolves, and ketones. Nontoxic.

FURTHER USEFUL READING/REFERENCES

General References

Berger, L. D., Jr., and Ivison, M. Thayer, "Water-Soluble Polymers," in R. L. Davidson and M. Sittig (ed.), *Water-Soluble Resins,* 1st ed., Reinhold Publishing Corp., New York, 1962, pp. 169–201; see also "Ethylene Oxide Polymers," by R. H. Harding and J. K. Rose from the 2d ed. (1968)

of the same title. Both editions of this volume also present good general discussions with considerable physical data.

Stone, F. W., and Stratta, J. J., "Ethylene Oxide Polymers," in *Encyclopedia of Polymer Science and Technology,* vol. 6, John Wiley & Sons, Inc., New York, 1967, pp. 103–145. Good general treatise on the subject with major emphasis on the preparation and properties of these polymers.

Bailey, F. E., Jr., and Kaleske, J. V., *Poly(Ethylene Oxide),* Academic Press, New York, 1976.

Numbered References

1. American Pharmaceutical Association, *Handbook of Non-Prescription Drugs,* 1971.
2. Barkman, J. H., Jr., et al. (to Shell Oil Co.), *Well Completion and Workover Fluid and Method of Use Thereof,* U.S. Patent 3,516,496 (June 23, 1970).
3. Beaumont, D. H., and Friedman, A. I. (to General Electric Co.), *Lamp Phosphor Adherence,* U.S. Patent 3,424,605 (Jan. 28, 1969).
4. Berger, L. D., Jr. and Ivison, M. T., "Water-Soluble Polymers," in R. L. Davidson and M. Sittig (eds.), *Water-Soluble Resins,* 1st ed., Reinhold Publishing Corp., New York, 1962, pp. 169–201.
5. Berger, L. D., Jr., and Ivison, M. T., "Water-Soluble Polymers," in R. L. Davidson and M. Sittig (eds.), *Water-Soluble Resins,* 1st ed., Reinhold Publishing Corp., New York, 1962, pp. 193–194.
6. L. D. Berger, Jr., and Ivison, M. T., "Water-Soluble Polymers," in R. L. Davidson and M. Sittig (eds.), *Water-Soluble Resins,* 1st ed., Reinhold Publishing Corp., New York, 1962, p. 199.
7. Colwell, C. E., and Miller, R. C. (to Union Carbide Corp.), *Coagulation,* U.S. Patent 3,020,231 (Feb. 6, 1962).
8. Duddy, J. C. (to Electric Storage Battery Co.), *Electrodes and Method for Making Same,* U.S. Patent 3,121,029 (Feb. 11, 1964).
9. Duddy, J. C. (to Electric Storage Battery Co.,), *Fuel Cell Electrode and Method of Making Same,* U.S. Patent 3,181,973 (May 4, 1965).
10. Eberle, J. F., "Fluid Jet Cutting," *Tappi,* **56**(10) (October 1973).
11. Ferry, R., "Glass Fiber Reinforced Concrete," *Concr. Constr.* (April 1975).
12. Goren M. B. (to Kerr-McGee Oil Industries, Inc.), *Process for Absorbing Solubilized Silica from Acidic Media with an Organic Containing Poly(Alkylene Oxide) Structure,* U.S. Patent 3,266,888 (Aug. 16, 1966).
13. Halko, J. J., Jr. (to Freeland Chemical Co.), *Lubrication,* U.S. Patent 3,699,057 (Oct. 17, 1972).
14. Hapff, H., et al. "Poly(Ethylene Oxides) as Protective Colloids," *Makromol. Chem.,* **84,** 282–5 (1965).
15. Herrett, R. A. and King, P. A. (to Union Carbide Corp.), *Plant Growth Medium,* U.S. Patent 3,336,129 (Aug. 15, 1967).
16. Jennings, J. W., *Method of Shaving,* U.S. Patent 3,811,349 (May 21, 1974).
17. Kelly, G. B., Jr. (to Union Carbide Corp.), *Ethylene Oxide Polymers Having Improved Stress Endurance,* U.S. Patent 3,154,514 (October 1964).
18. Kempster, E. (to National Research Developments), *Process for Dispensing Fiber in an Aqueous Mix,* U.S. Patent 3,716,386 (Feb. 13, 1973).
19. King, P. A. (to Union Carbide Corp.), *Disposable Absorbent Pads Containing Insoluble Hydrogels,* U.S. Patent 3,783,872 (Jan. 8, 1974).
20. Kingston, J. F., et al. (to The Glidden Co.), *Latex Webbing Finish Composition Containing Water-Soluble Poly(Ethylene Oxide) and Method of Applying,* U.S. Patent 3,117,942 (Jan. 14, 1964).
21. Long, W. D. (to International Minerals and Chemical Corp.), *High-Yield Bentonites,* U.S. Patent 3,687,846 (Aug. 29, 1972).
22. Manley, J. A. (to Nalco Chemical Co.), *Process of Improving Inorganic Filler Retention in Paper by Addition of Ethylene Oxide Homopolymer,* U.S. Patent 3,141,815 (July 21, 1962).
23. Markava, A. A., *Kinet. Katal.* **3,** 366 (1962); *Chem. Abstr.,* **58,** 579g (1963).
24. Marzocchi, A., and Ramel, G. E. (to Owens-Corning Fiberglas Corp.), *Yarns of Staple Glass Fibers and Compositions and Methods of Manufacturing Same,* U.S. Patent 3,042,544, (July 3, 1962).
25. McConnaughay, K. E., *Paving Composition and Method of Making It,* U.S. Patent 3,110,604 (Nov. 12, 1963).
26. Morrison C. R. W. (to Union Carbide Corp.), *Finish Removal Formulation,* U.S. Patent 3,179,609, (Apr. 20, 1965).
27. Nasca, S. *Poly(Ethylene Oxide) Dampening System for Lithographic Presses,* U.S. Patent 3,625,715 (Dec. 7, 1971).
28. Osipow, L. and Berger, L. D., Jr., "Poly(Ethylene Oxide) Gums in Toilet Goods," *Drug Cosmet. Ind.,* **82**(2), 166–167, 220 (February 1958).
29. Pilcher, W., and Eaton, S. L. (to Proctor and Gamble Co.), *Personal Use Detergent Lotion,* U.S. Patent 2,999,068 (Sept. 5, 1961).
30. Rankin, B. F. (to Burton Parsons Chemicals Inc.), *Ophthalmic Solution,* British Patent 1,340,516 (Dec. 12, 1973).
31. Radican, F. W., et al. (to Johns-Manville Corp.), *Manufacture of Asbestos-Cement Products,* U.S. Patent 3,219,467 (Nov. 23, 1965).

32. Richardson, D., et al. (to Union Carbide Corp.), *Suspension Polymerization of Vinyl Chloride,* French Patent 1,508,770 (Jan. 5, 1968).
33. Sandiford, B. B., and Keller, H. F., Jr. (to Union Oil Company of California), *Secondary Recovery of Petroleum,* U.S. Patent 3,116,791 (Jan. 7, 1964).
34. Simms, B. B., et al. (to U.S. Department of the Navy), *Method of Soil Stabilization,* U.S. Patent 3,696,621 (Oct. 10, 1972).
35. Stone, F. W., and Stratta, J. J., "Ethylene Oxide Polymers," in *Encyclopedia of Polymer Science and Technology,* vol. 6, John Wiley & Sons, Inc., New York, 1967, pp. 103–145.
36. Swafford, W. G., "Poly(Ethylene Oxide) Water Soluble Resin as a Suspending Agent for Calamine Lotion," *Am. J. Hosp. Pharm.* 19, 134–5 (March 1962).
37. Thompson, B. R. (to Union Carbide Corp.), *Coagulation of Dispersed Carbonaceous Material,* U.S. Patent 3,020,229 (Feb. 6, 1962).
38. Union Carbide Corp., *Better Fire Fighting with Ucar Rapid Water System,* Tech. Brochure F-43183B.
39. Union Carbide Corp., *Eliminating Airborne Mists During Cutting and Grinding Operations with Polyox® Water-Soluble Resins,* Tech. Service Bulletin TT-T1-729, April 1973.
40. Union Carbide Corp., *Forming Association Compounds,* Tech. Brochure F-43272.
41. Union Carbide Corp., *How to Dissolve Polyox Water-Soluble Resins,* Tech. Brochure F-42933.
42. Union Carbide Corp., *Polyox Packaging Film,* Tech. Brochure F-44703.
43. Union Carbide Corp., *Polyox Water-Soluble Resins Are Unique,* Tech. Brochure F-44029A.
44. Union Carbide Corp., *Polyox Water Soluble Resins Control Flow of Cleaning Solutions,* Tech. Brochure F-42934B.
45. U.S. Department of the Interior, Federal Water Pollution Control Administration, *Polymers for Sewer Flow Control,* Water Pollution Control Research Series, WP-20-22, August 1969.
46. Winslow, A. E., and Smith, K. L. (to Union Carbide Corp.), *Association Products of a Polymeric Ether and a Phenolic Compound,* U.S. Patent 3,125,544 (Mar. 17, 1964).

Chapter **20**

Polyvinyl Alcohol

T. W. Modi
E. I. du Pont de Nemours & Co., Inc.

General Information 20-2
  Chemical Nature 20-3
  Physical Properties 20-3
  Manufacture 20-15
  Physiological Properties 20-16
  Federal Drug Administration (FDA) Status 20-16
  Biochemical Oxygen Demand (BOD) 20-16
  Biodegradation 20-17
  Modifiers 20-17
  Handling and Storage 20-19
  Application Procedures 20-20
Commercial Uses: Compounding and Formulating 20-20
  Adhesives 20-20
  Paper and Paperboard Sizing 20-21
  Paper and Paperboard Coatings 20-21
  Pigmented Coatings 20-21
  Greaseproof Coatings 20-21
  Textile Finishing 20-22
  Binder Applications 20-22
  Cast Film 20-22
  Molded Articles 20-22
  Emulsions and Dispersions 20-22
  Cosmetics 20-23
  Chemical Derivatives 20-23
Commercial Uses: Processing Aids 20-23
  Textile Warp Sizing 20-23
  Temporary Binder 20-23
  Casting Slips 20-23
  Steel Quenchant 20-23
  Miscellaneous Coating Applications 20-24
  Materials Stabilization 20-24

Industries Using Polyvinyl Alcohol 20-24
   Textile Industry 20-24
   Paper Industry 20-24
   Adhesives Industry 20-24
   Cast-Film Industry 20-24
   Building Products Industries 20-25
   Packaging Industry 20-25
   Chemical Industry 20-25
   Cosmetics Industry 20-25
   Ceramics Industry 20-25
   Steel Industry 20-25
   Materials Binding 20-25
  Formulations 20-25
   Textile Warp Sizing: Slasher Operation 20-25
   Textile Warp Sizing: Size-Bath Formulas 20-26
   Adhesives 20-27
   Binder for Ceiling Tile Prime Coat 20-29
   Size for Printing and Writing Paper 20-29
   Pigmented Coating for Paperboard 20-29
   Strippable Coating 20-30
   Mold-Release Coating 20-30
   Concrete Adhesion Promoter 20-30
   Ready-Mix Cement Binder 20-30
  Laboratory Techniques 20-30
   Solution Viscosity 20-30
   Percent Hydrolysis 20-30
   Percent Volatiles 20-30
   Solution pH 20-31
   Percent Ash 20-31
   Screen Test: Particle-Size Distribution 20-31
  Product Tradename Glossary 20-31
  Further Useful Reading 20-31

GENERAL INFORMATION

Polyvinyl alcohol is a white, powdered synthetic resin which is manufactured by the hydrolysis of polyvinyl acetate. It is the only linear polyhydroxy polymer that is readily soluble in water. Polyvinyl alcohol is used in adhesives, in the manufacture of automotive safety glass, in textile sizes and finishes, in paper sizes and coatings, and as a binder for ceramics, foundry cores, nonwoven fabric, and various pigments. Polyvinyl alcohol is cast into films from water solutions, made into fibers, and molded or extruded into various sheets, tubes, etc., that are used where oil and solvent resistance is required. It may also be used in photosensitive coatings, steel quenching, protective coatings, cementitious products, production of emulsions, materials stabilization, cosmetics, oil-well drilling, and many other miscellaneous applications.

Worldwide demand for polyvinyl alcohol exceeded 295 metric tons (650 million pounds) in 1978. The United States market is comprised chiefly of textile, adhesive, and paper applications.

Polyvinyl alcohols are water-soluble synthetic polymers with excellent film-forming, adhesive, and emulsifying properties and outstanding resistance to oil, grease, and solvents. The first polyvinyl alcohol produced in the United States was by the Du Pont Co. in 1939. Du Pont and other producers have developed products of excellent quality and uniformity to meet industrial needs.

Chemical Nature

Chemically, polyvinyl alcohol can be described as a polyhydric alcohol with secondary hydroxyl groups on alternate carbon atoms. It is represented structurally as follows:

$$\left[\begin{array}{c} -CH_2-CH-CH_2-CH- \\ | | \\ OH OH \end{array} \right]_n$$

Although polyvinyl alcohol is usually used industrially without chemical modification, it undergoes reactions typical of long-chain polyhydric alcohols. Most important commercially is its reaction with aldehydes to form polyvinyl acetals. Hydroxyl groups along the chain can also be reesterified with acids, converted to cyanoethyl ether groups by reaction with acrylonitrile, or converted to hydroxyethyl groups using ethylene oxide. Polyvinyl alcohol can be crosslinked using difunctional or polyfunctional reagents.

TABLE 20.1
Polyvinyl Alcohol: Property Summary

Form	Granular or powder
Color	White to cream
Specific gravity	1.26–1.31
Specific volume, m^3/kg (in^3/lb)	7.95–7.62×10^{-4} (22.0–21.1)
Refractive index, n_D^{25}	1.49–1.53
Elongation, %, plasticized film	Up to 600
Tensile strength, dry, unplasticized, MPa (lb/in^2)	Up to 152 (22,000)
Hardness, Shore durometer, plasticized	10–100
Heat-sealing temperature, °C, dry, unplasticized	165–210
Compression-molding temperature, °C, plasticized	100–150
Heat stability	Darkens slowly above 100°C Darkens rapidly above 150°C Decomposes above 200°C
Storage stability	Excellent; no deterioration after storage for several years
Thermal coefficient of linear expansion, 0–45°C	7×10^{-5} to 12×10^{-5}
Specific heat, cal/(g·°C)	0.4
Flammability	Burns at about the rate of paper
Effect of light	Negligible
Effect of strong acids	Dissolves or decomposes
Effect of strong alkalies	Softens or dissolves
Effect of weak acids	Softens or dissolves
Effect of weak alkalies	Softens or dissolves
Effect of organic solvents	Resistant to most solvents

Physical Properties

Polyvinyl alcohol is sold as a white, granular powder. It is soluble in water, the only practical solvent for polyvinyl alcohol. Table 20.1 summarizes its properties. However, the temperature at which complete solution occurs varies with the grade, depending chiefly on percent hydrolysis. Polyvinyl alcohol grades in the 87 to 89% range show maximum solubility in that they dissolve readily in both cold and hot water. Above 89 to 90% hydrolysis, heating is required for complete solution. (See Tables 20.2 to 20.5.) Once dissolved in hot water, all grades of polyvinyl alcohol remain in solution when cooled.

Viscosity The viscosity of water solutions of polyvinyl alcohol varies with the grade, concentration, and temperature. These relationships are illustrated in Figs. 20.1 to 20.5. Grades within the same hydrolysis range can be combined to obtain intermediate solution viscosities for a given solids content.

Effect of Heating   Solutions of fully hydrolyzed polyvinyl alcohol show no apparent change in properties on heating for several days at temperatures up to 100°C (212°F). Partially hydrolyzed polyvinyl alcohols may show a gradual increase in percent hydrolysis on heating in aqueous solution.

Solubility   *In Water.*   All commercial grades of polyvinyl alcohol are soluble in water. The temperature at which the resin dissolves completely varies with the grade, depending chiefly on percent hydrolysis.   Grades in the 87 to 89% hydrolysis range show maximum solubility in that they dissolve in both cold and hot water.   Above 89 to 90% hydrolysis, heating is required for complete solution.   At 75 to 80% hydrolysis polyvinyl alcohols are soluble in cold water but tend to precipitate on heating.   Maximum practical concentrations for solutions prepared using conventional high-speed mixers are approximately 15 to 20% for high-viscosity grades, 20 to 25% for medium-viscosity grades, and 30 to 40% for low-viscosity grades.

TABLE 20.2
Fully Hydrolyzed, Medium-Viscosity Polyvinyl Alcohol: Typical Properties and Specifications

Form	Granular
Color	White
Hydrolysis, mol %*	99.0–99.8
Saponification number†	3–12
Residual polyvinyl acetate, wt. %	0.5–1.8
Viscosity,‡ Pa·s (cP)	0.028–0.032 (28–32)
Solution pH	5.0–7.0
Volatiles, % max.	5
Ash (as Na₂O), % max.*	1.0
Bulk density, kg/m³ (lb/ft³)	400–432 (25–27)
Specific gravity	1.30
Resin density, kg/m³ (lb/gal)	1294 (10.8)
Specific volume, m³/kg (in³/lb)	7.7×10^{-4} (21.3)
Refractive index, n_D^{25}	1.54
Hardness, Shore durometer, unplasticized	>100
Specific heat, cal/(g·°C)	0.4

* Dry basis.
† Milligrams potassium hydroxide per gram polymer.
‡ Viscosity of a 4% solids aqueous solution at 20°C, determined by Hoeppler falling-ball method.

In Water-Alcohol Mixtures.   Water is the only practical solvent for polyvinyl alcohol. Substantial quantities of the lower alcohols can be added to water solutions without causing precipitation, the proportion of alcohol increasing as percent hydrolysis of the polyvinyl alcohol decreases.   As shown in Fig. 20.6, polyvinyl alcohols with a percent hydrolysis less than approximately 73% are insoluble in water or alcohol alone, but are soluble in alcohol-water mixtures.

In Other Solvents.   Most of the common organic solvents have little or no effect on polyvinyl alcohol.   Sensitivity to organic solvents increases as percent hydrolysis is decreased.   However, there are only a few organic compounds in which any of the commercial grades of polyvinyl alcohol can be dissolved.   Most of these are polyhydroxy compounds such as glycerin, ethylene glycol, and some of the lower polyethylene glycols; amides such as formamide, dimethyl formamide, ethanol formamide, and ethanol acetamide; or amines including ethanolamines and ethanolamine salts.   Both partially and completely hydrolyzed grades of polyvinyl alcohol are also soluble in dimethyl sulfoxide.

Heat is required to dissolve even small amounts of polyvinyl alcohol in most of the above solvents.   For example, polyvinyl alcohol can be dissolved in glycerin at 120 to 150°C (248 to 302°F), but the mixtures gel on cooling to room temperature.   Dimethyl sulfoxide must be heated to dissolve polyvinyl alcohol, but the polymer remains in solution on cooling.   Diethylenetriamine and triethylenetetramine are among the few organic solvents that dissolve polyvinyl alcohol at room temperature.

TABLE 20.3
Commercially Available Grades

Elvanol® a Grade designations	Shipping specifications					Distinguishing properties and principal uses
	Vis-cosity b	Percent hydroly-sis c	Solu-tion pH	Vola-tiles, % max. d	Ash, % max.	
General purpose: HV	55–65	99.0–99.8	5.0–7.0	4	1.0	High-viscosity, fully hydrolyzed grade. Provides exceptionally high tensile and adhesive strength with strong resistance to grease and hydrocarbons. Used in adhesives, textile, and paper applications.
71–30	28–32	99.0–99.8	5.0–7.0	5	1.0	Medium-viscosity, fully hydrolyzed grade. Hot-water soluble, with excellent resistance to cold water. Used in textile, paper, and adhesive applications and for gastight, solvent-resistant film.
90–50	13–15	99.0–99.8	5.0–7.0	5	1.0	Medium-low viscosity, fully hydrolyzed grade designed to provide high film strength and binding power in relatively low-viscosity systems. Used in paper and paperboard coating and sizing, in adhesives, and as a pigment binder for highly filled systems.
Textile grades: T-25	25–31	99.0–99.8	5.0–7.0	5	1.0	Unique grade designed for textile warp sizing. Provides excellent weaving performance and desizes readily even after heat-setting sized fabric in the greige.
T-66	13–15	98.0–99.8	5.0–7.0	5	1.0	Unique grade for textile warp sizing. Designed for smooth slasher operation and easy desize, while still providing excellent weaving performance.
Specialty and adhesive grades: 85–60	25–57 e	—	5.9–6.6	—	—	Specialty grade for adhesives use in paper tube and core winding and solid fiber laminating. Useful in adhesive applications where a high degree of wet tack is desired.
85–82	25–31	99.0–99.8	5.0–7.0	5	1.0	Medium-viscosity grade which combines high tensile and adhesive strength with good gel resistance. Used for adhesives, films, and specialty applications.
75–15	13–15	98.0–99.8	5.0–7.0	5	1.0	Medium-low viscosity grade which imparts improved viscosity stability and gel resistance to aqueous solutions. For use in adhesives, films, paper applications and as an emulsifier for improved water resistance.

See page 20-6 for footnotes.

Table 20.3 *(Continued)*

Elvanol®ᵃ Grade designations	Shipping specifications					Distinguishing properties and principal uses
	Vis-cosityᵇ	Percent hydroly-sisᶜ	Solu-tion pH	Vola-tiles, % max.ᵈ	Ash, % max.	
BT	12–15	99.0–99.8	5.0–7.0	5	1.0	Specialty, borax-tolerant, fully hydrolyzed grade. Designed for use with starch in corrugating adhesive systems and other end uses requiring borax tolerance.

ᵃ Reg. U.S. Pat. & Tm. Off. for Du Pont's polyvinyl alcohol.
ᵇ Viscosity in mPa·s (cP) of a 4% solids aqueous solution at 20°C (68°F), determined by Hoeppler falling-ball method.
ᶜ Mole percent hydrolysis of acetate groups, dry basis.
ᵈ Dry basis; calculated as % Na₂O.
ᵉ Brookfield viscosity in mPa·s (cP) at 25°C (77°F) (Model LVF, #1 spindle, 60 rpm) of an aqueous solution containing 4% "Elvanol," as-is basis.

TABLE 20.4
Typical Partially Hydrolyzed Polyvinyl Alcohol Grades

Viscosity level	Typical properties				
	Viscosity,* cP (Pa·s)	Hydrolysis, mol %†	pH	Volatiles, % max.	Ash, % max.‡
High	35–45 (0.035–0.045)	87–89	5.0–7.0	5	0.5
Medium	21–25 (0.021–0.025)	86–89	5.0–7.0	5	0.5
Low	4–6 (0.004–0.006)	86–89	5.0–7.0	5	0.5

* Viscosity in cP (Pa·s) of a 4% aqueous solution at 20°C, determined by Hoeppler falling-ball method.
† Dry basis.
‡ Dry basis; calculated as % Na₂O.

Tolerance for Electrolytes   Solutions of polyvinyl alcohol show high tolerance for ammonium hydroxide, acetic acid, and most of the common inorganic acids, including hydrochloric, sulfuric, nitric, and phosphoric acids. Considerably lower concentrations of sodium hydroxide precipitate polyvinyl alcohol from solution.

Among the salts for which polyvinyl alcohol shows high tolerance are sodium nitrate, ammonium chloride, calcium chloride, zinc chloride, potassium iodide, and potassium thiocyanate. Salts acting as precipitants at low concentrations include sodium carbonate, sodium sulfate, and potassium sulfate.

Polyvinyl alcohol solutions are particularly sensitive to borax, which causes gelation. Equipment that has been used to handle borax or borated starches and dextrins should be washed thoroughly before using with solutions of polyvinyl alcohol. The amount of borax required to gel polyvinyl alcohol varies with the grade as well as the solution concentration. For example, as little as 0.1% of borax, based on the weight of solution, causes a 5% solution of fully hydrolyzed polyvinyl alcohol to gel, whereas up to about 1% of borax can be added to a 5% solution of partially hydrolyzed, low-viscosity polyvinyl alcohol before gelation occurs.

Table 20.6 illustrates the wide variation among electrolytes in their effect on polyvinyl alcohol solubility. The exact concentration required to precipitate polyvinyl alcohol from solution varies with the grade and concentration of polyvinyl alcohol as well as with the temperature and the test method.

TABLE 20.5
Correlation of Percent Hydrolysis, Saponification Number,* and Residual Polyvinyl Acetate Content of Polyvinyl Alcohols

Hydrolysis, mol %	Saponification number	Residual polyvinyl acetate, wt %	Hydrolysis, mol %	Saponification number	Residual polyvinyl acetate, wt %
100.0	00.0	0.0	49.0	436.9	67.0
99.0	12.6	1.9	48.0	442.7	67.9
98.0	25.0	3.8	47.0	448.3	68.8
97.0	37.1	5.7	46.0	453.9	69.6
96.0	49.1	7.5	45.0	459.4	70.5
95.0	60.8	9.3	44.0	464.9	71.3
94.0	72.3	11.1	43.0	470.2	72.1
93.0	83.6	12.8	42.0	475.6	73.0
92.0	94.7	14.5	41.0	480.8	73.8
91.0	105.6	16.2	40.0	486.0	74.6
90.0	116.3	17.8	39.0	491.1	75.3
89.0	126.8	19.5	38.0	496.2	76.1
88.0	137.1	21.0	37.0	501.2	76.9
87.0	147.3	22.6	36.0	506.1	77.7
86.0	157.3	24.1	35.0	511.0	78.4
85.0	167.1	25.6	34.0	515.8	79.1
84.0	176.3	27.1	33.0	520.6	79.9
83.0	186.3	28.6	32.0	525.3	80.6
82.0	195.7	30.0	31.0	529.9	81.3
81.0	204.9	31.4	30.0	534.5	82.0
80.0	213.9	32.8	29.0	539.1	82.7
79.0	222.8	34.2	28.0	543.6	83.4
78.0	231.6	35.5	27.0	548.0	84.1
77.0	240.2	36.9	26.0	552.4	84.8
76.0	248.7	38.2	25.0	556.8	85.4
75.0	257.1	39.4	24.0	561.1	86.1
74.0	265.3	40.7	23.0	565.4	86.7
73.0	273.5	42.0	22.0	569.6	87.4
72.0	281.4	43.2	21.0	573.7	88.0
71.0	289.3	44.4	20.0	577.8	88.7
70.0	297.1	45.6	19.0	581.9	89.3
69.0	304.7	46.8	18.0	585.9	89.9
68.0	312.2	47.9	17.0	589.9	90.5
67.0	319.7	49.0	16.0	593.9	91.1
66.0	327.0	50.2	15.0	597.8	91.7
65.0	334.2	51.3	14.0	601.6	92.3
64.0	341.3	52.4	13.0	605.5	92.9
63.0	348.3	53.4	12.0	609.3	93.5
62.0	355.2	54.5	11.0	613.0	94.1
61.0	362.0	55.5	10.0	616.7	94.6
60.0	368.7	56.6	9.0	620.4	95.2
59.0	375.3	57.6	8.0	624.0	95.7
58.0	382.0	58.6	7.0	627.6	96.3
57.0	388.4	59.6	6.0	631.1	96.8
56.0	394.7	60.6	5.0	634.7	97.4
55.0	401.0	61.5	4.0	638.2	97.9
54.0	407.2	62.5	3.0	641.6	98.4
53.0	413.3	63.4	2.0	645.0	99.0
52.0	419.3	64.3	1.0	648.4	99.5
51.0	425.3	65.2	0.0	651.8	100.0
50.0	431.1	66.1			

* Saponification number = mg KOH/g polymer.

Barrier Properties An outstanding property of polyvinyl alcohol is its high degree of impermeability to many gases. Continuous films or coatings of polyvinyl alcohol provide excellent barriers for such gases as oxygen, nitrogen, carbon dioxide, hydrogen, helium, and hydrogen sulfide. Table 20.7 shows the effect of humidity on oxygen permeability of a glycerin-plasticized film of fully hydrolyzed polyvinyl alcohol. Notable exceptions to the low permeability of polyvinyl alcohol to gases are ammonia and water vapor. The moisture vapor transmission rate of an unplasticized film of fully hydrolyzed polyvinyl alcohol at two different humidity gradients is shown in Table 20.8.

Moisture Absorption All grades of polyvinyl alcohol are hygroscopic. Table 20.9 shows typical weight-gain values for predried films of unplasticized polyvinyl alcohol

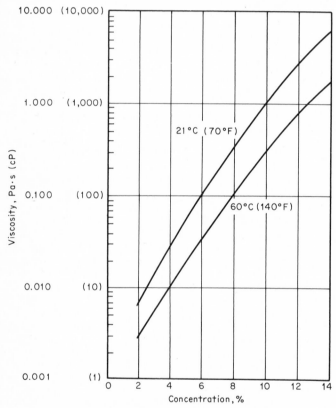

Figure 20.1 Viscosity of fully hydrolyzed, 30-cP-viscosity-grade polyvinyl alcohol vs. concentration. (Brookfield Viscometer, Model LVF, 60 rpm.)

at various humidities. The values shown in Table 20.9 are ranges for representative grades of polyvinyl alcohol.

Plasticizers commonly used with polyvinyl alcohol increase moisture absorption. Except for heavily plasticized compositions, which tend to "sweat," films and coatings of polyvinyl alcohol remain nontacky and dry to the touch even at high humidities.

Effect of Heat and Light Polyvinyl alcohols soften sufficiently to heat-seal at temperatures above 150°C, the required temperature increasing with degree of hydrolysis. All grades can be considered thermoplastic since, when plasticized, they can be molded and remolded repeatedly.

Films of polyvinyl alcohol show gradual discoloration, embrittlement, and reduction in water solubility on heating in air above 100°C (212°F), the rate increasing as the temperature is raised.   Rapid decomposition occurs at temperatures of 200°C and above. The discoloration of polyvinyl alcohol on heating can be prevented or retarded by incorporating a small amount of boric acid; 0.5 to 3% H_3BO_3, based on the weight of polyvinyl alcohol, is usually effective.   The boric acid can be added as a 5% solution. Polyvinyl alcohol shows excellent stability to sunlight and artificial light.

Optical Properties   Light transmission of a film prepared from completely hydrolyzed polyvinyl alcohol plasticized with 10% glycerin is given in Table 20.10 for several wavelengths in the ultraviolet band.   In the infrared region, polyvinyl alcohol is practically opaque from 15 to 6.6 μm and again at 3 μm.

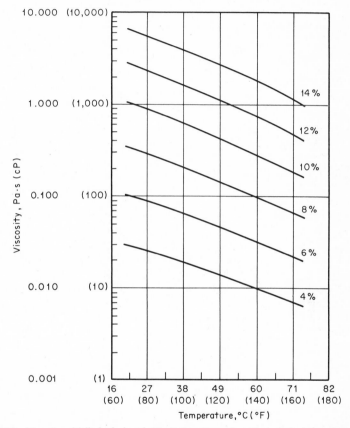

Figure 20.2   Viscosity of fully hydrolyzed, 30-cP-viscosity-grade polyvinyl alcohol vs. temperature. (Brookfield Viscometer, Model LVF, 60 rpm.)

Electrical Properties   The dielectric constants and power factors of polyvinyl alcohol are much higher than those of plastics commonly used as insulating materials.   Since all grades are hygroscopic, these properties vary widely with the humidity.   Table 20.11 illustrates the effect of changes in humidity on the electrical properties of films of polyvinyl alcohol.

Adhesive Properties   Polyvinyl alcohols form strong bonds to porous, water-absorbent surfaces such as paper, textiles, wood, and leather, and are effective binders for pigments and other finely divided solids.   Adhesion to smooth, nonabsorbent surfaces such as glass and metals improves as the percent hydrolysis decreases.   Adhesion of the partially hydrolyzed grades to nonporous surfaces can be further improved by adding 15% phosphoric acid, based on the weight of polyvinyl alcohol.

Adhesives based on polyvinyl alcohol are normally applied by wet bonding. Films can be joined by heat sealing. Temperatures of 160 to 200°C are required to thoroughly seal dried films of unplasticized polyvinyl alcohol, the temperature increasing with percent hydrolysis of the resin. Plasticizers, including moisture, lower the sealing temperature.

Emulsifying Properties   Polyvinyl alcohols function both as non-ionic emulsifiers and protective colloids in the preparation of oil-in-water–type emulsions and dispersions.

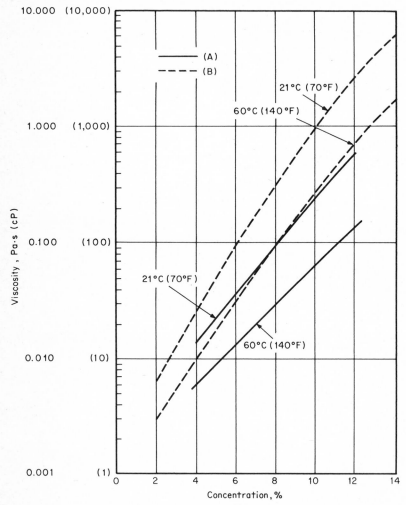

Figure 20.3   Viscosity vs. concentration for fully hydrolyzed polyvinyl alcohol: *(A)* = 13-cP-viscosity-grade; *(B)* = 30-cP-viscosity-grade. (Brookfield Viscometer, Model LVF, 60 rpm.)

The polyvinyl alcohol also retards creaming and settling by acting as a viscosity builder in the continuous phase. Both surfactancy and effectiveness as a protective colloid increase as the residual vinyl acetate content of the polyvinyl alcohol increases, i.e., with decreasing percent hydrolysis. Protective-colloid ability and viscosity-building effect increase with increasing molecular weight, while surfactancy is favored by decreasing molecular weight.

The partially hydrolyzed grades are generally preferred for emulsion applications.

The concentration of polyvinyl alcohol used to produce stable emulsions or dispersions containing 30 to 55% of the dispersed phase is generally from 2 to 5%, based on total weight of emulsion.   For initial trials, 3% polyvinyl alcohol or a combination of 2.5% polyvinyl alcohol and 0.5% anionic emulsifier is suggested.   In suspension polymerization systems, a polyvinyl alcohol concentration of 0.1% or less, based on the aqueous phase, is sufficient.

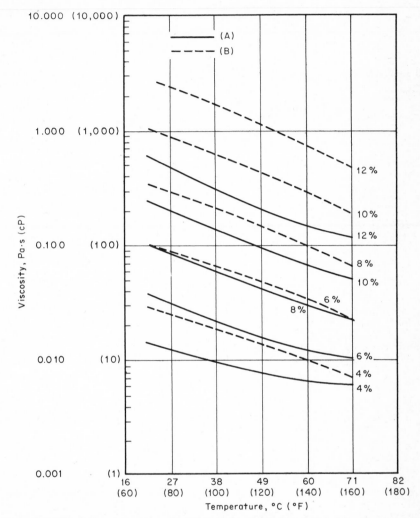

Figure 20.4   Viscosity vs. temperature for fully hydrolyzed polyvinyl alcohol: *(A)* = 13-cP-viscosity-grade; *(B)* = 30-cP-viscosity-grade. (Brookfield Viscometer, Model LVF, 60 rpm.)

Molding Properties   Plasticized polyvinyl alcohol can be compression-molded or extruded using conventional equipment.   Fully hydrolyzed grades of polyvinyl alcohol do not mold satisfactorily in the absence of plasticizer.   For a given plasticizer content, they require higher molding temperatures than the partially hydrolyzed grades and give harder compositions.   High-viscosity grades are generally used for compression molding, while medium-viscosity grades are favored for extrusion.

Mechanical Properties   Polyvinyl alcohol forms exceptionally tough, tear-resistant films with good abrasion resistance.   The tensile strength of polyvinyl alcohol is high in relation to plastics in general, and particularly impressive compared with that of other water-soluble polymers.   The tensile strength, elongation, tear resistance, and hardness of films and moldings vary with plasticizer content and relative humidity as well as with the grade of polyvinyl alcohol used.

Tensile Strength.   Assuming identical test conditions, the tensile strength of polyvinyl alcohol increases with degree of polymerization at constant percent hydrolysis, and

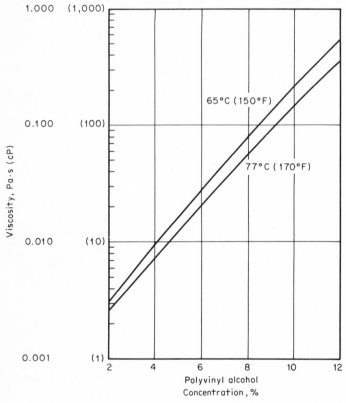

Figure 20.5   Effect of concentration and temperature on the viscosity of warp-size grade of polyvinyl alcohol.   (Brookfield Viscometer LVF, 60 rpm.)

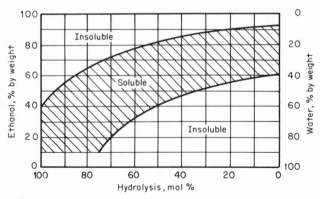

Figure 20.6   Solubility of polyvinyl alcohol in ethanol-water mixtures.

with percent hydrolysis at a given degree of polymerization. In comparing commercial grades of polyvinyl alcohol, there is a greater difference in strength between low- and medium-viscosity grades than between medium- and high-viscosity grades in the same hydrolysis range.

The tensile strength of films of polyvinyl alcohol increases markedly on stretching. Film having an average tensile strength of 69 MPa (10,000 psi) before orienting may show a tensile strength of more than 345 MPa (50,000 psi) after stretching to five times the original length.

Elongation. Elongation at break for previously unstretched films of polyvinyl alcohol may vary from less than 10% to more than 600%, depending primarily on degree of polymerization, plasticizer content, and humidity.

Tear Resistance. Tear resistance varies with the grade of polyvinyl alcohol in the same manner as does tensile strength, increasing with both percent hydrolysis and degree of polymerization or viscosity. Addition of a small amount of plasticizer substantially improves the tear resistance of films of polyvinyl alcohol.

TABLE 20.6
Maximum Salt Concentrations for Solubility of Fully Hydrolyzed, 0.030 Pa·s (30 cP) Viscosity, Polyvinyl Alcohol*

Salt	Salt concentration in water, %
NaCl	14
KCl	14
NH_4Cl	>25
Na_2SO_4	4
K_2SO_4	4
$(NH_4)_2SO_4$	6
$CuSO_4$	8
$Al_2(SO_4)_3 \cdot 18\ H_2O$	8
$K_2Al_2(SO_4)_4 \cdot 24\ H_2O$	8
$ZnSO_4 \cdot 7\ H_2O$	8
Na_2SO_3	5
$Na_2S_2O_3 \cdot 5\ H_2O$	10
Na_2CO_3	4
$NaHCO_3$	10
$Na_3PO_4 \cdot 12\ H_2O$	9
$Na_2HPO_4 \cdot 12\ H_2O$	8
$NaH_2PO_4 \cdot H_2O$	8
$NaNO_3$	23

* Determined by adding one drop of 10% solution of polyvinyl alcohol to 50-mL samples of salt solutions of various concentrations and observing the concentration above which precipitation of the polyvinyl alcohol occurred.

TABLE 20.7
Oxygen Permeability Through Polyvinyl Alcohol Film at 3°C (37°F)
(Fully hydrolyzed 0.03 Pa·s (30 cP) viscosity, plus 16% glycerin)

Relative humidity, %	Permeability at 1 atm	
	$cm^3/m^2/24\ h/\mu m$ ($cm^3/100\ in\ ^2/24\ h/mil$)	$cm^3/m^2/24\ h/\mu m$ ($cm^3/100\ in\ ^2/24\ h/mil$)
60	<0.244 (<0.04)	<0.0000347 (<0.000057)
60–80	<1.22 (<2)	<0.001769 (<0.0029)
100	32.94 ± 1.83 (54 ± 3)	0.047 ± 0.00244 (0.077 ± 0.004)

Hardness and Flexibility.   All grades of polyvinyl alcohol have a Shore durometer hardness above 100 when unplasticized.   Flexible compositions with a durometer hardness as low as 10 can be prepared by incorporating plasticizers.

Oil and Solvent Resistance   All grades of polyvinyl alcohol are unaffected by animal and vegetable oils, greases, and petroleum hydrocarbons.   Resistance to organic solvents

TABLE 20.8
Moisture Vapor Transmission Rate of Fully Hydrolyzed,
0.030 Pa·s (30 cP) Viscosity, Polyvinyl Alcohol at 22°C
(72°F)*

Film thickness, mils	Relative humidity at film faces, %	g H_2O/ 100 in²/ 24 h	g H_2O/ m²/24 h
3	0/50	0.45	7.0
3	50/72	9.5	147

* ASTM E 96-53T.

TABLE 20.9
Moisture Absorption of Unplasticized Films*

Relative humidity, %	Gain in weight, %
35	1–2
55	3–5
80	7–10
100	25–30

* Dried 10 min at 100°C (212°F) before conditioning at humidity indicated.

increases with the degree of hydrolysis.   There is no appreciable difference in the solvent resistance of low-, medium-, and high-viscosity grades within the same hydrolysis range.

Polyvinyl alcohol is substantially unaffected by most esters, ethers, ketones, aliphatic and aromatic hydrocarbons, and the higher monohydric alcohols. The lower monohydric

TABLE 20.10
Light Transmitted by Films of Polyvinyl Alcohol

Wavelength, nm	% transmitted
2536	77.5
3130	72.9
3650	81.1

TABLE 20.11
Effect of Humidity on Electrical Properties of Polyvinyl Alcohol

	Dried over P_2O_5	Conditioned at 50% relative humidity
60 cycles;		
Power factor	0.07–0.20	0.20–0.60
Dielectric constant	5–10	8–25
1.6 × 10⁶ cycles:		
Power factor	0.03–0.09	0.07–0.20
Dielectric constant	3–4	4–7

alcohols have some solvent or swelling action of the partially hydrolyzed grades, but the effect of these solvents on the completely hydrolyzed grades is negligible. (See Table 20.12.)

TABLE 20.12
Effect of Oils and Solvents on Polyvinyl Alcohol
Percent gain in weight of molded, unplasticized polyvinyl alcohol immersed for 10 days at 25–35°C (77–95°F)

Solvents	Partially hydrolyzed, 0.040 Pa·s (40 cP) viscosity	Fully hydrolyzed, 0.060 Pa·s (60 cP) viscosity
Alcohols:		
Methanol	30.0	0.4
Ethanol, 95%	9.6	<0.1
n-Butanol	< 0.1	<0.1
Esters:		
Ethyl acetate	< 0.1	<0.1
Amyl acetate	< 0.1	<0.1
Ethers:		
Ethyl ether (U.S.P.)	0.1	<0.1
Ketones:		
Acetone	0.2	<0.1
Hydrocarbons:		
Heptane	< 0.1	<0.1
Kerosene	< 0.1	<0.1
Toluene	0.1	<0.1
Turpentine	0.1	<0.1
Chlorinated hydrocarbons:		
Carbon tetrachloride	0.1	<0.1
Tetrachlorethane	1.0	<0.1
Ethylene dichloride	0.7	<0.1
Trichlorethylene	0.3	<0.1
Oils:		
SAE #10 oil	< 0.1	<0.1
Lard oil	0.1	<0.1
Cottonseed oil	< 0.1	<0.1
Raw linseed oil	< 0.1	<0.1
Miscellaneous:		
Oleic acid	0.9	0.9

Manufacture

Polyvinyl alcohol is manufactured by first producing polyvinyl acetate from vinyl acetate monomer and then following this with alcoholysis of the polyvinyl acetate to polyvinyl alcohol.

These reactions are illustrated as follows:

$$n \begin{bmatrix} CH{=}CH_2 \\ | \\ OCOCH_3 \end{bmatrix} \xrightarrow{\text{catalyst}} \begin{bmatrix} -CH-CH_2-CH-CH_2- \\ | \qquad\quad | \\ OCOCH_3 \quad OCOCH_3 \end{bmatrix}_n$$

Vinyl acetate   Polyvinyl acetate

$$\begin{bmatrix} -CH-CH_2-CH-CH_2- \\ | \qquad\quad | \\ OCOCH_3 \quad OCOCH_3 \end{bmatrix}_n + CH_3OH \xrightarrow{\text{NaOCH}_3}$$

Polyvinyl acetate   Methanol

$$\begin{bmatrix} -CH_2-CH-CH_2-CH-CH_2- \\ \qquad\quad | \qquad\qquad\quad | \\ \qquad\quad OH \qquad\qquad OH \end{bmatrix}_n + CH_3COOCH_3$$

Polyvinyl alcohol   Methyl acetate

Polymerization of vinyl acetate monomer is carried out by conventional techniques such as bulk, solution, or emulsion polymerization. The resulting polyvinyl acetate is then dissolved in a solvent, usually methanol, and alcoholized with either an acid or alkaline catalyst. Polyvinyl alcohol is insoluble in the methanol and the by-product methyl acetate and precipitates out. It is then filtered, washed, dried, and packaged. The polyvinyl alcohol is characterized by its solution viscosity or degree of polymerization and the percent alcoholysis or degree of saponification.

Both viscosity and alcoholysis may be separately controlled. The molecular weight, which is reflected in its solution viscosity, is a direct function of the molecular weight of the precursor polyvinyl acetate resin used in the alcoholysis reaction. This is usually controlled by the use of "chain-breakers" during the polyvinyl acetate polymerization. The average molecular weights of four types of polyvinyl alcohol are commercially characterized as:

Super-high viscosity	250,000–300,000
High viscosity	170,000–220,000
Medium viscosity	120,000–150,000
Low viscosity	25,000–35,000

The percent hydrolysis is controlled either by allowing the alcoholysis reaction to go essentially to completion or by stopping it at a desired level. There are two important levels of alcoholysis. These are the partially hydrolyzed grades which are alcoholized 87 to 89% and the fully hydrolyzed grades which are alcoholized 98 to 100%. Generally, when reference is made to polyvinyl alcohol without reference to the hydrolysis level, this means fully hydrolyzed polyvinyl alcohol.

Physiological Properties

Polyvinyl alcohol is not a primary skin irritant and does not produce skin sensitization. In any operation which produces dust from polyvinyl alcohol, a dust mask or respirator should be worn to avoid ingestion. If polyvinyl alcohol enters the body, it is not believed to be broken down by the tissues, but remains as a foreign material. Consequently, polyvinyl alcohol is not cleared for direct use in foods or other preparations that might be taken internally. For example, polyvinyl alcohol solutions must not be used as a substitute for blood plasma or as a component of preparations intended for injection into tissues or for spraying into the nose, throat, or lungs.

Federal Drug Administration (FDA) Status

Polyvinyl alcohol has been approved by the FDA for many indirect food additive uses such as in adhesives, paper sizes and coatings, and coatings for metal surfaces. Some of these uses require extraction tests on the finished product or place viscosity limitations on the polyvinyl alcohol. There has been no approval for direct addition of polyvinyl alcohol to food. Table 20.13 gives a summary of the status of polyvinyl alcohol under FDA regulations.

At present there is no FDA requirement for prenotification of use of polyvinyl alcohol in cosmetic formulas. However, each manufacturer must assume responsibility that the total formulation meets with FDA approval. It is understood that some manufacturers are using polyvinyl alcohol in cosmetic applications.

The use of polyvinyl alcohol in contact with most types of food classified as meats does not have approval of the Meat Inspection Division of the U.S. Department of Agriculture. Lack of approval has barred polyvinyl alcohol from meat-packaging applications.

Biochemical Oxygen Demand (BOD)

Measurement of the BOD of a given wastewater involves determining the oxygen uptake of the wastewater. The condition of the system tested influences the measured BOD for polyvinyl alcohol. Thus, a routine BOD test may not predict the oxygen requirement for biodegradation in waste treatment systems.

When polyvinyl alcohol is exposed to acclimated organisms, the BOD_{20} is about two-thirds of the chemical oxygen demand (COD). The COD for polyvinyl alcohol is 1.8 mg O_2/mg of polyvinyl alcohol. If polyvinyl alcohol is introduced into a sewage treatment system that has not been receiving polyvinyl alcohol, experience indicates that there is little biodegradation of the polyvinyl alcohol for several weeks. After about 5 weeks, activated sludge systems have been observed to acclimate to polyvinyl alcohol and over 90% degradation has resulted in 24 h. If polyvinyl alcohol were discharged into an environmental situation where acclimation did not occur, the BOD would be low (BOD_{20} would approximate 0.02 mg O_2/mg of polyvinyl alcohol).

Biodegradation

Prompted by concern over the fate of polyvinyl alcohol in textile-desize waste streams, considerable work has been carried out on the biodegradability of this polymer. These studies have established that the microorganisms in either domestic or textile mill–activated sludge waste treatment systems can adapt to polyvinyl alcohol.

Using acclimated organisms, polyvinyl alcohol removals in excess of 90% have been achieved in a properly designed and operated plant treating either type of waste. Since high polyvinyl alcohol removals are accompanied by correspondingly high reduc-

TABLE 20.13
Status of Polyvinyl Alcohol under FDA* Indirect Food Additive Regulations

FDA Regulation		Fully hydrolyzed grades	Partially hydrolyzed grades	Specialty grades
175.105	Adhesives	X	X	X
175.300	Resinous and polymeric coatings	X	X	
175.320	Resinous and polymeric coatings for poly-olefin films	X	X	
176.170	Components of paper and paperboard in contact with aqueous and fatty foods	X	X	
176.180	Components of paper and paperboard in contact with dry foods	X	X	
177.1200	Cellophane	X	X	
177.1670	Polyvinyl alcohol film	X	X	
177.2260	Filters, resin-bonded	X	X	
177.2800	Textiles and textile fibers	X	X	

* Subject to limitations specified in the applicable regulations.

tions in the COD of the waste, it appears that the polymer molecule is rapidly degraded after initial attack by microorganisms. The presence of polyvinyl alcohol in a waste stream does not inhibit the removal of other biodegradable materials normally present during waste treatment. As would be expected, molecular weight has no observable effect on the rate of polyvinyl alcohol biodegradation.

Modifiers

Plasticizers   The most effective plasticizers for polyvinyl alcohols are high-boiling, water-soluble organic compounds containing hydroxyl, amide, or amino groups. The ester-type, water-immiscible plasticizers are not compatible with polyvinyl alcohol.

Glycerin is the most widely used plasticizer for polyvinyl alcohol. It is compatible in substantial proportions with both partially and completely hydrolyzed grades and is colorless and odorless. It has good stability, is resistant to extraction by organic solvents, and has high softening effect per unit weight. Diglycerol has a lower vapor pressure but is less compatible with polyvinyl alcohol.

Ethylene glycol and some of the lower polyethylene glycols are also effective plasticizers for polyvinyl alcohol. Glycols, in general, are less compatible than glycerin with fully hydrolyzed grades. Tetraethylene, hexaethylene, and nonaethylene glycols can be used where lower volatility is required.

Formamide is compatible with polyvinyl alcohol in substantial proportions and has a high softening effect, but it is too volatile to serve as a permanent plasticizer. Ethanolamine salts, such as triethanolamine acetate or hydrochloride, can be used to plasticize both partially and fully hydrolyzed grades of polyvinyl alcohol. Except at very low humidities, sodium and ammonium thiocyanates function as plasticizers for polyvinyl alcohol because of their humectant properties.

Moisture alone is an effective plasticizer for polyvinyl alcohol. Since all the plasticizers discussed above are hygroscopic, their softening effect will vary to some extent with humidity.

Extenders　Polyvinyl alcohols may be extended with starch, dextrin, casein, urea, water-soluble amine-formaldehyde and phenol-formaldehyde resins, or with low-cost pigments such as clay and calcium carbonate to reduce cost or achieve certain properties.

Compatibility with starch varies with the grade of polyvinyl alcohol and the type of starch, as well as with the proportions in which they are combined. In general, compatibility with starches improves with increasing percent hydrolysis and increasing viscosity of the polyvinyl alcohol. Many commercial starches, particularly low-viscosity oxidized or hydroxyethylated starches, are sufficiently compatible with polyvinyl alcohol solutions for use in sizing and coating applications where film clarity is not important.

Only small proportions of most grades of dextrin are compatible with solutions of polyvinyl alcohol. Cold-water-soluble yellow potato dextrins show the best compatibility. This type of dextrin can generally be combined with polyvinyl alcohol in any desired proportion provided the total solids are less than 15%.

Insolubilizers　The water resistance of films and coatings of polyvinyl alcohol can be increased by heating to 100°C or above, incorporating an insolubilizer that reacts with the polyvinyl alcohol, or applying a protective coating. Practical insolubilizing agents include water-soluble amine-formaldehyde condensates such as dimethylolurea, trimethylolmelamine, dimethylolethylene urea, and triazone resins; dialdehydes; and polyvalent metal salts or complexes.

Commercial formaldehyde resins suitable for use as crosslinking insolubilizers for polyvinyl alcohol would be added, together with an acid catalyst, to the solution of polyvinyl alcohol before it is applied. Typically, 10% of the formaldehyde derivative and 2% of a catalyst such as ammonium nitrate, both based on polyvinyl alcohol, might be used. With most thermosetting resins, heat is required to develop maximum water resistance promptly. However, the insolubilizing reaction will take place slowly even at room temperature. Where heating to 100°C or above during drying is not feasible, aging for about 2 weeks will usually produce a fair degree of water resistance.

Metal compounds useful as insolubilizers for polyvinyl alcohol include bichromates, chromic nitrate, chrome complexes, cuprammonium hydroxide, and organic titanates. With most of these reagents, heat is required for maximum effectiveness.

Gelling Agents　Gelling agents can be used to control the penetration of polyvinyl alcohol solutions into porous substrates or to prevent run-off of coatings applied by dipping. Reagents capable of gelling polyvinyl alcohol solutions include certain dyes and aromatic hydroxy compounds, which form thermally reversible gels, and inorganic complexing agents, which form thermally stable gels. Organic compounds that have been reported* to form thermally reversible gels without imparting color include resorcinol, catechol, phloroglucinol, salicylanilide, gallic acid, and 2,4-dihydroxybenzoic acid.

Borax is a particularly effective gelling agent for polyvinyl alcohol solutions. Since gelation occurs immediately on contact and is not reversible on heating, borax is best utilized in two-stage treatments, where the polyvinyl alcohol and borax solutions are applied separately and complex formation occurs on the substrate. Alternately, boric acid can be incorporated in a polyvinyl alcohol solution to form a soluble complex which is converted to the gel form on contact with alkali. The reactions involved are as follows:

---

* U.S. Patents 2,249,536; 2,249,537; 2,249,538 (1941).

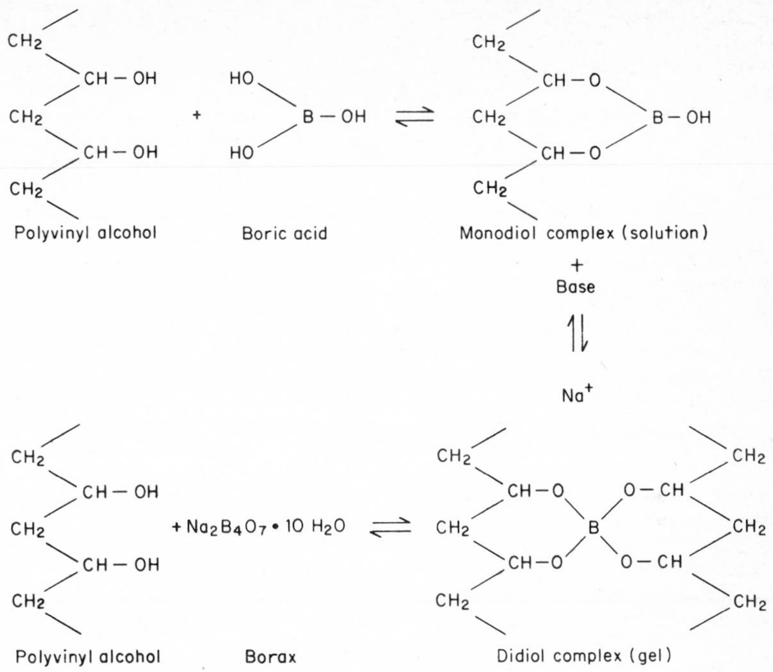

Polyvinyl alcohol   Boric acid   Monodiol complex (solution)

+
Base

Na⁺

Polyvinyl alcohol   Borax   Didiol complex (gel)

Precipitants Sodium carbonate can be used to precipitate polyvinyl alcohol from solution in a form that can be separated from the water. Sodium and potassium sulfates also act as precipitants at relatively low concentrations.

Antifoam Agents Foaming of polyvinyl alcohol solutions can be reduced by raising the temperature, reducing the amount of agitation, or adding an antifoam agent such as tributyl phosphate or a polyethylene glycol ether. If an antifoam is required, 0.1 to 0.5%, based on weight of solution, can be added during solution preparation. Excessive use of foam-control agents should be avoided since these products tend to impair the wetting properties of the solution.

Wetting Agents Practically any of the common wetting agents can be used in solutions of polyvinyl alcohol. The amount required varies with the application, and is usually between 0.05 and 0.20%, based on weight of solution.

Pigments and Dyes Pigments commonly used in paper-coating colors, i.e., clay, whiting, titanium dioxide, and calcium sulfate, can be incorporated in aqueous solutions of polyvinyl alcohol. A dispersing agent, for example, approximately 0.3% tetrasodium pyrophosphate based on pigment, is generally added to facilitate dispersion and decrease viscosity. Other pigments that can be used with polyvinyl alcohol in molding compositions or aqueous systems include lithopone, zinc oxide, carbon black, antimony oxide, and iron oxide. Finely divided silica can be used to reduce the gloss of clear films or coatings applied from solutions of polyvinyl alcohol.

Handling and Storage

Standard Package The standard shipping container for polyvinyl alcohol is a multiwall paper bag, net weight 22.7 kg (50 lb).

Shipping Polyvinyl alcohol may be shipped by freight, truck, express, and parcel post. Hopper trucks and hopper cars are used in bulk shipments.

Bulk Storage Major users of polyvinyl alcohol use silos for bulk storage. The silos should be constructed of epoxy-lined mild steel. Conveyor tubing should be of aluminum. All other parts of the system coming in contact with the product should be aluminum, stainless steel, or epoxy-coated.

Polyvinyl alcohol, as supplied, can be stored indefinitely without lumping, molding, or deteriorating. It should be kept dry to avoid caking.

Bulk transfer of polyvinyl alcohol from one area to another can be by mechanical or pneumatic means, or combinations of both, depending on the transfer distance and the related equipment in use at the plant. Mechanical systems are most economical for transfers over short distances. Proper safety procedures, similar to those used for handling bulk flour and other fine-particle-size products, should be followed to avoid dust explosions.

Safety Factors   Polyvinyl alcohol is not classified as a flammable material. However, it will burn at about the same rate as paper.

The chief hazards associated with the use of polyvinyl alcohol are dust inhalation and dust explosions. Normal handling of product as supplied in 22.7-kg (50-lb) bags entails no risk of encountering explosive mixtures. However, respirators and goggles should be worn in areas where work procedures produce dust from the polyvinyl alcohol. Spills can be swept up and disposed of by incinerating or burying.

Application Procedures

The following procedure is suggested for preparing solutions of all grades of polyvinyl alcohol.

1. Calculate level of finished solution in preparation vessel.
2. Free steam lines of any condensed water.
3. Add calculated amount of water to mixing tank, and begin agitation. Temperature of water should be no higher than 27°C (80°F). Live steam is recommended for heating. Therefore, sufficient water should be left out initially to compensate for steam condensate. Generally omitting 10 to 15% water is adequate.
4. Sift the polyvinyl alcohol directly into a vortex, if possible. Do not allow the polyvinyl alcohol to remain on the surface because it may cake or form lumps which are difficult to disperse.
5. Wash down any solids caught on the stirrer blades or sides of the mix tank. Continue stirring this mixture until a smooth slurry has been obtained.
6. Heat with live steam to 93°C (200°F), and stir at this temperature for at least 30 min. Turn the steam off once the temperature reaches 93°C (200°F). This will minimize dilution with condensate.
7. Continue agitation and allow to cool to usable temperature. If a jacketed tank is used, circulating cold water in the jacket will hasten cooling. This can also be accomplished by using cooling coils.
8. Add required water to bring total batch to required level determined in step (1). Cold water will help reduce the temperature rapidly.

COMMERCIAL USES: Compounding and Formulating

Adhesives

Polyvinyl alcohol is widely used in industrial adhesives for paper and paperboard, and in general purpose adhesives for bonding paper, textiles, leather, and porous ceramic surfaces. Completely hydrolyzed grades are used in formulating quick-setting, water-resistant adhesives, including bag-seam and bag-bottom pastes, and paper-laminating adhesives for use in the manufacture of solid fiberboard; linerboard; spiral-wound tubes, cans, and cores; and laminated specialties. Partially hydrolyzed grades are used in remoistenable adhesives. Polyvinyl alcohol is also used as a modifier for adhesives based on resin dispersions.

Water-Resistant (for Paper)   Combinations of fully hydrolyzed polyvinyl alcohol and starch are used as quick-setting adhesives for paper converting. Suitable starches include pearl corn starch, tapioca starch, and hydroxyethylated starches.

Polyvinyl alcohol/clay and polyvinyl alcohol-starch-clay mixtures are also used as paper adhesives. Simple binary mixtures of polyvinyl alcohol and clay are particularly effective laminating adhesives. They show high tack, set rapidly to form water-resistant bonds, and are economical since they give excellent mileage.

Borated polyvinyl alcohols, commonly called "tackified," are combined with clay and used in adhesive applications requiring a high degree of wet tack. They are used extensively to glue two or more plies of paper together to form a variety of tubes, cans, and cores.

Resin Dispersion Polyvinyl alcohols are used as viscosity modifiers in adhesives based on resin dispersions such as the polyvinyl acetate emulsions. The polyvinyl alcohol serves to maintain the desired adhesive viscosity at reduced solids without impairing adhesion. Partially hydrolyzed grades generally provide optimum stability.

Paper and Paperboard Sizing

The high film strength, toughness, abrasion resistance, and adhesive strength of polyvinyl alcohol, combined with outstanding resistance to oil and solvents, have led to its use for both internal and surface sizing. Polyvinyl alcohol can be applied from water solution by means of the size press, spray equipment, or water boxes on the calender stacks.

Internal Sizing Small amounts of polyvinyl alcohol, applied under conditions ensuring thorough penetration of the web, substantially increase the strength and flexibility of bond, ledger, index, blueprint, and other high-quality papers. In glassine paper, polyvinyl alcohol imparts improved strength, transparency, gloss, grease resistance, and dimensional stability. The polyvinyl alcohol can be applied, together with the usual plasticizers and transparentizing agents, at the size press after the sheet is formed. A 5% solution is typical.

Surface Sizing Polyvinyl alcohol is an effective surface size for paper or paperboard for high-gloss printing or other applications where oil or solvent holdout and surface smoothness are required. The polyvinyl alcohol can be used alone or as a modifier for improving the performance of starch.

Paper and Paperboard Coatings

Polyvinyl alcohol may be used as a binder in pigmented coatings for paper and paperboard, and in greaseproof and other specialty coatings, including solvent holdout coatings for reproduction-grade papers. By applying a continuous coating, free from pinholes or breaks, paper can be made completely greaseproof and highly impervious to oxygen and many other gases. (See Barrier Properties on page 20-8 and Tables 20.7 and 20.8.)

Pigmented Coatings

Polyvinyl alcohol is an exceptionally powerful pigment binder. Since it is effective at lower binder levels than other commonly used binders, it provides coatings of greater opacity and higher brightness. The improved optical properties obtained using polyvinyl alcohol make it possible to substitute a lower-cost pigment such as clay for all or part of the titanium dioxide used in some coatings to enhance brightness.

The use of 3 to 5 parts of polyvinyl alcohol in place of 14 to 20 parts of starch per 100 parts of clay allows lower-weight coatings with essentially equivalent pick resistance and hiding power. In substituting polyvinyl alcohol for a portion of the starch in a pigmented coating, 1 part polyvinyl alcohol generally replaces 4 to 6 parts starch. Pigmented coatings based on polyvinyl alcohol can be applied by size press, trailing blade, air knife, and roll coaters.

Greaseproof Coatings

In order to keep within practical viscosity limits, high-solids greaseproof coatings for on-machine application are formulated using polyvinyl alcohol as a modifier for a polyvinyl acetate emulsion rather than as the sole binder. Good grease resistance can be obtained at pigment/binder ratios of 1.5:1 or 2:1 using 5% polyvinyl alcohol based on polyvinyl acetate solids.

In commercial practice, these coatings are generally applied to paperboard at two coating stations, each applying approximately 9.77 to 14.66 g of coating solids/m² (2 to 3 lb/1000 ft²). Grease resistance can be further improved by applying a 5 to 7% solution of polyvinyl alcohol to the coated board at one or two water boxes on the calender stack.

Textile Finishing

Polyvinyl alcohol may be used to modify the hand of textile finishes based on thermosetting resins such as urea-formaldehyde, melamine-formaldehyde, dimethylolurea, and triazone resins. From 5 to 25% polyvinyl alcohol, based on the weight of thermosetting resin, is generally added to impart firmness and body to the finished fabric. Since the polyvinyl alcohol reacts chemically with the thermosetting resin during the curing step, the finish is washfast when properly applied.

Binder Applications

Polyvinyl alcohol is an efficient binder for solid particles, including pigments, ceramic materials, cement, plaster, and textile fibers. Polyvinyl alcohol is used as a binder for nonwoven fabrics and in ceiling tile prime coats. It can also be used as a binder in catalyst pellets, cork compositions, cementitious building products, and compressed waste products.

Nonwoven Fabrics and Ribbons   Polyvinyl alcohol shows excellent adhesion to cotton and rayon. As little as 3 to 5%, based on fiber weight, produces high-strength nonwoven fabrics. For use as a ribbon binder, polyvinyl alcohol is generally applied to 8 to 16% solids add-on.

Cementitious Binding Products   The addition of 1 to 2% polyvinyl alcohol improves the performance of many dry-mix cementitious products, including roof coatings, joint cements for drywall construction, thin-set mortars for ceramic tile work, and simulated acoustical ceiling coatings designed for spray application. A cold-water-soluble, partially hydrolyzed grade is preferred. Polyvinyl alcohol prolongs the open time of joint cements and mortars and improves both resiliency and adhesion.

Cast Film

Unsupported film cast from water solutions of polyvinyl alcohol and plasticizer is transparent, tough, tear-resistant, and puncture-resistant. It provides a unique combination of water solubility, gas barrier properties, and resistance to oils, grease, and solvents.

Use of water-soluble polyvinyl alcohol film for packaging preweighed quantities of materials such as dyes, detergents, bluing, powdered bleaches, and fungicides permits their addition to aqueous systems without breaking the package or removing the contents, thereby saving time and reducing material losses. Other suggested applications in the packaging field include container liners for oils, greases, and paints, and wrappers for protecting polished metals from gases such as oxygen and hydrogen sulfide.

Film made from polyvinyl alcohol may be used for hospital laundry bags that are added directly to the washing machine without the need for handling the contents, for oxygen tents, and for the manufacture of airtight bags for bag-molding operations. Since the film has little tendency to adhere to other plastics, it can be used to prevent sticking to molds.

When oriented by stretching, polyvinyl alcohol film will polarize light. This property can be utilized in the manufacture of sunglasses, desk lamps, automobile headlights and windshields, as well as in certain photographic techniques. A very strong, water-soluble thread can be prepared by stretching and twisting strips of film.

Molded Articles

Plasticized polyvinyl alcohol can be fabricated by compression or extrusion molding to produce flexible tubing, rod, sheeting, and miscellaneous molded articles. These molded products show high strength and flexibility; excellent resistance to oils, grease, and solvents; low permeability to gases; and good aging characteristics.

Flexible tubing has found use in lubricating equipment, for aircraft and automotive fuel lines, for oxygen and compressed-air lines, and in equipment for handling paints, lacquers, dry-cleaning solvents, fire-extinguishing fluids, and refrigerants.

Emulsions and Dispersions

Polyvinyl alcohol may be used as an emulsifier and protective colloid in the emulsion polymerization of vinyl acetate to form high-solids dispersions with outstanding stability and excellent adhesive properties. It is also effective in preparing stable polystyrene and styrene-butadiene copolymer latexes. In bead or suspension polymerization of vinyl monomers, including vinyl acetate, styrene, and vinyl chloride, polyvinyl alcohol can be used to control particle size and particle-size distribution. Partially hydrolyzed grades of polyvinyl alcohol are generally preferred for use as emulsifiers.

Cosmetics

The emulsifying, binding, film-forming, and thickening properties of polyvinyl alcohol have all found use in cosmetic preparations. Stable emulsions of natural fats, oils, and waxes may be readily prepared using partially hydrolyzed polyvinyl alcohol. Cold creams, cleansing creams, shaving creams, and facial masks may be formulated using polyvinyl alcohol.

The U.S. Food and Drug Administration does not require prenotification of cosmetic formulas. However, cosmetic manufacturers should clear with FDA the use of polyvinyl alcohol in their particular products.

Chemical Derivatives

The most important use for polyvinyl alcohol as a chemical intermediate is its reaction with aldehydes to form polyvinyl acetals. Polyvinyl butyral is used in the manufacture of automotive safety glass.

COMMERCIAL USES: Processing Aids

Textile Warp Sizing

The high film strength, flexibility, abrasion resistance, and good adhesion of polyvinyl alcohol, combined with its water solubility, make it an efficient warp size for filament and spun yarns. A specialty-grade polyvinyl alcohol is preferred for textile warp sizing, particularly for polyester/cotton blends.

Size-bath formulas based on polyvinyl alcohol are simple and easy to prepare, and run without difficulty on the slasher. Important performance advantages of polyvinyl alcohol as a warp size include the following:

1. *Low Add-On.* The high film strength, abrasion resistance, and excellent adhesion of polyvinyl alcohol permit its use at about one-half of the add-on required with starch formulas.

2. *High Weaving Efficiency.* Yarns sized with polyvinyl alcohol show excellent weaving performance with few loom stops.

3. *Low Weave-Room Humidity.* Although weaving efficiency remains high over a wide range of humidity, warps sized with polyvinyl alcohol do not require high humidity for good weaving performance.

4. *Size-Bath Stability.* Solutions of polyvinyl alcohol are noncorrosive and exhibit excellent resistance to spoilage. Unlike starch sizing materials, solutions of polyvinyl alcohol can be held at elevated temperatures for days without viscosity degradation.

5. *Ease of Desizing.* Polyvinyl alcohol dissolves readily in hot water, without the need for costly enzymes. It is easily removed, even from heat-set polyester/cotton fabrics.

Temporary Binder

Small amounts of polyvinyl alcohol are effective in improving the green strength of ceramic bodies. The polyvinyl alcohol burns out when the ware is fired. For example, addition of 2 to 3% polyvinyl alcohol, based on the dry weight of the clay, facilitates extrusion of steatite tubing and reduces breakage during handling and machining of the green ware.

Casting Slips

As little as 0.1% polyvinyl alcohol, based on dry weight of clay, improves the working characteristics of "short" casting slips in the manufacture of fine china flatware. Color and transparency of the finished ware are unaffected.

Steel Quenchant

Dilute water solutions of partially hydrolyzed polyvinyl alcohol give cooling rates and hardening responses intermediate between those obtained using oil and water, respectively. Concentrations as low as 0.05 to 0.3% have been found satisfactory for both spray and bath quenching.

Miscellaneous Coating Applications

Polyvinyl alcohol is useful in many specialized coating applications because of its toughness, oil and solvent resistance, gas barrier properties, and ease of application from water solution. Some of these are described briefly below.

Temporary Protection   Polyvinyl alcohol can be used as a temporary protective coating to prevent scratching of highly polished metals and plastic materials during fabrication and shipment. The coating can later be removed by stripping or washing with water. Polyvinyl alcohol coatings also protect metals against tarnishing.

Release Coatings   Polyvinyl alcohol may be used as a release coating to prevent sticking of plastics to molds. Smooth coatings of polyvinyl alcohol may also be used as film-casting surfaces.

Materials Stabilization

Dilution solutions of polyvinyl alcohol (1 to 3%) may be used for controlling soil erosion, for laying dust in playgrounds and parks, for controlling losses of fine particles of coal in transit, and for stabilizing a variety of other materials.

INDUSTRIES USING POLYVINYL ALCOHOL

The properties of polyvinyl alcohol and the role that these properties play in each major (and some minor) end-use applications are described in detail in other sections of this chapter, usually under the appropriate industry in which polyvinyl alcohol is used. For the convenience of the reader who wishes to refer quickly to the various industry usages of polyvinyl alcohol, this information has been summarized in the following paragraphs.

Textile Industry

The high film strength, flexibility, abrasion resistance, and good adhesion of polyvinyl alcohol, combined with its water solubility, make it an efficient warp size for filament and spun yarns.

Size-bath formulas based on polyvinyl alcohol are simple and easy to prepare, and run without difficulty on the slasher. Important performance advantages of polyvinyl alcohol as a warp size include low add-on, high weaving efficiency, low weave-room humidity, excellent size-bath stability, and ease of desizing.

Paper Industry

The high film strength, toughness, abrasion resistance, and adhesive strength of polyvinyl alcohol, combined with outstanding resistance to oil and solvents, have led to its use for both internal and surface sizing. Polyvinyl alcohol can be applied from water solution by means of the size press, spray equipment, or water boxes on the calender stacks.

Polyvinyl alcohol may be used as a binder in pigmented coatings for paper and paperboard, and in greaseproof and other specialty coatings, including solvent holdout coatings for reproduction-grade papers. By applying a continuous coating, free from pinholes or breaks, paper can be made completely greaseproof and highly impervious to oxygen and many other gases.

Adhesives Industry

Polyvinyl alcohol is widely used in the manufacture of industrial adhesives for paper and paperboard, and in the production of general-purpose adhesives for bonding paper, textiles, leather, and porous ceramic surfaces. Partially hydrolyzed grades are used in remoistenable adhesives, while completely hydrolyzed grades are used in formulating quick-setting, water-resistant adhesives, including bag-seam and bag-bottom pastes, and paper-laminating adhesives for use in the manufacture of solid fiberboard; linerboard; spiral-wound tubes, cans, and cores; and laminated specialties. Polyvinyl alcohol is also used as a modifier for adhesives based on resin dispersions.

Cast-Film Industry

Unsupported film cast from water solutions of polyvinyl alcohol and plasticizer is transparent, tough, tear-resistant, and puncture-resistant. It provides a unique combination of water solubility, gas barrier properties, and resistance to oils, grease, and solvents.

Building Products Industries

The addition of 1 to 2% polyvinyl alcohol improves the performance of many dry-mix cementitious products, including roof coatings, joint cements for drywall construction, thin-set mortars for ceramic tile work, and simulated acoustical ceiling coatings designed for spray application. A cold-water-soluble, partially hydrolyzed grade is preferred. Polyvinyl alcohol prolongs the open time of joint cements and mortars and improves both resiliency and adhesion.

Packaging Industry

The use of water-soluble polyvinyl alcohol for packaging preweighed quantities of materials such as dyes, detergents, bluing, powdered bleaches, and fungicides permits their addition to aqueous systems without breaking the package or removing the contents, thereby saving time and reducing material losses. Other suggested applications in the packaging field include container liners for oils, greases, and paints, and wrappers for protecting polished metals from gases such as oxygen and hydrogen sulfide.

Chemical Industry

The most important use for polyvinyl alcohol as a chemical intermediate is for its reaction with aldehydes to form polyvinyl acetals. Polyvinyl butyral is used in the manufacture of automotive safety glass.

Cosmetics Industry

The emulsifying, binding, film-forming, and thickening properties of polyvinyl alcohol have all found use in cosmetic preparations. Stable emulsions of natural fats, oils, and waxes may be readily prepared using partially hydrolyzed polyvinyl alcohol. Cold creams, cleansing creams, shaving creams, and facial masks may be formulated using polyvinyl alcohol.

Ceramics Industry

Small amounts of polyvinyl alcohol are effective in improving the green strength of ceramic bodies. The polyvinyl alcohol burns out when the ware is fired. For example, addition of 2 to 3% polyvinyl alcohol, based on the dry weight of the clay, facilitates extrusion of steatite tubing and reduces breakage during handling and machining of the green ware.

As little as 0.1% polyvinyl alcohol based on dry weight of clay improves the working characteristics of "short" casting slips in the manufacture of fine china flatware. Color and transparency of the finished ware are unaffected.

Steel Industry

Dilute water solutions of partially hydrolyzed polyvinyl alcohol give cooling rates and hardening responses intermediate between those obtained when using oil or water, respectively. Concentrations as low as 0.05 to 0.3% have been found satisfactory for both spray and bath quenching.

Materials Binding

Dilute solutions of polyvinyl alcohol (1 to 3%) may be used to stabilize soil to control erosion, to lay dust in playgrounds and parks, to control losses of fine particles of coal in transit, and to stabilize a variety of other materials.

Polyvinyl alcohol is an efficient binder for solid particles, including pigments, ceramic materials, cement, plaster, and textile fibers. Polyvinyl alcohol is used as a binder for nonwoven fabrics and in ceiling tile prime coats. It can also be used as a binder in catalyst pellets, cork compositions, cementitious building products, and compressed waste products.

FORMULATIONS

Textile Warp Sizing: Slasher Operation

Established slasher practices can be followed in applying sizes based on a special warp-size grade of polyvinyl alcohol.

Size-Box Temperature   Temperatures of 65 to 80°C (150 to 175°F) give best results with size formulations based on polyvinyl alcohol alone. This range is high enough to give satisfactory penetration and to allow stretching of the yarn when desired. At the same time, it is low enough to prevent skinning or film formation on the bath surface during "creep" operation of the slasher.

Squeeze-Roll Pressure   Normal squeeze-roll pressures are usually satisfactory. A 138-kPa (20-psi) air loading is generally used. The amount of size pickup can be varied by changing the squeeze-roll pressure.

Drying Cans   For spun yarns, the first drying can should be operated at a minimum temperature of 104°C (220°F) to prevent can sticking and film formation. Yarn temperatures above 138°C (280°F) should be avoided. If a double-size box is used, it is desirable to dry the sheds separately over at least two or three drying cans. This dries the yarn sufficiently to prevent cementing together when the sheds are joined. It is recommended that the first four or five cans be coated with TFE-fluorocarbon resin finish to prevent can sticking.

Stretch   Stretch should be kept at a minimum of 0.8 to 1.5% for most sizing requirements. An even tension should be maintained from section beam to section beam.

Warp Density   Proper spacing between individual warp yarns is important for optimum pickup of size. Adjacent ends should be separated by a space at least equal to the yarn diameter. It is preferable to use a double-size box, if available, to permit maximum separation.

Textile Warp Sizing: Size-Bath Formulas

Polyvinyl alcohol can be used alone or in combination with starches, depending on the yarn and weave and on individual mill preference. As a single-component size, this special grade of polyvinyl alcohol is effective at low add-on for use with the entire spectrum of spun yarns being woven today, including natural, synthetic, and blend yarns such as the increasingly popular polyester/cotton spun blends.

Tables 20.14 and 20.15 suggest concentration ranges for an initial trial in applying polyvinyl alcohol, alone or in combination with starch, to typical spun yarns. Wet size pickup will generally be in the range from 105 to 120%.

Preparing Warp-Size Solutions   Size baths based on polyvinyl alcohol alone or in combination with starch are easily prepared in all types of equipment now being used for size preparation.

TABLE 20.14
Guide for Sizing Spun Yarns with Special Warp-Size Grade of Polyvinyl Alcohol

Yarn	Polyvinyl alcohol, kg/m³ (lb/100 gal) of size*
Polyester/cotton, fine count, 40:1 to 50:1	5990–7787 (50–65)
Polyester/cotton, medium count, 20:1 to 30:1	4193–5990 (35–50)
Polyester/rayon, medium count, 20:1 to 30:1	4193–5990 (35–50)
Polyester/wool	6589–8386 (55–70)
Cotton, Fine count, 40:1 to 50:1	5990–7787 (50–65)
Cotton, medium count, 20:1 to 30:1	4193–5990 (35–50)
Acrylic, medium count	6589–8386 (55–70)
High-temperature-resistant Aramid fiber	8985–10,183 (75–85)

* The addition of 3–5% (on weight of polyvinyl alcohol) of a low-melting-point, emulsifiable wax is generally recommended.

TABLE 20.15
Guide for Sizing Spun Yarns with Special Warp-Size Grade of Polyvinyl Alcohol/Starch Mixtures

Yarn	kg/m³ (lb/100 gal) of size	
	Starch*	Polyvinyl alcohol
Synthetic/natural blends	11,980–17,970 (100–150)	1797–2995 (15–25)
Synthetic spun yarns	14,975–20,965 (125–175)	2396–4792 (20–40)

* The addition of 6–8% (based on the starch solids) of a mill wax is generally recommended.

Cleaning of Equipment. It is necessary to clean the dissolving equipment, pumps, pipelines, and size boxes thoroughly with hot water before dissolving and applying size formulas based on polyvinyl alcohol. Inorganic salts, soaps, alkalies, borax, and cleaning powders may increase solution viscosity and precipitate the polyvinyl alcohol.

Preparation Procedure:

1. Add cold water, about two-thirds of the total amount required, to the dissolving kettle.
2. Add the required amount of polyvinyl alcohol with agitation to form a slurry.
3. If the size formula includes a starch, add the required amount to the cold slurry, and stir for about 10 min.
4. Heat the mixture to 80 to 99°C (180 to 210°F), and stir at this temperature for 30 to 45 min. For polyvinyl alcohol–starch mixtures, check the final viscosity and cook subsequent kettles to the same viscosity.
5. Add water if necessary to bring the size up to the correct volume.

Wax-Containing Size Solutions Size solutions in which polyvinyl alcohol is used in combination with wax can be formulated as shown in the following.

Tight-Weave Fabrics. For fabrics such as broadcloth (50 count, 128 sley) and poplin (30 count, 96 sley), the following formula can be used:

Water	462 L (122 gal)
Polyvinyl alcohol	45.4 kg (100 lb)
Wax	2.27 kg (5 lb)
Total	598 L (158 gal)
Solids	8%

Print-Cloth Fabrics. For polyester/cotton print-cloth fabrics (38 singles, 78 sley), the following formula can be used:

Water	625 L (165 gal)
Polyvinyl alcohol	45.4 kg (100 lb)
Wax	2.27 kg (5 lb)
Total	776 L (205 gal)
Solids	6%

Preparation Procedure. Preparation of the preceding wax-containing warp-size formulas is by the following sequence of steps:

1. Add cold water, about two-thirds of the total amount required, to the dissolving kettle.
2. Add the required amount of polyvinyl alcohol, with agitation, to the kettle to form a slurry.
3. Add the required amount of wax to the slurry.
4. Heat the mixture to 80 to 99°C (180 to 210°F), preferably with live steam, and stir at this temperature for 30 to 45 min.
5. Add water if necessary to bring the size solution up to the correct total volume.

Note: Cooking by means of immersion heaters or direct flame should be avoided. Heating by circulating steam or hot water through coils or with a jacketed tank is acceptable, but will take longer.

Adhesives

The following are typical adhesive formulas for various applications:

Tubes and Cores: Spiral Winding

	Parts by weight
Fully hydrolyzed polyvinyl alcohol, tackified grade	6.3
Clay, kaolinite-type, aluminum silicate pigment having an average particle size of approximately 0.8 μm and a pH in the range of 3.8–5.5	15.7
Water	78
Adhesive viscosity, 26.7°C (80°F)	0.800–1.000 Pa·s (800–1000 cP)
Solids	22%

Tubes and Cores: Convolute Winding

	Parts by weight
Fully hydrolyzed polyvinyl alcohol, 0.030 Pa·s (30 cP) viscosity	15
Hydroxyethyl ether starch, low viscosity	37.5
Clay, kaolinite-type, aluminum silicate pigment having an average particle size of approximately 0.8 μm and a pH in the range of 3.8–5.5	7.5
Water	240
Adhesive viscosity, 26.7°C (80°F)	0.700–0.800 Pa·s (700–800 cP)
Solids	20%

Cans: Oils, Refrigerated Dough, Orange Juice

	Parts by weight
Fully hydrolyzed polyvinyl alcohol, 0.030 Pa·s (30 cP) viscosity	35
Polyvinyl acetate, 55% solids, viscosity 1.5–2.2 Pa·s (1500–2200 cP) at 26.7°C (80°F)	37
Clay, kaolinite-type, aluminum silicate pigment having an average particle size of approximately 0.8 μm and a pH in the range of 3.8–5.5	55
Water	267
Adhesive viscosity, 26.7°C (80°F)	3.500–4.000 Pa·s (3500–4000 cP)
Solids	28%

Laminating: Solid Fiber, Two-Ply

	Parts by weight
Fully hydrolyzed polyvinyl alcohol, 0.030 Pa·s (30 cP) viscosity	8.4
Clay, kaolinite-type, aluminum silicate pigment having an average particle size of approximately 0.8 μm and a pH in the range of 3.8–5.5	19.6
Water	72
Adhesive viscosity, 26.7°C (80°F)	2.100–2.400 Pa·s (2100–2400 cP)
Solids	28%

Multiwall Industrial Bags (Side Seam)

	Parts by weight
Fully hydrolyzed polyvinyl alcohol, 0.030 Pa·s (30 cP) viscosity	100
Thin-boiling starch, low or medium viscosity	100
Clay, kaolinite-type, aluminum silicate pigment having an average particle size of approximately 0.8 μm and a pH in the range of 3.8–5.5	200
Water	1950
Adhesive viscosity, 26.7°C (80°F)	1.000–1.300 Pa·s (1000–1300 cP)
Solids	17%

Multiwall Industrial Bags (Bottomer)

	Parts by weight
Fully hydrolyzed polyvinyl alcohol, 0.030 Pa·s (30 cP) viscosity	150
Thin boiling starch, high viscosity	200
Clay, kaolinite-type, aluminum silicate pigment having an average particle size of approximately 0.8 μm and a pH in the range of 3.8–5.5	200
Water	1950
Adhesive viscosity, 26.7°C (80°F)	35.000–38.000 Pa·s (35,000–38,000 cP)
Solids	22%

Preparation Procedure The following procedure should be followed for preparing all polyvinyl alcohol–clay adhesives:

1. Free steam lines of any condensed water.
2. Add calculated amount of water to mixing tank and begin agitation. Temperature of water should be no higher than 26.7°C (80°F).
3. Sift the polyvinyl alcohol directly into a vortex, if possible. Do not allow it to remain on the surface because it may cake or form lumps which are difficult to disperse.
4. After several minutes add starch, if the recipe calls for it, following the same procedure used for adding polyvinyl alcohol.
5. After several minutes add the clay, maintaining good agitation all the time. Wash down any solids caught on the stirrer blades or sides of the mix tank. Stir this mixture until a smooth dispersion has been obtained.
6. Heat with live steam to 93°C (200°F), and stir at this temperature for at least 30 min.
7. Turn off steam, continue agitation, and allow to cool to usable temperature. Circulating cold water in cooling coils or in a jacketed tank will hasten cooling.
8. Polyvinyl acetate emulsion can be added at this point, if required [the solution should be 40°C (100°F) or lower]. Continue stirring.
9. Add required water to make up the total batch. Continue stirring until the mixture is smooth and homogeneous. Cold water will help to reduce the temperature rapidly.

Binder for Ceiling Tile Prime Coat

	Parts by weight
Fully hydrolyzed polyvinyl alcohol, 0.013 Pa·s (13 cP) viscosity	1.6
Clay	48.4
Water	50
Viscosity of mixture, 26.7°C (80°F)	0.140–0.150 Pa·s (140–150 cP)
Solids in mixture	50%

To prepare this binder, use the same procedure as given for adhesives preparation. Only polyvinyl alcohol and clay are added to the water.

Size for Printing and Writing Paper

	Parts by weight
Fully hydrolyzed polyvinyl alcohol, 0.030 Pa·s (30 cP) viscosity	1.25
Oxidized starch	3.75
Water	95

This formulation is designed to give performance equivalent to a 10% starch size. To prepare, use the same procedure as given for adhesives preparation. Only polyvinyl alcohol and starch are added to the water.

Pigmented Coating for Paperboard

	Parts by weight
Fully hydrolyzed polyvinyl alcohol, 0.013 Pa·s (13 cP) viscosity	3
Styrene-butadiene latex, resin solids	12
Pigment (clay)	100
Glyoxal	0.5
Water (to give total solids of 45%)	As needed

Other additives as needed can be added for typical air-knife coating, e.g., dispersant, defoamer, etc. To prepare, cook the polyvinyl alcohol and clay in water as described previously. Add glyoxal and styrene-butadiene after the solution has cooled to 40°C (100°F). Stir for about 10 min. Total water is 59 kg (130 lb).

Strippable Coating

	Parts by weight
Fully hydrolyzed polyvinyl alcohol (10% solution), 0.030 Pa·s (30 cP) viscosity	10
Glycerin	5
Sulfonated castor oil	3
Water	90

To prepare, cook the polyvinyl alcohol as previously described. Add glycerin and sulfonated castor oil after the solution cools to about 40°C (100°F). Stir for about 10 min.

Mold-Release Coating

	Parts by weight
Fully hydrolyzed polyvinyl alcohol, 0.013 Pa·s (13 cP) viscosity	5
Ethanol	25
Methanol	15
Water	50

To prepare, cook the polyvinyl alcohol as previously described. Add the ethanol and methanol after the solution cools to about 32°C (80°F). Stir for about 10 min.

Concrete Adhesion Promoter

Use 3 to 12% fully hydrolyzed polyvinyl alcohol, 0.030 Pa·s (30 cP) viscosity. Coverage will be 0.020 to 0.146 kg/m² (0.4 to 3.0 lb/1000 ft²). To prepare, cook the polyvinyl alcohol at the required concentration, using the procedure described under Application Procedures on page 20-20.

Ready-Mix Cement Binder

Use a 3 to 12% solution of fully hydrolyzed polyvinyl alcohol, 0.030 Pa.s (30 cP) viscosity (based on cement). To prepare, use the same procedure as for the concrete adhesion promoter, preceding.

LABORATORY TECHNIQUES

Following are procedures and equipment for laboratory testing of polyvinyl alcohol.

Solution Viscosity

Follow standard procedures for viscosity determinations using Hoeppler viscometer or Brookfield viscometer, model LVF or RVF.

Percent Hydrolysis

Use standard method of determining saponification numbers. The saponification numbers are determined on a dry sample. Calculations are:

$$S = \frac{(A - B)(\text{normality of standard HCl solution})(56.1)}{\text{sample weight}}$$

where A = mL of standard HCl solution required for the blank
B = mL of standard HCl solution required for the sample

$$\text{Percent hydrolysis} = 100 - \frac{7.84\,S}{(100 - 0.075\,S)}$$

Percent Volatiles

A standard, constant-temperature oven or a forced-draft oven with a built-in balance may be used. Weigh sample initially, and dry to constant weight (usually 2 to 3 h). Reweigh and calculate volatiles.

$$\text{Percent volatiles} = \frac{A - B}{A} \times 100$$

where A = weight of sample before heating
B = weight of sample after heating

Solution pH

Standard pH meter measurement on a 4% solution.

Percent Ash

In this standard determination, polyvinyl alcohol is eliminated by slow burning, and then sulfuric acid is used to convert the ash to sodium sulfate. This sodium sulfate is then converted to ash by burning in a muffle furnace at 800°C.
Calculation:

$$\text{Percent ash as } Na_2O \text{ dry basis} = \frac{(\text{wt. of ash})(0.436)(100)(100)}{(\text{wt. of sample})(100 - \% \text{ volatiles})}$$

Screen Test: Particle-Size Distribution

Follow standard screen testing procedures using USS screens #10 to #325 plus pan and cover.
Calculation:

$$\text{Percent on screen} = \frac{\text{wt. of polyvinyl alcohol on screen}}{\text{sample weight}} \times 100$$

PRODUCT TRADENAME GLOSSARY

The following are the major products and sources associated with the commercial production of polyvinyl alcohol. These are trademarked products of their respective suppliers.

Elvanol	E. I. du Pont de Nemours & Co., Inc. 1007 Market Street Wilmington, Del. 19898
Vinol	Air Products and Chemicals Co. Five Executive Mall Swedesford Road Wayne, Pa. 19087
Gelvatol	Monsanto Chemical Company 800 N. Lindbergh Road St. Louis, Mo. 63166
Gohsenol	Nippon Gohsei Co. (Japan) Handled in the United States by: Marubeni America Corp. 14 Henderson Drive West Caldwell, N.J. 07006
Poval	Kuraray Co. (Japan) Handled in the United States by: Marubeni America Corp. (see Gohsenol, above)
Mowiol	Farbwerke Hoechst A.G. (Germany) Handled in the United States by: American Hoechst Corp. Chemicals & Plastics Div. Route 202–208 North Somerville, N.J. 08876

FURTHER USEFUL READING

The following two books are recommended for further reading about polyvinyl alcohol:

C. A. Finch (ed.), *Polyvinyl Alcohol Properties and Applications,* John Wiley & Sons, Inc., New York (1973).
- Functions of Water-Soluble Polymers, pages 13–15
- General Properties of Polyvinyl Alcohol, pages 64–65

■ Hydrolysis of Polyvinyl Acetate to the Alcohol, pages 115–120
■ Commercial Uses: Compounding and Formulating:
 Warp sizing and processing of textile fibers, pages 274–275
 In paper manufacture, pages 328–330
 Reactions with clay, page 338
 In emulsion polymerization, pages 454–460
 Photosensitized reactions for printing, pages 488–491
 In optical films, pages 513–521
 Miscellaneous applications, page 534
■ Commercial Uses: Processing Aids, same as immediately preceding
■ Physical and Chemical Data:
 Thermal properties, pages 180–181
 Chemical properties, pages 196–202
 Film properties, pages 388–389
 Compatibility with other water-soluble polymers, page 553
■ Laboratory Techniques, page 572

J. G. Pritchard, *Poly(Vinyl Alcohol) Basic Properties and Uses,* Gordon and Breach, Science Publishers, Inc., New York (1970).
■ Functions of Water-Soluble Resins, page 7
■ Molecular Structure, pages 35–37
■ Solid-State Properties, pages 54–56
■ Properties in Association with Water, pages 77–80
■ Chemical Properties, pages 99–102
■ Commercial Uses, Compounding and Formulating, pages 120–123
■ Commercial Uses, Processing Aids, pages 120–123
■ Physical and Chemical Properties, as listed above

In addition to the preceding references, the Du Pont Company has prepared and published the following technical literature, available from the Plastic Products and Resins Department, Ethylene Polymers Division of Du Pont.

Physiological Properties of Elvanol® Polyvinyl Alcohol
Long-Term Biochemical Oxygen Demand of Elvanol®
Biodegradation of Elvanol® Polyvinyl Alcohol
Increasing Water Resistance of Elvanol® Polyvinyl Alcohol
Bulk Handling and Storage of Elvanol® Polyvinyl Alcohol
Warp Sizing with Du Pont Elvanol® T-25 Polyvinyl Alcohol

Chapter 21

Polyvinylpyrrolidone

L. Blecher, D. H. Lorenz, H. L. Lowd, A. S. Wood, and D. P. Wyman
GAF Corporation

General Information 21-1
  Chemical Nature 21-2
  Physical Properties 21-2
  Manufacture 21-2
  Rheological Properties 21-4
  Toxicological Properties 21-5
  PVP Films 21-8
  Compatibilities 21-8
  Future Developments 21-8
Applications of PVP 21-11
  Pharmacy 21-12
  Medicine 21-13
  Beverages 21-14
  Cosmetics and Toiletries 21-14
  Textiles 21-15
  Paper 21-15
  Adhesives 21-15
  Detergents and Soaps 21-16
  Polymers and Polymerization 21-16
  Agricultural 21-16
  Photography and Lithography 21-16
References/Further Useful Reading 21-17
  References 21-17
  Further Useful Reading 21-17

GENERAL INFORMATION

Polyvinylpyrrolidone (PVP) is a polyamide that possesses unusual complexing and colloidal properties and is physiologically inert. Several grades are manufactured and sold in the United States under a variety of names: PVP, Plasdone, Polyclar AT, Peregal ST, Kollidon, and Albigen A.

PVP is offered in the form of a white, free-flowing powder and as aqueous solutions. It is used in a wide variety of applications, each based on its outstanding solution properties. Its film-forming and adhesive qualities are utilized in aerosol hair sprays, adhesives, and lithographic solutions. As a protective colloid, it is used in drug and detergent formulations, cosmetic preparations, polymerization recipes, and in pigment or dyestuff dispersions. The textile industry makes use of its dye-complexing ability to improve the dyeability of synthetic fibers and to aid in dye stripping.

Since it complexes with tanninlike compounds, crosslinked PVP is used as a clarifying agent for fermented products such as vinegar, wine, or beer. PVP forms a stable complex with iodine. The water-soluble complex retains the germicidal properties while reducing the iodine toxicity. For many of the above uses, the physiological inertness of PVP is an important property.

PVP was first developed in Germany by Dr. W. Reppe of the I.G. Farben Co. during the 1930s, and was widely used by the Germans as a blood plasma extender during World War II. In the United States, PVP was commercialized by GAF beginning in 1950. In early 1956, commercial production was started at GAF's Calvert City, Kentucky, plant. Production of PVP at GAF's second location in Texas City, Texas, was begun in 1969.

Chemical Nature

PVP is relatively inert to chemical modification. The dry polymer is quite stable when stored under normal conditions. Solutions, when protected from mold, are also stable. However, PVP may be crosslinked to make it water-insoluble by heating in air at 150°C, or by mixing with ammonium peroxydisulfate and heating at 90°C for 30 min.[21] Exposure of PVP to light in the presence of diazo compounds or oxidizing agents such as dichromates can cause the polymer to gel.[5,20,24] Aqueous solutions of PVP when heated with strong bases such as sodium metasilicate or trisodium phosphate form a precipitate. This results from opening of the pyrrolidone ring and subsequent reaction across different chains (crosslinking).

PVP forms complexes with many different substances. With iodine, for example, the complex is so stable that the iodine cannot be extracted with chloroform and there is no appreciable vapor pressure of iodine above the complex. The complex, however, retains the excellent germicidal properties of iodine but with greatly reduced toxicity and staining tendency.[8] Insoluble complexes are formed on addition of polybasic acids such as poly(acrylic acid), tannic acid, or the copolymer of methyl vinyl ether and maleic acid to aqueous solutions of PVP. Although these products are insoluble in water, alcohol, or acetone, the reaction can be reversed by neutralizing the polyacid with base. Spectral investigations indicate that these are hydrogen-bonding complexes, much as is observed in proteins.

In addition to iodine and polyacids, PVP forms complexes with phenolic materials, and these are often insoluble in water. The stability of such complexes is related to the acid strength of the phenolic material.[23] It has been reported that PVP complexes with toxins, drugs, and toxic chemicals to reduce their toxicity.[7,9] Some dyes also strongly complex with PVP, and this is the basis of the use of PVP as a dye-stripping agent.

Physical Properties

The glass temperature of PVP is 175°C.[25] This is reduced sharply by small amounts of absorbed moisture, and this had led to some confusion on the matter. Cast films of PVP, when dry, are very hard and clear. Like polyacrylonitrile, PVP is so strongly interacting via dipole-dipole bonding or attraction, the melt viscosity is impractically high for typical thermoplastic forming operations and, consequently, the chemistry of PVP does not include mechanical properties of molded parts.

PVP is water-soluble and hygroscopic. The equilibrium water content is equal to approximately one-third the relative humidity. It has been found that 0.5 mol of water is associated per monomer unit, much like the water of hydration of proteins.[17]

Manufacture

PVP is prepared by the polymerization of N-vinyl-2-pyrrolidone, a colorless liquid (with a freezing point of 13.80°C and a boiling point of 96°C[114 mm], 123°C[50 mm]), as follows:

The polymerization can be carried out in bulk, in solution, or in suspension and can be catalyzed cationically with, for example, BF_3, anionically with potassium amide, or with free-radical initiators such as hydrogen peroxide, benzoyl peroxide, or azobisbutyronitrile. The monomer, when sufficiently pure, combines with oxygen from the air to form peroxides which can serve to initiate polymerization.

The absolute rate constants for propagation and termination for the azobisbutyronitrile-initiated polymerization in the temperature range 40 to 70°C are found to be:

$$k_p = 1.87 \times 10^3 \times e^{-7100/RT} \quad \text{and} \quad k_t = 1.00 \times 10^9 \times e^{-1600/RT}$$

Woodhams studied the kinetics of polymerization of vinylpyrrolidone (VP), reporting the results in a 1954 thesis at Polytechnic Institute of Brooklyn. He found that the rate of the aqueous hydrogen peroxide–catalyzed polymerization is greatly affected by the concentration of ammonia, peroxide, and monomer as well as by the pH and temperature. The interrelation of these factors as expressed by Woodhams is:

$$\text{Rate} = K\,[\text{HOOH}]^{1/2}[\text{NH}_3]^{1/4}[\text{VP}]^{3/2}$$

The rate of polymerization is directly proportional to the monomer concentration up to about 50%. Beyond 60%, the rate falls off extremely rapidly. The rate of polymerization is insensitive to pH in the range of 7 to 12. Above a pH of 12 the rate slows, and above 13 it is completely inhibited. An induction period is observed in the hydrogen peroxide–catalyzed polymerization which can be reduced by excluding oxygen or by increasing the initiator concentration. The addition of ammonia (or an amine) greatly reduces the induction period.

The molecular weight of PVP in the ammonia–hydrogen peroxide system increases slightly with increasing ammonia concentration and is directly proportional to monomer concentration up to about 30%. Above 30% concentration, the molecular weight is dependent almost exclusively upon initiator (inversely proportional to initiator concentration), as expressed by:

$$\bar{M} \propto 1 + [I]$$

Typical polymerization recipes employing a variety of free-radical techniques are described in German Patent 2,439,187:

1. Vinyl pyrrolidone (500 parts) and isopropanol (214 parts) are mixed in a glass flask equipped with a stirrer and reflux condenser. After the addition of 5 parts of tert-butyl hydroperoxide, the contents of the flask are heated to boiling (96 to 88°C). After the boiling point is reached, 0.5 ppm of copper acetylacetonate, based on the weight of the vinylpyrrolidone and in the form of a highly dilute solution in 10 parts of anhydrous isopropanol, is added, and heating is continued until the residual vinylpyrrolidone content is <0.5%. Solid PVP with a K-value of 29.5 is then recovered by spray drying. (K-values are described in the rheology section of this chapter.)

2. A mixture of 1000 parts of toluene and 1000 parts of vinylpyrrolidone is freed of oxygen by vigorous degassing with nitrogen at 85°C (in a stirred flask provided with a reflux condenser); 200 parts of a solution of 25 parts of azobisisobutyronitrile in 475 parts of toluene is then added in one slug (polymerization ensuing). The temperature is held at 85°C, and the rest of the azobisisobutyronitrile solution is added dropwise over 4 h. Reaction is continued at 85°C for 1 h longer. Then 2000 parts of water is added, and the toluene is stripped with steam. When the temperature of the reaction mixture reaches 100°C, more steam is blown through until another 1000 parts of condensate is distilled. The clear aqueous solution which remains is adjusted to about 30% solids and dried in a spray dryer. A fine white powder with a solids content of 94.8% and a K-value of 38 is obtained.

PVP is manufactured in the United States in four viscosity grades identified by their K-value, which approximates K-15, K-30, K-60, and K-90. The number average of the molecular weights for these grades are about 10,000, 40,000, 160,000, and 360,000, respectively. (The exact K-value relationship is described later).

The K-value or molecular weight is a significant determinant in the properties of the product. The viscosity of a solution, obviously, increases at a fixed concentration with higher K-value resins. In addition, film and solution properties change with K-value. In selecting a product for a given application, the correct K-value range must be carefully established. PVP K-15, K-30, and K-90 are available as powders with a maximum of 5% water. PVP K-90 and K-60 are produced in aqueous solution with solids content of 20 and 45%, respectively. In addition, a pharmaceutical grade of PVP is available.

Rheological Properties

Intrinsic Viscosity   The practical rheology of PVP systems involves solution viscosities, since the melts are largely intractable.

Intrinsic viscosities have been measured in a number of solvents, but owing to the polar nature of PVP, water and methanol are particularly convenient. In these solvents the following Mark-Houwink relationships have been determined:[15]

$$[\eta] \text{ MeOH} = 2.3 \times 10^{-2} \times \bar{M}w^{0.65}$$
$$[\eta] \text{ H}_2\text{O} = 5.65 \times 10^{-2} \times \bar{M}w^{0.55}$$

It is of interest that the equations above, both of which were determined at room temperature (25°C), show that methanol is a better solvent for PVP than water. Indeed, water at room temperature is nearly a theta solvent.

Fikentscher's K-Values   Various molecular weight grades of PVP are often identified in commerce via K-values which are defined in the following equation:

$$\frac{\log h_{rel}}{c} = \frac{75K_0^2}{1 + 1.5K_0c} + K_0$$
$$K = 1000\ K_0$$

where c = concentration in g/100 mL solution

h_{rel} = viscosity of the solution compared with solvent

Other solvents with a broad range of chemical and rheological characteristics also dissolve PVP, as shown in Table 21.1. Neither pH (see Table 21.2) nor temperature (see Fig. 21.1) exert strong effects on the viscosities of aqueous solutions of PVP.

The relationship among $[\eta]$, K, $\bar{M}w$, and degree of polymerization are shown in Fig. 21.2. Viscosities of ethanol solutions are readily predictable and increase smoothly with increasing solute concentration, as shown in Fig. 21.3. Non-crosslinked PVP solutions are not especially thixotropic, unless they are very highly concentrated, and exhibit rapid relaxation times.

TABLE 21.1
Viscosities of PVP K-30 in Various Organic Solvents

Kinematic viscosities, cst			Kinematic viscosities, cst		
Solvent	2% PVP	10% PVP	Solvent	2% PVP	10% PVP
Acetic acid (glacial)	2	12	Glycerin	480	2,046
1,4-Butanediol	101	425	Isopropanol	4	12
Butyrolactone	2	8	Methyl cyclohexanone	3	10
Cyclohexanol	80	376	N-Methyl-2-pyrrolidone	2	8
Diacetone alcohol	5	22	Methylene dichloride	1	3
Diethylene glycol	39	165	Monoethanolamine	27	83
Ethanol (absolute)	2	6	Nitroethane	1	3
Ethyl lactate	4	18	Nonylphenol	3300	
Ethylene glycol	24	95	Propylene glycol	66	261
Monoethyl ether	3	12	Triethanolamine	156	666

NOTE:

$$\text{Kinematic viscosity, cst} = \frac{\text{absolute viscosity, cP}}{\text{density}}$$

$$\frac{cP}{1000} = \text{Pa·s}$$

TABLE 21.2
Effect of pH on Viscosity of 5% Aqueous PVP K-30 at 25°C

pH	10	9	7	4	2	1	0.1	Conc. HCl
Viscosity, cP	2.4	2.4	2.4	2.4	2.3	2.3	2.4	4.96
Pa·s	0.0024	0.0024	0.0024	0.0024	0.0023	0.0023	0.0023	0.00496

Toxicological Properties

General In common with most chemically inert large polymer molecules, soluble or insoluble in water, polyvinylpyrrolidone is also quite inert in physiological reaction. Higher molecular weights do not readily pass through most body membranes, such as skin and gut wall. The principal reactions, when noted in gross instances of storage (resulting from parenteral uses only), may be described as inert foreign-body reactions more disturbing to body mechanical functions than as a toxic reaction due to the chemical nature of the molecule.

Acute Toxicology The acute oral lethal dose (LD$_0$) of PVP (K-30) is reported as over 100 g/kg.[22] It is not a skin or eye irritant nor a skin-sensitizer, and it is used in

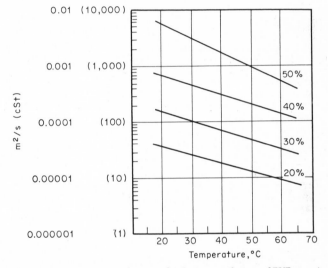

Figure 21.1 Effect of temperature on the viscosity of aqueous solutions of PVP at various concentrations.

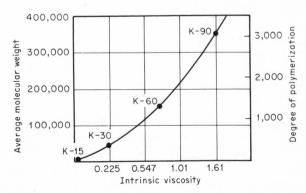

Figure 21.2 Relation of molecular weight, degree of polymerization, and intrinsic viscosity of PVP.

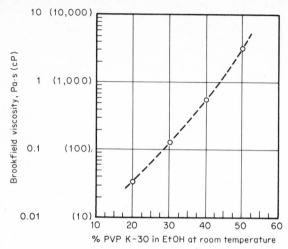

Figure 21.3 Viscosities of typical PVP K-30 ethanol solutions.

cosmetic and drug skin and eye preparations. By intraperitoneal (IP), intramuscular (IM), and intravenous (IV) routes, the material is also well tolerated, and parenteral uses including plasma volume expander solutions are based on this.

National Cancer Institute NCI-sponsored studies have investigated the acute effects of intravenously administered PVP K-15 (M.W. 12,500) in beagles, Rhesus monkeys, and other animals. In monkeys, the lethal IV dose of PVP K-15 (as 20% in saline) was observed to be greater than 5 g/kg, but less than 10 g/kg.[19] In dogs, doses of up to 10 g/kg (as 20% in saline) were not lethal.[18] In dogs, as contrasted with most other mammals (cattle, humans, monkeys, rats), a severe anaphylactic shock reaction to PVP of molecular weights of 12,500 to 40,000 is regularly noted by this route.

Subacute and Chronic *Oral.* Chronic oral feeding studies of up to 2 years have been reported[3] in dogs and rats. The same source describes the fate of single oral doses of ^{14}C-tagged PVP K-90 and PVP K-30 in rats. These studies are summarized in Table 21.3. There is no evidence for metabolism of PVP either orally or by any other route. Besides drug uses, certain food additive uses are established. PVP does not appear to be absorbed from the gastrointestinal tract (see Table 21.3).

Parenteral. Beginning in the early 1940s, observations on the parenteral effects of PVP have been reported exhaustively in the literature. Its original application as a human plasma volume expander in Germany during World War II involved an estimated 200,000 to 300,000 IV administrations. In 1947, a review[13] of this use considered critically various observed or suspected effects such as transient albuminuria, kidney and liver effects, and increased sedimentation rates.

The most well-established effect is storage, particularly in the reticuloendothelial system, and considerable attention has been given to the extent and significance of the presence of such a xenobiotic. Excretion rate and storage are dependent on molecular weight. Lower molecular weights up to around 20,000 appear to be readily excreted through the kidneys. It has been reported in the literature that various involved membranes as well as renal glomerular tubules are relatively impermeable to molecular weights of around 60,000 and over. The stored particles appear to be phagocytized principally by cells in the reticuloendothelial system and deposited in storage sites in the liver, spleen, lung, etc. (thesaurismosis).

Such storage is not ordinarily associated with pathologic changes. Potential for carcinogenic effect has been studied, particularly by Lindner[16] and Hueper.[10,11,12] Using the IP route in rats over an 18-month observation period, Lindner found no increased incidence of malignant tumor. In Hueper's first two studies, PVP was used, implanted as a powder or given IV variously to rats, rabbits, and mice. Results were most distinctly positive in rats, borderline in mice, and negative in rabbits.

Several sources and molecular weight grades were used in observations up to 2 years.

TABLE 21.3
Feeding Studies, Polyvinylpyrrolidone

Study	Maximum amount in diet, %	Toxic effects	Evidence of assimilation from GI tract	Carcinogenic effect
1. [14]C PVP K-30 rat study		Neg.	Less than 0.5%	
2. [14]C PVP K-90 rat study		Neg.	Less than 0.5%	
3. 90-day rat-feeding study with PVP K-90*	5	Neg.	Slight in surrounding lymph glands	None
4. 90-day dog-feeding study with PVP-90*	5	Neg.	No evidence found	None
5. 2-year rat-feeding study with PVP-30*	10	Neg.	No evidence found	None
6. 1-year dog-feeding study with PVP-30*	10	Neg.	Slight in surrounding lymph glands at 5% feeding level and over	None
7. 1-year dog-feeding study with PVP K-30†	10	Neg.	Slight in surrounding lymph glands at 5% feeding level and over	None
8. 2-year dog-feeding study with PVP K-30‡	10	Neg.	Slight in surrounding lymph glands at 5% feeding level and over	None
9. 90-day dog-feeding study with PVPP	10	Neg.	Neg.	None
10. Prenatal toxicity rats with K-25	10	Neg.	Neg.	None
11. Prenatal toxicity rats with K-90	10	Neg.	Neg.	None

* Study conducted by Industrial Biology Laboratories, 22 N. 36th Street, Phila., Pa., Dr. M. Shelanski, Director.
† Study conducted by Leberco Laboratories, 123 Hawthorne St., Roselle Park, N.J., Dr. I. Levenstein, Director.
‡ Study conducted by Chemo Medical Consultants, 431 N. Monroe St., Arlington, Va., Dr. V. Princiotta, Director.
NOTE: Based on the above studies as well as on extensive human experience as a tableting aid in many oral dosage forms, PVP has become well established as safe in oral uses involving such amounts.

Hueper's third study, designed to be definitive, used IP administration of four different PVP samples in rats and rabbits. No positive results were obtained in this study, and in rats, the tumor incidence in experimental animals was about half of that in the controls. No cancer effect in humans has been demonstrated by any route.

With respect to human use, some literature has developed since about 1964 with respect to gross storage effects (thesaurismosis), principally after prolonged intramuscular administration of PVP. Selected references are given.[2,4,6,14] Such clinical observations have been made principally in patients receiving long-term IM drug therapy in which PVP was incorporated for its depot effect. In some cases, amounts to over three kilograms of PVP were involved over a number of years. Storage effects were principally near, but not limited to, injection sites. Histopathological aspects are described by histiocytic and interstitial infiltration by the PVP and possibly with inflammatory response.

Inhalation. Because of its application as the first synthetic hair spray resin and its subsequent long-term use (now for the most part replaced by later ones), the inhalation toxicity of PVP has received considerable attention. An initial 1958 report[1] raises the question of the relation of hair spray resins to lung thesaurismosis (storage). An extensive hair spray inhalation literature has developed since then which includes epidemiological studies of hairdressers as well as animal experiments. The principal observation from the more detailed animal studies available is that certain of these resins, possibly the more insoluble ones, incite alveolar macrophage formation in susceptible individuals as a reaction to the foreign material (reversible). Such reaction continued to a further stage could be expected to produce granulomas (irreversible, scar tissue), but there is no evidence that this could normally result from less than impractically high and long-term exposure.

PVP Films

Dried, unmodified films of PVP are clear, transparent, glossy, and hard. Appearance does not vary when films are cast from different solvent systems, such as water, ethanol, chloroform, or ethylene dichloride.

Compatible plasticizers may be added without affecting clarity or luster of the film. Moisture taken up from the air by PVP can also act as a plasticizer. Among the several commercial modifiers that may be used in concentrations of 10 to 50% (based on PVP) to control tack and/or brittleness or to decrease hygroscopicity are:

Carboxymethylcellulose	Hyprin GP-25 (Dow)
Cellulose acetate	Igepal CO-430 (GAF)
Cellulose acetate propionate	Lorol (DuPont)
Dibutyl tartrate	Oleyl alcohol
Diethylene glycol	Resoflex R-363 (Cambridge)
Dimethyl phthalate	Santicizer 141, B-16, and E-15
Dytol B-35, J-68, and L-79 (Rohm & Haas)	(Monsanto)
	Santolite MHP (Monsanto)
2-Ethylhexanediol-1,3	Shellac
Glycerin	Sorbitol
Glyceryl monoricinoleate	

Carboxymethylcellulose, cellulose acetate, cellulose acetate propionate, and shellac effectively decrease tackiness. Dimethyl phthalate is less effective, and glycerin, diethylene glycol, and sorbitol increase tackiness. Films essentially tack-free over all ranges of relative humidity may be obtained with 10% arylsulfonamide-formaldehyde resin.

In comparative tests for plasticity at 33% relative humidity, PVP films containing 10% diethylene glycol show an elongation at break twice that of PVP films containing 10% glycerin, polyethylene glycol 400, sorbitol, or urea, and four times that of PVP films containing 10% ethylene glycol, dimethyl phthalate, or Santicizer E-15. At 50% relative humidity, 10% Santicizer E-15, 25% glycerin, and 25% diethylene glycol are effective plasticizers. At 70% relative humidity, 25% sorbitol and 25% dimethyl phthalate may be used successfully.

In PVP films used as hair fixatives, lanolin derivatives such as Acetulan and Amerchol L-101 (American Cholesterol), Ethoxylan 100 (Malstrom), Lanethyl (Croda), and Lanogel 41 (Robinson Wagner) are also effective plasticizers and humidity-control agents.

Compatibilities

PVP shows a high degree of compatibility, both in solution and film form, with most inorganic salt solutions and with many natural and synthetic resins, as well as with other chemicals.

A variety of specific examples of PVP-solution and PVP-solid (film) compatibilities are shown in Tables 21.4 and 21.5.

Future Developments

Because of its many unique properties, PVP lends itself to a wide variety of applications, only a few of which have been referenced here. New and expanded markets in cosmetics, especially skin care, can be foreseen. In pharmaceuticals, a biologically excretable molecular weight will open the door to widespread parenteral use. PVP's usefulness in paper manufacturing and adhesive formulations has only begun to be appreciated, and development is presently ongoing in these areas. It is expected that PVP's interesting ability to detoxify certain drugs and substances will be exploited.

TABLE 21.4
PVP Compatibility with Selected Solvents

Class	Compound tested	Solvent	% Total solids	Solution appearance	Film compatibility
Ether-alcohols	Polyethylene glycol,	Ethanol	5	S clear, colorless	C
	Ucon Oil 50-HB-5100 (polyalkylene glycol, Union Carbide)	Ethanol	5	S clear, colorless	I
	Ucon Oil 75-HB-90,000 (Union Carbide)	Ethanol	5	S clear, colorless	H
Ethers	Diethyl Carbitol (diethylene glycol diethyl ether, Union Carbide)	Ethanol	5	S clear, colorless	C
	Polyvinyl butyral	Chloroform	4	S clear, colorless	C
	Formvar (polyvinyl formal, Shawinigan)	Chloroform	4	PS cloudy, homogeneous	C(1:3), I(1:1, 3:1)
Gums	Arabic	Water	5	PS cloudy, heterogeneous	C
	Karaya	Water	4	I two-phase	H
	Tragacanth	Water	4	PS white, homogeneous	H
	Sodium alginate	Water	—	S	C
Glycerides	Olive oil	Chloroform	5	S clear, light yellow	I
	Castor oil	Ethanol	5	S clear, colorless	I
	Lanolin	Chloroform	6	S clear, light yellow	H
	Lecithin	Water	1	I opaque, homogeneous	I(1:3), H(1:1, 3:1)
Esters	Beeswax	Chloroform	5	S clear, colorless	H
	Diethylene glycol stearate	Propargyl alcohol	3	S clear, yellow	I
	Flexol 300 (triethylene glycol bis-alpha-ethylcaproate, Union Carbide)	Ethanol	5	S clear, colorless	I
	Cellulose acetate propionate	Chloroform	4	S clear, colorless	C
	Cellulose acetate	1:4 ethanol/ethylene dichloride	4	S clear, light yellow	C
Chlorinated hydrocarbons	Chlorinated mineral oil	Chloroform	5	S clear, colorless	I
	Arochlor 1262 (Monsanto)	Chloroform	4	S clear, colorless	I
Phenols	Actamer [2,2'-thiobis-(4,6-dichlorophenol), Monsanto]	Ethanol	6	S clear, colorless	
	Hexachlorophene [2,2'-methylene-bis(3,4,6-trichlorophenol)]	Ethanol	6	S clear, colorless	

TABLE 21.4 (continued)
PVP Compatibility with Selected Solvents

Class	Compound tested	Solvent	% Total solids	Solution appearance	Film compatibility
Miscellaneous resins	Shellac	Ethanol	5	S clear, yellow	C
	Ethylcellulose	Ethanol	4	PS cloudy, homogeneous	C
	Methylcellulose	Water	—	S	C
	Carboxymethylcellulose (low-viscosity grade)	Water	1	S clear, colorless	C
	Animal glue	Water	3	I solids settle out	I
	Bone glue	Water	—	S	C
	Corn dextrin	Water	—	S	C
	Piccolyte S-135 (terpene resin, Penn. Ind. Chemical)	Hot melt	—	S	C
Synthetic polymers	Vinylite VYHH (vinylchloride-vinylacetate copolymer, Union Carbide)	Ethylene dichloride	3	S clear, colorless	C
	Vinylite VYNW (vinylchloride-vinylacetate copolymer, Union Carbide)	Ethylene dichloride	3	S clear, colorless	I
	Polyethylene	Hot melt	—	S	C
	Polyacrylonitrile	N,N-Dimethylformamide	4	S clear, light yellow	C(1:3), H(1:1), I(3:1)
	Saran B-1155 (vinylidene chloride polymer, Dow)	N-Methyl-2-pyrrolidone	3	S yellow	I
	PVI-C poly(vinyl isobutyl ether)	Chloroform	4	I two liquid phases	I
	PVM poly(vinyl methyl ether)	Hot melt	—	S	C
	Polystyrene	Chloroform	—	S clear, colorless	I
	Polyvinyl chloride		—	S	C
	Polyvinyl alcohol	1:3 ethanol/methylethylketone	—	S	C
Surfactants	Alipal CO-436 [ammonium salt of sulfated nonylphenoxy-poly(ethyleneoxy)ethanol, GAF]	Water	3	S clear, colorless	C
	Duponol (sodium lauryl sulfate, DuPont)	Water	5	S clear, colorless	
	Igepon AP-78 (alkyl ester sulfonate, GAF)	Water	3	S hazy, light yellow	H
	Nacconol (alkylaryl sulfonate, Allied Chemical)	Water	5	S clear, colorless	
	Nekal BX-78 (sodium alkylnaphthalene sulfonate, GAF)	Water	3	S hazy, light yellow	I
Quaternary ammonium compounds	Hyamine 2389 (Rohm and Haas)	Water	2	S clear, colorless	
	Tetrosan (Onyx)	Water	10	S clear, colorless	
	BTC (alkyldimethylbenzyl ammonium chloride, Onyx)	Water	10	S clear, colorless	
	Isothan Q 15 (lauryl isoquinolinium bromide, Onyx)	Water	10	S clear, colorless	
	Ceepryn (cetylpyridinium chloride, Wm. S. Merrell)	Water	10	S clear, colorless	

TABLE 21.5
PVP Solubility in or Resistance to Various Solvents and Propellants

The following representative organic solvents will dissolve 10% or more PVP at room temperature:

Alcohols	*Acids*	*Ester*	*Amines*
Methanol	Formic acid	Ethyl lactate	Butylamine
Ethanol	Acetic acid		Cyclohexylamine
Propanol	Propionic acid	*Ketone*	Aniline
Isopropanol		Methylcyclohexa-	Ethylenediamine
Butanol	*Ether-Alcohols*	none	Pyridine
sec-Butanol	Glycol ethers		Morpholine
Amyl alcohol	Diethylene glycol	*Chlorinated*	2-Aminoethanol
2-Ethyl-1-hexanol	Triethylene glycol	*Hydrocarbons*	Diethanolamine
Cyclohexanol	Hexamethylene	Methylene	Triethanolamine
Phenol (50°C)	glycol	dichloride	Aminoethylethan-
Ethylene glycol	Polyethylene	Chloroform	olamine
Propylene glycol	glycol 400	Ethylene dichloride	2-Hydroxyethyl-
1,3-Butanediol	2,2¹-thiodiethanol		morpholine
1,4-Butanediol		*Lactams*	2-Amino-2-methyl-
Glycerin	*Lactone*	2-Pyrrolidone	1-propanol
	γ-Butyrolactone	N-Methyl-2-	
Ketone-Alcohol		pyrrolidone	*Nitroparaffins*
Diacetone alcohol		N-Vinyl-2-	Nitromethane
		pyrrolidone	Nitroethane

PVP is essentially insoluble in the following solvents under the same conditions of testing:

Hydrocarbons	Mineral oil	Isobutyl vinyl ether	*Ketones*
Benzene	Cyclohexane	Tetrahydrofuran	2-Butanone
Toluene*	Methylcyclohexane		Acetone
Xylene	Turpentine	*Chlorinated*	Cyclohexanone
Tetralin (DuPont)		*Hydrocarbons*	
Petroleum ether	*Ethers*	Carbon tetrachlo-	*Esters*
Hexane	Dioxane	ride	Ethyl acetate
Heptane*	Ethyl ether	Chlorobenzene	sec-Butyl acetate
Stoddard solvent*	Methyl ether		
Kerosene*	Ethyl vinyl ether		
Mineral spirits			

Anhydrous PVP is directly soluble in propellants containing a —CHX$_2$ group. It will dissolve in several other chlorofluoralkanes used in pressurized products if alcohol is used as a cosolvent.

Directly soluble in:

Freon 21 (DuPont)—dichloromonofluoro-
    methane

Freon 22
Genetron 141 } chlorodifluoro-
  (General Chemical) } methane

Soluble with cosolvent
(20–30% ethanol) in:

Freon 11
Genetron 11 } trichlorofluoromethane

Freon 12
Genetron 12 } dichlorodifluoromethane

Freon 113
Genetron 226 } 1,1,2-trichloro-1,2,2-trifluoroethane

Freon 114
Genetron 320 } 1,2-dichloro-1,1,2,2-tetrafluoroethane

Freon 142   1-chloro-1,1-difluoroethane

---

* PVP is soluble in these hydrocarbons in about 5% concentration when added to the solvent as a 25% butanol solution.

Looking ahead, an interesting aspect of PVP and vinylpyrrolidone copolymers is that their potential applications are so varied and are not limited to aqueous systems.

APPLICATIONS OF PVP

In view of its unique chemical and physical properties, PVP has found significant uses in pharmaceuticals, cosmetics and toiletries, textiles, paper, beverages, detergents and soaps, adhesives, polymerization aids, etc.

Because of an exceptionally low animal oral toxicity and very high parenteral lethal dose (LD$_{50}$), PVP has found extensive use in the pharmaceutical industry. Approximately 4000 papers have been published on the use of PVP in pharmacy and medicine.

The major use of PVP in pharmacy is as an excipient in oral dosage forms. PVP has been included under the name *povidone* in the *United States Pharmacopeia XIX*, as well as numerous other national compendia, and is accepted by health regulatory agencies in virtually all nations of the world.

Pharmacy

The compressed tablet continues to increase in popularity, especially in the United States. The dosage form offers several important advantages: precision of dosage, stability, long storage life, and convenience to both the pharmacist and patient.

A tablet is a blend of contradications. It must be strong enough to withstand the abuses of production, packaging, shipping, and dispensing. The tablet should be stable, free from chipped edges, cracks, etc., when it reaches the end user. On the other hand, it should break up as soon as it is in the stomach, and should release the drug quickly.

A key ingredient in every tablet is the binder, the adhesive that holds it all together. PVP has proven useful in this capacity, and, currently, this represents a large-volume use for the product (frequently K-30).

PVP may be utilized in both granulation and tableting. The binder must keep the tablet together until it reaches the stomach and then rapidly dissolve or disintegrate to release the active materials. PVP is soluble in water, alcohol, and many other organic solvents, thus giving the formulator flexibility. An example is an oral contraceptive formulation where PVP is used so that the tablet can be prepared from anhydrous solvents. In tableting, a normal use-level for PVP will run 2 to 5%.

In order to enhance the appearance of tablets and, in other cases, to contribute stability, a coating is often applied. The traditional coating has been multiple layers of sugar, usually colored with dyes. This procedure is complex and time-consuming. One improvement involves the use of lake colors instead of dyes (U.S. Patent 3,054,724), and PVP is used to disperse the finely divided lake, thus ensuring uniformly colored coatings.

In some cases a shellac coating is used, and in such formulations PVP may be used to modify the shellac. PVP also contributes to the shelf life of the product and to a stable disintegration time.

It has been shown that the solubility of water-insoluble drugs can be substantially increased when coprecipitated with PVP. Stupak and Bates investigated the in vitro and in vivo dissolution of reserpine-PVP mixtures and coprecipitate systems. They found that the coprecipitate yielded a threefold increase in reserpine excretion in the urine over a 48-h period following oral administration of the coprecipitant, as compared with oral dosage of a plain mixture. This technology is expected to prove useful in the development of new oral dosage forms.

PVP has also found extensive use in oral liquid dosage forms. In such products, the excellent dispersing properties of PVP permit production of uniform, stable products. PVP is used to retard crystal growth. An example which is well known involved cyclamate solutions. The early products tended to crystallize on the sides of the container and in the spout. PVP was incorporated to prevent this phenomenon.

Most solutions, suspensions, and emulsions are colored. PVP is used to achieve uniform dispersion of the coloring materials as well as to stabilize the solutions. In the case of suspensions and emulsions, PVP is used to stabilize the product as well as to disperse the colors (i.e., prevent agglomeration or precipitate formation).

Some formulators also use PVP as a hydrocolloid to achieve an acceptable mouthfeel. The thickening action of PVP provides an aesthetically appealing texture to the formulation.

A unique homopolymer of *N*-vinyl-2-pyrrolidone is produced by crosslinking, and the use of this crosslinked, insoluble form of PVP as a disintegrant has been reported. It was found that the mechanism of action is complex, but there are indications that the crosslinked PVP acts through capillary activity with a secondary swelling effect.

Another interesting application of crosslinked PVP is in the treatment of diarrhea (U.S. Patent 3,725,541). This use is more prevalent in European and Central American countries. A large, quickly disintegrable tablet, essentially all crosslinked PVP, is presently marketed in Europe and Mexico.

Medicine

The first use of PVP in medicine was as a blood plasma volume expander during World War II. At that time, it was found that a 3.5% solution of PVP in isotonic salt solution was effective in the treatment of shock. This application has recently been confirmed in a review by the NAS/NRC Drug Efficacy Study Group, where it was given an "effective" rating.

The K-25 to K-30 molecular weight range is preferred for this use because the effect lasts longer than with the more quickly eliminated lower molecular weights. The product was used in Germany on several hundred thousand wounded during World War II.

Long-term perservation of whole blood has been a challenge for many years. In a project sponsored by the Office of Naval Research, the Linde Division of Union Carbide Corporation and others investigated the protective influence of several materials on the cells in whole blood during quick (liquid nitrogen) freezing and subsequent thawing. Of the few efficacious materials found, PVP was the one of most interest for emergency use. In amounts of about 7% PVP based on blood weight, up to 97% viability of whole blood cells could be retained during the freezing, indefinite storage of up to 10 years, and subsequent thawing (liquid nitrogen temperatures).

In addition, of the few satisfactory materials found, such as glycerin, PVP was the only one which could be contemplated to be left in the thawed product. As an alternative, PVP or other protective agents can be removed by centrifuging, washing, and reconstituting the cells in fresh plasma. This has evolved as a procedure which is presently practiced using glycerin for long-term preservation as a supplement in usual periods where whole blood and plasma are regularly available and where such extra final steps can be afforded. Interestingly, the valuable oxygen-transport ability of frozen blood is retained, as contrasted with regularly stored blood where this is lost in about the first 5 or fewer days.

In this application, molecular weights in the K-25 to K-30 range are required for the proper cryoprotective effect. Lower molecular weights can be used, but at less practical concentrations.

In addition to complexing with and modifying the activity of certain therapeutic agents, PVP has also shown a valuable propensity to complex with a variety of toxic agents accidentally ingested, or with certain toxins bacterially generated. In this application, lower molecular weights are preferred since they complex as efficiently as higher ones but are eliminated through the kidney quicker, thereby decreasing the likelihood of storage. For example, PVP has proven to be a useful therapy in the treatment of toxic forms of acute gastrointestinal diseases of early childhood when administered IV in the same manner as a plasma expander. Typically, the manifestations of shock subside, vomiting and convulsions cease, arterial blood pressure returns, and pulse frequency, respiration, and diuresis return to normal.

The detoxifying effect of PVP has proven therapeutically useful in the treatment of burns, owing, apparently, to the inactivation and subsequent elimination of toxins by complex formation. Effect on tetanus toxins has also been studied, with the conclusion that PVP, if administered within 3 h of the toxin dose, has a very positive ameliorating effect.

A recent study of the potential of PVP with a molecular weight of 10,000 (K-15) as a detoxifying agent was conducted both in vitro and in vivo using salicylate as a model toxic drug. The in vitro studies established that PVP reversibly bound a significant quantity of salicylate, and that the binding of the drug was directly proportional to the molecular weight of the PVP as well as the concentrations of PVP and salicylate. The same detoxifying effect on salicylates has been observed using peritoneal dialysis instead of IV administration.

Rudy and Chernecki have studied PVP (molecular weight of 10,000, K-15) intravenously and also as an intraperitoneal lavage in the treatment of barbiturate intoxication in animals. It was concluded that the PVP binds both phenobarbital and secobarbital, promotes the renal excretion of the barbiturates, and effectively ameliorates the toxic effects.

PVP is known to retard the pharmacological action of a number of drugs. This action has been ascribed to formation of physiochemical combinations or complexes from which the drug may be more slowly liberated, to restricted diffusion of the drug due to the viscosity of the PVP medium at an IM injection site, or, possibly in other cases, to inactivation of the drug activity sites.   In the early years of penicillin, for example, PVP was a valuable aid in stretching the limited supply.   Other drugs which have been studied for this effect include sulfathiazole, sodium salicylate, chloramphenicol, insulin, phenobarbital, procaine, p-aminobenzoic acid, and many others.

PVP has also been observed to render water-soluble or water-dispersible certain materials which are ordinarily not soluble in water, thus offering a solution to otherwise difficult formulating problems.   As a special case, the formation by special procedures of coprecipitates as reported by Simonelli, Bates, and other workers indicates the utility of PVP in increasing the water solubility, sometimes by a few hundred times, of certain essentially insoluble but pharmacologically active materials such as certain ones of present interest in the National Cancer's Institute's continuing chemotherapy program.

The complexing ability of PVP is involved in the important PVP-iodine combination, useful as a microbicide.   The iodine, normally insoluble in water, is soluble in this complexed form, obviating the use of alcoholic tinctures.   The iodine is free, or available, for all its ordinary reactions, including its usual broad-spectrum antibacterial activity.   The complex appears in the *National Formulary* and is used extensively in surgical scrubs, antidandruff shampoos, and many related areas.   The complex also illustrates the detoxifying action of PVP, since the iodine in this complex is less toxic and less irritating than other free-iodine forms, such as Lugol's solution or tincture of iodine.

With regard to diagnostic aids, another PVP-iodine combination is widely reported in the literature.   In this case the product is a chemical reaction product, and the iodine is actually incorporated into the polymer and, in addition, is radioactive (usually ^{131}I).   It has been applied orally to study gastric emptying and small intestinal propulsion, ulcerative colitis, and cancer.   Introduced into the arterial system, it has been used, for example, in studying renal function and detection and location of brain tumors.   Interestingly, in confirmation of the toxicity studies described earlier, no absorption of this radioactive complex from the digestive tract has been observed with PVP of molecular weight of 35,000.

PVP has also been found useful in dispersing barium sulfate slurries used in x-ray examinations.

As an example of present special interest, both PVP and VP are cited in recent patent examples in combination with other polymers or monomers to produce soft contact lenses.   Other materials involved in the combinations include hydroxymethylmethacrylate and ethyleneglycoldimethyacrylate.   The PVP contributes hydrophilic properties.   Products are optically clear, water-insoluble, and machinable, and they absorb up to 90% by weight of water.   Other forms, incompletely crosslinked, may also be obtained, and these are fusible and moldable.

PVP, because of its unique stability in isotonic solutions and because of its ability to reduce eye irritation of many drugs, is commonly used in ophthalmic solutions.

Finally, in a readily recognized prosthetic device, the plaster cast, a recent U.S. patent describes casts of improved strength, water resistance, and physiological properties which incorporate a major amount of PVP along with other materials in the plaster of Paris.

Almost 4000 papers on PVP and its utility in medicine have appeared since the early 1940s. Papers continue to appear at the rate of about 150 per year, indicating an ongoing interest in this versatile molecule.   A representative, but not comprehensive, bibliography is listed at the end of this chapter.

Beverages

PVP forms insoluble complexes with polyphenolic structures of the type commonly referred to as *tannins* in beer, wine, vinegar, and many fruit and vegetable beverages. This property has found use in the colloidal stabilization of beer and in the stabilization of the color of white wines, rosé wines, and vinegar.

Cosmetics and Toiletries

In the early 1950s the formulation of cosmetic products, particularly those designed to control and condition women's hair, took a new direction.   Older hair fixatives, based on shellac or oils, were rapidly replaced by PVP-based sprays.   Substantivity to hair, formation of rewettable, transparent films, contributions to luster, and smoothness were reasons for the shift to a whole series of PVP-based hair-grooming products: shampoos, wave sets and creams, hair tints, and shaving products. Moreover, the broad spectrum of solubility of PVP in propellants used in pressurized products, and its anticorrosive nature, made it a natural for the rapidly developing aerosol industry.

Now recognized not merely as an additive but an integral part of many different kinds of cosmetics, PVP improves physical properties and contributes to performance in skin and eye makeup, lipsticks, deodorants, suntan lotions, and dentifrices, in addition to hair preparations. The polymer promotes emolliency, lubricity, and a natural moisture balance in skin cosmetics. It stabilizes creams, lotions, and foam products, disperses pigments, and aids in the reduction of toxicity and sensitivity problems.

In hair sprays, PVP today is useful to the formulator in nearly all the different propellant systems being investigated to replace the fluorocarbons. Where sensitivity to humid conditions presents a problem, copolymers with vinyl acetate are available, and these are less affected by moisture and humidity.

PVP is used in shampoos to improve foam stability and impart lustre to the hair. Even after a shampoo is rinsed away, enough PVP is adsorbed on the hair to give it sheen and manageability. Its use in antiseptic shampoos holds promise in view of its effectiveness in reducing the irritation of some types of bacteriostats.

The dye-complexing ability of PVP has led to its use in hair tints or color rinses. By adjusting the ratio of the PVP to dye, the tint can be made either more permanent or more transient, higher PVP concentrations favoring the latter.

PVP, which appears to impart emolliency and to reduce skin irritation, is being incorporated into preshave and after-shave lotions. Copolymers of vinylpyrrolidone can be made which are substantive to the hair and skin, and these have won formulator acceptance in conditioning shampoos, setting lotions, skin-care products, etc.

Textiles

PVP is used as a backbone for grafting monomers in the production of synthetic fibers with improved dye receptivity and antistatic properties. The polymer may also be incorporated by surface grafting or during spinning by impregnation of the wet-spun fibers. Similar results can be obtained, in other instances, by coating the polymer on the outside of a hydrophobic fiber. A related application is the modification of rubber latexes with a minor amount of PVP for use on coated fabrics which must be dyed uniformly.

Other dyeing and printing areas employing PVP are fugitive tinting, dye stripping, and scouring. The product also acts as a solubilizer for colors in the preparation of dye baths, as a protective colloid in oil-in-water emulsions, and as a dye scavenger in print washing.

In textile finishing, PVP is used in sizing, in delustering (because it is a specific dispersant for titanium dioxide), and in other finishing operations. In brightener-resin finishing baths containing metal-salt catalysts, PVP slows the exhaust rate and reduces tailing off during long runs on the padder. The polymer contributes outstanding adhesion to fiberglass sizes, and several compositions for this purpose have been patented.

Certain copolymers of the acrylates and PVP have found use for antisnag applications or polyester double-knit fabrics and have been patented.

Paper

PVP is employed in all types of paper manufacture, whether the stock is derived from cellulose fibers, rags, or inorganic materials. Its use covers a wide range: control of deposition of pitch, improvement of wet strength, solubilization of dyes and dispersing of pigments, and binding of inorganic flakes or fibers. It is a valuable adjunct in the de-inking of waste paper and in rag stripping, beating, and coloring.

PVP, moreover, is an economical fluidizer and antiblock agent in paper coating. It also contributes to gloss, printability, and grease resistance of the finished material.

Adhesives

PVP films are specifically adhesive to glass, metal, and plastic surfaces. In pressure-sensitive adhesives, the polymer imparts high initial tack, strength, and hardness. Water-remoistenable types cast on paper reduce the usual tendency to curl. Various compositions incorporating PVP have been used in glass-fiber sizes, PVA-based adhesives, wood glues, and abrasives.

In the coating field, PVP has been employed for such diverse applications as plastic packaging materials, graphite and lubricant compositions, inks, paints, waxes, and polishes. PVP is nonthixotropic, prevents gelation, and acts as a protective colloid in many formulations. It is an excellent dispersant for titanium dioxide, organic pigments, and latex polymers. In addition, it is an antiblock agent and fluidizer in paper coatings, and it forms films that are grease-resistant.

Detergents and Soaps

PVP is compatible in clear, liquid, heavy-duty detergent formulations. It prevents soil redeposition, particularly on synthetic fibers and resin-treated fabrics. It also acts as a loose-color scavenger during laundering because it has dye-binding power. Complete water solubility and superiority as a tablet binder make it ideal for detergent briquettes and tablet formulations. Common practice in the detergent industry is to incorporate carboxymethylcellulose (CMC) into synthetic detergent formulations to act as an antisoil redeposition agent.

Although CMC gives satisfactory results on cotton, the results on other fibers are, in general, unsatisfactory. With the increasing use of cotton–synthetic fiber blends, the need for more effective antisoil redeposition agents will become increasingly critical. Mixtures of PVP and CMC are worthy of investigation. PVP may be used in the presence of optical brightening agents and, in fact, aids in the pickup of the brightening agents onto the fabric. PVP and PVA have also been shown to have synergistic properties as soil antiredeposition agents (U.S. Patent 3,689,435).

PVP improves the emollient and detergent action of bars, soap sheets, and waterless hand cleaners. In the latter instance, it also helps to reduce skin irritation.

Polymers and Polymerization

In two-phase polymerization systems for styrene, vinyl chloride, vinyl ester, acrylics, and other monomers, PVP acts as a particle-size regulator, suspending agent, and viscosity modifier. Resins produced by this method exhibit improved strength, clarity, and color receptivity.

Alternatively, the polymer can be used as a postpolymerization additive to improve the dyeability and stability of other latexes. It is an equally effective pigment dispersant in plastic formulations. Incorporation of PVP has also been suggested for improving the heat and oxygen stability of sensitive polymers.

Agricultural

PVP is nonphytotoxic and therefore useful in agricultural sprays, fertilizing compositions, and wettable formulations for dusting and seeding. It has been recommended for use with mercurial herbicides in crabgrass eradication and has been shown to have some effect as a soil conditioner.

PVP films help protect foliage from wilting during transplanting and minimize windburn and frost damage on plant life. On seeds, the polymer reduces soaking injury and protects biological functions.

The PVP-iodine complex is an effective biocide with minimal plant toxicity. It is active against nematodes, meal worm larvae, some insects, and fungi-causing pathological conditions in plants.

Photography and Lithography

PVP is used in diazo and silver halide emulsions, etch coatings, gumming of lithographic plates, plate storage, and dampener roll solutions. It is a light-hardenable, water-soluble colloid that is reported to increase covering power, density, contrast, and effective speed of emulsions. In the case of metal plates, the use of PVP can obviate deep etching. Moreover, it offers uniform viscosity, temperature stability, tight adhesion in nonimage areas, and is nonthixotropic. The polymer also is an effective protective colloid in silver bromide emulsions. PVP films are grease-proof, water-receptive, and chemically inert to ink ingredients.

The patent literature describes the PVP-iodine complex as an antifogging and stabilizing agent in photosensitive emulsions and as a protective colloid for silver in development and fixation of films. PVP itself has also been used to prevent opacity in silver halide color transparencies.

REFERENCES/FURTHER USEFUL READING

References

1. Bergman, M., I. J. France, and H. T. Blumenthal, *N. Engl. J. Med.*, **258**, 471–476 (1958).
2. Bert, J. M., et al., *Sem. Hop. (Paris)*, **25**, 1809–1816 (1972).
3. Burnette, L. W., "A Review of the Physiological Properties of PVP," *Proc. Sci. Sect. Toilet Goods Assoc.*, **38**, 1–4 (December 1962).
4. Cabanne, F., et al., *Ann. Anat. Pathol. (Paris)*, **14**(4), 419–439 (1969).
5. Dorst, P. W., *P.B. Report 4116*, U.S. Department of Commerce, Office of Technical Services, 1945; *Bibliogr. Tech. Rep.*, U.S. Department of Commerce, **1**, 327 (1945).
6. Dupont, A., and J. M. Lachapelle, *Bull. Soc. Fr. Dermatol. Syphiligr.*, **71**, 508–509 (1964).
7. GAF Corporation, *PVP Bibliography, 1951–1966*, 1967, 124 pp.
8. GAF Corporation, *PVP-Iodine*, Bulletin, 7543–004.
9. GAF Corporation, *PVP—Preparation, Properties, and Applications in the Blood Field and Other Branches of Medicine*, 1951, 174 pp.
10. Hueper, W. C., "Bioassays on Polyvinylpyrrolidones with Limited Molecular-Weight Range," *J. Nat. Cancer Inst.*, **26**(1), 229–237 (1961).
11. Hueper, W. C., "Experimental Carcinogenic Studies in Macromolecular Chemicals I," *Cancer*, **10**, 8–18 (1957).
12. Hueper, W. C., *Arch. Pathol.*, **67**(6), 589–617 (1959).
13. Kleideser, I. C., et al., *P.B. Report No. 67620*, U.S. Department of Commerce, undated; *Bibliogr. Tech. Rep.*, U.S. Department of Commerce, **5**, 753 (1947).
14. Lachapelle, J. M., *Dermatologica*, **132**, 476–489 (1966).
15. Levy, J. B., and H. P. Frank, "Determination of Molecular Weight of PVP," *J. Polym. Sci.*, **17**, 247 (1955).
16. Lindner, J., "Experimental Investigations on Animals Regarding the Problem of So-called Polymer Cancers," *Trans. Ger. Soc. Path.*, 44th Meeting, Munich, June 8–10, 1960, pp. 272–283.
17. Miller, L. E., and F. A. Hamm, "Macromolecular Properties of Polyvinylpyrrolidone: Molecular Weight Distribution," *J. Phys. Chem.*, **57**, 110–122 (1953).
18. National Institutes of Health, *Report No. 256*, NIH Contract No. 69–2067, May 27, 1970.
19. National Institutes of Health, *Report No. 258*, NIH Contract No. 69–2067, Oct. 14, 1970.
20. Rose, C. E., and R. D. White, *P.B. Report 1308*, U.S. Department of Commerce, Office of Technical Services, 1945; *Bibliogr. Tech. Rep.*, U.S. Department of Commerce, **1**, 223 (1945).
21. Schildknecht, C. E. (to GAF Corp.), U.S. Patent 2,658,045 (Nov. 3, 1953); *Chem. Abstr.*, **48**, 2413 (1954).
22. Shelanski, H., M. V. Shelanski, and A. Cantor, *J. Soc. Cosmet. Chem.*, **5**, 129–132 (1954).
23. Siggin, S., A. Carpenter, and S. Carter, *Anal. Chem.*, **48**, 225 (1976).
24. Slifkin, S. C., *P.B. Report 78256*, U.S. Department of Commerce, Office of Technical Services, 1947; *Bibliogr. Tech. Rep.*, U.S. Department of Commerce, **7**, 616 (1947).
25. Tan, Y. Y., and G. Challa, *Polymer*, **17**, 739 (1976).

Further Useful Reading

There is a sizable listing of technical literature that deals with the various aspects of PVP and its applications. The following listing, in addition to the preceding numbered references, is a representative, not comprehensive, sampling of this literature.

General Chemistry

Kline, G. M., "Polyvinylpyrrolidone," *Mod. Plast.*, p. 157 (November 1957).
Miller, L. E., and Hamm, F. A., "Macromolecular Properties of Polyvinylpyrrolidone: Molecular Weight Distribution," *J. Phys. Chem.*, **57**, 110–122 (1953).

Medicine

Luchtrach, H., Freibel, H., and Graevenitz, H., "Chronic Application of Periston in Experimental Silicosis," *Arch. Gewerbepathol. Gewerbehyg.*, **16**, 361–374 (1958).
Stern, K., Sabet, L., and Gleason, M., "Effect of PVP on Incidence of Spontaneous Mouse Mammary Cancer," *Proc. Am. Assoc. Cancer Res.*, **2**, 150 (1956).
Chevallier, A., Manuel, S., and Chambron, J., "Action of PVP on Experimental Cancer Induced by 3,4-Benzopyrene," *Compt. Rend.*, **155**, 895–897, 916–918 (1961).
Lederman, E., "Ophthalmic Irritation Following Gaseous Anesthesia," *Eye, Ear, Nose, Throat Mon.*, **35**, 785–787 (1956).
Calandra, J., and Kay, J. A., "Inhalation of Aerosol Hair Sprays," *Drug Cosmet. Ind.*, **84**, 174–175, 252–254 (1959).

Draize, J. H., Nelson, A. A., Newburger, S. H., and Kelley, E. A., "Aerosol Hair Sprays," *Soap Chem. Spec.*, **35**, 91–94 (1959).

Shelanski, M. V., "Is PVP in Hair Sprays a Potential Hazard?," *Soap Chem. Spec.*, **34**, 64–66, 87 (1958).

Bergmann, M., Flance, I. J., and Blumenthal, H. T., "Thesaurismosis Following Inhalation of Hair Spray," *N. Engl. J. Med.*, **258**, 471–476 (1958).

Bergmann, M., Flance, I. J., Cruz, P. T., Klam, N., Aronson, P. K., Joshi, K. A., and Blumenthal, H. T., "Thesaurosis Due to Inhalation of Hair Spray," *N. Engl. J. Med.*, **266**, 750–755 (1962).

Schepers, G. W. H., "Thesaurosis vs. Sarcoidosis," *J. Am. Med. Assoc.*, **181**, 635–637 (1962).

Brunner, M. J., Giovacchini, R. P., Wyatt, J. P., Dunlap, F. E., and Calandra, J. C., "Pulmonary Disease and Hair Spray Polymers: A Disputed Relationship," *J. Am. Med. Assoc.*, **184**, 851–857 (1963).

"Storage Disease," *J. Am. Med. Assoc.*, **184**, 888 (1963).

Adhesives

USA Sec. Commerce, *Pressure-Sensitive Adhesive Composition Comprising Polyvinylpyrrolidone and Polyethylene Polyamine*, U.S. Patent 3,028,351 (Apr. 3, 1963).

National Starch and Chemical, *Remoistenable Adhesive Compositions*, U.S. Patent 2,978,343 (Apr. 4, 1961).

GAF, *Composition Containing Vinylpyrrolidone Polymer Stabilized with Alkylated Phenol*, U.S. Patent 2,941,980 (June 21, 1960).

Owens-Corning Fiberglass Corp., *Plastics and Laminates of Glass Fibers and Epoxy Resins*, U.S. Patent 2,931,739 (Apr. 5, 1960).

GAF, *Polyvinyl Halide Resin Compositions Containing N-Vinylpyrrolidone Polymers and Process of Sizing Glass Therewith*, U.S. Patent 2,853,465 (Sept. 23, 1958).

GAF, *Amine-Formaldehyde Resin Compounds Modified with Vinylpyrrolidone Polymers and Process of Sizing Glass Surfaces Therewith*, U.S. Patent 2,813,844 (Nov. 19, 1957).

Agriculture

Kellog, H. B., "PVP-Iodine in Agricultural Pest Control," *Farm Chem.*, **119**, 41–43 (1956).

D. Milner, *Protection of Plant Life from Frost Damage*, U.S. Patent 3,045,394 (July 24, 1962).

Barton, L. V., "Relation of Different Gases to the Soaking Injury of Seeds," *Contrib. Boyce Thompson Inst.*, **17**, 7–34 (1952).

Kozlov, Y. P., and Tarusov, B. N., "Radiation Grafting of Polymers on Biological Objects," *Vysokomolek. Soedin.*, **3**, 1265–1276 (1962).

Beverage Clarification

Baxter Labs, *Stabilizing of Malt Beverages*, U.S. Patent 3,061,439 (Oct. 30, 1962).

Baxter Labs, *Stabilized Malt Beverages*, U.S. Patent 2,943,941 (July 5, 1960).

Canadian Breweries, *Treatment of Brewer's Wort*, U.S. Patent 2,939,791 (June 7, 1960).

N. S. Berntsson, *Process of Preventing Haze, Formation in Beverages*, U.S. Patent 2,919,193 (Dec. 29, 1959).

Canadian Breweries, *Process for Clarifying and Stabilizing Vegetable Beverages*, U.S. Patent 2,688,550 (Sept. 7, 1960).

Clemens, R. A., and Martinelli, A. J., "PVP in the Clarification of Wines and Juices," *Wines and Juices*, **39**, 55–58 (1958).

GAF, *Making Coffee and Tea Brews and the Resultant Product*, U.S. Patent 3,022,173 (Feb. 20, 1962).

Coatings

Printing Arts Research Labs, *Ultraviolet Absorbing Watercolor Paint Compositions*, U.S. Patent 2,944,912 (July 12, 1960).

Surface Chemical Development Corp., *Aqueous Suspensions of Colloidal Graphite and Their Preparation*, U.S. Patent 2,978,428 (Apr. 4, 1961).

Badische Anilin- & Soda-Fabrik, *Production of Pigmented Prints and Coatings on Fibrous Material*, U.S. Patent 2,719,831 (Oct. 4, 1955).

Cosmetics

Prescott, F. J., Hahnel, E., and Day, D., "Cosmetic PVP," *Drug. Cosmet. Ind.*, **93**, 443–445, 540–541, 629–630, 702, 739 (1963).

Freifeld, M., Lyons, J. R., and Martinelli, A. J., "PVP in Cosmetics," *Am. Perfum.*, **77**, 25–27 (1962).

Shelanski, H. A., Shelanski, M. V., and Cantor, A., "PVP—Useful Adjunct in Cosmetics," *Am. Perfum. Essent. Oil Rev.*, **64**, 267–268 (1954).

La Mauer Inc., *Sprayable Water-Free Alcoholic Polyvinylpyrrolidone Hair Preparation*, U.S. Patent Re. 25,022 (reissued Aug. 8, 1961; original no. 2,871,161, Jan. 27, 1959).

Cifelli, T., "PVP Aerosol Spray Patent," *Drug. & Cosmet. Ind.*, **90**(4), 467–468 (April 1962).

Detergents and Soaps

USA Sec. Agriculture, *Use of Polyvinylpyrrolidone as Soil-Suspending Agent*, U.S. Patent 3,000,830 (Sept. 19, 1961).

Goldschmiedt, H., "Soap Sheets," *Soap Chem. Spec.*, **33**, 47–49, 54 (1957).

Electrical

Electric Storage Battery, *Sintered Plate for Alkaline Storage Batteries*, U.S. Patent 2,964,582 (Dec. 13, 1960).

Bjorksten Research Labs, *Battery Constituents*, U.S. Patent 2,902,530 (Sept. 1, 1959).

Polytechnic Inst. Brooklyn, *Porous Membrane Materials and Process for Producing Same*, U.S. Patent 2,884,387 (Apr. 28, 1959).

McGraw-Edison, *Alkaline Storage Cells with Cadmium Type Negative Electrodes*, U.S. Patent 2,870,234 (Jan. 20, 1959).

Fibers and Textiles

Hansen, E. C., Bergman, C. A., and Witwer, D. B., "PVP—A Versatile Compound," Proc. Am. Assn. Text. Chem. Col., *Am. Dyest. Rep.*, 43, 72–76 (1954).

Henglein, H., and Schnabel, W., "Surface Grafting of Polyvinylpyrrolidone with Acrylonitrile under the Influence of Co^{60}-Gamma Radiation and Swelling of Resultant Product," *Makromol. Chem.*, 25, 119–133 (1957).

Eastman Kodak, *Mixtures Comprising Acrylonitrile Polymers with Polyvinylpyrrolidone and Fibers Thereof*, U.S. Patent 2,790,783 (Apr. 30, 1957).

Chemstrand Corp., *Solutions of Polyvinylpyrrolidone in Aqueous Phytic Acid and Process for Making Same*, U.S. Patent 2,980,641 (Apr. 18, 1961).

Dow Chemical Co., *Graft Copolymers of Mixtures of Vinyl Pyridine Monomers and Monomeric Sulfonic Acids Upon Polyvinyllactams, Acrylonitrile Polymer Compositions Obtainable Therewith and Method of Preparation*, U.S. Patent 3,036,033 (May 22, 1962).

Dow Chemical Co., *Graft Copolymers of Mixtures of Monomeric Bisacrylamide and Monomeric Organic Sulfonic Acid Compounds upon Polyvinyllactams*, U.S. Patent 3,036,032 (May 22, 1962).

Dow Chemical Co., *Graft Copolymers of Monomeric Acrylates and Monomeric Organic Sulfonic Acid Compounds with Polyvinyllactams*, U.S. Patent 3,035,009 (May 15, 1962).

Dow Chemical Co., *Graft Copolymers of Monomeric Aminoethyl Acrylates and Methacrylates on Polyvinyllactams*, U.S. Patent 3,029,220 (Apr. 10, 1962).

Dow Chemical Co., *Graft Copolymers Comprised of Monomeric Vinyl Benzyl Polyglycol Ethers on Polyvinyllactams*, U.S. Patent 3,029,219 (Apr. 10, 1962).

Dow Chemical Co., *Graft Copolymers Comprised of Monomeric Diacrylate Esters of Polyglycols on Polyvinyllactams*, U.S. Patent 3,029,218 (Apr. 10, 1962).

Dow Chemical Co., *Graft Copolymers of Certain Monomeric Sulfonic Acids on Polyvinyllactams*, U.S. Patent 3,026,291 (Mar. 20, 1962).

Chemstrand Corp., *Acid Pretreatment of Polyacrylonitrile-Type Fibers and the Treated Fibers*, U.S. Patent 2,932,550 (Apr. 12, 1960).

Dow Chemical Co., *Process for Preparation of Permanently Stiffened Textiles*, U.S. Patent 3,011,917 (Dec. 15, 1961).

Olpin, H. C., and Wesson, A. J., "The Application of Identification Tints with Particular Reference to Acetate Rayon Staple," *J. Soc. Dyers Colour.*, 69, 357–362 (1953).

British Celanese, *Fugitive Tinting of Textile Materials*, U.S. Patent 2,802,713 (Aug. 13, 1957).

E. A. Murray, *Fugitive Tints for Natural and Synthetic Fibers*, U.S. Patent 2,959,461 (Nov. 8, 1960).

Geigy, *Compositions for Printing and Dyeing Textiles*, U.S. Patent 3,002,939 (Oct. 3, 1961).

Dow Chemical Co., *Method for Preparing Acrylonitrile Graft Copolymers-Fiberforming Systems*, U.S. Patent 2,949,435 (Aug. 16, 1960).

American Cyanamid, *Vat Dye Dispersions Containing Polyvinylpyrrolidone*, U.S. Patent 2,971,813 (Feb. 14, 1961).

GAF, *Dyeing Polyacrylonitrile Fibrous Material*, U.S. Patent 2,955,008 (Oct. 4, 1960).

American Cyanamid, *Soluble Vat Dye Printing Compositions with Heterocyclic Solubilizing Agents*, U.S. Patent 2,970,880 (Feb. 7, 1961).

GAF, *Dyestuff Compositions*, U.S. Patent 2,955,011 (Oct. 4, 1960).

GAF, *Stable Solution for Producing Ice Colors*, U.S. Patent 2,953,422 (Sept. 20, 1960).

Celanese Corp., *Vat Dyeing of Cellulose Ethers and Esters*, U.S. Patent 2,798,788 (July 9, 1957).

Ferro Corp., *Glass Fiber Size Comprising an Aqueous Dispersion of a Film Forming Polymer, a Coupling Agent, and Polyvinylpyrrolidone*, U.S. Patent 3,082,183 (Mar. 19, 1963).

Lithography and Photography

DuPont, *Photographic Elements Exhibiting Reduced Haze*, U.S. Patent 3,058,826 (Oct. 6, 1962).

DuPont, *Photographic Gelatin-N-Vinyllactam Halide Emulsions Containing Phenolic Antifoggants*, U.S. Patent 3,043,697 (July 10, 1962).

GAF, *Stabilization of Photographic Emulsions Sensitized with Alkylene Oxide Polymers*, U.S. Patent 2,995,444 (Aug. 8, 1961).

DuPont, *Photographic-Stripping Film*, U.S. Patent 2,976,147 (Mar. 21, 1961).

Powers Chemco Inc., *Etching Bath*, U.S. Patent 2,890,943 (June 16, 1959).

Azoplate, *Process for Fixing Lithographic Diazotype Printing Foils having been Exposed to Light*, U.S. Patent 2,739,889 (Mar. 27, 1956).

Azoplate, *Light-Sensitive Photographic Element and Process of Producing Printing Plates*, U.S. Patent 2,702,243 (Feb. 15, 1955).

Azoplate, *Lithographic Plates*, U.S. Patent 2,692,826 (Oct. 26, 1954).

Azoplate, *Light-Sensitive Layers for the Printing Industry*, U.S. Patent 2,687,958 (Aug. 31, 1954).
Azoplate, *Process for Producing Positive Photolithographic Printing Foils*, U.S. Patent 2,667,415 (Jan. 26, 1954).
Keuffel and Esser, *Photographically Light-Sensitive Silver Halide-Diazide Colloid Layers*, U.S. Patent 2,663,640 (Dec. 22, 1953).
S. D. Warren, *Photosensitive Printing Plate Having a Light-Sensitive Coating Consisting of a Nonprotein Colloid, a Phosphate and a Chromium Compound*, U.S. Patent 2,624,673 (Jan. 6, 1953).
DuPont, *Polyvinyllactam Photographic Silver Halide Emulsions*, U.S. Patent 2,495,918 (Jan. 31, 1950).

Paper

Callinan, T. P., "Powdered Resin Improves Mechanical Properties of Inorganic Specialty Papers," *Rep. NRL Prog.*, 6 pp. (September 1955).
GAF, *Inorganic Papers and Methods of Making Same*, U.S. Patent 3,008,867 (Nov. 14, 1961).
GAF, *Inorganic Papers and Methods of Making Same*, U.S. Patent 3,005,745 (Oct. 24, 1961).
GAF, *Inorganic Papers and Methods of Making Same*, U.S. Patent 2,901,390 (Aug. 24, 1959).
Burgess Cellulose, *Process for Incorporating Resins into Paper*, U.S. Patent 3,036,950 (May 29, 1962).
Rohm and Haas, *Prevention of Deposition of Pitch in Papermaking*, U.S. Patent 3,081,219 (Mar. 21, 1963).

Pharmacy and Veterinary

GAF Corp., *Plasdone in the Manufacture of Pharmaceutical Dosage Forms*, AP-123 (August 1962), 24 pp. A review with 37 references.
Prescott, F. J., Lane, F., and Hahnel, E., "Pharmaceutical Dispersions (featuring Plasdone)," *Texas J. Pharm.*, 4, 300–308 (1963).
Lehrman, G. P., and Skauers, D. M., "A Comparative Study of Polyvinylpyrrolidone and Other Binding Agents in Tablet Formulations," *Drug Stand.*, 26, 170–175 (1958).
Ahsan, S. S., and Blaug, S. M., "A Study of Tablet Coating Using Polyvinylpyrrolidone and Acetylated Monoglyceride," *Drug Stand.*, 26, 29–33 (1958).
McAuliff, J. L., Phillips, W. V., and Steele, J. R., "Polyvinylpyrrolidone (HPX): A New Treatment for Bovine Mastitis," *J. Am. Vet. Med. Assoc.*, 133, 169–171 (1958).
Dawe's Labs., *Feed Containing an Arsenical and Polyvinylpyrrolidone*, U.S. Patent 3,015,564 (Jan. 2, 1962).
Pabst Brewing Co., *Poultry and Swine Feed*, U.S. Patent 2,767,094 (Oct. 16, 1956).
Rinfret, A. P., "Research on Procedures for the Low-Temperature Preservation of Blood," *U.S. Gov. Res. Rep.* 38: 79 AD-286,754, 286,744 (1963).
Persidsky, M., and Richards, R., "Mode of Protection of Polyvinylpyrrolidone in Freezing of Bone Marrow," *Nature*, 196, 586–6 (1962).
Bricka, M., and Bessis, M., "Preservation of Erythrocytes by Congealing at Low Temperature in the Presence of Polyvinylpyrrolidone or Dextran," *Comp. Rend*, 149, 875–877 (1955).
Burstone, M. S., "Polyvinylpyrrolidone as a Mounting Medium for Stains for Fat and for Azo-Dye Procedures," *Am. J. Clin. Pathol.*, 28, 429–430 (1957).
Gude, W. D., and Odell, T. T., "Vinisil as a Diluent in Making Bone Marrow Smears," *Stain Technol.*, 30, 27–28 (1955).
Hill, G. A., Fung, J., and Marius, S., "Polyvinylpyrrolidone as an Indicator for Measuring Intercellular Space in Packed Cell Pellets," *J. Bacteriol.*, 84, 191–192 (1962).
Sullivan, M. F., "Polyvinylpyrrolidone Labelled with I^{131} as an Agent for the Diagnosis of Radiation Injury in Rats," *Int. J. Radiat. Biol.*, 2, 393–398 (1960).
Sukiasian, G. V., Dzhavadian, N. S., Novikova, M. N., Beliayeva, B. F., Probatova, N. A., and Shitikova, M. G., "A Study of the Effect of the Transfusion of Polyvinylpyrrolidone on the Course of Acute Radiation Sickness," *Probl. Gematol. Pereliv. Krovi*, 4, 48–55 (1959).
Lesser, M. A., "PVP," *Drug & Cosmet. Ind.*, 75, 32–33, 126–131 (1954).
Schmutzler, W., and Giertz, H., "Influence of Polyvinylpyrrolidone on the Anaphylactoid PVP Effects on the Dog," *Arch. Expl. Pathol. Pharmakol.*, 243, 363–364 (1962).
Hanna, C., "Use of Polyvinylpyrrolidone in the Assay of Coronary Drugs," *Arch. Int. Pharmacodyn.*, 119, 305–310 (1959).
Perlmutt, J. H., Parkins, W. M., and Vars, H. M., "Response of Dogs to I.V. Administration of PVP and its Monomer," *Proc. Soc. Exp. Biol. Med.*, 83, 146–150 (1953).
Besse, H., and Patay, R., "On the Hypotensive Action of Polyvinylpyrrolidone in the Dog," *Compt. Rend.*, 147, 1721–1722 (1953).
Halpern, B. N., and Briot, M., "Histaminic Mechanism of Action of Polyvinylpyrrolidone in the Dog," *Compt. Rend.*, 147, 643–647 (1953).
Hahn, F., and Wellman, A., "Experimental Study of Histamine Liberation by Synthetic Blood Substitutes," *Klin. Wochenschr.*, 30, 998 (1952).

Polymerization Additive

Diamond Alkali, *Polymerization Employing Polyvinylpyrrolidone in Combination with Secondary Dispersants as Suspension Stabilizers*, U.S. Patent 2,890,199 (June 9, 1959).

Diamond Alkali, *Polymerization Process Utilizing Dextran as a Primary Dispersing Agent,* U.S. Patent 2,857,367 (Oct. 21, 1958).

Badische Anilin- & Soda-Fabrik, *Production of Porous Shaped Articles from Thermoplastic Substances,* U.S. Patent 2,787,809 (Apr. 9, 1957).

Badische Anilin- & Soda-Fabrik, *Production of Porous Shaped Articles from Thermoplastic Substances,* U.S. Patent 2,744,291 (Apr. 16, 1957).

Badische Anilin- & Soda-Fabrik, *Copolymerization of Styrene and Acrylonitrile in the Presence of an Organic Solvent having at least One Hydroxyl Group,* U.S. Patent 2,895,938 (July 21, 1959).

Schulz, G., "The Use of Surface Active Agents in Plastics and Coatings," *Farbe & Lack,* **66,** 621–624 (1960).

Badische Anilin- & Soda-Fabrik, *Production of Poly-N-vinylcarbazole,* U.S. Patent 2,877,216 (Mar. 10, 1959).

Chapter **22**

Starch and Its Modifications

M. W. Rutenberg
National Starch and Chemical Corporation

General Information 22-1
   Structure and Properties 22-2
   Starch Supplies 22-21
   Manufacture of Starch 22-22
   Starch Modification 22-23
   Applications of Starches 22-54
Laboratory Techniques 22-70
Tradenames of Starch Products 22-71
Acknowledgment 22-73
Further Reading 22-73
References 22-73

Glossary of Terms

AGU = Anhydroglucose unit
D.P. = Degree of polymerization
D.S. = Degree of substitution
M.S. = Molar substitution
SEM = Scanning electron microscopy
TEM = Transmission electron microscopy

GENERAL INFORMATION

Starches are used as components and/or processing aids in the manufacture of products such as adhesives, textiles, paper, food, pharmaceuticals, and building materials. The use of this natural polymeric material is based on its thickening, gelling, adhesive, and film-forming properties, as well as its low cost, controlled quality, and ready availability.

The characteristics of a starch can be modified by chemical, physical, and/or enzyme treatment to enhance or repress its intrinsic properties, or to impart new ones. This capability for modification has been a necessary factor in developing new uses for starch and in maintaining old markets.

There are five prime factors that determine the properties of starches: (1) starch is a polymer of glucose (dextrose); (2) the starch polymer is of two types: linear and branched; (3) the linear polymeric molecules can associate with each other, giving insolubility in water; (4) the polymeric molecules are organized and packed into granules which are insoluble in water; and (5) disruption of the granule structure is required to render the starch polymer dispersible in water. The modification of starch takes into account these factors.

Structure and Properties

Starch is widely distributed as the reserve carbohydrate in the leaves, stems, roots, and fruits of most land plants. The commercial sources are the seeds of cereal grains (corn, sorghum, wheat, rice), certain roots (potato, tapioca or cassava, arrowroot), and the pith of the sago palm. Since the growth conditions are different in each plant, the starch from each plant source will vary somewhat in appearance, composition, and properties. Consequently, the starch is described by its plant source as corn starch, potato starch, tapioca starch, rice starch, wheat starch, etc.

Chemical Composition Starch is composed of carbon, hydrogen, and oxygen in the ratio of $6:10:5$ ($C_6H_{10}O_5$), placing it in the class of carbohydrate organic compounds. It can be considered to be a condensation polymer of glucose and yields glucose when subjected to hydrolysis by acids and/or certain enzymes (Fig. 22.1). The glucose units in the starch polymer are present as anhydroglucose units (AGU), the linkage between the glucose units being formed as if a molecule of water is removed during a condensation polymerization. Since the starch is formed in the plant by a biosynthetic process, the polymerization process is a complex one involving enzymes. The glucose units are connected through an oxygen atom attaching carbon atom 1 of one glucose unit to carbon atom 4 of the next glucose unit, forming a long chain or polymer of interconnected glucose units (see Fig. 22.1 for the numbering system). This linkage of one glucose to another through the C-1 oxygen is known as a *glucoside bond.*

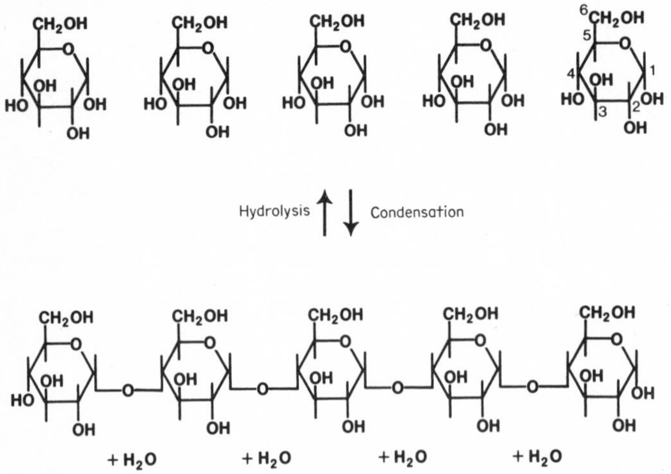

Figure 22.1 Starch considered as a condensation polymer of glucose. Note numbering of carbon atoms in the glucose and the spatial relationships of the hydroxyl groups with respect to the six-membered ring containing 5 carbon atoms and 1 oxygen atom.

From a chemical modification standpoint, the significant structural features of the starch are the glucoside oxygen linkages between glucose units, the primary hydroxyl group on carbon atom 6, and the secondary hydroxyl groups on carbon atoms 2 and 3 on almost every anhydroglucose unit in the starch molecule. The glucoside linkage is an acetal, stable under alkaline conditions and hydrolyzable under acid conditions.

Thus, acid hydrolysis leads to depolymerization of the starch, which if carried to completion results in the release of glucose (also known as dextrose).   The hydroxyl groups can react to form ethers and esters and can be oxidized to aldehyde, ketone, and carboxyl groups.

It should be noted that the glucose unit at one end of the polymeric chain is not tied to another glucose unit through its carbon atom 1.   This unit has a latent aldehyde group present as a hemiacetal and is known as a reducing end group since the aldehyde group can act as a reducing agent in tests with Fehling's or Tollen's solution or alkaline ferricyanide.

Depending upon the variety, commercial starches contain small amounts of proteins, fatty materials, phosphorus-containing materials or phosphate ester groups, and traces of inorganic materials.

Molecular Structure   Most starches consist of a mixture of two polysaccharide types: amylose, an essentially linear polymer, and amylopectin, a highly branched polymer. The relative amounts of these starch fractions in a particular starch are a major factor in determining the properties of that starch.

Amylose.   This linear polymer consists of a chain of glucose units connected to each other by 1–4 linkages.   The glucose units are in the *"alpha*-D-glucopyranose" form. This means that the glucose is arranged in the form of a six-membered ring in such a way that the hydroxyl group on carbon atom 1 is on the same side of the ring as the hydroxyl group on carbon atom 2 (Fig. 22.1).   This spatial configuration of the *alpha*-glucose units in amylose differs from that of *beta*-glucose units in cellulose where the *beta* configuration indicates that the hydroxyl group on carbon atom 1 is on the opposite side of the glucopyranose ring from the hydroxyl group on carbon atom 2.   Thus the link connecting the glucosidic oxygen atom to carbon atom 1 of the anhydroglucose units in the amylose chain has a different spatial relationship to the glucopyranose ring than the analogous link in cellulose (Fig. 22.2).   This seemingly minor difference in

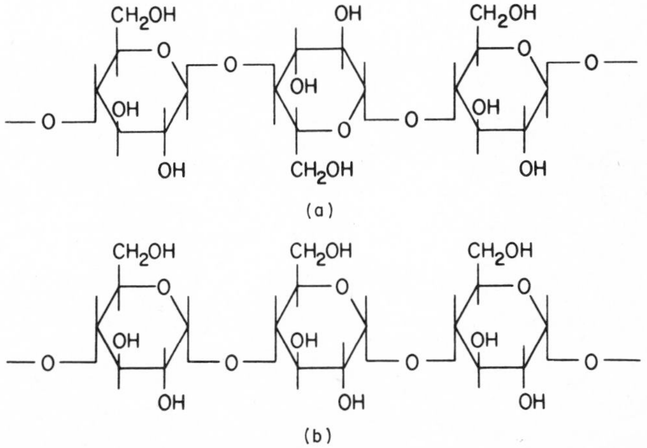

Figure 22.2   The *beta*-glucoside linkage of amylose *(a)* and the *alpha*-glucoside linkage of cellulose *(b)*.   *(Ref. 79.)*

the spatial arrangement of the interconnected glucose units causes a considerable difference in the properties of the amylose and cellulose, which are both linear polymers of D-glucose connected through 1–4 linkages.

The amylose chain has a single reducing end group and a single nonreducing end group.   Enzyme studies with *beta*-amylase, which converts amylose to maltose, indicate a trace amount of branching in the amylose.   The amylose polymer fraction of a starch will show a distribution of molecular sizes, and the average degree of polymerization (D.P.) will vary with the plant variety from which the starch is obtained.   Depending upon the type of starch, the D.P. will range from about 250 to 4000 AGU per amylose molecule, corresponding to a molecular weight of approximately 40,000 to 650,000. The amylose from potato and tapioca starches has a higher molecular weight than that from corn starch. The relative severity of the treatment used to isolate the starch from its plant source and to fractionate the amylose from the starch will affect the molecular size of the amylose and the amylopectin.[1,2,3,4,5]

As noted earlier, starch molecules have a multitude of hydroxyl groups which impart hydrophilic, or water-loving, properties to the starch. In addition to their affinity for water, these hydroxyl groups also tend to attract each other, forming hydrogen bonds. Since amylose is a linear polymer containing hydroxyl groups, it shows special properties when it is dispersed or dissolved in water. The linear amylose molecules can readily align themselves next to each other and form interchain hydrogen bonds through the hydroxyl groups. When sufficient interchain hydrogen bonds are formed, the individual amylose molecules are associated to form molecular aggregates with reduced hydration capacity and hence, lower solubility.

In dilute solutions (less than 1%), the amylose precipitates. In more concentrated dispersions, the aggregated amylose entraps the aqueous fluid in a network of partially associated amylose molecules, forming a gel (Fig. 22.3). This process of alignment,

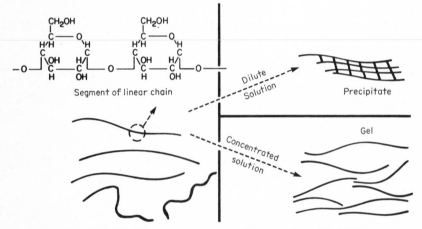

Figure 22.3   Mechanism of starch retrogradation (amylose).   *(Ref. 71b.)*

association, and precipitation is essentially a crystallization process and is known to starch chemists as *retrogradation*.[3,6] The process takes place at room temperature. The rate of retrogradation depends upon the molecular size and concentration of the amylose, the temperature, and the pH.[7,8,9,10,11] The lower the temperature, the more rapid the retrogradation rate. Any process or additive that interferes with the molecular alignment or intermolecular hydrogen bond formation will tend to prevent or reverse the retrogradation.

The higher the temperature [80 to 150°C (176 to 302°F)], the lower the degree of retrogradation and the greater the tendency to destroy the retrograded crystalline aggregates. Large amylose molecules have a slower rate of retrogradation (as for potato and tapioca amylose), presumably because the alignment process is more difficult. As the molecular size decreases, the retrogradation rate increases (as for corn amylose). The same effect can be seen with mild acid hydrolysis of the amylose (potato amylose, for example). The retrogradation rate increases to a maximum and then declines as the molecular size of the amylose is reduced. The retrogradation rate is fastest at pH 5 to 7, decreasing at higher and lower pH's. Retrogradation does not occur at pH above 10 and is slow below pH 2.

Retrogradation is retarded by the salts of monovalent anions and cations.[10,12] Listed in decreasing order of retardation: iodide > thiocyanate > nitrate > bromide > chloride > fluoride; potassium > ammonium > sodium > lithium. Calcium nitrate exhibits a very strong retardation effect on retrogradation. Other compounds such as formaldehyde, urea, and dimethylsulfoxide also act to retard retrogradation.

The tendency of amylose in aqueous dispersion to gel is the most significant property of this fraction and, in many cases, determines the utility or nonutility of a particular starch in a specific application. Many of the modifications of starch are directed toward repressing or eliminating this retrogradation tendency of amylose.

Amylose forms inclusion complexes with iodine[13,14] and polar organic compounds, such as butanol, fatty acids, nitroparaffins, and phenols, as well as with nonpolar compounds such as aliphatic and cyclic hydrocarbons.[2,6,15,16,17] These complexes are essentially insoluble in water. The iodine complex is formed by interaction with iodine in the presence of iodide ions to give a blue color. The color is related to chain length, with a chain length in excess of 40 AGU needed to give the blue color. Shorter chains give purple, red, brown, or yellow complexes.[13,14]

The blue iodine color is useful in detecting the presence of amylose-containing starch and in determining the amount of amylose in a starch. The formation of complexes with polar and nonpolar organic compounds has been used to fractionate starch into amylose and amylopectin and would have an effect on the properties imparted by starch to complex systems containing these materials, such as food systems containing fats and fatty acids.[18]

Because amylose is a linear molecule with a marked tendency to associate, it can form strong, unsupported films similar to those of cellulose. These films become brittle at low relative humidities and must be plasticized with water-holding humectant compounds such as glycerol or sorbitol to remain flexible.

Amylopectin. Amylopectin has a highly branched structure consisting of short linear amylose chains, with a D.P. ranging from 12 to 50 AGU and an average chain length of about 20 AGU, connected to each other by *alpha*-1,6-linkages (Fig. 22.4). These

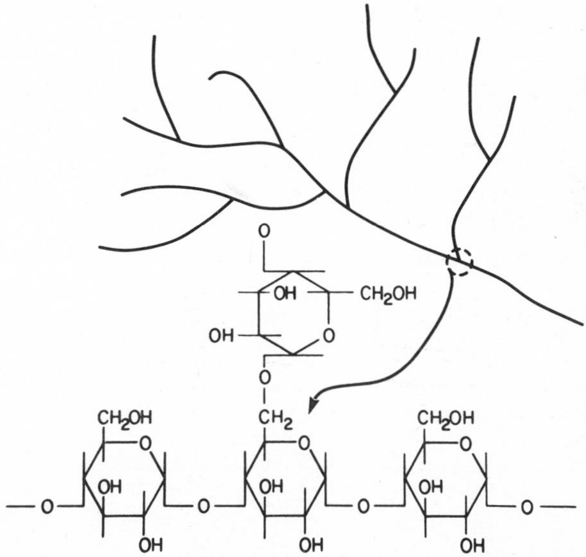

Figure 22.4 Branched structure of amylopectin and chemical configuration at the branch point. *(Ref. 71b.)*

alpha-1,6-linked AGU make up about 4 to 5% of the total AGU in amylopectin and are the branch points of the molecule. There is no single definite structure for amylopectin. The structure is therefore described in the statistical picture of its structural features and details.[19,20,21,22]

The average length of the outer chains of amylopectin before reaching a branch point is estimated from enzymatic analysis to be about 12 AGU, and the average chain length of the inner branches is estimated at about 8 AGU. K. H. Meyer, in 1940, proposed a bush or treelike structure for amylopectin that was generally accepted. Later work with enzymes showed that a significant proportion of the branches are separated by no more than a single glucose unit, suggesting the presence of regions of dense branching. On the basis of these results, French proposed an elongated cluster formula for amylopectin to account for the high viscosity of amylopectin, which would require an asymmetric structure, and high crystallinity in the starch granule, which would require that a high proportion of the branches be able to run parallel to each other (Fig. 22.5).

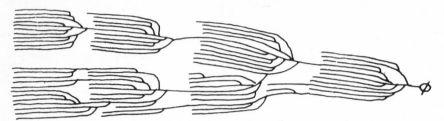

Figure 22.5 Amylopectin structure according the cluster hypothesis of D. French. *(Refs. 21 and 22.)*

Amylopectin is a very large molecule, one of the largest natural polymers. Light-scattering studies indicate a molecular weight (weight-average) of the order of 10^7 or 10^8. Subjecting the amylopectin molecule to shear forces results in a marked decrease in molecular weight.[1b,2,23,24] Amylopectin, particularly from roots and tubers, contains bound phosphate, estimated to be about one phosphate ester group per 400 or more AGU.

Amylopectin gives a characteristic red-to-purple color with iodine, presumably because the outer chains available for complexing with iodine are not long enough (before the branch points) to give the blue color seen with amylose.

Because the amylopectin is highly branched, it cannot readily undergo the retrogradation or crystallization phenomenon that amylose so easily does. Hence, in contrast to amylose, isolated amylopectin is easily dispersed in water and does not readily gel. However, when these amylopectin dispersions are stored at refrigerator temperatures or under freezing conditions, they show a decrease in clarity, a lower water-holding capacity, and a tendency to gel. These effects are attributed to the association of the outer chains of the amylopectin molecule. Prevention of this phenomenon in food starches is important and will be discussed further. It is also noted that amylopectin does not form strong, unsupported films because the highly branched molecule cannot readily orient itself in parallel alignment with other amylopectin molecules to form the multitude of associative hydrogen bonds necessary to produce strong films.

The Starch Granule Land plants synthesize starch and accumulate it in the form of minute, discrete particles or granules which are distinctive to the plant in which they are formed, varying in size, shape, and relative proportions of amylose and amylopectin (Table 22.1). Thus, the source of a starch can be determined by microscopic examination of the granules (Fig. 22.6)[37,45−49,52,53] and the qualitative and quantitative analysis of the iodine color reaction via spectrophotometry or potentiometric titration or visual observation. The amylose and amylopectin molecules are organized and packed into these granules, which are insoluble in water at room temperature. The granules range in size from 1 to 100 micrometers (μm), the smallest being found in leaves. Most cereal and root starches contain 17 to 28% amylose. However, special varieties of plants that produce starch containing virtually no amylose (waxy maize, waxy or glutinous rice) or a much higher amylose content (amylomaize varieties with 50 to 75% amylose in the starch) have been developed through scientific breeding. The presence of some crystalline organization in the starch granules is indicated by X-ray diffraction patterns and birefringence in polarized light. Microscopic examination of starch granules under polarized light will show the presence of a polarization cross

TABLE 22.1
Starch Granule Properties

Starch	Type	Size (diameter,[a]) μm	Shape[a] (light microscope)	Gelatinization temperature,[b] °C	Amylose content,[c] %
Corn	Cereal	5–26(15)	Round, polygonal	62–72	22–28
Waxy corn (waxy maize)	Cereal	5–26(15)	Round, polygonal	63–72	<1
Tapioca (Thailand)	Root	5–25(20)	Truncated, round, oval	62–73	17–22
Potato	Root	15–100(33)	Oval, spherical	59–68	23
Sorghum	Cereal	6–30(15)	Round, polygonal	68–78	23–28
Wheat	Cereal	2–35[d]	Round, lenticular	58–64	17–27
Rice	Cereal	3–8(5)	Polygonal, angular	68–78	16–17
Sago	Pith	15–65	Oval, truncated	—	26
Amylomaize (high-amylose corn)	Cereal	3–24(12)	Round, elongated, deformed	63–92[e]	50–80

[a] References 44–49. Values in parentheses are average sizes.
[b] References 44, 47, and 48. Determined by loss of birefringence while heating aqueous suspension on Kofler hot stage and observing under microscope. Temperature of first observed disappearance of polarization cross in initially swelling granules and temperature when almost all birefringence crosses have disappeared.[37]
[c] References 4, 6, 44, 50, and 51.
[d] Wheat starch has 2 populations: large granules (20–35 μm) and small granules (2–10 μm).
[e] High-amylose corn starches are not completely gelatinized in boiling water.

if the granule is undamaged (Fig. 22.6B, D). If the granule has been partially or completely disorganized, perhaps by the milling process or by gelatinization, there will be only a part or no polarization cross visible.[1c,20]

Transmission (TEM) and scanning electron microscopy (SEM) have provided insights of the fine structure of the starch granule. Both techniques have been used in conjunction with the action of acids, enzymes, and other reagents which modify the starch and affect the granule structure.[54–63] The scanning electron microscope provides a picture of the starch granules which is almost three-dimensional in character and certainly reveals dramatically the physical nature of the granule (Fig. 22.7).

The SEM has also been used to study the effects of gelatinization on the granule and to interpret these effects in terms of the physical properties of the starch pastes, such as viscosity.[42,43,64] The SEM studies have shown that the surface of the starch granule is relatively smooth with indentations in some cases, presumably produced by the pressure of other granules or the protein matrix during growth in the plant. There may be surface damage on some granules, resulting from the milling separation of the starch from the rest of the plant material. Fissures and cavities may be formed in the granule because of stresses developed during drying; the more severe the drying conditions, the more cavities formed.

Granule Swelling and Gelatinization   The properties of the starch granule in water are the major factors in its commercial utility. Since the granule is insoluble in water below about 50°C (122°F), the starch can be extracted from its plant source in aqueous systems, purified and modified in suspension in water, and recovered by filtration and drying. It can be handled as an aqueous slurry (30 to 40% solids) which is pumpable until its thickening, adhesive, binding, and film-forming properties are needed. Then the aqueous slurry is heated to a temperature [about 55 to 80°C (131 to 176°F)] where, depending on the type of starch and concentration, the intermolecular hydrogen bonds holding the granule together are weakened and the granule undergoes a rapid, irreversible swelling.

The critical temperature at which this occurs is known as the *pasting* or *gelatinization temperature.*[25] The granules take up the water, swelling to many times their original volume, rupturing and collapsing as the heating and agitation of the mixture continues, and releasing concurrently some of the starch molecules, particularly amylose, into solution. The viscosity increases to a maximum that corresponds to the largest hydrated, swollen volume of the granule before it bursts apart, and then the viscosity declines

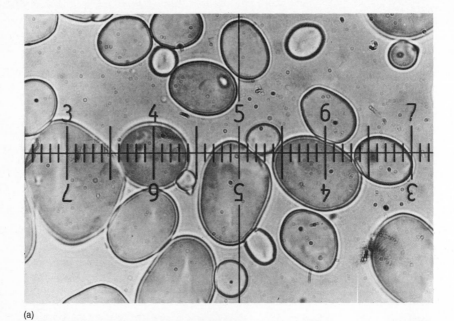

(a)

(b)

Figure 22.6   Photomicrographs of starch granules. Transmitted light. Smallest scale division equals 3 μm. *(R. Holzer, National Starch and Chemical Corporation).* *(a)* Potato starch. *(b)* Potato starch, polarized light.

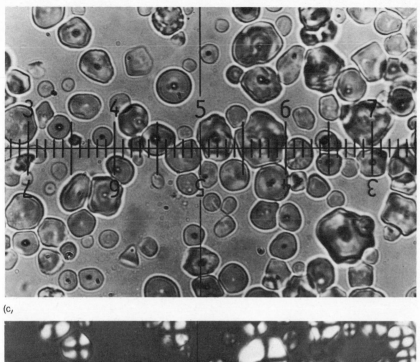

(c)

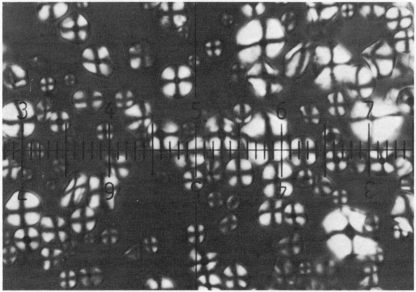

(d)

Figure 22.6 (cont.) *(c)* Corn starch. *(d)* Corn starch, polarized light.

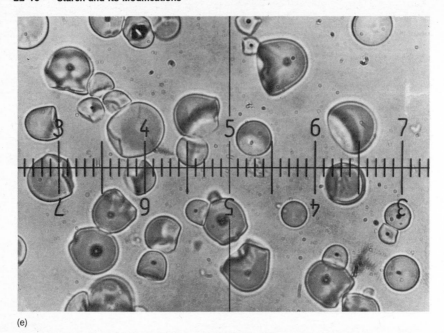

(e)

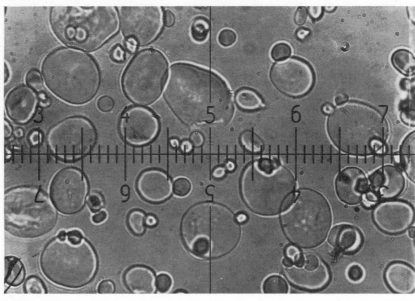

(f)

Figure 22.6 (cont.)   *(e)* Tapioca starch.   *(f)* Wheat starch.

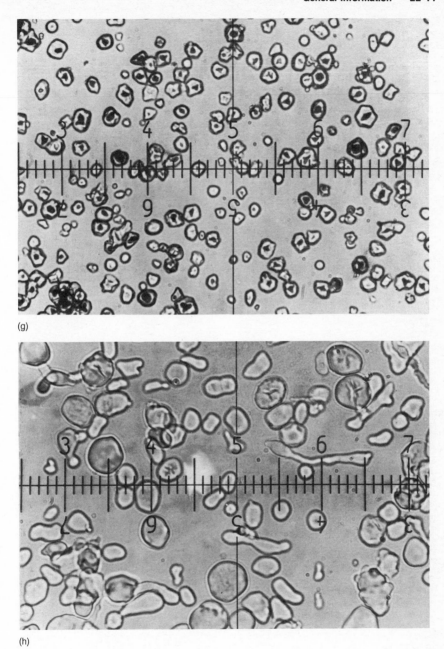

(g)

(h)

Figure 22.6 (cont.)   *(g)* Rice starch.   *(h)* Amylomaize, class 7.

as the swollen granules disintegrate. The result is a viscous colloidal dispersion which is a complex mixture of residual, swollen granule masses, hydrated molecular aggregates, and dissolved molecules.

Although the starch granule is insoluble in water below the gelatinization temperature, there is a limited absorption of water when the granule is wetted or exposed to high humidity.[25,26,27,28] This results in a slight swelling of the granule which is reversible on drying. The birefringence and X-ray diffraction patterns displayed by the granules

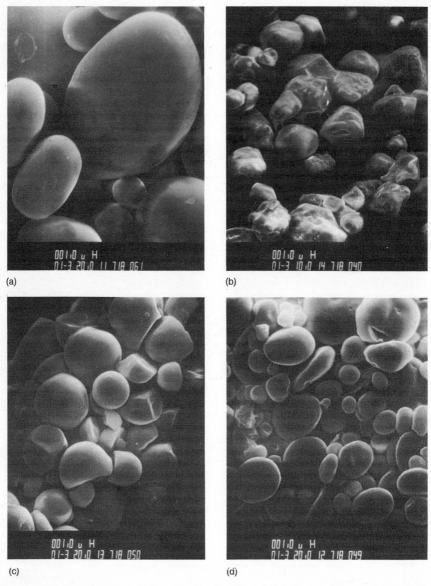

(a)

(b)

(c)

(d)

Figure 22.7 Scanning electron photomicrographs. H symbol designates distance of 1 μm. *(R. Holzer, National Starch and Chemical Corporation.)* *(a)* Potato starch. *(b)* Corn starch. *(c)* Tapioca starch. *(d)* Wheat starch.

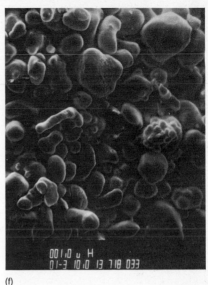

(e)   (f)

Figure 22.7 (cont.)   *(e)* Rice starch.   *(f)* Amylomaize, class 7.

remain.   Considering the granule to consist of crystalline and amorphous or gel-like regions, it has been suggested that, during the reversible swelling, water enters the gel or amorphous regions of the granule and causes swelling of these portions while leaving the crystalline regions unchanged.

On drying, the water is then released without major effect on the granule properties. When the starch granule is heated in water to a temperature above the gelatinization temperature, the water also enters the crystalline regions, which are disrupted.   This results in loss of birefringence and X-ray diffraction pattern.   Observation of the polarization cross disappearance can be used to determine the gelatinization temperature.

Starch exposed to the atmosphere establishes a moisture equilibrium with the moisture in the air, sorbing or desorbing moisture depending on the temperature and relative humidity (Table 22.2).   Under normal atmospheric conditions, most commercial starches contain 10 to 17% moisture.

Thermal Gelatinization.   One of the most important properties of starch is the formation of a viscous paste on cooking in water.   Thus, the measurement of the changes in viscosity or consistency as an aqueous suspension of starch is heated is of practical value in predicting the useful properties of a particular starch or modified starch.

TABLE 22.2
Water Sorption and Granule Swelling as a Function of Relative Humidity[27,28]

	Percent relative humidity							
	100*		75*		31*		8*	
Starch	H_2O sorption, %†	Diameter increase, %‡	H_2O sorption, %†	Diameter increase, %‡	H_2O sorption, %†	Diameter increase, %‡	H_2O sorption, %†	Diameter increase, %‡
Corn	39.9	9.1	17.4	5.4	10.1	2.6	5.0	1.5
Waxy corn	51.4	22.7	18.6	13.5	10.4	7.0	4.5	1.5
Potato	50.9	12.7	19.4	10.3	10.0	5.4	4.4	1.9
Tapioca	42.9	28.4	17.6	12.5	10.4	5.5	5.2	3.6

* Equilibration in evacuated desiccator over saturated solutions of various salts at 25°C.
† Moisture content as % of dry starch.
‡ Increase over vacuum-dry diameter (corn, 16.96 μm; potato, 38.23 μm; tapioca, 20.16 μm; waxy corn, 10.59 μm).

The Brabender amylograph or visco-amylograph may be used to follow these changes.[36,37] These instruments provide a continuous record of viscosity changes over a controlled heating and cooling cycle. Initially, no viscosity effect is noted as the suspension of starch is heated until the pasting or gelatinization temperature is reached. Then there is a marked and rapid rise in viscosity, as seen in Fig. 22.8. It can be seen that potato starch gelatinizes at a lower temperature than waxy maize starch, which gelatinizes at a lower temperature than regular corn starch. The peak viscosity while hot is greatest for potato starch, followed by waxy maize and then corn starch. The potato and waxy maize starches break down in viscosity or thin out with continued heating and stirring much more rapidly and drastically than does corn starch.

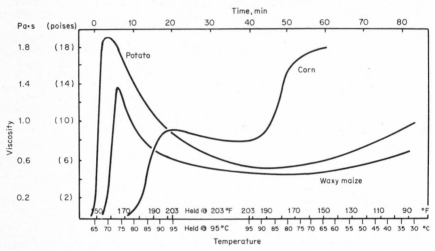

Figure 22.8 Viscosity changes as starch is cooked in water and then cooled. Concentration of corn starch is 35 g in 450 cm³ of water; potato and waxy maize starches are at 25 g in the 450 cm³ of water. *(Ref. 80.)*

On cooling, the corn starch dispersion increases in viscosity more rapidly and reaches a higher viscosity than either the potato or waxy maize starches. In fact, the corn starch forms a gel, whereas the other two starches do not. Note that the pasting temperature shown by the initial increase in viscosity is higher than the gelatinization temperature measured by loss in birefringence. Usually the granules have lost their birefringence before the granules have swollen sufficiently to develop enough viscosity to be detectable in the Brabender viscosity-temperature curve. The pasting temperature, as measured by the development of initial viscosity, is related to the starch concentration in the aqueous suspension. The higher the concentration, the lower the pasting temperature, since the swelling of more granules per unit volume will be evident more quickly (at a lower temperature) by a detectable viscosity increase.

The gelatinization temperature of the starch granules varies with the particular starch type. However, not all the granules of a given starch start to swell at the same temperature; rather, there is a temperature range. Since the granules are held together by intermolecular hydrogen bonding and gelatinization results from a weakening and disruption of these bonds, the gelatinization temperature and the rate of swelling can be considered a measure of the strength and character of the associative bonds. As starch granules swell, there is a corresponding increase in clarity, starch solubility, and viscosity. Each species of starch has a characteristic swelling and solubility pattern.[38,39,25]

Potato starch undergoes a very rapid and unrestricted swelling at relatively low temperature, indicative of weak bonding forces of approximately uniform strength. Tapioca starch starts to swell at about the same temperature as potato starch, but the swelling continues at a much slower rate. This suggests that the internal bonding forces in the tapioca granule have a wider range of bonding strengths than those in potato granules. Cereal starches such as corn show a characteristic two-stage swelling pattern as related to temperature, presumably indicative of two types of internal bonding forces. These cereal starches exhibit restricted swelling patterns, whereas the waxy varieties of these starches are less restricted. Even though the potato starch granule swells much more readily than even the waxy starches, it gives much less solubles at any given degree of swelling than the cereal starches. This suggests that the potato starch has a more extensive and uniform molecular network.

The swelling patterns of the various starches and the results of X-ray diffraction studies and examination of granules by optical and electron microscopy have suggested a picture of the possible structure and organization of the starch granule.[10,20,25,34,39,40] It appears that the amylopectin is the main factor in the crystallinity and structural integrity of the starch granule. The granule is composed of crystalline and amorphous regions, and the transition between these regions is a gradual one. The large amylopectin molecules take part in both the highly organized crystalline regions as well as in the less-organized amorphous regions. The crystalline regions are formed by a large number of intermolecular hydrogen bonds holding the linear chain portions of the branched amylopectin molecules together. The number of these bonds and their concentration along a molecular association of chains decreases on passing from a crystalline, micellular region to an amorphous region (Fig. 22.9).

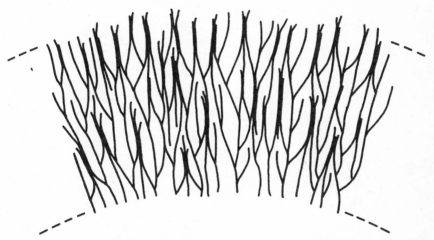

Figure 22.9 Structure of molecules in a layer of a starch granule. Thickened areas represent micelles produced by association of chains. *(Ref. 80.)*

The crystallites initially radiate from the hilum. (The hilum is a spot on the granule believed to be the nucleus around which the granule has grown. As observed by microscopy, under ordinary light the hilum usually appears at the intersection of two or more creases or, under polarized light, at the intersection of the arms of the polarization cross.) In the peripheral parts of large asymmetric granules, the crystallites are oriented perpendicularly to the granule surface. The results suggest that the starch molecules are deposited in an orientation perpendicular to the growing surface of the starch granule.

The linear amylose fraction apparently does not contribute much to the crystalline nature of the granule in normal starches with an amylose content of 15 to 30%. When these starches are leached with water at or slightly above their gelatinization temperatures, the shorter-chain amylose molecules are solubilized and leached from the granule, indicating that the amylose is not tightly bound in the crystalline areas of the granule. The longer amylose molecules are more intimately entangled in the granule structure.

In the case of high-amylose starches (50 to 80% amylose content), the amylose is more tightly bound in the granule, reinforcing the structural integrity. These starches have a more restricted swelling and require higher temperatures for gelatinization.

Waxy starches, with little or no amylose, have a structural integrity and show the X-ray and birefringence patterns indicative of crystallinity. The gelatinization tempera-ture ranges of regular and waxy varieties of the same species are the same [corn, 62 to 72°C (144 to 162°F); sorghum, 67.5 to 74°C (153 to 165.2°F)], indicating that the amylose does not affect the gelatinization temperature. However, since the regular varieties show more restricted swelling than the waxy varieties, the linear amylose mole-cules must be involved in the structural network, providing some reinforcement. This is most marked in the high-amylose starches.

The initial swelling during gelatinization takes place in the amorphous regions of the granule, disrupting the weak hydrogen bonding between the molecules and hydrat-ing them. As the temperature of the aqueous starch mixture rises, more hydration occurs in the amorphous region and the hydrogen bonding in the crystalline regions begins to be disrupted. The granule swells to give an expanded network of hydrated amylopectin molecules (and some amylose molecules) held together by residual orga-nized micellular areas which have not been disrupted (Fig. 22.10). It is this network

Figure 22.10 Structure of molecules in a cooked starch paste. Thickened portions represent persistent micellar associations which are part of the highly hydrated, highly swollen, residual granule structure. *(Ref. 80.)*

that gives the swollen granule its elastic character, which determines the viscosity and textural properties of the cooked starch paste. The higher the molecular weight of the amylopectin and the stronger the bonds in the persistent micelles, the greater the swollen volume of the granule before breaking apart and hence the higher the peak viscosity. The greater the swelling of the granule, the more susceptible it is to rupture and breaking apart on continued cooking.

Hence, starches that are capable of swelling to a high degree are susceptible to break-down in viscosity to a greater extent on cooking (see Fig. 22.8). Potato and waxy corn starch pastes tend to be rubbery and cohesive, whereas corn starch pastes tend to be salve-like before gelling. It should be noted that the organized micelles are quite persis-tent, and starches cooked at boiling water bath temperature for 1 h still contain highly swollen, hydrated starch aggregates. These are present in potato starch even after the paste is autoclaved at 121°C (250°F) for almost 1 h.[41] To obtain complete disaggrega-tion of the starch molecules in solution, it is necessary to heat aqueous suspensions of the common starches to 150 to 160°C (302 to 320°F). Heating at these temperatures for prolonged periods will, however, cause degradation of the starch. It has also been proposed that the starch molecules exuded from the granule on cooking are the major factor in viscosity development.[42,43]

The initial viscosity of the gelatinizing starch suspension is primarily the result of the swollen granule. After the peak viscosity, the starch paste viscosity is determined by the rapidity and extent of the breaking apart of the swollen granules with continued heating and agitation, as well as by the solubilization of the starch. The final viscosity is caused by the mixture of hydrated granule fragments and dissolved molecules, and is related to the molecular weight of the dissolved molecules and molecular aggregates and the amount of amylose present. The retrogradation of the amylose causes the paste to gel on cooling. The molecules of starch exuded from the granule on cooking have been considered as a major factor in the development of viscosity.[42,43]

Chemical Gelatinization. Granular starches can be gelatinized at room temperature by aqueous solutions of alkalies, certain salts, and organic compounds. In aqueous alkali, granular starches adsorb alkali in accordance with the well-known Freundlich adsorption equation, the amount adsorbed being proportional to the alkali solution concentration at equilibrium.[29] Under similar conditions, there is relatively little difference in the amount of alkali adsorbed by the various types of native starches despite differences in granule structure, granule size, or content of linear fraction.

For example, after 30 min, a suspension of 10 g (dry basis) starch in 200 mL 0.0494 N sodium hydroxide solution at 30°C (86°F), potato, corn, waxy corn, high-amylose corn (61% amylose), and rice starch adsorbed 0.23 to 0.24 milliequivalents (meq) of sodium hydroxide per gram. The following hydroxides are adsorbed in decreasing degrees: calcium and barium > potassium > sodium > lithium > benzyltrimethylammonium > tetramethylammonium. Gelatinization takes place when the sorbed alkali exceeds a certain critical concentration, which is dependent upon the type of alkali and the starch.

For sodium hydroxide, the critical adsorption concentrations of corn, waxy corn, high-amylose corn, and potato starches are 0.40, 0.43, 0.41, and 0.32 meq/g, respectively. With calcium hydroxide, no granule swelling is noted in an almost saturated solution. In the presence of sodium sulfate or sodium chloride, the alkali adsorption capacity and critical adsorption concentration are higher. These salts also increase the pasting temperature and are often added to prevent swelling during the preparation of starch derivatives. It is interesting to note that native starch adsorbs sodium hydroxide slowly and continuously from ethanol solution without reaching any apparent equilibrium. Presumably, the adsorbed alkali is in the form of sodium ethylate. A method for predicting the conditions for reduced filterability of an alkaline starch slurry has been reported.[30]

The gelatinization of starch is affected by the kind and concentration of salts present (Table 22.3). Sodium chloride and sodium sulfate repress the gelatinization of starch; i.e., the gelatinization temperature in aqueous solutions of these salts is raised. Thus, starch will not gelatinize in boiling saturated sodium sulfate solution. These salts are used in the preparation of modified starches where the strongly alkaline reaction conditions or level of substitution would result in swelling or gelatinization if the salt were not present. On the other hand, certain salts lower the gelatinization temperature. For example, starch will gelatinize at room temperature in sodium iodide solution.

A study of the swelling and gelatinization of wheat starch at 30°C (86°F) in varying concentrations of salt solutions showed that the swelling effect follows the Hofmeister lyotropic series.[31] In decreasing order of starch granule swelling power, the anions are listed as follows: hydroxide > salicylate > thiocyanate > iodide > bromide > chloride. The relative activity of the cations indicated some effect other than a lyotropic series. It has been suggested, in work with 24 different chloride salts, that both water and hydrated salts can act as gelatinizing agents, and the interplay of these two factors results in a complex relationship between gelatinization temperature and salt concentration.

As salt concentration increases, the gelatinization temperature goes through a maximum and then a minimum.[31a] The swelling power of the reagents, in general, increased with concentration. Starch, at 2% concentration, after 3 h in the salt solution showed appreciable swelling with sodium hydroxide at 0.09 mol/L and complete gelatinization at about 0.14 mol/L; with potassium hydroxide, at 0.17 and 0.21; with sodium salicylate, at 0.8 and 2.8; potassium and ammonium thiocyanates, at about 2.0 and 4.2, 3.8 and

TABLE 22.3
Gelatinization Temperatures of Corn Starch in Aqueous Solutions[47,48]

Solution, %	Temperature,* °C	Solution, %	Temperature,* °C
Water	62–72		
Sucrose:		Ammonium nitrate:	
5	60.5–72.5	10	62–75
10	60–74	20	54.5–70.5
20	65.5–78	30	52–66.5
30	69.5–81	40	44.5–63
40	72–85	50	42–61
50	76–90.5	60	Room temp.–57
60	84–96.5		
		Ammonium sulfate:	
Urea:		10	78–90
10	51–64	20	88–99
20	39–55	30	No gelatinization
30	Room temp.–47		on boiling
40	Room temp. complete		
		Sodium carbonate:	
Sodium hydroxide:		5	64–75
0.2	55.5–69.5	10	67–76
0.3	49–65	20	77.5–87
		30	92–103

* Suspension of starch (0.1–0.2% concentration) observed on Kofler hot stage under polarized light microscope. Temperatures correspond to loss of birefringence by 2% of the granules in the field and by 98% of the granules, indicative of the start and end of the gelatinization range.[37,47]

4.5, respectively; with sodium and potassium iodides, at 2.5 and 4.0; with sodium bromide, at 4.5, but doesn't completely gelatinize. Sodium chloride (5 M) solutions cause no swelling of starch even after a week at 30°C (86°F). Calcium chloride solutions above 2.4 M cause pronounced swelling if allowed to act for a sufficiently long time. Strontium and magnesium chlorides are less effective.

Increased calcium chloride concentration will decrease the time required for complete gelatinization. Urea, a nonelectrolyte, also causes swelling of starch at 30°C (86°F), appreciable swelling being noted at about 6 M concentration with complete gelatinization at about 8 to 10 M urea solution (Table 22.3). The effects of these salts on the swelling characteristics of wheat and corn starches as measured by the Brabender amylograph and swelling power determinations confirm these observations.[18,32] An aqueous solution containing lithium bromide (2 M concentration) and guanidinium chloride (4 M) will dissolve starch in 4 to 6 h at room temperature, whereas neither LiBr (6 to 8 M) nor guanidinium chloride (4 to 8 M) alone will completely dissolve starch.[33]

Granular starches dissolve in cold anhydrous dimethyl sulfoxide without swelling, the rates of solution varying with the type of starch.[34,35] Waxy corn and waxy sorghum starches dissolve almost completely in about 24 h; corn and sorghum starches require about 48 h. Tapioca and arrowroot starches show an initial lag and then slowly dissolve, tapioca solubilizing to a slightly lesser extent than corn starch in 48 h, while arrowroot starch becomes only 55% solubilized in the same time. High-amylose corn starch and rice starch are completely dissolved in 40 h. In 48 h, wheat starch is about 95% solubilized; potato starch, 68%. If a small amount of water (3 to 15%) is added to the dimethyl sulfoxide, the solubilization of the granular starches is increased, potato starch becoming completely soluble.

Factors Affecting Starch Paste Properties In addition to the variety of starch and/ or the type of modification, the properties of the starch paste are determined by the cooking procedure and the presence of other materials. Thus, careful attention must be given to controlling the conditions of cooking in order to obtain the desired properties consistently. The factors involved are concentration, temperature, time, intensity of agitation, pH, and the type of additive or impurity present.

The inherent swelling capability of the starch will determine the selected concentration. There is a critical concentration at which the swollen granules occupy virtually the entire volume, entrapping all the available water so that there is no "free" water between the granules. This value, determined at 95°C (203°F), in grams of anhydrous starch per 100 mL of water, is 0.1 for potato starch, 1.4 for tapioca, 1.6 for waxy maize, 4.4 for corn, and 20.0 for high-amylose corn starch.[25,39]

In most applications, the concentration is substantially greater than the critical concentration, but the relative amounts of starch used to obtain a given viscosity will generally be in line with these values. The objective of cooking is to disrupt the starch granules and hydrate the starch. When the starch concentration is high enough to inhibit the complete hydration of the starch granules, the rheological properties of the paste will be different from those at a lower concentration. At concentrations higher than the critical concentration, the granules become more susceptible to rupture by shearing forces. The higher the concentration, the higher the viscosity of the cooled paste and the greater its tendency to thicken and gel.

Once the peak viscosity of a starch paste is reached during cooking, the swollen granules begin to disintegrate and the viscosity decreases with further cooking, even if the temperature is kept constant. A corn starch slurry in water heated to 90°C (194°F) and held constant will take longer to reach peak viscosity and to break down in viscosity than one cooked to 95°C (203°F).

Faster and more effective agitation will provide faster and more uniform heat transfer and a more rapid cooking process. However, the increased shear produced by the faster agitation will also hasten granule rupture, causing faster breakdown in viscosity and a lower final viscosity. Agitation during the cooling interferes with the retrogradation of amylose and yields softer gels. Faster cooling, because of better heat transfer, tends to result in weaker gels because of a shortening of the time for orientation and association of the amylose molecules.

Most starches gelatinize more quickly at higher and lower pH's. At the intermediate pH range 4 to 7, gelatinization takes place more slowly and the start and rate of breakdown are retarded. Figure 22.11 shows the early gelatinization at pH 10 and pH 2.5, as well as the early peak and breakdown in viscosity. The low peak viscosity and rapid breakdown when cooking under high-acid conditions is probably fostered by hydrolysis during the cooking process. This would be of importance, for example, when starch is used in the preparation of salad dressings and is cooked in the presence of vinegar.

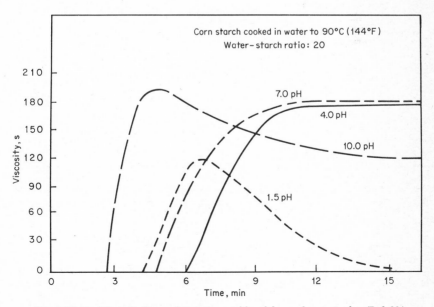

Figure 22.11 Effect of pH on gelatinization and breakdown of corn starch. *(Ref. 80.)*

When starch is cooked in water containing substantial amounts of sugars or other water-soluble hydroxyl-containing compounds, the rate of gelatinization and hydration is retarded and there is less viscosity breakdown because of a decreased tendency of the granules to rupture and collapse. Sucrose hinders the swelling of the starch granule presumably because it binds water and withholds the water from the granule by an osmotic effect. Increasing concentrations of sucrose cause correspondingly greater retardation of granule swelling and increases in the gelatinization temperature (Table 22.3). The peak viscosity of corn starch pastes increases somewhat up to a maximum of 20% sucrose on the weight of the paste. At 50% sucrose, the gelatinization temperature is noticeably higher, the peak viscosity is markedly reduced, and the rate of gelatinization is also substantially reduced, as is the breakdown.

Other sugars such as glucose, fructose, maltose, lactose, and sorbitol have similar, but quantitatively different effects. Gel strength of starch pastes decreases with increasing sugar content, presumably because of incomplete cooking of the granules and decreased solubility of the amylose.[65] Thus, where high concentrations of sugar are needed in a product containing cooked starch, the starch should be cooked alone or with a decreased amount of sugar and the remainder of the sugar added to the cooked paste. The effect of sugar on the gelatinization temperature of corn starch as measured by loss of birefringence is seen in Table 22.3.

Long-chain fatty acids and certain other surfactants complex with the linear amylose fraction of starch. The amylose forms a helical structure in the presence of the complexing agent, and these materials are enclosed in the interior of the helix. Hence, the dimensions and steric configuration of the complexing material must enable it to fit into the helix. Monoglycerides of saturated fatty acids are good complexing agents. Polar surfactants which complex strongly with amylose restrict the swelling and solubilization of corn, potato, and waxy sorghum starches over the pasting range of 60 to 95°C (140 to 203°F).[66]

If no complex is formed, or if a waxy starch which doesn't contain amylose is used, then the surfactant should act as a wetting agent to assist hydration of the granule, causing increased swelling and solubilization. Sodium lauryl sulfate (Du Pont "Duponol ME") represses the swelling and solubilization of corn and potato starches below 85°C (185°F), but shows a reverse action above this temperature, presumably indicating dissociation of the complex. With waxy sorghum starch, which does not contain amylose, the swelling and solubilization are enhanced. The sodium lauryl sulfate acts as a wetting agent over the entire temperature range with the waxy sorghum starch, but only above 85°C (185°F) with the corn and potato starches.

A non-ionic polyoxyethylene graft polymer on polyoxypropylene (Wyandotte Chemical, "Pluronic F-68"), which has no hydrophobic group capable of complexing with amylose, causes a moderate increase in the hydration of corn and waxy sorghum starches. Stearic acid (Armour, "Neofat 18"), glycerol monostearate-monopalmitate mixture (Distillation Products, "Myverol 18–07"), and polyoxyethylene monostearate with 8 ethylene oxide groups or 50 ethylene oxide groups (Atlas Chemical, "Myrj" 45 or 53, respectively) reduce the swelling and solubilization of corn, potato, and waxy sorghum starches. Since the waxy sorghum was affected, it would appear that the outer branches of the amylopectin molecules can also participate in some helical complex formation with a long hydrocarbon chain to restrict expansion of the molecule by hydration, thereby repressing granule swelling.

Although these surfactants were merely added to the starch suspension in water (the stearic acid being added as the sodium salt and the pH adjusted to 6 to regenerate the free acid), the most effective method for complexing monoglycerides or fatty acids is by heating a starch slurry in water containing the monoglyceride or potassium salt of the fatty acid at 50°C (122°F) for 1 to 16 h (the longer the better) to allow penetration of the granule by the agent. Then, the pH is adjusted to 5.0 and the starch recovered by filtration, washing, and air-drying. Increasing the stearic acid content of potato starch in this way to 2.5% on the starch progressively decreased the swelling power of the starch [at 85°C (185°F)] from 170 to 12 and that of the solubles from 35 to 0.7%, and produced a marked increase in the pasting temperature, a decrease in the pasting peak viscosity, and an increase in the cold paste viscosity. The stearic acid complex is thought to dissociate at 120°C (248°F).[66]

When oil or melted fat is blended with starch and the mixture slurried in water containing sufficient sodium hydroxide to bring the cooked pH to 6.5, it was found that the temperature of maximum hot viscosity decreased with increasing amounts of the fat (soybean oil) up to 12% on the starch paste (6% starch solids) without a decrease in the maximum viscosity. The source of the fat or the degree of unsaturation had no effect. However, when a number of non-ionic surfactants were tested in this starch-fat-water system, it was found that they generally raised the temperature of maximum hot viscosity and also changed the cooling curve of viscosity.[67]

Some of the most effective amylose-complexing agents among the food-grade emulsifiers are saturated distilled monoglycerides with a minimum monoester content of 90%. Increasing unsaturation of the fatty acid chains of these materials results in decreasing the amylose-complexing ability. Acetylation of the remaining hydroxyl groups of the monoglycerides greatly reduces complex formation. Esters of propylene glycol with fatty acids have a comparatively low complexing effect. Molecules which have a large polar group, such as sucrose and sorbitan esters of fatty acids, lecithin, and citric acid esters of monodiglycerides, are poor amylose-complexing agents. Monoglycerides, the stearoyl-2-lactylates and the stearoyl fumarates, which have a straight hydrocarbon chain and a small polar group, are very effective complexing agents.[68]

In distilled water, the distilled monoglycerides increase the pasting temperature of wheat, tapioca, corn, and potato starches (to a lesser extent with the latter two starches). The sodium and calcium stearoyl-2-lactylates act similarly but to a lesser extent. The peak viscosity (hot) of wheat starch was increased by these emulsifiers, whereas the peak viscosity of the tapioca and potato starches was decreased and the corn starch was relatively unchanged. These emulsifiers increased the temperature at which corn starch paste started to gel. In hard tap water, the pH is higher, which affects the wheat starch. In this case, only the distilled monoglyceride shows a higher pasting temperature and a higher peak viscosity.

When the starch begins to gelatinize, water is absorbed into the amorphous area. The formation of complexes with the amylose at this stage will delay the swelling process and increase the pasting temperature. Complex formation will also make the structure of the swollen granule more rigid and hence more resistant to breakdown and loss in viscosity.[69] When wheat or tapioca starch is suspended in water containing monoglycerides, part of the monoglycerides is irreversibly bound by the starch granules at 30°C (86°F). During gelatinization, the quantity of monoglyceride-bound starch increases to reach a higher equilibrium level at 90°C (194°F). Presumably the monoglyceride molecules penetrate the swollen starch granules and form an insoluble complex with the amylose.[70]

Starch Supplies

The starches of commerce are corn, tapioca, potato, wheat, sorghum, rice, sago, and arrowroot. Corn starch is by far the major starch produced and used in the United States. A total of nearly 375 million bushels of corn were processed by wet milling in the United States in 1977 to obtain corn starch. This is equivalent to about 5.5 billion kg (12 billion lb) of starch. About 900 million kg (2 billion lb) were sold as unmodified starch and about 700 million kg (1.5 billion lb) were sold as modified starches. The rest was processed to make corn syrups, maltodextrins, high-fructose corn syrups, and dextrose.[71a]

Tapioca starch is imported from Thailand and Brazil, amounting to approximately 43 million kg (94 million lb) in 1976. Smaller amounts of potato starch are used in the United States, some produced domestically [approx. 7 million kg (15 million lb) in 1976], and some imported [8.5 million kg (18.7 million lb) in 1976]. Wheat starch, rice starch, arrowroot starch, and sago starch are used in relatively small quantities. Approximately 3.8 million pounds of the latter two were imported in 1976.

The paper and paperboard industry is the largest consumer of industrial starches, taking an estimated 1 billion kg (2.1 billion lb) in 1974 or approximately 60% of the starch sold. The textile industry used an estimated 161 million kg (355 million lb) of starch in 1974. In the same year, miscellaneous usage, including building materials, adhesives, pharmaceuticals, mineral refining, explosives, and drilling muds, consumed about 260 million kg (570 million lb). The food applications of starches took about 230 million kg (500 million lb).[72] Estimates of the usage of various types of starches in the paper industry for 1976 are given in Table 22.4. The distribution of corn starch

TABLE 22.4
Starch Used in Paper Industry, 1976[73]

Starch	Millions of kg	Millions of lb
Corn	1,080*	2,380*
Tapioca	48	105
Potato	38	81
Wheat	33	72
Total	1,199	2,628

* Includes 295 million kg (650 million lb) used in corrugating.

TABLE 22.5
U.S. Markets, Corn Starch, 1976[73]

Industry	Millions of kg	Millions of lb
Paper/board/corrugating	1,080	2,380
Food processing	158	350
Brewing	107	235
Textiles	107	235
Pharmaceutical	50	110
Building products	50	110
Adhesives	43	95
Mining	29	65
Briquetting	18	40
Miscellaneous	36	80
Total	1,678	3,700

TABLE 22.6
Sales of Corn Starch Products, 1972[74]
(Non-food uses)

Starch type	Millions* of kg	Millions* of lb
Unmodified	824	1,817
Acid-converted	154	340
Oxidized	75	166
Dextrins	68	150
Cationic	45	100
Pregelatinized	45	100
Others†	190	420
Total	1,403	3,093

* Includes sorghum, waxy corn, and high-amylose corn starches.
† Includes hydroxyethyl starch and starch acetates.

in U.S. markets is estimated in Table 22.5. A rough breakdown of the type of corn starches sold for industrial uses in 1972 is shown in Table 22.6.

Manufacture of Starch

The separation of starch from the seed, root, or pith is generally carried out in a milling and grinding process in the presence of water as a transporting, processing, and washing medium.[75,76] Laboratory procedures for the isolation of corn, wheat, and potato starches have been described.[37] For corn starch, the procedure involves steeping the clean corn kernels (300 g) in a dilute aqueous solution of sodium bisulfite (1.22 g/750 mL water) at 50°C (122°F) for about 24 h, draining off the steep water, and grinding the corn in a mill or Waring blender (in water); the ground slurry is then passed successively through a series of screens (80 to 400 mesh) to remove hull and germ particles; and finally the fine gluten and starch particles in the aqueous suspension are separated by settling since the starch is of greater density than the gluten protein (1.5 vs. 1.1 g/

cm³). The upper layer of protein can be scraped off the sediment, and the reslurrying and the settling or centrifuging repeated until the residual starch sediment is white. A bushel of dent corn (56 lb) will normally yield about 61% starch, 3.8% oil, 19.2% feed products, and 16% water from the commercial wet milling process. As shown in Fig. 22.12, the process separates the component hull, germ, and endosperm of the corn kernel. The endosperm is separated into starch and gluten, and the oil is removed from the germ. The oil and starch are sold as separate products, whereas the rest of the components are combined and sold as animal feeds.

The shelled corn is cleaned and then steeped in a dilute solution of sulfurous acid for 30 to 50 h at 49 to 54°C (120 to 130°F). This is a controlled fermentation which produces lactic acid and results in leaching of solubles from the corn and softening of the kernel to permit separation of the germ, as well as facilitating the starch-gluten separation. The grain is separated from the steep water, then milled to crack open the kernel to liberate the germ as intact as possible, and the germ is separated centrifugally. The remainder of the kernel is finely ground to release the hull, starch, and gluten. The slurry of these components is then screened to remove the ground hull or bran, and the starch is separated from the gluten in centrifuges or hydrocyclones. The starch is washed and either pumped to reaction tanks for modification or is dried for use or sale as such.

Starch Modifications

Modifications of native starches are carried out to provide products with the properties needed for specific uses. In many cases, the properties of the native starches preclude their use in a particular application or process. The modifications are designed to change the gelatinization characteristics, solids-viscosity relationships, gelling tendency of starch dispersions, hydrophilic character, moisture content, water-holding power of dispersions at low temperature, resistance of dispersions to breakdown in viscosity by acids and mechanical shear, and to introduce ionic character. The modifications may involve merely a reduction in moisture content, a change in physical form, a controlled degradation, a change in the amount of amylose or amylopectin content, a molecular rearrangement, or the introduction of chemical groups not normally present (Table 22.7).

Modifications Changing Amylose/Amylopectin Content As noted previously, the linear amylose is the starch fraction that causes gelling of starch dispersions and is a better film-former than the branched amylopectin. The amylopectin gives stable starch sols which do not gel. Thus, starches containing one or the other fraction would have special utility.

Hybrid Breeding. The most economical way to obtain this end is through breeding specific plant strains that can be grown readily and which yield starch with only one component. The waxy varieties of corn and milo maize (sorghum) do yield a starch that is almost entirely amylopectin. These waxy starches have been available commercially since 1942. The paste stability properties are characteristic of the modified waxy starches. These products are used in foods, in textile finishing and printing, in adhesives, and in the manufacture of paper and paper products.

The attempt to develop a hybrid with 100% amylose starch has not been as successful. Although experimental strains with starch containing as much as 85% amylose have been seen, the commercially developed high-amylose corn hybrids contain starch of 50 to 75% amylose (Class 5 and 7 amylomaize).[44] The agronomic properties of these amylomaize hybrids are not as good as those of dent and waxy corn, giving lower yields per acre. This, in conjunction with the necessity of segregating the crop during growth, transport, and storage to prevent contamination (as is also required for the waxy corn crop), raises the cost of the corn grain and hence of the starch. Further, the milling yields of starch are also lower.

The high-amylose corn starch does not disperse readily by cooking in a boiling water bath as does corn and waxy corn starch, even though birefringence is lost at 92°C (198°F) for almost all the granules. It should be noted, however, that many of the granules do not show any birefringence even though they have not been damaged. This is particularly seen in Class 7 starch.[49] To disperse the high-amylose corn starch, heating the aqueous suspension at about 150 to 160°C (302 to 320°F) under pressure is required. Commercially, continuous cookers utilizing high-pressure steam injection can be used to disperse the amylomaize starch. Chemical gelatinization with sodium

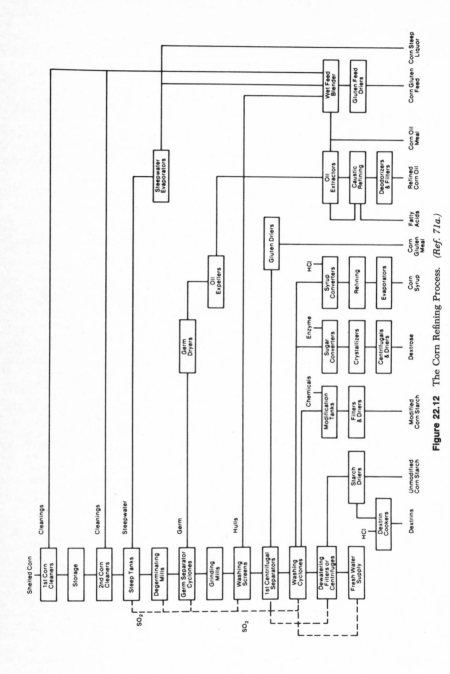

Figure 22.12 The Corn Refining Process. *(Ref. 71a.)*

TABLE 22.7
Modification of Starch

Type	Objectives	Treatment
1. Hybrid	Stability of sols (amylopectin) High gel strength (amylose) High film strength (amylose)	Plant breeding
2. Amylose Amylopectin	Same as 1.	Fractionation
3. Acid fluidity	Lower viscosity High gel tendency	Acid hydrolysis
4. Oxidized	Lower viscosity with improved sol stability	Oxidation (hypochlorite)
5. Dextrins	Lower viscosity Range of sol stability Range of solubility in cold water	Heat treatment, dry (may be in presence of acid)
6. Inhibited	Modify cooking characteristics	Crosslinking
7. Stabilized	Improve sol stability (resistance to gelling) Lower gelatinization temperature	Esterification Etherification
8. Functional groups	Change colloidal, hydrophobic, hydrophilic, cationic, anionic character	Esterification Etherification
9. Changed physical character	Improve moisture absorption, molding flow characteristics, cold-water-dispersibility	Redrying to low moisture additives (oil, MgO) drum-drying

or potassium hydroxide (0.5 N or higher) solutions, or 80 to 95% aqueous dimethyl sulfoxide can be used.

The high amylose content, which confers strong gelling tendencies, makes these starches useful in the preparation of gum candies and snack foods. The strong film-forming capability suggests its use as a coating on foods and confections and in non-food uses such as sizing and coating of paper, warp sizing, and fiberglass forming size. Since the amylose will retrograde and cause gelling of the high-amylose starch dispersion as it cools, aqueous dispersions must be maintained at elevated temperature [above 60 to 70°C (140 to 158°F)] until use, or some other means of stabilizing the dispersion must be used, such as alkali or chemical derivatization of the starch. The temperature of retrogradation will be related also to the concentration of the starch dispersion.

Fractionation. Although amylose complexes with butanol, pentasol, and other polar compounds can be used to selectively precipitate the linear amylose from a starch dispersion, the only successful commercial fractionation process was developed based on a liquid-liquid phase separation using magnesium sulfate to salt out the amylose.[4,6,8,12,15,16,17]

The process is carried out on a commercial scale by the AVEBE company of Veendam, Holland.[77] The process involves suspending potato starch in a 13% magnesium sulfate solution, buffered at pH 6.5 to 7.0, heating rapidly to 160°C (320°F) to disperse the starch, holding at this temperature for 15 min, then cooling to 80°C (176°F) and holding about 30 to 60 min for the amylose to separate out and retrograde. The amylose is separated by centrifuging, and the solution containing amylopectin is cooled to 20°C (68°F). The amylopectin precipitates and becomes water-insoluble if it is allowed to remain in contact with the magnesium sulfate solution for several hours. It is then recovered by filtration, washed free of magnesium sulfate with water, and dried. The amylose is also washed with water and dried. The retrograded amylose is insoluble in water and a suspension in water must be heated to 135 to 150°C (275 to 302°F) to redisperse. The amylopectin becomes water-soluble after it is dried at temperatures about 60°C (140°F) and can be dispersed in water at room temperature. Since it is not in the native starch granule, the amylopectin viscosity for a given solids is much less than that of waxy corn starch.

Although considerable work was done on unsupported films of amylose and derivatives of amylose (acetates, hydroxypropyl ethers) in analogy to cellulosic films and some semicommercial film-forming processes were developed, full commercial production was not carried out.[81-85] The films are impermeable to air, oxygen, and nitrogen and have approximately the same tensile strength as cellophane films. The films must be highly plasticized with humectants to prevent brittleness at low relative humidity. The isolated amylose would be useful wherever the gelling and film-forming properties are needed.

Although the isolated amylose is insoluble in water, it will absorb up to four times its weight of water. In addition to being soluble in sodium and potassium hydroxide solutions and in aqueous dimethyl sulfoxide, the amylose is also soluble in aqueous formaldehyde and in formic acid. In the latter case, the concentrated formic acid forms an ester with the amylose. The ester is labile, and dilution will allow the amylose to regenerate. In the case of the formaldehyde, hemiacetals are formed with the amylose. This is an equilibrium which can be shifted to cause retrogradation of the amylose and precipitation or gelling by decreasing the concentration of formaldehyde. The viscosity of the solutions of amylose will increase with increasing concentration.

Pregelatinized Starches   When a starch that is swellable in cold water without cooking is required, pregelatinized starches are used. These starches are prepared by cooking and drying starch slurries on heated drums. Since the granule structure has been disrupted by this procedure, the pulverized dried film from the drum dryer will hydrate and thicken in water. In essence, the starch manufacturer has cooked the starch for the user. The pregelatinized starch will have less thickening power and less gelling tendency than the corresponding granular starch from which it is made. The loss in thickening and gelling potential is related to the destruction of the hydrated granular structure on the drum and, in the case of an amylose-containing starch, to some retrogradation of the wet film on the drum while drying.

Aging of the pregelatinized starch in dry storage will also result in some decrease in thickening power and slowing of rehydration in water. Presumably some association of both amylose and amylopectin takes place under these conditions. The way in which a pregelatinized starch behaves on addition to water is important. The hydration rate may be slow, or if initially fast, the particles may form lumps because the initial wetting produces a sticky surface on the particles, which then clump together. Sometimes, on initial wetting, a particle forms a swollen shell which prevents further diffusion or causes very slow penetration of the water into the particle, so that it is not completely wetted and dispersed. This leads to "fisheyes."

If the drum dryer is adjusted to produce a thick, dense, horny film, the particles, after grinding, will not hydrate so rapidly in dispersing and will give a low paste viscosity and a grainy paste which may be desirable or undesirable depending on the application. A thin film, with rapid drying on the drum, will give a thin flake that hydrates rapidly, giving a high paste viscosity without appreciable graininess.[78] The grinding to give a suitable particle size distribution is also critical to the dispersibility properties in cold water.

Many starches, such as corn, sorghum, and wheat, contain small quantities of unsaturated fatty acids. In the intact granule they are protected from oxidation, but in the drum-dried starch they may develop rancid flavors on storage of the pregelatinized starch. Possibly the fatty acids are deposited on the particle surfaces where oxygen has easy access.

Pregelatinized starches can be made from almost any native starch, modified starch, or starch derivative. The market for pregelatinized starches in prepared foods is expanding rapidly, particularly in the development of "convenience foods" such as instant soups, instant puddings, instant beverages, cake mixes, and salad dressing mixes. Pregelatinized tapioca starches are of particular value here.

The dry mixes are usually added to water or milk at room temperature, and the pregelatinized starch ingredient is required to hydrate and thicken the mixture to give a palatable texture and to hold the other components in a uniform suspension. Pregelatinized starches have been used as a binder for charcoal briquettes and as foundry core binders. They are useful as adhesives, such as in cold-water-dispersible wallpaper pastes.

Modification by Crosslinking   The gelatinization and swelling properties of the starch granule can be changed considerably by a minimal amount of crosslinking which reinforces the associative hydrogen bonds holding the granule together. The toughening of the granule causes some restriction in the swelling during cooking and results in less viscosity breakdown under shear and low-pH conditions. The swollen granules that are crosslinked are not as easily disrupted and fragmented. Hence, a higher working viscosity is reached and maintained.[75,78a,94]

The crosslinking reaction is carried out commercially by adding the required amount of crosslinking agent to the aqueous starch suspension (35 to 41% solids) at the required temperature [20 to 50°C (68 to 122°F)] and pH (usually 8 to 11, depending upon the reagent, adjusted with 3% sodium hydroxide solution). After the reaction is completed, the pH is adjusted to 4.5 to 6 with dilute mineral acid, filtered, washed with water, and dried. The reaction conditions are chosen to yield an ungelatinized granular starch. The reagents that are effective are those containing two or more functional groups able to react with at least two hydroxyl groups, such as phosphorus oxychloride,[78b] soluble trimetaphosphates,[78c] epichlorohydrin,[78d,e] linear dicarboxylic acid anhydrides (e.g., adipic),[100] N,N'-methylenebisacrylamide,[78f] 2,4,6-trichloro-*s*-triazine, etc.[78g] (equations 1–5, Fig. 22.13). In most cases, the treatment level is in the range of 0.005 to 0.1%

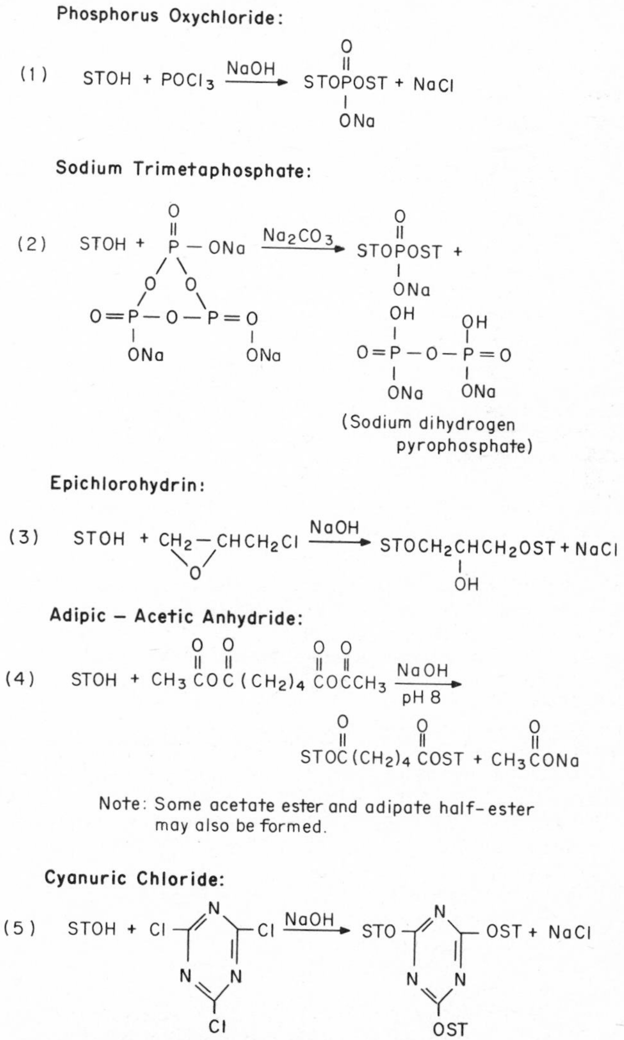

Phosphorus Oxychloride:

(1)   STOH + POCl$_3$ $\xrightarrow{\text{NaOH}}$ STOPOST + NaCl

Sodium Trimetaphosphate:

(2)   STOH + (sodium trimetaphosphate) $\xrightarrow{\text{Na}_2\text{CO}_3}$ STOPOST + ...

(Sodium dihydrogen pyrophosphate)

Epichlorohydrin:

(3)   STOH + CH$_2$—CHCH$_2$Cl $\xrightarrow{\text{NaOH}}$ STOCH$_2$CHCH$_2$OST + NaCl

Adipic — Acetic Anhydride:

(4)   STOH + CH$_3$COC(CH$_2$)$_4$COCCH$_3$ $\xrightarrow[\text{pH 8}]{\text{NaOH}}$ STOC(CH$_2$)$_4$COST + CH$_3$CONa

Note: Some acetate ester and adipate half-ester may also be formed.

Cyanuric Chloride:

(5)   STOH + (cyanuric chloride) $\xrightarrow{\text{NaOH}}$ STO—(triazine)—OST + NaCl

Figure 22.13 Crosslinking reactions on starch.

reagent, to give a relatively low degree of crosslinking (one crosslink per 200 to 1000 AGU). The reaction is followed by the physical properties of the starch, such as viscosity, and the reaction is stopped at the desired level of crosslinking.

The Brabender viscosity curve shows that increasing levels of crosslinking result in changes in the hot peak viscosity and the degree of breakdown in viscosity after the peak with continued heating and stirring. With corn starch (Fig. 22.14), mild crosslinking has little effect on the initial rate of gelatinization but does cause an increase in the maximum viscosity during the 30-min holding period at 95°C (203°F), eliminating the viscosity breakdown seen with the native starch.

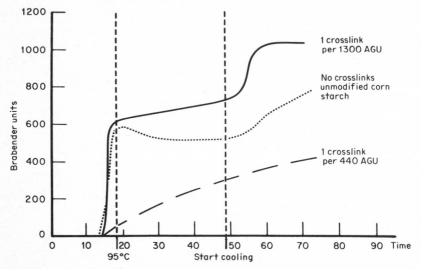

Figure 22.14 The effect of crosslinking on the gelatinization and viscosity of corn starch. Starch at 6% solids, pH 6, heated at 1.5°C/min in Brabender Visco-amylograph (350 cm-gm cartridge) from 30 to 95°C, held at 95°C for 30 min and then cooled at 1.5°C/min to 25°C. Number of crosslinks are estimates. (*Ref. 79.*)

As the level of crosslinking is increased, the reinforcing effects of the chemical covalent bonding, which is not as susceptible to rupture during cooking as are the hydrogen bonds, cause the granules to resist the swelling, and hence the viscosity tends to decrease proportionately. If the level of crosslinking is carried far enough, the granules will not gelatinize at all, even when heated at superatmospheric pressure, as in autoclave sterilization. In the latter case, these types of starches have been used as dusting powders for surgical gloves.[78h,i] The gelatinization temperature is not affected by the crosslinking. Because the crosslinking inhibits granule swelling, these cross-bonded starches are also called *inhibited* starches.

The crosslinking of waxy corn starch and tapioca starch is of particular importance because of the effects on viscosity and texture when these starches are used in food applications. As shown in Fig. 22.15, the effect of crosslinking on lowering the peak viscosity and minimizing or eliminating the drastic drop in viscosity on continued cooking is dramatic. The sensitivity of the starch to pH variations during cooking can also be minimized or overcome as shown by the viscosity curves at pH 5 (neutral) and pH 3 (acid). Here again the effect of toughening the granule by crosslinking so that the swollen granules do not rupture and fall apart so readily leads to a much higher working viscosity under neutral or acid conditions.

As mentioned previously, higher temperatures of cooking accelerate the rate of gelatinization and increase the tendency toward viscosity breakdown. If the temperature is high enough, a lightly crosslinked starch will also tend to show this viscosity breakdown. A higher degree of crosslinking must then be used to overcome the breakdown in viscosity resulting from the particular cooking conditions to obtain the highest final viscosity needed in a particular application.

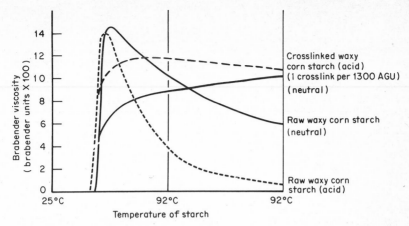

Figure 22.15   The effect of mild crosslinking on viscosity of waxy corn starch.   *(Ref. 80.)*

Since crosslinking toughens the swollen granule so that it is not disintegrated as rapidly as an unmodified starch, it will be less sensitive to the mechanical forces applied during processing of mixtures containing the starch such as the pumping and mixing involved in the commercial manufacture of foodstuffs.   In Fig. 22.16, the viscosity of waxy corn starch and a mildly crosslinked waxy corn starch (both cooked at 5% solids to the same viscosity) was measured at 25°C (77°F), using different rates of shear.   In going from 50 to 200 rpm, the crosslinked starch shows little change in viscosity, while the native waxy starch undergoes a marked drop in viscosity because the shearing action causes the fragile swollen granules to break apart.

Waxy corn, tapioca, and potato starches cook to give cohesive, rubbery, elastic pastes with a clinging character that is unpalatable in foods.   The viscoelasticity is produced by the interaction of the highly swollen, fragile granules with each other so that it is difficult for them to rub past each other freely and instead they become entangled and interpenetrating.   A low level of crosslinking strengthens the granule, restricting

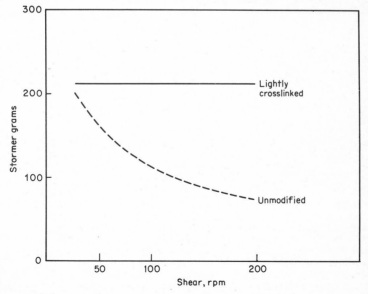

Figure 22.16   Effect of shear on swollen granules of native and crosslinked waxy corn starch. *(Ref. 71b.)*

its swelling, limiting the disintegration, and thereby minimizing interpenetration of granules rubbing against each other. This then yields a non-cohesive, salvelike texture which is non-clinging in the mouth and is palatable in a food.

When acidic canned foods with pH's usually below 4.5 are heated at temperatures up to 99°C (210°F), or retorted at temperatures up to 118°C (245°F), with local temperatures reaching as high as 132°C (270°F), there can be an extreme drop in viscosity if an unmodified starch is used as the thickener. Thus, to obtain a final viscosity that is high enough in the finished product, the concentration of unmodified starches must be increased to the extent that excessive viscosity is produced in the initial cooking or processing stage. This creates handling problems, interferes with heat penetration during the sterilization retorting, and may adversely affect product quality. However, the use of a crosslinked starch will give a higher final viscosity at a lower concentration because it will not lose viscosity excessively and will therefore not give rise to these processing and quality problems.

Where a crosslinked granular starch is needed, the extent of the crosslinking will be determined by the conditions of use such as the pH and the processing temperature, time, and techniques of agitation, mixing, and pumping. Usually, the level of crosslinking can be chosen to give the highest level of viscosity for the conditions employed. Crosslinked starches are found in fruit pie fillings, in canned soups, gravies, and sauces, in baby foods, and in canned cream-style corn. Crosslinked starches are also used in corrugating, in adhesives, in oil well drilling muds, and in binders for charcoal briquettes.

Modification by Controlled Degradation Low-viscosity starches are needed when a high-solids starch dispersion with a pumpable and workable viscosity is required. Usually, the application involves the properties of a dried starch film, and it is necessary to deposit a higher level of starch solids to be effective, as, for example, in an adhesive film on a gummed tape. With a native starch, the high viscosity of the cooked dispersion requires a large amount of water (low starch concentration) to provide a workable viscosity for pumping, mixing, and applying to a substrate. This low-solids dispersion would then require large quantities of water to be removed to obtain the dried film (slower drying and requires more energy), reducing machine speeds and yielding thin films.

The commercial degraded starches are known as *converted starches* and comprise the "thin-boiling" acid-converted starches and oxidized starches, as well as the dextrins. These products accounted for somewhat less than half of the sales of modified corn starches in 1977. In 1972, these products accounted for about 51% of the sales of modified corn starches, or 300,000,000 kg (656,000,000 lb) (Table 22.6).[74]

The degradation of starches by conversion processes involves mainly a scission of the starch molecules to lower-molecular-weight fragments. When this is carried out on a granular starch, the granule structure is weakened so that the granules tend to disintegrate more readily and rapidly during the gelatinization process, leading to a lower hot paste viscosity. For those starches that do not contain amylose and do not retrograde (waxy starches), a lower cold paste viscosity also results. The lower viscosity produced by the disintegration of the swollen granule before it can reach its maximum swelling results because the amylopectin network in the granule is cut in a number of places so that the hydrated network holding the swollen granule together is smaller and not as extensible.

The commercial conversion process is carried out by the action of acids, oxidizing agents, and/or heat (Table 22.8). Other agents that can be used are alkalies and enzymes. Most of these conversions are run on dry starch or aqueous suspensions of starch which are then recovered by filtration, washing, and drying. Conversion can also be run on dispersed starch and the product recovered by drum drying or spray drying. However, in most cases, the dispersed conversion reaction products are used at the point of preparation, as in the preparation of a paper-coating adhesive or surface size by enzyme conversion.

The widespread use of continuous cookers and continuous thermal converters allows *in situ* conversion of dispersed starch in the presence of acids or oxidizing agents or enzymes in the user's plant. The conversion of dispersed starch may also be used to prepare liquid adhesives, which are sold as such or drum-dried to give cold water–soluble products.

The comparative viscosity ranges of the native and converted starches in terms of parts of water per part of starch to give roughly the same hot viscosity are shown in Fig. 22.17.

TABLE 22.8
Starches: Modification by Conversion

Type	Preparation				Major reactions			Properties		
		Acid	Heat	Moisture (Dry starch)	Hydrolysis	Rearrang.	RePolym	Viscosity	Sol Stability	Misc. uses
Acid fluidity (thin-boiling)	Acid hydrolysis—slurry (0.2% H_2SO_4 or HCl 50–55°C, about 12–14 hs)				Hydrolytic breakdown (chain scission at glucoside bond)			Lower-range (10–90 fluidity)	Poor-firm gels (waxy-stable)	Warp sizes Confectionery Paper sizing
Oxidized (chlorinated)	Alkaline oxidation—slurry NaOCl				Oxidative breakdown (form COOH, $\diagdown$C=O)			Lower-range (10–85 fluidity)	Increased over acid fluidity	Paper and warp sizing; paper coating; adhesives
Dextrinization: White Dextrin		High	Low (95–120°C)	Moisture (5–12%)	Main	—	—	Lower-range	Unstable–pasty; poor	0–90% cold water solubility (CWS), light color—Paper coating, textile finishes, adhesives
British gum		Low to none	High (170–195°C)	Low moist. (<5%)	Minor	Major	—	High to low	Intermed.	High CWS; dark color—adhesive
Yellow dextrin (canary dextrin)		Intermed.	High (150–180°C)	Low moist. (<5%)	Early stages	Major (later stages)	Major	Low viscosity	Excellent	>95% CWS; Intermed. to dark color— tacky, sticky— adhesives

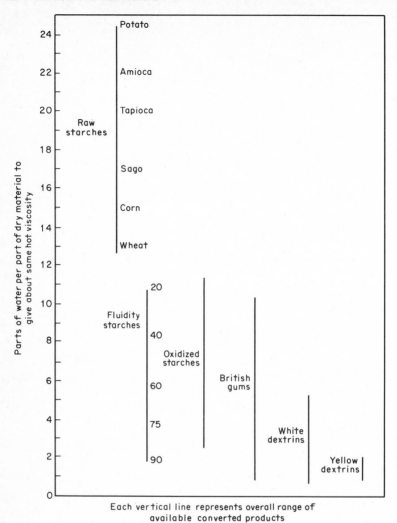

Figure 22.17 Comparative viscosity ranges of different types of starches. The vertical lines span the normal viscosity ranges for the more common starch types. *(Ref. 80.)*

Thin-Boiling, Acid-Converted Starches. These acid-modified starches are manufactured by controlled acid-catalyzed hydrolysis of granular starches in aqueous suspension at temperatures below the gelatinization temperature, so that the products can be recovered by filtration, washing, and drying.[86] The reaction is controlled by monitoring the viscosity during the reaction and stopping the reaction by neutralization (sodium hydroxide, sodium carbonate, calcium hydroxide) when the required viscosity is reached.

The rate of hydrolysis is determined by the acid concentration and temperature, and these factors are taken into account in predicting the time to stop the hydrolysis. Since the product is recovered by filtration, the degree of hydrolysis must be limited so that the starch granule is not degraded to the point where it becomes solubilized. During the hydrolysis, starch fragments of various sizes are split off the molecules and dissolve in the reaction medium; they are lost when the acid-converted starches are filtered and washed.

From a commercial point of view, the hydrolysis must be stopped before the level of losses becomes economically unjustified in terms of product recovery and pollution in effluents. The acid-modified starch is still birefringent, and the acid presumably attacks the starch in the amorphous areas of the granule more rapidly than in the crystalline areas. Initially, the amylopectin in the granule is degraded more rapidly than the amylose.

When an aqueous suspension of acid-converted starch is heated to gelatinize the starch, the modified granules swell less than those of the native starch and tend to fragment after a limited swelling. More starch is dissolved or dispersed and the peak viscosity is lower than that of the unmodified starch in the initial cooking cycle. The fluidity starch disperses to a clear, fluid sol. On cooling, these sols retrograde to form firm gels, particularly at the lower range of conversion. The low viscosity has also been attributed to the greater solubilization of the acid-modified starch and hence a smaller proportion of a discontinuous gel phase (the residual, hydrated, highly swollen granular masses).

Acid-converted starches, sometimes known as *fluidity starches,* are classified commercially in terms of fluidity, an inverse viscosity measurement. The fluidity number is the number of milliliters of a standard alkaline starch dispersion that will flow through a standard funnel in the time required for 100 mL of water to flow through the funnel at 25°C (77°F), usually 40 to 70 s.[45] The higher the fluidity number, the more degraded the starch and the thinner the viscosity. Thus, in Fig. 22.17, the 75 fluidity starch is less viscous than the 20 fluidity, and therefore, the 75 fluidity takes only 4 parts of water to give the same dispersion viscosity on cooking as the 20 fluidity in 11 parts of water.

Unlike the fluidity starches from amylose-containing native starches, the acid-converted waxy starches are more resistant to gelling, in keeping with the inherent stability of the parent starch. At high concentrations, the waxy fluidity starches show gelling tendencies. These fluidity starches are used where stable, low-viscosity, high-solids dispersions are needed, as in paper surface sizing, liquid laundry starches, and adhesive applications. The high gel strength and film-forming properties of the corn and tapioca fluidity starches are useful in textile warp sizing and in the manufacture of gum candy.

As starch is hydrolyzed, there will be an increasing number of reducing end groups formed, reflecting the increasing number of shorter molecules. A relative measure of the reducing end groups is found through the determination of the "alkali number." When starch is digested in hot alkali, the aldehyde end groups undergo an enediol rearrangement which then leads to a slow progressive destruction of the starch chains, producing simple organic acids such as acetic and lactic acids.

Under standard conditions the amount of acid produced provides a relative measure of the number of aldehyde end groups.[8,36,37] The alkali number is the milliequivalents of sodium hydroxide consumed per 10 grams of dry starch during a 1 hour digestion in 0.1 N aqueous sodium hydroxide at boiling water bath temperature. A better measure of reducing value for evaluation of the number of reducing end groups is the ferricyanide number or the 3,5-dinitrosalicylic acid method. (See Reference 37 for procedure.)

Oxidized Starches. The commercial production of thin-boiling oxidized starches involves treating an aqueous starch suspension (35 to 44% solids) with sodium hypochlorite solution (containing 5 to 10% available chlorine) at pH 8 to 10 and 21 to 38°C (70 to 100°F).[87] These starches are sometimes called *chlorinated starches* because of the reagent used, even though no chlorine is introduced into the starch. The reaction is neutralized to pH 5.0 to 6.5 when the required level of oxidation (degradation) is reached and excess oxidant is destroyed by addition of sodium bisulfite solution or sulfur dioxide.

The reaction product is washed to remove impurities, solubilized starch, and by-products of the reaction either on continuous vacuum filters or in hydrocyclones, recovered by filtration, and dried. The hypochlorite oxidizes a limited number of hydroxyl groups to aldehyde, ketone, and carboxyl groups with concomitant cleavage of the glucoside bonds. The introduction of carboxyl groups into the linear amylose molecules reduces the tendency to retrograde and gel. The degree of substitution (D.S.), defined as the average number of substituent groups per anhydroglucose unit (moles substituent per mole of AGU), is in the range of 0.01 to 0.03 for the carboxyl group and 0.005 to 0.01 for carbonyl groups.[87,88]

The oxidized starches are manufactured in the same fluidity ranges as the acid-modified starches (Fig. 22.17), but dispersions of the oxidized starches will generally not gel.

The oxidized starches will be whiter because of the bleaching action of the hypochlorite and the removal of most proteinaceous impurities. Since carboxyl groups are present, the oxidized starches will have an anionic character at pH above 5 and will therefore stain with cationic dyes such as methylene blue.[47] This technique is particularly useful when mixtures of starches are encountered. Any anionic starch will stain similarly, such as carboxymethyl starch or a phosphate derivative.

The oxidized starches gelatinize at lower temperatures than the parent starches, giving a faster rate of gelatinization with a lower peak viscosity. At the higher oxidation levels, the granules disintegrate more readily and yield very clear sols. Films from these dispersions also are clear and adhesive. Since the acid-converted waxy starches are quite stable with less tendency to gel than their amylose-containing counterparts, there is less need to use oxidation to prepare stable, thin-boiling products from waxy starches.

The oxidized starches are used in laundry finishing, textile warp sizing, and paper surface sizing, and as a clay binder in paper coating.

Dextrins. There are four major groups of dextrins:

1. Products obtained by hydrolysis of dispersed starch by action of liquifying enzymes such as the amylases

2. Degradation products by acid hydrolysis of dispersed starch

3. Schardinger dextrins' formed from dispersed starch by the action of *Bacillus macerans* transglycosylase

4. Pyrodextrins produced by the action of heat alone or in combination with acid on dry, granular starch.

The first two products are generally prepared by the user at the site of the application, as in paper surface sizing. The Schardinger dextrins are cyclic molecules consisting of rings of 6, 7, 8, or more 1–4 linked *alpha*-D-glucopyranose units which are called *alpha-, beta-, gamma-,* or *delta*-Schardinger dextrins, respectively. They are of interest because of their ability to form inclusion complexes with organic molecules[89] but are not presently of major commercial interest.

The pyrodextrins are manufactured in large quantities and are commercially important. A large number of dextrin grades and types are manufactured by controlling the parent starch source, the moisture, catalyst, temperature, and time of the heat reaction.[90] Three types of chemical transformations take place during the dextrinization process.[91-93] In the presence of acid and moisture, degradation by hydrolysis takes place (Fig. 22.18). At high temperature and low moisture content, there is rearrangement involving breaking the glucoside linkage at one point in the molecule and reattaching the severed part of the chain to another point in the same or a different molecule by formation of another glucoside bond attached at the 2, 3, 4, or 6 position of the anhydroglucose unit at the attachment point. At high temperature in the presence

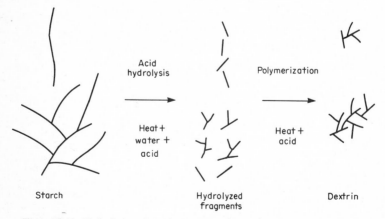

Acid hydrolysis

Heat + water + acid

Polymerization

Heat + acid

Starch

Hydrolyzed fragments

Dextrin

Figure 22.18 Hydrolysis and repolymerization during dextrinization. *(Ref. 80.)*

of acid and anhydrous conditions, the small fragments repolymerize to form larger, highly branched molecules.

Thus, the conditions of the dextrinization determine the properties of the dextrin product. If hydrolysis predominates, the dextrin obtained initially will tend to show retrogradation properties. If repolymerization and rearrangement predominate, the dextrin will be more stable to retrogradation and more soluble in cold water. Although it is more accurate to designate aqueous dextrin systems as *dispersions, sols, hydrosols,* or *hydrogels,* terms appropriate to the colloidal nature of the system, it is common practice to refer to dextrin dispersions as solutions or pastes having solubility, viscosity, and fluidity.

The pyrodextrinization process is carried out in four steps: acidification, predrying, dextrinization, and cooling. Uniformity of distribution of the acid catalyst is of primary importance and a volatile acid is usually used to facilitate this. The dilute acid is generally sprayed or atomized into the starch as it is mixed. The starch contains at least 5% moisture to minimize local specking. Usually, hydrochloric acid (0.05 to 0.15%) is used, the amount added being determined by the degree of conversion required and the nature of the native starch raw material. Time is allowed for equilibration distribution of the sprayed acid throughout the starch. The starch can also be acidified by suspension in an aqueous acid medium followed by filtration and drying. Gaseous hydrogen chloride can also be added to dry starch.

The starch may be dried prior to entering the dextrinizer, depending on the type of dextrin to be made. Where a hydrolysis reaction is desired initially, moisture must be present and the starch at normal moisture (10 to 18%) may be added directly to the dextrin cooker. The predrying would bring the moisture content down to 1 to 5%. The predrier is designed to remove moisture rapidly without overheating the starch. Sometimes an air stream is passed over the starch in the dextrinizer during the dextrinization to remove volatiles and, particularly in the initial heating period, to remove moisture more rapidly. Because dry, powdered starch is highly combustible and constitutes an explosion hazard, the usual bulk dextrin cooker is heated indirectly with high-pressure steam or hot oil in jacketed walls or by coils or by passage through hollow agitators. In a bulk dextrin cooker, containing as much as 4,500 kg (about 10,000 lb) of starch, it is essential that the starch be heated uniformly and that large temperature gradients throughout the starch be avoided. The heating rate must also be adjusted so that the starch does not char in contact with the heating surfaces. Some continuous processes have also been employed via heated, endless stainless steel belts or rotary drier type units.

The dextrin conversion is carried out with close attention to the rate of heating (temperature schedule), the maximum temperature, and total time. Hydrolysis is the major reaction in the initial stage of cooking, and the viscosity is reduced rapidly in this stage to near the level of the finished dextrin. If the moisture content is high and the rate of increase in temperature is slow, the resulting low-viscosity products have a high reducing sugar content and generally have retrogradation tendencies. If the moisture content is low and the rate of temperature increase is fast, hydrolysis is minimized and the rearrangement and repolymerization reactions are predominant, yielding low-to high-viscosity dextrins.

These products tend to be more stable (less gelling). The cold-water-solubility of the starch is minimal at the initial stage of conversion but rapidly approaches 100% when the temperature reaches the range of 130 to 145°C (266 to 293°F). Reducing sugar values, indicative of low-molecular-weight saccharides and oligosaccharides, rise to a maximum in the initial hydrolytic stage of the conversion and then decrease as higher temperatures, which initiate the repolymerization reaction, are reached. The longer the conversion time, the higher the acidity, and the higher the temperature, the darker the color of the dextrin, ranging from white to brown.

The dextrinization is continued until the product shows the desired characteristics of viscosity, solubility, and color. To stop the dextrinization reactions, the dextrin must be cooled rapidly. This is accomplished by discharging the hot dextrin from the converter to a cooling mixer or conveyor, cooled by water passing through a jacket. Sometimes the acid is neutralized at this point by the addition of ammonia or other alkaline material. However, dextrin is frequently packed without neutralization of the residual acid, and the user should keep this in mind. The dextrin is screened before packing for shipment to remove charred material and hard lumps from the dextrinization process. When the dextrin leaves the dextrinizer, it is dry and contains adsorbed gases, mainly carbon dioxide. Thus, when dispersed in water, the release of this gas causes foaming and the rapid uptake of water on the surface of the dry dextrin particle results in lumps which are very difficult to disperse. Therefore the dextrin may be rehumidified by exposure to moist air after the dextrinization process or may be allowed to pick up atmospheric moisture during the normal storage of the packed product. In the latter case, the adsorbed gases are released as the water is adsorbed.

Three types of dextrins are manufactured: white dextrins, yellow or canary dextrins, and British gums (Table 22.8). The white dextrins have a high reducing sugars value and a low viscosity, being used at the level of 1 part of dextrin to 1 to 5 parts of water (Fig. 22.17). They generally require cooking to disperse in water. The yellow dextrins range in color from cream to dark yellow or brown. They are highly branched in structure and do not retrograde, yielding stable pastes at high solids (1 part of dextrin per ¾ to 1½ parts of water; Fig. 22.17). These products do not require cooking to disperse in water. The dispersions of corn canary dextrins show a thixotropic character, tending to be less fluid or setting up when not agitated.

British gums are made by heating with no acids or in the presence of small amounts of alkaline buffers, such as sodium phosphate, sodium bicarbonate, or sodium acetate, to neutralize acids formed at higher temperature. There is little hydrolytic degradation in the absence of acid, and since the rearrangement reactions are slow, the British gum conversion requires longer times (up to 18 h) to produce the highly branched, high-molecular-weight products. The British gums range in color from light tan to dark brown. The viscosity of British gums covers a broad range from the low viscosities of white dextrins up to the high viscosity of fluidity starches. The British gums exhibit better sol stability than white dextrins or fluidity starches at the same viscosity.

Dextrins can be made from all the commercial starches, the ease of conversion and quality of the dextrins varying with the type and quality of the raw starch. To obtain a high-quality dextrin requires a high-quality base starch with a low level of impurities, protein, and buffering components, which make acidification difficult. Tapioca starch gives high-quality dextrins, converting easily to yield products that give dispersions of excellent clarity, stability, and bland taste, a major factor in food applications.

Potato starch also converts easily to give dextrins of good solution clarity, stability, and adhesiveness. There is an inherent problem in that potato dextrins tend to have disagreeable taste and odor characteristic of potato starch. However, the odor has been minimized or eliminated in certain imported dextrins which have been given special treatment. Waxy corn and waxy sorghum starches yield high-quality dextrins almost comparable to tapioca dextrins with good stability and excellent adhesion properties. Corn dextrins have good adhesion properties, but the dispersions do not show the clarity of tapioca and potato dextrins and tend to thicken rapidly on storage.

When dextrins are used as the major component in an adhesive or binder system, additives may be added to modify the properties of the dextrin. Solubilizers such as urea, dicyandiamide, thiocyanates, sodium nitrate, and iodides may be added to decrease viscosity and setback of the pastes as well as hold the dextrin in solution in the film. Humectants such as sorbitol, glycerol, sugar, corn syrup, and other polyhydroxy compounds retain moisture in the dextrin film and prevent it from becoming brittle at low humidities. However, the quantity of these materials must be limited to provide flexibility without stickiness or blocking. Fatty compounds such as sulfonated castor oils or soluble soaps may be used as film lubricants.

Borax (sodium tetraborate), added in amounts up to 20% on the dextrin, increases the viscosity and stability of the dispersion and makes it more cohesive, hence, tackier. Too much borax will make the dextrin difficult to spread as an adhesive film and will cause rubberiness and a loss in tackiness. Addition of sodium hydroxide (0.5 to 1.5% on the dextrin) will enhance the borax effect by converting the borax to metaborate, increasing viscosity, stability, and tack and increasing the penetration and bite of the dextrin adhesive into the substrate. To improve water-resistance of dextrin adhesive films, resins such as urea-formaldehyde are added (10 to 20%), and the paste adjusted to pH 4.5 to 6 with acids to foster the crosslinking in the film. Preservatives are necessary in aqueous dextrin dispersions to prevent spoilage in the stored paste as well as in the film.

Modification by Derivatization   In addition to the crosslinking effects produced by very low levels of derivatization by multifunctional agents, derivatization by monofunctional reagents can impart resistance to retrogradation and gelling, lower gelatinization temperature, increase viscosity, impart improved colloidal properties such as emulsifying capability, and introduce functional groups that change the polarity of the starch as well as the hydrophilic or hydrophobic character (Table 22.7).

To fully describe a starch derivative the following factors must be included:

1. Nature of the substituent group
2. Degree of substitution
3. Molar substitution
4. Physical form
5. Plant source
6. Starch composition in terms of amylose/amylopectin content
7. Presence of associated materials such as protein, fatty acids, and phosphorus compounds or substituents
8. Molecular weight distribution
9. Prior treatment, such as dextrinization, acid modification, or oxidation

Usually, not all this information is available.

As mentioned before, derivatives are usually esters or ethers or formed by oxidation or graft polymerization. Although a large number of derivatives of starch have been prepared, only a few are produced commercially. This section will deal only with those of commercial interest. Information on the multitude of other starch derivatives may be obtained from the literature.[94,95] It should be noted that commercial interest is generated when a need for a product becomes economically justifiable in a cost-benefit relationship. Thus, derivatives which may be only of academic or scientific interest could become commercially interesting if an inexpensive manufacturing method were developed, if the material were required for an essential property that no other material could match at the price, or the starch derivative had a unique characteristic.

Interest in starch derivatives generally centers on the water-dispersible products, and since starch is inherently water-dispersible, it is not necessary to impart water-dispersibility by high degrees of substitution as is required for cellulose. Most commercial starch derivatives are in the range of D.S. 0.02 to 0.1 substituents per AGU. By derivatizing converted starches or converting the starch derivative where possible, a wide range of viscosity grades can be made. The sequence of treatments is determined by the stability of the substituent groups to the treatment, facilities for handling and treatment, and the physical and chemical properties of the intermediates. The properties of the derivatives can also be modified somewhat by the choice of raw starch to be derivatized. The use of combination treatments is also of value in obtaining the desired properties. Crosslinking in combination with other modifications, such as esterification or etherification, is often employed. As mentioned previously, most derivatives can be put in cold-water-dispersible form by pregelatinization on heated rolls.

Process of Derivatization.   The derivative reactions are usually carried out by treating the starch in aqueous suspension (35 to 45% solids) with the reagent, maintaining a pH of 7 to 12, depending upon the type of reaction. Conditions are chosen so that the starch does not gelatinize and the reaction product can be isolated in granule form, washed, and dried. The reaction may be run by controlling the pH by metered addition of a 3% sodium hydroxide solution controlled by a pH-sensing device, as in acetylation with acetic anhydride, or all the required alkali may be added initially, as in etherification with ethylene oxide. The swelling of the starch under strongly alkaline conditions or because of derivatization can be controlled by the presence of high concentrations (10 to 30%) of sodium chloride or sulfate in the reaction mixture.[95]

Since there is a limit to the repression of the swelling of starch, the gelatinization temperature being lowered as the D.S. increases, the degree of modification that can be made in aqueous slurry while retaining the granular form is also limited. Higher D.S. products containing hydrophilic groups tend to become cold-water-dispersible because the introduction of the substituents weakens the granule structure, so that the granules swell on contact with water. To obtain higher D.S. cold-water-dispersible products in granular form, derivatization reactions may be carried out in a nonswelling solvent like isopropanol (or mixtures with water) or by blending the reagents with "dry" starch (5 to 20% moisture) without a solvent and heating [up to 150°C or (115°F)].

The reagents can also be impregnated on the starch by spraying a wet filter cake prior to drying or by suspending the starch in a solution of the reagents, filtering and drying prior to heat reaction. Derivatization reactions can also be run on dispersed starch and the products recovered by drum drying or spray drying either directly or after dialysis to remove low-molecular-weight by-products. Alternatively, hydrophobic products will precipitate from the aqueous reaction mixture and can be isolated by filtration.

Stabilization. Aqueous dispersions of starches that contain amylose tend to increase in viscosity and/or gel on cooling or aging because of the tendency of the linear amylose molecules to associate and become insoluble. This aggregation tendency can be minimized or eliminated by introducing substituent groups along the amylose molecule to interfere with the alignment and association process of retrogradation. This is accomplished via the introduction of a few ester or ether groups by reacting some of the hydroxyl groups with monofunctional reagents (Fig. 22.19).

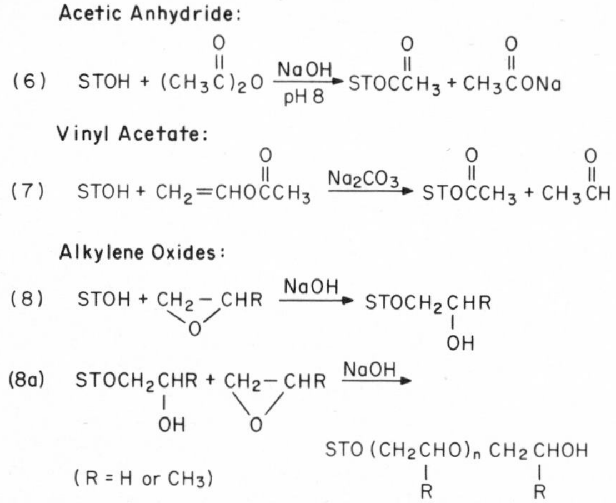

Acetic Anhydride:

$$(6) \quad STOH + (CH_3C)_2O \xrightarrow[pH\,8]{NaOH} STOCCH_3 + CH_3CONa$$

Vinyl Acetate:

$$(7) \quad STOH + CH_2{=}CHOCCH_3 \xrightarrow{Na_2CO_3} STOCCH_3 + CH_3CH$$

Alkylene Oxides:

$$(8) \quad STOH + CH_2{-}CHR \xrightarrow{NaOH} STOCH_2CHR$$
$$\underset{O}{\diagdown\diagup} \qquad\qquad\qquad \underset{OH}{|}$$

$$(8a) \quad STOCH_2CHR + CH_2{-}CHR \xrightarrow{NaOH}$$
$$\underset{OH}{|} \qquad\qquad \underset{O}{\diagdown\diagup}$$

$$STO\,(CH_2CHO)_n\,CH_2\,CHOH$$
$$(R = H\ or\ CH_3) \qquad \underset{R}{|} \qquad\qquad \underset{R}{|}$$

Figure 22.19   Stabilization reactions. Reaction of monofunctional reagents with starch.

Although dispersions of the waxy corn and sorghum starches, which contain no amylose, are stable to gelling, some instability is noticed when these dispersions are subjected to low-temperature storage or freezing. This is shown by the tendency of these starch dispersions to lose clarity and hydration ability, exuding water (syneresis), and to gel during storage at or below freezing temperature. Since these waxy starches are used to a considerable extent in food products, these changes are detrimental to the texture, taste, and eye appeal of the food, and must be prevented. This is accomplished by low D.S. derivatives which are cleared under the appropriate Food and Drug Administration regulations for "Food Starch—Modified."[96]

Low D.S. (0.05 to 0.1) granular corn starch acetates and hydroxyethyl or hydroxypropyl ethers are manufactured in large quantities. These products have the noncongealing dispersion properties useful in such applications as paper sizing and coating and textile warp sizing. The acetates and hydroxypropyl ethers of corn, tapioca, waxy corn, and waxy sorghum starches are used in foods as thickeners and stabilizers. These products are also usually crosslinked to provide the desired processing and texture characteristics. The tendency of a corn starch dispersion to gel decreases with increasing level of substitution and is eliminated under normal conditions when the substitution level for acetylation reaches a D.S. of 0.05 (Fig. 22.20). Thus, the viscosity of a corn starch acetate cook does not increase as much as that of the unmodified corn starch when it is cooled.

The introduction of substituents into the starch molecules in the granule interferes with the associative hydrogen bonding that holds the granule together. Hence, the weakened granule will gelatinize at a lower temperature. The gelatinization temperature is progressively lowered as the level of substitution is increased, and if enough substituents are introduced, the gelatinization temperature is lowered to room temperature and the starch derivative becomes "soluble" in cold water. Thus, as the D.S. of

Figure 22.20 Effect of acetyl levels on tendency of corn starch to form gels. *(National Starch and Chemical Corporation.)*

a corn starch acetate increases from 0.04 to 0.08 to 0.12, the gelatinization temperature, as measured by initiation and completion of loss of granule birefringence on a Kohler hot stage, goes from 56 to 63°C (133 to 145°F) to 48 to 56°C (118 to 133°F) to 41 to 51°C (106 to 124°F). Similarly, hydroxyethyl corn starch at D.S. 0.04 showed a gelatinization temperature of 57.5 to 67.5°C (135.5 to 153.5°F), which dropped to 45.5 to 54.5°C (114 to 130°F) at 0.12 D.S. The raw corn starch showed a gelatinization temperature of 62 to 72°C (144 to 162°F).[47] It should be noted that the birefringence loss gelatinization temperature of a crosslinked starch is not significantly different from the raw starch. Uniformity of substitution of a starch ester or ether can be qualitatively estimated by observing the loss of birefringence on the microscope hot stage. Any granules which remain unswollen (birefringent) when most of the others have lost birefringence are obviously not derivatized to any significant extent.[47]

This phenomenon of decreasing gelatinization temperature with increasing D.S. is used to lower the high gelatinization or dispersion temperatures of high-amylose starches and isolated amylose to permit dispersion by cooking at the usual boiling-water temperatures and normal cooking procedures instead of requiring temperatures in the 135 to 160°C (273 to 320°F) range. The amylose so treated does not retrograde as rapidly or, if enough substituent is introduced, at all. Thus, a D.S. of about 0.05 to 0.15 in acetate or hydroxyalkyl ether is sufficient to yield an amylose derivative that will disperse at boiling-water bath temperature.[97 98] The treatment does decrease the gel strength and film strength of the amylose or high-amylose starch.

The starch acetates are manufactured by treatment of the aqueous starch suspensions with acetic anhydride or vinyl acetate.[37,75,99-101] The acetic anhydride reaction is carried out at a pH controlled at 8 to 8.4 at room temperature with slow addition of the anhydride. The vinyl acetate reaction is run at about 35 to 40°C (95 to 104°F) at pH 9 to 10 (sodium carbonate, trisodium phosphate). Reaction efficiency of both reactions is about 70%. The products are recovered by neutralization to pH about 5 with dilute mineral acids (hydrochloric, sulfuric), filtration, washing, and drying. Granular starch acetates can also be made by heating dry starch in glacial acetic acid at 90 to 120°C (194 to 248°F) for 8 to 12 h. Products of this type were sold in 1910.

Acetic anhydride alone or in combination with glacial acetic acid can be used to make starch acetates, usually with strong acid or salt catalysts, reaching D.S. levels of 2 to 2.5. These products tend to be degraded. These acid processes are not currently commercial. High D.S. acetates including amylose and amylopectin acetates have not been developed commercially (D.S. 1.0 = 21.1% acetyl; D.S. 2 = 35% acetyl; D.S. 3 = 44.8% acetyl; D.S. 0.1 = 2.5% acetyl). The solubility characteristics of the starch acetates depend upon the D.S., the degree of polymerization (D.P.), and the type of treatment given for acetylation. Whole starch acetates up to about 15% acetyl tend to be soluble in hot water [50 to 100°C (122 to 212°F)]. The higher D.S. derivatives are insoluble in water. Degradation tends to increase water solubility.

The hydroxyalkyl starches are ethers and the substituent group is resistant to cleavage by acids, alkalies, and mild oxidizing agents. Hence, the hydroxyalkyl derivatives can be subjected to conversion by acid, oxidation or dextrinization to produce lower-viscosity products. The starch esters, on the other hand, are subject to cleavage by alkaline hydrolysis, which is relatively rapid. Deacetylation of a granular starch acetate (1.8% acetyl) is complete within 4 h at pH 11, maintained constant by addition of 3% sodium hydroxide solution. Thus, starch acetates should not be used under alkaline conditions and should not be subjected to conversions.

Low D.S. granular hydroxyalkyl starch ethers are prepared by treating aqueous starch suspensions with alkylene oxides (ethylene oxide, propylene oxide) in the presence of alkalies (sodium hydroxide, calcium hydroxide) at temperatures up to 50°C (122°F). Sodium sulfate or sodium chloride are required in high concentrations to prevent swelling of the starch so that the granular starch remains filterable. At higher degrees of substitution, the hydroxyalkyl starch will tend to swell as the salts are washed out of the cake during recovery and purification.[37,94,95,102,103]

Since ethylene and propylene oxides are flammable and form explosive mixtures with air over a broad composition range, care must be taken in running large-scale reactions. For safety, the reaction should be run in a closed-system pressure reactor under a nitrogen blanket and steps taken to ensure that there is no residual epoxide before further processing. Generally there have been few problems in large-scale processing over the years of production.

Low and high D.S. products can also be made by reacting commercially dry starch with alkylene oxide vapors.[103,104] Highly substituted, ungelatinized hydroxyalkyl starch ethers can be prepared by reaction of alkylene oxides with starch suspended in solvents such as isopropanol.[105] Since the alkylene oxide can react further with the hydroxyalkyl ether group already attached to the starch to form polyoxyalkylene graft polymers, the substitution may be reported as molar substitution (M.S.) in place of D.S. The M.S. is defined as the average number of moles of alkylene oxide combined per mole of anhydroglucose unit.

The starch acetates and the hydroxyethyl and hydroxypropyl starch ethers are nonionic and their dispersions are therefore not subject to the solubility and viscosity effects dissolved electrolytes have on polyelectrolyte polymers. Hydroxyethyl and hydroxypropyl starches are cold-water-swelling at M.S. 0.4 to 0.5, with higher substitution levels giving increasingly rapid cold water gelatinization. The cold-water-dispersibility properties of starch derivatives of high D.S. in granule form differ from those of the pregelatinized starches in that the granular starch derivatives tend to wet out easily and disperse and swell readily and gradually without the lumping so noticeable in the pregelatinized starches. As the hydroxyalkyl M.S. becomes greater than 1.0, the starch ether derivative becomes soluble in the lower alcohols, methanol, ethanol, etc., while still retaining solubility in water. The higher D.S. hydroxyethyl and hydroxypropyl starches begin to show thermoplasticity above D.S. about 1.

The distribution of hydroxyethyl groups in a sample of commercial hydroxyethyl corn starch, D.S. approximately 0.1, was found to be about 85% on carbon atom 2 and about 15% on carbon atom 6 of the anhydroglucose units substituted, with only a trace substitution on carbon atom 3.[105,106] Similar results were obtained with an hydroxyethyl starch of M.S. 0.6 (D.S. estimated at 0.5) with an indicated 82% hydroxyethyl on carbon atom 2, 15% on carbon atom 6, and 2.9% on carbon atom 3 with very little polysubstitution (more than 1 hydroxyethyl group on an AGU) and little polyoxyethylene substituent (reaction of the ethylene oxide on an hydroxyethyl group already substituted on the starch).[107]

However, there is some evidence that at M.S. = 0.45, complex polyoxyethylene substitution reaches a measurable level, and at M.S. 0.8, some 15% of the hydroxyethyl groups are present as polyoxyethylene groups. For hydroxyethyl amylose, at M.S. values of 0.6, 0.9, and 1.25, the corresponding D.S. values were 0.56, 0.69, and 0.77. Thus, depending on preparative method, the hydroxyethyl starch can have varying M.S. and D.S. values and this must be taken into account in evaluating and comparing samples. This is of importance in considering these derivatives for use as blood plasma volume expanders.[108]

New Functional Properties. Stabilization effects can be produced with reagents that also introduce other chemical groups which affect the viscosity, dispersibility, and other functional properties of the starch (Figs. 22.21 and 22.22). Starch derivatives containing

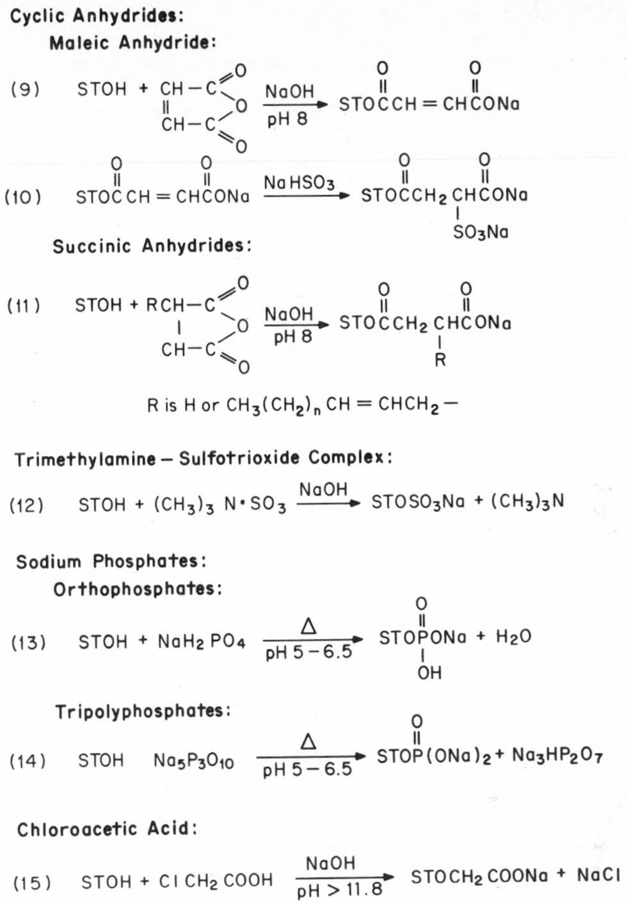

Cyclic Anhydrides:
Maleic Anhydride:

(9) $\quad$ STOH $+$ $\begin{array}{c} CH-C \\ \| \\ CH-C \end{array}$ $\overset{O}{\underset{O}{<}}$ O $\xrightarrow[\text{pH 8}]{\text{NaOH}}$ $STO\overset{\overset{O}{\|}}{C}CH=CH\overset{\overset{O}{\|}}{C}ONa$

(10) $\quad$ $STO\overset{\overset{O}{\|}}{C}CH=CH\overset{\overset{O}{\|}}{C}ONa$ $\xrightarrow{NaHSO_3}$ $STO\overset{\overset{O}{\|}}{C}CH_2\underset{\underset{SO_3Na}{|}}{CH}\overset{\overset{O}{\|}}{C}ONa$

Succinic Anhydrides:

(11) $\quad$ STOH $+$ $\begin{array}{c} RCH-C \\ | \\ CH-C \end{array}$ $\overset{O}{\underset{O}{<}}$ O $\xrightarrow[\text{pH 8}]{\text{NaOH}}$ $STO\overset{\overset{O}{\|}}{C}CH_2\underset{\underset{R}{|}}{CH}\overset{\overset{O}{\|}}{C}ONa$

R is H or $CH_3(CH_2)_n\,CH=CHCH_2-$

Trimethylamine – Sulfotrioxide Complex:

(12) $\quad$ STOH $+$ $(CH_3)_3\,N\cdot SO_3$ $\xrightarrow{\text{NaOH}}$ $STOSO_3Na + (CH_3)_3N$

Sodium Phosphates:
Orthophosphates:

(13) $\quad$ STOH $+$ NaH_2PO_4 $\xrightarrow[\text{pH 5}-6.5]{\Delta}$ $STO\overset{\overset{O}{\|}}{\underset{\underset{OH}{|}}{P}}ONa + H_2O$

Tripolyphosphates:

(14) $\quad$ STOH $\quad$ $Na_5P_3O_{10}$ $\xrightarrow[\text{pH 5}-6.5]{\Delta}$ $STO\overset{\overset{O}{\|}}{P}(ONa)_2 + Na_3HP_2O_7$

Chloroacetic Acid:

(15) $\quad$ STOH $+$ $Cl\,CH_2\,COOH$ $\xrightarrow[\text{pH} > 11.8]{\text{NaOH}}$ $STOCH_2\,COONa + NaCl$

Figure 22.21 Functional group introduction: anionic groups.

hydrophilic groups such as carboxylate, sulfonate, sulfate, and phosphate in the form of sodium, potassium, and ammonium salts have a high affinity for water and give high-viscosity, clear, non-gelling dispersions. Of course these ionic derivatives are affected by the presence of other electrolytes in the solution which modify their solubility and viscosity.

By using cyclic dibasic acid anhydrides, such as succinic and maleic anhydrides, in the esterification of granular starch under conditions similar to those used with acetic anhydride, the derivative is a half-ester of the dibasic acid with a free carboxylate group.[100] Starch esters of this type include starch succinate, phthalate, and maleate half-esters. The latter can be treated further with sodium bisulfite to yield the sulfosuccinate half-ester with the enhanced water-taking sodium sulfonate group, which yields a much higher viscosity dispersion for given solids.[109] The succinate half-ester of starch has been cleared for use in foods.[96]

Starch sulfates have been made using a number of complexes of sulfur trioxide with tertiary amines to treat starch dispersed in dimethylformamide, pyridine, or suspended in aqueous alkali, yielding a D.S. of up to 0.38 with starch in granular form and D.S. 1.4 to 2 for amylose and amylopectin sulfates.[37,94,95,110,111] Sulfur trioxide complexes with dimethylformamide or dimethylsulfoxide have also been used.[112,113] The starch sulfates are usually recovered as the sodium salts and are readily dispersible in water,

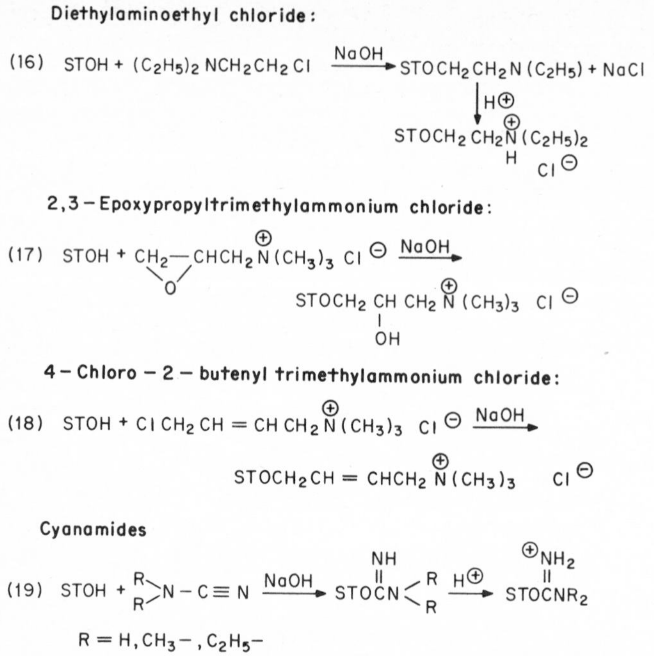

Diethylaminoethyl chloride:

(16) $STOH + (C_2H_5)_2 NCH_2CH_2 Cl \xrightarrow{NaOH} STOCH_2 CH_2 N (C_2H_5) + NaCl$

$$\downarrow H^{\oplus}$$

$$STOCH_2 \overset{\oplus}{CH_2 N} (C_2H_5)_2$$
$$\underset{H \quad Cl^{\ominus}}{}$$

2,3 − Epoxypropyltrimethylammonium chloride:

(17) $STOH + CH_2\!-\!\!-\!CHCH_2 \overset{\oplus}{N}(CH_3)_3 \; Cl^{\ominus} \xrightarrow{NaOH}$

$$STOCH_2 \underset{\underset{OH}{|}}{CH} CH_2 \overset{\oplus}{N} (CH_3)_3 \; Cl^{\ominus}$$

4 − Chloro − 2 − butenyl trimethylammonium chloride:

(18) $STOH + Cl CH_2 CH = CH CH_2 \overset{\oplus}{N}(CH_3)_3 \; Cl^{\ominus} \xrightarrow{NaOH}$

$$STOCH_2 CH = CHCH_2 \overset{\oplus}{N}(CH_3)_3 \quad Cl^{\ominus}$$

Cyanamides

(19) $STOH + \underset{R'}{\overset{R}{>}} N - C \equiv N \xrightarrow{NaOH} STOC\underset{\overset{||}{N}}{\overset{NH}{<}}\underset{R}{R} \xrightarrow{H^{\oplus}} STOC\underset{}{\overset{\overset{\oplus}{NH_2}}{\overset{||}{N}}}R_2$

$R = H, CH_3-, C_2H_5-$

Figure 22.22 Functional group introduction: cationic groups.

showing good dispersion stability under neutral or alkaline conditions. Under acid conditions, the ester is hydrolyzed rapidly, liberating sulfuric acid, which degrades the starch. This is an autohydrolytic process. Starch sulfate inhibits pepsin and other enzymes[95] and is reported to be nontoxic when taken orally.[114]

In contrast to the crosslinked phosphate esters produced by treating starch with phosphorus oxychloride or sodium trimetaphosphate, non-crosslinked starch monophosphate esters are made by heating starch with the sodium acid salts of ortho-, pyro-, and tripolyphosphoric acids.[37,75,94,115-117] The starch is impregnated with the phosphate salt solution by suspending the starch in the solution and then filtering, by spraying the solution on the dry starch, or by mixing the solution with the wet filter cake. The phosphate solutions are about pH 5 to 8.5, depending upon the particular phosphate used. The phosphate-impregnated starch is dried and then heated at temperatures of 120 to 175°F (248 to 347°F) for 0.5 to 3 h or more. The phosphation generally runs to a D.S. 0.02 to 0.15. The phosphates can be made in a range of fluidities by adjusting pH, temperature, and time of reaction.

Although the starch phosphates give a higher-viscosity dispersion than the parent starch, the viscosity of these starches is drastically reduced in the presence of salts. The starch phosphates tend to become cold-water-dispersible at D.S. greater than about 0.07. Analysis of a granular corn starch phosphate with a D.S. of 0.016 (0.3% P), prepared by heating with sodium tripolyphosphate at 150°C (302°F), indicated that the disodium phosphate ester groups were distributed as 63% on carbon atom 6, 28% on carbon atom 2, and 9% on carbon atom 3.[118]

It is interesting to note that potato starch contains phosphate ester groups (about 0.07% P) which are primarily attached to the amylopectin (0.07 to 0.09% P vs. 0.002 to 0.004% P in the amylose). The phosphate is on carbon atom 6 of the anhydroglucose unit. The D.S. is about 0.0025 to 0.005 in amylopectin and about 0.0002 in amylose. Since the potato starch and amylopectin are polyelectrolytes, dispersion viscosity will be decreased in the presence of salts.[118a,118b]

Highly water-soluble carboxymethyl starch ether, usually obtained as the sodium salt, is manufactured by the reaction of starch with chloroacetic acid or sodium chloroacetate in the presence of aqueous sodium hydroxide. Low D.S. products (up to D.S. 0.1) can be obtained in granular form by reaction in aqueous alkaline suspension in the presence of sodium sulfate or in aqueous alcohol. By crosslinking before the carboxymethylation reaction, a slightly higher D.S. granular product can be obtained.[94,95,119]

Higher D.S. carboxymethyl starch derivatives can be made by reaction in water-miscible solvents (isopropanol) in the presence of small amounts of water.[119a,120] Alternatively, an ungelatinized, cold-water-dispersible carboxymethyl starch can be made by blending the starch (12% moisture or less) with alkali and sodium chloroacetate and reacting at room temperature or higher [up to 85°C (185°F)] for a time determined by the reaction temperature.[121] The higher D.S. cold-water-dispersible products will contain a high sodium chloride content, a by-product of the reaction, unless they are purified by alcohol extraction or dialysis.

Carboxymethyl starch can be insolubilized by reaction with polyvalent ions, such as aluminum, ferric, chromic, and cupric ions, leading to precipitation or gelling of dispersions or insolubilization of films.

Cationic starch ether derivatives are manufactured in large quantities and are used mainly by the paper industry as strength and retention aids.[75,122] These derivatives are prepared by treating starch with reagents that contain tertiary amino or quaternary ammonium groups such as 2-diethylaminoethylchloride hydrochloride,[123-125] 2,3-epoxypropyltrimethylammonium chloride,[126] 4-chloro-2-butenyltrimethylammonium chloride,[127] chloropropyltrimethylammonium chloride,[128] and 2,3-epoxypropyldiethylamine.[123]

Cationic starch has also been made by reaction with cyanamide[129] or dialkylcyanamides,[130] yielding pseudourea or imino disubstituted carbamates. These products and the cationic ether derivatives are usually made by treating the starch in aqueous alkaline suspension with the reagent at temperatures up to about 50°C (122°F) for 8 to 16 h and then adjusting the pH to about 3 to 7 with dilute mineral acid, filtering, washing, and drying. The products can also be made in organic solvents or as "semi-dry" reactions. The reactions can also be run on gelatinized starch.

These products are generally low D.S. granular products (D.S. 0.05 or less) which are available in a range of viscosities, yielding stable, clear, fluid dispersions. The quaternary ammonium ion is very effective in lowering gelatinization temperature, cold water swelling of the granule beginning at a D.S. of about 0.07 vs. a D.S. of 0.4 for hydroxyethyl starches.

Some primary amino derivatives of starch have been prepared on a semicommercial scale using the reaction of ethyleneimine vapor with dry starch at temperatures up to 120°C (248°F). The products were in granular form and had value as pigment retention aids in paper-making.[131,131a,131b]

Because the cationic starches have a positive charge, they are attracted to, and retained by, anionic substrates, making them useful for pigment flocculation and retention in the paper-making process as well as for increasing the fiber-to-fiber bonding strength of paper. They also find use in paper sizing and coating, as well as in textile warp sizing.

The introduction of hydrophobic groups at low D.S. levels (0.01 to 0.1) imparts some hydrophobic properties to the starch without destroying the water-dispersibility. The hydrophobic-hydrophilic balance imparts useful emulsifying and emulsion stabilization properties. Thus, alkenylsuccinate half-esters of starch are made by esterification procedures involving cyclic dibasic acid anhydrides.[132] If these granular starch esters are treated with water-soluble salts of polyvalent ions, such as copper, aluminum, iron, chromium, and calcium, they become free-flowing and water-repellent. These free-flowing powders are useful as no-offset dry sprays for graphic arts work and other dusting applications.[133]

To disperse these products in water, it is necessary to moisten them with a water-miscible solvent such as ethanol or to use a high-shear mixer to suspend the starch uniformly in the water in which the starch is to be heated to disperse it. These metal ion-treated starch esters can be used as encapsulating agents for flavoring agents and perfumes.[134] Films from aqueous dispersions of this starch have some degree of water-repellency. The octenylsuccinate half-ester of starch and the corresponding reaction product with aluminum sulfate have been cleared as a "Food Starch—Modified" under FDA regulations for use in foods.[96]

A starch benzyl ether, substituted to a sufficient degree to prevent gelatinization when cooked in water at temperatures up to 100°C (212°F) (D.S. about 0.2), develops some interesting properties when dispersed in water at superatmospheric pressure in a continuous steam-injection cooker at 150°C (302°F). The resulting dispersion is a fluid sol containing submicron-size particles (most of the particles are in range of 0.03 to 0.05 μm). The sol is opaque and there is no settling on standing for several months. When cast as a film and dried, a clear, continuous, water-resistant film results. The benzyl starch is made by treating starch, in aqueous alkaline suspension containing sodium sulfate swelling repressant, with benzyl chloride (10% on starch) at 50°C (122°F).[135] This material was used as a sizing for glass fiber.

Dialdehyde starch is prepared by treatment with periodic acid which selectively oxidizes the adjacent hydroxyl groups on carbon atoms 2 and 3 to aldehyde groups (Fig. 22.23).[37,136] All levels of oxidation up to 100% of the anhydroglucose units in the starch can be made.

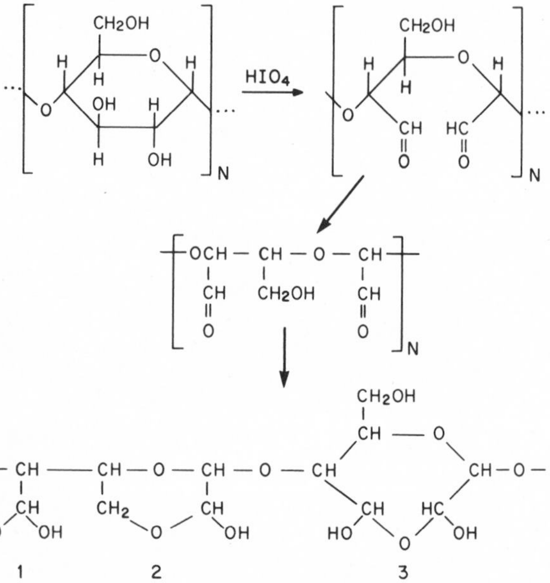

1 = Hydrated carbonyl
2 = Hemiacetal form (intramolecular; can be intermolecular)
3 = Hemialdal form

Figure 22.23 Dialdehyde starch.

The commercial process for manufacturing dialdehyde starch involves a two-stage process in which periodic acid oxidizes the starch in a separate reaction vessel and the spent oxidant solution is recycled through an electrolytic cell to regenerate the periodic acid, which is then pumped back to the reaction tank.[75,136-139] In the starch oxidation, the periodic acid is reduced to iodic acid, which is electrolytically oxidized back to the periodic acid. The starch oxidation is carried out at pH 0.7 to 1.5. The major oxidation is completed in about 1 h, but considerably more time is necessary to reach levels greater than 95%. Hence, the commercial products generally are at a maximum of about 90% of the AGU oxidized.

A chemical process has been suggested that might be suitable for small-scale production of dialdehyde starches in a range of oxidation levels without requiring expensive electrolytic equipment for regeneration of periodic acid. In this process, the spent oxidant is converted by alkaline hypochlorite to insoluble sodium paraperiodate which is recovered in high yield (97%) by filtration for recycling.[140]

Dialdehyde starch is insoluble in water but swells to some extent on cooking at 90°C (194°F). The higher the degree of oxidation, the more difficult to disperse by cooking in water. Mild alkalinity helps dispersion of the dialdehyde starch by controlled degradation, but strongly alkaline conditions cause rapid depolymerization which is excessive. In the presence of small amounts of sodium bisulfite, the dialdehyde starch can be dispersed to give clear, fluid sols of the degraded polymer.

The aldehyde groups in the dialdehyde starch are not present as such, but are in the form of hydrated hemialdal or gem-dihydroxy structures, as well as in hemiacetal structures. This probably is the cause of the difficulty in dispersing the starch. The derivative behaves as a polyaldehyde since the hydrated and hemiacetal forms are easily cleaved by bisulfite, amines, hydrazines, hydroxyl groups, and other agents that react with aldehydes. Chlorous acid will oxidize the aldehyde groups to carboxylic acids, yielding a polycarboxylate polymer. Conversely, reduction with sodium borohydride yields a polyalcohol. Hydrolysis yields glyoxal and a sugar, erythrose.

~The major uses of dialdehyde starch reflect the reactive nature of polymeric polyaldehyde in crosslinking substrates containing amino, hydroxyl, and imino groups. It has utility in hardening gelatin, tanning leather, imparting water-resistance to adhesives, and in developing wet strength in paper. A cationic dialdehyde starch was prepared for use in making wet-strength paper by treating the dialdehyde starch with (carboxymethyl)-trimethylammonium chloride hydrazide, Girard T Reagent, which reacted with about 2% of the carbonyl groups. This derivative disperses more easily than the parent dialdehyde starch.[141] Unfortunately, the cost of dialdehyde starch was high, discouraging really large-scale use and the single manufacturer in the United States has discontinued offering the product.

Starch xanthates are prepared by the reaction of starch with carbon disulfide in the presence of strong alkali, e.g., sodium or potassium hydroxide (Fig. 22.24).[75,94,95] A

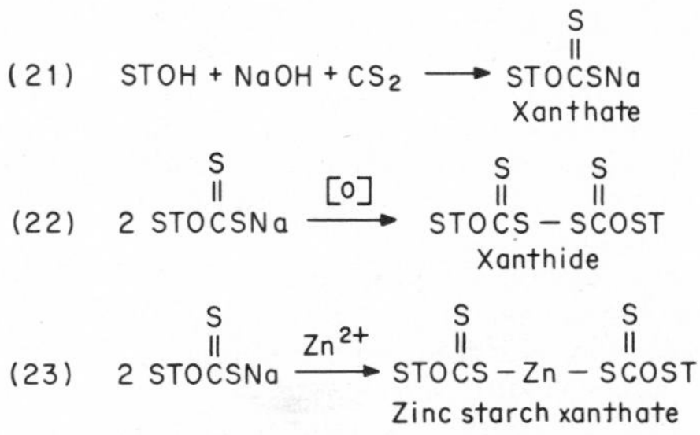

$$(21) \quad STOH + NaOH + CS_2 \longrightarrow STOC\overset{\overset{\displaystyle S}{\|}}{}SNa$$
Xanthate

$$(22) \quad 2\ STOC\overset{\overset{\displaystyle S}{\|}}{}SNa \xrightarrow{\ [O]\ } STOC\overset{\overset{\displaystyle S}{\|}}{}S - SC\overset{\overset{\displaystyle S}{\|}}{}OST$$
Xanthide

$$(23) \quad 2\ STOC\overset{\overset{\displaystyle S}{\|}}{}SNa \xrightarrow{\ Zn^{2+}\ } STOC\overset{\overset{\displaystyle S}{\|}}{}S - Zn - SC\overset{\overset{\displaystyle S}{\|}}{}OST$$
Zinc starch xanthate

Figure 22.24 Starch xanthate.

continuous process for large-scale preparation of starch xanthate, D.S. 0.07 to 0.47, has been developed.[142] The process involves metering granular starch, carbon disulfide, and sodium hydroxide solution separately and sequentially into a high-shear, screw-type mixer-reactor and discharging the viscous paste product (53 to 61% solids) after a reaction time of about 2 min.

Conversion of the carbon disulfide to xanthate is about 80 to 87% complete for D.S. 0.17 to 0.07 within 10 min of discharge and, upon standing 1 h after discharge, 90 to 93% complete for a D.S. of 0.10 to 0.29. Maximum reaction efficiency was increased by raising the reaction temperature from 26 to 38°C (79 to 100°F), by introducing the carbon disulfide before the sodium hydroxide, by increasing the $NaOH/CS_2$ mole ratio, and by using a smaller discharge orifice. The xanthation appears to be uniform with virtually no unreacted starch present and the amylose and amylopectin fractions substituted to the same degree.[143] In concentrated alkali metal hydroxide solutions, the xanthation of starch takes place initially on carbon atoms 2 and 6, with further xanthation taking place rapidly. The C_2 xanthate substitution reaches a maximum and then decreases, while the C_6 substitution continues to increase, finally leveling off and then decreasing at a slower rate.

Xanthation at C_3 is very much slower. Starch xanthates prepared by the continuous process at D.S. 0.12 and 0.33 showed a $C_6/C_2/C_3$ xanthate substitution of 56/44/0 and

67/27/6 ratio, respectively. This shows that the primary hydroxyl group on carbon atom 6 is most readily xanthated under the continuous reaction conditions used.[144,145] Considerable degradation of the starch occurs during the short-time continuous reaction, presumably because of the high mechanical shearing action. Potassium hydroxide is more effective than sodium hydroxide, the xanthation proceeding more rapidly with the former alkali.[146]

To produce a starch xanthate (D.S. 0.12 maximum) suitable for use at the site of application without the need of an expensive high-shear mixer-reactor, a continuous process was developed which discharges a low-viscosity starch xanthate solution that can be used directly in the paper-making process.[147] A 10% starch slurry is cooked to disperse the starch and the dispersion is metered into a receiving tube where it is mixed with carbon disulfide and a 40% sodium hydroxide solution, also metered in, and the reaction mixture is passed through a polyethylene holding tube to give a reaction time of about 10 min. The reaction is run at ambient temperature, the exit temperature being about 35 to 38°C (95 to 100°F).

Starch xanthates are water-soluble and can be isolated by alcohol precipitation. Aqueous solutions of starch xanthates are unstable, undergoing hydrolysis and oxidative reactions. Isolated xanthates usually have sufficient moisture to be unstable also. Storage under low-temperature or freezing conditions has been recommended. This instability has been a major drawback in developing commercial usage, and the preparation of the material at the site of use has been recommended. By decreasing the alkali content of the xanthate, bringing the pH below 14, and spray drying, it is claimed that a stable xanthate powder can be obtained.[148]

For a D.S. less than 0.2, the starch xanthate can be spray-dried without neutralization, since the alkali used in preparation will not be too high. At higher than D.S. 0.2, larger amounts of alkali are used in the preparation and this should be decreased by neutralization with dilute or weak acids, dialysis, or ion exhange. The moisture content of the spray-dried powder should be less than 5% for stability.

Starch xanthates can be oxidized to crosslinked starch xanthides by chlorine, iodine, nitrous acid, sodium tetrathionate, nitrogen tetroxide, etc. In contrast to the starch xanthates, the xanthides are quite stable. They can be dried at 100°C (212°F) for 7 h *in vacuo* without losing sulfur. Starch xanthides are insoluble in boiling water, boiling 5% acetic acid, and boiling 1% sodium sulfite solution. In 3% aqueous sodium hydroxide, the rate of swelling and solution is slow.

Treating xanthates with heavy metal salts such as zinc chloride precipitates insoluble xanthates. Cupric ions, acting partly as oxidizing agents, precipitate black cuprous xanthate and starch xanthide. Addition of the soluble starch xanthate (D.S. 0.05 to 0.75) to a pulp in the paper-making process and then forming the insoluble xanthide or metal xanthate increased dry tensile, burst and fold strength, and wet tensile strength of the paper when added up to a level of 20% on the pulp. Tear strength is reduced. Drainage time of greaseproof pulps is reduced by addition of starch xanthates.[149,150]

Powdered elastomers (rubber) suitable for fabrication by powder processing techniques, such as extrusion and injection molding, can be made by mixing starch xanthate and elastomer latexes, and then coprecipitating by simultaneously crosslinking the starch (as xanthide or zinc xanthate) and destabilizing the latex. The precipitated curd is dried to give a friable, starch xanthide–encased rubber crumb. When milled, the rubber phase coalesces to form a continuous phase, and the starch xanthate provides a reinforcement action for the rubber. The starch xanthide products are light in color.[72,151-156] The starch xanthate is not currently being used in rubber on a large scale. This is not likely to change unless the users are willing to prepare the xanthate on the application site or starch xanthate can be prepared in a form stable enough for shipment and storage under usual conditions.

Another area of great interest in starch xanthates is their use in removal of heavy metal ions from wastewaters. This is proposed using a highly crosslinked, insoluble starch as the base for xanthation. The insoluble starch xanthate then attracts and holds the heavy metal ions (copper, mercury, cadmium, silver, cobalt, chromium, lead, iron, nickel, zinc) for removal by filtration. The insoluble starch xanthate can be used as a precoat on a filter through which the metal-containing effluent can be passed. This approach has attracted the interest of metal-plating companies who have difficulty meeting Environmental Protection Agency effluent limits.

The stability of the xanthate is claimed to be improved by preparing the magnesium salt of the insoluble starch xanthate. The magnesium starch xanthate had good room temperature storage stability for several months if the xanthate was completely in the magnesium salt form. A mixed sodium-magnesium salt (Na/Mg ratio 1.7) decomposes significantly on storage at 28°C (82°F) after 6 weeks but is quite stable at 0°C (32°F). A moisture content below 2% is critical for storage stability.[157-159] The insoluble starch xanthate is effective in lowering metal concentrations of solutions below stringent discharge limits for many industrial effluents.

The starch xanthates have also been suggested for encapsulating chemical pesticides within a starch matrix to improve safety and reduce loss of the active ingredient.[72,160] The active agent is dispersed in an aqueous starch xanthate solution, which is then crosslinked by oxidation, multivalent metal ions, or a difunctional agent such as epichlorohydrin. The entire mass becomes a gel which, on continued mixing, becomes a particulate solid. This can be dried with minimal loss of the entrapped chemical. Storage stability is good with no loss of pesticide noted in closed containers. The active agent is released from the matrix when the product is placed in water or soil.

In many products, the desired properties are obtained by combining two or more derivatizing treatments. As mentioned previously, crosslinking is often used in conjunction with stabilization treatments. The introduction of both cationic and anionic groups in various proportions is also of commercial interest.[161-163]

Modification by Graft and Block Polymerization For the past twenty years, there has been considerable interest in block and graft copolymers of starch.[75,94,95,164,165] Most of the work has been based on the initiation of free radicals on the starch and then allowing the radical to react with polymerizable vinyl and acrylic monomers, thereby producing a polymeric chain attached to the starch. In many cases, free radicals are also formed on monomer or polymer not attached to the starch, resulting in the initiation of polymerization to produce homopolymer instead of graft polymer. Thus, a mixture of starch graft polymer and homopolymer results. The aim of graft polymerization is to obtain a high "grafting efficiency," that is, to obtain a high percentage of the total synthetic polymer as grafted polymer on the starch. The ultimate goal is 100% grafting efficiency.

Free radical initiating methods may be considered in terms of physical methods or chemical methods. Physical initiation includes irradiation, as with cobalt 60 gamma rays or an electron beam, as well as mechanochemical techniques such as mastication, ball-milling, and freezing and thawing starch dispersions. When starch is subjected to shear, the starch molecules are broken apart, producing free radical sites at the break points. If this application of shear is done in the presence of monomers, copolymerization is initiated and block copolymer is produced, attached to the starch at the site of the free radical formation. Masticating mixtures of starch with methyl methacrylate, styrene, vinyl acetate, and acrylonitrile resulted in grafting.[166,167]

Irradiation can be applied to mixtures of the starch and monomer which will yield some homopolymer but will also provide for the reaction of short-lived free radicals. On the other hand, irradiation of the starch alone and then allowing the activated starch to react with monomer will produce less homopolymer. Low temperature, low moisture content, and the absence of oxygen will favor increased stability of the free radicals. Free radicals generated by this pre-irradiation technique in starch may remain in the starch for several days at room temperature.

The most used method of chemical initiation is probably the reaction of starch with ceric salts such as ceric ammonium nitrate dissolved in dilute nitric acid (Fig. 22.25).[168] The ceric ion forms a complex with the starch, probably at the C_2 and C_3 hydroxyls, which then causes one of the hydrogen atoms on C_2 or C_3 to be oxidized while the ceric ion is reduced to cerous ion and a free radical is formed on the starch at the C_2 or C_3 carbon atom.[164,165] The bond between C_2 and C_3 is broken with the formation of an aldehyde group on the C_2 or C_3, which does not have the free radical form. The free radical may then initiate polymerization with monomer or react further with ceric ion to form another aldehyde group with destruction of the free radical. Thus, in the absence of monomer, the ceric ion will oxidize the C_2 and C_3 hydroxyl groups of the AGU to aldehyde.

Other types of initiating systems have been used, such as hydrogen peroxide–ferrous ion redox system or ozone.[169,170] In the hydrogen peroxide–ferrous ion redox system,

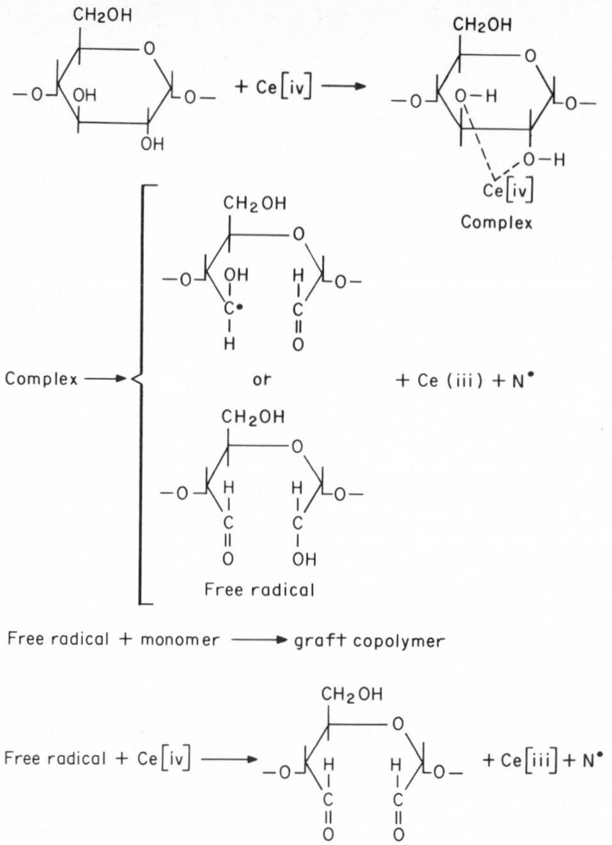

Figure 22.25   Ceric ion initiation: graft polymerization.   *(Ref. 164.)*

hydroxyl radicals are produced.   These radicals then abstract hydrogen atoms from a monomer, such as methyl methacrylate, or from starch, producing a free radical on the starch or on the monomer (Fig. 22.26).   The monomer free radical or starch free radical can react with more monomer, producing polymeric or starch graft polymeric free radicals which eventually terminate to form homopolymer or starch graft polymer. In this method, significant amounts of homopolymer are usually formed.   Starch xanthate has been used in conjunction with hydrogen peroxide as a redox system to graft onto the starch a number of monomers.[171]   Another initiating system based on manganic pyrophosphate was claimed to give less homopolymer in starch grafting.[172]

Monomers that give water-soluble starch graft polymers are acrylic acid, acrylamide, and acrylate esters containing quaternary ammonium groups.   Water-insoluble starch grafts are obtained with acrylonitrile, alkyl acrylate esters, and styrene.   Grafting has been carried out on granular and dispersed starch.

Polyacrylonitrile.   Acrylonitrile graft-copolymerizes onto starch quite easily with a high grafting efficiency by free radical initiation.   Although the starch-polyacrylonitrile graft copolymer (SPAN) is insoluble in water, it can be converted to a water-soluble form by alkaline hydrolysis of the pendant nitrile groups to form carboxylate and carboxamide groups.   The hydrolyzed starch-polyacrylonitrile graft copolymers (H-SPAN) have generated considerable interest because of their high viscosities and high water-taking properties.

The graft polymerization is carried out on a laboratory scale by gelatinizing starch (wheat, 42.4 g) in water (400 mL) by heating at 85°C (185°F) for an hour while sparging

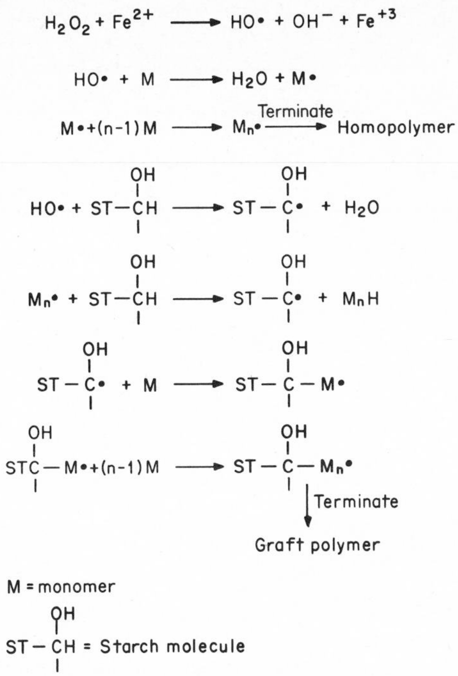

$$H_2O_2 + Fe^{2+} \longrightarrow HO\bullet + OH^- + Fe^{+3}$$

$$HO\bullet + M \longrightarrow H_2O + M\bullet$$

$$M\bullet + (n-1)M \xrightarrow{\text{Terminate}} M_n\bullet \longrightarrow \text{Homopolymer}$$

$$HO\bullet + ST\overset{OH}{\underset{|}{-}}CH \longrightarrow ST - \overset{OH}{\underset{|}{C}}\bullet + H_2O$$

$$M_n\bullet + ST\overset{OH}{\underset{|}{-}}CH \longrightarrow ST - \overset{OH}{\underset{|}{C}}\bullet + M_nH$$

$$ST - \overset{OH}{\underset{|}{C}}\bullet + M \longrightarrow ST - \overset{OH}{\underset{|}{C}} - M\bullet$$

$$ST\overset{OH}{\underset{|}{C}} - M\bullet + (n-1)M \longrightarrow ST - \overset{OH}{\underset{|}{C}} - M_n\bullet$$

$$\downarrow \text{Terminate}$$

Graft polymer

M = monomer

$$ST\overset{OH}{\underset{|}{-}}CH = \text{Starch molecule}$$

Figure 22.26 Chain transfer graft polymerization (hydrogen peroxide–ferrous ammonium sulfate initiation-activator system). *(Ref. 169.)*

with nitrogen to remove air (oxygen). After cooling to 25°C (77°F), ceric nitrate reagent (14.6 mL of 0.1 M ceric ammonium nitrate in 1 N nitric acid) is added and, after a 20-min initiation period, acrylonitrile (61.6 g, redistilled to remove polymerization inhibitors) is added to the reaction mixture, still kept in a nitrogen atmosphere. The polymerization is exothermic and the temperature is held at 30 to 33°C (86 to 91°F) by cooling. After a 2- to 3-h reaction period, the product is recovered by filtration and washed with water to remove acid and unreacted acrylonitrile (yield of air-dried solid, 73.5 g).[173,174,175]

Granular starch can also be used and may first be heated in water at 50°C (122°F) for an hour to swell before cooling to 25°C (77°F) for the polymerization reaction.[174] The polymerization reaction may be stopped by the addition of hydroquinone or by dilution with water and separation of the product. A continuous process for SPAN preparation with ceric ion catalysis was developed.[176] The process involves simultaneous addition of deoxygenated acrylonitrile-starch dispersions and aqueous ceric ammonium nitrate solution at controlled rates to the top of a stirred reactor column and, after a few minutes retention (8 to 10 min), discharging to a holding tank prior to filtration, washing, drying, and grinding.

The polymerization reaction conditions influence the molecular weight of the grafted chains and the grafting frequency, which is the average number of anhydroglucose units (AGU) separating each grafted chain (AGU/graft). The weight of the grafted polymer per mole of AGU divided by the number average molecular weight of the grafted polymer gives the moles of grafted side chain per mole of AGU. This is equivalent to the D.S. calculated for starch derivatives. The reciprocal of D.S. is equivalent to the number of AGU per substituent or the grafting frequency. The average molecular weight of the grafted polymeric side chain is determined after removal of the starch (by acid and/or enzymic hydrolysis with or without periodic acid oxidation).[169,170,174,177–179,183,185] The weight of the grafted polymer/AGU can be obtained from the percent synthetic polymer in the starch graft copolymer (add-on percentage).

More PAN chains of lower molecular weight are grafted to the starch at higher dilution of the reactants in water.[180]   Using granular wheat starch at 0.27 mol/L and acrylonitrile at 1.20 mol/L with 1.5×10^{-3} mol/L of ceric ion in the aqueous reaction medium, the SPAN obtained after 3 h reaction had 44% grafted PAN, a grafting frequency of 600 AGU/graft, and the grafted PAN had a molecular weight of 122,000.   Under the same conditions, using lower concentrations of starch and acrylonitrile (0.023 and 0.235 mol/L of water, respectively), the grafting frequency was lowered to 280 AGU/graft and the molecular weight of the grafted PAN was reduced to 36,000.   The grafting frequency and molecular weight are not significantly affected by the variety of starch (rice, waxy corn, wheat, 70% high-amylose corn starches), the amylose content, or granule size.

Although the percentage of polyacrylonitrile in SPAN does not vary much with the degree of swelling of the granule prior to the polymerization reaction, the molecular weight of the grafted PAN increases and the number of grafted chains decreases (grafting frequency increases) with the degree of swelling of the starch, heated in water to 60 or 80°C (140 or 175°F) instead of 25°C (77°F).[181,182]   These results apply to both ceric ion initiation and hydrogen peroxide–ferrous ion initiation.

Graft polymerization of acrylamide also gave higher molecular weight and less frequent grafts with swollen starch than with unswollen starch.[183]   On the other hand, methyl methacrylate, N,N-dimethylaminoethyl methacrylate nitric acid salt, and 2-hydroxy-3-methacryloyloxypropyltrimethylammonium chloride yielded less frequent grafts of higher molecular weight when the starch was not swollen.   With acrylic acid, the molecular weight of the graft was independent of the swelling of the starch, although grafting was less frequent when swollen starch was the substrate.[183]

Scanning electron micrographs of wheat starch polyacrylonitrile graft copolymers, which contained 22% and 44% grafted PAN, suggested that the PAN graft was located mainly on the surface of the granules with the lower add-on and throughout most of the granule interior with the higher add-on graft.   The observed locations are reasonable if the rate of diffusion of monomer and/or initiator through the starch granule matrix is slow relative to the polymerization rate.   In the low add-on polymer, the limited amount of monomer would become exhausted through polymerization at the granule surface before diffusion throughout the granule could take place.   In the high add-on graft copolymer, where the high add-on is obtained by a higher concentration of acrylonitrile in the reaction mixture, there is sufficient monomer to graft throughout most of the granule.[184]

Ceric ammonium nitrate is a more effective catalyst in grafting acrylonitrile to starch than is ceric ammonium sulfate.   The concentrations of acrylonitrile and catalyst are the major factors influencing the grafting.   Using a 1:1 water-dimethylformamide solvent system, with 0.5 mol/L of starch and 1.0 mol/L of monomer and reaction at 30°C for an hour, it was noted that the degree of grafting increased with increasing ceric ammonium nitrate concentration up to 5.0×10^{-3} mol/L and then decreased slightly.

The amount of homopolymer increased slightly (10 to 14% of original acrylonitrile present) as catalyst concentration increased from 2.5×10^{-3} to 1.0×10^{-2} mol/L. Increasing the acrylonitrile concentration increased the add-on, and although total conversion of monomer to polymer was low (24 to 35%), the amount of homopolymer produced was always lower than the amount grafted, the grafting efficiency being always higher than 50%.   Grafting efficiency increased with increasing acrylonitrile concentration, reaching a maximum of 62.5% at 1 mol/L with 2.5×10^{-3} mol/L of ceric ammonium nitrate and 87% with 5×10^{-3} mol/L of catalyst.   Acrylonitrile grafts at a faster rate when water alone is the solvent.[185]   The rates of graft polymerization were directly related to the square root of the ceric ion concentration and the 1.3 power of the monomer concentration.[186]

Hydrolyzed Starch Polyacrylonitrile Graft Copolymers.   Saponification of starch polyacrylonitrile graft copolymer (SPAN) in aqueous alkali converts the nitrile groups into a mixture of carboxamide and carboxylate groups, and the insoluble SPAN becomes soluble in water.[164,165,173-175,187,188]   The alkaline hydrolysis may be carried out on an aqueous slurry (10% solids) of SPAN by heating in the presence of potassium or sodium hydroxide (0.5 part of NaOH/part SPAN) at 80 to 100°C (176 to 212°F) for 2 to 3 h. The hydrolysis mixture becomes a viscous, relatively clear dispersion which could cause stirring difficulties and require a heavy-duty mixer.   After cooling to 25°C (77°F), the hydrolysis product is precipitated by acidification of the reaction mixture to pH 3 with hydrochloric or sulfuric acid.

The product (H-SPAN) is washed with water to remove salts and dried. During the saponification the reaction mixture develops a deep red color, presumably caused by the formation of partially hydrogenated naphthyridine ring structures, which gradually fades to a straw color as the saponification reaches completion. The saponification has been carried out in aqueous alcoholic alkali (methanol, ethanol) which prevents the H-SPAN from solubilizing, thereby preventing the mixing problems of high-viscosity dispersion as the hydrolysis proceeds.[189] In cases where the presence of acrylamide in the H-SPAN would be undesirable and detrimental, the residual acrylonitrile monomer is removed from the SPAN by distillation before the saponification.

Saponification conditions can be adjusted in terms of alkali concentration, temperature, and time to give products with varying amounts of nitrogen, but a nitrogen-free product is never obtained. Infrared spectra indicate that the nitrile is not present and is hydrolyzed to the carboxamide and carboxylic acid.[173] Maximum conversion of nitrile to carboxyl is about 65%.[174] Table 22.9 shows that a range of products with different

TABLE 22.9
Alkaline Hydrolysis of Starch-Polyacrylonitrile Graft Copolymer:
Composition of Hydrolyzed Copolymer[174]
(S/PAN = 1/1.90)

	KOH, wt %	KOH/ PAN	Temp., °C	Time, h	N, %	COOH, %	Acrylic acid, %	Acrylamide, %	Starch, %
1	8.6	3.6	100	2	5.34	27.4	44	27	29
2	8.6	3.6	80	3	6.10	26.6	42	31	27
3	17.2	7.2	80	2	6.33	24.8	40	32	28
4	3.4	1.4	80	2	7.97	20.2	32	40	28

* Starch determined by difference. Theory: 28%.

nitrogen and carboxyl contents can be obtained. Assuming that only poly(acrylic acid-acrylamide) is obtained from the original PAN graft, the composition of the hydrolyzed H-SPAN can be calculated.

A 1:1 starch-polyacrylonitrile graft copolymer (molecular weight of PAN approximately 800,000), hydrolyzed to H-SPAN and dried in the acid-precipitated form, cannot be dispersed in water, although it will hydrate to some extent. However, after sifting the H-SPAN into water and raising the pH above 4.5 to convert some of the carboxylic acid groups to carboxylate salt form, a smooth, viscous dispersion can be obtained. Maximum viscosity of a 1% dispersion is found at the neutralization point, pH 8.5, with 90% of maximum viscosity of 19 Pa·s (19,000 cP) retained between pH 5.5 and 11.[173,190] Table 22.10 shows viscosity changes of one sample with concentration and addition of salts. Trivalent cations suppress viscosity more than the divalent cations which are more effective than monovalent cations. The initial increase in viscosity at 0.1% CaCl₂ may be caused by a few crosslinks from carboxylic acid salt formation.

At 1% concentration by weight, the H-SPAN dispersion is a stiff paste showing a yield point. At stresses above the yield point, the viscous dispersion is non-Newtonian and very elastic. These dispersions of H-SPAN in carboxylate form appear to consist mainly (80% or more) of highly swollen gel particles, which can be centrifuged down in the presence of salt. The remainder of the dispersion is soluble material. Within the concentration range where major thickening occurs, the dispersion appears to consist of highly swollen, deformable gel particles closely packed in contact with the surrounding particles and little, if any, free solvent. If the dispersion is diluted to a level where swollen gel particles are no longer in close contact, the viscosity and elasticity decrease markedly. Thus the dispersions are not solutions, and at the high concentration levels where thickening occurs, all the solvent is absorbed by the swelling particles.[191]

Further dilution or increasing the ionic strength leads to excess solvent and a consequent drop in viscosity.[191] Sonication can lower the viscosity of a 1% H-SPAN dispersion from 30 to 0.02 Pa·s (30,000 to 20 cP) where the H-SPAN no longer consists of gel particles but is in solution. The solution now exhibits Newtonian behavior and

TABLE 22.10
Viscosities of H-SPAN (1:1)[a] Dispersions[b173]

Dispersion concentration, %	Viscosity,[c] cP	Salt	Viscosity[c] (cP) at salt concentration of:			
			1%	2%	4%	8%
0.032	950					
0.063	2,000					
0.125	3,350					
0.25	5,450	KCl	765	180	100	—
0.5	10,000	KCl	1,960	1,000	520	240
1.0	19,000	KCl	11,000	8,900	7,000	5,000
2.0	—	KCl	—	—	—	15,300
1.0	19,000	NaCl	10,000	7,300	6,300	3,900
		NaH_2PO_4	12,400	7,000	5,800	4,200
		NaOAc	10,500	9,400	6,450	3,240
		$Na_2B_4O_7 \cdot 10H_2O$	13,000	10,700	9,500	9,200
		Salt Conc.	0.1%	0.2%	0.3%	0.4%
		$CaCl_2$	22,100	10,800	6,250	520
		$AlCl_3$	15,500[d]	12,100[e]	10,300[f]	

[a] Wheat starch polyacrylonitrile graft copolymer, PAN molecular weight 794,000, hydrolyzed at 100°C for 2 h at KOH/PAN 5.4 by wt. H-SPAN had 26.6% COOH, 4.40% N.
[b] Initial pH 8.5
[c] Brookfield Synchro-Lectric viscosimeter, LVT, #4 spindle, 30 ppm, 25°C; multiply by 0.001 to convert to Pa · s.
[d] $AlCl_3$ concentration 0.06%.
[e] $AlCl_3$ concentration 0.12%.
[f] $AlCl_3$ concentration 0.18%.

the low viscosity is independent of shear rate. Films cast from these low-viscosity solutions are soluble in contrast to the insoluble, highly swellable films cast from the dispersions.[192]

The soluble films, prepared from the sonified polymer solution and dried at 30°C (86°F) in a forced-air oven, can be rendered insoluble by heating in a vacuum oven at 165°C (329°F) for 30 min, by irradiation with a cobalt 60 source, or by exposure at ambient temperatures for 2 weeks to relative humidities of at least 75%. Apparently crosslinking occurs, resulting in films which, though insoluble, absorb up to 1800 times their weight in water, with most values of the order of 500 times.[192] Scanning electron microscopy of saponified corn starch–polyacrylonitrile graft polymer shows gel particles which retain the outward appearance of graft-polymerized starch granules. This suggests some crosslinking during grafting or saponification.[192a]

The aqueous dispersions of H-SPAN in the potassium carboxylate form will yield films which, on drying, are hard and brittle. Small amounts of water will plasticize the film so that it becomes pliable and translucent. In excess water, the film will swell rapidly to give a clear, self-supporting gel sheet, which retains the gross shape of the original film. Water absorption was in the range of 300 times the dry film weight, and expansion up to 33 times the original surface area. The swelling of the film is reversible, decreasing to its original size at pH 2.3. By alternate changes from alkaline to acid pH, the film can undergo repeated expansion and contraction. Electrolytes will also cause contraction of the film.[190]

Other forms of the absorbent polymer can be obtained by alternate drying methods. Alcohol precipitation gives a powdery product. Drum drying yields flakes. When the insoluble film is redispersed in water by applying mechanical shear, such as with a Waring blender or colloid mill, and, if the shearing has not been too extensive, the film cast from this dispersion and dried will regain the absorbency and insolubility of the original film.[187,188,190] This insoluble, swellable gel product may be useful in absorbable soft goods such as diapers, bandages, hospital bed pads, and sanitary napkins. These products must compete with other materials such as cellulose, carboxymethylcellulose, and synthetic polymers in terms of absorption of body fluids (blood, urine) and holding fluids under pressure so that the absorbed fluid cannot be squeezed out.[175,193] The same types of products have been made from flour.[188,194]

Other Monomers for Graft Polymerization. Free radical initiation has been used to produce starch graft copolymers with methyl methacrylate,[169,170,183] acrylate esters,[164,165] vinyl acetate,[164,165] styrene,[164,165] acrylamide,[164,165,179,183,185,195] and acrylic acid.[164,165,195] The water-insoluble polyacrylate ester graft copolymers can be saponified to give dispersible starch-polyacrylate sodium salts. Cationic starch graft copolymers can be made by using amine-containing monomers, usually quaternary ammonium or amine acid salts.[165] The types of monomers used are N,N-dimethylaminoethyl methacrylate and N-tertiary-butylaminoethylmethacrylate nitric acid salts,[165,183] 2-hydroxy-3-methacryloyloxypropyltrimethylammonium chloride,[165,183,196] and 4-vinylpyridine.[165]

Graft polymerization may also be carried out with mixtures of monomers to form graft polymers containing these monomers.[196a] Thus, mixtures of acrylamide and the nitric acid salt of dimethylaminoethyl methacrylate have been grafted onto wheat starch using hydrogen peroxide–ferrous ammonium sulfate initiation since ceric ion is not effective with dimethylaminoethyl methacrylate.[179]

Corn starch graft copolymers have been made using acrylonitrile in combination with 2-acrylamido-2-methylpropanesulfonic acid (AASO$_3$H), vinylsulfonic acid, acrylic acid, methyl acrylate, methyl methacrylate, and styrene monomers initiated with ceric ammonium nitrate.[194] The resulting graft copolymers were subjected to alkaline saponification to hydrolyze the nitrile groups and the products dispersed, cast as films, and dried to yield absorbent polymers. The saponification time was shortened by the incorporation of the 2-acrylamido-2-methylpropanesulfonic acid, acrylamide, and acrylic acid. Graft polymerization on gelatinized starch gave much higher fluid absorbencies than the products made with granular starch: an H-SPAN (granular) with 55% add-on absorbed 300 g deionized water or 28 g synthetic urine per gram of polymer vs. 1000 g and 40 g, respectively, for the H-SPAN (59% add-on) made with gelatinized corn starch. The AASO$_3$H comonomer increased the absorbency considerably (5300 g deionized water or 73 g synthetic urine per gram of polymer for gelatinized starch-based graft copolymer).

Graft Polymers on Modified Starches. Commercial converted corn starches (acid fluidity, hypochlorite-oxidized, dextrinized) and an *alpha*-amylase-treated wheat starch were subjected to graft polymerization with acrylonitrile using ceric ammonium nitrate initiator.[197] With highly soluble starches, much of the carbohydrate is not grafted and the grafted material consists mainly of dimethylformamide-soluble product with a high polyacrylonitrile (PAN) content, presumably because grafting took place on small carbohydrate fragments of the converted starches. The converted starches which were not highly water-soluble and remained in granule form in aqueous suspension at room temperature gave grafted copolymers that were similar to those obtained with unmodified starch.

Starch methacrylate esters (D.S. 0.12) yield graft copolymers with methyl methacrylate using hydrogen peroxide–ferrous ammonium sulfate alone or in combination with ascorbic acid as initiator-activator catalyst systems, with high grafting efficiency (92 to 97%). The grafting efficiency is higher on the starch methacrylate than on an unmodified starch (72 to 88%), presumably because peroxide causes grafting directly on the starch as well as on the pendant methacrylate ester groups on the starch.[198] When azo-bis-isobutyronitrile (AIBN) is used as the initiator, grafting efficiencies of 54 to 67% were found on the starch methacrylate ester, whereas this initiator gives a high conversion to poly(methyl methacrylate) homopolymer and no grafting with unmodified starch. With corn starch maleate half-esters (D.S. 0.13), AIBN gives little or no grafting of methyl methacrylate; with peroxide initiation, there is no difference in the grafting efficiency between the starch maleate and unmodified starch. This indicates that the maleate half-ester does not participate in the polymerization.

Starch derivatives containing pendant acrylamide groups, which can be copolymerized with other vinyl monomers by free radical initiation, can be prepared by treatment of starch or a modified starch with N-methylolacrylamide in the presence of an acid catalyst (ammonium chloride, ammonium dihydrogen phosphate) and a polymerization inhibitor (hydroquinone).[199] Starch methacrylate, crotonate and itaconate esters, allyl carbonate starch esters, allyl starch ethers, and 2-hydroxy-3-methacryloyloxypropyl starch ethers have been claimed as the starting materials for copolymerization with vinyl monomers.[200] These products were suggested as being useful as binders in pigmented paper coating.[200]

Acrylonitrile can be graft-polymerized onto a gelatinized cationic starch containing diethylaminoethyl ether substituents by ceric ion initiation to give stable latex-like copolymer dispersions.[201] The opaque, aqueous polymerization reaction mixtures (up to 8% solids) are agitated in a blender at pH 7 to give smooth creams with Brookfield viscosities of 1.5 to 3 Pa·s (1500 to 3000 cP). More fluid dispersions with viscosities of 0.015 to 0.04 Pa·s (15 to 40 cP) are obtained by sonification. These sonified dispersions dry to give clear, hard adhesive films, formed by the coalescence of particles measuring 0.05 to 0.15 μm in diameter. (Same diameter of particles in nonsonified dispersions.) The sonified cationic starch-polyacrylonitrile copolymer (1:1) dispersions did not form sediments on aging 6 months at room temperature but sometimes increased in viscosity to about 0.1 to 0.2 Pa·s (100 to 200 cP).

Graft Polymers by Derivatizing Reactions. Graft polymers on starch can be made without free radical initiation. Reaction of ethyleneimine with tertiary aminoalkyl ethers of starch yields a cationic starch with short polyethyleneimine side chains.[202] The interest in cationic starches for use in the paper industry as wet-end additives to impart pigment retention and strength, as sizing, and as coating binders has resulted in the use of various techniques to introduce amino-containing groups and side chains on the starch molecule. Starch polyethyleneiminothiourethane derivatives have been prepared by reaction of sodium starch xanthate (D.S. 0.12 to 0.35) with polyethyleneimine (molecular weight about 100,000). The soft gel formed during the 3-h reaction period dissolved on dilution with water and, when added to a pulp suspension in the paper-making process, increased the burst and tear strength of the resulting paper.[203,204]

Cationic starch derivatives have been prepared by treating gelatinized starch dispersions with a preformed polymer made by condensing epichlorohydrin and ammonia in such a way that there are residual chlorohydrin groups available for reaction with starch under alkaline conditions.[205] Another analogous starch graft copolymer is prepared by coupling a starch (preferably converted starch) with a polyamideamine made by the condensation polymerization of a polyalkyleneamine (diethylenetriamine) and a dibasic acid or ester (adipic acid, dimethylsuccinate) such that the ratio of total amine equivalents to total carboxylic acid equivalents is 1.25:1 to 3.0:1. The coupling agent is preferably epichlorohydrin, which is added to the starch dispersion containing the polyamideamine condensate polymer in a starch-polyamideamine ratio of 1:9 to 9:1 and heating the aqueous mixture (5 to 60% solids) at 40 to 100°C (104 to 212°F). The product dispersions are stable at pH 4 to 6 and are useful as wet-strength additives in paper. The polyamideamine can be prepared incorporating other compounds such as lactams, acrylates, or lactones.[206]

Other types of polymers, made by reaction of polyepichlorohydrin and an amine (e.g., trimethylamine, dimethylamine) or condensation polymerization of epichlorohydrin and dimethylamine, are reacted with dispersed starch (gelatinized) through residual chlorohydrin groups under alkaline conditions to give a cationic starch graft copolymer in aqueous dispersion.[207] These cationic starch graft copolymer dispersions are useful wet-end additives.

Whenever granular starch is used in grafting reactions, there is the possibility of crosslinking, which would have an adverse effect on the dispersibility of the starch product and its utility.

Applications of Starches

The Paper Industry The paper and paperboard industry is by far the largest industrial consumer of starches in the United States, using almost 900 million kg (2 billion lb) in 1976 (Table 22.4).[73] It was estimated that in 1972, out of a total of 590 million kg (1.3 billion lb) used by the paper industry, 52 million kg (115 million lb) were used in wet-end addition in the paper-making process, 354 million kg (780 million lb) were used in surface sizing and 200 million kg (440 million lb) as coating adhesives.[74]

Wet-End Addition. The principle of making paper is deceptively simple. A very dilute suspension (0.1 to 0.5% solids) of cellulose pulp in water is filtered on a fine wire screen to drain off most of the water, leaving a wet fibrous mat which is pressed to remove more water and then dried. The cellulosic pulp must be prepared by mechanical treatment in beaters or refiners which fibrillate and hydrate the fibers. This increases the surface area available for interfiber bonding. The increased hydration of the fiber resulting from the beating operation causes slower, less complete drainage on the wire screen. However, the increased hydration produces stronger paper because of the greater area and intimacy of fiber-to-fiber bonding when the sheet is formed, pressed, and dried. The additional hydration with the longer beating time causes the paper to shrink more on drying and to become dense, translucent, less compressible, and less oil-receptive, properties not desirable in a paper for printing. The beating processes require considerable energy. Therefore, ways are sought to develop strength with a minimum of beating.

Although the principles of paper-making appear simple, the application to commercial production has led to complex machine systems to manufacture paper. Thus, additives are used to improve drainage, to retain fillers and pigments, to provide wet strength and dry strength, and to meet pollution abatement requirements. Further, the additives must not have an adverse effect on the repulping and recycling of the paper. Unmodified starch and starch derivatives are used as wet-end additives to the pulp slurry, called the *furnish*, which is fed to the wire of the Fourdrinier paper-making machine or to the wire-covered cylinder of the cylinder machine.[74,208,209,210]

Starch-based, wet-end additives are used primarily to increase the dry strength of the paper. They may also be used to increase the retention of fillers or pigments, as well as to minimize or overcome the strength losses generated by the incorporation of fillers in the paper. The starch is usually added as a dispersion produced by cooking the starch or using a pregelatinized starch. The starch can be cooked in high-shear, continuous, steam-injection cookers that give a well-dispersed starch. This type of equipment can also be used to cook and derivatize at the site of application.

The starch may be added at any point in the stock delivery system, such as at the headbox, fan pump, stock chest, or beater, depending upon the characteristics of the starch, the type of paper being made, and the operating characteristics of the machine. The unmodified starches used are corn, potato, tapioca, or sorghum. Modified starches include oxidized, phosphate ester, hydroxyethyl, and cationic starches. In order for the starch to be effective in increasing strength, it must be retained in the wet sheet as it is formed on the wire.

Cationic starches, because of their attraction to the anionic cellulosic fiber, are virtually 100% retained on the pulps at the levels of addition usually found (0.2 to 2.0% on the fiber).[211] The strongly adsorbed cationic starch is not removed from the fiber by the usual paper-making processes and repulping. Thus, it is a very effective agent and does not contribute to pollution in the paper-making process. Unmodified starches or non-ionic starch derivatives are not retained as well; hence they must be used at higher levels of addition (2 to 3% on dry pulp), and do contribute to the build-up of solubles in the recycling water system as well as to pollution problems directly and on repulping.

The starch phosphates are effectively retained in the wet sheet in the presence of alum. They are useful in acid paper-making systems, which are the usual systems. In some mills where heavy grades of paper are made, ungelatinized starch is added to the pulp slurry so that the starch is retained by physical entrapment as the wet sheet is formed on the wire, and the starch is then gelatinized as the sheet is dried. Low gelatinization temperature is important here, and derivatives such as oxidized starch or hydroxyethyl starch are useful.

Ampholytic starches containing both anionic and cationic groups have been effective in both strength and pigment retention properties under a wide range of conditions. Acceptable performance can be obtained in terms of pigment retention and strength improvement over a wide range of pH, alum concentration, and water hardness by introducing anionic phosphate ester groups into corn starch, together with cationic ether substituents containing tertiary amino, or quaternary ammonium groups.[161,162]

For most effective results, there must be a suitable balance of anionic to cationic groups, preferably 0.07 to 0.18 moles of anionic groups per mole of cationic group. The presence of both anionic and cationic groups extends the effective range of pigment retention over the pH range 4.6 to 7.6. The anionic phosphate starch is not as effective at the neutral pH and the cationic diethylaminoethyl starch ether is less effective at pH 4.6. Cationic potato starches which normally contain a small amount of natural phosphate act in a similar way, although the effects may be enhanced by introducing more phosphate to give a better anionic/cationic balance. The cationic, anionic, and ampholytic starches are much more effective than the unmodified or non-ionic starches in development of the internal strength of the paper.   The anionic starches are used in the presence of alum to attain maximum efficacy.

Fillers or pigments such as clay, calcium carbonate, or titanium dioxide are used to increase opacity and to improve the brightness and printing qualities of the paper sheet. It is important to retain these fillers in the wet sheet as it is formed on the wire to realize the full potential of the added material and minimize the economic loss of not holding the filler in the sheet.[212]   Further, the cleanliness of the water system in the paper-making process and the pollution load of the effluent wastewater from the paper mill are affected by the efficiency of the filler retention.   Filler or pigment retention is the percentage of the pigment/filler added to the pulp slurry (furnish) that is retained in the finished paper.   The filler particles generally have a negative surface charge in water suspension, and positively charged cationic polymers which promote flocculation of the pigment particles are effective retention aids.

Synthetic polymers such as polyethylenimines, polyamideamines, and polyacrylamides are used as retention aids.   The cationic starches also act as retention aids, as well as increasing the strength of the sheet.   Since the retention of fillers decreases the internal strength of the sheet, this capability of maintaining internal bonding strength in the presence of these fillers is a most important property.   The cationic starches also act to retain the fiber fines in the wet sheet, reducing the amount carried away in the water draining from the sheet during formation on the wire as well as using the fiber fines to contribute to the strength of the sheet.   Cationic starch graft polymers are also of interest for filler retention.[213]

To obtain best results in retention of fillers and fines, the starch dispersion should be added to the diluted pulp stock slurry after the points of high-shear mixing and working and at a point where the added starch will be thoroughly mixed into the slurry.   Generally this will be close to the headbox of the Fourdrinier machine, which feeds the furnish to the moving wire screen.   The objective is to obtain maximum interaction of the starch with the pigments and fines to form flocs and not to break up the flocculated agglomerates by shearing forces.   The point of addition of the starch dispersion for best results should be determined by trial on the specific machine.

When used as combination strength and retention aid, the cationic starch is generally added at 0.25 to 1% on the pulp.   The level required depends upon the system and the character of the pulp being used, including the level of anionicity, the level of alum, and the presence of other materials in the furnish.[214,215,216]   The cationic demand can be assessed by a titration with cationic polymer as measured by the zeta potential, which is the electrical potential between the particle surface and the bulk of the medium (water) and is related to the charge on the particle.[215]   A test for "cationic efficiency" to predict the retention efficiency of cationic starches involves optical measurement of the amount of anionic blue pigment remaining in suspension after addition of cationic starch-pulp complexes.[217]

If an excess of cationic starch is added to the pulp-filler-water system, the negative surface potential of the pigment particles will change, becoming positive, and the filler particles will again repel each other, thereby reducing retention.   The cationic starches should be fully cooked (dispersed) for maximum retention efficacy.[218]   The drainage of the water from the wet sheet on the moving wire of the paper-making machine is improved by the addition of cationic starch and polymers which act as retention aids and also through flocculation and redistribution of the fines throughout the wet sheet or web, which increase the permeability of the wet web.[212]

As concern with reducing pollution in mill effluents becomes a major factor in planning and operations, the paper mills are closing up their water recycling systems to minimize solids and solubles in the wastewater discharge.   This leads to increased solubles in the water used to make up the furnish and problems in the process and the quality of product that must be overcome.   Additives such as the cationic starches and other polymers will be employed to solve these problems in the paper-making process. Similarly, with the cost of energy rising precipitously, any means that will minimize energy requirements is in demand.   The use of additives to remove water more rapidly and easily from the wet sheet is of interest.   Cationic starches make up a major portion of the starch wet-end additives used at present.

Anionic polyelectrolytes would act as dispersants for the anionic fibers and fillers if they were not used in combination with alum (aluminum sulfate), which develops the cationic charge.   Starch phosphate acts as an effective retention aid in the presence of alum, as well as developing dry strength.[219]   Oxidized starches have a strong detrimental effect on filler retention, even in the presence of alum.[220]   Since oxidized starch, which is used as a wet-end additive and in sizing and coating, may return to the paper-making process through the "broke" (trimmings and other waste paper that is recycled through repulping), means of nullifying its dispersant effect on pigment must be used.   Cationic starches are effective in the presence of small amounts of oxidized starch.[220]

When it was available, dialdehyde starch could be used to impart a temporary wet-strength to paper.[141]   It had the advantage over the synthetic resins normally used (urea-formaldehyde, melamine-formaldehyde) of allowing for easy broke recovery with dilute alkali since the dialdehyde starch is degraded by alkali.   It is easily disintegratable and would therefore be useful in disposable paper towels and tissues.   The high cost of producing dialdehyde starch on a small scale (under 454,000˙kg, or 1 million lb, annually) had limited its use which never exceeded 340,000 kg (750,000 lb) per year.[74]

Surface Sizing.   After the paper is formed and dried, it may be given a surface sizing treatment to improve strength and surface properties for printing and writing.   The surface properties desired are improved appearance, surface strength, scuff resistance, printability, erasability, and resistance to the penetration of fluids, inks, and coating into the sheet.   The size is employed to anchor the surface fibers and particles so that dusting and linting do not occur.   The size is applied on the paper machine on the size press or at the calender stack.[74,208,209,210]

The starch paste is applied on both sides of the sheet by a shower and/or by a roll revolving in a starch dispersion, and the excess paste is squeezed out by the pressure of the press rolls which also drives the size into the paper.   The sheet is then dried. In calender sizing, the sheet passes around a series of steel rolls where the starch paste is applied and is then wound on reels without further drying.   Thus the removal of water must be through evaporation by the latent heat in the sheet as well as by absorption of moisture into the sheet.   Because of this, heavy sheets such as linerboard and boxboard are the only types sized at the calender stack.   The operating variables of the equipment will determine the type of starch, pickup, and penetration into the sheet.

Starch is the most used surface sizing agent.   Corn starch, being most readily available, is the type generally used, although tapioca, potato, and wheat starches may be used wherever and whenever the economics are favorable.   The starches used in the sizing operation are generally converted starches since the native starches are too high in viscosity to provide high enough solids in the sizing solution to obtain enough pickup on the sheet.   The equipment used and the required properties of the size will determine the type of starch employed.   Generally the starch dispersion is used at 2 to 12% solids, although in tub sizing the solids may reach 20%.   (Tub sizing involves immersing the sheet in a tub of starch dispersion to saturate the sheet and then passing the sheet through press rolls and drying.   This process may be used by older mills for making rag bond paper.)[208,210]

The degree of conversion (the viscosity level) needed varies, and therefore a range of converted products is used.   The higher the viscosity, the lower the solids in the size solution and, hence, the lower the amount of starch picked up by the sheet.   The less-converted starch has a higher bonding strength and will also penetrate to a lesser degree into the sheet.   The lower-viscosity products can be used at higher solids in the size, giving higher pickup.   The lower-viscosity products will also penetrate further into the sheet, giving a greater effect in improving internal strength.[208,209]   Oxidized starches, hydroxyethyl starch ethers, starch acetates, and cationic starches have been the main products used for paper sizing.   Other derivatives such as carboxymethyl, hydroxypropyl, and cyanoethyl starch ethers have been used.   These products are based on converted starches and have the dispersion stability and film-forming capabilities of particular value in surface sizing.

The enzyme-conversion of native starches to the desired viscosity has been used by many mills to meet their surface size requirements.   The conversions are carried out in the mill, using *alpha*-amylases, either in batch or continuous processes with automatic equipment to control the conversion-cooking temperature cycle for close control of the viscosity.   The conversion is carried out as the starch is being heated (to gelatinize it) so that the high viscosity of the native starch is not reached and the agitation can be maintained in the high-solids mixture (18 to 35%). The enzyme conversion is completed at about 80 to 85°C (176 to 185°F) and the enzyme is inactivated, usually by heating to about 90 to 100°C (194 to 212°F) for a short period. The converted paste is then diluted with cold water to about 12 to 15% solids for storage and is used after further dilution. The enzyme conversion has given flexibility and economic advantages in the paper mills' sizing operations. Specially buffered starches are sold for this enzyme-conversion process.

A recent development is the thermochemical conversion process, which is a continuous process for degradation of the starch carried out in the paper mill. Ammonium persulfate (0.15 to 0.3% on starch) is the oxidation reagent used to effect the conversion at high temperature, about 150°C (302°F) for up to 5 min, and low pH. The conversion is more like an acid hydrolysis since carboxyl and aldehyde groups are not detected and minimal carbonyl groups are introduced, the product displaying the dispersion instability expected of an acid-converted starch.[209,221] In some cases the thermal converter is used without the oxidant, depending upon low pH to effect the conversion at the high temperature.

The thermal converter system consists of a positive displacement pump supplying starch slurry to a steam-injection jet cooker to which is attached a retention coil. A back pressure is maintained on the coil to ensure complete condensation of the steam and full retention volume so that the starch is thoroughly dispersed and the oxidant is completely consumed. From the back-pressure valve, the starch dispersion is flashed into a steam separation chamber and finally into a holding tank. Instrumentation provides temperature and feed control so that automatic operation is feasible.[221] Viscosity control is primarily accomplished by controlling the level of ammonium persulfate with constant temperature and retention time. If the latter two parameters are also varied, then a range of conditions is available.

Although there is a trend toward the use of lower-cost enzyme and thermochemical conversion of native starches, these are not without functional disadvantages. When significant amounts of starch are included in the paper web, opacity may be harmed. Size-press starches have been a major contributor to the pollution problems of paper mill effluents because of the relatively large amounts of starch that are solubilized when the sized paper is recycled to the pulping system as broke. This includes converted starches, non-ionic, and oxidized starches. Further, when oxidized starches are used for sizing and are returned to the paper-making system as broke, they adversely affect the retention of fines and fillers as mentioned previously. Consequently, there has been some movement away from the use of oxidized starches in sizing because of the loss of fillers and increase in suspended solids in the mill effluent as well as the costs of increased useage of filler and higher production costs.

Cationic starches are tenaciously held on the fibers by electrostatic attraction and, when used as surface sizes, remain substantially with the cellulosic fiber and the filler during the repulping of the broke, thereby reducing pollution (measured by the BOD, or biochemical oxygen demand) and providing more effective use of the broke.[222-224] Depending upon the type of paper mill, the finished paper can contain from 15 to 35% broke. Some reports have indicated that as much as 50 to 60% of the starch that comes back with the broke goes through the paper-making process and out as effluent.[223] Anionic starch esters and ethers may be bound to the fiber through the alum at pH up to 6, and may be retained in the pulp when broke is recycled.

The cationic starches have other advantages in surface sizing. Since there is an electrochemical attraction between the positively charged starch and the negatively charged fiber, the size is attracted to the surface fibers and retained on or near the surface of the sheet. Consequently, less starch is needed to maintain surface strength and quality. Less starch is therefore used, which improves opacity and often increases drying rates and machine speeds, thereby improving economics. The printing quality of the paper sized with the cationic starches also is improved.[222,225]

Sometimes acid-converted starches are used at the calendar stack as a surface size for heavyweight kraft papers and liner boards to improve scuff resistance and printability. The requirement is for the size to set up rapidly, remaining on the surface. The retrogradative tendency of the acid-fluidity starch is put to good use here.[86,208]

Paper coating. Coating is the application of one or more layers of non-cellulosic material to the paper surface either to improve the printability of the surface or to provide certain functional properties, such as barriers for water vapor, grease, fats, solvents, etc. Starches are not the primary material for use in functional coatings. The pigmented coatings are used to improve the printing characteristics of the paper surface, to increase brightness, and to increase opacity. The aqueous coating system consists of a dispersion of pigments, an adhesive to bind the pigment particles to each other and to the paper surface, and auxiliary agents to disperse the pigments and modify the flow properties (rheology) and the characteristics of the finished coating. This aqueous mixture is known as the *coating color.*

The factors to be considered in formulating coating colors are the coating application equipment to be used, the characteristics of the paper or paperboard to be coated, and the use requirements for the finished, coated sheet, including the type of printing to be used on it.[74,208,209,226,227,228] The pigments used are mainly clay, calcium carbonate, and titanium dioxide, while hydrated alumina, barium sulfate, synthetic silicas and silicates, zinc oxides, diatomaceous silica, and plastic pigments based on polystyrene solid particles are used to a lesser extent, sometimes in special applications.[229] Detailed procedures for preparing the coating colors with starch have been published.[227,228]

The main purpose of the binder in the coating color is to act as a carrier for the pigment, to regulate the degree of water retention, and to impart suitable flow properties, and to bind the pigment to itself and the paper. The binder dispersion must have high adhesive strength and suitable rheological properties (controlled, reproducible viscosity that is stable during storage, good flow, and good water-holding properties that minimize penetration into the sheet being coated).[208,209,228c,228d,228f,228g] The binder should not adversely affect the pigment dispersion. The application of the coating to the paper has been done through a variety of machines, including the roll coater, the brush coater, the air knife coater, and various types of blade coaters, all, except the air knife coater, designed to give a smooth coating of a given thickness. The trailing blade coater appears to be the one of choice. Size press coating also is used to a considerable extent.[208,209,230]

The starches used in coating colors are enzyme-converted starches, thermochemically converted starches, oxidized starches, dextrins, hydroxyethyl starch ethers, and starch acetates. In the latter case, since this is an ester, care must be taken that the pH is not so high (above 8) as to saponify the ester and cause instability in the color. Depending upon the equipment used, the coating colors range in solids from about 45% (air knife coater) to about 70% (blade coaters). Coating solids for the size press may run about 30 to 45% solids.

The starch is generally used at a level of about 5 to 15% on the clay pigment. The lower levels are used in coatings for papers to be printed by rotogravure, and the higher binder levels for letterpress papers. Since the coating color must have viscosity stability, and thermochemically converted starches tend to retrograde on aging, the use of fatty acid derivatives was suggested as inhibitors of retrogradation. A modified polyethylene glycol laurate additive appeared to be effective.[231]

To improve water-resistance of coatings containing starch binders, required in the offset process, crosslinking agents such as urea-formaldehyde or melamine-formaldehyde resin or glyoxal may be added to the coating color. Cobinders such as polyvinyl acetate, styrene-butadiene, and polyacrylic latexes may also be added to meet more demanding water-resistance and flexibility requirements.

The surface strength of the coated paper is manifest by its resistance to picking or rupture in the printing process. The resistance to picking is dependent upon the base sheet and the pigments used, as well as upon the strength of the binder. The cationic starches give high binding strength because of the previously mentioned affinity for negatively charged particles and substrates which results in adsorption onto and bonding to these anionic materials. Thus the cationic starches could be used at lower binder levels than the other starches to obtain equivalent pigment binding, leading to increased opacity, gloss, brightness, and equal ink holdout.[232-234]

In aqueous systems the cationic starches can act to flocculate the anionic pigment particles, a highly undesirable effect. It was possible to develop cationic starches that could be used without undue viscosity and flocculation effects in the coating color. These starches had a specific level of cationic derivatization and an adjusted viscosity. It was also found that formulation of the coating required some changes to eliminate the flocculation effect. Non-ionic surfactants are preferable as defoamers. The commonly used inorganic clay dispersants, sodium hexametaphosphate and tetrasodium pyrophosphates, are preferable to organic dispersants and should be used at higher levels than usual to minimize pigment flocculation and maintain good viscosity control. The starch can be cooked in the usual way, but the pH should be at 2.5 to 3.5 and the starch dispersion cooled to 60 to 85°C (140 to 185°F) before being added to the pigment slurry. To help eliminate pigment flocculation and control final color pH, alkali (ammonia, sodium hydroxide, sodium carbonate) is added to the pigment slurry to adjust the pH to 9.0 to 10 before the addition of the cationic starch binder dispersion (pH 2.5 to 3.5). This generally gives a coating color pH of 6.5 to 8.5. Where a synthetic resin cobinder is normally used, this can be incorporated with the cationic starch by adding it at the appropriate time in making up the color; styrene-butadiene latex is added before the cationic to the pigment slurry, and polyvinyl acetate or acrylic latexes are added immediately after the cationic.[234]

A higher degree of water-resistance can be obtained by the addition of 5% glyoxal on the weight of the starch binder using the cationic starch as compared to oxidized and hydroxyethyl starches. The coatings containing the cationic starch have high binding strength, excellent ink holdout, high gloss, and desirable optical properties.[233]

Starch binders for paper and paperboard coatings comprise the largest volume of all the binders used, exceeding the combined totals of all the other binders (proteins, synthetic polymers). However, with the development of alkali-sensitive polymer latexes, significant inroads are being made into completely replacing the natural binders (proteins) in coating formulations because the water-holding properties and color viscosities are similar to the starch-based binders.

These synthetic latexes have a number of advantages which may enable them to capture a large share of the market.[235] However, the significant economic advantage of the low-cost starch binders may override the functional advantages of an all-synthetic system as a replacement for starch binders alone or in combination with synthetic latexes. It is interesting to note that the cationic starches, which are about twice the cost of hydroxyethyl starches, have not captured a significant share of the starch binder market at this time.

FDA Indirect Food Additive Regulations. The federal Food and Drug Administration (FDA) regulates and defines the types of starch derivatives that may be "used as a component of articles intended for use in producing, manufacturing, packing, processing, preparing, treating, packaging, transporting or holding food." These starches are classified as "Industrial Starch—Modified." [96a] The full regulation should be consulted before using a starch derivative in paper and paperboard for the uses defined above.

However, a listing of the permissible reagents for treatment of starch under "Industrial Starch—Modified" allows: ammonium persulfate; (4-chlorobutene-2)-trimethylammonium chloride; *beta*-dimethylaminoethylchloride hydrochloride; dimethylol ethylene urea; 2,3-epoxypropyltrimethylammonium chloride; ethylene oxide; phosphoric acid and urea; starch irradiated with cobalt 60 or an electron beam source to produce free radicals for subsequent graft polymerization with acrylamide and [2-(methacryloyloxy)ethyl] trimethylammonium methyl sulfate. The full regulation defines the limitations on the levels of treatment and/or derivatization and, in some instances, the use of the derivative and the level of addition in the paper or paperboard intended for contact with foods.

The starch derivatives that are allowed as direct additives in foods, classified as "Food Starch—Modified,"[96] are also permitted in those applications defined for "Industrial Starch—Modified." There is also provision for using starches that combine treatments acceptable under "Industrial Starch—Modified" and "Food Starch—Modified." Pregelatinized starch, unmodified starch, acid-modified starch, and dextrins, which are classified as GRAS (Generally Recognized As Safe), may also be used in paper and paperboard for food packaging.[236]

The Food Industry   Starch is a major component of the human diet, being the primary carbohydrate energy source (1 gram of starch supplies about 4 calories). The use of starches as food additives is based, however, not on their nutritional value, but on their functional value in facilitating the processing of foods and/or imparting to the food system certain desirable properties such as texture or "mouthfeel," thickening, gelling, binding, and stability.

In the United States the consumer is supplied with an abundance and variety of foods without regard to season or distance. This is done through large-scale preparation, transportation, storage, and distribution. This has brought to a high development the concept of "convenience foods" which require a minimum of preparation while providing a high quality in nutrition, eye appeal, and palatability. The starch additives play a significant role in making this possible at reasonable costs. The food industry and its suppliers have developed a high level of technical sophistication to enable it to meet the stringent demands of this complex preparation-distribution system.

The food industry consumes about 272 million kg (600 million lb) of starch and modified starches annually in a variety of applications such as pie fillings, pudding mixes, salad dressings, canned foods, frozen foods, dry food mixes, confections, bakery products, beverages, snack foods, and pet foods. Corn starch and waxy corn starch make up the major share of this usage in the United States with some sorghum, waxy sorghum, tapioca, wheat, potato, and arrowroot starches also used. Tapioca, arrowroot, and some of the potato starches are imported. Economic considerations as well as functional properties dictate which starch will be used. In Australia and New Zealand, wheat starch is produced in larger quantities than corn or sorghum starches and is used to a greater extent.[237a,238-241]

To obtain the starch products that meet the specific property requirements for a particular food process or product, manipulation of the inherent properties of starch described previously is required: gelatinization, thickening power, gelation, viscosity stability, binding and adhesive characteristics, emulsion stabilization properties, and film-forming abilities. By choosing the type of starch (corn, waxy corn, tapioca, potato, etc.) and the modification (conversion, crosslinking, derivatization, pregelatinization), a starch product can usually be developed to meet the requirements of a particular application. The use of starch in the food industry is readily considered in terms of the properties of the intact, dry granule, the starch dispersion or paste, and the starch film.[238] Reference should be made to the preceding sections on *Structure and Properties* and *Starch Modifications* to relate properties to the uses described in the following sections.

Dry Granular Starch. Starch powders are used as diluents, bulking agents, molds, moisture-adsorbing, and fluidifying agents. The most commonly used starches are those from corn and waxy corn. The granules range in diameter from 5 to 25 μm. The surface area of the corn starch granules calculated from nitrogen sorption is 0.7 m^2/g (roughly 3400 ft^2/lb). However, on the basis of water sorption, the calculated surface area of corn starch is 280 m^2/g (about 1.4 million ft^2/lb). Water sorption is not solely a surface phenomenon, and, as mentioned previously, the granule is permeable to water so that the water molecules are attracted to, and held by, the hydroxyl groups in the starch molecules. Photomicrographic measurements indicate that there are 1.27 $\times$ 10^9 granules per gram of corn starch and that potato starch has 6.20 $\times$ 10^7 granules per gram.[28]

By drying the starch to a level below its equilibrium moisture content (Table 22.2), it will preferentially draw moisture from other materials in close contact which have less affinity for the moisture. Thus, starch redried to 1.5 to 7.5% moisture is used as a diluent and moisture adsorbent to protect active ingredients in powders from the detrimental effects of moisture. Powdered corn starch and redried corn starch are used (at 15 to 40% of total) in baking powder to physically separate the sodium bicarbonate and the acidic ingredient (e.g., and sodium acid tartrate), and to preferentially absorb moisture to prevent premature interaction and loss of the carbon dioxide-producing activity. Similarly, redried starch is used in powdered sugar at about 3% to prevent caking by preferentially taking up moisture. Since whiteness is important, a bleached starch may be used.

In some cases, the starch is added as a diluent or bulking agent for active ingredients such as enzymes, flavors, dyes, or pigments to permit measurement of the material on a larger unit scale and to facilitate standardization of potency. If the additive is used in cold or warm water and a clear solution is required, a soluble dextrin or pregelatinized starch may be used.

Powdered starch may be used as a dusting powder. Blending with a small amount of tricalcium phosphate will coat the starch granules and increase the lubricity so that they flow better. A starch modified to contain hydrophobic ester groups and a polyvalent metal ion exhibits a very mobile, free-flowing property and is water-repellent, even at

16% moisture.[133] This is of value as a dusting powder and as a fluidizing and anticaking agent.

Powdered corn starch, usually sprayed with small amounts of mineral oil or acetylated monoglycerides, is used in molding beds in the manufacture of gum candies and other cast confectionery products. The starch is formed into molds with a die, and the high-solids fluid (usually sugar, corn syrup and water and, perhaps, some acid-converted starch) is poured into the starch molds and allowed to set by cooling and loss of moisture. The starch is usually redried to about 5 to 8% moisture to absorb some moisture from the cast piece.

Starch Dispersions. As mentioned in the discussion of the properties of the starch granule and its behavior in water, the starch dispersion is a complex mixture of swollen hydrated granules and granule fragments, colloidal hydrated molecular aggregates, and dissolved molecules, mainly amylose. These starch sols provide viscosity, texture, and gelling effects.[245] The selection of the particular starch or modified starch for a particular food system will be determined by the processing requirements and the characteristics required in the finished food product. The processing requirements will depend upon the shear to be applied, the temperatures, and the time factors. The other ingredients in the food system may affect the gelatinization, viscosity, and stability properties of the starch, and be affected by the starch. The type of texture, flavor, appearance, and viscosity will also be influenced by the starch. Native starches are used in food systems for thickening and gelling. However, they do have limitations such as the opacity and retrogradative tendencies of corn, sorghum, and wheat starches, and the syneresis effects of their gels (release water upon standing). Many of the cereal starches (corn, sorghum, wheat) have a characteristic flavor which may become less apparent on cooking, or be masked by other flavors. The flavor has been attributed to oxidation of the lipids in the starch catalyzed by trace amounts of iron or copper.[237a]

Tapioca and waxy cereal starches have a less pronounced starchy flavor. The flavor of potato starch and flour is apparently caused by a non-lipid, steam-volatile compound.[237a] Potato, waxy corn, and tapioca starches are used to provide temporary viscosity during high-speed, can-filling operations to prevent splashing and to keep the solids in uniform suspension. Since these starches lose viscosity on continued cooking (Fig. 22.8), the viscosity contributed by the starch will be substantially decreased, as desired, when the cans undergo heat sterilization. An improved product based on crosslinked waxy corn starch acetate has been suggested as a can-filling starch.[242]

Crosslinked starches are generally used for thickening foods to impart texture, particularly where the viscosity and/or texturizing effect might be lost because of exposure to high acidity (low pH), high temperature, longer heating times, strong agitation, or homogenization. The more severe the conditions of preparation and processing, the higher the level of crosslinking needed. Crosslinked corn and sorghum starches are used where clarity is not needed and gelling or setback is desirable or not objectionable. Crosslinked potato, waxy corn, and tapioca starches will give greater thickening powder and clarity. The tapioca starches tend to have a bland taste, which may provide a better perception of the food flavor. The waxy starches would provide less tendency for viscosity change and gelling during storage at room temperature.

Where the food storage will be at low temperatures, crosslinked starches stabilized by acetate, hydroxypropyl ether, succinate, or phosphate substituents will be useful. Low temperatures cause increased tendencies toward association and retrogradation, even with the waxy starches, so that clarity and hydration capacity of the starches are decreased. This results in syneresis or weeping and loss of eye appeal. The crosslinked, stabilized products show strong resistance to these effects and are therefore used in foods that will be refrigerated or frozen.

Starches are used as thickeners in canned foods such as puddings, pie fillings, soups, sauces, and gravies, as well as specialty products like cream-style corn, chow mein, spaghetti, and stews. The canning of foods involves preparation, can filling, can sealing, and sterilization by heating or retorting to destroy bacteria and molds which may be present in the raw foods, followed by cooling. The temperature and time of the sterilization heat treatment is set according to the type of food. A low-pH food (less than pH 4.5) may only require a temperature of about 93°C (200°F). At pH's above 4.5, temperatures as high as 127°C (260°F) may be used. The time required for sterilization temperature to be reached depends upon the rate of heat transfer (determined by the type of food), the viscosity during heating, and the amount of agitation.

Since exposure to high temperature may be detrimental to the flavor, the processor is interested in shortening the time of heating and cooling while being sure that the sterilization has been effective. Thus, special starch derivatives have been developed which have been crosslinked sufficiently to delay full gelatinization so that little viscosity is developed in the early stages of the heat treatment. The low viscosity facilitates heat transfer; hence the sterilization temperature is reached more quickly. The crosslinked starch thickens at the higher temperature of the later stage of heating and gives a higher final product viscosity since it does not break down.[243,244]

The National Canners Association has recommended standards for thermophile counts on the low-thermophile starches used in certain canning applications. For these uses, the native starch or modified starch is treated with heat, hydrogen peroxide, or peracetic acid to destroy thermophile organisms or spores in the starch.

Crosslinked tapioca and waxy corn starch acetates are used as thickeners and stabilizers in baby foods. These products impart excellent low-temperature stability and resist high-temperature and low-pH breakdown during processing.

Crosslinked waxy starches and crosslinked stabilized starches (phosphates, acetates, hydroxypropyl ethers), including the pregelatinized versions, are used as thickeners in fruit pie fillings, fruit fillings, imitation jellies, icing stabilizers, cream fillings, and whipped topping stabilizers. Corn starch is used in custard and cream fillings where a gel texture is desired.

It is instructive to review the requirements for a starch to be used in a fruit pie filling as indicative of the types of factors to be considered in deciding on a starch for any food application. The starch must resist viscosity breakdown during cooking in the presence of the fruit acids (low pH). It must bring out the fresh fruit appearance through good clarity and sheen. The starch thickener must give a short, smooth, heavy, non-rubbery texture with a pleasant "mouthfeel." The starch must not set to an opaque, rigid gel (as would be expected with native corn starch), so that it maintains the clarity and smooth texture. However, the starch must provide sufficient thickening so that the filling does not flow out when the pie is cut. The starch must have bland flavor which does not mask the fruit flavor nor impart an off-flavor. The starch should be resistant to overcooking, so that the juices do not boil out during baking.

When the pie filling or pie is stored at room temperature or lower (refrigerator or freezer), the texture should remain the same with no syneresis or weeping of liquid from the filling. Examination of the gelatinization characteristics and body and textural characteristics of various starches would lead one to select a crosslinked waxy corn or tapioca starch stabilized with acetate, phosphate, or hydroxypropyl ether substituents. The D. S. and level of crosslinking would be selected on the basis of test data.

Starch pastes serve as stabilizers for oil emulsions in mayonnaise-type salad dressings. Since these are generally of low pH, the starch must resist the viscosity breakdown by acid cooking (vinegar) and also the high shear used in homogenizing the formulation. Crosslinked starches are employed here also. In some cases, a starch containing a hydrophobic group for better emulsion stabilization may be of value. The starch must not gel or exhibit syneresis.

The candy industry uses a large amount of starch to impart texture, particularly in gum types such as jelly beans, gumdrops, orange slices, and hard gums. The starch is cooked with the sugar and syrup to a fluid sol (to which is added color and flavor), which is cast in starch molds where setting and drying occur. The essential contribution of the starch is to form a rigid gel on cooling. Acid-hydrolyzed starch (about 60 fluidity) is commonly used because of its tendency to retrograde and gel, and because it can be cooked at high solids to give a workable viscosity.

Soft gums use about 10 to 14% starch in a sol cast at 72 to 80% solids, requiring 1 to 4 days in the molds to set and dry to about 18% moisture.[238] Tapioca dextrins are used in hard gums at high solids (up to 40% of the total gum). High-amylose starches are of value since they gel much faster than do the fluidity starches, thus giving firmer gels. Time in the mold can be reduced drastically. However, the high-amylose starch must be cooked at 160°C (320°F) in a continuous cooker, to disperse it for confectionery formulations. The resulting sol must be kept hot to prevent premature gelling.

Pregelatinized starches and pregelatinized, crosslinked, stabilized starches are used in many kinds of dry mixes which the consumer mixes with cold or warm water or milk. Instant puddings are an important outlet for these materials, as are soups, cake mixes, and gravy and sauce bases.

Starch Films. When the water evaporates from a starch dispersion, a film is formed. If the dispersion is supported on a surface for which the starch has an affinity, it may bond to the surface, forming a coating. The strength of the film is determined by the size and shape of the molecules and the conditions of formation. Although high-molecular-weight starches give the strongest films, the viscosity of these starches is so high that only low-solids dispersions can be cooked and handled. Therefore, converted starches, which yield workable, pumpable dispersions at relatively high solids, are used, even though a weaker film is deposited, because a thicker film is formed. The ability of starch to form films is used to produce protective and decorative coatings, to provide a matrix for carrying various food ingredients, and to act as a binder.

Pan coatings on chewing gum and other confections may also contain fluidity starches or dextrins along with the sugar or syrup to protect the hard coating from chipping and to give a smooth, even surface. To prevent migration of oils on coatings for nuts and chocolate candies, starch or modified starches or dextrins are included in coatings.

Water-insoluble flavor oils and clouding agents can be encapsulated in a matrix of dextrins and special starch derivatives to obtain dry powders suitable for incorporation in dry mixes, providing protective action against loss of flavor through oxidation or volatilization. The process involves dispersing the dextrin or starch in water (30% solids) and adding the oil (20 to 40% on the starch). The mixture is subjected to strong agitation to form an emulsion which is spray-dried to a dry powder containing entrapped oil. This process has many applications.[238]

Starches are used as binders in emulsion meats such as bologna and sausage to keep meat, fat, and juices together. They are used as binders in extruded pet foods and snack foods.

Governmental Regulations. The U.S. Food and Drug Administration has issued regulations specifying the reagents and amounts of these reagents that are cleared in treating starch to make products suitable for direct addition to foods. These starches are designated "Food Starch—Modified" (Table 22.11).[96] The regulation should be consulted for details. Additions and changes are published in the *Federal Register*. Native starches and dextrins, not covered by this section of the regulations, are considered as GRAS on the basis of long usage and are permitted as direct additives in foods.

In Europe, each country has its own regulations on food additives, and these vary considerably. (See Reference 240 for a brief summary.) The Joint FAO/WHO Expert Committee on Food Additives (JEC) has recently reported a list of modifications of starch which are considered as permissible in food ingredients according to available information.[246,247] For all the major modified food starches, the JEC report states that these are toxicologically safe and may be used in foods without limitation by ADI values (acceptable daily intakes). The starch modifications covered include most of those listed in the FDA regulations on "Food Starch—Modified" (Table 22.11). Further studies were recommended by the European Economic Community Scientific Committee for Food.[248,249] Additional information is being obtained and reviewed. The Committee was most concerned with hydroxypropyl starches and starches treated with epichlorohydrin, and recommended that these not be used in infant foods.

The Textile Industry Starches are used in the textile industry in sizing warp yarn, in fabric finishing, in color printing, and as a component in finishes for glazing and polishing sewing thread. Only minor amounts of starch are used in the latter two applications. Warp sizing takes the major share of the starch used in the textile industry. Considerable amounts of starch are used in fabric finishing.[73,237b,250,251]

Warp Sizing. Warp sizing is a process used in the manufacture of fabrics woven from yarn. It is not applied to yarns used for knitted or non-woven fabrics. Weaving produces fabric by interlacing two sets of yarns at right angles to each other. The warp yarn runs lengthwise in the fabric and the shuttle, carrying the filling or weft yarn, runs back and forth between the warp yarns during the weaving process on the loom. Thus, the stresses and strains of the weaving process are imposed on the warp yarns, while the weft yarns receive little stress and are not sized.

The warp yarn is sized to provide a protective coating, probably discontinuous, which

TABLE 22.11
Food Starch—Modified[96] Starch Treatment[a]

A. Acid-modification:
   Hydrochloric and/or sulfuric acid

B. Bleaching:
   Hydrogen peroxide and/or peracetic acid[b]
   Ammonium persulfate (0.075) and sulfur dioxide (0.05)
   Sodium hypochlorite[c]
   Potassium permanganate (0.2)[d]
   Sodium chlorite (0.5)

C. Oxidation:
   Sodium hypochlorite[e]

D. Esterification:
   Acetic anhydride[f]
   Acetic anhydride[f] and adipic anhydride (0.12)
   Monosodium orthophosphate[g]
   1-Octenylsuccinic anhydride (3)
   1-Octenylsuccinic anhydride (2) and aluminum sulfate (2)
   Phosphorus oxychloride (0.1)
   Phosphorus oxychloride (0.1) and either acetic anhydride (8)[f] or vinyl acetate (7.5)[f]
   Sodium trimetaphosphate[h]
   Sodium tripolyphosphate and sodium trimetaphosphate[g]
   Succinic anhydride (4)
   Vinyl acetate[f]

E. Etherification:
   Acrolein (0.6)
   Epichlorohydrin (0.3)
   Epichlorohydrin (0.1) and propylene oxide (10)[i]
   Epichlorohydrin (0.1) and propylene oxide (25)[i]
   Propylene oxide (25)[i]

F. Esterification and Etherification:
   Acrolein (0.6) and vinyl acetate (7.5)[f]
   Epichlorohydrin (0.3) and acetic anhydride[f]
   Epichlorohydrin (0.3) and succinic anhydride (4)
   Phosphorus oxychloride (0.1) and propylene oxide (10)[i]

G. Miscellaneous:
   Sodium hypochlorite[e] and hydrogen peroxide[b] and propylene oxide (25)[i]
   Sodium hydroxide (1)

H. Combination Treatments[j]

---

[a]Numbers in parentheses give maximum allowable treatment level in percent on starch. Consult regulation.
[b]Not to exceed 0.45% active oxygen.
[c]Not to exceed 0.0082 lb chlorine/lb dry starch.
[d]Residual Mn not to exceed 50 ppm in starch.
[e]Not to exceed 0.055 lb chlorine/lb dry starch.
[f]Not to exceed 2.5% acetyl in starch.
[g]Residual phosphate in starch not to exceed 0.4% calculated as phosphorus.
[h]Residual phosphate in starch not to exceed 0.04% calculated as phosphorus.
[i]Residual propylene chlorohydrin not more than 5 ppm in starch.
[j]Food starch may be modified by a combination of treatments prescribed by A and/or B and any one of the treatments prescribed by C, D, E, F, or G, subject to limitations shown.

increases the tensile strength and improves abrasion resistance, thereby minimizing breaks and shedding that result in costly decreases in weaving efficiency and defects in the fabric. The fabric actually produced as a percentage of the fabric that could be produced if the loom ran continuously without stops is the weaving efficiency.

The warp sizing, or "slashing" operation, involves passing the strands of yarn, lined up side by side, into a hot size solution, then through squeeze rolls to remove the excess size, and then finally over steam-heated rolls or "cans" to dry the sized yarn. After the fabric is woven, the size is removed since it may interfere with the subsequent processes of bleaching, dyeing, or printing. The warp sizing process requires a polymer that is easy to disperse in water, giving non-congealing, non-foaming size solutions which maintain uniform viscosity and deposit non-tacky, flexible, tough, strongly adhering, and non-shedding films on the yarn. The films must be resistant to damage by heat since they may be overdried on the drying rolls.

If the size gives tacky films during drying, the sized yarn will stick to the drying cans, and/or adjacent yarns will stick to each other. If the size congeals or gels, "hard size" formation will occur with resulting "breakouts" at the slasher. The hard size consists of globs of gelled starch which dry on a section of the fibers, glueing them together. Then, on separating the yarns, the fibers are broken because of the strength of the glob of gelled starch. This is the "breakout." If insufficient size is placed on the fiber, this "soft warp" will not weave because the fiber will abrade and there will be too many breaks. If the size film does not adhere strongly and is not flexible enough, it will be shed from the fiber with consequent loss of its protective function. It also must be tough enough to withstand the abrasive action on the loom.

Since starch films are plasticized by moisture, weaving rooms are usually kept at high humidity when starch is the sizing agent. Further, the sizing film, even when overdried to low moisture, must be readily solubilized by enzymatic degradation or alkaline scouring solutions during the desizing operation.

The amount of size picked up by the yarn in the slasher, the penetration into the yarn, and the adhesion of the size to the yarn are influenced by the viscosity and solids content of the sizing solution and the pressure of the squeeze rolls. Close control of the temperature of the sizing solution is necessary to control the viscosity of the solution.

Two types of yarn are sized: synthetic continuous-filament yarns; spun yarns composed of natural fibers, synthetic fibers, or their blends. The spun yarns consist of short staple fibers which are spun and twisted into a continuous strand. The natural fibers occur as short staple fibers, but the synthetics are produced as continuous filaments which must be chopped into the short staple fiber lengths.

Since the natural-fiber spun yarns have good cohesion between the fibers in the yarn bundle, the sizing is applied as a protective coating at the surface to cement the upright surface fibers to the main body of the yarn to minimize abrasion and prevent shedding. Because spun yarns of synthetic fibers have poor interfiber cohesion and the fiber surface is smooth and slippery, penetration of the size into the yarn is needed to cement the fibers together, as well as to provide the usual coating at the surface for abrasion resistance. About 10 to 15% size on the weight of the spun yarn is usually added.

In sizing continuous-filament yarns, in addition to surface coating, the size must penetrate into the filament bundle to bond the individual filaments together, preventing their separation from the yarn bundle. This prevents fraying and entanglement with adjacent yarns when the yarn breaks. The amount of size added to filament yarns is about 3 to 5% of the yarn weight.

Starch performs well on hydrophilic yarns such as the natural fibers, but does not perform as well on the hydrophobic synthetics in terms of penetration and adhesion. Thus, the starches are widely used for sizing natural fibers and the blends of synthetic with natural fibers. Modified starches with suitable additives can be used to size synthetic spun yarns. However, starches are unsuitable for sizing synthetic filament yarns.[251]

The cost of the polymer used in sizing is a major consideration. The size is used as an aid to weaving and is then removed and discarded. It has no effect on the consumer's choice of fabric or the price he or she is willing to pay. Hence, the manufacturer will choose the size on the basis of cost and performance. On this basis, starch has held its own against synthetic polymers in warp sizing since it is inexpensive and generally works well.

Unmodified starch is often used by the manufacturer instead of modified starches because of its lower cost. However, since the viscosity of the unmodified, native starch is sometimes too high, it must be converted in the textile mill to the required viscosity. This can be done, as mentioned previously in connection with paper coating and sizing, by using heat, shear, oxidizing agents, enzymes, or acids. Open kettles or various types of continuous cookers can be employed to prepare the starch dispersion for sizing. Generally, however, for convenience, converted starches are purchased.

Lubricants, softeners, preservatives, antifoam agents, and other materials may be added to the size solution. The more important ingredients are the film formers, binders, and lubricants or waxes. Typically, about 6% on the starch of wax or other lubricant is used. To obtain better film properties and increased adhesion, particularly for spun yarns containing synthetic fibers, polyvinyl alcohol or the sodium salts of polyacrylic acid copolymers may be used as binders with starch or modified starches (typically, 20 to 50% on starch). Sodium carboxymethylcellulose and alginates have also been used as adjuncts with starch sizes.[250,251]

Acid-converted fluidity starches are used in warp sizing, mainly for cotton goods and cotton-synthetic blends, because they are low in cost and are available in a range of usable viscosities.[86] By maintaining the temperature of the size solution above 90°C (194°F), the tendency of these starches to thicken and congeal can be reduced or eliminated. Oxidized starches,[87] hydroxyethyl starch ethers,[102] starch acetates,[99] and derivatives containing cationic[122,252] and hydrophilic anionic groups are used, depending upon the type of yarn and the requirements of the particular mill.

The derivatives are generally easier to disperse, have lower gelatinization temperatures, give stable viscosities, are non-congealing, and tend to give better film properties; they are also more readily removed in the desizing operation. The high-amylose corn starch derivatives are also of interest because of their increased film strength which allows lower size add-on.

To remove the starch from the woven fabric, agents that break down the starch molecule to more soluble fragments and sugars are employed. Acids, acid salts, or oxidizing agents can be used, but may be deleterious to the fabric. Diastatic enzymes, amylases, which liquefy and saccharify the starch are preferred for desizing. This treatment is usually followed by washing and scouring.

The pollution of streams by effluent wastewaters is a major problem of textile mills, and the desizing operation is a prime contributor to this. Environmental Protection Agency regulations are requiring a drastic reduction in the actual suspended and dissolved solids discharged into the waterways. Initially, the biochemical oxygen demand (BOD) tests, which measure the amount of oxygen removed from the water by bacterial decomposition of the organic waste, indicated that highly modified starches and synthetic polymers have a lower BOD in 5 days than native starch. There was a movement to replace starch with low BOD carboxymethylcellulose, polyvinyl alcohol, and other synthetic polymeric materials. However, it would appear that the installation of wastewater treatment plants is the necessary solution to the problem.

The use of more efficient sizing agents which permit lower add-on to obtain the necessary abrasion resistance properties would be useful in lowering the pollution potential of desizing wastes. Another method is the recovery and reuse of polyvinyl alcohol sizes by ultrafiltration. This affects the starch market.

Cationic starches, dextrins, and high-amylose corn starches are used as forming sizes for glass fibers. This size is applied at the time the fibers are formed at the spinnerets to protect from abrasion and to improve handling of the bobbins. The rapid setting of the high-amylose starch reduces migration of the size as the yarn is wound at high speed on spools (high centrifugal force) and during the drying.[253] The glass forming size requires high film strength, low penetration, high abrasion resistance of the film, and low conductivity. The latter is required because the fiberglass is used in many electronic applications for insulating function (e.g., circuit boards).

Specially washed high-amylose starches with low salt content are used. The size is usually removed from the fiber by burning it off in an oxidizing atmosphere which can transform inorganic constituents present in the size to undesirable water-soluble compounds with high conductivity. Phosphoric acid or a variety of phosphate compounds or silicon-containing compounds may be added to the starch where conductive salts are present. With the use of pyrolysis to remove the size, these phosphorus or silicon compounds insolubilize the inorganic components of the ash and prevent them from contributing to conductivity.[254]

Textile Finishing. Starch is traditionally used as a finishing agent for textile fabrics to add weight, smoothness, stiffness, and strength to the cloth by filling in the interstices of the weave, thereby improving its "hand," body, and general appearance. To accomplish this, the last stage of the finishing operation involves passing the fabric through a dilute starch solution, squeezing to remove the excess (as well as to force the size into the weave), and then drying on steam-heated rolls or tenter frames which are hot air driers. The starch finish is non-durable since it doesn't resist washing and is used for inexpensive fabrics, work clothing, and white goods, as well as on fabrics such as book cloth, window shades, and interlinings. Corn starch, dextrins, oxidized, and acid-converted starches are used.

In backfilling operations, starch and a filler, such as clay or talc, are applied to the back of a fabric to fill in the interstices of the weave to increase opacity and stiffness. Fabrics of this type are used in book cloth and window shades. Generally, high-viscosity products are used. Starch acetates and hydroxyethyl starch ethers of appropriate viscosity may also be used.

The advent of new finishing demands involving wash-and-wear, durable press, waterproofing, flameproofing, mildew-proofing, and other types of functional properties has required finishes based on synthetic resins which are better able to meet these demands than starch. Although in some areas the starch derivatives such as hydroxyethyl starch ethers may be combined with insolubilizing resins such as thermosetting melamine-formaldehyde or urea-formaldehyde to give laundry-resistant finishes, starch has lost ground to synthetic resins in the finishing market.

Printing. The printing of colored patterns on cloth using thickened dye pastes was a good market for special starch derivatives which competed against natural gums and other materials. The thickener was used to provide good dispersion of the expensive dyestuff and help to give sharp transfer of the dye pattern to the cloth as well as good color value. Modified starches, such as dextrins, and crosslinked hydroxyethyl and carboxymethyl starch ethers, have been recommended for this purpose and were used in large quantities alone or in combination with other thickeners.[237b] However, another style of printing using insoluble pigments which are bound to the surface by insolubilizing binders (synthetic resins) has become predominant. Hence, the market for starch print paste thickeners has decreased to a major extent. Carpet printing still uses natural gum thickeners for dye pastes, but the starch derivatives have not been able to make inroads because of higher cost-effectiveness factors.

Adhesive Applications   Paper and paperboard converting processes comprise the major market for starch-based adhesives, including fabrication of corrugated board, paper bags, paper boxes, laminated paperboard, spiral-wound tubes, gummed labels and tapes, and other gumming applications. It was estimated that 295 million kg (650 million lb) of corn starch were consumed in corrugating in 1976.[73] In the same year, other adhesives uses were estimated to take 43 million kg (95 million lb).

Starch adhesives are manufactured to meet a variety of use requirements and may be supplied as dry powders or liquids that are ready to use.[74,237c,255,256,257] These products are classified as glues, which exhibit good mobility, viscosity stability, and pumpability through pipe lines, and pastes, which have little or no mobility. The dry powders may be cold-water-dispersible, or they may require warm water or cooking to disperse. The advantage of the liquid adhesives is that they are ready for use and have a wider range of properties since they can be made with a large number of additives and other modifications not readily done with the dry powders. Although the shelf lives of the liquid adhesives may be more limited, and they are subject to spoilage by microbiological attack, the products are adequate with proper protection. However, freight costs are higher because of the water in the formulation.

Corn, waxy corn, tapioca, potato, and wheat starches are used in the manufacture of adhesives. Each imparts its own characteristics to the final adhesive. Corn starch has strong setback or gelling characteristics unless modified. It is used mainly in the manufacture of corrugated board. Wheat starch gives a thick paste and is often used in pastes for posting on billboards and paper bag making. Potato, tapioca, and waxy corn starches can be made into fluid, stable glues giving films of good clarity. Modifications of the starch are made in the preparation of liquid glues, and/or modified starches are used in the preparation of adhesives. Dextrins, oxidized starch, acid-converted starches, hydroxyethyl starch ethers, and starch acetates can be used.

Numerous additives are included in starch adhesives. Plasticizers such as glycerol, sorbitol, glycols, glucose, corn syrup, and sucrose are used to regulate drying speed and impart some flexibility to the glue film. Humectants and liquefying agents such as urea, calcium chloride, sodium nitrate, dicyandiamide, thiocyanate salts, and acetamide, may be used to reduce viscosity or control speed of drying and "open time" (where the adhesive remains active before the substrates have to be joined). Some are also effective in dispersing the starch (e.g., urea, calcium chloride).

Mineral fillers may be added to lower costs or control penetration into porous substrates by acting as sealants or barriers. These include calcium carbonate, china clay, and titanium dioxide, and may be used at 2 to 50% on the starch base. Defoamers may also be used (sulfonated oils, etc.) and are usually chosen by trial. Preservatives are most important. Copper sulfate, benzoates, formaldehyde, and other agents are used.

When choosing a preservative, consideration must be given to toxic effects in preparation and use, the regulations of the FDA, Environmental Protection Agency (EPA), and Toxic Substances Control Act (TOSCA), as well as the efficacy in protecting the adhesive.

Sodium tetraborate (borax) is used to modify starch-based adhesives, particularly dextrins, to increase viscosity, stability, and cohesive nature of the dispersion, as mentioned in the section on dextrins. Borax is used widely in starch adhesives. By adding borax to a highly soluble British gum, it is possible to prepare an adhesive at 20% solids that is equivalent in tack and viscosity to the base dextrin at the 50% solids necessary to give a workable hydrosol.[256]

Urea acts to reduce the viscosity of starch adhesives, the opposite of the effect of borax. This is important because it permits the use of a high concentration of a less converted starch in an adhesive, giving greater flexibility and bonding strength. Note that the higher the molecular weight of the starch, the stronger the adhesive bond. Conversely, the more converted the starch, the tackier the adhesive. When a low solids is necessary to obtain a workable viscosity, the adhesive is slow to dry because more water has to be removed and poor initial tack is found. The urea allows these higher-molecular-weight starches to be used with smaller amounts of water, thereby giving more rapid development of tack. Thiocyanates, dicyandiamide, and other solubilizing chemicals act similarly.[257]

The ingredients used in adhesives that are applied to substrates which will come in contact with foods must conform to FDA regulations.[260]

Liquid Adhesives. The simplest liquid adhesives are made by cooking starch or converted starches in water with only preservatives added, yielding pastes suitable for bag making and bill posting. Addition of inorganic salts (e.g., calcium or magnesium chloride) improves the paste, and the incorporation of borax increases tack and viscosity. Wheat and corn starches are used in this application for making cigarette seams.[255]

Alkaline gelatinization of starch yields stringy, rubbery, viscous dispersions. Tapioca, waxy corn, and potato starches give fluid, transparent adhesives, whereas corn and wheat starches yield thick pastes. These dispersions may be neutralized for use at 18 to 26% solids as such, or formulated with resins to obtain water resistance. These products are used for aluminum foil–to–paper laminations and bottle labeling.

Pastes. Pastes are made from heavy-bodied starches or dextrins that have a short, non-stringy character with an appreciable amount of thixotropy. They may be neutral, alkaline, or acid, and light to dark in color. They are made from dextrins and native or slightly converted oxidized or acid-modified starches. These pastes generally are excellent adhesives for paper stock. Library pastes are made with white dextrins.

Dextrins. Borated dextrins give high tack with good aging characteristics, good adhesion, and machining properties. They are used in case sealing, carton sealing, tube winding, and laminating applications. Non-borated, highly converted dextrins (yellow dextrins and highly converted British gums) are used in high-solids, low-viscosity adhesives requiring good viscosity stability. The solids content is 50 to 65%. Formulations of these dextrins are used for remoistening adhesives, for stamps, labels, envelopes, and wherever high initial tackiness is required. They are also used in bottle labeling. British gums have filminess or spreadability of value in adhesives.

Cold-Water-Soluble Adhesives. These products are made by drum drying and the resulting flakes are usually coarsely ground and fines removed to minimize lump formation when they are redispersed. Long storage life and low shipping costs are their main advantages in the marketplace. The products are formulated to give adhesives at pH 6 to 14 at 2 to 40% solids. When water-resistance is required, they may be formulated with additives or resins during manufacture, or the user may add urea-formaldehyde or similar resin during the makeup of the liquid adhesive, controlling the pH to obtain a proper cure. These cold-water-dispersible adhesives are used for applications such as multiwall paper bags, laminated board, wallpaper paste, case and carton sealing, and tube winding.

Water-Resistant Adhesives. Starch adhesives can be made more water-resistant by formulating with thermoplastic or thermosetting resins. Mixtures with polyvinyl acetate or polyacrylate emulsions can be made using tapioca, potato, or waxy corn starch adhesives which give an intermediate level of water-resistance. As mentioned above, urea-formaldehyde or resorcinol-formaldehyde resins can be added with control of pH. Blends of starch and polyvinyl alcohol will also give some water-resistance.

Corrugating Adhesives. Starch is used as the primary adhesive in the fabrication of corrugated board in the United States.[237c,255,256,257] The adhesive consists of raw ungelatinized starch suspended in a diluted starch paste containing sodium hydroxide and borax to lower gelatinization temperature and increase tack and adhesive strength. Corn starch is the predominant starch but other starches (tapioca, potato, wheat) could be used, as dictated by economics and local availability. The adhesive is placed on the tips of the flutes and does not develop adhesive properties until the flutes and the paper liner in contact with the adhesive-coated tips are passed through the heated fluting rolls. The raw starch is instantly gelatinized, taking up the water to give an extremely tacky adhesive at 60 to 70°C (140 to 158°F).

Starch solids are in the 20 to 30% range. This two-component system of caustic-gelatinized carrier starch and raw ungelatinized starch suspended in it is known as the "Stein-Hall" system.[258] When a water-resistant bond is required, the formula includes heat-curing resins such as resorcinol-formaldehyde, acetone-formaldehyde, or urea-formaldehyde. High-amylose corn starches, used in the carrier portion, are also able to produce a formula with a high degree of water-resistance. These starches are able to be run at high speeds on the corrugator. Formulations and handling directions for corrugating adhesives are given in a new monograph.[259]

Miscellaneous Adhesive Applications. Multiwall bag adhesives are of three types: seam, cross-pasting of the plies, and bottoming. For seams and plies, the same adhesive formulation may be used, and it usually contains less water than the bottom adhesive. Thin-boiling corn starches and dextrins are used, with urea-formaldehyde if water-resistance is required. Bottom pastes contain native starch, acid-converted starches, or dextrins, formulated with borax, clay, and a lubricant (and resins if water-resistance is desired). Laminating adhesives are based on dextrins or acid-converted starches, with resins for water-resistance. Starch tube-winding adhesives include liquid adhesives, acid-converted starches, and dextrins with and without borax and caustic.

The adhesive is selected on the basis of the type of paper stock and the type of tube. Potato, tapioca, and waxy maize thin-boiling starches or combinations with dextrins have been used as the water-remoistenable adhesive on gummed tapes. The starches are modified to give low viscosity and non-congealing properties at about 50% solids. They may be used in combination with protein or resin fortifiers to try to match the animal glue tapes. Envelope seals are made from pastes of highly soluble dextrins.

Users should consult suppliers for adhesive recommendations for the specific substrate. This is a specialized field.

Miscellaneous Uses Starch is used in pharmaceutical tablets as a binder, filler, and disintegrant. Corn starch, cooked or pregelatinized, is used as a binder. Uncooked raw or bleached corn starch is used as a carrier or diluent in the tablet. Modified corn or potato starch derivatives, such as carboxymethyl starch ethers, are useful as tablet disintegrants. Starches are used as body powders. Highly crosslinked starches that resist gelatinization at sterilization temperatures are used as dusting powders for surgeons' gloves.[78f,78i]

Starch is used in tape joint cements as a binder where it must have cold-water-dispersibility, low viscosity, and stable viscosity, as well as non-lumping, low tack, and strong binding properties. Acid-converted and oxidized corn starches, starch acetates, and hydroxyethyl corn starch are used.

Corn starch and tapioca starch in ungelatinized form are used as ceiling tile binders. They are incorporated in the forming of the tile from a slurry of wood fiber or mineral wool (silicate fibers). When the wet, formed tile is dried by heating, the starch gelatinizes and binds the tile together. The starch is used at a 7 to 10% level on the mineral wool.

Corn starch derivatives are used as binders in sand molds used in foundries. High-viscosity, low-temperature gelatinization is needed.

LABORATORY TECHNIQUES

There are a number of sources of information on analytical procedures for the detection and determination of starch and modified starch components of materials, as well as for the characterization and analysis of starch and its various modifications. The interested reader should refer to these for direction to specific procedures (Refs. 36, 37, 45, 71c, 249, 261, 262, 263, 264).

TRADENAMES OF STARCH PRODUCTS

The following list is not all-inclusive. Suppliers should be called wherever a starch is to be selected for a specific use; they can suggest the starch product most likely to be effective. Since product lines are continually changing, some of the products may not be available, or others substituted with or without changing the name. The numbers in parentheses in the following tabulation refer to the subsequent list of manufacturers, while the letters refer to the notes giving the type of starch, treatment, and possible application area. The designations of base starch and treatment were obtained from company literature and an interpretation of the information given. Since this information is an interpretation, it may not be completely accurate.

Abinco Gums (4, h, i, P)
Amaizo Amioca Starch (2, a, F)
Amaizo Amylomaize VII Starch (2, j, F)
Amaizo 400 Stabilizer (2, b, c, F)
Amaizo W-13 (2, b, c, F)
Amaizo W-11 (2, b, c, F)
Amaizo 721A and 721 AE (2, b, e, F)
Amidex (7, h, t, A)
Amioca (10, a, F)
Amioca-50, 85 (10, b, r, P)
Apollo 15, 20 (11, o, u, P)
Astro Gum 21, 3010, 3020 (11, h, w, P)
Astro X-100 (11, o, u, P)
Binasol-15 (12, e, m, F)
Binasol-81 (12, c, d, e, m, F)
Bondcor-11 (5, g, C)
Capsul (10, b, F)
Cardinal Gums (4, h, q, P)
Cato (10, h, u, P)
Cato-Kote (10, u, dd, P)
Cato-Size (10, h, u, dd, P)
Ceri-Gel 300 (9, b, e, F)
Cerioca 100B (9, b, c, d, F)
Charge (5, u, P)
Claro 5541, 5581, 5591 (7, h, i, P, T)
Clinco (6, h, i, P)
Clineo (6, h, q, r, P, T)
Clinivert (t, g, n, P)
Clinsize (6, h, q, r, P)
Col-Flo 67 (10, b, c, d, f, F)
Consista Starch (12, b, c, d, F)
CPC International Starch 6448, 6485, 6486 (7, b, r, P, A, T)
CPC Internation Starch 3372 (7, g, n, P, T)
Crisp Film (10, d, j, F)
Crystal Gum (10, t, cc, F)
Cut/Las (7, h, r, T)
Delta 7444 (4, d, h, F)
Delta Gums (4, h, v, P)
Delta SD 7390, 7391 (4, c, h, F)
Douglas E, EA, EC, E 3607 (11, g, n, P)
Douglas Clearsol Gums (11, h, i, P)
Dry-Flo (10, h, bb, E, F, D)
Dura-Jel HV Starch (12, b, F)
Dura-Jel Starch (12, b, c, F)
Electra Size (4, h, r, u, P)
Electra Starch-AB 7970 (4, h, u, P)
Emulsol-Maize 488, 522, 608 (2, a, b, c, g, h, k, F)
Essex Gum (11, o, q, P)
Ethylex (12, h, q, P, T)
Express A/C Starch, Carrier Starch (4, C)
Fibersize (10, h, i, P)

Flojel Series (10, h, r, F, P, T)
Flokote 64 (10, h, i, P)
Fluftex (2, h, F)
Freezist M Starch (12, c, d, f, m, F)
Fruitfil-1 (12, c, m, F)
Hamaco 267, 277 (12, h, d, B)
Hi-Fi Starch A-B 7385 (4, b, r, A)
Hi-Fi Starch A-B 738 (4, p, A. P)
Hubstar 110 (9, b, c, F)
Hydropruf (z, j, C)
Hylon (10, j, F)
Instant Keogel 30 (9, b, c, d, e, f, F)
Instant Clearjel (10, b, c, d, e, l, F)
Interbond (12, u, P)
Jetsize 16 (9, h, P)
Keo Chlor 1075 (9, h, i, P)
Keogel 97 (9, e, h, P, A)
Keogum Starches (9, h, q, P, T)
Keojel 28, 30 (9, b, c, d, f, F)
Keokor (9, g, k, C)
Keosize 46B (9, d, h, r, P, A, T)
Kofilm (10, h, aa, P)
Kol Guard-A-B 7370, 7373, 7410 (4, b, c, d, F)
Loksize (12, u, P)
Maxi-Gel 7690, 7710, 7720 (4, b, c, d, F)
Maxi-Gel 7750 (4, c, d, h, F)
Melojel (10, g, E, F)
Miracleer-340 (12, c, d, f, h, F)
Mira-Quik C (12, j, F)
Nabond (10, j, p, C)
Nadex 772 (10, g, t, F)
National Frigex (10, c, d, m, f, F)
Nu-Col 231, 326 (12, d, h, l, F)
OK Brand (9)
OK Stabilizer PK 44 (9, b, r, F)
OK Stabilizer 2825 (9, b, c, h, k, F)
Pen-Cor (11, h, q, C)
Pen-Cote (11, h, q, P)
Penford Gum (11, h, q, P)
Perma-Flo Starch (12, b, c, d, f, F)
Polar-Gel 5, 8, 10, 15 (2, b, c, d, f, F)
Polymer 35-A-100 (8, y, z, M)
Pure-Flo (10, b, c, d, F)
Purity 21 (10, h, F)
Purity-69 A (10, c, d, m, F)
Purity Gum BE (10, b, F)
Redisol-4 (12, c, d, e, f, m, F)
Redisol 313 (12, e, o, F)
Redi-Tex Starch (12, c, e, h, F)
Rezista Starch (12, b, c, d, f, F)
Secap 7400 (4, b, d, t, F)
S G P Absorbent Polymer (13, y, z, M)
Shur-Fil 327 Starch (12, b, d, F)
Shur-Fil CS500 Starch (12, b, c, d, F)

Snow Flake Starch 4408 (7, a, F)
Snow Flake Starch 4814 (7, b, d, h, k, F)
Snow Flake Starch 4816 (7, b, d, F)
Sta-Lok (12, u, P)
Sta-O-Paque Starch (12, k, p, F)
Starpol-100 (12, h, x, P)
Tenderfil-8 (12, c, d, f, m, F)
Textaid (10, h, F)

Thin-N-Thik 35 (12, c, h, F)
Thin-N-Thik 65 (12, c, d, m, F)
Thin-N-Thik 99 (12, b, c, d, F)
Ultra-Fast 3820 (7, h, C)
X-Tra Gel 7 and 9 (2, b, c, F)
Vinamyl-II (10, j, p, C)
Vulca-90 (10, c, h, M)

Type of Starch, Probable Treatment, and Possible Applications (for the preceding tabulation of tradenames)

Applications
A. Remoistenable Adhesives
B. Building/Construction
C. Corrugating
D. Cosmetics
E. Pharmaceuticals
F. Foods
G. Adhesives
L. Laundry
M. Miscellaneous
P. Paper
T. Textiles

Types and Treatments
a. Unmodified waxy corn starch
b. Modified waxy corn starch
c. Crosslinked
d. Stabilizing modification
e. Pregelatinized
f. Low temperature stable
g. Corn starch
h. Modified corn starch
i. Oxidized
j. High-amylose corn starch
k. Blend
l. Cold-water-swelling
m. Modified tapioca starch
n. Buffered, enzyme-converted starch
o. Potato starch
p. Modified
q. Hydroxyethyl starch ether
r. Acid-modified
s. Redried starch
t. Dextrin or enzyme-converted starch
u. Cationic starch
v. Esterified starch
w. Anionic starch
x. Acrylamidomethyl ether
y. Starch graft polymer
z. Saponified starch–polyacrylonitrile graft polymer
aa. Acetylated
bb. Hydrophobic
cc. Tapioca
dd. Converted

U.S. Manufacturers of Starches (for the preceding tabulation of tradenames)
1. ADM Corn Sweeteners (Division of Archer-Daniels-Midland, Co.)
2. American Maize-Products Co.

3. Amstar Corporation
4. Anheuser-Busch, Inc.
5. Cargill, Inc.
6. Clinton Corn Processing Co. (A Division of Standard Brands, Inc.)
7. CPC International, Inc.
8. Grain Processing Corp.
9. The Hubinger Co.
10. National Starch and Chemical Corp.
11. Penick Ford, Ltd., Division of Pacific Resins and Chemicals (a subsidiary of Univar Corp)
12. A. E. Staley Manufacturing Co.

Manufacturer of special starches:
13. General Mills Chemicals, Inc.

ACKNOWLEDGMENT

The author acknowledges with thanks the following individuals who reviewed parts or all of the manuscript and offered suggestions for clarification and accuracy: T. Agin, R, Boettger, K. Bristol, R. Forsythe, L. Gaspar, M. Hasuly, M. McKenna, B. Nappen, H. Olsen, P. Reitano, R. W. Reiter, W. Sederlund, K. Sickafoose, P. Smith, M. Tessler, O. B. Wurzburg, G. Zwiercan. The author also thanks Maureen Szabelak and Shirley Laggini, who typed the manuscript and were very patient.

FURTHER READING

R. L. Whistler and E. F. Paschal (eds.), *Starch: Chemistry and Technology,* Academic Press, New York, vol. I: Fundamental Aspects, 1965; vol. II: Industrial Aspects, 1967. Very good, comprehensive compendium of starch chemistry, properties, uses with good bibliographies at end of each chapter.
R. W. Kerr (ed.), *Chemistry and Industry of Starch,* Academic Press, New York, 1950. Good review of fundamental and applied knowledge available prior to 1950. Contains information not necessarily found in later books.
J. A. Radley (ed.), *Starch and Its Derivatives,* 4th ed., Chapman and Hall, Ltd., London, 1968. Review of starch chemistry with a number of chapters on the action of enzymes.
W. Banks and C. T. Greenwood, *Starch and Its Components,* Halsted Press, New York, 1975. Review of fundamentals of starch chemistry and structure.
J. A. Radley (ed.), *Starch Production Technology,* Applied Science Publishers, Ltd., London, 1976.
J. A. Radley (ed.), *Industrial Uses of Starch and Its Derivatives,* Applied Science Publishers, Ltd., London, 1976. Good review of applications.
J. A. Radley (ed.), *Examination and Analysis of Starch and Starch Products,* Applied Science Publishers, Ltd., 1976.
R. W. James (ed.), *Industrial Starches,* Noyes Data Corp., Park Ridge, N.J., 1974. Compilation of U.S. patents issued from the early 1960s to 1974. Covers modifications and applications of starch.
R. L. Whistler (ed.), *Methods in Carbohydrate Chemistry,* vol. IV: "Starch," Academic Press, New York, 1964. Excellent compilation of methods for preparation of starch, starch fractions, modifications, and derivatives; analytical procedures; physical methods of analysis; microscopy. The series (volumes I to VII) is an excellent reference for all types of carbohydrates, covering preparation and analysis. Some procedures applicable to starch will be found in other volumes.
W. Jarowenko, "Starch," pp. 787–862, in N. M. Bikales (ed.), *Encyclopedia of Polymer Science and Technology,* vol. 12, John Wiley & Sons, 1970. Brief review of starch chemistry and technology with good bibliography.
N. P. Badenhuizen, *The Biogenesis of Starch Granules in Higher Plants,* Appleton-Century-Crafts, New York, 1969. A review of knowledge on growth of starch granules.

Periodicals of interest are:
1. *Die Staerke* (Starch). An international journal with articles on all aspects of starch in English, German, or French. Includes patent abstracts (in German).
2. *Cereal Chemistry,* published by the American Association of Cereal Chemists. Publishes original research papers on starch and other cereal products.

REFERENCES

1. W. Banks and C. T. Greenwood, *Starch and Its Components,* Halsted Press, New York, 1975.
1a. Reference 1, pp. 140–168, 196.

1b. Reference 1, pp. 44–66.
1c. Reference 1, pp. 242–273.
2. C. T. Greenwood, "Aspects of the Physical Chemistry of Starch," pp. 335–393, in M. L. Wolfrom and R. S. Tipson (eds.), *Advances in Carbohydrate Chemistry*, vol. 11, Academic Press, New York, 1956.
3. J. F. Foster, "Physical Properties of Amylose and Amylopectin in Solution," pp. 349–391, in R. L. Whistler and E. F. Paschall (eds.), *Starch: Chemistry and Technology*, vol. I, Academic Press, New York, 1965.
4. C. T. Greenwood and J. Thomson, *J. Chem. Soc., (London)*, 222–229 (1962). The Fractionation and Characterization of Starches of Various Plant Origins.
5. R. Geddes, C. T. Greenwood, and S. MacKenzie, *Carbohydr. Res.*, 1, 71–82 (1965). Properties of Components of Starches from the Growing Potato Tuber.
6. D. P. Langlois and J. A. Wagoner, "Production and Use of Amylose," pp. 451–497, in R. L. Whistler and E. F. Paschall (eds.), *Starch: Chemistry and Technology*, vol. II, Academic Press, New York, 1967.
7. R. L. Whistler and C. Johnson, *Cereal Chem.*, 25, 418–424 (1948). Effect of Acid Hydrolysis on the Retrogradation of Amylose.
8. S. Lansky, M. Kooi, and T. J. Schoch, *J. Am. Chem. Soc.*, 71, 4066–4075 (1949). Properties of the Fractions and Linear Subfractions from Various Starches.
9. M. Ott and E. E. Hester, *Cereal Chem.*, 42, 476–484 (1965). Gel Formation As Related to Concentration of Amylose and Degree of Starch Swelling.
10. F. A. Loewus and D. R. Briggs, *J. Am. Chem. Soc.*, 79, 1494–1497 (1957). A Potentiometric Study of the Change in Iodine Binding Capacity of Amylose While Retrograding in Dilute Solution.
11. A. J. Kalb and C. Sterling, *J. Appl. Polym. Sci.*, 6, 571–574 (1962). Role of Hydrogen Ion Concentration in Retrogradation of Starch Sols.
12. P. Hiemstra, W. C. Bus, and J. M. Muetgeert, *Die Staerke*, 8, 235–241 (1956). Fractionation of Starch.
13. J. M. Bailey and W. J. Whelan, *J. Biol. Chem.*, 236, 969–973 (1961). Physical Properties of Starch. I: Relation Between Iodine Stain and Chain Length.
14. J. A. Thoma and D. French, *J. Am. Chem. Soc.*, 82, 4144–4147 (1960). The Starch-Iodine-Iodide Interaction. Part I. Spectrophotometric Investigation.
15. D. French, A. O. Pulley, and W. J. Whelan, *Die Staerke*, 15, 349–354 (1963). Starch Fractionation by Hydrophobic Complex Formation.
16. J. Muetgeert, "Fractionation of Starch," pp. 299–333, in M. L. Wolfrom and R. S. Tipson (eds.), *Advances in Carbohydrate Chemistry*, vol. 16, Academic Press, New York, 1961.
17. R. L. Whistler and G. E. Hilbert, *J. Am. Chem. Soc.*, 67, 1161–1165 (1945). Separation of Amylose and Amylopectin by Certain Nitroparaffins.
18. E. M. Osman, "Starch in the Food Industry," pp. 163–215, in Whistler and Paschall (eds.), *Starch: Chemistry and Technology*, vol. II, op. cit.
19. S. Erlander and D. French, *J. Polym. Sci.*, 20, 7–28 (1956). A Statistical Model for Amylopectin and Glycogen. The Condensation of A-R-B$_{f-1}$ Units.
20. D. French, "Chemistry and Biochemistry of Starch," pp. 267–335, in W. J. Whelan (ed.), *Biochemistry of Carbohydrates*, Biochemistry Series one, vol. 5, University Park Press, Baltimore, 1975.
21. D. French, *J. Anim. Sci.*, 37, 1048–1061 (1973) Chemical and Physical Properties of Starch.
22. D. French, *Denpun Kagaku J. Jpn. Soc. Starch Sci.*, 19, 8–25 (1972). Fine Structure of Starch and Its Relationship to the Organization of Starch Granules.
23. W. Banks, R. Geddes, C. T. Greenwood, and I. G. Jones, *Die Staerke*, 24, 245–251 (1972). The Molecular Size and Shape of Amylopectin.
24. S. R. Erlander and D. French, *J. Am. Chem. Soc.*, 80, 4413–4420 (1958). Dispersion of Starch Granules and the Validity of Light Scattering Results on Amylopectin.
25. H. W. Leach, "Gelatinization of Starch," pp. 289–307, in Whistler and Paschall (eds.), *Starch: Chemistry and Technology*, vol. I, op. cit.
26. L. Sair and W. R. Fetzer, *Ind. Eng. Chem.*, 36, 205–208 (1944), Water Sorption By Starches. 36, 316–319 (1944), Water Sorption by Corn Starch and Commercial Modifications of Starches.
27. N. N. Hellman, T. F. Boesch, and E. H. Melvin, *J. Am. Chem. Soc.*, 74, 348–350 (1952). Starch Granule Swelling in Water Vapor Sorption.
28. N. N. Hellman and E. H. Melvin, *J. Am. Chem. Soc.*, 72, 5186–5188 (1950). Surface Area of Starch and Its Role in Water Sorption.
29. H. W. Leach, T. J. Schoch, and E. F. Chessman, *Die Staerke*, 13, 200–203 (1961). Adsorption of Alkalies by the Starch Granule.
30. E. B. Lancaster and H. F. Conway, *Cereal Sci. Today*, 13, 248–249 (1968). Alkali Sorption and Swelling of Starch.
31. C. E. Mangels and C. H. Bailey, *J. Am. Chem. Soc.*, 55, 1981–1988 (1933). Relation of Concentration to Action of Gelatinizing Agents on Starch.
31a. B. M. Gough and J. N. Pybus, *Die Staerke*, 25, 123–130 (1973). Effect of Metal Cations on the Swelling and Gelatinization Behavior of Large Wheat Starch Granules.

32. D. G. Medcalf and K. A. Gillies, *Die Staerke*, 18, 101–105, (1966). Effect of a Lyotropic Ion Series on the Pasting Characteristics of Wheat and Corn Starches.
33. S. R. Erlander, R. Tobin, and R. J. Dimler, Abstracts of Papers, 144th ACS Meeting, 3/31–4/5/63, p. 18c. Dispersion of Starch Granules at Room Temperature with a New Solvent System and Molecular Weights of the Resulting Components.
34. H. W. Leach and T. J. Schoch, *Cereal Chem.*, 39, 318–327 (1962). Solubilities of Granular Starches in Dimethyl Sulfoxide.
35. R. H. Kurtzman, Jr., F. T. Jones, and G. F. Bailey, *Cereal Chem.*, 50, 312–322 (1973). Dissolution of Starches in Dimethyl Sulfoxide and Variations in Starches of Several Species, Varieties and Maturities.
36. R. J. Smith, "Characterization and Analysis of Starches," pp. 569–635, in Whistler and Paschall (eds.), *Starch: Chemistry and Technology*, vol. II, op. cit.
37. R. L. Whistler (ed.), *Methods in Carbohydrate Chemistry*, vol. IV, "Starch," Academic Press, New York, 1964. This volume covers a wide range of analytical and preparative methods used in starch chemistry and is highly recommended.
38. F. E. Kite, T. J. Schoch, and H. W. Leach, *Baker's Dig.*, 31, 42–46 (1957). Granule Swelling and Paste Viscosity of Thick Boiling Starches.
39. H. W. Leach, L. D. McCowen, and T. J. Schoch, *Cereal Chem.*, 36, 534–544 (1959). Structure of the Starch Granule. I: Swelling and Solubility Patterns of Various Starches.
40. B. Zaslow, "Crystalline Nature of Starch," pp. 279–287, in Whistler and Paschall (eds.), *Starch: Chemistry and Technology*, vol. I, op. cit.
41. H. B. Wigman, W. W. Leathen, and M. J. Brackenmeyer, *Food Technol.*, 10, 179–184 (1956). Phase Contrast Microscopy in the Examination of Starch Granules.
42. B. S. Miller, R. I. Derby, and H. B. Trimbo, *Cereal Chem.*, 50, 271–280 (1973). A Pictorial Explanation for the Increase in Viscosity of a Heated Wheat Starch–Water Suspension.
43. R. D. Hill and B. L. Dronzek, *Die Staerke*, 25, 367–372 (1973). Scanning Electron Microscopy Studies of Wheat, Potato and Corn Starch during Gelatinization.
44. F. R. Senti, "High Amylose Corn Starch," pp. 499–552, in Whistler and Paschall (eds.), *Starch: Chemistry and Technology*, vol. II, op. cit.
45. R. W. Kerr, *Chemistry and Industry of Starch*, 2d ed., Academic Press, New York, 1968, pp. 14–17, 133, 679.
46. G. E. Moss, "The Microscopy of Starch," pp. 1–31, in J. A. Radley (ed.), *Examination and Analysis of Starch and Starch Products*, Applied Science Publishers, Ltd., London, 1976.
47. T. J. Schoch and E. C. Maywald, "Microscopic Examination of Modified Starches," *Anal. Chem.*, 28, 382–387 (1956).
48. T. J. Schoch and E. C. Maywald, "Industrial Microscopy of Starches," pp. 637–647, in Whistler and Paschall (eds.), *Starch: Chemistry and Technology*, vol. II, op. cit.
49. M. J. Wolf, H. L. Seckinger, and R. J. Dimler, *Staerke*, 16, 375–380 (1964). Microscopic Characteristics of High Amylose Corn Starches.
50. R. L. Whistler and P. Weatherwax, *Cereal Chem.*, 25, 71–75 (1948). Amylose Content of Indian Corn Starches from North, Central and South American Corns.
51. W. L. Deatherage, M. M. MacMasters, and C. E. Rist, *Trans. Am. Assoc. Cereal Chem.*, 13, 31–42 (1955). A Partial Survey of Amylose Content in Starch from Domestic and Foreign Varieties of Corn, Wheat and Sorghum and from Some Other Starch-Bearing Plants.
52. O. A. Sjostrum, *Ind. Eng. Chem.*, 28, 63–74 (1936). Microscopy of Starches and Their Modifications.
53. G. P. Wivinis and E. C. Maywald, "Photographs of Starches," pp. 649–683, in Whistler and Paschall (eds.), *Starch: Chemistry and Technology*, vol. II, op. cit.
54. D. J. Gallant and C. Sterling, "Electron Microscopy of Starch and Starch Products," pp. 33–59, in J. A. Radley (ed.), *Examination and Analysis of Starch and Starch Products*, Applied Science Publishers, Ltd., London, 1976.
55. M. S. Buttrose, *Die Staerke*, 15, 85–92 (1963). Electron Microscopy of Acid-Degraded Starch Granules.
56. R. L. Whistler, J. D. Byrd, and W. L. Thornburg, *Biochim. Biophys. Acta*, 18, 146–147 (1955). Surface Structure of Starch Granules.
57. R. L. Whistler and E. S. Turner, *J. Polym. Sci.*, 18, 153–156 (1955). Fine Structure of Starch Granule Sections.
58. R. L. Whistler, W. W. Spencer, J. L. Goatley, and Z. Nikuni, *Cereal Chem.*, 35, 331–336 (1958). Effect of Drying on the Presence of Cavities in Corn Starch Granules.
59. R. L. Whistler, J. L. Goatley, and W. W. Spencer, *Cereal Chem.*, 36, 84–90 (1959). Effect of Drying on the Physical Properties and Chemical Reactivity of Corn Starch Granules.
60. A. D. Evers, *Die Staerke*, 21, 96–99 (1969). Scanning Electron Microscopy of Wheat Starch.
61. D. M. Hall and J. G. Sayre, *Text. Res. J.*, 39, 1044–1052 (1969); 40, 256–266 (1970); 41, 880–894 (1971). A Scanning Electron Microscope Study of Starches.
62. D. M. Hall and J. G. Sayre, *Text. Res. J.*, 40, 147–157 (1970); 41, 404–414 (1971). Internal Architecture of Potato and Canna Starch. Crushing and Swelling Studies.
63. D. M. Hall and J. G. Sayre, *Die Staerke*, 25, 119–123 (1973). A Comparison of Starch Granules as Seen by Scanning Electron and Ordinary Light Microscopy.

64. J. F. Chabot, L. F. Hood, and J. E. Allen, *Cereal Chem.*, 53, 85–91 (1976). Effect of Chemical Modifications on the Ultra Structure of Corn, Waxy Maize and Tapioca Starches.

65. M. L. Bean and E. M. Osman, *Food Res.*, 24, 665–671 (1959). Effects of Different Sugars on the Viscosity and Gel Strength of Starch Paste.

66. V. M. Gray and T. J. Schoch, *Die Staerke*, 14, 239–246 (1962). Effects of Surfactants and Fatty Adjuncts on the Swelling and Solubilization of Granular Starches.

67. E. M. Osman and M. R. Dix, *Cereal Chem.*, 37, 464–475 (1960). Effects of Fats and Nonionic Surface-Active Agents on Starch Pastes.

68. N. Krog, *Die Staerke*, 23, 206–210 (1971). Amylose Complexing Effect of Food Grade Emulsifiers.

69. N. Krog, *Die Staerke*, 25, 22–27 (1973). Influence of Food Emulsifiers on Pasting Temperature and Viscosity of Various Starches.

70. H. Van LonKhuysen and J. Blankenstijn, *Die Staerke*, 26, 337–342 (1974). Interaction of Monoglycerides with Starches.

71. "Products of the Corn Refining Industry in Food," Seminar Proceedings, May 9, 1978, Corn Refiners Association, Inc., Washington, D.C.

71a. J. Harness, Wet-Milling Industry, in Reference 71.

71b. O. B. Wurzburg, Starch, in Reference 71.

71c. K. M. Brobst, Analysis, in Reference 71.

72. W. M. Doane, *J. Coating Technol.*, 50, 88–98 (1978). Starch: Renewable Raw Material for the Chemical Industry.

73. J. E. Maryanski, *Pulp Pap.*, 104–105 (May 1977). A Look at Trends in Starch Usage for Paper and Container Industries.

74. C. R. Russell, "Industrial Use of Corn Starch," pp. 262–284, in *Industrial Uses of Cereals*, Symposium Proceedings, 58th Annual Meeting (Y. Pomeranz, Chairman), American Association of Cereal Chemists, St. Louis, Mo., 1973. (Obtainable from the Association, 3340 Pilot Knob Road, St. Paul, MN 55121.)

75. J. A. Radley (ed.), *Starch Production Technology*, Applied Science Publishers, Ltd., London, 1976.

76. S. A. Watson, "Manufacture of Corn and Milo Starches," pp. 1–51; R. A. Anderson, "Manufacture of Wheat Starch," pp. 53–63; J. T. Hogan, "The Manufacture of Rice Starch," pp. 65–78; R. H. Treadway, "Manufacture of Potato Starch," pp. 87–101; and L. Shipman, "Manufacture of Tapioca, Arrowroot and Sago Starches," pp. 103–119, in Whistler and Paschall (eds.), *Starch: Chemistry and Technology*, vol. II, op. cit.

77. W. C. Bus, J. Muetgert, and P. Hiemstra, U.S. Patents 2,829,987 to 2,829,990 (1958). Covering the AVEBE Process for Starch Fractionation.

78. E. L. Powell, "Production and Use of Pregelatinized Starches, pp. 523–536, in Whistler and Paschall (eds.), *Starch: Chemistry and Technology*, vol. II, op. cit.

78a. C. H. Hullinger, "Production and Use of Cross-Linked Starch," pp. 445–450, in Whistler and Paschall (eds.), *Starch: Chemistry and Technology*, vol. II, op. cit.

78b. G. F. Felton and H. H. Schopmeyer, U.S. Patent 2,328,537 (1943) (to American Maize Products Co.). Cross-linking granular starch by treatment with phosphorus oxychloride.

78c. R. W. Kerr and F. C. Cleveland, Jr., U.S. Patent 2,801,242 (1957) (to Corn Products Refining Co.). Cross-linking granular starch by treatment with sodium trimetaphosphate.

78d. M. Konigsberg, U.S. Patent 2,500,950 (1950) (to National Starch and Chemical Corp.). Cross-linking of granular starches by treatment with epichlorohydrin and other difunctional halogen compounds.

78e. G. E. Hamerstrand, B. T. Hofreiter, and C. L. Mehltretter, *Cereal Chem.*, 37, 519–524 (1960). Determination of the Extent of Reaction Between Epichlorohydrin and Starch.

78f. D. Trimnell, C. P. Patel, and J. F. Johnston, U.S. Patent 3,035,045 (1962) (to Union Starch and Refining Corp.). Reaction of Starch with N,N'-methylene-bis-acrylamide in presence of an oxidation-reduction catalyst.

78g. T. S. W. Gerwitz, U.S. Patent, 2,805,220 (1957) (to Anheuser-Bush, Inc.). Cross-linking granular starch with cyanuric chloride.

78h. C. G. Caldwell, T. A. White, W. L. George, and J. J. Eberl, U.S. Patent 2,626,257 (1953) (to National Starch and Chemical Corp.). Medical dusting powder made by cross-linking granular starch with epichlorohydrin or dihalohydrins.

78i. R. W. Kerr and F. C. Cleveland, Jr. U.S. Patent 2,938,901 (1960) (to Corn Products Co.). Surgical dusting powder made by crosslinking granular starch with sodium trimetaphosphate.

79. O. B. Wurzburg, "Corn Starch and Modified Starch," in Symposium Proceedings, Products of the Wet-Milling Industry in Food, Corn Refiners Association, Inc., Washington, D.C., 1970.

80. "Corn Starch," 3d ed., Corn Industries Research Foundation, Inc. (now Corn Refiners Association, Inc.), Washington, D. C., 1964.

81. I. A. Wolff, H. A. Davis, J. E. Cluskey, L. J. Gundrum, and C. E. Rist, *Ind. Eng. Chem.*, 43, 915–919 (1951). Preparation of Films from Amylose.

82. I. A. Wolf, D. W. Olds, and G. E. Hilbert, *Ind. Eng. Chem.*, 43, 911–914 (1951). Triesters of Corn Starch, Amylose and Amylopectin.

83. W. B. Roth, and C. L. Mehltretter, *Food Technol.*, **21**, 72–74 (1967). Some Properties of Hydroxypropylated Amylomaize Starch Films.
84. J. C. Rankin, I. A. Wolfe, H. A. Davis, and C. E. Rist, *Ind. Eng. Chem., Chem. Eng. Data Ser.*, **3**, 120–123 (1958). Permeability of Amylose Film to Moisture Vapor, Selected Organic Vapors and the Common Gases.
85. C. H. Hullinger, *Cereal Sci. Today*, **10**, 508–510 (1965). Starch Film and Coating.
85a. G. A. Politis and J. J. McCabe, *Package Eng.*, **16**(4), 58–60 (1971). Edible Water-Soluble Films Break the Machinability Barrier. (High Amylose Starch Films.)
86. P. Shildneck and C. E. Smith, "Acid-Modified Starch: Production and Use," pp. 217–235, in Whistler and Paschall (eds.), *Starch: Chemistry and Technology*, vol. II, op. cit.
87. B. L. Scallet and E. A. Sowell, "Production and Use of Hypochlorite-Oxidized Starches," pp. 237–251, in Whistler and Pachall (eds.), *Starch: Chemistry and Technology*, vol. II, op. cit.
88. J. Schmorak, D. Mejzler, and M. Lewin, *Die Staerke*, **14**, 278–290 (1962). A Study of the Mild Oxidation of Wheat Starch and Waxy Maize Starch by Sodium Hypochlorite in the Alkaline pH Range.
89. D. French, A. O. Pulley, and W. J. Whelan, *Die Staerke*, **15**, 280–284 (1963). Preparation of Schardinger Dextrins on a Larger-than-Laboratory Scale.
90. R. B. Evans and O. B. Wurzburg, "Production and Use of Starch Dextrins," pp. 253–278, in Whistler and Paschall (eds.), *Starch: Chemistry and Technology*, vol. II, op. cit.
91. D. Horton, "Pyrolysis of Starch," pp. 421–437, in Whistler and Paschall (eds.), *Starch: Chemistry and Technology*, vol. I, op. cit.
92. B. Brimhall, *Ind. Eng. Chem.*, **36**, 72–75 (1944). Structure of Pyrodextrins.
93. H. C. Srivastava, R. S. Parmar, and G. B. Dave, *Die Staerke*, **22**, 49–54 (1970). Studies on Dextrinization. Part I. Pyrodextrinization of Corn Starch in the Absence of Any Added Catalyst.
94. H. J. Roberts, "Nondegradative Reactions of Starch," pp. 439–493, in Whistler and Paschall (eds.), *Starch: Chemistry and Technology*, vol. I, op. cit.
95. H. J. Roberts, "Starch Derivatives," pp. 293–350, in Whistler and Paschall (eds.), *Starch: Chemistry and Technology*, vol. II, op. cit.
96. Code of Federal Regulations, Title 21, Chapter I, Part 172, Food Additives Permitted in Food for Human Consumption, Section 172.892, Food Starch—Modified, U.S. Government Printing Office, Washington, D.C., 1978.
96a. Code of Federal Regulations, Title 21, Chapter I, Part 178, Indirect Food Additives, Section 178.3520, Industrial Starch—Modified, U.S. Government Printing Office, Washington, D.C., 1978.
97. M. W. Rutenberg and W. Jarowenko, U.S. Patent 3,038,895 (1962) (to National Starch and Chemical Corp.). Water-dispersible, partially substituted derivatives of amylose.
98. R. B. Evans, U.S. Patent 3,130,081 (1964) (to National Starch and Chemical Corp.). Limited stability amylose dispersions and method of preparation. Esterification of amylose.
99. L. H. Kruger and M. W. Rutenberg, "Production and Uses of Starch Acetates," pp. 369–401, in Whistler and Paschall (eds.), *Starch: Chemistry and Technology*, vol. II, op. cit.
100. C. G. Caldwell, U.S. Patent 2,461,139 (1949) (to National Starch and Chemical Corp.). Reaction of granular starch in aqueous suspension with carboxylic acid anhydrides to form ungelatinized starch esters.
101. J. V. Tuschhoff and C. E. Smith, U.S. Patents 2,928,828 (1960) and 3,022,289 (1962) (to A.E. Staley Mfg. Co.). Acetylation and esterification of starch by transesterification with vinyl esters.
102. E. J. Hjermstad, "Production and Uses of Hydroxyethyl Starch," pp. 423–432, in Whistler and Paschall (eds.), *Starch: Chemistry and Technology*, vol. II, op. cit.
103. C. C. Kesler and E. T. Hjermstad, U.S. Patent 2,516,633 (1950) (to Penick and Ford, Ltd.). Preparation of hydroxyalkyl starch by wet process.
104. C. C. Kesler and E. T. Hjermstad, U.S. Patents 2,516,632 (1950) and 2,516,634 (1950) (to Penick and Ford, Ltd.). R. W. Kerr and W. A. Faucette, U. S. Patent 2,733,238 (1956) (to Corn Products Refining Co.). Reaction of Starch with Alkylene Oxide Vapors to Prepare Hydroxyalkyl Starch.
105. C. C. Kesler, E. T. Hjermstad, U.S. Patent 2,845,417 (1958) (to Penick and Ford, Ltd.). Hydroxyalkylation of ungelatinized starches and dextrins in aqueous, water-miscible alcohols with alkylene oxides.
106. H. C. Srivastava, K. V. Ramalingam, and N. M. Dushi, *Die Staerke*, **19**, 295–300 (1967); **21**, 181–193 (1969). Distribution of hydroxyethyl Groups in Commercial Hydroxyethyl Starch.
107. G. N. Bollenbach, R. S. Golick, and F. W. Parrish, *Cereal Chem.*, **46**, 304–309 (1969). Distribution of Hydroxyethyl Groups in Hydroxyethyl Starch.
108a. W. Banks, C. T. Greenwood, and D. D. Muir, *Die Staerke*, **24**, 181–212 (1972). A review of the chemistry of hydroxyethyl starch, with reference to its use as a blood plasma volume expander.
108b. W. Banks, C. T. Greenwood, and D. D. Muir, *Br. J. Pharmacol.*, **47**, 172–178 (1973). The Structure of Hydroxyethyl Starch.

109. C. G. Caldwell, U.S. Patent 2,825,727 (1958) (to National Starch and Chemical Corp.). Preparation of granular starch derivatives containing sulfonate groups.
110. O. B. Wurzburg, M. W. Rutenberg, and L. J. Ross, U.S. Patent 2,786,833 (1957) (to National Starch and Chemical Corp.); R. W. Kerr, E. F. Paschall, and W. H. Minkema, U.S. Patent 2,967,178 (1961) (to Corn Products Co.); L. H. Kruger, O. B. Wurzburg, U.S. Patent 3,441,558 (1969) (to National Starch and Chemical Corp.). These patents cover the sulfation of starch and starch fractions using tertiary amine-sulfur trioxide complexes.
111. H. E. Smith, C. R. Russell, and C. E. Rist, Cereal Chem., 39, 273–281 (1962). Preparation and Properties of Sulfated Wheat Flour.
112. R. G. Schweiger, U.S. Patent, 3,401,160 (1968) (to Kelco Co.). Sulfation of Starch with Sulfur trioxide–Dimethylformamide Complex in DMF or Dimethylsulfoxide.
113. R. L. Whistler, U.S. Patent 3,507,855 (1970) (to Purdue Research Foundation). Sulfation of Starch with Sulfur trioxide–Dimethylsulfoxide Complex in Dimethylsulfoxide.
114. T. Astrup, I. Galsman, and M. Volkert, Acta Physiol. Scand., 8, 215–226 (1944); Chem. Abstr., 39, 4976 (1945). Polysaccharide Sulfuric Acid (Esters) as Anticoagulants.
115. R. M. Hamilton and E. F. Paschall, "Production and Uses of Starch Phosphates," pp. 351–368, in Whistler and Paschall (eds.), Starch: Chemistry and Technology, vol. II, op. cit.
116. R. W. Kerr and F. C. Cleveland, U.S. Patents 2,884,413 (1959) and 2,961,440 (1960) (to Corn Products Co.). Starch orthophosphate esters prepared by heating starch with sodium orthophosphate, sodium tripolyphosphate, sodium trimetaphosphate, sodium pyrophosphate and sodium polymetaphosphate.
117. H. Neukom, U.S. Patents 2,865,762 (1958) and 2,884,412 (1959) (to International Minerals and Chemical Corp.). Starch phosphate prepared by heating starch with mixture of mono- and di-sodium orthophosphate.
118. R. E. Gramera, J. Heerema, and F. W. Parrish, Cereal Chem., 43, 104–111 (1966). Distribution and Structural Form of Phosphate Ester Groups in Commercial Starch Phosphates.
118a. G. Harris and I. C. MacWilliams, Die Staerke, 15, 98–101 (1963). Phosphate in Starches and Glycogens.
118b. C. T. Greenwood and A. W. MacGregor, Die Staerke, 21, 199–202 (1969). The Action of Phosphatase on Potato Amylose and Amylopectin.
119a. W. F. Filbert, U.S. Patent 2,599,620 (1952) (to E. I. du Pont de Nemours and Co.). Treatment of starch with chloroacetic acid in aqueous alcohol.
119b. E. T. Hjermstad and C. C. Kesler, U.S. Patents 2,773,057 (1956), 3,062,810 (1962) (to Penick and Ford, Ltd.); L. O. Gill and J. A. Waggoner, U.S. Patent 3,014,901 (1961) (to A. E. Staley Mfg. Co.). Preparation of granular carboxymethyl starch.
120. J. W. Sloan, C. L., Mehltretter, and F. R. Senti, J. Chem. Eng. Data, 7, 156–158 (1962). Carboxymethyl High Amylose Starch.
121. C. G. Caldwell and I. Martin, U.S. Patent 2,802,000 (1957) (to National Starch and Chemical Corp.). Ungelatinized cold-water-soluble starch ethers including carboxymethyl, hydroxypropyl, benzyl by dry heat reaction.
122. E. F. Paschall, "Production and Uses of Cationic Starches," pp. 403–422, in Whistler and Paschall (eds.), Starch: Chemistry and Technology, vol. II, op. cit.
123. C. G. Caldwell and O. B. Wurzburg, U.S. Patents 2,813,093 (1957) and 2,917,506 (1959) (to National Starch and Chemical Corp.). Preparation of cationic starch ethers by reaction with diethylaminoethyl chloride and other tertiary amino-containing reagents.
124. C. H. Hullinger and N. H. Yui, U.S. Patent 2,970,140 (1961) (to American Maize Products Co.). Treatment of starch with dialkylaminoalkyl halides.
125. E. T. Hjermstad and L. C. Martin, U.S. Patent 3,315,739 (1964) (to Penick and Ford, Ltd.). Etherification of starch in aliphatic ketone medium.
126. E. F. Paschall, U.S. Patent 2,876,217 (1959) (to Corn Products Co.); P. A. Shildneck and R. J. Hathaway, U.S. Patent 3,346,563 (1967) (to A. E. Staley Mfg. Co.); G. C. Harris and R. A. Leonard, U.S. Patent 3,070,594 (1962) (to Hercules Powder Co.); T. A. McGuire and C. L. Mehltretter, U.S. Patent 3,558,501 (1971) (to U.S. Dept. of Agriculture). Treatment of starch with 2,3-epoxypropyltrimethylammonium chloride.
127. C. P. Patel, M. A. Jaeger, and R. E. Pyle, U.S. Patent 3,378,547 (1968) (to Union Starch and Refining Co.); W. G. Hunt, U.S. Patent 3,624,070 (1971) (to Anheuser-Busch, Inc.). Cationic starch ethers made by reaction with 4-halo-2-butenyltrimethylammonium halide.
128. K. W. Kirby, U.S. Patent 3,336,292 (1967) (to Penick and Ford, Ltd.). Cationic starch prepared by reaction with haloalkylenetrimethylammonium halide.
129. L. H. Elizer, C. G. Glasscock, and J. M. Seitz, U.S. Patents 3,051,700 (1962) and 3,136,646 (1964) (to Hubinger Co.). Cationic starch made by reaction with cyanamide.
130. E. F. Paschall, U.S. Patent 2,894,944 (1959) (to Corn Products Co.). Cationic starch made by reaction with disubstituted cyanamides.
131. J. C. Rankin and C. R. Russell, U.S. Patent 3,522,238 (1970) (to U.S. Dept. of Agriculture); J. C. McClendon and E. L. Berry, U.S. Patent 3,725,387 (1973) (to Dow Chemical Co.). Reaction of starch with ethyleneimine.
131a. J. C. Rankin, M. M. Holzapfel, C. R. Russell, and C. E. Rist, TAPPI, 52, 82–86 (1969). Preparation and Properties of Cationic Cereal Flours.

131b. H. D. Heath, A. J. Ernst, B. T. Hofreiter, and J. C. Rankin, *TAPPI*, 52, 1647–1651 (1969). Cationic Aminoethyl Cereal Flours as Wet-End Additives.

132. C. G. Caldwell and O. B. Wurzburg, U.S. Patent 2,661,349 (1953) (to National Starch and Chemical Corp.). Esterification of starch with alkenylsuccinic anhydride.

133. C. G. Caldwell and O. B. Wurzburg, U.S. Patent 2,613,206 (1952) (to National Starch and Chemical Corp.). Treatment of starch alkenylsuccinate half ester with polyvalent metal salts.

134. O. B. Wurzburg, W. Herbst, and H. M. Cole, U.S. Patent 3,091,567 (1963) (to National Starch and Chemical Corp.). Encapsulating agents with controlled water repellency (hydrophobic starch esters).

135. E. T. Hjermstad, L. C. Martin, and K. W. Kirby, U.S. Patent 3,462,283 (1969) (to Penick and Ford, Ltd.). Hydrophobic starch by reaction with benzyl chloride.

136. C. L. Mehltretter, "Production and Use of Dialdehyde Starch," pp. 433–444, in Whistler and Paschall (eds.), *Starch: Chemistry and Technology*, vol. II, op. cit.

137. C. L. Mehltretter, *Die Staerke*, 18, 208–213 (1966). Recent Progress in Dialdehyde Starch Technology.

138. V. F. Pfeifer, V. E. Sohns, H. F. Conway, E. B. Lancaster, S. Dabic, and E. L. Griffin, Jr., *Ind. Eng. Chem.*, 52, 201–206 (1960). Two-Stage Process for Dialdehyde Starch Using Electrolytic Regeneration of Periodic Acid.

139. J. Slager, U.S. Patent 3,086,969 (1963) (to Miles Laboratories, Inc.). Periodic acid oxidation of starch.

140. T. A. McGuire and C. L. Mehltretter, *Die Staerke*, 23, 42–44 (1971). Chemical Process for Making Dialdehyde Starch.

141. C. L. Mehltretter, T. E. Yeates, G. E. Hamerstrand, B. T. Hofreiter, and C. E. Rist, *TAPPI*, 45, 750–752 (1962). Preparation of Cationic Dialdehyde Starches for Wet Strength Paper.

142. C. L. Swanson, T. R. Naffziger, C. R. Russell, B. T. Hofreiter, and C. E. Rist, *Ind. Eng. Chem., Prod. Res. Dev.*, 3, 22–27 (1964). Xanthation of Starch by a Continuous Process.

143. W. M. Doane, C. R. Russell, and C. E. Rist, *Die Staerke*, 17, 77–81 (1965). Uniformity of Starch Xanthates Prepared by a Rapid, Continuous Process.

144. W. M. Doane, C. R. Russell, and C. E. Rist, *Die Staerke*, 17, 176–179 (1965). Location of Xanthate Groups in Starch Xanthate.

145. E. G. Adamek and C. B. Purves, *Can. J. Chem.*, 38, 2425–2432 (1960). The Location of Xanthate Groups in Partially Substituted Starch Xanthates.

146. E. G. Adamek and C. B. Purves, *Can. J. Chem.*, 35, 960–968 (1957). The Divergent Behavior of the Hydroxides of Lithium, Sodium, Potassium, Rubidium and Cesium in the Xanthation of Cellulose and Starch.

147. E. B. Lancaster, L. T. Black, H. F. Conway, and E. L. Griffin, Jr., *Ind. Eng. Chem., Prod. Res. Dev.*, 5, 354–356 (1966). U.S. Patent 3,385,719 (1968) (to U.S. Dept. of Agriculture). Xanthation of Starch in Low Concentration Pastes.

148. D. J. Bridgeford, U.S. Patents 3,291,789 (1966), 3,339,069 (1968) and 3,484,433 (1969) (to Tee-pack, Inc.). Decausticized, spray-dried polymeric alcohol xanthates and starch xanthate stable to room temperature storage.

149. C. R. Russell, R. A. Buchanan, C. E. Rist, B. T. Hofreiter, and A. J. Ernst, *TAPPI*, 45, 557–566 (1962). Preparation and Application of Cross-Linked Cereal Xanthates in Paper Products.

150. G. E. Lauterbach, E. J. Jones, J. W. Swanson, B. T. Hofreiter, and C. E. Rist, *Die Staerke*, 26, 58–66 (1974). Starch Xanthide in Handsheets and Machine-Made Papers.

151. R. A. Buchanan, W. F. Kowek, H. C. Katz, and C. R. Russell, *Die Staerke*, 23, 350–359 (1971). Starch in Rubber. Influence of Starch Type and Concomitant Variables in Reinforcement of Styrene-Butadiene Rubbers.

152. R. A. Buchanan, O. E. Weislogel, C. R. Russell, and C. E. Rist, *Ind. Eng. Chem., Prod. Res. Dev.*, 7, 155–158 (1968). Starch in Rubber. Zinc Starch Xanthate in Latex Masterbatching.

153. T. P. Abbott, C. James, W. M. Doane, and C. R. Russell, *Rubber World*, 169, 40–42 (1974). Elastomers Encased in a Little Starch Could Put Starch in Powdered Rubber Mart.

154. R. A. Buchanan, H. C. Katz, C. R. Russell, and C. E. Rist, *Rubber J.*, 153(10), 28, 30, 32, 35, 88–91 (1971). Starch in Rubber. Powdered Elastomers from Starch-Encased Latex Particles.

155. T. P. Abbott, W. M. Doane, and C. R. Russell, *Rubber Age*, 105(8), 43–49 (1973). Starch Xanthide-Encased Rubbers.

156. H. F. Conway, and V. E. Sohns, *J. Elastomers Plast.*, 7, 365–371 (1975). Starch-Encased Powdered Rubber: Pilot Plant Production.

157. R. E. Wing, C. L. Swanson, W. M. Doane, and C. R. Russell, *J. Water Pollut. Control Fed.*, 46, 2043–2047 (1974). Heavy Metal Removal with Starch Xanthate-Cationic Polymer Complex.

158. R. E. Wing, W. M. Doane, and C. R. Russell, *J. Appl. Polym. Sci.*, 19, 847–854 (1975). Insoluble Starch Xanthate: Use in Heavy Metal Removal.

159. R. E. Wing, B. K. Jasberg, and L. L. Navickis, *Die Staerke*, 30, 163–170 (1978). Insoluble Starch Xanthate: Preparation, Stabilization, Scale-Up and Use.

160. B. S. Shasha, W. M. Doane, and C. R. Russell, *J. Polym. Sci., Polym. Lett. Ed.*, 14, 417–420 (1976). Starch-Encapsulated Pesticides for Slow-Release.

161. C. G. Caldwell, W. Jarowenko, and I. D. Hodgkin, U.S. Patent 3,459,632 (1969) (to National

Starch & Chemical Corp.). Method of making paper containing starch derivatives having both anionic phosphate and cationic tertiary amino or quaternary ammonium groups.

162. K. B. Moser and F. Verbanac, U.S. Patent 3,562,103 (1971) (to A. E. Staley Mfg. Co.). Quaternary ammonium ether starch phosphates and phosphonates and their use as beater additives.

163. R. M. Powers and R. W. Best, U.S. Patent 3,649,624 (1972) (to A. E. Staley Mfg. Co.). Oxidized starch containing carboxyl groups etherified with quaternary ammonium-containing reagent.

164. G. F. Fanta, "Synthesis of Graft and Block Copolymers of Starch," pp. 1–27, and "Properties and Applications of Graft and Block Copolymers of Starch," pp. 29–45, in R. J. Ceresa (ed.), *Block and Graft Copolymerization,* vol. I, John Wiley & Sons, London, 1973.

165. G. F. Fanta and E. B. Bagley, "Starch Graft Copolymers," pp. 665–699, in N. M. Bikales (ed.), *Encyclopedia of Polymer Science and Technology,* Supplement Volume 2, Interscience, New York, 1977.

166. D. J. Angier, R. J. Ceresa, and W. F. Watson, *J. Polym. Sci.,* **34,** 699–708. (1959). Mastication of Rubber VIII. Preparation of Block Polymers by Mechanical Shear of Polymer-Monomer Systems (contains data on starch copolymerization).

167. B. H. Thewlis, *Die Staerke,* **16,** 279–285 (1964). The Mechanical Mastication of Wheat Flour and Wheat Starch.

168. G. Mino and S. Kaizerman, *J. Polym. Sci.,* **31,** 242–243 (1958). A New Method for the Preparation of Graft Copolymers. Polymerization Initiated by Ceric Ion Redox Systems.

169. C. E. Brockway and K. B. Moser, *J. Polym. Sci.,* **A1,** 1025–1039 (1963). Grafting of Poly(methyl Methacrylate) to Granular Corn Starch.

170. C. E. Brockway, *J. Polym. Sci.,* **A2,** 3721–3731 (1964). Efficiency and Frequency of Grafting of Methyl Methacrylate to Granular Corn Starch.

171. R. W. Faessinger and J. S. Conte, U.S. Patent 3,359,224 (1967) (to Scott Paper Co.). Thioated cellulosic/amylaceous substrate-ethylenically unsaturated monomer graft copolymer.

172. R. Mehrotra and B. Ranby, *J. Appl. Polym. Sci.,* **22,** 2991–3001, 3003–3010 (1978). Grafting of acrylonitrile on gelatinized potato starch or methyl methacrylate on granular potato starch by manganic pyrophosphate initiation.

173. L. A. Gugliemelli, M. O. Weaver, C. R. Russell, and C. E. Rist, *J. Appl. Polym. Sci.,* **13,** 2007–2017 (1969). Base-Hydrolyzed Starch-Polyacrylonitrile (SPAN) Graft Copolymer. SPAN-1:1, PAN M. W. 794,000.

174. M. O. Weaver, L. A. Gugliemelli, W. M. Doane, and C. R. Russell, *J. Appl. Polym. Sci.,* **15,** 3015–3024 (1971). Hydrolyzed Starch-Polyacrylonitrile Graft Copolymers: Effect of Structure on Properties.

175. M. O. Weaver, E. B. Bagley, G. F. Fanta, and W. M. Doane, U.S. Patents 3,935,099 (1976), 3,981,100 (1976), 3,985,616 (1976), and 3,997,484 (1976) (to U.S. Dept. of Agriculture). Water-insoluble, highly absorbent, base-hydrolyzed starch polyacrylonitrile graft polymer for soil conditioning, reducing water content of emulsions, immobilizing enzymes and incorporation in sanitary goods.

176. Z. Reyes, C. F. Clark, F. Dreier, R. C. Philips, C. R. Russell, and C. E. Rist, *Ind. Eng. Chem., Process Des. Dev.,* **12,** 62–67 (1973). Continuous Production of Acrylonitrile-Starch Graft Copolymers by Ceric Ion Catalysis.

177. L. A. Gugliemelli, M. W. Weaver, and C. R. Russell, *J. Polym. Sci., Polym. Lett. Ed.,* **6,** 599–602 (1968). New Method for Isolation of Grafts of Starch Vinyl Graft Copolymers.

178. C. E. Brockway and P. E. Seaberg, *J. Polym. Sci.,* **A1,** 5, 1313–1326 (1967). Grafting of Polyacrylonitrile to Granular Corn Starch.

179. G. F. Fanta, R. C. Burr, W. M. Doane, and C. R. Russell, *J. Appl. Polym. Sci.,* **16,** 2835–2845 (1972). Graft Copolymers of Starch with Mixtures of Acrylamide and the Nitric Acid Salt of Dimethylaminoethyl Methacrylate.

180. G. F. Fanta, R. C. Burr, C. R. Russell, and C. E. Rist, *J. Appl. Polym. Sci.,* **13,** 133–140 (1969). Copolymers of Starch and Polyacrylonitrile: The Dilution Effect.

181. R. C. Burr, G. F. Fanta, C. R. Russell, and C. E. Rist, *J. Macromol. Sci. Chem.,* **A1,** 1381–1385 (1967). Influence of Swelling and Disruption of the Starch Granule on the Composition of the Starch-Polyacrylonitrile Copolymer.

182. G. F. Fanta, R. C. Burr, C. R. Russell, and C. E. Rist, *J. Macromol. Sci. Chem.,* **A4,** 331–339 (1970). Copolymers of Starch and Polyacrylonitrile. Influence of Granule Swelling on Copolymer Composition under Various Reaction Conditions.

183. G. F. Fanta, R. C. Burr, W. M. Doane, and C. R. Russell, *J. Appl. Polym. Sci.,* **15,** 2651–2660 (1971). Influence of Starch Granule Swelling on Graft Copolymer Composition. A Comparison of Monomers.

184. G. F. Fanta, F. L. Baker, R. C. Burr, W. M. Doane, and C. R. Russell, *Die Staerke,* **25,** 157–161 (1973). Polyacrylonitrile Distribution in Grafted Starch Granules Determined by Scanning Electron Microscopy.

185. Z. Reyes, C. E. Rist, and C. R. Russell, *J. Polym. Sci.,* **A1,** 4, 1031–1043 (1966). Grafting Vinyl Monomers to Starch by Ceric Ion. I. Acrylonitrile and Acrylamide.

186. L. A. Gugliemelli, C. L. Swanson, and W. M. Doane, *J. Polym. Sci., Polym. Chem. Ed.,* **11,** 2451–2462 (1973). Kinetics of Grafting Acrylonitrile onto Starch.

187. L. A. Gugliemelli, M. O. Weaver, and C. R. Russell, U.S. Patent 3,425,971 (1969) (to U.S. Dept. of Agriculture). Salt-resistant thickeners comprising base-saponified starch-polyacrylonitrile graft copolymers.

188. G. F. Fanta and W. M. Doane, U.S. Patent 4,045,387 (1977) (to U.S. Dept. of Agriculture). Highly absorbent saponified flour-polyacrylonitrile graft copolymers.

189. T. Smith, U.S. Patent 3,661,815 (1972) (to Grain Processing Corp.). Water-absorbing alkali metal carboxylate salts of starch-polyacrylonitrile graft copolymers.

190. M. O. Weaver, E. B. Bagley, G. F. Fanta, and W. M. Doane, *J. Appl. Polym. Sci., Appl. Polym. Symp.,* **25,** 97–102 (1974). Gel Sheets Produced by Hydration of Films from the Potassium Salt of Hydrolyzed Starch-Polyacrylonitrile Graft Polymer.

191. N. W. Taylor and E. B. Bagley, *J. Appl. Polym. Sci.,* **18,** 2747–2761 (1974). Dispersions or Solutions? A Mechanism for Certain Thickening Agents.

192. E. B. Bagley and N. W. Taylor, *Ind. Eng. Chem., Prod. Res. Dev.,* **14,** 105–107 (1975). Starch-Polyacrylonitrile Copolymers. Properties of Hydrogels.

192a. G. F. Fanta, F. L. Baker, R. C. Burr, W. M. Doane, and C. R. Russell, *Die Staerke,* **29,** 386–391 (1977). Scanning Electron Microscopy of Saponified Starch-g-Polyacrylonitrile.

193. M. O. Weaver, G. F. Fanta, and W. M. Doane, "Highly Absorbent Starch-Based Polymer," pp. 169–177, in *Proceedings of the Technical Symposium on Non-Woven Product Technology,* International Non-Wovens & Disposables Association, New York, 1974.

194. G. F. Fanta, R. C. Burr, W. M. Doane, and C. R. Russell, *Die Staerke,* **30,** 237–242 (1978). Absorbent Polymers from Starch and Flour Through Graft Polymerization of Acrylonitrile and Comonomer Mixtures.

195. Z. Reyes and C. R. Russell, U.S. Patent 3,518,176 (1970) (to U.S. Dept. of Agriculture). Graft polymerization of starch with vinyl monomers in alcohol medium.

196. D. A. Jones, G. F. Fanta, and R. C. Burr, U.S. Patent 3,669,915 (1972) (to U.S. Dept. of Agriculture). Flocculants from starch graft copolymers made from monomers containing cationic quaternary ammonium substituents.

196a. G. F. Fanta, R. C. Burr, W. M. Doane, and C. R. Russell, *J. Appl Polym. Sci.,* **18,** 2205–2209 (1974). Graft Copolymers by Simultaneous Cobalt-60 Irradiation of Starch, Acrylamide and *N,N,N*-Trimethylaminoethyl Methacrylate Methyl Sulfate.

197. G. F. Fanta, R. C. Burr, C. R. Russell, and C. E. Rist, *J. Polym. Sci.,* **A1, 7,** 1675–1681 (1969). Copolymers of Modified Starches with Acrylonitrile.

198. C. F. Brockway, *J. Polym. Sci.,* **A, 3,** 1031–1036 (1965). Copolymerization of Methyl Methacrylate with Unsaturated Esters of Granular Starch.

199. F. Verbanac, U.S. Patent 4,060,506 (1977) (to A. E. Staley Mfg. Co.). Starch acrylamides, preparation and polymerization.

200. A. H. Young and F. Verbanac, U.S. Patent 4,079,025 (1978) (to A. E. Staley Mfg. Co.). Copolymerization of starch derivatives containing ethylenic unsaturated substituent groups with vinyl monomers.

201. L. A. Gugliemelli, C. L. Swanson, F. L. Baker, W. M. Doane, and C. R. Russell, *J. Polym. Sci., Polym Chem. Ed.,* **12,** 2683–2692 (1974). Cationic Starch-Polyacrylonitrile Graft Copolymer Latexes.

202. W. Jarowenko, U.S. Patent 3,331,833 (1967) (to National Starch & Chemical Corp.). Graft polymerization of ethylenimine onto tertiary amino starch.

203. G. C. Maher, U.S. Patent 3,436,305 (1969) (to Secretary of Agriculture). Starch polyethyleniminothiourethane additives for paper.

204. G. C. Maher, A. J. Ernst, H. D. Heath, B. T. Hofreiter, and C. E. Rist, *TAPPI,* **55,** 1378–1384 (1972). Starch Polyethyleniminothiourethane: A Wet-and Dry-Strength Agent.

205. W. Jarowenko and M. W. Rutenberg, U.S. Patents 3,666,751 (1972) and 3,737,370 (1973) (to National Starch & Chemical Corp.). Liquid cationic starch derivatives by reaction with a condensate of epichlorohydrin and ammonia and their use in paper-making.

206. D. K. Ray-Chaudhuri and C. P. Iovine, U.S. Patent 3,962,159 (1976) (to National Starch & Chemical Corp.). Graft copolymers of a polyamideamine substrate and starch.

207. T. Aitken, D. R. Anderson, and M. J. Jursich, U.S. Patent 3,674,725 (1972); T. Aitken, U.S. Patents 3,854,970 (1974) and 3,930,877 (1976); and T. Aitken and W. D. Pote, U.S. Patent 4,097,427 (1977) (to Nalco Chemical Co.). Starch grafted with epichlorohydrin-amine condensate polymers or amine-modified-polyepichlorohydrin polymers.

208. E. K. Nissen, "Starch in the Paper Industry," pp. 121–145, in Whistler and Paschall (eds.), *Starch: Chemistry and Technology,* vol. II, op. cit.

209. A. H. Zijderveld and P. G. Stoutjesdijk, "The Paper Industry," pp. 199–228, in J. A. Radley (ed.), *Industrial Uses of Starch and Its Derivatives,* Applied Science Publishers, Ltd., London 1976.

210. James P. Casey, *Pulp and Paper,* vol. 2: "Paper-Making," 2d Ed., Interscience, New York, 1961. Covers all facets of the paper-making operation including wet-end additives, filler properties and retention, and surface sizing.

211. H. W. Moeller, *TAPPI,* **49,** 211–214 (1966). Cationic Starch as a Wet End Strength Additive.

212. W. A. Foster, "Water-Soluble Polymers as Flocculants in Paper-Making," pp. 3–19, in N. M. Bikales (ed.), *Water-Soluble Polymers,* Plenum Press, New York, 1973.

213. H. D. Heath, B. T. Hofreiter, A. J. Ernst, W. M. Doane, G. E. Hamerstrand, and M. I. Schulte, *Die Staerke,* 27, 76–82 (1975). Cationic and Nonionic Graft Polymers for Filler Retention.
214. J. G. Penniman, *Pap. Trade J.,* 162(5), 52, 54 (March 1, 1978). Maximizing Retention/Drainage in Pulp Slurries.
215. J. G. Penniman, *Pap. Trade J.,* 162(7), 36, 37 (April 1, 1978). The Importance of Eliminating Anionic Trash.
216. D. D. Halabisky, *TAPPI,* 60(12), 125–127 (1977). Wet-End Control for the Effective Use of Cationic Starch.
217. C. L. Mehltretter, F. B. Weakley, M. L. Ashby, D. W. Herlocker, and C. E. Rist, *TAPPI,* 46, 506–508 (1963). Spectrophotometric Method for Determining Cationic Efficiency of Pulp-Cationic Starch Complexes.
218. L. A. Gaspar, "Starch, Starch Derivatives and Related Materials," pp. 83–88, in "Retention of Fine Solids during Paper Manufacture," C. A. Report #57, September 1, 1975, Technical Association of the Pulp and Paper Industry (One Dunwoody Park, Atlanta, GA, 30341).
219. R. W. Kerr and F. C. Cleveland, Jr., U.S. Patent 3,132,066 (1964) (to Corn Products Co.). Process of Adding Starch Phosphates to Paper Pulp Containing Titanium Dioxide to Improve Retention of the Pigment.
220. H. C. Brill, *TAPPI,* 38, 522–526 (1955). An Evaluation of Various Beater Retention Aids for Titanium Dioxide Filler in the Presence of Chlorinated Corn Starch.
221. D. A. Brogly, *TAPPI,* 61(4), 43–45 (1978). Thermal-Chemical Starch Conversion for Size-Press Application.
222. K. J. Bristol, G. H. Brown, *Pap. Trade J.,* 153, 42–44 (1969). New Tool for Use at the Size Press—Cationic Starch.
223. K. J. Bristol, *Pulp Pap.,* 48(10), (September 1974). How to Reduce BOD Levels from Size-Press Starches. (Reprint available from author, National Starch and Chemical Corp.)
224. T. Walsh, *Am. Pap. Ind.,* April 1974. Cationic Additives Benefit Ecology. (Reprint available from author, National Starch and Chemical Corp.)
225. G. H. Brown and E. D. Mazzarella, U.S. Patent 3,671,310 (1972) (to National Starch & Chemical Corp.). Sizing Paper and Paperboard with Cationic Amylose Derivatives.
226. J. P. Casey, *Pulp and Paper,* vol. 3: "Paper Testing and Converting," 2d ed., Interscience, New York, 1961. In addition to excellent descriptions of paper properties and testing, this volume covers in detail the pigment coating of paper, laminating and corrugating, and other paper converting processes.
227. "Starch and Starch Products in Paper Coating," *TAPPI Monogr. Ser.,* no. 17, Technical Association of the Pulp and Paper Industry, New York, 1957. (See reference 218 for current address.)
228. A. H. Nadelman and G. H. Baldauf, *Pap. Trade J.,* 149 (1965). This is a series of articles appearing over several months.
228a. "Coating Formulation-Principles and Practices," 72–73, Reference 228 (Aug. 9).
228b. "Coating Methods Dictate Coating Formulations," 74–78, Reference 228 (Aug. 9).
228c. "What Do Coating Adhesives Have in Common," 57–62, Reference 228 (Sept. 6).
228d. "Coating Adhesives—Their Use in Formulations," 46–51, Reference 228 (Oct. 4).
228e. "The ABC's of Coating Additives," 43–50, Reference 228 (Nov. 1).
228f. "What the Formulator Needs to Know about Coating Color Preparation Systems," 66–70, Reference 228 (Nov. 15).
228g. "Coating Colors Must Be Formulated to Fit the Application Equipment," 82–88, Reference 228 (Dec. 6).
229. C. W. Cairns, *TAPPI,* 57(5), 85–87 (1974). Evolution of Raw Materials.
230. F. Kaulakis, *TAPPI,* 57(5), 80–84 (1974). Evolution of Coating Machinery.
231. R. D. Harvey and L. J. Welling, *TAPPI,* 59(12), 92–95 (1976). Viscosity Stabilizers for High Solids Thermal-Chemically Converted Starch Pastes Used as Coating Adhesives.
232. D. S. Greif, *TAPPI,* 43, 254–260 (1960). Cationic Starches as Paper Coating Binders.
233. E. D. Mazzarella and L. J. Hickey, *TAPPI,* 49, 526–532 (1966). Development of Cationic Starches as Paper Coating Binders.
234. J. Kronfeld, U.S. Patent 3,052,561 (1962) (to National Starch and Chemical Corp.). Paper coating compositions containing cationic starch.
235. E. J. Heiser, *Pap. Trade J.,* 162(9), 30–32 (May 1–15, 1978). Study of New Developments and Trends in Coating Formulations and Binders.
236. Code of Federal Regulations, Title 21, Chapter I, Part 182, Substances Generally Recognized As Safe, Section 182.90, Substances Migrating to Food from Paper and Paperboard Products.
237. J. A. Radley, *Industrial Uses of Starch and Its Derivatives,* Applied Science Publishers, Ltd., London, 1976.
237a. "The Food Industry," pp. 51–115, in Reference 237.
237b. "The Textile Industry," pp. 149–197, in Reference 237.
237c. "Adhesives from Starch and Dextrin," pp. 1–50, in Reference 237.
238. O. B. Wurzburg, "Starch in the Food Industry," pp. 361–395, in T. E. Furia (ed.), *Handbook of Food Additives,* vol. I, 2d ed., CRC Press, Cleveland, Ohio, 1972.

239. E. M. Osman, "Starch in the Food Industry," pp. 163–215, in Whistler and Paschall (eds.), *Starch: Chemistry and Technology*, vol. II, op. cit.

240. W. M. Marrs, "The Properties and Uses of Natural and Modified Starches," Scientific and Technical Surveys No. 85, British Food Manufacturing Industries Research Association, Randalls Road, Leatherhead, Surrey, England, June 1975.

241. N. B. Petersen, *Edible Starches and Starch-Derived Syrups*, Noyes Data Corporation, Park Ridge, N.J., 1975. A citation and abstract compilation of U.S. Patents issued since 1968.

242. J. E. Eastman, U.S. Patent 3,959,514 (1976) (to A. E. Staley Mfg. Co.). Acetylated potato, tapioca and waxy corn starches give thick viscosity at normal cooking temperatures and thin viscosity after retorting.

243. C. D. Szymanski, M. M. Tessler, and H. Bell, U.S. Patents 3,804,828 (1974) and 3,857,976 (1974) (to National Starch & Chemical Corp.). Epichlorohydrin-cross-linked, hydroxypropyl starch ethers which have thin viscosity at normal cooking temperatures and develop high viscosity at retort temperatures.

244. J. V. Tuschoff, G. L. Kessinger, and C. E. Hanson, U.S. Patent 3,422,088 (1969) (to A. E. Staley Mfg. Co.). Phosphorus oxychloride cross-linked, hydroxypropyl cereal starch ethers which are thin on cooking at normal temperature and thicken at retort temperatures.

245. L. H. Kruger and R. Murray, "Starch Texture," pp. 427–444, in J. M. deMan, P. W. Voisey, V. F. Rasper, and D. W. Stanley, (eds.), *Rheology and Texture in Food Quality*, Avi Publishing, Westport, Conn., 1976.

246. "Toxicological Evaluation of Some Food Additives Including Anticaking Agents, Antimicrobials, Antioxidants, Emulsifiers and Thickening Agents," WHO Food Additives Series, No. 5, World Health Organization, Geneva, 1974. Also issued by Food and Agricultural Organization of the United Nations, Rome, as "FAO Nutrition Meetings Report Series, No. 53A." This reference includes summaries of the biological data (biochemical and toxicological studies) on which the evaluation and ADI limits are based.

247. G. Graefe, *Die Staerke*, 26, 145–153 (1974). Modified Starches as Ingredients of Food. (In German)

248. European Economic Community, "Report of the Scientific Committee for Food on Some Chemically Modified Starches," 1976 (Paper 1421/VI/76-E). Reproduced in full in Special Circular No. 55/76, published by the British Food Manufacturing Industries Research Association, Randalls Road, Leatherhead, Surrey, England.

249. C. J. Blake, "The Detection and Determination of Starch in Foods—A Literature Review," Scientific and Technical Surveys No. 109, British Food Manufacturing Industries Research Association. Randalls Road, Leatherhead, Surrey, England, July 1978.

250. J. Compton and W. H. Martin, "Starch in the Textile Industry," pp. 147–162, in Whistler and Paschall (eds.), *Starch: Chemistry and Technology*, vol. II, op. cit.

251. "An Economic Evaluation of Starch Use in the Textile Industry," U.S. Department of Agriculture, Economic Research Service, Agricultural Economic Report No. 109, U.S. Government Printing Office, Washington, D.C., 1967.

252. H. C. Olsen, U.S. Patent 2,946,705 (1960) (to National Starch and Chemical Corp.). Warp sizing with cationic starches.

253. D. H. Griffiths, U.S. Patent 3,227,192 (1966) (to Pittsburgh Plate Glass Co.). Sizing glass fiber with high amylose starch.

254. C. E. Rix and K. J. Mysels, U.S. Patent 3,682,685 (1972) (to Penick & Ford, Ltd.). Treatment of starch to be used in sizing glass fibers, with phosphorus-or silicon-containing compounds to reduce conductivity after heat-cleaning the fiber.

255. E. F. W. Dux, "Production and Uses of Starch Adhesives," pp. 538–552, in Whistler and Paschall (eds.), *Starch: Chemistry and Technology*, vol. II, op. cit.

256. W. Jarowenko, "Starch Based Adhesives," pp. 192–211, in I. Skeist (ed.), *Handbook of Adhesives*, 2d ed., Van Nostrand Reinhold Co., New York, 1977.

257. A. Frieden, "Starch Adhesives," pp. 589–626, in Ralph W. Kerr (ed.), *Chemistry and Industry of Starch*, 2d ed., Academic Press, New York, 1950.

258. J. V. Bauer, U.S. Patents 2,051,025 (1936), 2,102,037 (1937), and 2,212,557 (1940) (to Stein-Hall Co.). A corrugating system using starch.

259. W. O. Kroeschell (ed.), *Preparation of Corrugating Adhesives*, Tappi Press, Atlanta, Ga. 1977.

260. Code of Federal Regulations, Title 21, Chapter I, Part 175, Indirect Food Additives: Adhesives Coatings and Components, Subpart B, Substances for use only as components of adhesives, Section 175.105, Adhesives, U.S. Government Printing Office, Washington, D.C., 1978.

261. J. A. Radley (ed.), *Examination and Analysis of Starch and Starch Products*, Applied Science Publishers, Ltd., 1976.

262. *Standard Analytical Methods of the Member Companies of the Corn Industries Research Foundation, Inc.*, 1001 Connecticut Avenue, N.W., Washington, D.C., 20036.

263. *Cereal Laboratory Methods*, 7th ed., American Association of Cereal Chemists, St. Paul, Minn., 1962.

264. W. Horwitz (ed.), *Official Methods of Analysis of the Association of Official Analytical Chemists*, 12th ed., Association of Official Analytical Chemists, Washington, D.C., 1975.

Chapter **23**

Tamarind Gum

Thomas Gerard
Eastern Research Laboratory, Stauffer Chemical Company

General Information 23-1
    Chemical Nature 23-2
    Physical Properties 23-3
    Manufacture 23-3
    Biological/Toxicological Properties 23-3
    Electrochemical Properties 23-4
    Rheological Properties 23-4
    Additives/Extenders 23-6
    Handling 23-7
    Applications 23-7
    Application Procedures 23-7
    Future Developments 23-7
Commercial Uses 23-8
    Processing Aids 23-8
Industries Using Tamarind Gum 23-8
Formulations 23-8
    Size for Jute Yarn 23-8
    Size for Cotton Warps 23-9
    Latex Manufacture 23-9
    Other Uses 23-9
Laboratory Procedures 23-9
    Viscosity Method 23-9
    Acid-Insoluble Residue 23-10
    Fat Content 23-10
Term Glossary 23-10
References 23-11

GENERAL INFORMATION

Tamarind gum, obtained from the endosperm of the seeds of the tamarind tree, *Tamarindus indica*, came into commercial production in 1943 as a replacement for starch in cotton sizing used in Indian textile mills. The tamarind seed, which is a by-product of the tamarind pulp industry, has a flat, irregular shape, being round, ovate, or obtusely four-sided. The length of a side is 1.5 cm, and the thickness is 0.75 cm, with the edge broadly keeled and more often slightly furrowed. The seed has about 70% kernel or endosperm enclosed by about 30% testa, which is a rich brown color.[3] The powdered tamarind gum can be distinguished from locust bean gum and guar gum under a microscope or by determination of the proportions of sugars released by hydrolysis.[62]

The composition of tamarind kernels resembles the cereals:[6,15,19,23,38,41] 15.4 to 22.7% protein, 3.0 to 7.4% oil, 0.7 to 8.2% crude fiber, 65.1 to 72.2% nonfiber carbohydrates, 2.45 to 3.3% ash, all measured on a dry basis. The oil resembles peanut oil and is obtained in a yield of 3.89 to 6.38% of the weight of the dry seed[2,18,19,22,23,38] and 6 to 8% of the weight of the kernels. The testa is always removed before the kernel is pulverized to produce the tamarind seed powder, or more specifically, since the kernel or endosperm is only part of the seed, the tamarind kernel powder (TKP). This tamarind kernel powder, which is commercially available, is called *tamarind gum*, and will be so designated in the remainder of this chapter.

Tamarind gum is a convenient starting material for the preparation of a more purified tamarind seed polysaccharide which is available in limited amount. This latter carbohydrate fraction of the kernel, which readily dissolves in water, typically contains 15.3% moisture, 0.5% nitrogen, 0.11% phosphorus, 1.0% fibrous matter, and a small amount of oil. Aqueous solutions of it are slightly acidic, nonreducing toward Fehling's solution, and have $[\alpha]_D = +71.4°$ (c 0.25, water). When pure, this polysaccharide is a free-flowing, pale, creamy white powder without taste or odor; the commercially available tamarind gum is creamy white to light tan, with a fatty taste and odor and a sticky appearance, and it has a tendency to lump.

Chemical Nature

Structure A proposed structure[25] of the carbohydrate, shown below, consists of a main chain or β-D-(1→4)-linked glucopyranosyl units, with a side chain consisting of a single xylopyranosyl unit attached to every second, third, and fourth D-glucopyranosyl unit through an α-D-(1→6) linkage.

$$
\left[
\begin{array}{l}
4-\beta-\text{D}-\text{Glc}p\ (1\!\rightarrow\!4)-\beta-\text{D}-\text{Glc}p\ (1\!\rightarrow\!4)-\beta-\text{D}-\text{Glc}p\ (1\!\rightarrow\!4)-\beta-\text{D}-\text{Glc}p\ (1\!\rightarrow\!4) \\
\qquad\quad 6 \qquad\qquad\qquad\quad 6 \qquad\qquad\qquad\quad 6 \\
\qquad\quad \uparrow \qquad\qquad\qquad\quad \uparrow \qquad\qquad\qquad\quad \uparrow \\
\qquad\quad 1 \qquad\qquad\qquad\quad 1 \qquad\qquad\qquad\quad 1 \\
\qquad \alpha-\text{D}-\text{Xyl}p \qquad\quad \alpha-\text{D}-\text{Xyl}p \qquad\quad \alpha-\text{D}-\text{Xyl}p \\
\qquad\qquad\qquad\qquad\qquad 2 \\
\qquad\qquad\qquad\qquad\qquad \uparrow \\
\qquad\qquad\qquad\qquad\qquad 1 \\
\qquad\qquad\qquad\quad \beta-\text{D}-\text{Gal}p
\end{array}
\right]_n
$$

One D-galactopyranosyl unit is attached to one of the xylopyranosyl units through a β-D-(1→2) linkage. The exact sequential distribution of branches along the main chain is uncertain. Another study proposes an L-(1→6) linkage of an L-arabinosyl unit on every fourth D-glucopyranosyl unit.[60]

Molecular Weight The molecular weight of the purified polysaccharide, based on measurements of viscosity,[24] osmometry,[8] copper number,[8] and reduction of 3,5-dinitrosalicylate,[25] has been calculated as 52,350, 54,600, 55,600, and 115,000, respectively. The fact that the molecular weight using 3,5-dinitrosalicylate is twice as great as that using copper number is rather surprising, since both are based on the reducing power of the polysaccharide.

Derivatives Metal complexes are obtained when an aqueous solution of the polysaccharide is treated with a solution of a metal hydroxide or an alkaline salt. Thus, the copper(II) compound is formed by the addition of either Fehling's solution or copper ammonium sulfate; the lead complex, by the addition of potassium plumbate; the barium complex, by the addition of a saturated solution of barium hydroxide; and the strontium complex, by the addition of strontium hydroxide.

These compounds separate as gels, and their purification is difficult because they have a tendency to adsorb some of the unreacted polysaccharide. Boiling water cannot be used to remove the adsorbed polysaccharide because the compounds decompose when heated. Therefore, they are purified by a thorough washing with a dilute solution of the complexing agent.[60] Other metal salts have also been reported.[52] The polysaccharide also forms an insoluble complex with tannic acid.[43]

Tamarind gum has been hydrolyzed with fungal enzymes, as well as with acids.[59] Dry roasting in the presence of acid (dextrinization)[39] also reduces viscosity and increases penetration of the tamarind gum sols into fibers. The alkyl and hydroxyalkyl forms of tamarind gum have improved swelling properties in water. Alkyl and carboxyalkyl ethers of tamarind gum are useful as emulsifiers and in textile printing paste.[55] Alkoxyl tamarind gum yields highly viscous dispersions.

Miscellaneous Iodine stains tamarind seed polysaccharide with a stain that is characteristically different from that of stained starch. Tamarind gum forms mucilaginous sols with water, is insoluble in alcohol, and with sodium tetraborate forms a semisolid gel that can be thinned by heating.

Physical Properties

Commercial tamarind gum, which is creamy white to light tan, is a mixture of substances. The presence of a small amount of fat causes a tendency for the gum to lump and to have a fatty odor. It must be stored dry to avoid lumping. Tamarind gum disperses instantly, even in cold water; however, a uniform sol is obtained only on heating the mixture. It is insoluble in most organic solvents. The molecular weight of the polysaccharide is above 50,000 (see Molecular Weight).

Manufacture

Tamarind gum preparation is a relatively simple operation. It consists of decortication of the seeds and pulverization of the creamy white endosperms or the kernels. First the seeds are thoroughly washed with water to free them from attached pulp. Hollow seeds, which float, are discarded.

Decortication of the seeds is the most difficult step because the testae are tenaciously held to the endosperms and cannot be removed easily. Generally the seeds are heated to above 150°C (302°F) for at least 15 min to render the testae brittle and friable. These testae are then removed by simultaneous decortication of the seeds and pulverization of the endosperms. Advantage is taken of the fact that the testae and endosperms have different degrees of pulverizability. The finely divided testae are blown off in an air stream, while the crushed endosperms fall out or remain undischarged. These endosperms are pulverized to form commercial tamarind gum, sometimes called tamarind seed or kernel powder.[14,17,26,56]

This powdered tamarind gum is a convenient starting material for the preparation of a more purified tamarind seed polysaccharide. The gum is gradually added to 30 to 40 times its weight of boiling water containing 0.2% organic acid. The mixture is stirred vigorously, and the boiling is continued for 30 to 40 min. The resulting thin, whitish liquid is then discharged into a tall settling tank where it is left overnight in order to allow most of the proteins and fibers to precipitate out. The supernatant liquor is then siphoned into a steam-heated concentration pan and reduced to half the original volume.

After it is mixed with infusorial earth (0.5% of the volume of the liquor) and stirred, the mixture is passed through a filter press. The resulting translucent solution is then evaporated to dryness on a drum drier. The flake product is pulverized in a ball mill to yield a powder of the required mesh. The yield of purified tamarind seed polysaccharide is a little more than 50% of the starting tamarind gum.[42]

Biological/Toxicological Properties

It is essential that the outer testa be completely removed before pulverizing the endosperm, since ingestion of the testa may cause depression, constipation, diarrhea, and inflammation. However, the analysis of the endosperm does not indicate any toxic principles. Rather, it is regarded as having medicinal effects and is said to function as a stomachic. The gum is also recommended as a valuable remedy in diarrhea and dysentery. Purified tamarind seed polysaccharide has many properties similar to those of fruit pectins, and therefore might find use in the treatment of colitis, diarrhea, dysentery, and other intestinal disorders.

Because the tamarind gum can contain from 15 to 23% protein, 4 to 7% oil, and 65 to 72% carbohydrate, it is susceptible to microbiological degradation in sols as well as under high-humidity conditions. Under the high-humidity conditions prevalent in the weaving departments of textile mills, mildew growth could first attack the size, and then the fiber. Wheat flour sizes are similarly attacked.[12]

It is essential that the preservative used with tamarind gum have both antifungal and antibacterial properties. Pentachlorophenol, β-naphthol, p-chloro-m-cresol, combinations of salicylanilide and alum, sodium pentachlorophenate and alum, and also sodium silicofluoride and zinc chloride are effective preservatives.[30]

Electrochemical Properties

The tamarind seed polysaccharide is non-ionic, although chemical derivatives of it may be either cationic or anionic.

Rheological Properties

Tamarind gum disperses instantly, even in cold water. However, a uniform sol is obtained only upon heating the mixture. The maximum viscosity is attained after boiling for 20 to 30 min.[9,39] These sols exhibit a non-Newtonian flow; that is, the rate of flow in the viscometer increases more rapidly than the rate of increase of the pressure applied.[29] Hence, sols of tamarind gum are pseudoplastic. In this respect, they resemble those of gum tragacanth but differ from those of gum arabic.

As with starches, the viscosity of a tamarind gum dispersion depends on the purity of the material, the concentration of gum in the sol, the method of sol preparation, the pH, and the addition of other chemicals. For example, oven-roasted or microbiologically degraded gums yield less viscous sols. Similarly, the viscosity of dispersions or pastes prepared from unwashed kernels decreases rapidly upon prolonged boiling. However, if the kernels are thoroughly washed to remove the hydrolyzing agents, the viscosity does not fall, even after boiling for 4 h.[29]

In general, the viscosities of tamarind gum sols are much higher than those of starch at equal concentrations. A 5% (w/v) paste of tamarind gum has a peak viscosity of

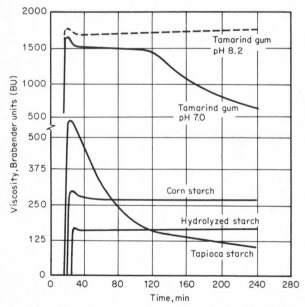

Figure 23.1 Viscosity as a function of heating of 5% paste solutions of tamarind gum and other common sizing materials at 95°C. (*Ref. 39.*)

more than 1600 Brabender units (BU), whereas a paste of corn starch of the same concentration has a viscosity of only 300 BU (see Fig. 23.1).[39] The viscosity of tamarind gum sols increases gradually as the concentration of gum is increased up to 0.5%, then more rapidly at higher gum concentrations. For gum above 1%, a slight increase in concentration results in a considerable increase in viscosity, and a very rapid rise occurs between 3 and 4% concentrations. Above 5% concentration, a paste without any flow is formed (see Fig. 23.2).[39]

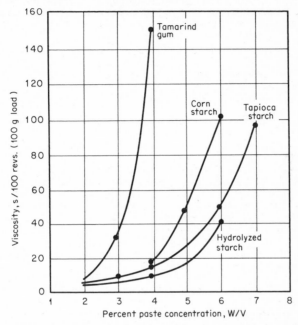

Figure 23.2 Viscosity as a function of concentration of sols of tamarind gum and other common sizing materials. Viscosity measurements were made with a Stormer viscometer at 70°C. (*Ref. 39.*)

Boiling has a profound effect upon the viscosity of tamarind gum dispersions or pastes. As with starches, the viscosity first reaches a maximum, then begins to fall. As stated earlier, the maximum viscosity is attained after boiling for 20 to 30 min, and falls to nearly one-half of that value in about 5 h.[9] The change in the viscosities of dispersions of tamarind gum, sago, and corn starch when boiled are presented in Table 23.1. Even prolonged heating at a lower temperature, at 95°C (203°F), brings about similar changes (see Fig. 23.2). However, the viscosity is stabilized if the paste is made slightly alkaline.

TABLE 23.1
Effect of Continued Boiling on Viscosity of Tamarind Gum and Starch Dispersions[9]

Boiling time, min	Relative viscosities		
	Tamarind gum 0.5% solution	Sago starch 1.0% solution	Corn starch 1.5% solution
10	4.7	5.2	4.9
20	5.0	4.0	4.6
30	4.3	3.3	4.5
60	3.8	2.9	4.0
180	2.9	2.8	2.7
300	2.5	2.5	2.4

A tamarind gum sol attains its maximum viscosity at pH 7.0 to 7.5. Beyond these limits, whether on the acid or alkaline side, the viscosity falls. This decrease is more pronounced in acid media. Boiling tamarind gum with mineral acids, even with low acid concentrations, brings about a rapid decrease in the sol viscosity, whereas boiling with organic acids does not have such a marked effect.[29] The fall in the sol viscosity at pH levels above and below 7.0 to 7.5 is caused by depolymerization, whereas the maximum viscosity in that range is attributed to uncoiling of the molecules. Sodium hydroxide, sodium carbonate, sodium bicarbonate, or even sodium tetraborate (borax) may be used to attain a sol pH of 7.0 to 7.5. These substances are added so as to form 0.001 M solutions in the final size pastes, which do not subsequently become thin, even after boiling for as long as 5 h.[9]

Vigorous stirring has an adverse effect on the ultimate viscosity of starch sols. However, tamarind gum sols do not become thin on agitation. This dissimilarity of the two types of sols has been attributed to a basic difference in their dispersion mechanisms. Tamarind gum possibly undergoes a straightforward dispersion, rather than an initial swelling of particles followed by disintegration and dispersion.[29] Measurement of the particle size of tamarind gum suspensions at various temperatures indicates no evidence of extreme distension of the particles at any stage when tamarind gum is heated in water from 30°C (86°F) to 98°C (208°F). The mean diameter, taken at the long and short axes of a number of random particles in a sample, varies from 44 μm at 33°C (91°F) to 25 μm at 92°C (198°F). Also, tamarind gum particles appear to have no characteristic shape or size, such as starch particles have. The size of tamarind gum particles depends on the extent and type of grinding. This lack of swelling of tamarind gum sols allows a constant sol viscosity to be maintained throughout the sizing operation.

The viscosity of the tamarind gum paste increases as the temperature decreases; but, in contrast with most starches, the tamarind gum pastes do not gel or retrograde upon cooling and are easily redispersed, even after storage for several days. However, microbial degradation must be prevented during storage.

Variation of the viscosity of tamarind gum sols with temperature is not completely reversible. This behavior cannot be explained by a change in molecular structure occurring with the change in temperature. The decrease in viscosity at higher temperatures is attributed mainly to the molecules' coiling, although depolymerization may make some contribution.[9]

The use of preservatives such as copper(II) sulfate and zinc(II) chloride is not recommended, since they reduce the viscosity of size pastes. However, soluble calcium and magnesium salts have a minimum viscosity effect on size pastes, and are recommended. Although a slight initial drop in the consistency of tamarind gum paste occurs on the first addition of these latter salts, further addition does not have further significant effect on the paste viscosity.[29]

The purified tamarind seed polysaccharide disperses easily in cold water, forming viscous, mucilaginous sols, even in low concentrations. The viscosity increase with increasing concentration is so rapid that it is difficult to prepare a mobile sol of more than 2%. The viscosity is not affected by pH of the sol, or by the addition of sodium, calcium, or iron salts. Addition of sucrose, D-glucose, starch syrup, and other oligosaccharides increases viscosity, whereas addition of hydrogen peroxide decreases it.[61]

Additives/Extenders

Sodium hydroxide, sodium carbonate, sodium bicarbonate, or even sodium tetraborate (borax) can be used with advantage to adjust the pH of a tamarind gum sol to 7.0 to 7.5 to minimize depolymerization and to maintain the viscosity of the sol.

Plywood Industry Quality plywood for tea chests is produced using urea-formaldehyde resin adhesive extended with 50% tamarind gum.[31] In lower-grade commercial plywood, only 40% of tamarind gum is used. It is estimated that about 15 to 20 million square meters of plywood are manufactured annually in India using tamarind gum–extended glue.

Textile Industry The purified tamarind polysaccharide has been recommended in the textile industry for sizing, finishing, and printing cotton and artificial silk.[47,48,49]

Tamarind seed polysaccharide, either alone or in combination with gum tragacanth, can be used economically with most colors in printing pastes. With some dyestuffs, it does not produce a suitable paste because it gels upon addition of the dye and then becomes ropy and insoluble. The formulations and lists of dyes which can be used with tamarind seed polysaccharide, as well as those which cannot, have been determined.[51] These and other aspects of the industrial utilization of purified tamarind seed polysaccharide have been reviewed in detail.[49,50,53]

Handling

Tamarind gum should be packed in moisture-proof containers for storage. Excessive heat and humidity should be avoided. The gum may deteriorate through enzymic action during long storage, particularly in moist areas. To avoid this, mix the powder with 0.5% by weight of sodium bisulfite and pack it in moisture-proof containers,[20,21] or store it in a dry place and fumigate it with sulfur dioxide at frequent and regular intervals.[44]

Applications

Typical commercial and industrial applications of tamarind gum are summarized in the following:

By Result

Binding. Sizing cotton and jute yarns; in adhesives, sawdust briquettes, plywood adhesives, building bricks.

Coagulation. As a creamer for latex.

Thickening. For yarn sizes; in explosives, gypsum, latex.

Gelation. In borax-printing paste, explosives, sugar-jellies, jams, marmalades.

Dispersion. In paper manufacture.

Suspension. In paper manufacture.

Stabilization. Of soil, ice cream, mayonnaise.

Film forming. For the sizing of yarns.

Emulsions. In textile paste.

Water retention. In explosives combinations.

By End Product
Tamarind gum is used as a component of cotton textile sizes, jute yarns, stabilized soils, building bricks, sawdust briquettes, adhesives, paper products, explosives, plywood, gypsum boards, jellies, ice creams, mayonnaise.

By Industry
Industries making use of tamarind gum include textile, construction, lumber, adhesive, paper, explosive, food, pharmaceutical, and synthetic rubber manufacture.

By Process
Tamarind gum is used as a manufacturing processing agent for the sizing and printing of fabrics, the concentration of latex by creaming agent, and in paper making.

Application Procedures

When tamarind gum is treated with water, it disperses instantly, even in cold water, but a uniform sol is obtained only upon heating the mixture. The pH should be adjusted to 7.0 to 7.5 with dilute sodium hydroxide, sodium carbonate, or sodium bicarbonate. The sol should be boiled 20 to 30 min to attain maximum viscosity.

In preparing fruit gels with purified tamarind seed polysaccharide, the gel strength increases tremendously with additional boiling time. Thus, the gel strengths after boiling for 5, 7, and 10 min are, respectively, 420, 540, and 650.[54]

Future Developments

The uses of commercial tamarind gum and purified tamarind seed polysaccharide promise to expand from their present major use in textiles and foods and go into the cosmetic industry to produce emulsions of fatty oils, shaving creams, and dentrifices, and into the pharmaceutical industry as a binder, an excipient, and an emulsifier.

Tamarind products may also find more widespread use in the treatment of colitis, diarrhea, dysentery, and other intestinal disorders. Modified forms of the gum may find application in the paper and other industries.

COMMERCIAL USES

By far the largest consumer of tamarind gum is textile sizing. The powdered gum, when heated with water, forms thick, viscous, colloidal dispersions that yield strong, transparent, elastic films upon evaporation. Its dispersions also have sufficient gluing power, but their fluidity and consequently their power of penetration into the yarn is rather poor when compared with thin-boiling starches. The viscosity characteristics can easily be modified by hydrolysis with acids or enzymes. It has completely replaced starch in jute sizing and to a large extent in cotton sizing in India (see Formulations).

Tamarind seed polysaccharide may also be used as a stabilizer in ice cream and mayonnaise-type dressings.

Processing Aids

The largest use of tamarind gum is as a processing aid in the weaving of cotton and jute textiles. It is also a good creaming agent for the concentration of rubber latex.

Random flocculation of the fibers in suspension is important in paper and rayon manufacture. Tamarind gum has been found to be a good dispersing agent for such fiber suspensions. In this respect, it is superior to starch ethers and pectins, but inferior to locust bean gum, guar gum, and cellulose ethers.

INDUSTRIES USING TAMARIND GUM

Tamarind seed powder is used in the following industries:

Textiles. It is used as a sizing and weaving agent for jute and cotton fabrics because it has excellent binding, film, and, in a modified form, thickening properties. In India, it has replaced starch in many formulations because of its lower cost as well as its equivalent or better weaving efficiency.[4,5,29,34,35,58]

Latex. It is used as a creaming agent for the concentration of latex. The latex creaming time is also drastically reduced because of tamarind gum's hydrophilic character and its ability to form sols of high viscosity at low concentrations.[27,46]

Construction. As a binder and stabilizer, tamarind gum enables the molding of sand, soil, clay, calcium hydroxide, and water into building bricks.[13,28,40]

Lumber. Tamarind gum binds sawdust into briquettes for use as a fuel.[57] The powder is extensively used as a 50% extender for urea-formaldehyde resin adhesive in the production of quality plywood for tea chests, as well as lower-grade commercial plywood.[31]

Adhesive. Label pastes containing tamarind gum with boric acid have good strengths and are suitable for the bonding of paper, glass, and metal.[37,45]

Explosives. The gum is used as a thickener in slurry explosives and as a water barrier in blasting explosives because of its ability to absorb moisture.[7,16]

Papermaking. Tamarind gum promotes random flocculation of the fiber suspension in the formation of paper sheet because of its dispersing and suspending actions. Locust bean gum and guar gum are superior to tamarind gum in this application.[11] Oxidized tamarind gum is reported to increase the wet strength of bleached kraft paper by 20%.[63]

FORMULATIONS

Tamarind gum has completely replaced starches in the sizing of jute yarns, because no blending is necessary. It is also being used extensively as a size for cotton warps. Typical formulations for size mixes for jute and cotton warps are as follows:

Size for Jute Yarn[21]

	1.5 to 2% w/v
Tamarind gum	30–40 kg
Sodium carbonate	1 kg
Salicylic acid (or β-naphthol)	1 kg (0.5 kg)
Sizing auxiliaries	As required
Water	To make 2,000 L

Size for Cotton Warps[1,36]

	Size for bleached goods (7% w/v)	Heavy size for grey goods (14% w/v)
Tamarind gum	67.0 kg	50.5 kg
Guar gum	—	1.0 kg
Sodium silicate	6.7 kg	—
China clay	—	63.0 kg
Magnesium chloride	—	12.6 kg
Mutton tallow	1.3 kg	15.2 kg
Paraffin wax	—	5.0 kg
Sodium silicofluoride	0.2 kg	0.3 kg
Water	To make 1000 L	To make 1000 L

The warp breakages during weaving of jute and cotton yarns are of the same order of magnitude whether tamarind gum, starch, or thin-boiling starch is used. It is a general practice to use 5 to 8% of a softener such as mutton tallow in starch-based size mixes meant for cotton warp. The softener makes the starch film smooth, soft, and pliable, and prevents excessive abrasion against the loom parts.

Since tamarind gum already contains 6 to 8% of fatty material, little or no softener need be added. Furthermore, to prevent the sized yarn from getting excessively soft, as well as to prevent the peeling of the size film, it has been found that addition of sodium silicate (10% on the weight of the tamarind gum) improves the weaving efficiency. Sodium silicate also makes the size paste slightly alkaline, thus stabilizing the paste viscosity during sizing operations. Homogenization of tamarind gum reduces its paste viscosity, and thus slightly more size can be put onto the yarn.[34]

Desizing by boiling with 1% sodium carbonate for 3 h is satisfactory for white goods. In the case of colored goods, the same treatment at a lower temperature, for example, 70°C (158°F), is recommended.[34] Boiling for 3 h with a solution containing 0.15% of ammonium sulfate and 0.3% of sodium chlorite, which gives simultaneous bleaching, is used for desizing grey yarn.[4]

Latex Manufacture

Because of its hydrophilic character and its ability to form high-viscosity solutions even in low concentrations, tamarind gum serves as a good creaming agent for the concentration of rubber latex. The addition of a 3% solution of tamarind gum to the ammoniated latex in an amount calculated to give a tamarind gum concentration of 0.3% of the weight of the water phase brings about an effective concentration. After creaming for nearly 38 h, the dry rubber content of the latex increases to 58.3%.[46]

Other Uses

Tamarind gum stabilizes soil[13,28,40] when 2 to 4 parts of it are thoroughly mixed with 94 to 97 parts of dry sand, soil, clay, or earth along with 1 to 4 parts of calcium hydroxide and 10 to 25 parts of water. This stabilized material can be molded into strong, compact bricks for building purposes.

The addition of 0.5 to 15% of tamarind gum to blasting explosives that contain water-soluble salts makes them water-resistant.[7] The gum is also recommended as a thickener for aqueous slurry explosives.[16]

Tamarind gum is used in the plywood industry as an extender for urea-formaldehyde resin adhesives, as discussed earlier.

LABORATORY PROCEDURES

Viscosity Method[33]

1. Weigh 8.000 g of tamarind seed gum into a 600-mL tared beaker.
2. Add 10 mL of alcohol to wet the gum evenly.
3. Add 390 mL of water to the gum-alcohol mixture, and stir rapidly with a thermometer and thick glass stirring rod to make the solution uniform. (If lumps form, repeat the above steps.)

4. Place the glass beaker in a hot-water bath (87 to 88°C) or into a steam bath where the temperature of the solution is maintained at 85° ± 1°C for exactly 10 min.
5. Then transfer the beaker to a water bath containing running water at 15°C, or to an ice bath, to cool the solution until the temperature reaches about 25°C. This cooling takes from 20 to 35 min and depends on the viscosity of the solution. (The time necessary for cooling varies for the same gum by varying the intensity of the flow of running water, by varying its temperature, and also by varying the stirring of the solution.) During the heating and the cooling periods, stir the solution slightly now and then with the thermometer and stirring rod to make the temperature uniform or even and to forego the possibility of drying on the surface or the forming of a film.
6. Then add distilled water to bring the weight of the solution to a total of 400 g. Stir slightly and adjust the temperature of the solution to 25°C (exactly).
7. Determine the viscosity at exactly 45 min with a Brookfield viscometer Model RVF using spindle #3 at 20 rpm. Take a reading at 1 min.
8. Read pH on electric meter.
9. Record characteristics, e.g., color and uniformity.
10. Remove the thermometer and stirring rod. Cover the beaker with parafilm, and place in 25°C water bath.
11. Repeat steps (7), (8), and (9) after 24 h.

Acid-Insoluble Residue (AIR)[33]

1. Weigh 1.0000 g of tamarind seed gum into a 250-mL beaker.
2. Add 150 mL of 1% sulfuric acid to the gum.
3. Stir with thin, rubber-tipped stirring rod, and cover with a watch glass, leaving the stirring rod in the beaker.
4. Place beaker into the steam bath for 6 h.
5. Stir occasionally (once an hour) and maintain level of liquid by adding distilled water.
6. Filter through a coarse, fritted glass crucible. If the filter is too fine, then use a Gooch crucible. If still too fine or difficult to filter, add 0.5000 g filter aid (Celite 545 or 535, dried for 2 h at 105°C).
7. Wash several times with hot distilled water (about 200 mL).
8. Dry in air oven at 105°C for 2 h.
9. Report as percent AIR.

Fat Content[33]

1. Weigh accurately 10.0000 g of the gum.
2. Extract it completely by subjecting it to the action of solvent (carbon tetrachloride) in a continuous extraction apparatus for 20 h.
3. Transfer the carbon tetrachloride solution to a tared evaporating dish, and allow it to evaporate over a steam bath with gentle heat.
4. Dry the residue in a desiccator for 18 h, and weigh.
5. Calculate the percentage of this extractive from the original sample, and report as percent fat content.

TERM GLOSSARY

Tamarind gum is also known as tamarind seed powder, tamarind kernel powder (TKP), and tamarind endosperm powder.

Tamarind gum is a high-molecular-weight, non-ionic, carbohydrate polymer obtained from the endosperm of the seeds of the tamarind tree, *Tamarindus indica*, an evergreen tree, one of the most important and common trees of India.

Purified tamarind seed polysaccharide is prepared from tamarind gum by removing most of the impurities. The pure polysaccharide is composed of simple sugars. It has a long main chain of β-D-(1→4)-linked glucopyranosyl units and small side chains of xylopryanosyl, galactopyranosyl, and arabinosyl units.   The exact structure is not certain.

REFERENCES

1. Anonymous, *A.T.I.R.A., Tech. Dig.,* **2,** 219 (1968).
2. Anonymous, *Olien Vetten,* **30,** 387 (1920); *Chem. Abstr.,* **14,** 1230 (1920).
3. Anonymous, *Pharmacographic Indica,* 1890.
4. Bhat, S. G., *J. Sci. Ind. Res., Sect. A,* **16,** 563 (1957).
5. Bhathena, N. S., *Tamarind Powder Industry,* Din. Products, Ltd., Bombay, India, 1953.
6. Bose, S. M., and N. Subramanian, *Bull. Cent. Food Technol. Res. Inst.,* **3,** 66 (1954).
7. Boyd, G., Australian Patent 229,190 (1960); *Chem. Abstr.,* **55,** 25255 (1961).
8. Chakraverti, I. B., S. Nag, and W. G. Macmillan, *J. Sci. Ind. Res., Sect. D,* **20,** 380 (1961).
9. Das, D. B., and K. K. Basak, *J. Indian Chem. Soc.,* **27,** 115 (1950).
10. Datta, R., and S. Basu, *Text. Res. J.,* **40,** 728 (1970).
11. deRoos, A. J., *Tappi,* **41,** 354 (1958).
12. Desai, N. F., A. Sreenivasan, and K. Venkataraman, *J. Sci. Ind. Res., Sect. B,* **44** (1946).
13. Dun, Dehra (President, Forest Res. Inst. & Colleges), Indian Patent 31,362 (1944).
14. Dutt, K. G., and U. Chatterji, Indian Patent 30,321 (1943).
15. Esh, G. C., and K. N. Bagchi, *J. Proc. Inst. Chem., Calcutta,* **24,** 11 (1952).
16. Fee, H. R., J. D. Gerguson, and R. W. Lawrence, U.S. Patent 3,350,246 (1967); *Chem. Abstr.,* **68,** 14630W (1968).
17. Ghose, T. P., and S. Krishna, *Indian Text. J.,* **53,** 236 (1943).
18. Grindley, D. N., *J. Soc. Chem. Ind., London,* **65,** 118 (1946).
19. Hooper, D., *Agr. Ledger,* **2,** 13 (1907).
20. Indian Standards Institution, *Indian Standard 189* (1951).
21. Indian Standards Institution, *Indian Standard 511* (1954).
22. Kafuku, K., and C. Hata, *Nippon Kagaku Kaishi,* **55,** 369 (1934).
23. Kehar, N. D., and B. Sahai, *Sci. Cult.,* **14,** 534 (1949).
24. Khan, N. A., and B. D. Mukherjee, *Chem. Ind. (London),* 1413 (1959).
25. Kooiman, P., *Rec. Trav. Chim. Pays-Bas,* **80,** 849 (1961).
26. Krishna, S., Indian Patent 29,620 (1943).
27. Krishna, S., Indian Patent 29,870 (1943).
28. Krishna, S., Indian Patent 32,035 (1947); *Chem. Abstr.,* **42,** 7506 (1948).
29. Macmillan, W. G., and J. B. Chakravetti, *J. Sci. Ind. Res., Sect. B,* **10,** 270 (1951).
30. Macmillan, W. G., J. B. Chakravetti, and P. N. Pal, *J. Sci. Ind. Res., Sect. B,* **11,** 438 (1952).
31. Maity, P., *Indian Wood Panels,* **2,** 30 (1969).
32. Meer Corp., *Product Guide.*
33. Meer Corp., *Analytical Procedures.*
34. Mehta, P. C., P. N. Bhatt, S. S. Trivedi, and C. C. Shah, *J. Sci. Ind. Res., Sect. A,* **16,** 194 (1957).
35. Mehta, P. C., and C. C. Shah, *A.T.I.R.A. Res. Notes,* **3,** 2 (1953).
36. Modi, J. R., personal communication.
37. Narayanamurti, D. P., and P. R. Rao, *Res. Ind.,* **6,** 127 (1961).
38. Padilla, P., and F. A. Soliven, *Philipp. Agric.,* **22,** 408 (1933).
39. Patel, K. F., M. M. Patel, S. N. Harshe, J. R. Modi, and H. C. Srivastava, *Proc. Tech. Conf., A.T.I.R.A., B.T.R.A.,* 9th, 1967, 219 (1968).
40. Puntambekar, S. V., *Bull. Nat. Inst. Sci. India,* **6** (1955).
41. Rajnarian and S. Dutt, *Indian J. Agric. Sci.,* **15,** 209 (1945).
42. Rao, P. S., *Indian Export Trade J.,* **10,** 120 (1956).
43. Rao, P. S., T. P. Ghose, and S. Krishna, *J. Sci. Ind. Res.,* **4,** 705 (1945–1946).
44. Rao, P. S., and S. Krishna, *Indian Text. J.,* **59,** 314 (1949).
45. Rao, P. S., and A. K. Misra, *Indian For.* **89,** 686 (1963).
46. Rhodes, E., and K. C. Sekar, Brit. Patent 474,651 (1937); *Chem. Abstr.,* **32,** 3665 (1938).
47. Sakai, Kazuyoshi, Japanese Patent 30,073 (1970).
48. Savur, G. R., *Curr. Sci.,* **24,** 235 (1955); *Chem. Ind. (London),* 212 (1956); *Sci. Cult.,* **21,** 744 (1956); *J. Chem. Soc.,* 2600 (1956).
49. Savur, G. R., *Indian Text. J.,* **65,** 418, 547 (1955).
50. Savur, G. R., *Indian Text. J.,* **66,** 33 (1956).
51. Savur, G. R., *Indian Text. J.,* **66,** 309 (1956).
52. Savur, G. R., *J. Indian Chem. Soc., Ind. News Ed.,* **19,** 67 (1956).
53. Savur, G. R., *Melliand Textilber.,* **37,** 588 (1956); *Chem. Abstr.,* **50,** 15091 (1956).
54. Savur, G. R., and A. Sreenivasin, *J. Soc. Chem. Ind., London,* **67,** 190 (1948).
55. Schiavio, A., and C. Maderno, Italian Patent 574,928 (1958); *Chem. Abstr.,* **54,** 25911 (1960).
56. Schroff, M. L., Indian Patent 30,487 (1944).
57. Shekar, A. C., P. L. Taneja, V. K. Sood, and M. G. Karnik, *Res. Ind.,* **8,** 291 (1963).
58. Srivastava, H. C., S. N. Harshe, M. M. Gharia, and G. P. Mudia, *J. Text. Assoc.,* **33,** 139–147 (1972).

59. Srivastava, H. C., S. N. Narsche, and M. M. Gharia, unpublished data.
60. Srivastava, H. C., and P. P. Singh, *Carbohydr. Res.,* **4,** 326 (1967).
61. Tamura, A., K. Wada, and O. Shoji, *Nippon Nogei Kagaku Kaishi,* **38,** 300 (1964); *Chem. Abstr.,* **63,** 6247 (1965).
62. Tseuneo, N., T. Itiro, and T. Shintaro, *Shoyakugaku Zasshi,* **21,** 47 (1967); *Chem. Abstr.* **69,** 53042w (1968).
63. Vemuri, K. P., Germ. Offen. Pat. 1,949,401 (1970).

Chapter **24**

Xanthan Gum

**Ian W. Cottrell, Kenneth S. Kang, and
Peter Kovacs**

Kelco, Div. of Merck & Co., Inc.

General Information 24-2
   Chemical Structure 24-2
   Physical Properties 24-3
   Solution Properties 24-4
   Suspensions 24-13
   Emulsions 24-13
   Dispersions 24-16
   Application Procedures 24-16
   Handling and Storage 24-16
   Reactions with Galactomannans 24-16
   Toxicology and Safety 24-19
Commercial Uses: Food 24-20
   Xanthan Gum 24-20
   Xanthan Gum with Locust Bean Gum 24-21
Commercial Uses: Industrial 24-21
   Xanthan Gum 24-21
   Xanthan Gum with Locust Bean Gum 24-22
Formulations 24-22
   Dessert Soufflés 24-22
   Bakery Jellies 24-23
   Salad Dressings 24-24
   Dry Sauce Mixes 24-26
   Frozen Pizzas 24-27
   Animal Feeds (Liquid) 24-27
Laboratory Techniques 24-27
   Viscosity (Food Grade) 24-28
   Viscosity (Industrial Grade) 24-28
   Moisture Content 24-28
   Powder Color 24-28
   Determination of Gum in Mixtures 24-29
Product Tradename Directory. 24-29
Further Reading/References 24-30

GENERAL INFORMATION

Xanthan gum is a high-molecular-weight natural carbohydrate. It is a polysaccharide produced in a pure culture fermentation by the microorganism *Xanthomonas campestris,* an organism originally isolated from the rutabaga plant.

In the fermentation process, the microorganism is cultured in a well-aerated medium containing carbohydrates, such as glucose, a suitable nitrogen source, dipotassium hydrogen phosphate, and appropriate trace elements. The microorganism itself is grown in several stages with associated identification tests conducted prior to introduction into the final fermentation medium. Following fermentation, xanthan gum is recovered by precipitation in isopropyl alcohol, then dried and milled.

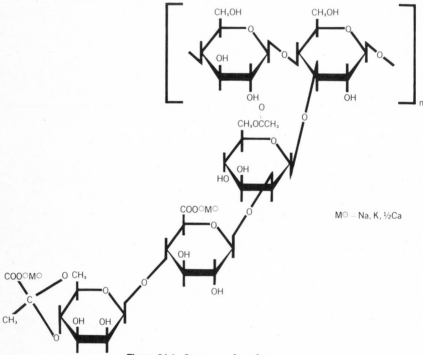

Figure 24.1 Structure of xanthan gum.

Chemical Structure

Figure 24.1 illustrates the repeating-unit structure of xanthan gum; the structure is based on the most recent experimental evidence. Although its molecular weight is probably on the order of 2 million, it has been reported to be as high as 13 to 50 million. These differences in measurement are probably caused by association phenomena occurring between polymer chains.

Xanthan gum contains three different monosaccharides: mannose, glucose, and glucuronic acid (as a mixed potassium, sodium, and calcium salt). Each repeating block of the polymer chain has five sugar units (two glucose, two mannose, one glucuronic acid). The polymer's main chain is made up of β-D-glucose units linked through the 1- and 4-positions; thus, the chemical structure of the main chain is identical to that of cellulose.

Two mannose units and the glucuronic acid unit make up the side chain. The terminal β-D-mannose unit is glycosidically linked to the 4-position of β-D-glucuronic acid, which in turn is glycosidically linked to the 2-position of α-D-mannose. This side chain is linked to the 3-position of every other glucose residue on average in the polymer main chain. Roughly half of the terminal D-mannose residues carry a pyruvic acid residue linked ketalically to the 4- and 6-positions. The nonterminal D-mannose unit on the side chain has an acetyl group at the 6-position.

The shielding of the backbone of xanthan gum by its side chains could explain its extraordinary resistance to enzymes. Also unique among the natural gums are the unvarying chemical structure and the uniformity of chemical and physical properties of xanthan gum. The structural rigidity of xanthan gum produces several unusual properties. Contrary to the behavior expected from a typical anionic polysaccharide, the addition of salts to a salt-free xanthan gum solution causes the viscosity to increase when the gum concentration is greater than 0.15%. Also, whereas most polysaccharide solutions show a decrease in viscosity when heated, salt-free xanthan gum solutions in

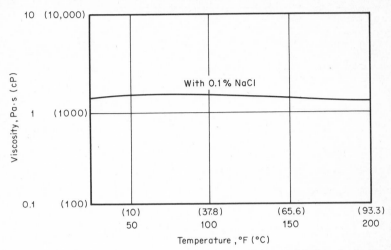

Figure 24.2 Effect of temperature on the viscosity of a 1% solution of food-grade xanthan gum.

deionized water show an increase in viscosity after an initial viscosity decrease. However, in the presence of a small amount of salt, moderate increases in solution temperature, that is, to ≃ 90°C, have very little effect on xanthan gum solution viscosities (see Fig. 24.2).

Physical Properties

Table 24.1 gives the physical properties of a typical food-grade xanthan gum, whereas Table 24.2 gives the physical properties of a typical industrial grade of xanthan gum.

TABLE 24.1
Typical Physical Properties of a Food-Grade Xanthan Gum

Physical state	Dry, cream-colored powder
Moisture content, %	11
Ash, %	9
Color (white enamel standard, 75% reflectance)	70
Specific gravity	1.5
Bulk density, kg/m³ (lb/ft³)	836 (52.2)
Browning temperature, °C	165
Charring temperature, °C	240
Ashing temperature, °C	470
Ignition temperature, °C	*
As a 1% solution in distilled water:	
Heat of solution (cal/g soln)	0.080
Refractive index at 20°C	1.3338
pH	7.0
Surface tension, dyn/cm	75
Freezing point, °C	0.0
1% viscosity with 1% electrolyte added,† Pa·s (cP)	1.4 (1400)
Typical mesh sizes (Tyler standard)	80,200

* Spontaneous combustion did not occur in an air environment.
† Measured with Brookfield LVF viscometer, 60 rpm.

Solution Properties

Viscosity/Concentration Relationships Figure 24.3 shows the viscosity/concentration relationship of food-grade xanthan gum solutions as measured with a Brookfield Model LVF viscometer at 60 rpm. These solutions have an apparent high viscosity at low-solids concentration and ~ 11 s^{-1} shear rate. Pouring shear would approximate this shear rate.

For industrial-grade xanthan gum solutions, Fig. 24.4 details the viscosity/concentration relationship. Like the food-grade solutions, again there appears high viscosity at low-solids concentration and a shear rate of approximately 11 s^{-1}.

Effect of Salts on Viscosity The effect of salts on the viscosity of food-grade xanthan gum solutions depends upon the solution concentration of the xanthan gum. Where gum concentrations are low, below 0.15%, monovalent salts, e.g., sodium chloride, cause a slight decrease in viscosity. At higher concentrations of salt the viscosity is increased. At a monovalent salt concentration level of about 0.08% the peak viscosity is reached; additional salt has no effect on viscosity, as shown in Fig. 24.5.

TABLE 24.2
Typical Physical Properties of an Industrial-Grade Xanthan Gum

Physical state	Dry, cream-colored powder
Moisture content, %	12
Ash, %	10
Color (white enamel standard, 75% reflectance)	70
Specific gravity	1.6
Bulk density, kg/m³ (lb/ft³)	839 (52.4)
Browning temperature, °C	160
Charring temperature, °C	270
Ashing temperature, °C	470
Ignition temperature, °C	*
Heat of combustion (cal/gm)	3.48
Nitrogen, %	1.2
As a 1% solution in distilled water:	
Heat of solution (cal/g soln)	0.055
Refractive index at 20°C	1.3332
pH	7.0
Surface tension, dyn/cm	75
Viscosity,† Pa·s (cP)	0.85 (850)
Freezing point, °C	0.0
Mesh size (Tyler standard)	40

* Spontaneous combustion did not occur in an air environment.
† Measured with Brookfield LVF viscometer, 60 rpm.

This same viscosity-peaking effect also occurs with salts of most divalent metals, e.g., calcium and magnesium. The extent of viscosity change that will occur in formulated systems depends upon the pH and other aspects of the system. In order to develop optimal rheological properties and a uniformity of the solution properties, some type of salt should be present in the system. Often, the metallic salts found naturally in tap water are sufficient for this purpose.

The effect of salts on solution viscosity of the industrial-grade xanthan gum also depends on the concentration of the gum in the solution. For low gum concentrations (below 0.25%) monovalent salts, such as sodium chloride, cause slight viscosity decreases, whereas higher concentrations of the salt will increase the solution viscosity. When monovalent salt is at a concentration level of about 0.08%, it will, as for the food-grade gum, produce the ultimate solution viscosity; addition of more salt will have no effect on the viscosity. This salt-effect phenomenon for industrial-grade xanthan gum is illustrated in Fig. 24.6. As is the case for food-grade xanthan gum, this effect also occurs with most divalent metal salts, and the amount of viscosity change depends on pH and other systems dynamics.

Effect of Temperature on Viscosity The unique temperature/viscosity relationship for food-grade xanthan gum is shown in Fig. 24.7. A solution that contains 0.1% sodium chloride is examined. At temperatures up to 93°C (200°F) there is almost no change in viscosity of the solution. To the food industry this means that the viscosity contributed to the solution by the xanthan gum will remain constant, whether or not the foods are stored in a refrigerator, on a kitchen shelf, or on a steam table. In addition flow characteristics of the foods will be the same in all climate zones.

Figure 24.8 illustrates these same properties for the industrial-grade xanthan gum. At a 1% concentration of xanthan gum, there is practically no decrease in viscosity as temperature is increased. However, at lower concentrations, the solution viscosity does decrease slightly as the temperature increases.

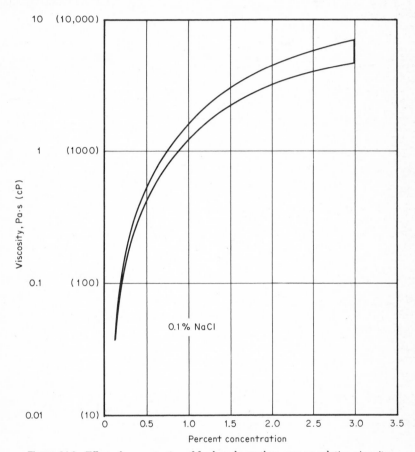

Figure 24.3 Effect of concentration of food-grade xanthan gum on solution viscosity.

Effect of pH on Viscosity For food-grade xanthan gum solutions that contain a minimal amount of salt, pH has very little effect on viscosity, as shown in Fig. 24.9. Similarly, pH has little effect on the viscosity of industrial-grade xanthan gum solutions, as shown in Fig. 24.10.

Gelation with Metals Industrial-grade xanthan gum generally shows unusually good compatibility with divalent ions, and it forms smooth solutions in their presence. The solution conditions that govern the reaction of the gum with the metallic ions are mainly the pH level and the salt content. Figure 24.11 illustrates the pH ranges in which different metallic ions react in solutions of industrial-grade xanthan gum. The data were obtained using 1.0% solutions of the gum containing 0.5% metallic salt concentration; pH was adjusted with dilute sodium hydroxide or hydrochloric acid as needed. There are no other additives present. Gelation occurred within the pH boundaries noted on the figure for each ion.   Where concentrations of monovalent salts are high, gelation may be minimal, or sometimes it may be entirely prevented. Most xanthan gum solutions that contain divalent ions require a high pH for gelation to occur, but trivalent ions may produce gels in neutral or even acidic conditions, contingent upon the reactivity of the metal ion present.   The reactivity with xanthan gum is specific for the metallic ion involved.   In general, the anion does not have any influence on reactivity.

Since sequestrants normally have a greater affinity for both divalent and trivalent cations than does the xanthan gum itself, sequestrants can be used to prevent gelation.

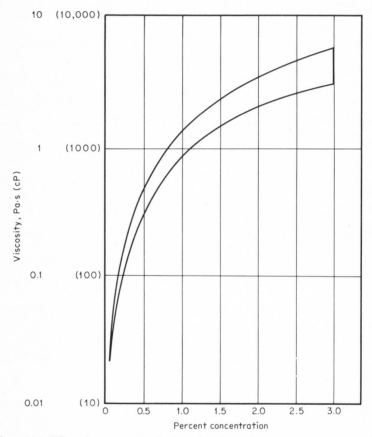

Figure 24.4 Effect of concentration of industrial-grade xanthan gum on solution viscosity.

In cases where there are low concentrations of metallic ions in solution, the viscosities can be increased by a simple increase of pH.   The viscosity-increase reaction at high pH with chromium ions has been successfully utilized in oil-well drilling fluids.   If the pH of such a solution is raised quickly with cessation of stirring, firm, brittle gels are formed.   However, these gels are unstable, and they tend to exhibit syneresis.

Rheological Properties An important property of xanthan gum is the control of aqueous rheological properties.   Water solutions of xanthan gum are extremely pseudoplastic.   When shear stress is applied, the viscosity is reduced in proportion to the amount of shear, once the yield point of the solution has been exceeded.   When the shear is released, total viscosity recovery occurs almost instantly.

The relationship between viscosity and shear rate for xanthan gum over a shear-rate range of 0.1 to 40,000 s⁻¹ is given in Fig. 24.12. This figure clearly shows the pseudoplasticity of xanthan gum for this range of shear rates. The chart at the bottom of the figure shows "application" shear rates and shear-rate ranges covered by various types of viscometers.

The suspending ability of xanthan gum is predictable and uniform over a wide range of conditions because of the uniformity of the viscosity/shear-rate relationship and because of the known yield point. At very low shear rates suspended particles remain stationary because of the very high apparent viscosity of xanthan gum solutions below the yield point. Shear rates that are encountered during pouring or spreading operations are usually sufficient to materially reduce apparent viscosity. Under high-shear conditions, such as pumping, xanthan gum imparts very little apparent viscosity to the solution.

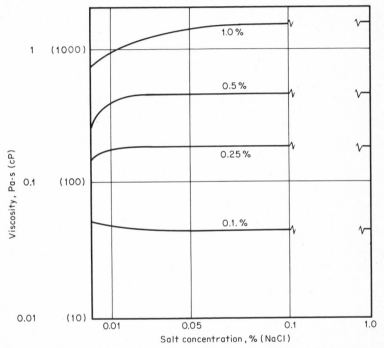

Figure 24.5 Effect of sodium chloride concentration on the viscosity of food-grade xanthan gum solutions.

Xanthan gum solutions are usually more resistant to degradation resulting from prolonged shearing than other thickeners. A 1% xanthan gum solution that was sheared at 46,000 s⁻¹ (comparable to that encountered in a homogenizer or colloid mill) exhibited no viscosity loss after 1 h.

It is possible, if desired, to effectively modify the pseudoplasticity of xanthan gum solutions by combining a refined sodium alginate with the xanthan gum. Figure 24.13 compares the flat slope of the viscosity versus rpm curve for a refined sodium alginate solution in distilled water (curve D) with those of xanthan gum solutions (curves A and B). The solution of xanthan gum has its rheological properties altered by the presence of the sodium alginate, as shown in curve C. The result is a solution that has a significantly lower degree of pseudoplasticity.

Applications Based on Rheological Properties. The effects of rheological properties are present, if not always apparent, in most of the fluids that we use in our daily lives.

One obvious application based on rheological properties is latex paints. Paints require low viscosity at high shear rates, i.e., brushing, but high viscosity at low shear so that the coating clings to the surface and does not run. Thickening agents such as xanthan gum provide these properties. Among the many other applications where the control of rheological properties by xanthan gum enhances the overall properties of products are:

Salad dressings. In these products xanthan gum gives emulsion stability and pleasing mouthfeel, and imparts cling to the dressing.

Gravies. In gravies xanthan gum produces cling and uniform viscosity over wide temperature ranges.

Paper coatings. Coatings for papers are improved, because xanthan gum gives the coatings thixotropy during application, controlled water release during drying of the coating, and migration control.

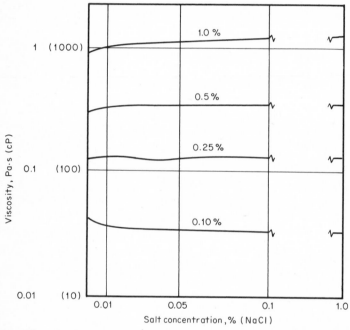

Figure 24.6 Effect of sodium chloride concentration on the viscosity of industrial-grade xanthan gum solutions.

Textile printing. In the process of printing textiles xanthan gum aids in the suspension of dye pigments and the control of the rheology of the dye solution during the printing operation.

Cleaning compounds. Household and other cleaning solutions are improved by the addition of xanthan gum, which gives low viscosity to the solution during pumping and spraying, high at-rest viscosity for retention and run-off control on vertical surfaces, and provides shear thinning with water from a hose to assist in easy wash-off.

Compatibility   *Acids.* Xanthan gum dissolves directly in many acidic solutions. The best results are obtained, however, when acid is added to a xanthan gum solution, rather than when the gum is added to the acid solution. Xanthan gum stability is excellent in the presence of most organic acids. The compatibility with mineral acids depends upon the type of acid and its concentration in the solution. However, at elevated temperatures acid hydrolysis of the polysaccharide is accelerated, and the lower viscosities may result.

Bases. Xanthan gum has compatibility with many basic compounds, including concentrated ammonium hydroxide. Sodium hydroxide concentrations greater than

12.0% will cause gelation or precipitation of the xanthan gum in the solution. Basic salts, such as sodium carbonate, phosphate, or metasilicate, may also produce gelation after prolonged storage of xanthan gum solutions, if their concentrations in the solutions are greater than 5%.

Salts. In the presence of many salts, xanthan gum solutions have unusually good compatibilities and stabilities. In some cases compatibility is limited only by the solubility of the salt. There exists, however, incompatibility between xanthan gum and polyvalent metal ions for solutions with high pH. This often can be controlled or prevented by the addition of sequestrants, such as polyphosphates.

Very low levels of borates, generally less than 300 ppm as boron, can cause gelation when other soluble salts are present in xanthan gum solutions. Usually, this gelation can be avoided by increasing the boron ion concentration to greater than 300 ppm or by lowering the pH to ~5.0 or less.

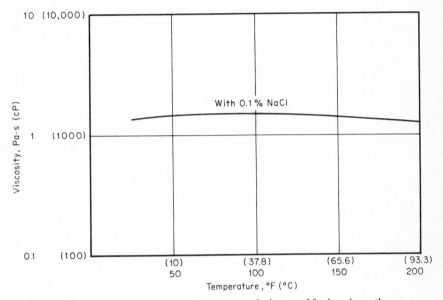

Figure 24.7   Effect of temperature on the viscosity of solutions of food-grade xanthan gum.

Substances that contain vicinal hydroxyl groups also prevent gelation by forming soluble complexes with the borate ion. Ethylene glycol and mannitol are useful for this purpose.

Oxidizing Agents. Persulfates, peroxides, hypochlorites, and other strong oxidizers degrade xanthan gum. High temperature accelerates this reaction.

Reducing Agents. In the presence of reducing agents xanthan gum is generally stable. Even so, care is needed to prevent the formation of free radicals that may cause degradation of long-chain polymers. Such degradation can also occur if oxidizers are present.

Thickeners. Xanthan gum has good compatibility with sodium alginate and the starches. With dextrin, guar, and locust bean gums, xanthan gum exhibits synergistic viscosity increases when mixed.

Compatibility is good with the synthetic, water-soluble resins, with the exception of high concentrations of Carboset 525, which will precipitate xanthan gum. Gum arabic forms complexes with xanthan gum at medium to high concentrations. And it is generally not recommended that xanthan gum be used along with cellulose derivatives.

Solvents. Aqueous xanthan gum solutions tolerate up to a 50 to 60% concentration of water-miscible solvents, such as methanol, ethanol, isopropanol, or acetone. Higher alcohol concentrations will produce gelation or precipitation of the xanthan gum.

For the most stable results the xanthan gum must first be dissolved completely in the water, and then the solvent added slowly under continuous agitation.

Xanthan gum is not generally soluble in organic solvents, although it will dissolve directly in glycerol or ethylene glycol if the solution is heated to 65°C or higher. Xanthan gum solutions in glycerol have long, stringy flow characteristics.

Enzymes. Those enzymes that are commonly met as microbial by-products or as commercially available products, e.g., protease, cellulase, pectinase, amylase, have no effect on xanthan gum. This resistance to enzymes is attributed to the nature of the sugar linkages in the polymer chains, as well as the nature of the side chain substituents on the polysaccharide backbone.

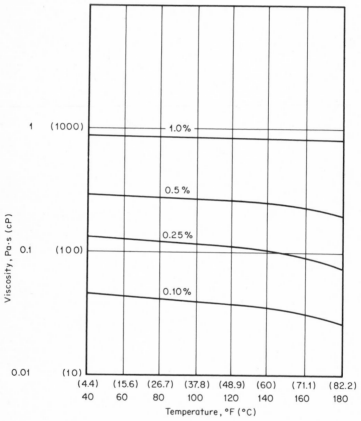

Figure 24.8 Effect of temperature on the viscosity of solutions of industrial-grade xanthan gum.

Surfactants. The compatibility of xanthan gum solutions with non-ionic surfactants is quite good for surfactant concentrations of 20% or lower. At surfactant concentrations of about 15% or higher, anionic and amphoteric surfactants tend to salt out the xanthan gum. Below this concentration level, stability is good.

Preservatives. Although xanthan gum is compatible with most of the commonly used preservatives, quaternary ammonium compounds should not be used as preservatives unless there is a suitable salt present to serve as a shielding agent. As is true for other polysaccharides, xanthan gum solutions will support microbial growth, even though it is a poorer supporter of growth than most other polysaccharides. Therefore, an antimicrobial preservative is recommended if xanthan gum solutions are to be stored for periods of 24 h or longer.

Latex Emulsions. For the manufacture of paints, foams, coatings, or adhesives, xanthan gum is compatible with the common types of latex emulsions, making it effective as a stabilizer, thickener, and rheological-properties modifier.

Cationic Dyes. Although xanthan gum is incompatible with cationic dyes, it may be stabilized to prevent reaction by lowering the solution pH to 1.5, or by adding a shielding agent (e.g., a soluble monovalent or divalent salt). The salt needed depends upon the type of dye and the cation involved and which of these must be shielded, but the salt concentration used is generally in the 2.0 to 3.0% range.

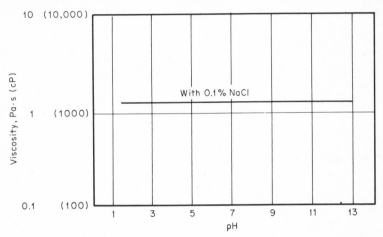

Figure 24.9   Effect of pH on the viscosity of a 1% solution of food-grade xanthan gum.

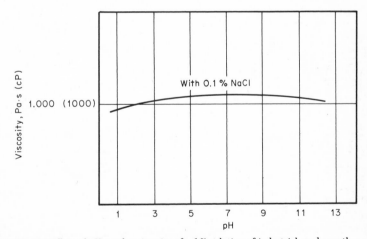

Figure 24.10   Effect of pH on the viscosity of a 1% solution of industrial-grade xanthan gum.

In the presence of many other chemicals xanthan gum solutions have unusually good compatibility and stability. The data shown in Table 24.3 illustrate the compatibilities of industrial-grade xanthan gum with a selection of natural and synthetic materials. The concentrations given in the table are not necessarily the maximum limits for compatibility for any given material. Food-grade xanthan gum exhibits compatibility and stability that are comparable to those given for industrial-grade xanthan gum in Table 24.3.

The compatibilities given in the table were determined by preparing the xanthan gum solutions first, then adding the proper amount of salt or other material to produce the percentage shown (based on total concentration). After mixing, the solutions were placed in sealed glass bottles, and the viscosities were determined at 60 rpm with a Brookfield LVF viscometer. The viscosities were again determined after standing at ambient temperature for 90 days. All the solutions tested were preserved with 0.2% formaldehyde.

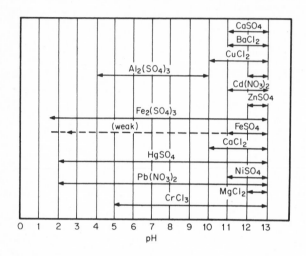

Figure 24.11   Reactivity of industrial-grade xanthan gum with metallic ions.

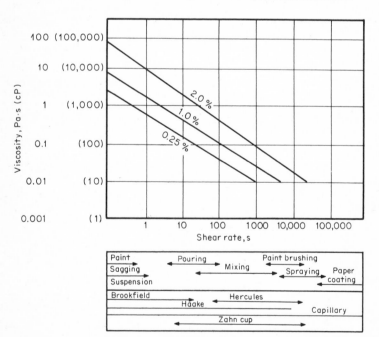

Figure 24.12   Effect of shear rate on the viscosity of xanthan gums solutions. Chart at bottom shows the shear rates of various applications for such solutions, and the ranges of measurement for various analytical procedures.

Suspensions

Xanthan gum is a nearly ideal suspension stabilizer because it not only has a yield value, but can maintain a viscosity which is almost independent of temperature and pH. It is also chemically inert with respect to most of the ingredients that may also be in the system.

For nonattracting solid particles, the stability that xanthan gum can impart to a suspension depends upon the rheological properties and yield point of the xanthan gum solution itself. Xanthan gum is particularly functional as a suspending agent for a wide variety of materials because of its high viscosity at low shear rates. (It has been shown that 1% concentration of xanthan gum in solution gives the solution an apparent yield value of 20 to 50 dyn/cm².)

The properties that render food-grade xanthan gum a funtional suspending agent have provided applications in many areas, such as in the suspension of fruit and pulp in drinks and concentrates; in suspension of solids in calamine lotion; in suspension of

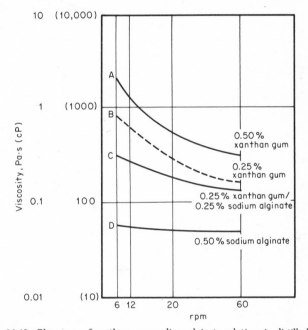

Figure 24.13   Rheogram of xanthan gum–sodium alginate solutions in distilled water.

nutrients in molasses; for liquid in animal feeds; and in suspension of solids in nonsettling toppings and syrups. Stable barium sulfate and bismuth suspensions have also been made with xanthan gum. Industrial-grade xanthan gum has been used to suspend the solids in ceramic glazes; this enables a more uniform application of the glaze to the ceramic body. In paints, xanthan gum prevents settling of the pigments, whereas the oil-well drilling industry utilizes it as a viscosifier for cuttings removal and as a lubricating agent at the bit to improve penetration rate. Dyes and tinting colors can be suspended indefinitely in textile print pastes and in paints. The formulation of stable flowable pesticides for agricultural applications is possible with the use of xanthan gum.

Emulsions

Among emulsion stabilizers, xanthan gum is unique because of the many desirable characteristics that it imparts to the final emulsion. Most of the functionality of xanthan gum is undoubtedly due to its particular rheological properties. It is nearly ideal as an emulsion stabilizer, since it not only has a yield value, but it also maintains, as was

TABLE 24.3
Compatibility of Xanthan Gum with Various Materials

Material	Percent material	Percent xanthan gum*	Viscosity† Initial cP	Initial Pa·s	After 90 days cP	After 90 days Pa·s
1. Organic and inorganic acids:						
Acetic acid	10.0	2.0	4200	4.200	4050	4.050
	20.0	2.0	4300	4.300	4400	4.400
Citric acid	10.0	1.0	1200	1.200	900	0.900
	20.0	1.0	1400	1.400	1030	1.030
Hydrochloric acid	5.0	2.0	3400	3.400	2750	2.750
	10.0	2.0	3500	3.500	1250	1.250
Phosphoric acid	30.0	2.0	4850	4.850	5250	5.250
	40.0	2.0	5600	5.600	7000	7.000
Sulfuric acid	5.0	2.0	3610	3.610	3150	3.150
	10.0	2.0	3800	3.800	3100	3.100
Tartaric acid	10.0	1.0	1210	1.210	900	0.900
	20.0	1.0	1430	1.430	1010	1.010
2. Organic and inorganic salts and bases:						
Aluminum sulfate	0.5	1.0	2860	2.860	Gel	
	5.0	1.0	1460	1.460	1330	1.330
Ammonium chloride	10.0	1.0	1120	1.120	980	0.980
	30.0	1.0	1380	1.380	1050	1.050
Diammonium phosphate	5.0	1.0	1100	1.100	1100	1.100
	10.0	1.0	1190	1.190	1230	1.230
Barium chloride	5.0	1.0	1240	1.240	1180	1.180
	15.0	1.0	1330	1.330	1130	1.130
Calcium chloride	10.0	1.0	1260	1.260	1300	1.300
	20.0	1.0	1580	1.580	1610	1.610
Chrome alum	5.0	1.0	1430	1.430	Gel	
Cobalt chloride	5.0	1.0	1280	1.280	1110	1.110
	15.0	1.0	1600	1.600	1160	1.160
Cobalt sulfate	5.0	1.0	1300	1.300	1120	1.120
	15.0	1.0	1170	1.170	1170	1.170
Cupric chloride	5.0	1.0	1180	1.180	970	0.970
	15.0	1.0	1500	1.500	2000	2.000
Ferrous sulfate	5.0	1.0	1410	1.410	300	0.300
	15.0	1.0	1600	1.600	Gel	
Magnesium chloride	5.0	1.0	1130	1.130	1070	1.070
	15.0	1.0	1210	1.210	1180	1.180
Potassium chloride	5.0	1.0	1240	1.240	1200	1.200
	15.0	1.0	1200	1.200	1115	1.115
Sodium bisulfite	5.0	1.0	1210	1.210	1160	1.160
	15.0	1.0	1540	1.540	1485	1.485
Sodium tetraborate	5.0	1.0	1140	1.140	1140	1.140
	15.0	1.0	1630	1.630	1520	1.520
Sodium carbonate	5.0	1.0	1130	1.130	1050	1.050
Sodium chloride	5.0	1.0	1110	1.110	1090	1.090
	15.0	1.0	1330	1.330	1200	1.200
Sodium chromate	5.0	1.0	1620	1.620	Gel	
Sodium citrate	5.0	1.0	1240	1.240	1170	1.170
	15.0	1.0	1390	1.390	1240	1.240
Sodium hydroxide	5.0	1.0	1360	1.360	810	0.810
	10.0	1.0	1390	1.390	115	0.115
Disodium hydrogen phosphate	5.0	1.0	1310	1.310	1100	1.100
	10.0	1.0	1190	1.190	1230	1.230
Sodium sulfate	5.0	1.0	1070	1.070	1100	1.100
	15.0	1.0	1460	1.460	1550	1.550
Zinc chloride	5.0	1.0	1200	1.200	1090	1.090
	15.0	1.0	1460	1.460	1320	1.320

Note: See page 24-16 for footnotes.

3. Thickeners:						
Sodium alginate	0.5	0.5	330	0.330	275	0.275
	1.0	1.0	1430	1.430	1390	1.390
Guar gum	1.0	1.0‡	4300	4.300	3030	3.030
Gum tragacanth	0.5	0.5	1020	1.020	1230	1.230
	1.0	1.0	4570	4.570	4900	4.900
4. Organic compounds:						
Acetone	10.0	1.0	890	0.890	980	0.980
	20.0	1.0	1000	1.000	1050	1.050
Methanol	10.0	1.0	940	0.940	970	0.970
	20.0	1.0	980	0.980	1050	1.050
Isopropanol	10.0	1.0	990	0.990	1000	1.000
	20.0	1.0	1040	1.040	1080	1.080
Benzyl alcohol	10.0	1.0	1060	1.060	1100	1.100
Urea	25.0	1.0	980	0.980	1200	1.200
	50.0	0.5	500	0.500	510	0.510
5. Preservatives:						
Dowicide® A	0.1	1.0	970	0.970	980	0.980
	0.5	1.0	1030	1.030	940	0.940
Dowicide B	0.1	1.0	1040	1.040	1040	1.040
	0.5	1.0	1100	1.100	1080	1.080
Dowicide G	0.05	1.0	900	0.900	970	0.970
	0.1	1.0	990	0.990	990	0.990
	0.5	1.0	1090	1.090	1100	1.100
Formaldehyde	0.1	1.0	860	0.860	990	0.990
	1.0	1.0	860	0.860	890	0.890
Methyl p-hydroxybenzoate	0.5	1.0	880	0.880	890	0.890
6. Enzymes:						
Alkalase®	1.0	1.0	1140	1.140	1150	1.150
Cellulase	1.0	1.0	1030	1.030	1030	1.030
Papain	1.0	1.0	1150	1.150	1060	1.060
Rhozyme® A-4	1.0	1.0	1150	1.150	1130	1.130
Gumase® HP 150	1.0	1.0	1085	1.085	1075	1.075
Pectinol® 10M	1.0	1.0	1040	1.040	920	0.920
7. Surfactants:						
Anionic:						
Stepanol® WAT	1.0	0.5	350	0.350	340	0.340
	10.0	0.5	370	0.370	340	0.340
Duponal® C	10.0	0.5	430	0.430	340	0.340
Non-ionic:						
Igepal® CO 610	20.0	0.5	1330	1.330	1130	1.130
Tween® 80	20.0	0.5	590	0.590	860	0.860
Tergitol® NPX	20.0	0.5	710	0.710	690	0.690
Amphoteric						
Miranol® 2MCA	1.0	0.5	360	0.360	340	0.340
	10.0	0.5	370	0.370	130	0.130
8. Plasticizers (glycols):						
Glycerol	50.0	0.5	690	0.690	620	0.620
Propylene glycol	50.0	0.5	610	0.610	550	0.550
Triethanolamine	50.0	0.5	720	0.720	640	0.640
Hexylene Glycol	50.0	0.5	1590	1.590	1500	1.500
Kromfax®	50.0	0.5	630	0.630	540	0.540
Ethylene glycol	50.0	0.5	470	0.470	430	0.430
9. Water-soluble resins:						
Vinol® PA-20	1.0	1.0	1090	1.090	1270	1.270
	5.0	1.0	930	0.930	1670	1.670
Carboset® 525	1.0	1.0	960	0.960	950	0.950
	10.0	1.0	36	0.036	36	0.036
Carbopol® 934	0.5	0.5	1310	1.310	900	0.900
	1.0	0.5	1930	1.930	1580	1.580

TABLE 24.3
Compatibility of Xanthan Gum with Various Materials *(Continued)*

Material	Percent material	Percent xanthan gum*	Initial cP	Initial Pa·s	After 90 days cP	After 90 days Pa·s
10. Latex emulsions:						
Rhoplex® AC 490	60.0	0.5	950	0.950	740	0.740
UCAR® 360	60.0	0.5	1080	1.080	1200	1.200
Dow Latex 460	60.0	0.5	800	0.800	760	0.760
Dow Latex 307	60.0	0.5	800	0.800	730	0.730
Geon® 652	60.0	0.5	720	0.720	580	0.580
Airflex® 500	60.0	0.5	890	0.890	800	0.800

* 0.5% xanthan gum (preserved with 0.1% formaldehyde): initial viscosity, 0.290 Pa·s (290 cP); after 90 days, 0.285 Pa·s (285 cP).

1.0% xanthan gum (preserved with 0.1% formaldehyde): initial viscosity, 0.860 Pa·s (860 cP); after 90 days, 0.990 Pa·s (990 cP).

2.0% xanthan gum (preserved with 0.1% formaldehyde): initial viscosity, 2.350 Pa·s (2350 cP); after 90 days, 2.500 Pa·s (2500 cP).

† Brookfield Model LVF viscometer at 60 rpm, appropriate spindle.

‡ KELZAN DR grade.

mentioned earlier, a viscosity that is virtually independent of both temperature and pH, along with a general inertness to other chemicals that might be encountered in emulsified systems.

The pseudoplasticity of xanthan gum solutions, coupled with the yield value, results in immediate viscosity response to shear. Because of this, emulsions stabilized with xanthan gum exhibit very high viscosities at rest and are unusually stable under the low-shear conditions frequently met during shipment and storage. Low-shear conditions, of which pouring is typical, are adequate to make the emulsion flow. A typical product that uses this unique rheological property is salad dressing. Such oil-in-water emulsions are very stable in the container, but pour easily and show good cling characteristics on lettuce and other salad ingredients.

The direct result of the high degree of shear thinning exhibited by xanthan gum solutions is the excellent quality and pleasing mouthfeel, as well as the lack of gumminess on the palate.

Dispersions

Xanthan gum solutions, because of their yield value and rheological properties, have received wide application for the stabilization of solid, liquid, and gaseous dispersions. For instance, by the use of xanthan gum, solids (suspension), liquids (emulsion), or gases (foam) may be suspended in water. In some products, such as ice cream, liquids, solids, and gases are suspended in water.

Application Procedures

The application and mixing procedures and methods that are appropriate to xanthan gum for dispersion and mixing are discussed in detail under Application Procedures in Chap. 2. They will not be repeated in this chapter.

Handling and Storage

The dry storage stability of xanthan gum is excellent at moderate temperatures, that is, 25°C (77°F) or less. Like any hydrophilic colloid, xanthan gum should be stored in a cool, dry place.

Compared to other polysaccharides, xanthan gum is unusual in that long-term storage at elevated temperatures, that is, about 49°C (120°F), causes only a slight increase in viscosity, and this can be reversed by subjecting the solution to shear and/or heat.

Reaction with Galactomannans

There is a synergistic increase in solution viscosity when xanthan gum is combined at ambient temperature with the galactomannans such as guar gum or locust bean gum.

In addition, the combination of xanthan gum and locust bean gum forms a thermally reversible gel when the solution is heated to higher than 54.4°C (130°F) and then cooled. This gel contains a small amount of polysaccharide and a large amount of water. It is differentiated from a viscous solution, such as that produced by high-shear solubilization of locust bean gum, by its bulk properties, one of which is gel strength.

Physical Properties Typical physical properties of an industrial-grade xanthan gum–guar gum combination are given in Table 24.4. The gel properties of the xanthan gum–

TABLE 24.4
Typical Physical Properties of a Xanthan Gum–Guar Gum Combination

Physical state	Dry, cream-colored powder
Moisture, %	11
Ash, %	6
Color (white enamel standard, 75% reflectance)	70
Specific gravity	1.56
Bulk density, kg/m³ (lb/ft³)	790 (49.32)
Browning temperature, °C	190
Charring temperature, °C	270
Ashing temperature, °C	490
Ignition temperature, °C	*
Heat of combustion, cal/g	3.46
As a 1% solution:	
Refractive index at 20°C	1.3335
pH	7.0
Surface tension, dyn/cm	88
Viscosity,† 0.4%, sheared, Pa·s (cP)	2.0+(2000+)
Heat of solution, cal/g soln	0.070
Freezing point, °C	0.0
Mesh size (Tyler standard)	40

* Spontaneous combustion did not occur in an air environment.
† Measured with Brookfield LVF viscometer, 60 rpm.

locust bean gum combination are affected by certain variables, such as colloid concentration and ratio, and hydrogen ion concentration. Figure 24.14 illustrates the synergistic increase in viscosity with locust bean gum.

The effect of colloid ratio and pH on the gel strength of the xanthan gum–locust bean gum combination in distilled water is shown in Figures 24.15 and 24.16, respectively. Wherever it was necessary to make pH adjustments, dilute solutions of sodium hydroxide or hydrochloric acid, as needed, were used. A Bloom gelometer was used to check gel strength.

Changes in the ratios of xanthan gum to locust bean gum affect the gel system, as shown in Figure 24.15. Maximum gel strength occurs at a ratio of approximately 1:1. Two pH levels (6.0 and 3.0) were tested to demonstrate the effect of reducing solution pH on gel strength.

In Fig. 24.16, it can be seen that the greatest gel strengths arise from solutions that are neutral or slightly alkaline, regardless of the ratio of xanthan gum to locust bean gum.

Time also affects the gel strengths of the xanthan gum–locust bean gum combination. A marked increase in gel strength is noted after preparation. The strength gradually increases until the maximum strength is reached in about 7 days. Afterward, a slight decrease in gel strength is observed until such time as the gel stabilizes.

Figure 24.17 illustrates the variation in viscosity for a combination of xanthan gum and locust bean gum put into solution under shear. The figure shows two shear rates. The data for the lowest of the two curves were developed under normal laboratory stirring conditions with a propeller agitator running at 800 rpm for 2 h. Viscosity data for the other curve were obtained after the solution was sheared for 1 min in a Waring blender at high speed. The resultant solutions were semigelatinous.

Compatibilities For combinations of xanthan gum with locust bean gum or guar gum, the compatibility with other common additives will be the same as that for each gum by itself in solution. Borate ions will not be tolerated, for instance. For complex formulations it is recommended that the user conduct laboratory studies to be sure of the solution compatibilities, as well as to ensure adequate shelf life of the final formulation.

Suspensions   Because of the high apparent viscosities of such solutions, solids suspension can be obtained with very low gum concentrations when combinations of xanthan gum and locust bean gum are used.

Preservatives   Both the galactomannan gums and xanthan gum are, as polysaccha-

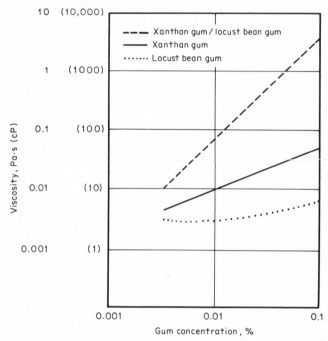

Figure 24.14   Viscosity as a function of concentration for (1) a combination of xanthan gum and locust bean gum, (2) xanthan gum, and (3) locust bean gum.   Solutions prepared by diluting at 140°C from 1% concentration.

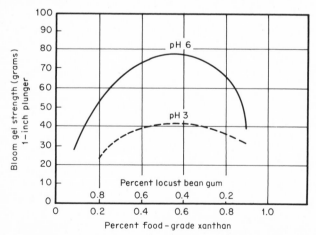

Figure 24.15   Gel strength of 1% solution of xanthan gum–locust bean gum combination as a function of colloid ratio.

rides, subject to bacterial action if storage is prolonged. Part 5 of Table 24.3 gives a list of suitable preservatives.

Toxicology and Safety

In short-term, subacute toxicological studies of xanthan gum, it has not been possible to determine acute oral toxicity (LD_{50}) to rats and dogs. The amount of xanthan gum that could be fed to the test animals within a 24-h period caused no fatalities, no signs

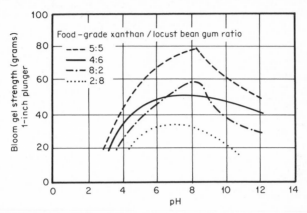

Figure 24.16 Gel strength of 1% solution of xanthan gum–locust bean gum combinations as a function of pH.

of toxicity, and no changes in the internal organs of rats and dogs that were given doses as great as 45 g/kg of body weight for rats and 20 g/kg of body weight for dogs.

Histopathological observations of rats that had been subjected to long-term oral administration (104 to 105 weeks) of xanthan gum at levels of 0.25, 0.5, and 0.1 g/kg of body weight showed that there were no distinguishable differences between the control and the test groups. Studies of three-generation reproduction by albino rats indicated that xanthan gum at levels of 0.25 and 0.5 g/kg of body weight had absolutely no

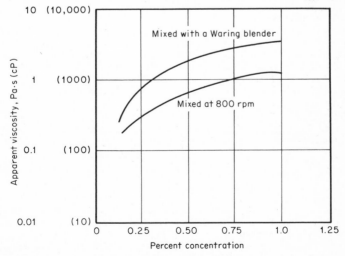

Figure 24.17 Viscosity development of xanthan gum–locust bean gum (1:1) solutions produced by various shear inputs. Data for top line by mixing in Waring blender for one minute at high speed. Bottom line is data for laboratory mixer at 800 rpm.

adverse effect. Long-term (107-week) oral administration of xanthan gum at levels of 0.25, 0.37, and 1.0 g/kg of body weight to dogs also failed to produce any observable adverse effects. Skin- and eye-irritation studies of xanthan gum indicated that the gum is not an irritant. In addition, recent sensitization studies have shown that xanthan gum is not a sensitizer.

Regulatory Status The general use of xanthan gum in foods has been cleared by the United States Food and Drug Administration where Standards of Identity do not preclude such use (21 CFR 172.695). The Canadian Governor-in-Council has also formally approved the general use of xanthan gum in foods. Xanthan gum is on Annex II of the European Economic Community emulsifier/stabilizer list. The Joint Expert Committee of the Food and Agriculture Organization/World Health Organization of the United Nations (FAO/WHO) has issued an acceptable daily intake (ADI) for xanthan gum. Many other countries have approved xanthan gum for various food uses.

In the United States, xanthan gum is permitted by Food and Drug Administration Regulations in many standardized foods, such as cheeses, cheese products, milk and cream products, mellorine, food dressings, table syrups, and vegetables in butter sauce. In the standards for these products xanthan gum is either mentioned by name or the phrase "safe and suitable stabilizers" appears in the list of approved ingredients. The Food and Drug Administration has also approved xanthan gum as a component of paper and paperboard intended to contact food.

United States Department of Agriculture Regulations permit the use of xanthan gum in sauces, gravies, and breadings used with meat and poultry products. The United States Environmental Protection Agency has exempted xanthan gum from the requirement of a tolerance when used as an inert ingredient in pesticide formulations.

The Code of Federal Regulations should be consulted for details on conditions of use and permitted use levels in the above products.

COMMERCIAL USES: Food

Xanthan Gum

Bakery Products Xanthan gum allows cold makeup of bakery fillings. In addition, the filling has excellent texture, mouthfeel, and flavor release, and the pastry does not absorb the filling.

In the preparation of bakery flavor emulsions with xanthan gum, excellent stability along with a smooth-textured, pourable body are obtained. When compared with bakery flavor emulsions that are conventional stabilizers, those using xanthan gum as the stabilizer give to the user both savings in preparation time and in overall costs.

Dressings In the production of pourable dressings, it takes a smaller quantity of xanthan gum than other commerical gums to give the dressing superior stability, easy pumpability, better mouthfeel and cling, excellent pourability at refrigerated temperatures, and better flavor release. These advantages are the result of pseudoplasticity of xanthan gum.

For spoonable dressings, mouthfeel is less pasty and flavor release is better if there is a partial replacement of starch with xanthan gum.

Foods and Drinks For the canning of foods, xanthan gum provides excellent viscosity control under processing conditions. Its shear-thinning properties make pumping and filling operations easier. If xanthan gum is used for partial starch replacement, heat penetration is faster. Foods maintain nutritional quality and appearance.

In dry mixes, xanthan gum provides rapid, high-viscosity build-up in both cold and hot systems and permits easy preparation of salad dressings, milkshakes, sauces, gravies, and beverages that have excellent texture, mouthfeel, and flavor release.

The syneresis control and temperature stability of xanthan gum provide for emulsion and suspension stability and viscosity maintenance of dressings, sauces, and gravies during both freeze-thaw and heating cycles. Small amounts of xanthan gum will improve the freeze-thaw stability of starch-thickened products.

In juice drinks low concentrations of xanthan gum are effective for suspending fruit pulp for long periods of time. This gives the drink more uniformity of flavor, body consistency, and better mouthfeel.

Other Products The presence of xanthan gum in syrups improves pourability and cling of syrup to ice cream, fruits, pancakes, and other foods. Penetration and runoff are also controlled.

When xanthan gum is used in the preparation of relishes, it requires no cooking in preparation; therefore, the loss of liquor during filling operations is eliminated. The relish and its liquor cling better to hot dogs and hamburgers. This reduces the sogginess of rolls and buns.

In addition to the previously mentioned role that xanthan gum plays in improving sauces and gravies, the viscosities of these products will remain constant over wide temperature ranges for long periods of time when xanthan gum is incorporated into their systems.

Xanthan gum is used to thicken and suspend molasses-based liquid feed supplements and provides temporary suspension of insoluble proteins in calf-milk replacer systems.

Xanthan Gum with Locust Bean Gum

When xanthan gum is combined with locust beam gum (and, in some cases, starch), the mixture allows easy gel formation necessary for the preparation of puddings and pie fillings, bakery jellies, chip dips, imitation sour cream, tomato aspic, and liver paté. When used for partial starch replacement, the combination of xanthan gum and locust bean gum gives the food a less-pasty mouthfeel, a better texture, and improved flavor release.

In dairy products the combination of gums gives heavier consistency and improved texture to pasteurized processed cheese spreads. It also allows the preparation of instant flavored milkshakes.

Xanthan gum blended with locust bean gum is also used for the gelling of canned pet foods where the thermal-reversible nature of the system aids the reprocessing of production batches.

When xanthan gum is combined with both locust gum and guar gum, the mixture provides for viscosity control of cottage cheese dressings. Drainage of dressing from the curd is minimized, giving improved uniformity of the product. The combination of the three gums is also effective for the stabilization of ice cream and ice milk processed by batch or high-temperature, short-term (HTST) systems. The three-gum blend is also effective in mixes with high whey-solids concentrations.

COMMERCIAL USES: Industrial

Xanthan Gum

Suspensions The effectiveness of xanthan gum as a suspending agent accounts for a large proportion of its industrial applications. Examples are:

Agricultural Chemicals. Xanthan gum is used as a suspending agent for herbicides, pesticides, fertilizers, and fungicides. Also, because of its excellent control of drift, a precisely defined area can be sprayed during spraying operations. Also, xanthan gum imparts cling, which increases the contact time between the pesticide and the crop. Xanthan gum is also useful as a stabilizer for flowable agrochemicals.

Paints. Xanthan gum provides excellent suspension in thixotropic paints. It also has outstanding enzyme resistance and good compatibility with other thickeners that might be used in paint systems.

Polishes. Xanthan gum suspends solids in shoe polishes, abrasives in silver and brass polishes, and emulsions in wax polishes.

Pigments. Slurried pigments that are being shipped or stored can be held in suspension by addition of xanthan gum, thus discouraging reagglomeration.

Textiles. Dye pigments can be suspended in solution by the use of xanthan gum, thereby controlling dye application in space printing and Kuester dyeing. The gum also acts as a flow modifier during the printing operation.

Inks. Used in both water-based and emulsion inks, xanthan gum is a suspending agent and stabilizer, providing controlled penetration and water release under uniform gloss.

Ceramics. The ingredients of ceramic glazes are suspended by the addition of xanthan gum.

Wallpaper. Xanthan gum in wallpaper printing systems is used as a flow modifier and suspending agent.

Viscosity Control The low viscosity of xanthan gum under high shear, its unique rheological properties, and other flow characteristics make it useful for a number of viscosifying and flow-control applications.

Abrasives. Low viscosity under high shear aids abrasive machining by providing rapid grinding action and fast removal of cuttings; meanwhile, it suspends the abrasives.

Adhesives. The rheological properties of xanthan gum permit controlled penetration of substrates by the adhesive. During application, the adhesive will have a low viscosity, thus pumping easily, yet will have high viscosity under the low-shear conditions on the surface of the substrate. Water release during drying is fast.

Paper. Xanthan gum has several important functions in the manufacture of paper products. It acts as a modifier of rheological properties for high-solids size-press and roll coatings, as a wet-end formation aid, for suspension of raw starch for jet cookers, and for dewatering control of air-knife coatings.

Oil Production. Xanthan gum is used as a viscosifier in oil-well drilling fluids. It also finds use in fluids used in reservoir flooding, well completion and workover, and formation fracturing.

Other Applications Xanthan gum assists in the manufacture of gel-type acid and alkali cleaners used industrially. It also promotes cling of the cleaner to surfaces to give longer contact times and better cleaning action.

In explosives, xanthan gum is a gelling agent. And the gum lubricates and provides green strength during the extrusion of welding rods.

During ore beneficiation processing, xanthan gum controls the settling rate of ores during the sedimentation phase. It also acts as a flocculant during flotation and provides foam stabilization. In the slurry pumping of pulverized ore, xanthan gum reduces drag by up to 65%, especially under low-shear conditions.

In general, the stability of xanthan gum to pH, temperature, salts, acids, alkalis, and shear changes allows it to produce stable emulsions with uniform particle sizes.

Xanthan Gum with Locust Bean Gum

Agricultural Sprays This combination produces stable emulsions of oil-soluble ingredients in herbicides, fungicides, and similar products, providing end products with flowability, controlled droplet size, improved cling to sprayed surfaces, and controlled spray drift. In agricultural foams, the gum combination aids in the production of fine bubble structures and, by adding viscosity, adds stability to the continuous phase.

Gelled Products Deodorant gels made with the gum combination are thermally reversible; when filled hot, they gel upon cooling. The gels are firm and rubbery and slowly release the deodorants.

Slurried Explosives Xanthan gum and guar gum blends are used to suspend and gel slurried explosives systems containing saturated ammonium nitrate solutions.

Fire Fighting Because of high viscosity at low concentrations, the gum combination is able to improve the drop pattern and the cling of fire-fighting fluids which are dropped from the air and used to fight forest fires.

Paper Sizing The paper industry uses the gum combination to enhance the efficiency of rosin-alum sizes, to increase Mullen readings, and to improve internal water resistance.

Photographic Processing The gum combination is quite compatible with photoprocessing solutions, and it is thermally reversible within the temperature range encountered during photoprocessing. Use of the gum pair in photoprocessing solutions results in smooth surfaces with low synersis.

FORMULATIONS

Dessert Soufflés

Xanthan gum is used in easy-to-prepare dessert soufflés which have the advantages of convenience of preparation, stability during service, and a fat content and total caloric value lower than found in home-prepared recipes.

Vanilla Soufflé

	Weight percent
Baker's sugar	69.2
Redisol No. 4 Starch (A. E. Staley)	14.7
Clearjel Starch (National Starch & Chemical)	14.7
Xanthan gum	0.59
Table salt	0.49
Artificial Vanilla Flavor No. 59.290 (Firmenich)	0.29
F. D. & C. Yellow No. 5	0.005
F. D. & C. Yellow No. 6	0.001

Dry-blend the ingredients, then add the blend to 180 mL (6 oz) of cold milk in a small mixing bowl. With an electric mixer, blend at a slow speed (to prevent air bubbles) for 3 min, then set aside.

Next, prepare an egg white mix by adding 15.1 g of egg white solids (P-20, Henningsen Foods) to 120 mL (4 oz) of cold water. Whip at high speed until soft peaks are formed, about 2 min. Sprinkle on 14 g (1 tbsp) of sugar, and continue to whip until stiff peaks are formed, approximately 1 min.

Stir one-fourth of the beaten egg white mix into the soufflé mix. Fold this mixture into the rest of the beaten egg whites. Place the product into a 1.4 L (about 6-cup) ceramic soufflé dish which has been greased with butter and coated with sugar. The product should be 2.5 cm (1 in) below the rim for best results. Bake at 177°C (350°F) for about 30 min. If a metal container is used, place it into a slightly larger second container with 2.5 cm (1 in) of water. Serve the soufflé with a favorite sweet sauce.

Chocolate Soufflé

	Weight percent
Baker's Sugar	61.4
Cokay No. 35 Cocoa (R. A. Johnston Co.)	14.2
Redisol No. 4 Starch (A. E. Staley)	11.8
Clearjel Starch (National Starch & Chemical)	11.8
Xanthan gum	0.47
Table salt	0.39

Dry-blend the ingredients, then add the blend to 180 mL. (6 oz) of cold milk in a small mixing bowl. With an electric mixer, blend at slow speed (to prevent air bubbles) for 3 min. Set aside.

Prepare an egg white mix the same as for the vanilla soufflé, except use 30 g (2 tbsp) of sugar. Mixing and baking procedures are the same as for the vanilla soufflé, except baking should be for 1 h.

Lemon Soufflé

	Weight percent
Baker's sugar	66.6
Redisol No. 4 Starch (A. E. Staley)	13.9
Clearjel Starch (National Starch & Chemical)	13.9
Lemon Powder No. 14110 (Borden, Inc.)	2.8
Adipic acid	1.9
Xanthan gum	0.56
Table salt	0.46
F. D. & C. Yellow No. 5	0.02

The procedure is the same as for the vanilla soufflé.

Bakery Jellies

Xanthan gum permits the formulation of bakery jellies which are both economical and easy to prepare. Used in cold applications, these jellies exhibit minimal syneresis, so that the jelly is not absorbed by the doughnut or other pastry. Texture, mouthfeel, flavor release, and appearance are excellent. Additionally, the jellies are easily pumped and can withstand several pumpings without change in consistency. Two such formulations follow:

	Weight percents	
	55% solids	35% solids
Water	45.00	65.00
Sugar	—	32.02
Dextrose	27.14	—
Sucrose	25.00	—
Colflo 67 Starch (National Starch & Chemical)	2.00	2.00
Anhydrous citric acid (fine, granular)	0.40	0.40
Xanthan gum	0.14	0.20
Locust bean gum	0.14	0.20
Sodium citrate (powder)	0.10	0.10
Raspberry color	0.05	0.05
Raspberry flavor	0.03	0.03

To prepare the 55% solids jelly, blend the dry ingredients, with the exception of the sugar, and add to the water while stirring with a mechanical mixer. Heat to 93°C (200°F), and continue stirring. Add the sugar, and dissolve. Package and cool. For the 35%-solids jelly, blend the dry ingredients and add to the water while stirring with a mechanical mixer. Heat to 93°C (200°F) with continued stirring. Stir 2 to 3 min longer, package, and cool. For firmer gel bodies for either jelly, increase the concentration of xanthan and locust bean gum, keeping the original ratio.

Salad Dressings

Salad dressings made with xanthan gums have excellent emulsification properties with superior stability. These dressings are smooth in texture, pour easily, and look appealing. They also have excellent flavor release and clean mouthfeel.

Green Goddess

	Weight percent
Vegetable oil	38.00
Water	27.70
Vinegar (50 grain)	14.70
Sugar	12.00
Salt	4.00
Egg yolk (fresh)	2.00
Anchovy paste	0.60
Onion powder	0.30
Garlic powder	0.30
Xanthan gum	0.25
Shredded green onions	0.10
Chopped chives	0.05

To prepare, dry-blend the xanthan gum with one-half of the sugar, and hydrate with water, vinegar, shredded green onions, and chives with vigorous agitation for 15 min. Add egg yolk and then the remaining dry ingredients. Next, add oil, slowly at first, then at normal rate. Emulsify with a colloid mill at a setting of 0.05 cm. (0.02 in).

Creamy Russian

	Weight percent
Vegetable oil	35.00
Water	24.75
Vinegar (50 grain)	13.50
Sugar	12.00
Pickle relish	7.00
Salt	4.00
Tomato paste (26%)	2.50
Mustard	1.00
Xanthan gum	0.25

To prepare, dry-blend the xanthan gum with one-half of the sugar, and hydrate with water, vinegar, and pickle relish liquor under vigorous agitation for 15 min. Add a blend of the remaining sugar, salt, and mustard; then add the tomato paste. Next, add oil slowly at first, then at normal rate. Emulsify with a colloid mill at 0.05 cm (0.02 in). Add pickle relish solids.

French Dressing

	Weight percent
Vegetable oil	38.00
Water	34.65
Sugar	11.50
Vinegar (100 grain)	9.00
Salt	4.00
Paprika (powdered)	1.35
Mustard (powdered)	1.25
Xanthan gum	0.25

To prepare, dry-blend the xanthan gum with one-half of the sugar, and hydrate with water and vinegar under vigorous agitation for 15 min. Add a blend of all the remaining solids. Next, add oil, slowly at first, then at normal rate. Emulsify with a colloid mill at 0.05 cm (0.02 in).

Creamy Italian

	Weight percent
Oil	55.0
Water	17.6
Cider vinegar (100 grain)	11.8
Salt	4.0
Wine vinegar (55 grain)	4.0
Lemon juice (reconstituted)	3.0
Egg yolks	2.0
Sugar	1.5
Garlic powder	0.4
Onion powder	0.3
Oregano powder	0.2
Xanthan gum	0.2

To prepare, add a blend of the xanthan gum with sugar to the vinegars with agitation. Mix until hydrated. After the xanthan is hydrated, add a blend of salt and spices, and mix thoroughly. Next, slowly add oil and mix until homogenous. Emulsify with a colloid mill at a setting of 0.05 cm (0.02 in). In separating-type Italian dressings, the xanthan gum helps to hold the oil and water phases together for uniform pouring.

Italian Dressing

	Weight Percent
Oil	56.00
Water	21.50
Cider vinegar (100 grain)	14.00
Lemon juice (reconstituted)	5.00
Salt	3.00
Minced garlic	0.20
Minced onion	0.16
French Celery Seed (Schilling's)	0.04
Java Black Pepper (cracked)	0.04
Italian Seasoning (Schilling's)	0.02
Crushed Red Pepper (Schilling's)	0.02
Xanthan gum	0.02

To prepare, add the xanthan gum to the vinegar with agitation. Add all spices, salt, and lemon juice. Fill oil and water phases separately by normal two-stage fill operation. (*Note:* The three Schilling's products are made by McCormick & Co., Inc.)

Dry Sauce Mixes

The following formulations produce dry sauce mixes with outstanding heat and freeze-thaw stability. Finished products have excellent flavor release and mouthfeel, and the viscosity remains constant over a wide temperature range.

Cheese Sauce Mix

	Weight Percent
Cheeztang (Kraft Foods)	39.68
Milk solids (nonfat)	32.20
Beatreme 743 (Beatrice Foods)	15.22
Rokatang (Kraft Foods)	6.41
Xanthan gum	2.58
Adipic acid	1.79
Salt	1.76
Garlic salt	0.18
Onion salt	0.18

To prepare, disperse 56.70 g (2 oz) of dry mix in 237 mL (1 cup) of cold milk. Slowly heat to a boil, stirring constantly.

Barbecue Sauce Mix

	Weight percent
Tomato solids	39.60
Sugar	20.80
Barbecue Spice (Spice Islands)	19.80
Salt	9.24
Spaghetti Sauce Seasoning (Spice Islands)	6.60
Onion powder	2.64
Xanthan gum	1.32

To prepare, place 237 mL (1 cup) of water in a mixing bowl. Add sauce mix (75.5 g) and blend at low speed. Mixing time for an electrical mixer is approximately 3 min; for an egg beater, 1.5 to 2 min. Just before the end of the mixing period, add 5 mL (1 tbsp) of vinegar.

Spaghetti Sauce Mix

	Weight percent
Tomato solids	60.99
Cheeztang (Kraft Foods)	15.20
Salt	6.30
Maggi 3H-1 Powder (Nestle Co.)	6.30
Minced onion	5.06
Xanthan gum	4.10
Monosodium glutamate	0.79
Pepper	0.63
Garlic powder	0.47
Thyme	0.16

To prepare, disperse 63.60 g (2.25 oz) in 474 mL (2 cups) of cold water. Heat slowly to a simmer, stirring constantly; continue to simmer for 5 min.

White Sauce Mix

	Weight percent
Whey solids	50.00
Colflo 67 Starch (National Starch)	25.00
Salt	15.00
Xanthan gum	7.50
Onion powder	1.25
White pepper	1.00
Paprika	0.25

To prepare, disperse 20.0 g (0.7 oz) in 237 mL (1 cup) of cold milk. Heat slowly to a boil, stirring constantly.

Frozen Pizzas

Xanthan gum helps to produce a tomato-based sauce for frozen pizza with improved viscosity and freeze-thaw stability. Even with reduced tomato solids, the sauce will appear rich and thick without gumminess. Because of xanthan gum's high-temperature stability, the gum will not boil out during baking. And because of its exceptional freeze-thaw stability, xanthan gum contributes excellent shelf life to the pizza.

To obtain the optimum functional benefits, the recommended level of xanthan gum is 0.3% by weight of the sauce. This level may be adjusted to obtain the desired consistency. Following are two formulations for spice mixes for use in the pizza sauce.

	Weight percent	
Xanthan gum	9.6	2.6
Nu-Col 231 Modified Food Starch (A. E. Staley)	—	24.7
Salt	60.0	48.2
Garlic powder	9.9	8.0
Onion powder	7.3	5.9
Sweet basil powder	6.6	5.3
Oregano powder	4.4	3.5
Black pepper powder	2.2	1.8

To prepare either formulation, dry-blend the ingredients well. Then add the spice mix to an agitated 9%-solids tomato puree (or diluted tomato paste) in the following ratio:

For the first spice mix: add 88.04 g (3.1 oz) to 2.84 kg (100 oz) of puree. For the second spice mix: add 113.6 g (4.0 oz) to 2.84 kg (100 oz) of puree. Fill and spread 170.4 g (6 oz) of prepared sauce onto a 36-cm (14-in) round pizza crust. Top with mozzarella cheese, and, if desired, pepperoni, sausage, ground beef, shrimp, mushrooms, peppers, anchovies, olives, etc.

Animal Feeds (Liquid)

Xanthan gum solutions are highly effective thickening and suspending agents in liquid feed systems, exhibiting stability over the entire pH range and with its viscosity unaffected by temperature variations. Four typical formulations are shown in the following:

	Percents in compositions			
Standard Brix molasses	67.50	67.50	35.00	40.00
50% ammonium lignosulfonate	—	—	—	8.80
100% urea	—	22.00	—	22.00
50% urea liquor	20.00	—	10.00	—
Salt	5.00	—	10.00	—
Water	0.67	3.67	24.77	22.37
Trace mineral sulfates	0.20	0.20	0.20	0.20
Preservative	0.20	0.20	—	0.20
75% phosphoric acid	6.40	6.40	—	6.40
Vitamin ADE emulsion	0.03	0.03	0.03	0.03
Ground limestone	—	—	20.00	—
Xanthan gum	0.075	0.1	0.15	0.1

In the preparation of these formulations, best results are obtained if the xanthan gum is dissolved in the free water available in the mixture. If there is no free water (or if it accounts for less than 5% of the total), add urea liquor to make up the difference. The other liquid-feed supplement ingredients may be added first to the xanthan gum solution, which can then be added to the overall formulation. To prepare the xanthan gum solution, it is necessary to have uniform particle wetting and dispersion of the powdered gum as described in the section Application Procedures in Chap. 2.

LABORATORY TECHNIQUES

Following are several of the more frequently needed analytical techniques required for the laboratory testing and evaluation of xanthan gums. Other analytical methods, such as analysis for trace elements, residual isopropyl alcohol, and others may be found in the monograph for xanthan gum in the *Food Chemicals Codex.*

Viscosity (Food Grade)

1. Dry-blend 2.00 g of xanthan gum and 2.00 g of potassium chloride.
2. Add blend slowly to 196 mL of distilled water while stirring the solution at approximately 800 rpm. Continue stirring for 2 h.
3. Remove the solution from the stirrer, adjust temperature to 25°C, and stir vigorously by hand to eliminate any possible thixotropic effect.
4. Determine the viscosity immediately with a Brookfield Model LVF viscometer at 60 rpm and with a No. 3 spindle.

Viscosity (Industrial Grade)

1. Prepare a 1% solution of xanthan gum by pouring 297 mL of distilled water into a 500-mL jar (about 1 pt) and adding 3.0 g of the gum slowly, while stirring the solution at approximately 800 rpm. Continue stirring for 2 h.
2. Remove the solution from the stirrer, adjust temperature to 25°C, and stir vigorously by hand to eliminate any possible thixotropic effect.
3. Read viscosity immediately with a Brookfield Model LVF viscometer at 60 rpm and with a No. 3 Spindle.

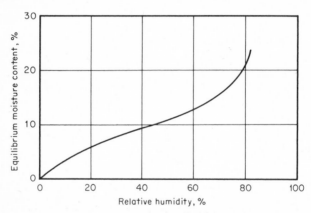

Figure 24.18 Xanthan gum moisture content–humidity relationship.

Moisture Content

Like the other hydrophilic polysaccharides, xanthan gum absorbs moisture from the atmosphere. The resulting equilibrium moisture content is a function of the relative humidity, as shown in Figure 24.18. Mixtures of xanthan and locust bean gums behave similarly. The procedure for determining the moisture content of xanthan gum is:

1. Weigh 1.00 g of xanthan gum into a wide (50 mm inside diameter), low-form weighing bottle.
2. Dry in an oven at 105°C for 3 h.
3. Replace the top of the weighing bottle, cool in a desiccator, and reweigh.
4. Calculate moisture content as the percentage of weight lost during the drying operation.

Powder Color

1. Place a flat-bottomed 90-mm petri dish or optically flat glass container on top of the search unit of a Photovolt reflectometer.
2. Spread powdered xanthan gum into the glass container to give a powder layer at least ¼ in deep over the entire top of the search unit. Do not shake the glass container, as this would result in sifting and would give erroneous readings.

3. Standardize the search unit against a white enamel working standard of about 75% reflectance. Record the reflectometer reading while using the green tristimulus filter in the search unit.

Determination of Gum in Mixtures

Because of the great number of products in which xanthan gum is used, it is not possible to give a general method of analysis that would not be subject to interferences from one or more of the materials present. The method which follows is valid only if xanthan gum is known to be the sole carbohydrate present.

High levels of proteins or other organic materials that are charred by contact with hot concentrated sulfuric acid also interfere with the analysis. High levels of salts will cause slight changes in the standard curve. For high levels of accuracy in analyses, it is suggested that the standard solutions used have salt contents similar to those of the sample being analyzed.

Among the procedures suggested for removing interferences are dialysis to remove salts and lower sugars, precipitation of proteins with trichloroacetic acid, and precipitation of the gum with alcohol. A method for the determination of xanthan gum in some food products is given by Graham (see reference in Further Reading/References, given later in this chapter).

The quantitative procedure for determination of xanthan gum is:

1. Pipette a 1.0-mL sample solution (5 to 250 ppm xanthan gum) into a 10-mL Erlenmeyer flask.
2. Add 1.0 mL resorcinol solution, and mix. (Resorcinol solution is a 4% solution in water, prepared fresh daily.)
3. From a Mohr pipette, rapidly add 6 mL of concentrated sulfuric acid to the mixture from step (2). *CAUTION:* The use of safety goggles and gloves is recommended, since the heat generated by the addition of the sulfuric acid is sufficient to bring the mixture close to the boiling point, and there is a possibility of spattering.
4. Place the flask in a shallow ice-water bath until the solution is near room temperature.
5. Remove from bath and let stand at room temperature for 10 to 25 min.
6. Next, carefully transfer the solution to a 10-mm spectrophotometer cell, avoiding the formation of air bubbles in the cell.
7. Read the absorbance of the peak at about 494 nm against a water blank.
8. Repeat with a water blank.
9. Determine the concentration of xanthan gum by comparison of the corrected absorbance with a standard curve prepared with solutions of known concentration ranging from 5 to 250 ppm. Typical absorbance for different concentrations in a 10-mm cell are 25 ppm, 0.14 nm; 50 ppm, 0.28 nm; 100 ppm, 0.55 nm; and 200 ppm, 1.10 nm. Sensitivity of the analysis may be increased by using longer-path-length cells, or by using a spectrophotometer with scale expansion.

PRODUCT TRADENAME DIRECTORY

(Reference taken from TSCA 1977 CANDIDATE LIST, provided by Chemical Abstracts; Kelco tradenames supplemented)

CAS REGISTRY NUMBER: 11138–66–2
FORMULA: UNKNOWN
REPLACES CAS REGISTRY NUMBER(S): 9088–32–8–12771–06–1–37189–49–4–37279–85–9–37383–52–1–39393–27–6–39444–54–7–54511–23–8–
CODE DESIGNATION: R261–9467
CA NAME (1): HP-9CI Xanthan gum
SYNONYMS: Xanthan; Xanthomonas gum: Polysaccharide B 1459: Biopolymer XB 23: Biopolymer XB-23:

KELZAN® (technical grade); KELZAN® D (technical grade); KELZAN® M (technical grade); KELZAN XC® Polymer (technical/drilling fluid); XANFLOOD® (technical grade); XANCO-FRAC® (technical grade); KELTROL® (food grade); KELTROL® F (food grade); GFS® (food grade); KELFLO®

FURTHER READING/REFERENCES

Ahmed, Z. F., and Whistler, R. I., "The Structure of Guaran," *J. Am. Chem. Soc.*, **72**, 2524 (1950).

Andrew, T. R., "Applications of Xanthan Gum in Foods and Related Products," in *Extracellular Microbial Polysaccharides*, American Chemical Society Symposium Series, No. 45, American Chemical Society, Washington, D.C., 1977.

Anonymous, *Fed. Regist.*, **34**, 5376 (March 19, 1969).

Anonymous, Canada Gazette, Part II, **105** (January 27, 1971).

Anonymous, *Fed. Regist.*, **36**, 17333 (August 28, 1971).

Anonymous, *Fed. Regist.*, **36**, 24217 (December 22, 1971).

Anonymous, *Food Chemicals Codex*, 2d ed., National Academy of Sciences, Washington, D.C., 1972, p. 856.

Anonymous, *Fed. Regist.*, **38**, 1218 (January 10, 1973).

Anonymous, *Fed. Regist.*, **38**, 6883 (March 14, 1973).

Anonymous, *Fed. Regist.*, **39**, 4466 (February 4, 1974).

Araujo, O. E., "Emulsifying Properties of New Polysaccharide Gum," *J. Pharm. Sci.*, **56**, 1141–1145 (1967).

Booth, A. N., Hendrickson, A. P., and De-Eds, F., "Physiologic Effects of Three Microbial Polysaccharides on Rats," *Toxicol. Appl. Pharmacol.*, **5**, 478–484 (1963).

Charm, S. E., and McComis, W., "Physical Measurements of Gums," *Food Technol.*, **19**, 948–953 (1965).

Cottrell, I. W., and Kang, K. S., "Xanthan Gum: A Unique Bacterial Polysaccharide for Food Applications," in *Developments in Industrial Microbiology 19*, American Institute of Biological Sciences, Washington, D.C., vol. 19, chap. 10, pp. 117–131 (1979).

Courtois, J. E., and Le Dizet, P., "Action de l'α-galactosidase du café sur quelques galactomannanes," *Carbohydr. Res.*, **3**, 141 (1966).

Courtois, J. E., and P. Le Dizet, "Galactomannans. VI. Action of Some Mannanases on Various Galactomannans," *Bull. Soc. Chem. Biol.*, **52**(1), 15–22 (1970) (Fr.).

Dea, I. C. M., McKinnon, A. A., and Rees, D. A., Tertiary and Quaternary Structure in Aqueous Polysaccharide Systems Which Model Cell Wall Cohesion. Reversible Changes in Conformation and Association of Agarose, Carrageenan, and Galactomannas," *J. Mol. Biol.*, **68**(1), 153–172 (1972).

Dintzis, F. R., Babcock, G. E., and Tobin, R., Studies on Dilute Solutions and Dispersions of the Polysaccharide from *Xanthomonas campestris* NRRL B-1459," *Carbohydr. Res.*, **13**, 257–267 (1970).

Graham, H. D., "Microdetermination of KELTROL (Xanthan Gum)," *J. Dairy Sci.*, **54**, 1622–1628 (1971).

Hui, P. A., and Nekom, H., "Some Properties of Galactomannas," *Tappi*, **47**, 39–42 (1964).

Jansson, P. E., Keene, L., and Lindberg, B., "Structure of the Extracellular Polysaccharide from *Xanthomonas campestris*," *Carbohydr. Res.*, **45**, 275–282 (1975).

Jeanes, A., Pittsley, J. E., and Senti, F. R., "Polysaccharide B-1459: A New Hydrocolloid Polyelectrolyte Produced from Glucose by Bacterial Fermentation," *J. Appl. Polym. Sci.*, **5**, 519–526 (1961).

Kang, K. S., and Cottrell, I. W., "Polysaccharides," in *Microbial Technology*, 2d ed., Academic Press, Inc., New York, in press.

Kovacs, P., "Useful Incompatibility of Xanthan Gum with Galactomannans," *Food Technol.*, **27**(3), 26–30 (1973).

Kovacs, P., "Xanthan Gum, A New and Unique Colloidal Stabilizer for the British Food Industry," *Food Trade Rev.*, 17–22 (1973).

Kovacs, P., and Igoe, R. S., "Xanthan Gum/Galactomannan System Improves Functionality of Cheese Spreads," *Food Prod. Dev.*, 37–39 (1976).

Kovacs, P., and Kang, K. S., "Xanthan Gum," in *Food Colloids*, Avi Publishing Co., Westport, Conn., pp. 500–521.

Matsumoto, T., Takashima, A., Masuda, T., and Onogi, S., "A Modified Casson Equation for Dispersions," *Trans. Soc. Rheol.*, **14**(4), 617–620 (1970).

McNeely, W. H. and Kang, K. S., "Xanthan and Some Other Biosynthetic Gums," in *Industrial Gums*, 2d ed., Academic Press, Inc., New York, pp. 473–497 (1973).

McNeely, W. H. and Kovacs, P., "The Physiological Effects of Alginates and Xanthan Gum," in *Physiological Effects of Food Carbohydrates*, American Chemical Society Symposium Series, No. 15, American Chemical Society, Washington, D.C., 1975.

Moorhouse, R., Walkinshaw, M. D., Winter, W. T., and Arnott, S., "Solid State Conformations and Interactions of Some Braided Microbial Polysaccharides," in *Cellulose Chemistry and Technology*, American Chemical Society Symposium Series, No. 48, American Chemical Society, Washington, D.C., 1977.

Moorhouse, R., Walkinshaw, M. D., and Arnott, S., "Xanthan Gum: Molecular Conformation and Interactions," in *Extracellular Microbial Polysaccharides*, American Chemical Society Symposium Series, No. 45, American Chemical Society, Washington, D.C., 1977.

Rees, D. A., "Double Helix Structure in Food," *Sci. J.*, **6**(12), 47–51 (1970).

Rees, D. A., "Polysaccharide Gels: A Molecular View," *Chem. Ind.*, 630–636 (1972).

Rees, D. A., "Shapely Polysaccharides," *Biochem. J.*, **126**, 257–273 (1972).

Rees, D. A., Biophysical Society Winter Meeting, London (1973).

Rocks, J. K., "Xanthan Gum," *Food Technol.*, **25**(5), 22–31 (1971).

Sloneker, J. H., and Jeanes, A. R., Exocellular Bacterial Polysaccharides from *Xanthomonas campestris* NRRL B-1459. Part I: Constitution, *Can. J. Chem.*, **40**, 2066–2071 (1962).

Smith, F., and Montgomery, R., *The Chemistry of Plant Gums and Mucilages*, Reinhold Publishing Corporation, New York, 1959.

Szczesniak, A. S., and Farkas, E., "Objective Characterization of the Mouthfeel of Gum Solutions," *J. Food Sci.*, **27**, 381–385 (1962).

Woodard, G., Woodard, M. W., McNeely, W. H., Kovacs, P., and Cronin, T. I. I., "Xanthan Gum: Safety Evaluation by Two-Year Feeding Studies in Rats and Dogs and a Three-Generation Reproduction Study in Rats," *Toxicol. Appl. Pharmacol.*, **24**, 30–36 (1973).

Index

NOTE: Page numbers in italics followed by the letter F indicate formulations.

Abinco Gums, 22-71
Abrasive suspensions, 5-18, 5-23
ABS terpolymer, 19-7
Acacia decurrens, 8-15
Acacia gum, 8-2
Acacia mucilage, 8-2
Acacia senegal, 8-2
Accofloc, 16-25
Accrostrength, 16-25
Accurac, 16-25
Acetamide, 22-68
Acetanilide, 18-12
Acetic acid, 4-11, 4-13; 8-7; 11-9; 13-5, 13-6; 16-3; *19-30F;* 20-6; 21-4, 21-11; 24-14
Acetone, 2-9, 2-12, 2-21; 4-6; 7-4; 12-5, 12-21; 13-5, 13-6; 18-11; 19-9; 20-15; 21-11; 24-9, 24-15
Acetone-formaldehyde resin, 22-70
Acetonitrile, 19-9
Acetophenetidin, 18-12
Acetulan, 21-8
N-Acetyl ethanolamine, 3-12; 12-4
Acetyleugenol, 8-8
Acetylsalicylic acid, 18-12
Acetyltriethyl citrate, 19-7
Acids:
  and agar, 7-4
  and carboxymethylcellulose, 4-13
  and gum karaya, 10-10
  and polyethylene glycols, 18-10
  and poly(ethylene oxide), 19-3, 19-15, *19-30F*
  and polyvinyl alcohol, 20-6
  and polyvinylpyrrolidone, 21-11
  and starches, 22-5, 22-20
  and tamarind gum, 23-6
  and xanthan gum, 24-8, 24-14
  (See also specific acids)
Acrylamide, 22-48, 22-53
Acrylate esters, 22-48, 22-53
Acrylic acid, 16-3; 17-3; 22-48, 22-53
Acrylic ester polymers, 12-13
Acrylic polymers, 21-16
Acrylonitrile, 22-48

Acrylonitrile-butadiene rubber, 19-7
Acrylonitrile-styrene copolymer, 19-7
Acrysol, 17-18
Actamer, 21-9
Adhesives:
  in abrasives, 21-16
  alginates in, 2-25
  for aluminum foil, 8-13
  animal glue, 18-19
  for artificial flowers, 8-13
  for bandages, 17-17
  British gum in, 22-69
  carboxymethylcellulose in, 4-24
  casein in, 18-19
  for casting cements, 8-13
  for ceramics, 3-14, 3-18; 20-20, 20-24
  cold-water soluble, 22-69
  for colostomy rings, 10-9
  in confections, 11-21
  for construction, 19-16
  corrugating, 2-35; 22-30, 22-69, 22-70, 22-72
  cured, 16-16
  dental, 17-14; 19-16
  denture (*see* Denture adhesives)
  dextrin in, 18-19; 22-36
  for envelopes, 8-13; 22-69
  for facial masks, 8-12
  and the FDA, 22-69
  for film, photographic, 17-17
  flock, 12-14
  and food, 20-16
  for food packaging, 12-6
    (*See also* Packaging)
  general, 3-18
  for glass-bottle coatings, 17-17
  for glass fiber binding, 3-18
  for glassine paper, 8-13
  gum arabic in, 8-16
  gum karaya in, 10-5, 10-9, 10-10, *10-11F, 10-12F*
  hot-melt, 18-19
  hydroxypropylcellulose in, 13-11, 13-12
  hydroxypropylmethylcellulose in, 3-14, 3-18

Adhesives *(Cont.):*
  for labels, 8-13; 17-17; 18-19; 22-68, 22-69;
    23-8
  laminating, 22-70
  latex, 18-19; 24-11
  for leather products *(see* Leather products,
    adhesives for)
  liquid, 22-30, 22-68, 22-69
  Lunel's paste, 8-13
  methylcellulose in, 3-14, 3-18
  mounting, 8-13
  multiwall bag, 22-70
  and paper products, 2-35; 3-19; 8-13; 10-9; 19-
    16, 19-26; 20-9, 20-16, 20-20, 20-21, 20-
    24, *20-27 to 20-29F;* 22-30, 22-68 to 22-
    70
  pastes, 22-69
  in personal care products, 19-16
  phenolformaldehyde, 3-18
  photographic film, 17-17
  for plywood, 3-19; 12-11, 12-15, 12-16; 17-17;
    23-6, 23-7, 23-9
  polyacrylamide, 16-10
  polyacrylate, 17-9, 17-11, 17-17
  polyethylene glycols in, 18-16, 18-18, 18-23,
    18-24
  poly(ethylene oxide) in, 19-5, 19-16
  polyvinyl alcohol in, 20-2, 20-5, 20-6, 20-20
  polyvinylpyrrolidone in, 21-2, 21-10, 21-15, 21-
    16, 21-18
  for postage stamps, 8-13
  pressure-sensitive, 7-9; 12-6; 18-19; 19-14
  quickset, 19-14
  remoistenable, 22-69, 22-72
  for rubber-to-steel, 17-17
  for stamps, 8-13; 22-21, 22-22, 22-69
  starch in *(see* Starches, and adhesives)
  stationery, 3-18
  steel, rubber onto, 17-17
  for tablets *(see* Pharmaceuticals, tablets)
  and tamarind gum, 23-6, 23-8
  for tapes, 22-68
  for textiles *(see* Textiles, adhesives for)
  tile, 3-18
  for tobacco products, 3-15, 3-20; 11-19, 11-21
  and TOSCA, 22-69
  wallpaper, 3-14, 3-18; 4-21; 8-13; 12-15, 12-16;
    22-26, 22-69
  water-proof, 22-45, 22-69, 22-70
  water-remoistenable, 22-70
  water-soluble, 16-16; 19-13, 19-14
  for wood products, 17-14, 17-17; 19-14, 19-26;
    20-9; 21-16
  xanthan gum in, 24-11, 24-22
Advacide, 2-21
Aerofloc, 16-25
Aerosols, 3-18
Aerosol toppings *(see* Foods, toppings for)
Aerotex, 13-2, 13-16
Agar, 7-1 to 7-19
  and acids, 7-4
  additives for, 7-5
  and alcohols, 7-4
  applications of, 7-6
  bacteriological grade, 7-3
  and battery electrodes, 7-9
  biochemistry of, 7-7

Agar *(Cont.):*
  biological properties of, 7-4
  and carboxyethylcellulose, 7-5
  and cationic exchange, 7-9
  chemical nature of, 7-2
  and cholesterol, 7-9
  and chromatography, 7-9
  commercial uses of, 7-7 to 7-9
  compatibilities of, 7-6
  constituents of, 7-3
  in cosmetics, 7-9
  and criminology, 7-8
  and culture media, 7-9
  and dentistry, 7-7, 7-8
  and denture molds, 7-6, *7-10F*
  derivatives of, 7-6
  and detergents, 7-9
  and dextrin, 7-5
  and electrolytes, 7-4
  electromechanical properties of, 7-4
  and electrophoresis, 7-9
  and explosives, 7-9
  extenders for, 7-5
  and the FDA, 7-2, 7-4
  fetotoxicity of, 7-4
  films of, 7-9
  and filtration, 7-9
  in foods, 7-2, 7-6, 7-7, 7-9; 8-18
  formulations with, 7-9
  in fuels, 7-9
  and gelation, 7-5 to 7-7
  gelling curves for, 7-16
  gelling temperatures of, 7-17
  gel strengths of, 7-17
  grades of, 7-17
  and GRAS, 7-2, 7-4
  and gum karaya, 7-5
  and gum kino, 7-6
  and gum tragacanth, 7-7
  handling of, 7-4, 7-5
  and Iceland moss, 7-5
  and immunodiffusion, 7-9
  and isinglass, 7-5
  laboratory applications of, 7-8
  laboratory techniques for, 7-13, 7-14, 7-18
  in laxatives, 8-12
  and lichenin, 7-5
  and locust bean gum, 7-5, 7-7
  manufacture of, 7-4
  melting curves for, 7-16
  melting temperatures of, 7-17
  and microbiology, 7-6 to 7-8
  microbiology specifications for, 7-6
  and microorganisms, 7-3
  and microtomy, 7-8
  molds of, 7-8
  in paraffin emulsions, 8-12
  and pharmaceuticals, 7-7, 7-8
  and photographic products, 7-9
  physical properties of, 7-2, 7-3
  precipitation of, 7-6
  and proteins, 7-6, 7-9
  rheology of, 7-4, 7-5
  and Russian isinglass, 7-5
  and salts, 7-4, 7-5
  and sodium alginate, 7-5
  solubilities of, 7-2

Agar *(Cont.):*
  and solvents, 7-4, 7-5
  specifications for, 7-2, 7-6
  and starch, 7-5, 7-7
  storage of, 7-4
  and tool making, 7-8
  toxicological properties of, 7-4
  viscosity of, 7-4, 7-5
  and water-soluble gums, 7-6
Agaran, 7-5
Agaroids, 7-3
Agarose, 7-3, 7-5 to 7-7, 7-9
Agricultural products:
  for aquaculture, 5-18
  dust stickers, 3-24
  fertilizers, 21-16; 24-21
  films, protective, 3-24
  fungicides, 3-14; 9-5; 12-7; 24-21, 24-22
  herbicides, 5-24; 18-19; 21-16; 24-21, 24-22
  insecticides, 3-18; 8-16; 11-3, 11-19, 11-21; 18-19
  nicotine sprays, 7-9
  orchid culture, 7-9
  packaging for, 19-25
  pesticides, 2-18; 5-24; 12-7; 18-14; 22-47; 24-13, 24-20, 24-21
  seed coatings, 3-18, 3-24; 12-16
  seed tapes, 19-16, 19-20, 19-25
  soil stabilization, 19-16, 19-21, 19-25
  sprays, 3-18, 3-24; 12-15, 12-16; 18-19, 18-23; 19-16, 19-24, 19-25
  suspensions, 5-18
  weed killers, 3-18
  wettable powders, 3-24
Airflex, 2-21; 24-16
Air fresheners, 2-25, 2-26; 5-18, 5-19, 5-21, 5-22, 5-24, 5-25
AKU-CMC, 4-23
Albigen, 21-2
Alcohols:
  and alginates, 2-9
  and carrageenans, 5-10, 5-18
  and guar gum, 6-4
  and gum tragacanth, 11-16
  and hydroxyethylcellulose, 12-5
  and hydroxypropylmethylcellulose, 3-11
  and locust bean gum, 6-7
  and methylcellulose, 3-11
  and pectins, 15-3
  and polyethylene glycols, 18-9
  and polyvinyl alcohol, 20-15
  and polyvinylpyrrolidone, 21-11
  and starches, 22-40
  and tamarind gum, 23-3
  (See also specific alcohols)
Aldehydes, 18-10
Algin, 2-1 to 2-43
  (See also Alginates)
Alginate, sodium *(see* Sodium alginate)
Alginates, 2-1 to 2-43
  in air fresheners, 2-25, 2-26
  in beer, 2-24, 2-25
  in beverages, 2-24, 2-25
  BOD of, 2-23
  in boiler compounds, 2-25, 2-26
  in ceramic glazes, 2-25, 2-26
  in cleaners, 2-25, 2-26

Alginates *(Cont.):*
  COD of, 2-23
  compatibilities of, 2-15 to 2-17, 2-21, 2-22
  complexes with, 2-13
  concentration effect, 2-9, 2-13
  derivatives from, 2-2, 2-3
  dry-blending of, 2-20, 2-22
  in explosives, 2-25, 2-26
  films from, 2-23
    insolubilization of, 2-15, 2-23, 2-24
    plasticization of, 2-24
  food uses for, 2-24 to 2-26
  formulations with, 2-26 to 2-34
  gelation of, 2-10, 2-13 to 2-15
  industrial uses for, 2-25, 2-26
  laboratory techniques for, 2-37, 2-38
  in latex emulsions, 2-25, 2-26
  manufacture of, 2-3 to 2-5
  mixing of, 2-20, 2-22, 2-23
  in paper manufacture, 2-25, 2-26
  pH effect on, 2-9, 2-10
  physical properties of, 2-4
  in polishes, 2-25, 2-26
  powdered, 2-5, 2-6
  reactions of, 2-10, 2-13, 2-14
  rheology of, 2-6, 2-7
  solution properties of, 2-6 to 2-15
  and solvents, 2-9, 2-12
  and starches, 22-67
  storage of, 2-6
  in textile manufacture, 2-25, 2-26
  toxicology of, 2-17, 2-18, 2-20
  viscosities of, 2-7, 2-8, 2-13 to 2-15
  in welding rod coatings, 2-25, 2-26
Alginic acid, 2-3, 2-4, 2-18, 2-20
Algipon, 2-41
Alipal, 21-10
Alkalase, 2-22; 24-15
Alkalies, 10-10
Alkali metals, 4-4
Alkaloids, 8-7, 8-12
Alkanolamines, 7-5
Alkox, 19-31
Alkyl celluloses, 3-1 to 3-25
  (See also Methylcellulose)
Alligator (gum karaya), 10-13
Allyl alcohol, 19-3
Aloin, 18-12
Alum, 23-4
Aluminum, 4-3
Aluminum chloride, 10-4, 10-6
Aluminum foil adhesive, 8-13
Aluminum oxide, 16-4
Aluminum plating, 8-16
Aluminum sulfate, 3-10; 10-4, 10-6; 13-7; 16-2; 24-14
Amaizo, 22-71
Amerchol, 21-8
Amides, 20-4
Amidex, 22-71
Amines, 18-10; 20-4; 21-11
2-Aminoethanol, 21-11
Aminoethylethanolamine, 21-11
2-Amino-2-methyl-1-propanol, 21-11
Aminopyrine, 8-8
Amioca, 22-71

Ammonia, 20-8; 22-54
  manufacture of, 16-16
Ammonium alginate, 2-4 to 2-6, 2-12, 2-13, 2-
  17 to 2-20, 2-23
Ammonium/calcium alginate, 2-34
Ammonium chloride, 2-22; 20-6; 24-14
Ammonium compounds, 21-10; 24-10
Ammonium fluoride, 19-4, 19-5
Ammonium hydroxide, 8-7, 8-8; 19-17; 20-6;
  24-8
Ammonium nitrate, 13-7; 22-18; 24-22
Ammonium peroxydisulfate, 21-2
Ammonium polyacrylate, 17-18
Ammonium salts, 4-3; 22-4
Ammonium sulfate, 2-22; 7-4; 9-2; 13-7; 22-18
Ammonium thiocyanate, 19-4, 19-5; 20-18; 22-
  17
Amnucol, 2-41
Amoloid, 2-39
Amyl acetate, 20-15
Amyl alcohol, 21-11
Amylase, 24-10
Amylomaize starch, 22-7
Amylopectin, 22-5, 22-6
  breeding for, 22-23
  structure of, 22-6
Amylose:
  breeding for, 22-23
  change in, 22-25
  films of, 22-5, 22-25
  solubilities of, 22-26
  structure of, 22-3 to 22-5
Analgine, 2-39
Anesthetics, 8-12
Anhydrides, 18-10
Aniline, 21-11
Animal feed, 24-13, 24-27F
  (See also Foods, pet)
Animal glue, 18-19
Anise oil, 18-12
Anisole, 19-9
Annealing, 18-21
Anogeissus leiocarpus gum, 9-2
Antacids (see Pharmaceuticals, antacids)
Antibiotic suspensions (see under Pharma-
  ceuticals)
Antifoam, 2-25; 13-16
Antifogging coatings, 17-16
Antimigrant, 2-41
Antimony oxide, 20-19
Antioxidants, 18-14
Antiperspirants (see Cosmetics/toiletries, anti-
  perspirants)
Antipyrine, 18-12
Antiseptics, 8-12
Antisoil agents, 4-17; 21-16
Antistatic finishes (see under Textiles)
Apollo, 22-71
Apomorphine, 8-8
Arabic (see Gum arabic)
Arabic acid, 8-2, 8-3
Arabinose production, 9-6
Archaeological finds, 18-25
Arochlor, 21-9
Arrowroot starch, 22-18, 22-21, 22-61
Asbestos-cement products, 16-12, 16-17; 17-14
Ascophyllum nodosum, 2-2

Asphalt products, 3-18; 12-15
Asphalt release coating, 3-18
Aspirin, 8-7
Astro Gum, 22-71
Atlantic, 15-20
Aubygel, 5-28
Aubygum, 5-28
Automotive polishes (see Polishes)
Azobisbutyronitrile, 21-3

Bakery products (see Foods, bakery products)
Baking powder, 22-61
Banana gel base, 2-31F
Bandages, 22-52
  adhesives for, 17-17
Barbecue sauces (see Foods, barbecue sauces)
Barbitol, 18-12
Barbiturates, 8-11
Barium, 4-3
Barium chloride, 24-14
Barium hydroxide, 22-17; 23-2
Barium production, 16-16
Barium sulfate (see Pharmaceuticals, barium sul-
  fate suspensions)
Bases, 24-8, 24-14
Bassora Gum, 11-30
Batteries, 7-9; 8-16; 17-16; 18-22; 19-16, 19-21,
  19-26
Beauty aids (see Cosmetics/toiletries)
Beer (see Beverages, beer)
Beeswax, 18-11; 21-9
Benzaldehyde, 18-12
Benzene, 3-12; 13-5, 13-6; 19-9; 21-11
Benzoates, 22-68
Benzocaine, 18-12
Benzoic acid, 10-5; 11-13; 18-12
Benzoyl peroxide, 21-3
Benzyl alcohol, 2-21; 18-12; 24-15
Bermocoll, 3-25
Betz, 16-25
Beverages:
  and agar, 7-9
  and alginates, 2-24, 2-25
  apple, 11-17, 11-21
  beer, 2-24, 2-25; 5-18, 5-22; 8-6, 8-8, 8-11, 8-
    16, 8-19F; 19-16; 21-2, 21-14; 22-22
  carbonated, 8-6
  and carboxymethylcellulose, 4-20
  and carrageenans, 5-18, 5-20 to 5-22
  and cellulose gum, 4-20
  ciders, 7-9
  clarification of, 17-13; 21-2, 21-18
  clouding agents for, 22-64
  dry mix, 11-18
  flocculants for, 7-9
  foam stabilization of, 13-10; 19-12
  fruit, 2-24, 2-25; 4-20; 5-21, 5-22; 8-6, 8-8, 8-
    11, 8-16, 8-18F, 8-19F; 11-12, 11-17, 11-
    21, 11-24F; 15-8, 15-9, 15-13, 15-14, 15-
    18F; 17-11; 21-14; 24-13, 24-20, 24-21
  and guar gum, 6-12
  and gum arabic, 8-6, 8-8, 8-11, 8-16, 8-18F, 8-
    19F
  and gum tragacanth, 8-18F; 11-12, 11-17 to 11-
    19, 11-21, 11-24F
  instant, 4-8, 4-20; 22-26

Beverages *(Cont.)*:
  juices, 6-12; 7-9; 11-17, 11-21
  and locust bean gum, 14-14
  malts, 19-12
  milk-based, 5-20, 15-13, 15-14, *15-18F, 15-19F*
    (*See also under* Foods)
  nectar, 11-17, 11-18, 11-21
  papaya juice, 11-17, 11-21
  and pectins, 15-8, 15-9, 15-13, 15-14, *15-18F,
    15-19F*
  and poly(acrylic acid), 17-11, 17-13
  and poly(ethylene oxide), 19-12, 19-16
  and polyvinylpyrrolidone, 21-2, 21-14, 21-18
  protein, 4-20
  soft drinks, 11-18, 11-21; 14-14
  vegetable juices, 6-12; 21-14
  wines, 5-18, 5-22; 7-9; 21-3, 21-14
  and xanthan gum, 24-13, 24-20, 24-21
Bicarbonate, 15-7
Bichromates, 20-18
Binasol, 22-71
Binders:
  adhesive, 23-7
  battery electrode, 19-16, 19-21, 19-26
  briquettes, charcoal, 22-22, 22-26, 22-30;
    23-7
  building materials, 20-22
  burn-out, 13-12
    (*See also* Ceramics, green strength for)
  catalyst pellet, 20-25
  ceiling tile, 22-70
  ceramic glaze, 12-11, 12-15, 12-16
  for ceramics (*see* Ceramics, binders for)
  clay, 22-34
  color, 22-59
  cork product, 20-25
  cough drop, 11-3
  and dextrins, 22-36
  explosives, 19-21
  fluorescent lamp, 19-16, 19-21, 19-26
  foundry core, 20-2, 22-26
  fuel cell electrode, 19-26
  ink, 13-11
  meat, 22-64
  metal-foundry cores, 12-11, 12-16
  paint, *20-29F*
  paper coating, 22-53, 22-60
  pencil, 12-11, 12-15
  pharmaceutical (*see under* Pharmaceuticals)
  pigment, 20-9, 20-22, 20-25
  polish, 11-21
  poly(ethylene oxide) as, 19-16
  polyvinyl alcohol as, 20-22
  for refractory shapes, 12-15
  ribbon, 20-22
  roof coating, 20-22
  sand mold, 22-70
  sawdust briquettes, 23-8
  seed coating, 12-16
  soil stabilization, 19-16; 20-25
  tablets, 11-21, 11-22; 19-21; 21-12; 22-70
  tape joint cement, 22-70
  textile (*see* Textiles, binders for)
  tobacco product, 19-21

Binders *(Cont.)*:
  waste product, 20-25
  welding rod flux, 2-25, 2-26; 4-21; 12-11, 12-
    15
Biochemical oxygen demand (*see* BOD)
Biological properties:
  of agar, 7-4
  of carboxymethylcellulose, 4-7, 4-8, 4-27
  of carrageenans, 5-11
  of guar gum, 6-9, 6-10
  of gum arabic, 8-4
  of gum ghatti, 9-5
  of gum karaya, 10-8
  of gum tragacanth, 11-14
  of hydroxyethylcellulose, 12-6, 12-7, 12-16
  of hydroxypropylcellulose, 13-2
  of locust bean gum, 14-6, 14-8
  of pectins, 15-6, 15-7
  of polyacrylamide, 16-7, 16-9
  of polyethylene glycols, 18-14
  of poly(ethylene oxide), 19-7, 19-8
  of tamarind gum, 23-3, 23-4
Biscon, 4-23
Bismuth chloride, 8-8
Bismuth suspensions, 24-13
Blanose, 4-23
Blood plasma extenders, 21-13; 22-40
BOD (biochemical oxygen demand):
  of alginates, 2-20, 2-23
  of carboxymethylcellulose, 4-19
  of hydroxyethylcellulose, 12-7, 12-11
  of hydroxypropylmethylcellulose, 3-6
  of methylcellulose, 3-6
  of polyethylene glycols, 18-14
  of poly(ethylene oxide), 19-8
  of polyvinyl alcohol, 20-16, 20-17
  of starches, 22-67
Boiler compounds, 2-25, 2-26; 18-14
Bomud, 4-23
Bonaril, 16-25
Bondcor, 22-71
Bone glue, 21-10
Boniadril, 4-23
Boniasol, 4-23
Bonracel, 4-23
Borated dextrin, 12-5
Borates, 6-9; 24-9, 24-17
Borax, 8-3, 8-7; 14-5; 16-12; 20-18; 22-36, 22-69,
    22-70; 23-6
Boric acid, 20-18
Boxboard, 6-16; 14-12
Brass polishes, 24-21
Bread products (*see under* Foods)
Breakfasts (*see under* Foods)
Brines, 5-18; 16-16, 16-17; 17-16
Briquettes, charcoal, binders for, 22-22, 22-26,
    22-30; 23-7
British gum, 22-31, 22-32, 22-69
Bromides, 22-4, 22-17
Bromine, 19-4, 19-5
Bronchography, 3-15, 3-23
BTI, 16-25
BTC, 21-10
Bubble baths, 12-14
Buckeye CMC, 4-23
Buffing compounds (*see under* Metal products)

Building products:
  asbestos board, 17-14
  asbestos cement, 19-16, 19-25, 19-26
  bricks, 23-7
  caulking, 3-18; 12-14, 12-16
  cements, 3-14, 3-15, 3-18, 3-21; 4-24; 8-13; 12-
    11, 12-14, 12-16; 16-11, 16-12, 16-17, *16-
    24F;* 17-12, 17-14; 20-2, 20-22, 20-25, *20-
    30F*
  concrete, 12-11, 12-16; 19-16, 19-22, 19-26
  crack fillers, 12-14, 12-16
  drywall finishes, 3-14, 3-18
  insulation, sound, 3-14
  matrix board, 13-10, 13-11, 13-12
  mortars (*see* Mortars)
  patching compounds, 3-14
  paving, 19-16, 19-17, 19-26
  plaster, 3-14, 3-21; 12-14, 12-16; 20-22, 20-25;
    21-14
  plaster of Paris, 21-14
  plywood (*see* Adhesives, for plywood)
  refractories, 3-14, 3-18, 3-21
  roof coatings, 20-22, 20-25
  slurry trenching, 19-16, 19-24
  stuccos, 3-14, 3-18
  tile, 18-16; 20-22, 20-25
  (*See also* Wood products)
Bulking agents, 10-9
Bulmer, 15-20
Burn therapy, 3-23
Butadiene (*see* Petroleum processing)
Butadiene-styrene rubber, 19-7
1,2-Butanediol, 21-11
1,4-Butanediol, 21-4, 21-11
Butanol, 2-12; 13-5, 13-6; 20-15; 21-11; 22-5, 22-
  25
sec-Butanol, 21-11
2-Butanone, 21-11
Butter sauces (*see under* Foods)
Butyl acetate, 13-5, 13-6; 19-9; 21-11
Butylamine, 21-11
Butyl Cellosolve, 2-12; 13-5, 13-6; 19-9
Butyl phthalylbutyl glycolate, 19-7
Butyrolactone, 21-4, 21-11

CA 33, 2-39
Cadmium chloride, 19-4, 19-5
Caffeine, 8-8; 18-12
Caffeotannic, 8-8
Cakes (*see under* Foods)
Calamine (*see* Pharmaceuticals, calamine lotions)
Calcium, 2-9; 4-3, 4-6; 6-9; 10-4; 15-5
Calcium alginate, 2-17, 2-18
Calcium carbonate, 19-22; 22-68
Calcium carboxymethylcellulose, 4-22
Calcium chloride, 6-8; 8-7; 10-6; 13-7; 19-11; 20-
  6; 22-18, 22-68, 22-69; 24-14
Calcium hydroxide, 22-17
Calcium nitrate, 6-8; 22-4
Calcium salts, 18-12; 23-6
Calcium-sodium alginate, 2-19
Calcium sulfate, 20-19
Calendar sizes, 4-18
Calginate, 2-39
Calgon Coagulant Aid, 16-25

Callatex, 2-39
Camphor, 18-12
Cancer research, 17-17
Candy (*see* Foods, candies)
Capsul, 22-71
Capsules, pharmaceutical, 7-8; 8-7
Caramels, 8-6
Carastay, 5-28
Carbitol solvent, 19-9
Carbocel, 4-23
Carbohydrates, 8-7; 10-9
Carbohydrate sweeteners, 15-9
Carbon black, 20-19
Carbon dioxide, 20-8
Carbon tetrachloride, 19-9; 20-15; 21-11
Carbopol, 2-21; 24-15
Carbose, 4-23
Carboset, 2-21; 24-15
Carbowax, 13-4, 13-10, 13-16; 18-4, 18-5, 18-30
Carboxal, 4-23
Carboxyethylcellulose, 7-5
Carboxymethyl alginate, 2-3
Carboxymethylcellulose (CMC), 4-1 to 4-28
  in adhesives, 4-24
  and alkali metal ions, 4-3
  analysis of, 4-24
  and antisoil agents, 4-17
  applications of, 4-17
  aqueous solutions of, 4-5
  bag handling for, 4-14
  in bakery products, 4-20
  in beverages, 4-20
  biological properties of, 4-7, 4-8, 4-27
  BOD of, 4-19
  bulk handling of, 4-14
  and calendar stock, 4-18
  in cement, 4-24
  in ceramic glazes, 4-21
  in ceramics, 4-24
  chemical nature of, 4-2 to 4-5
  in coatings, 4-2, 4-21
  compatibilities of, 4-4
  in convenience foods, 4-20
  in cosmetics, 4-2, 4-7, 4-13, 4-17, 4-21, 4-24,
    4-25
  in dehydrated foods, 4-20
  in desserts, 4-20
  and detergency, 4-16
  in detergents, 4-2, 4-7, 4-13, 4-14, 4-17, 4-18,
    4-26, 4-27
  in drilling muds, 4-18
  emulsions with, 4-25
  and FAO/WHO, 4-8, 4-21
  and the FDA, 4-1, 4-8
  film from, 4-6
    insolubilization of, 4-6
  in fire fighting, 4-21
  in foods, 4-2, 4-7, 4-13, 4-17, 4-20, 4-25; 11-
    16
    grade definition for, 4-8
  in fruit juices, 4-20
  GRAS, 4-8
  and gum arabic, 8-8
  and gum tragacanth, 11-12
  handling of, 4-11 to 4-17
  and heavy metals, 4-6
  and hydroxyethylcellulose, 12-5

Carboxymethylcellulose (CMC) *(Cont.)*:
  and hydroxypropylcellulose, 4-12; 13-4, 13-5
  in ice cream, 4-20
  industrial uses for, 4-7
  and latex paints, 4-2, 4-21
  in low-calorie foods, 4-20
  manufacture of, 4-6, 4-7
  and metal ions, 4-3
  microbiological stability of, 4-7
  in mining, 4-2, 4-7
  moisture content of, 4-5
  molecular weights of, 4-5
  and multivalent salts, 4-6
  and organic solvents, 4-6
  packaging of, 4-7, 4-12 to 4-14
  in paints, 4-26
  in paperboard, 4-18
  in paper manufacture, 4-2, 4-13, 4-14, 4-18
  in paper products, 4-2, 4-19, 4-26
  in paper sizing, 4-18
  in pet foods, 4-20
  and petroleum recovery, 4-2, 4-7
  in pharmaceuticals, 4-7, 4-13, 4-17, 4-21, 4-26
  physical properties of, 4-4 to 4-6
  in polishes, 4-22
  in pollution control, 4-2
  and polyvinylpyrrolidone, 21-8, 21-10, 21-16
  powdered, 4-16
  in processed foods, 4-20
  and proteins, 4-3
  purities of, 4-7
  rheology of, 4-9
  in rubber goods, 4-26
  shipping of, 4-15 to 4-17
  in soaps, 4-26, 4-27
  and sodium chloride, 4-11
  solubilities of, 4-6
  in soup powders, 4-20
  source of, 1-2
  vs. starch BOD, 4-19
  and starches, 22-67
  storage of, 4-11 to 4-15
  and stream pollution, 4-19
  in synthetic detergents, 4-2
  in textiles, 4-2, 4-4, 4-7, 4-13, 4-14, 4-19, 4-20,
    4-27
  toxicological properties of, 4-8, 4-27
  tradenames for, 4-23, 4-24
  with vegetable juices, 4-20
  viscosity of, 4-4, 4-10 to 4-13
  in wallpaper adhesives, 4-21
  and welding rod coatings, 4-21
  world production of, 4-23
  worldwide uses for, 4-2
Carboxymethyl dextran, 19-4
Carcinogenicity, 5-11
Cardinal Gums, 22-71
Cards (*see* Paper products)
Carnauba wax, 18-11
Carpets (*see* Textiles, carpets)
Carrageenans, 5-1 to 5-30
  acid degradation of, 5-9
  and additives, 5-17
  and agricultural products, 5-18, 5-24
  in air fresheners, 5-19, 5-21, 5-22, 5-24, 5-25
  and alcohols, 5-18
  and antacids, 5-22, 5-23, 5-26

Carrageenans *(Cont.)*:
  applications for, 5-17 to 5-26
  and beer processing, 5-22
  and beverages, 5-17, 5-21 to 5-24
  biological properties of, 5-11
  and brines, 5-18
  and casein, 5-8
  and ceramics, 5-23
  chemical nature of, 5-2 to 5-9
  compatabilities of, 5-18
  composition of, 5-6
  and cosmetics, 5-21 to 5-23, 5-25
  and dairy products, 5-8, 5-16, 5-17, 5-19 to 5-
    21, 5-23, 5-24
  depolymerization of, 5-9
  extenders for, 5-17
  and foods, 5-4, 5-8, 5-18, 5-21 to 5-26
  in formulations, 5-24 to 5-26
  gelation of, 5-12 to 5-18
  and germicides, 5-18
  and guar gum, 5-17
  and gum tragacanth, 11-20
  handling of, 5-17
  hydrolysis of, 5-9
  in infant formulas, 5-8, 5-23
  and locust bean gum, 5-17, 5-22
  manufacture of, 5-10, 5-11
  and metal products, 5-23
  molecular weight of, 5-4, 5-6
  and paints, 5-21 to 5-23
  and pectins, 15-8
  in pet foods, 5-21 to 5-23
  and pharmaceuticals, 5-18, 5-22, 5-23
  physical properties of, 5-3, 5-9, 5-10
  and proteins, 5-7, 5-18, 5-23
  reactivities of, 5-6 to 5-9
  and relishes, 5-21 to 5-23
  rheology of, 5-11 to 5-17
  and sauces, 5-21 to 5-23
  shipping of, 5-17
  solubilities of, 5-9, 5-10
  source of, 5-2
  stability of, 5-18
  structure of, 5-4, 5-5, 5-14
  toxicological properties of, 5-11
  viscosity of, 5-11, 5-12
  water gels of, 5-12 to 5-16, 5-21, 5-22
  and wine processing, 5-22
Carvol, 8-8
Casein, 4-3; 5-8; 12-5; 15-8; 18-11; 20-18
Casting cements, 8-13
Casting of metal products, 18-14, 18-16, 18-22,
  18-24
Castor oil, 3-12; 18-11; 21-9
Catalyst pellets, 20-22, 20-25
Catalyst systems, 17-14, 17-16, 17-17
Catamenial devices, 19-16, 19-19, 19-27
  (*See also* Sanitary napkins)
Catechol, 20-18
Catechol tannins, 19-4, 19-5, 19-14
Cathodoluminescent coatings, 17-12
Cato, 22-71
Cato-Kote, 22-71
Cato-Size, 22-71
Caulking, 3-18; 12-14, 12-16
Caustics, 19-15; 22-70
Caviar, imitation, 2-33

Cecalgine, 2-41
Ceepryn, 21-10
Cekol, 4-23
Celacol, 3-4, 3-25
Cellofas, 4-23
Cellofix, 4-23
Cellogen, 4-23
Cellophane film, 18-16, 18-19, 18-24; 20-17
Cellosize, 12-3, 12-71
Cellosolve, 13-5, 13-6
Cellosolve acetate, 19-9
Cellosolve solvent, 19-9
Cellucol, 4-23
Cellujel, 4-23
Cellulase, 2-22; 24-10, 24-15
Cellulose acetate, 21-8, 21-9
Cellulose acetate butyrate, 19-7
Cellulose acetate phthalate, 13-5
Cellulose acetate propionate, 21-8, 21-9
Cellulose gums, 4-20
    definition of, 4-1
    in foods, 4-20
    types of, 4-3
Cellulosics, 1-2; 3-2; 17-15; 18-21; 24-9
Cel-Pro, 4-23
Cements (*see* Building products, cements)
Ceramics:
    adhesives for, 3-14, 3-18; 20-20, 20-24
    binders for, 13-10, 13-11; 18-14; 19-16, 19-20,
        19-21, 19-25; 20-2, 20-21 to 20-23, 20-25
    carboxymethylcellulose in, 4-24
    core washes for, 5-23
    enamels for, 18-21, 18-23
    extrusion of, 18-16
    formability of, 18-21, 18-23
    glazes for, 2-25, 2-26; 4-21; 5-23; 8-16; 12-11,
        12-15, 12-16; 13-10, 13-13; 17-11, 17-14;
        18-16, 18-21, 18-23; 24-13, 24-22
    green strength for, 11-19 to 11-21; 12-16; 13-
        10, 13-11, 13-13; 17-14; 18-21, 18-23
    and gum tragacanth, 11-3
    molded, 18-14
    mortars for, 20-22
    patching compounds for, 17-11
    plastic mixes for, 3-15, 3-18
    tiles of, 12-16; 18-16
Cereals, breakfast, 6-12; 7-7; 11-18
Cereal starches, 22-15, 22-62
Ceri-Gel, 22-71
Cerioca, 22-71
Cesium polyacrylate, 17-2
Cetavlon, 11-4
Cetyl alcohol, 18-12
Charcoal briquettes, 22-22, 22-26, 22-30; 23-7
Charge, 22-71
Cheeses (*see* Foods, cheese products)
Chemical oxygen demand (COD), 2-23; 20-17
Chemicals, fine, 16-16
Chewing gum, 8-9
China clay, 22-68
Chinaware, 20-23, 20-25
Chloral hydrate, 18-12
Chlorides, 12-5; 22-4, 22-17
Chlorinated hydrocarbons, 20-15; 21-11
Chlorinated mineral oil, 21-9
Chlorinated phenols, 10-5
Chlorinated phenol salts, 12-5

Chlorine manufacture, 17-16
Chloroacetic acid, 16-3
Chlorobenzene, 21-11
Chlorobutanol, 11-13; 18-12
p-Chloro-*m*-cresol, 23-4
Chloroform, 13-5, 13-6; 21-9 to 21-11
Chlorothymol, 18-12
Chocolate products (*see* Foods, chocolate products)
Cholesterol, 7-9
Chrome alum, 24-14
Chromic nitrate, 20-18
Chromium, 4-3; 19-3
Ciders, 7-9
Cigarette packages, 18-19
Cigars (*see* Tobacco products)
Cinnamon oil, 18-12
Citrates, 15-7
Citric acid, 4-11, 4-13; 18-12; 24-14
Citroflash, 15-20
Citrus products (*see under* Beverages; Foods, citrus products)
Clarification of beverages, 17-13; 21-2, 21-18
Claro, 22-71
Clarocel, 16-25
Clays, 13-7; 16-16; 19-22, 19-26; 20-19 to 20-21;
    22-67, 22-68, 22-70
Clay slimes, 16-13
Cleaning materials, 2-25, 2-26; 3-14, 3-18; 12-15;
    13-10 to 13-12; 17-11, 17-13; 18-20, 18-23;
    19-15, 19-16, 19-26, *19-27F*; 24-8, 24-22
Cleansing creams, 18-16
Clearjel, 22-71
Clinco, 22-71
Clineo, 22-71
Clinivert, 22-71
Clinsize, 22-71
Clouding agents for beverages, 22-64
Clove oil, 18-12
CMC (*see* Carboxymethylcellulose)
Coal, 19-22, 19-26
Coal fines, 16-12, 16-17
Coating color, 22-59
Coatings:
    and agar, 7-9
    antifogging, 17-16
    and carboxymethylcellulose, 4-2, 4-21
    cathodoluminescent, 17-12
    for confections, 13-10; 22-25, 22-64
    cured, 16-16
    electrically conductive, 17-16
    electrophotographic, 17-17
    for fiber glass, 19-16
    for fluorescent lamps, 19-21, 19-26
    for food (*see* Foods, coatings for)
    for glass, 17-17; 18-20; 19-16
    and hydroxypropylcellulose, 13-10 to 13-13,
        13-15F
    latex, 4-21; 24-11
    for metals, 7-9
    for optical products, 17-16
    packaging, 21-16
    for paper products (*see* Paper products, coatings for)
    permanent, 17-9, 17-15
    photographic, 17-17, 17-18; 20-2
    and polyethylene glycols, 18-24

Coatings *(Cont.)*:
  and poly(ethylene oxide), 19-16, 19-21, 19-26, 19-27
  and polyvinyl alcohol, 20-2, 20-24, *20-29F, 20-30F*
  and polyvinylpyrrolidone, 21-11, 21-15, 21-16, 21-18
  release, 3-18; 12-16; 17-9, 17-15, 17-16; 20-24, *20-30F*
  removable, 12-15; 17-9, 17-14
  shellac, 13-5; 21-8, 21-9, 21-12
  soil-release, 17-16
  and starches, 22-25, 22-31, 22-34, 22-38, 22-43, 22-53, 22-58 to 22-60, 22-64
  strippable, *20-30F*
  tablets, 13-10, 13-12, *13-15F;* 19-16, 19-21, 19-27; 21-11
  textile *(see* Textiles, coatings for)
  texturized, 12-14, 12-16
  (See also under Paints; Paper coatings; *under* Pharmaceuticals)
Cobalt chloride, 24-14
Cobaltous salts, 18-12
Cobalt sulfate, 24-14
Cobra (gum karaya), 10-13
Cocoloid, 2-40
COD (chemical oxygen demand), 2-23; 20-17
Cohasal, 2-41
Col-Flo, 22-71
Collatex Arm Extra, 2-39
Collowell, 4-24
Colostomy rings, adhesives for, 10-9
Combinace, 2-39
Combretaceae, 9-4
Complex, 16-25
Compound 20M, 18-17, 18-18, 18-30
Concrete *(see* Building products, concrete)
Confections *(see* Foods, confections)
Consista Starch, 22-71
Constipation control, pharmaceuticals, 3-23
Construction products *(see* Building products)
Convenience foods, 4-20, 4-22; 5-18
Copagen, 4-24
Copper ammonium sulfate, 23-2
Copper pellets, 17-11
Copper plating, 8-16
Copper processing, 16-12, 16-14
Copper sulfate, 22-68; 23-6
Core washes, 13-10
  for ceramics, 5-23
Cork products, 18-21; 20-22
Corn dextrin, 21-10
Corn starches:
  acetates of, 22-38, 22-62, 22-63
  and acetyl levels, 22-39
  acid-catalyzed, 22-22
  and adhesives, 22-68 to 22-70
  amylopectin in, 22-23
  amylose in, 22-23
  bleached, 22-70
  cationic, 22-22
  cooked, 22-16
  crosslinked, 22-28, 22-62
  derivatives of, 22-38, 22-67
  dextrins in, 22-22, 22-36
  dispersions of, 22-62

Corn starches *(Cont.)*:
  in foods, 22-29, 22-38, 22-61
  gelation of, 22-16, 22-20, 22-23, 22-25
    chemical, 22-17
    and pH, 22-19
    temperature for, 22-39
  and glass fibers, 22-67
  graft polymers of, 22-53
  granule properties of, 22-7
  and humidity, 22-13
  hydration of, 22-20
  hydroxyethyl ether of, 22-38, 22-40
  markets for, 22-22
  modified, 22-30, 22-72
  and mold binders, 22-70
  oxidized, 22-22
  and paper products, 22-22, 22-55, 22-57
  pastes of, 22-19, 22-20
  pasting temperature of, 22-21
  phosphate esters of, 22-42, 22-55
  polymerization of, 22-50
  powdered, 22-62
  pregelatinized, 22-22, 22-26
  solution of, 22-18
  stabilized, 22-63
  and starches, 22-68
  supplies of, 22-21
  swelling of, 22-15, 22-18, 22-20
  in tablets, 22-70
  and textile finishing, 22-67
  viscosity of, 22-14, 22-19, 22-29; 23-5
Corn syrup, 22-36
Corrosion inhibitors, 7-9; 8-16; 18-22, 18-25
Cosmetics/toiletries:
  aerosols, 13-12
  antiperspirants, *12-21F;* 13-10, 13-12, *13-13F, 13-14F*
  binding of, 8-7
  bubble baths, 12-14
  cakes, makeup, 8-12, 8-20
  cleansing creams, 18-16
  colognes, 13-10, 13-12
  creams, 3-14, 3-18; 5-18, 5-22, 5-23; 7-9; 8-12, 8-16; 11-3, 11-19, 11-21; 12-11, 12-14; 13-10, 13-12; 18-16, 18-20, 18-23, *18-27F, 18-28F;* 19-16; 20-23, 20-25
  dentifrices, 21-15
  deodorants, 3-14, 3-18; *12-21F;* 18-16, 18-20, 18-23; 21-15; 24-22
  depilatory creams, 3-14
  detergents, 18-20; 19-16, 19-18, 19-27, *19-28F*
  face creams, 3-14, 3-18
  face powders, 8-16
  facial masks, 8-12, 8-16; 11-3, 11-19, 11-21; 20-23, 20-25
  gels, 10-11; 12-11, 12-14; 18-20
  grooming aids, 13-10, 13-12
  hair dressings, 3-14, 3-18; 11-3, 11-19, 11-21; 12-11, 12-14, 12-16; 13-10, 13-12, *13-13F, 13-14F;* 18-16, 18-20, 18-23
  hair finishing, 8-13; 17-16
  hair preparations, 4-21; 11-16; 17-17; 21-2, 21-8, 21-14, 21-15
  hair rinses, 18-20
  hand creams and lotions, 3-14, 3-18; 4-21; 5-19; *18-27F*
  jellies, 12-11, 12-14; 18-20

Cosmetics/toiletries *(Cont.):*
  lotions, 3-14, 3-18; 5-18, 5-19, 5-21 to 5-23;
    7-9; 8-12, 8-16; 11-3, 11-16, 11-19, 11-21;
    12-11, 12-14; 13-10, 13-12, *13-14F;* 18-20,
    18-23, *18-27F;* 19-16
  makeup, 4-21; 11-19; 12-14; 18-23; 21-15
  mascara, 18-20
  ointments, 11-19; 12-14
  perfumes, 13-7, 13-10, 13-12; 18-20, *18-28F;*
    22-43
  pomades, 18-20
  powders, 18-16, 18-20
  protective creams, 8-12, 8-16; 20-24
  rouge, 8-7, 18-12; 18-20, *18-28F*
  salves, 11-19
  shampoos, 3-14, 3-19; 4-21; 12-11, 12-14, *12-
    21F;* 13-12, *13-14F;* 15-13; 18-20, 18-23;
    21-2, 21-14, 21-15, 21-18
  shaving products, 17-11; 18-20, 18-23, *18-27F;*
    19-16, 19-19, 19-26; 20-23, 20-25; 21-15
  skin cleaners, 18-20
  skin conditioners, 18-23; 21-15
  soaps, 8-12; 12-14; 19-16, 19-27
  sticks, 18-16, 18-20, 18-23, *18-28F*
  suntan lotions, 18-20; 21-15
  toothpastes, 3-14, 3-19; 4-7, 4-13, 4-21; 5-18,
    5-21, 5-22, 5-23, *5-25F;* 11-3, 11-16, 11-19,
    11-21; 12-14; 17-11; 18-16, 18-20, 18-23;
    19-16, 19-19, 19-26, *19-31F*
  vanishing creams, 18-20
  wave sets, 10-12, *10-13F*
Cottonseed oil, 20-15
Cough drops *(see* Pharmaceuticals, cough drops)
Courlose, 4-24
Crayons, 11-3, 11-19, 11-20
Creams *(see under* Cosmetics/toiletries; *under*
    Foods; Pharmaceuticals, creams)
Cresols, 8-8
Criminology, 7-8
Crisp Film, 22-71
Crude oil *(see* Petroleum recovery)
Crystal Gum, 22-71
Culminal, 3-4, 3-25
Culture media, 16-11
Cuprammonium hydroxide, 20-18
Cupric salts, 18-12; 24-14
Cut/Las, 22-71
Cutting fluids, metal products, 19-16
Cyanamer, 16-25
Cyclohexane, 3-12, 3-14; 21-11
Cyclohexanol, 13-5, 13-6; 21-4, 21-11
Cyclohexanone, 13-5, 13-6; 21-11
Cyclohexylamine, 21-11
Cymel, 13-2, 13-16
Cystoscope jellies, 3-15

Dacron, 17-15
Dairy products *(see under* Foods)
Dariloid, 2-39, 2-40
Decane, 17-3
Defoaming agents *(see under* Beverages)
De-inking of wastes, 16-13
Delta Gums, 22-71
Demulcents, 8-11
Denture adhesives, 10-5, 10-9 to 10-11, *10-12F;*
    17-14; 19-16, 19-19, 19-27, *19-28F*

Denture molds, 7-6 to 7-9, *7-10F*
Deodorants *(see* Cosmetics/toiletries, deodo-
    rants)
Depilatory creams, 3-14
Desserts *(see under* Foods)
Detergents, 4-2, 4-7, 4-13, 4-14, 4-16 to 4-18, 4-
    26, 4-27; 7-9; 12-15; 13-7; 17-13; 18-16, 18-
    20, 18-23; 19-7, *19-28F;* 21-2, 21-16, 21-18
Dewatering, 18-16
Dextrins, 22-34 to 22-36, 22-72
  in adhesives, 22-36, 22-68 to 22-70
  and agar, 7-5
  in binders, 22-36
  borated, 22-69
  canary, 22-31, 22-36
  and confections, 22-64
  films from, 22-36
  in foods, 22-61
  and glass fibers, 22-67
  and gum arabic, 8-13
  humectants for, 22-36
  and hydroxyethylcellulose, 12-5
  and paper products, 22-59
  and polyvinyl alcohol, 20-6, 20-18
  and polyvinylpyrrolidone, 21-10
  preservatives for, 22-36
  pyro-, 22-34
  and resins, 22-36
  solubilizers for, 22-36
  storage of, 22-36
  tapioca, 22-63
  and textiles, 22-68
  types of, 22-34
  white, 22-31, 22-32, 22-36
  yellow, 22-31, 22-32, 22-36, 22-69
  (See also British gum)
Dextrose, 15-7
Diabetic syrups, 8-11
Diacetone alcohol, 21-4, 21-11
Dialdehydes, 12-6; 20-18
Dialysis solutions, 18-16, 18-21
Diammonium phosphate, 2-22; 24-14
Diapers, 19-16, 19-19, 19-27; 22-52
Diarrhea control, pharmaceuticals, 3-23; 21-12
Dibutyl phthalate, 18-11
Dibutyl tartrate, 21-8
Dichloroethyl ether, 18-11
Dichlorophene, 19-4, 19-5
Dichromates, 21-2
Dicyandiamide, 22-36, 22-68, 22-69
Dietetic foods *(see* Foods, dietetic)
Diethanolamine, 3-11; 21-11
Diethoxin, 18-12
Diethyl Carbitol, 21-9
Diethyl Cellosolve, 19-9
Diethylene glycol, 21-4, 21-8, 21-11
Diethylene glycol diethyl ether, 21-9
Diethylene glycol ethers, 18-11
Diethylene glycol stearate, 21-9
Diethylenetriamine, 12-4; 20-4
Digamon, 2-39
Diglycerol, 20-17
2,4-Dihydroxybenzoic acid, 20-18
Diisopropyl sebacate, 19-7
Dimethylaminoethyl methacrylate, 17-8, 17-9
Dimethyl Cellosolve, 19-9

Dimethylformamide, 3-12; 6-4; 12-4; 13-5, 13-6; 14-3; 17-3; 19-9; 20-4; 21-10
Dimethylolethylene urea, 20-18
Dimethylolurea, 13-2; 20-18, 20-22
Dimethyl phthalate, 18-11; 21-8
Dimethyl sulfoxide, 3-12; 5-10; 6-4; 12-4; 13-5, 13-6; 14-3; 20-4; 22-5, 22-18, 22-25, 22-26
Dioxane, 13-5, 13-6, 13-10, 13-16; 17-3; 19-9; 21-11
Dipropyl glycol, 19-7
Disodium hydrogen phosphate, 24-14
Disodium phosphate, 13-7
DMU, 13-16
Douglas Gums, 22-71
Dowicide, 2-21; 13-16; 24-15
Dowicil, 13-16
Dow Latex, 2-21
Drag reduction, 16-5, 16-7, 16-15 to 16-17; 19-3, 19-16, 19-23, 19-26
Drewfloc, 16-25
Dricoid, 2-39; 2-40
Drinks (see Beverages)
Drugs (see Pharmaceuticals)
Dry-cleaning solvents, 17-13
Dry-Flo, 22-71
Duponal, 21-10; 22-20; 24-15
Dura-Jel, 22-71
Dust stickers, 3-24
Dye pigments, 10-11; 21-2; 22-61; 24-8, 24-11, 24-21
Dytol, 21-8

Edifas, 4-24
EHEC (see Ethylhydroxyethylcellulose)
Elastomers, 3-15; 22-46
(See also Rubber products)
Electra Size, 22-71
Electra Starch, 22-71
Electrical products, 21-19
(See also Batteries)
Electrode separators, 17-16
Electrolytes, 5-10; 6-8; 7-4; 8-5, 8-7; 10-4, 10-5, 10-10; 20-6
Electrophoresis, 7-9; 17-12, 17-17
Electroplating, 12-11, 12-15; 17-13; 18-16, 18-21; 18-23
Electropolishing, 18-21, 18-23
Electrorefining, 16-12
Electrowinning, 12-15; 18-23
Elephant (gum karaya), 19-13
Elvanol, 20-31
Emcol, 13-16
Emulsol-Maize, 22-71
Encapsulation:
of clouding agents, 22-64
and hydroxypropylcellulose, 13-11, 13-12
of pesticides, 22-47
and starch derivatives, 22-43
(See also Microencapsulation)
Engine coolants, 17-11
Environmental Protection Agency (EPA), 12-7; 24-20
Enzymes, 2-16, 2-22; 12-7; 17-17; 22-42, 22-61; 24-10, 24-15
Eosinophils, 9-5

EPA (Environmental Protection Agency), 12-7; 24-20
Epichlorohydrin, 22-54
Epinephrine, 8-8
Eserine, 8-8
Essex Gum, 22-71
Esters, 18-9, 18-11; 20-14, 20-15; 21-11
Ethane, 13-5
Ethanol, 2-12; 4-6; 7-4; 9-2; 13-4 to 13-6; 17-3; 19-9; 20-5; 21-4, 21-9, 21-11; 22-40; 24-9
(See also Ethyl alcohol)
Ethanol acetamide, 20-4
Ethanolamines, 12-6; 20-4, 20-18
Ethanol formamide, 20-4
Ether-alcohols, 18-9; 21-11
Ethers, 15-3; 20-14, 20-15; 21-11
Ethoxylan, 21-8
Ethyl acetate, 13-6; 18-11; 19-9; 20-15; 21-11
Ethyl alcohol, 18-11; 19-3
(See also Ethanol)
Ethyl carbamate, 18-12
Ethylcellulose, 13-5; 21-9
Ethylene chlorohydrin, 12-4; 13-5
Ethylenediamine, 12-4; 19-9; 21-11
Ethylene dichloride, 19-9; 20-15; 21-9 to 21-11
Ethylene glycol, 2-12, 2-22; 3-11; 8-3; 16-3; 17-3; 18-11; 19-3; 20-4, 20-17; 21-4, 21-8, 21-11; 24-9, 24-10, 24-15
Ethyl ether, 18-11; 20-15; 21-11
Ethylex, 22-71
2-Ethylhexanediol-1,3, 21-8
2-Ethyl-1-hexanol, 21-11
Ethylhydroxyethylcellulose (EHEC):
additives for, 3-11
manufacture of, 3-5
physical properties of, 3-3, 3-13
as a processing aid, 3-15, 3-16
and salts, 3-11
solubility of, 3-3, 3-7
and sucrose, 3-11
tradenames for, 3-25
uses for, 3-3, 3-16
viscosities of, 3-7, 3-10
Ethyl lactate, 21-4, 21-11
Ethyl vinyl ether, 21-11
Euftotal, 16-25
Eugenol, 8-8
Exchange, 15-20
Explosives, 2-25, 2-26; 6-10, 6-13, 6-18; 7-9; 9-5; 16-16F, 16-18F, 16-19F; 19-21; 22-21; 23-7 to 23-9; 24-22
Express Starch, 22-71
Extropol, 4-24

Fabricated foods (see Foods, fabricated)
Fabrics (see Textiles)
FAO/WHO (Food and Agriculture Organization/World Health Organization), 2-20; 4-1, 4-8, 4-21; 15-6; 22-64; 24-20
Fat barriers, 13-11, 13-13
Fats, 6-12; 8-6; 22-20
Fatty acids, 22-5, 22-20
FDA (Federal Drug Administration), 3-14; 4-1, 4-8; 6-9, 6-13; 7-2, 7-4; 8-4; 9-5; 10-8; 11-2; 12-6, 12-7; 13-2, 13-8; 14-6; 17-18; 18-14;

FDA (Federal Drug Administration) *(Cont.)*:
19-7, 19-12; 20-16, 20-17; 22-38, 22-43, 22-60, 22-64, 22-69; 24-20
Fehling's solution, 23-2
Ferric chloride, 8-3, 8-7, 8-8; 11-4; 13-7
Ferric ions, 4-3
Ferric salts, 18-12
Ferrocyanides, 12-5
Ferrous ions, 4-3
Ferrous sulfate, 24-14
Fertilizers, 21-16; 24-21
Fiberglass *(see* Textiles, fiberglass)
Fibers *(see* Textiles)
Fibersize, 22-71
Films:
  from agar, 7-9
  from alginates, 2-23
  from amylose, 22-5, 22-26
  from carboxymethylcellulose, 4-6
  from dextrins, 22-36
  from gum arabic, 8-6, 8-7
  from gum karaya, 10-5
  from gum tragacanth, 11-15
  from hydroxyethylcellulose, 12-5, 12-6, 12-10
  from hydroxypropylcellulose, 13-6, 13-8, 13-12, 13-13
  from hydroxypropylmethylcellulose, 3-11 to 3-13
  from methylcellulose, 3-11 to 3-13
  photographic, 17-18
  from poly(acrylic acid), 17-8, 17-16, 17-18
  from poly(ethylene oxide), 19-9, 19-19, 19-20
  from polyvinyl alcohol, 20-2, 20-5, 20-8, 20-9, 20-13, 20-22, 20-24, 20-25
  from polyvinylpyrrolidone, 21-2, 21-8
  for printing plates, 8-6, 8-7
  from propylene glycol alginate, 2-24
  protective, 3-24
  from sodium alginate, 2-24
  from starch *(see* Starches, films from)
  water-soluble, 16-16
    for packaging, 12-15
  from zinc alginate, 2-24
Filtration aids, 16-11; 17-13
Fines removal, clarification of beverages, 17-13; 21-2, 21-18
Finnfax, 4-24
Fire fighting, 4-21; 12-16; 16-10; 19-16, 19-23, 19-26; 24-22
Fish products *(see* Foods, fish products)
Flanogen, 5-28
Flans *(see* Foods, flans)
Flavors *(see* Foods, flavorings)
Flexol, 21-9
Flocculation, 16-4, 16-7; 17-9, 17-13, 17-14
Flojel, 22-71
Flokote, 22-71
Flotation, 6-12; 14-13, 14-14; 16-14
Flour, 8-13; 12-15
Flowers, artificial, 8-13
Flue-dust recovery, 16-12
Fluftex, 22-71
Fluid-jet cutting, 19-23, 19-26
Fluorescent lamps, 12-15; 19-16, 19-21, 19-26
Fluorides, 22-4
Fluorocarbons, 21-11
Fluxes, 18-14, 18-16

Food and Agriculture Organization *(see* FAO/WHO)
Foods:
  additives to, 14-6, 14-8; 19-12
  and adhesives, 18-14
  agar in, 7-2, 7-6, 7-7, 7-9; 8-18
  alginates in, 2-17, 2-18, 2-21, 2-24
  animal, 24-13
  baby, 22-30, 22-63
  bakery products, 2-24, 2-25, 2-29, 2-30, 2-32; 3-14, 3-19, 3-22; 4-20; 5-19, 5-20, 5-23, *5-26F;* 6-12, *6-19F;* 7-7, 7-9, *7-10F;* 8-6, 8-10, 8-11, 8-16; 11-2, 11-18; 14-14, *14-15F;* 15-12, 15-14; 22-26, 22-61, 22-64; 24-20, 24-21, *24-23F, 24-24F*
  barbecue sauces, 2-27; 5-21, 5-22; 11-2, 11-17, 11-21, *11-24F*
  binders for, 13-10 to 13-12; 18-14
  bologna, 10-11; 22-64
  breakfast cereals, 6-12; 7-7; 11-18
  breakfast powders, 5-19, 5-20, 5-23
  buttermilk, *6-19F;* 14-14, *14-15F*
  candies, 7-6, 7-7, 7-18; 8-6, 8-9, *8-17F;* 11-3, 11-18, 11-21; 15-9; 22-25, 22-33, 22-61 to 22-63
  canned, 22-30, 22-62; 24-20
  caramels, 8-6
  carboxymethylcellulose in *(see* Carboxymethylcellulose, in foods)
  catsup, 11-18
  cellulose gum in, 4-20
  cheese products, 2-30; 5-17, 5-20, 5-23; 6-11, 6-12; 7-6, 7-7, 7-9; 10-10, 10-11; 11-18; 14-8, 14-14; 24-20, 24-21, *24-26F*
  chewing gum, 22-64
  chip dips, 11-19, *11-26F;* 24-21
  chocolate products, 2-30, 2-32; 5-17, 5-19, 5-20, 5-23, 5-24; 15-14, *15-19F, 15-20F;* 24-23F
  chop suey, 2-28
  chow mein, 22-62
  citrus oils, 11-2, 11-18
  citrus products, 8-6, 8-8, 8-11; 15-13; 18-14
  coatings for, 2-18; 3-22; 13-10 to 13-12; 18-14; 22-25
  cocktail sauces, 8-18
  cod liver oil, 8-6, 8-11, 8-12; 11-12
  coffee cream, imitation, 5-21
  cole slaw dressing, 2-32
  condiment carrier, 3-22
  confections, 4-20; 7-6, 7-7, 7-9, 7-18; 8-9, 8-16, *8-17F;* 11-3, 11-18, *11-25F;* 13-10, 13-12; 15-8 to 15-10, 15-12, 15-14, *15-16F, 15-17F;* 22-25, 22-31, 22-61 to 22-64
  convenience, 4-20, 4-22; 5-18
  cookies, 7-7
  corn: canned, 22-30
    cream style, 22-62
  corn starch in, 22-29
  cottonseed oil, 20-15
  cream, imitation, coffee, 5-21
  cream products, 5-17, 5-20, 5-23; 8-9; 10-11; 14-14, *14-15F*
  cream shells, 7-7
  custards, 5-19, 5-20, 5-23, *5-26F;* 22-63
  dairy products, 2-29; 3-22; 5-23; 6-12; 7-7, 7-9; 8-9, 8-10, 8-16; 10-10; 11-13, 11-17, 11-

Foods, dairy products *(Cont.)*:
    18, 11-21, *11-25F, 11-26F;* 14-14, *14-15F;* 15-
    13, 15-14; 24-21
defoamers for, 18-14
dehydrated, 4-20
dessert products, 2-24 to 2-26, 2-28 to 2-32,
    2-34; 3-22; 4-20; 5-17, 5-19, 5-21 to 5-24,
    5-25F, 5-26F; 6-19F; 7-6, 7-7, 7-9; 8-9;
    10-10, 10-11, *10-13F;* 11-2, 11-17 to
    11-21, *11-25F;* 14-8, 14-13, *14-14F, 14-*
    15F; 15-4, 15-8, 15-14, *15-19F, 15-20F;*
    22-30, 22-61 to 22-63; 23-7, 23-8; *24-22F,*
    24-23F
diabetic, 8-11
dietetic, 3-14, 3-19, 3-22, 3-23; 5-17; 8-16, *8-*
    17F
dressings, 2-24, 2-25, 2-32, 2-33; 3-22; 8-18; 10-
    10; 11-2, 11-9, 11-16, 11-18, 11-21, *11-21F*
    to *11-24F, 11-27F;* 14-8; 22-26, 22-61, 22-
    63; 24-20, 24-21, *24-24F, 24-25F*
dry mix, 2-24; 22-61
edible paper, 7-18
eggs, frozen, 4-20
emulsions, 8-11, 8-17, 8-18; 18-19, 18-23; 22-
    21
evaporated milk, 5-8, 5-17, 5-20, 5-23
extruded, 4-20
fabricated, 2-24, 2-25, 2-33; 4-20, 4-22; 13-10,
    13-12
fats, 6-12; 8-6; 22-20
fish oils, 11-3, 11-16, 11-19, 11-21
fish products, 3-22; 5-17, 5-21 to 5-23; 7-7;
    16-9
flans, 5-19, 5-20, 5-23, *5-26F;* 15-12, *15-17F,*
    15-18F
flavorings, 3-22; 8-6, 8-8, 8-10, 8-11, 8-16, 8-
    18; 9-5; 22-42, 22-61; 24-20
flour, 3-22
French dressing, 2-32; 10-10; *24-25F*
fried, 3-14
frozen, 2-24 to 2-27; 3-14, 3-22; 5-22, 5-23; 8-
    9; 11-2, 11-18, 11-21; *14-15F;* 22-61; *24-*
    27F
fruit juices (*see under* Beverages)
fruit products, 2-18, 2-24 to 2-27, 2-31, 2-33,
    2-34; 3-14, 3-19; 5-21, 5-22; 7-7; 8-6, 8-8,
    8-11, *8-18F, 8-19F;* 15-4, 15-7 to 15-9, *15-*
    10F, 15-12 to 15-14, *15-15F, 15-19F;* 16-
    9, 16-17; 18-14; 23-4; 24-13, 24-21
gelatin, 14-14
glazes, 3-22; 7-7, 7-9; 8-9, 8-10; 13-12
GRAS (*see* GRAS)
gravies, 2-24, 2-25; 3-22; 4-20; 6-12; 8-18; 11-
    18; 22-30, 22-62, 22-64; 24-8, 24-20, 24-
    21
and guar gum, 6-2, 6-3, 6-11, 6-13, 6-18
    levels of usage, 6-12
gum arabic in, 8-6, 8-8 to 8-11, 8-16, *8-17F,*
    8-18F
gum candy, 22-25
gum drops, 8-9; 11-3, 11-18, 11-21
gum karaya in, 10-10, 10-11
ham, 7-8
hydroxyethylcellulose in, 12-6, 12-7
hydroxypropylcellulose in, 13-7, 13-8, 13-12
hydroxypropylmethylcellulose in, 3-6, 3-14

Foods *(Cont.)*:
ice cream, 4-20; 5-17, 5-19, 5-20, 5-23; 6-11,
    6-18F; 8-9; 11-18, 11-19, 11-21; 14-13, *14-*
    14F; 17-11; 23-7, 23-8; 24-16, 24-21
ice milk, *14-15F;* 24-21
ices, 7-6, 7-7, 7-9; 8-9; 10-10, 10-11; 11-18, 11-
    20, 11-21
icings, 2-24, 2-25; 7-6, 7-7, 7-9, *7-10F;* 8-10
infant, 5-8, 5-17, 5-20, 5-23; 22-64
instant, 2-29, 2-30; 5-23; 15-14
Italian dressing, 2-32; *24-25F*
jams, 6-12; 14-14; *15-10F;* 23-7
jellies, 2-30; 5-17, 5-21, 5-22; 6-12, *6-19F;* 14-
    14, *14-15F;* 15-4, 15-10, 15-12, 15-14, *15-*
    15F, 15-17F, 15-18F; 23-7; 24-21, *24-23F,*
    24-24F
jujubes, 8-9; 11-3, 11-19, 11-21
liver paté, 24-21
locust bean gum in, 14-8, 14-13, 14-14; 24-21
low-calorie, 4-20
lubricants, equipment, 18-21
margarine, 8-12
marmalades, 15-2; 23-7
marshmallows, 7-7; *8-17F*
mayonnaise, 8-18; 11-17, *11-27F;* 23-7, 23-8
meat products, 3-22; 5-18, 5-23; 7-7 to 7-9; 10-
    10, 10-11; 20-16; 22-64; 24-20
mellorine, 24-20
meringues, 2-24, 2-25, 2-31; 3-22; 4-20; 7-7; 10-
    10, 10-11; 11-2, 11-17, 11-18, 11-21
milk products, 2-24, 2-25; 3-14, 3-19; 5-2, 5-3,
    5-8, 5-10, 5-16, 5-17, 5-19 to 5-21, 5-23,
    5-24, *5-25F, 5-26F;* 6-12; 8-9; 11-17, 11-18;
    15-4, 15-7 to 15-9, 15-13, 15-14, *15-18F,*
    15-19F, 15-24; 18-14, 18-23; 24-20, 24-21
milkshakes, 11-17, 11-18
minerals in, 18-14
molasses, 24-13, 24-21
mustard, 8-18
mustard sauce, 11-2, 11-17, 11-21, *11-28F*
Neufchâtel cheese, 7-7
nuts, *8-19F;* 13-10, 13-12
oils, 6-12; 8-6, 8-11; 18-11; 21-9
olive oils, 8-11; 18-11; 21-9
packaging for, 2-18; 12-6; 18-14, 18-21, 18-23;
    19-12; 20-16, 20-17; 22-60
pancakes, 24-21
pastilles, 8-9; 11-3, 11-19, 11-21
pastries (*see* bakery products, *above*)
pectins in, 15-2, 15-5, 15-10
    (*See also under* Pectins)
pet, 4-20; 5-18, 5-21 to 5-23; 6-11; 7-8; 14-8,
    14-14; 22-61, 22-64; 24-21
pickle juice, 2-32; *8-18F*
pickle relish, 11-17, 11-18, *11-28F*
pickling oils, *8-18F*
pies, 2-24 to 2-26, 2-29, 2-30, 2-34, 2-35; 3-14,
    3-19, 3-22; 4-20; 5-19, 5-20, 5-23, 5-26; 7-
    7; 11-2, 11-18, 11-21; 22-30, 22-61 to 22-
    63; 24-21
piping gels, 7-7
pizzas, 11-2, 11-7, 11-21; *24-27F*
pizza sauces, 2-28; 5-21, 5-22; *24-27F*
polyethylene glycols in, 18-14, 18-16, 18-19,
    18-21, 18-23
polyvinyl alcohol in, 20-16
potato products, 3-22

Foods *(Cont.)*:
  potato starch in, 22-29
  poultry products, 7-7; 24-20
  preserves, 15-5, 15-7, 15-12
  processed, 4-20
  processing of, 22-22
  puddings, 2-24, 2-25, 2-29; 5-17, 5-19 to 5-21,
    5-23; 11-18; 14-14; 15-8, 15-14, *15-19F, 15-
    20F;* 22-61, 22-62, 22-64; 24-21
  relishes, 5-17, 5-21, 5-22, 5-23; 24-21
  salad dressings, 2-24, 2-25, 2-32, 2-33; 3-14, 3-
    19; 6-11; 8-18; 22-26, 22-61, 22-63; 24-8,
    24-20, *24-24F, 24-25F*
  salad spreads, *11-27F*
  sandwich spreads, 11-17
  sauces, 2-24, 2-25, 2-27, 2-28; 3-22; 4-20; 5-17,
    5-18, 5-21 to 5-23; 6-11, 6-12; 8-18; 11-2,
    11-17, 11-18, 11-21, *11-24F, 11-28F;* 14-8;
    22-30, 22-62, 22-64; 24-20, 24-21, *24-26F,
    24-27F*
  sausages, 7-8; 10-11; 22-64
  sherbets, 5-19; *6-18F;* 7-6, 7-7, 7-9; 8-9; 10-10,
    10-11, *10-13F;* 11-18, 11-21; *14-15F*
  snack, 3-14, 3-19; 4-20; 15-9; 22-25, 22-61, 22-
    64
  soufflés, 2-31, 2-32; *24-22F, 24-23F*
  soups, 4-20; 5-23; 6-12; 22-30, 22-62, 22-64
  sour cream, *6-19F*
    imitation, 24-21
  soy products, 5-21; 22-21
  spaghetti, 2-28, 2-33; 8-18; 11-2; 22-62
  spaghetti sauces, 8-18; 11-17, 11-21, *11-28F;
    24-26F*
  specialty, 4-20
  starches in, 22-28 to 22-30, 22-38, 22-41, 22-
    60 to 22-64, 22-72
  stews, 22-62
  sweeteners, 15-9; 18-14, 18-21
  sweet jelly, 7-18
  syrups, 2-24; 4-20; 5-21; 6-12; 9-5, 9-6; 11-18,
    11-25F; 24-13, 24-20, 24-21
  tablets, 18-14, 18-21
  and tamarind gum, 23-4
  and tapioca starch, 22-29, 22-36
  tomato aspic, 24-21
  toppings for, 2-24, 2-25; 3-19, 3-22; 4-20; 5-20,
    5-21, 5-23; 6-12; 8-10; 10-11; 11-18; 13-10,
    13-12; 22-63; 24-13
  TV dinners, 3-22
  vegetable oils, 3-12; 18-10; 20-14
  vegetable products, 2-33; 4-20; 5-23; 6-12, 6-
    17; 8-11; 16-9, 16-17
  vinegar, 21-2, 21-14
  vitamins, 3-23; 8-6, 8-11, 8-12; 9-5; 11-13; 18-
    14, 18-21
  whey solids, 24-21
  xanthan gums in (*see* Xanthan gum, in foods)
  yogurt, 5-17, 5-20, 5-21, 5-23; *6-19F;* 7-7; 14-
    14, *14-15F*
  (*See also* Beverages)
Formaldehyde, 2-21; 10-5; 12-5; 13-10; 16-3; 22-
    5, 22-26; 24-15
Formaldehyde resins, 20-18
Formamide, 20-4, 20-18
Formic acid, 3-12; 12-4; 13-5, 13-6; 21-11; 22-26
Formulations, 2-26 to 2-37; 3-16 to 3-24; 5-24
    to 5-26; 6-18, 6-19; 7-9 to 7-13; 8-17 to 8-

Formulations *(Cont.)*:
    20; 9-6, 9-7; 10-10 to 10-13; 11-21 to 11-28;
    12-18 to 12-21; 13-13 to 13-16; 14-14, 14-15;
    15-10, 15-11, 15-15 to 15-20; 16-16, 16-18,
    16-19; 18-19 to 18-21, 18-25 to 18-29; 19-27
    to 19-31; 20-25 to 20-30; 23-8, 23-9; 24-22
    to 24-27
Formvar, 21-9
Foundry cores for metal products, 20-2; 22-26
FR-14, 16-25
Freezist Starch, 22-71
Freons, 21-11
Friction reduction (*see* Drag reduction)
Fructose, 22-20
Fruitfil, 22-71
Fuel-cell electrodes, 19-26
Fuels, 7-9
Fumaric acid, 4-13
Fungicides (*see* Agricultural products, fungi-
    cides)
Furcellaran, 15-8
Furfuryl alcohol, 3-12
Furnace-wall repair, 3-16
Furose, 5-28

Galactomannans, 24-16 to 24-19
Gallic acid, 8-8; 20-18
Gelation, 2-10, 2-13 to 2-15, 2-24, 2-25; 3-8 to
    3-10; 4-3; 5-12 to 5-17; 7-5 to 7-7; 8-3, 8-5,
    8-7; 9-2; 12-5; 13-4; 18-11; 19-4, 19-5; 22-45;
    24-5, 24-6, 24-12, 24-17, 24-19
Gelcarin, 5-28
Gelgard, 16-25
Gelloid, 5-28
Gelogen, 5-28
Gelvatol, 20-31
Generally Recognized As Safe (*see* GRAS)
Genetron, 21-11
Genflo, 2-21
Gen Floc, 16-25
Genu, 15-20
Genugel, 5-28
Genulacta, 5-28
Genuvisco, 5-28
Geon, 2-21; 24-16
Germicides, 5-18
Gigartinaceae, 5-2
Glacial acetic acid, 3-12
    (*See also* Acetic acid)
Glass:
    coatings for, 17-17; 18-20; 19-16
    safety, 20-2
    (*See also* Textiles, fiberglass)
Glassine paper, 20-21
Glazes, 2-23, 2-25, 2-26; 3-15, 3-18; 4-21; 8-16;
    12-11, 12-15, 12-16
    (*See also* Ceramics, glazes for)
Glove coating, 17-10
Glucose, 22-68, 22-70; 23-6
Glues (*see* Adhesives)
Glycerides, 13-4
Glycerin, 3-12; 12-6; 21-4, 21-8, 21-11
    (*See also* Glycerol)
Glycerol, 2-12, 2-22, 2-24; 7-5; 8-3; 19-19; 22-5,
    22-36, 22-68; 24-10
    (*See also* Glycerin)

Glycerol diacetate, 19-7
Glycerol monoacetate, 19-7
Glycerol monostearate, 13-7, 13-10
Glycerol triacetate, 19-7
Glyceryl diacetate, 3-11
Glyceryl monoricinolate, 21-8
Glycol ethers, 21-11
Glycols, 2-9, 2-22; 12-5, 12-6; 18-9; 20-17; 22-68
Glyoxal, 12-14; 22-59
Goats Thorn, 11-30
Gog's Gum, 11-30
Gohsenol, 20-31
Gold recovery, 16-14
Graphic arts, 22-43
Graphite, 12-11, 12-15
GRAS (Generally Recognized As Safe), 2-17, 2-
    18; 3-6; 4-8; 5-11; 6-9; 7-2, 7-4; 8-4; 9-1, 9-
    5; 10-2, 10-8; 11-2, 11-14; 14-6, 14-8; 22-60,
    22-64
Greases, 20-14
Green-liquor clarification, 16-13
Grinding media, 17-12, 17-14
Guaiacol, 8-8
Guanidmium chloride, 22-18
Guar gum, 6-1 to 6-19
    and alcohols, 6-4
    applications of, 6-10 to 6-12
    biodegradability of, 6-8
    biological properties of, 6-9, 6-10
    borax reaction of, 6-9
    and calcium chloride, 6-8
    and calcium nitrate, 6-8
    carboxymethylation of, 6-8
    and carrageenans, 5-17
    chemical properties of, 6-3, 6-8 6-9
    commercial applications of, 6-12 to 6-17
    compatibilities of, 6-8
    derivatives of, 6-3, 6-8
    and dimethylformamide, 6-4
    and dimethylsulfoxide, 6-4
    esterification of, 6-8
    etherification of, 6-1, 6-8
    in explosives, 6-10, 6-13, 6-18; 24-22
    food grade, 6-2, 6-3
    in foods, 6-2, 6-9, 6-11 to 6-13, 6-18
    formulations with, 6-18
    gels of, 6-9
    grades of, 6-2, 6-3
    GRAS, 6-9
    and gum karaya, 10-13
    and gum tragacanth, 11-12, 11-20
    handling of, 6-10
    hydroxyalkylation of, 6-8
    and hydroxyethylcellulose, 12-5
    and hydroxypropylcellulose, 13-4
    in ice cream, 6-18F
    industrial grade of, 6-3
    and locust bean gum, 14-11
    manufacture of, 6-2, 6-3
    and metals recovery, 6-18
    and minerals recovery, 6-16 to 6-18
    and mining, 6-12
    in paper manufacture, 6-11, 6-14 to 6-16, 6-
        18; 14-11
    in petroleum recovery, 6-10, 6-13, 6-17, 6-18
    physical properties of, 6-3 to 6-8
    reactivities of, 6-8

Guar gum (Cont.):
    rheology of, 6-4 to 6-8
    salt reactions of, 6-9
    shear response of, 6-5, 6-6
    and sodium alginates, 2-21
    and sodium chloride, 6-8
    solubility of, 6-3, 6-4
    solutions from, 6-10
    specifications for, 6-2, 6-3
    storage of, 6-10
    structure of, 6-3, 6-4
    vs. tamarind gum, 23-8
    with textiles, 6-10, 6-11, 6-18
    usage levels for, 6-12
    viscosity of, 6-4 to 6-7
    worldwide consumption of, 6-2
    and xanthan gum, 24-9, 24-15 to 24-17, 24-22
Guluronic acid, 2-3
Gum arabic, 8-1 to 8-24
    additives for, 8-5, 8-13
    in adhesives, 8-13, 8-16
    in agricultural products, 8-16
    and alkaloids, 8-12
    and aluminum, 8-16
    in batteries, 8-16
    for beer processing, 8-6, 8-8, 8-16, 8-19F
    in beverages (see Beverages, and gum arabic)
    biological properties of, 8-4
    in building products, 8-12, 8-13
    in candies and confections, 8-6, 8-9, 8-16, 8-
        17F
    and carboxymethylcellulose, 8-8
    chemistry of, 8-2, 8-3
    in chewing gum, 8-9
    and cod liver oil, 8-6, 8-11, 8-12
    compatibilities of, 8-7, 8-8, 8-13
    in cosmetics, 8-7, 8-12, 8-13, 8-16
    in dairy products, 8-9, 8-10, 8-16
    and dextrins, 8-13
    in dressings, 8-18
    extenders for, 8-6
    and flour, 8-13
    in foods (see Foods, gum arabic in)
    formulations with, 8-17 to 8-20
    and glazes, 8-16
    grades of, 8-2
    and gum ghatti, 8-6; 9-1, 9-2, 9-5
    and gum karaya, 8-11; 10-10
    and gum tragacanth, 8-6, 8-7, 8-11 to 8-13; 11-
        12, 11-16, 11-17
    handling of, 8-6
    hydrolyzation of, 8-3
    and hydroxyethylcellulose, 12-5
    in inks, 8-6, 8-7, 8-13 to 8-16
    and kaolin, 8-11
    laboratory tests for, 8-20, 8-21
    and lithography, 8-6, 8-14, 8-15, 8-17
    manufacture of, 8-3
    and metal products, 8-16
    and metals protection, 8-16, 8-17
    and oils, 8-11, 8-12
    and paints, 8-6, 8-13
    and paper products, 8-13, 8-15, 8-17F
    and pectins, 8-11
    and petroleum recovery, 8-16
    and pharmaceuticals, 8-6, 8-7, 8-9, 8-11, 8-12,
        8-16

Gum arabic *(Cont.):*
and photographic chemicals, 8-16
physical properties of, 8-3
and polyvinylpyrrolidone, 21-9
rheology of, 8-4, 8-5
and salts, 8-7, 8-12
in sauces, 8-18
and soaps, 8-6, 8-12
solubility of, 8-3, 8-4
and starch, 8-6, 8-13
and steroids, 8-11
and textiles, 8-14, 8-15, 8-17
toxicological properties of, 8-4
and vinyl emulsions, 8-6, 8-13, 8-17
viscosity of, 8-4 to 8-6
and vitamins, 8-6, 8-11, 8-12
and xanthan gum, 24-9
Gumase HP, 2-22
Gum drops, 8-9; 11-3, 11-18, 11-21
Gum ghatti, 9-1 to 9-8
and agricultural products, 9-5
and alcohol, 9-2
biological properties of, 9-5
chemical nature of, 9-2
and explosives, 9-5
and foods, 9-5, 9-6
formulations with, 9-6, 9-7
and gelation, 9-2
grades of, 9-4
and gum arabic, 8-6; 9-1, 9-2, 9-5
and gum karaya, 9-1, 9-2
handling of, 9-5
laboratory techniques for, 9-5 to 9-7
manufacture of, 9-4
and mining, 9-6
in paints, 9-5
and paper products, 9-5
and petroleum recovery, 9-5, 9-6
and pharmaceuticals, 9-5
physical properties of, 9-2 to 9-4
and pigments, 9-5
in polishes, 9-2, 9-5, *9-6F*
as polymerization aid, 9-5, 9-6
and proteins, 9-5
and salts, 9-2
and soil analysis, 9-6
toxicological properties of, 9-5
viscosity of, 9-6
Gum karaya, 10-1 to 10-14
in adhesives, 10-5, 10-9, 10-10, *10-11F, 10-12F*
and agar, 7-5
biological properties of, 10-8
chemistry of, 10-2
compatibilities of, 10-5, 10-9, 10-10
in cosmetics, 10-11, *10-12F, 10-13F*
in dairy products, 10-10
and dyes, 10-11
and electrolytes, 10-4, 10-5
films of, 10-5
in foods, 10-10, 10-11, *10-13F*
formulations with, 10-12, 10-13
grades of, 10-6 to 10-8
and guar gum, 10-13
and gum arabic, 8-11; 10-10
and gum ghatti, 9-1, 9-2
and gum tragacanth, 11-2
handling of, 10-9

Gum karaya *(Cont.):*
hydrolysis of, 10-5
and hydroxyethylcellulose, 12-5
manufacture of, 10-6
and paper products, 10-9, 10-10
and petroleum recovery, 10-11
and pharmaceuticals, 10-8 to 10-11, *10-12F*
physical properties of, 10-3 to 10-6
and polyvinylpyrrolidone, 21-9
preservatives for, 10-5
and psyllium seeds, 10-10
rheology of, 10-8
in sealing pads, 10-11, *10-12F*
solubilization of, 10-11
and textiles, 10-9, 10-11
toxicological properties of, 4-9; 10-8
tradenames for, 10-13
viscosity of, 10-3 to 10-7
and xanthan gum, 24-15
Gum kino, 7-6
Gum tragacanth, 11-1 to 11-31
additives for, 11-15
in adhesives, 8-13; 11-19
and agar, 7-7
in agricultural products, 11-3, 11-19
and alcohols, 8-12; 11-16
and beverages (*see* Beverages, and gum tragacanth)
biological properties of, 11-14
and carboxymethylcellulose, 11-12
and carrageenan, 11-20
and ceramics, 11-3, 11-19, 11-20
chemical nature of, 11-3
and cod liver oil, 8-12
commercial uses of, 11-16 to 11-20
compatibilities of, 11-12
in cosmetics, 11-3, 11-18, 11-19, 11-21
electromechanical properties of, 11-5
extenders for, 11-15
and the FDA, 11-2
films from, 11-15
in foods, 8-18; 11-2, 11-18, 11-21
formulations with, 11-21
functions of, 11-18
gelation of, 11-4
GRAS, 11-2, 11-14
and guar gum, 11-12, 11-20
and gum arabic (*see* Gum arabic, and gum tragacanth)
and gum karaya, 11-2
handling of, 11-15
hydration of, 11-7
hydroxyethylcellulose, 12-5
interfacial tension of, 11-11
laboratory techniques for, 11-29
and locust bean gum, 11-12, 11-20
manufacture of, 11-14, 11-15
modified, 11-13
and paper products, 11-19, 11-20
in pharmaceuticals, 8-12; 11-3, 11-16, 11-18, 11-19, 11-21
physical properties of, 11-7
and polyvinylpyrrolidone, 21-9
as a preservative, 11-13
and printing, 11-19
and propylene glycol alginate, 11-12
reactivities of, 11-4

Gum tragacanth *(Cont.):*
  rheology of, 11-7
  and sodium alginate, 2-21
  and starches, 11-12, 11-17
  structure of, 11-3
  surface tension of, 11-10, 11-12, 11-13
  and tamarind gum, 23-6
  and textiles, 11-3, 11-19, 11-20
  toxicological properties of, 11-14
  tradenames for, 11-29
  viscosity of, 11-5 to 11-11
  and xanthan gum, 11-12; 24-15
Gypsum filtration, 16-17

Hair products *(see under* Cosmetics/toiletries)
Halogen removal *(see* Pollution control)
Hamaco, 22-71
HBMC (Hydroxybutylmethylcellulose), 3-4
Headlights, 20-22
Heat-transfer fluids, 18-16, 18-21, 18-25
HEC *(see* Hydroxyethylcellulose)
Helium, 20-8
HEMC (Hydroxyethylmethylcellulose), 3-16
Heptane, 18-11; 20-15; 21-11
Herbicides *(see* Agricultural products, herbicides)
Hercofloc, 16-25
Hexachlorophene, 19-4, 19-5; 21-9
Hexaethylene glycol, 20-18
Hexamethylene glycol, 21-11
Hexamethylenetetramine, 18-12
Hexane, 17-3; 21-11
1,2,6-Hexanetriol, 7-5
Hexylene glycol, 2-22; 24-15
Hi-Fi Starch, 22-71
Hi-moloc, 16-25
Hi-Poly, 15-20
Histological specimens, 18-24
Hodag Flocs, 16-25
Horsil, 4-24
Hoses, 18-22
HPC *(see* Hydroxypropylcellulose)
HPMC *(see* Hydroxypropylmethylcellulose)
H-SPAN, 22-50 to 22-52
Hubstar, 22-71
Humectants, 22-36
Hyamine, 21-10
Hydraid, 16-25
Hydraulic fluids, 17-10; 18-21, 18-24
Hydrazine gel, *16-18F;* 17-11
Hydrocarbons, 15-3; 17-3; 18-9; 20-14, 20-15; 21-11; 22-5
Hydrochloric acid, 8-7; 10-7; 17-3; *19-30F;* 24-14
Hydrogen, 20-8
Hydrogen peroxide, 21-3; 23-6
Hydrogen sulfide, 20-8
Hydromulching, 2-25
Hydropruf, 22-71
Hydroxides, 22-17
Hydroxyalkylalkylcelluloses, 3-1 to 3-25
  (See also specific compounds)
p-Hydroxybenzoate, 8-12
Hydroxybutylmethylcellulose (HBMC), 3-4
2-Hydroxyethyl alginate, 2-3

Hydroxyethylcellulose (HEC), 12-1 to 12-22
  with acrylic ester polymers, 12-13
  additives for, 12-10
  in adhesives and binders, 12-11, 12-14 to 12-16
  in agricultural products, 12-7, 12-15, 12-16
  applications of, 12-10, 12-11
  as a binder, 12-2, 12-11
  biological properties of, 12-6, 12-7, 12-16
  blends of, 12-5
  BOD of, 12-7, 12-11
  in building products, 12-11, 12-14 to 12-16
  and carboxymethylcellulose, 4-12
  in ceramics, 12-11, 12-15, 12-16
  chemical nature of, 12-2, 12-3
  and chlorides, 12-5
  in cleaners, 12-15
  in coatings, 12-15
  commercial uses of, 12-13 to 12-18
  compatibilities of, 12-5
  compounding uses of, 12-13 to 12-15
  and copper recovery, 12-11, 12-15
  in cosmetics, 12-11, 12-14, 12-16, *12-21F*
  in detergents, 12-15
  and dyes, 12-11, 12-14, 12-18
  and enzymes, 12-7
  and the EPA, 12-7
  and the FDA, 12-6, 12-7
  films from, 12-5, 12-6, 12-10, 12-15
  in fire fighting, 12-16
  for flow modification, 12-2, 12-11, 12-16
  and fluorescent lights, 12-15
  in foods, 12-6, 12-7
  formulations using, 12-13 to 12-15
  and fungi, 12-7
  in gels, 12-11, 12-14
  and glass products, 12-15
  grades of, 12-3
  with gums and resins, 12-5
  handling of, 12-10
  and hydroxypropylcellulose, 13-4
  and inks, 12-15
  interactions of, 12-5
  laboratory techniques for, 12-21
  and laundry products, 12-15
  manufacture of, 12-6
  and metal products, 12-11, 12-15, 12-16
  and microorganisms, 12-7
  molecular weight of, 12-3
  in packaging, 12-11, 12-15
  in paints and coatings, 12-11 to 12-14, 12-17 to 12-19, *12-18F, 12-19F*
  in paper products, 12-11, 12-14, 12-17
  and petroleum recovery, 12-11, 12-15, 12-17, *12-20F*
  and photographic film, 12-6
  physical properties of, 12-3, 12-6
  plasticizers for, 12-6
  and poly(ethylene oxide), 19-17
  as polymerization aid, 12-13, 12-16, 12-17
  with polymers, 12-13, 12-14
  and printing, 12-16, 12-18, *12-20F*
  as a processing aid, 12-15, 12-16
  rheology of, 12-7 to 12-10
  and salts, 12-5
  solubility of, 12-4, 12-5
  and solvents, 12-4, 12-5, 12-14

Hydroxyethylcellulose (HEC) *(Cont.):*
  storage of, 12-10
  and sucrose, 12-9
  and textiles, 12-11, 12-14, 12-15, 12-17, 12-18,
    12-20F
  and tobacco products, 12-11, 12-15
  toxicological properties of, 12-6, 12-7
  tradenames for, 12-21
  viscosity of, 12-3 to 12-9
  and waste treatment, 12-11
  and welding-rod flux, 12-11, 12-15
  and wheat flour, 12-15
Hydroxyethylmethylcellulose (HEMC), 3-16
2-Hydroxyethylmorpholine, 21-11
Hydroxypropylcellulose (HPC), 13-1 to 13-17
  additives for, 13-8
  and adhesives, 13-12
  applications of, 13-2, 13-10
  biological stability of, 13-2
  and building products, 13-10
  and Carbowax, 13-4
  and carboxymethylcellulose, 4-12; 13-4, 13-5
  and cellulose acetate phthalate, 13-5
  and ceramics, 13-10, 13-11, 13-13
  chemical nature of, 13-2
  chemical stability of, 13-2
  commercial uses for, 13-11 to 13-13
  compatibilities of, 13-4, 13-7, 13-11
  and cosmetics, 13-7, 13-11, 13-12, *13-13F, 13-
    14F*
  and defoamers, 13-10
  and detergents, 13-7
  and ethylcellulose, 13-5
  and the FDA, 13-2, 13-8
  fillers for, 13-7
  films from, 13-6, 13-13
  and foods, 13-8, 13-11, 13-12
  and foodstuffs, 13-7
  formulations using, 13-13 to 13-16
  and gelation, 13-4
  grades of, 13-3
  ang guar gum, 13-4
  handling of, 13-10
  and hydroxyethylcellulose, 13-4
  insolubilization of, 13-2
  laboratory techniques for, 13-16
  and locust bean gum, 13-4
  manufacture of, 13-7
  and methylcellulose, 13-4
  molding of, 13-9
  and paper products, 13-10, 13-13
  and pharmaceuticals, 13-10, 13-12, *13-15F*
  physical properties of, 13-2
  and plasticizers, 13-10
  and poly(ethylene oxide), 13-4
  and poly(vinyl alcohol), 13-4
  preservatives for, 13-10
  properties of, 13-9
  and resins, 13-8
  rheology of, 13-3
  and salts, 13-5, 13-7, 13-8, 13-11
  and shellac, 13-5
  and sodium alginate, 13-4, 13-5
  and sodium caseinate, 13-4
  solution properties of, 13-3
  and solvents, 13-3, 13-5, 13-6
  and starches, 13-7

Hydroxypropylcellulose (HPC) *(Cont.):*
  and textiles, 13-10, 13-13
  toxicological properties of, 13-8
  tradenames for, 13-16
  viscosities of, 13-5, 13-6
  and waxes, 13-4
  and zein, 13-5
Hydroxypropyl guar gum, 6-8
Hydroxypropylmethylcellulose (HPMC), 3-1 to
    3-25
  and acids, 3-12
  additives for, 3-10
  and alcohols, 3-11
  BOD of, 3-6
  compatibilities of, 3-11
  in cosmetics, 3-6
  films from, 3-11 to 3-13
  in foods, 3-6
  formulations using, 3-14, 3-15
  and hydroxyethylcellulose, 12-5
  insolubilization of, 3-11
  laboratory techniques for, 3-24
  manufacture of, 3-5, 3-6
  and oils, 3-12
  in paints, 3-16, 3-17
  in pharmaceuticals, 3-6
  physical properties of, 3-3, 3-13, 3-14
  powders from, 3-11, 3-12
  as a processing aid, 3-15, 3-16
  rheology of, 3-7, 3-8
  and salts, 3-10, 3-12
  solution properties of, 3-3, 3-6 to 3-11
  and solvents, 3-11, 3-12
  storage of, 3-6
  and sucrose, 3-10
  thermal gelation of, 3-8 to 3-10
  toxicological properties of, 3-6
  tradenames for, 3-4, 3-25
  viscosities of, 3-4, 3-6 to 3-10
Hylon, 22-71
Hypneaceae, 5-2
Hypochlorites, 24-9
Hyprin, 21-8

Ice cream *(see* Foods, ice cream)
Iceland moss, 7-5
Igepal, 2-22; 21-8; 24-15
Igepon, 21-10
Immunodiffusion, 7-9
Indian gum, 9-1
Infant food *(see* Foods, infant)
Initial, 10-13
Inks:
  ballpoint pen, 18-20, 18-24, *18-28F*
  carbon paper, 8-15
  duplicating, 12-15
  electrically conductive, 8-14
  emulsion, 8-14; 24-21
  encapsulation of, 8-7
  for fabrics, 8-14
  gloss-finish, *8-19F, 8-20F*
  hectographic, 8-6, 8-14
  and hydroxypropylcellulose, 13-10 to 13-12
  laundry-marking, 8-14
  lithographic, 8-6
  microencapsulated, 19-16, 19-18, 19-27

Inks *(Cont.)*:
  pigmented, 5-18; 8-14
  polyvinylpyrrolidone in, 21-16
  printing, 11-19; 13-15; 17-12; 18-20, 18-24
  quick-drying, 8-6, 8-14
  record, 8-14
  soluble, 8-6, 8-14
  stamp pad, 18-20, 18-24
  steam-set, 18-20, 18-24
  stencil, 18-20, 18-24
  textile, 11-19
  textile-marking, 8-14
  typographic, 8-6, 8-14
  watercolor, 8-14
  wood-grain, *8-20F*
Insecticides (*see* Agricultural products, insecticides)
Instant Clearjel, 22-71
Instant Keogel, 22-71
Insulation, 3-14; 13-10, 13-13; 16-12; 20-9
Interbond, 22-71
Iodides, 22-4, 22-6, 22-17, 22-36
Iodine, 19-4, 19-5; 21-2, 21-14; 22-6; 23-3
Ion exchange, 17-16
Iota-carrageenan, 5-28
Irish moss extract, 8-18
Iron, 19-3
Iron analyses, 9-6
Iron arabate, 8-12
Iron corrosion, 8-16
Iron ore concentration, 16-13
Iron ore tailings, 16-17
Iron oxide, 20-19
Iron rust removal, 17-16
Iron salts, 23-6
Irrigation, 16-10; 19-23
Isinglass, 7-5
Isobarbaloin, 8-8
Isobutyl vinyl ether, 21-11
Isopropanol, 2-12, 2-21; 11-16; 19-9; 21-4, 21-11; 24-9, 24-15
  (*See also* Isopropyl alcohol)
Isopropyl alcohol, 13-5, 13-6; 19-3
  (*See also* Isopropanol)
Isopropyl ether, 18-11
Isothan, 21-10

Jams (*see* Foods)
Jeffox, 18-4, 18-5, 18-30
Jellies (*see* Cosmetics, jellies; Foods, jellies; *under* Pharmaceuticals)
Jetsize, 22-71
Jet-Sol, 15-20
Joint cements, 3-18; *16-24F*

Kappa-carrageenan, 5-29
Karaya gum (*see* Gum karaya)
Kaserose, 4-24
Kelacid, 2-39
Kelco Gel, 2-40
Kelcoloid, 2-39, 2-40
Kelco Pac, 2-40
Kelcosol, 2-40
Kelflo, 24-29
Kelgin, 2-40

Kelmar, 2-39
Kelp, 2-2
Kelset, 2-40
Keltex, 2-40
Keltone, 2-40
Keltose, 2-39
Keltrol, 24-29
Kelzan, 24-29
Keo Chlor, 22-71
Keogel, 22-71
Keogum Starch, 22-7
Keojel, 22-71
Keokor, 22-71
Keosize, 22-71
Kerosene emulsions, 9-5
Kerosine, 20-15; 21-11
Ketone-alcohols, 21-11
Ketones, 18-9; 20-14, 20-15; 21-11
Kiccolate, 4-24
Klucel, 13-2, 13-16
Kofilm, 22-71
Kol Guard, 22-71
Kollidon, 21-2
Konafloc, 16-25
Kraft products (*see* Paper products, Kraft)
Kromfax, 2-22; 24-15
Kromflex, 13-6

Labels (*see* Adhesives, for labels)
Laboratory techniques:
  for agar, 7-13, 7-14
  for alginates, 2-37, 2-38
  for carrageenans, 5-27, 5-28
  for gum arabic, 8-20, 8-21
  for gum ghatti, 9-7
  for gum tragacanth, 11-29
  for hydroxyethylcellulose, 12-21
  for hydroxypropylcellulose, 13-16
  for hydroxypropylmethylcellulose, 3-24
  for methylcellulose, 3-24
  for pectins, 15-20
  for polyacrylamide, 16-19 to 16-24
  for polyethylene glycols, 18-29
  for poly(ethylene oxide), 20-30, 20-31
  for polyvinyl alcohol, 20-30, 20-31
  for starch, 22-70
  for tamarind gum, 23-9, 23-10
  for xanthan gum, 24-27 to 24-29
Lacquers, 12-17
Lactams, 21-11
Lactase purification, 17-17
Lactic acid, 4-11, 4-13; 13-5, 13-6
Lactones, 21-11
Lactose, 22-20
L'Algiline, 2-41
Lambda-carrageenan, 5-29
Lamitex, 2-41
Landalgine, 2-39
Lanethyl, 21-8
Lanogel, 21-8
Lanolin, 21-8, 21-9
Larch gum, 8-6
Lard oil, 20-15
Latex, 2-16, 2-21, 2-25, 2-26; 3-15, *3-16F, 3-17F,* 3-19; 4-2, 4-21, 4-26; 5-21 to 5-23; 8-13, 8-17, 8-24; 12-11 to 12-14, 12-17, *12-18F,* 12-

Latex *(Cont.):*
  19F; 17-10, 17-12; 18-16, 18-19, 18-20, 18-24, *18-26F;* 19-16 to 19-18, 19-26; 23-7 to 23-9; 24-8, 24-11, 24-16, 24-21
Laundry products, 17-13; 21-16, 21-18; 22-33, 22-34, 22-72
  marking inks for, 8-14
  soil suspension, 18-20
  (See also Antisoil agents; Detergents; Soaps)
Laxatives *(see* Pharmaceuticals, laxatives, bulk)
Lead, 4-3
Lead acetate, 8-3; 9-2; 11-4
Lead analyses, 9-6
Lead subacetate, basic, 8-7
Leaf Gum, 11-30
Leather products, 3-16, 3-19, 3-23; 17-15, 17-16; 18-16, 18-22, 18-24; 22-45
  adhesives for, 3-16, 3-18, 3-19, *3-23F;* 11-20; 17-16; 20-9, 20-20, 20-24; 22-30, 22-68, 22-69
  polishes for, 7-9
Lecithin, 21-9; 22-21
Lemon oil, 18-12
Levamid, 16-25
Library paste, 3-14
Lichenin, 7-5
Lignin sulfates, 19-4, 19-11
Lignin sulfonates, 19-5
Lime-sulfur filtration, 16-12, 16-17
Linseed oil, 8-12; 20-15
Lithium bromide, 22-18
Lithium chloride, 19-11
Lithium hydroxide, 22-17
Lithium manufacture, 16-16
Lithium salts, 22-4
Lithographic inks, 8-6
Lithopone, 20-19
Locust bean gum, 14-1 to 14-16
  and agar, 7-5, 7-7
  and agricultural products, 24-22
  and ammonium nitrate, 24-22
  analysis of, 14-2
  in bakery jellies, 24-21
  biological properties of, 14-6, 14-8
  borax reaction with, 14-5
  and carrageenans, 5-17, 5-22
  in cheese spreads, 24-21
  in chip dips, 24-21
  commercial uses for, 14-3, 14-8
  in dairy products, 24-21
  in deodorants, 24-22
  derivatives of, 14-3, 14-5, 14-6, 14-8, 14-9
  in desserts, 24-21
  in dressings, 24-21
  and the FDA, 14-6
  in fire fighting, 24-22
  in foods, 14-6, 14-8, 14-13, 14-14, *14-15F;* 24-21
  formulations with, 14-14, 14-15
  in fungicides, 24-22
  grades of, 14-2, 14-3
  GRAS, 14-6, 14-8
  and guar gum, 14-11
  and gum tragacanth, 11-12, 11-20
  handling of, 14-8
  and herbicides, 24-22
  and hydroxypropylcellulose, 13-4

Locust bean gum *(Cont.):*
  in ice cream, 24-21
  in jellies, 24-21
  manufacture of, 14-2, 14-3
  in milk products, 24-21
  and mining, 14-13, 14-14
  and ore processing, 14-13, 14-14
  in paper manufacture, 6-11, 6-14, 6-15
  and paper products, 14-11 to 14-14; 24-22
  in pet foods, 24-21
  in photographic products, 24-22
  properties of, 14-3 to 14-6
  reactivities of, 14-3, 14-5, 14-6
  rheology of, 14-3 to 14-7
  and salts, 14-6
  shear responses of, 14-3, 14-5, 14-6
  and sodium alginate, 2-21
  solubilities of, 14-3
  structure of, 14-3
  vs. tamarind gum, 23-8
  viscosity of, 14-3 to 14-7
  and white paper manufacture, 6-16
  and WHO, 14-8
  and xanthan gum, 14-5, 14-6; 24-9, 24-16 to 24-19
Loksize, 22-71
Lorol, 21-8
Lotions *(see* Cosmetics/toiletries, lotions; Pharmaceuticals, lotions)
Lozenges, candy, 8-9
Lubricants, 21-16
  food handling, 18-21, 18-24
  low-temperature, 18-21
  metal-working, 18-16, *18-28F*
  rubber, 18-24; 19-26, *19-31F*
    tire mounting, 19-16, 19-17, *19-31F*
  water-soluble, 16-16; 18-24
Lucel, 4-24
Lunel's paste, 8-13
Lygomme, 5-29

Macaroni and cheese, 2-28
Machine oil, 3-12
Magnafloc, 16-24, 16-25
Magnesia production, 16-12
Magnesium, 4-3, 4-6
Magnesium chloride, 2-22; 3-10, 3-11; 22-18, 22-69; 24-14
Magnesium salts, 18-12
Magnesium sulfate, 7-4; 19-11; 22-25
Maize starch:
  pastes of, 22-19
  viscosity of, 22-14
Majol, 4-24
Maleic acid, 21-2
Maltose, 22-20
Manganous chloride, 19-4, 19-11
Manganous salts, 18-12
Mannitol, 24-9
Mannuronic acid, 2-3
Manucol, 2-41
Manugel, 2-41
Manutex, 2-41
Margel, 2-40
Marpolose, 3-25
Matches, 11-19, 11-20

Maxi-Gel, 22-71
Mayonnaise, 8-18
MC (*see* Methylcellulose)
Meat products (*see* Foods, meat products)
Meat sauces, 2-24, 2-25
Medicinal materials (*see* Pharmaceuticals)
Melamine-formaldehyde resins, 12-6;
    22-59
Melojel, 22-71
Membranes, semipermeable, 17-16
Menthol, 18-12
Mercuric acetate, 18-12
Mercuric chloride, 8-8
Mercuric halides, 19-4, 19-5
Mercuric nitrate, 8-3
Mercuric salts, 10-5
Mesquite gum, 8-6
Metal corrosion, 8-17
Metal fabrication, 5-23
Metal foundry cores, 20-2; 22-26
Metal plating, 12-15
Metal products, 17-11, 17-13; 19-16
   and antimist fluids, 19-24, 19-26
   casting of, 18-14, 18-16, 18-22, 18-24
   cleaners for, *19-27F*
   coatings for, 7-9; 20-16
   copper-plating of, 7-9
   corrosion-inhibition for, 7-9
   finishing of, 18-22, 18-24
   fluid-jet cutting of, 19-16
   forming of, 18-16, 18-22, 18-24
   foundry cores for, 20-2; 22-26
   lubrication of, 18-22, 18-24, 18-28; 19-16
   machining of, 18-22, 18-24; 24-22
   mold binders for, 22-20
   plating of, 7-9
   purge dams, 19-16, 19-17, 19-26
   quenching of, 20-2, 20-25
   soldering fluxes for, 18-22, 18-24
   welding of, 19-16, 19-17, 19-26
Metasol, 2-21
Methacrylic acid, 17-3
Methanol, 2-12, 2-21; 4-6; 6-7; 13-5, 13-6; 17-3;
    18-11; 19-9; 20-15; 21-11; 22-40; 24-9, 24-15
   (*See also* Methyl alcohol)
Methocel, 2-21; 3-4, 3-25
β-Methoxyethanol, 17-3
Methoxypolyethylene glycols, 18-30
Methyl acetate, 13-5, 13-6
Methyl alcohol, 18-11
   (*See also* Methanol)
Methylan, 4-24
Methyl Cellosolve, 13-5, 13-6
Methylcellulose (MC), 3-1 to 3-25
   and acids, 3-12
   additives tolerance of, 3-10
   in adhesives, 3-14
   and alcohols, 3-11, 3-12
   BOD of, 3-6
   compatibilities of, 3-11
   in cosmetics, 3-6
   films from, 3-11 to 3-13
     plasticizers for, 3-12
     water resistance of, 3-13
   in foods, 3-6; 11-16
   formulations with, 3-14, 3-15
   and hydroxyethylcellulose, 12-5

Methylcellulose (MC) *(Cont.):*
   and hydroxypropylcellulose, 13-4
   insolubilization of, 3-11
   laboratory techniques for, 3-24
   for leather processing, 3-23
   in machine oil, 3-12
   manufacture of, 3-4 to 3-6
   and oils, 3-12
   in pharmaceuticals, 3-6
   physical properties of, 3-3, 3-13, 3-14
   and polyethylene glycols, 18-11
   and polyvinylpyrrolidone, 21-9
   powders from, 3-11, 3-12
   as a processing aid, 3-15, 3-16
   rheology of, 3-7, 3-8
   and salts, 3-10 to 3-12
   solubility of, 3-3
   solution properties of, 3-6 to 3-11
     and foaming, 3-10
     stability of and pH, 3-10, 3-11
   and solvents, 3-11, 3-12
   storage of, 3-6
   and sucrose, 3-10
   surface activity of, 3-10
   thermal gelation of, 3-8 to 3-10
   thickening properties of, 3-6, 3-7
   toxicological properties of, 3-6
   tradenames for, 3-4, 3-25
   uses for, 3-3
   and vegetable oils, 3-12
   viscosity of, 3-4, 3-6 to 3-10
     blending chart for, 3-9
Methylcyclohexane, 21-11
Methylcyclohexanol, 21-4, 21-11
Methylene bisacrylamide, 16-16; 17-14
Methylene chloride, 13-5, 13-6
Methylene dichloride, 19-9; 21-4, 21-11
Methyl ether, 21-11
Methylethylketone, 13-5, 13-6; 19-9
Methyl-*p*-hydroxybenzoate, 24-15
Methylisobutylketone, 19-9
Methylmethacrylate:
   polymerization of, 17-12
   and starches, 22-53
Methyl Parasept, 2-21
Methylphthalylethyl glycolate, 19-7
N-Methyl-2-pyrrolidone, 21-4, 21-10, 21-11
Methyl salicylate, 3-12; 18-12
Methyl vinyl ether, 21-2
Metolose, 3-4, 3-25
Meypralgin, 2-41
Microballoons, 17-16
Microbiology, 7-6 to 7-9, *7-10F to 7-12F*
Microencapsulation, 8-7, 8-11; 19-5, 19-16, 19-18,
    19-27, *19-29F*
   (*See also* Encapsulation)
Microorganisms, 12-7
Microtomy, 7-8
Milk products (*see* Foods, milk products)
Millon's reagent, 8-3; 9-2; 11-4
Mill tailings disposal, 16-15
Mill water clarification, 16-17
Milo maize, 22-23
Mimosa extract powder, 19-14
Mineral oil, 3-12; 18-11; 21-11
   (*See also* Pharmaceuticals, mineral oil)
Mineral spirits, 21-11

Minerals processing, 6-12; 19-16, 19-22, 19-26; 22-21, 22-22
(*See also* Mining; Ore processing; *under* Pharmaceuticals)
Mine water treatment, 16-13, 16-17
Mining, 4-2, 4-7; 6-16, 6-17, 6-18; 9-6; 14-13, 14-14; 16-10, 16-13 to 16-15, 16-17, 16-18; 19-16, 19-26
(*See also* Minerals processing)
Minus, 2-41
Miracleer, 22-71
Miranol, 2-22; 13-16; 24-15
Mira-Quik, 22-71
Mitsumame, 7-18
Modocoll, 3-13
Moisture barriers, 9-5
Molding:
of candies, 22-62
and criminology, 7-8
and dentistry, 7-6, 7-8, 7-9, *7-10F*
and elastomer extrusion, 22-46
of structural forms, 13-9, 13-12, *13-15F, 13-16F;* 20-22
Mold-release agents, 3-15, 3-19, 3-20; 17-14, 17-16; 18-16, 18-22
(*See also under* Coatings)
Monoethanolamine, 21-4
Monoethyl ether, 21-4
Monoglycerides, 13-4; 22-20, 22-21
Monosodium citrate, 7-7
Morphine, 8-8
Morpholine, 13-5; 21-11
Mortars, 12-16
concrete-block, 3-14, 3-18
masonry, 3-14, 3-18
Portland cement, 3-21
refractory, 3-15, 3-18, 3-21
thin-set, 20-22
tile, 3-18
Mowiol, 20-31
Muriatic acid, thickened, *19-30F*
Myrj, 22-20
Myvacet, 13-4, 13-16
Myverol, 13-4, 13-16; 22-20

Nabond, 22-71
Nacconol, 21-10
Nadex, 22-71
Nalco, 2-21; 16-25
Nalcolyte, 16-25
Naphtha, 13-5
Naphthol, 8-8; 19-4, 19-5; 23-4
National Frigex, 22-71
Natrosol, 12-4, 12-12, 12-21; 13-16
Natural gas recovery, 10-11
Nekal, 21-10
Neofat, 22-20
Nickel, 19-3
Nickel briquettes, 17-14
Nickel plating, 17-11
Nisso HPC, 13-2, 13-16
Nitrates, 12-5; 22-4
Nitric acid, 20-6
Nitrocellulose, 18-11
Nitroethane, 21-4, 21-11
Nitrogen, 20-8

Nitroglycerin, *16-16F*
Nitromethane, 21-11
Nitroparaffins, 18-9; 21-11; 22-5
Nonaethylene glycol, 20-18
Nonwoven fabrics, 12-4; 20-2, 20-22, 20-25
Nonylphenol, 21-4
Nonylphenol polyglycol, 19-10
Nopco NDW, 13-16
Norbak, 16-25
Norgine, 2-39
Nouralgine, 2-41
Novolak resins, 19-5, 19-14
Nucleosides (*see under* Pharmaceuticals)
Nu-Col, 22-71
Nut coating, *8-19F*
Nutricol, 5-29
Nylon, 17-8, 17-15; 18-16, 18-23
Nymcel, 4-24

Obipektin, 15-20
Oblate, 7-18
OG, 2-41
Oil barriers, 13-11, 13-13
Oil reclamation, 16-12, 16-17
(*See also* Petroleum recovery)
Oils, 9-5; 20-15; 22-20
(*See also under* Foods)
Ointments (*see* Cosmetics/toiletries, ointments; *under* Pharmaceuticals)
OK Brand, 22-71
Oleic acid, 20-15
Oleyl alcohol, 21-8
Olive oils, 8-11; 18-11; 21-9
Omacide, 2-21
Onyxide, 13-16
Ophthalmic medicinals, 3-15, 3-19; 17-7
Opraspray Lake Pigment, 13-16
Optical coatings, 17-16
Optics, 20-9
Orchids, 7-9
Ore processing, 14-13, 14-14; 16-13, 16-17; 17-13; 24-22
(*See also* Minerals processing; Mining)
Organic acids, 4-13
Oxidizers, 21-2; 24-9
Oxygen, 20-8

Packaging, 2-18; 3-15; 12-6, 12-11, 12-15; 18-16, 18-24; 19-12, 19-16, 19-20; 20-16, 20-17, 20-22, 20-25; 22-60
Paint removers, 3-15, 3-17, 3-19 to 3-21; 13-10 to 13-12, *13-14F, 13-15F;* 18-20; 19-16 to 19-18, 19-26, *19-29F*
Paints, 2-23; 17-11; 21-16; 24-13
artists, 18-20, 18-24
binders for, 20-25, *20-29F*
cement-based, 3-14, 3-19
enamels, 12-17
lacquers, 12-17
latex, 3-15, *3-16F, 3-17F,* 3-19; 4-2, 4-21, 4-26; 5-21 to 5-23; 8-13, 8-17, 8-24; 12-11 to 12-14, 12-17, *12-18F, 12-19F;* 17-10; 18-20, 18-24, *18-26F;* 19-16 to 19-18, 19-26; 24-8, 24-11, 24-21
linseed-oil based, 11-12

Paints *(Cont.):*
  multi-color, 3-19
  paste, 17-12
  primer, 17-15
  shellac, 18-20
  spatter finish, 19-16 to 19-18, 19-26
  texture, 3-19
  watercolor, 18-20, 18-24
Pamak, 13-16
Papain, 2-22; 9-2; 24-15
Paper coatings, 2-23, 2-25, 2-26; 3-15, 3-19; 9-5;
    12-11, 12-14, 12-17; 17-15; 18-22, 18-24, *18-28F;* 24-8, 24-22
Paper manufacture, 3-15; 10-10; 16-10, 16-11, 16-13; 17-13, 17-15, 17-18; 18-16; 19-22, 19-23; 22-43, 22-54 to 22-57; 23-7, 23-8
Paper products:
  additives for, 14-10 to 14-14
  adhesives for *(see* Adhesives, for paper products)
  and alginates, 2-25, 2-26
  aluminum lamination with, 22-69
  antistat, 19-16
  board, 4-18; 22-68
  boxboard, 6-16; 14-12
  calendar stocks, 4-18
  carbonless, 19-18
  and carboxymethylcellulose, 4-2, 4-13, 4-14, 4-18, 4-26
  and cationic starch ethers, 22-42
  coatings for, 7-9; 13-10, 13-12, 13-13; 19-16; 20-2, 20-5, 20-16, 20-21, 20-24, *20-29F;* 21-12, 21-15, 21-16; 22-25, 22-31, 22-34, 22-38, 22-43, 22-53, 22-58 to 22-60
  corrugated, 6-15; 14-12; 22-30, 22-68, 22-72
  cutting of, 19-23
  dry strength for, 4-19
  glassine, 8-13
  gloss finishing of, 7-9
  and guar gum, 6-11, 6-14 to 6-16, 6-18
  and gum arabic, 8-13, 8-15, 8-17
  and gum ghatti, 9-5
  and gum karaya, 10-9, 10-10
  and gum tragacanth, 11-19, 11-20
  and hydroxyethylcellulose, 12-14, 12-17
  and hydroxypropylmethylcellulose, 3-19
  Kraft, 2-35; 6-15; 14-12; 16-13; 18-24; 22-58
  labels, 22-68
  laminated, 22-68
  light-sensitive, 9-5
    (See also Photographic products)
  linerboard, 6-15; 14-12; 22-58
  and locust bean gum, 6-11; 14-11 to 14-13; 24-22
  and methylcellulose, 3-15, 3-19
  offset news stock, 6-16
  pigment coatings for, 22-43, 22-53, 22-59
  and polyacrylamide, 16-11, 16-13
  and poly(acrylic acid), 17-13, 17-16, 17-17
  and polyethylene glycols, 18-16, 18-21, 18-22, 18-24, *18-28F*
  and poly(ethylene oxide), 19-16
  and polyvinyl alcohol, 20-2, 20-5, 20-21, 20-24, *20-29F*
  and polyvinylpyrrolidone, 21-15, 21-20
  printing on, 11-19
  and sodium alginate, 2-35

Paper products *(Cont.):*
  and sodium polyacrylate, 17-12
  and starches, 22-21 to 22-23, 22-25, 22-30, 22-31, 22-33, 22-34, 22-38, 22-43, 22-45, 22-46, 22-53, 22-54 to 22-60, 22-72
  and tamarind gum, 23-8
  tapes, 22-68
  typing, 8-15
  wet strength, 22-45, 22-46
  white, 6-16; 14-12, 14-13
  white-water effluent from, 16-13
  and xanthan gum, 24-22
Paraffin emulsions, 8-6
Paraffin wax, 18-11
Paraldehyde, 18-12
Parenteral solutions *(see under* Pharmaceuticals)
Pastilles, 8-9; 11-3, 11-9, 11-21
Patching compounds, 3-4
Peanut oil, 3-12
Pectalgine, 2-41
Pectans, 15-14
Pectinase, 24-10
Pectinol, 24-15
Pectins, 15-1 to 15-21
  and additives, 15-7
  and alcohols, 15-3
  applications of, 15-7, 15-8, 15-12 to 15-14
  in bakery products, 15-12, 15-14
  in beverages *(see* Beverages, and pectins)
  biological properties of, 15-6, 15-7
  and carrageenans, 15-8
  and casein, 15-8
  chemical nature of, 15-2, 15-3
  in confections, 15-3, 15-8 to 15-10, 15-12, 15-14, *15-16F, 15-17F*
  in convenience foods, 15-14
  in cosmetics, 15-10, 15-13
  in dairy products, 15-4, 15-7 to 15-9, 15-13, 15-14, *15-18F, 15-19F*
  in desserts, 15-4, 15-8, 15-14, *15-19F*
  and dextrose, 15-7
  dissolving of, 15-8
  and ethers, 15-3
  extenders for, 15-7
  and FAO/WHO, 15-6
  in foods, 15-2, 15-5, 15-10
  formulations with, 15-10
  with fruits, 15-4, 15-7, 15-10, 15-12 to 15-14, *15-15F, 15-18F*
  and furcellaran, 15-8
  gelation of, 15-2, 15-7, 15-9
  and gum arabic, 8-11
  handling of, 15-7
  and hydrocarbons, 15-3
  in jams, 15-2, 15-4, 15-5, 15-7, 15-10 to 15-12, *15-11F,* 15-14, *15-17F, 15-18F*
  in jellies, 15-2, 15-4, 15-7, 15-9, 15-10, 15-12, 15-14, *15-15F, 15-17F, 15-18F*
  laboratory techniques for, 15-20
  manufacture of, 15-5, 15-6
  in marmalades, 15-2
  nomenclature for, 15-3
  in pharmaceuticals, 15-10, 15-13
  and phosphates, 15-7
  physical properties of, 15-3 to 15-5
  in preserves, 15-5, 15-7, 15-12
  in puddings, 15-8, 15-14, *15-19F, 15-20F*

Pectins *(Cont.):*
  rheology of, 15-5
  and salts, 15-5, 15-7 to 15-9
  in snack foods, 15-9
  and solvents, 15-3
  storage of, 15-7
  and sucrose, 15-7
  vs. tamarind gum, 23-8
  and tartrates, 15-7
  tradenames for, 15-20
  viscosity of, 15-3, 15-5, 15-8
PEGs *(see* Polyethylene glycols)
Pencils, 12-11, 12-15
Pen-Cor, 22-71
Pen-Cote, 22-71
Penford Gum, 22-71
Pentachlorophenol, 23-4
Pentasol, 22-25
PEO, 19-31
Peppermint oil, 18-12
Pepsin, 22-42
Percol, 16-25
Peregal ST, 21-2
Perma-Flo Starch, 22-71
Peroxides, 24-9
Persulfates, 24-9
Pesticides *(see* Agricultural products, pesticides)
Pet foods *(see* Foods, pet)
Petrolatum emulsions, 8-11
Petroleum ether, 21-11
Petroleum processing, 18-22, 18-25
Petroleum recovery, 4-2, 4-7, 4-18; 6-10, 6-13,
  6-17, 6-18; 7-9; 8-16; 9-5, 9-6; 10-11; 12-11,
  12-15, 12-17, *12-20F;* 16-10, 16-14, 16-15, 16-
  18; 17-10, 17-11; 18-22, 18-25; 19-16, 19-24,
  19-25, 19-27; 20-2; 22-21, 22-30; 24-13, 24-
  22
Pharmaceuticals (including drugs and medicinals):
  alkaloids, 8-7, 8-12
  anesthetics, 8-12
  antacids, 5-19, 5-22, 5-23, 5-26; 11-19
  antibiotics, 7-8
  anticoagulants, 7-8
  antilipemic, 7-8
  antiseptics, 8-12
  antitumor drugs, 9-5
  aspirin, 8-7
  barbiturates, 8-11
  barium sulfate suspensions, 5-18, 5-23; 7-8; 21-
    14; 24-13
  benzalkonium chloride, 8-12
  bismuth suspensions, 24-13
  bleeding control, 3-23
  blood plasma extenders, 21-13
  blood pressure, 8-12
  bronchography, 3-15, 3-23
  burn therapy, 3-23
  calamine lotions, 4-21; 8-11; 19-16, 19-18, 19-
    27, *19-28F;* 24-13
  capsules, 7-8; 8-7
  and carboxymethylcellulose, 4-7, 4-13, 4-17, 4-
    21, 4-26
  chlorbutanol, 8-12
  colostomy rings, 10-11, *10-12F*
  constipation control, 3-23

Pharmaceuticals *(Cont.):*
  cough drops, 8-9, 8-12, 8-16; 11-3, 11-18, 11-
    21
  creams, 3-15, 3-19; 11-16; 12-11, 12-14; 18-20
  crystal modification, 3-23
  culture media, 7-9
  cystoscope jellies, 3-15
  demulcents, 8-11
  dental medicinals, 3-23
  denture adhesives *(see* Denture adhesives)
  detoxification, 21-12
  diagnostic aids, 21-14
  diarrhea control, 3-23; 21-12
  drug carriers, 18-21
  dusting powders, 22-28, 22-70
  elixirs, 13-12
  emulsions, 8-11, 8-12; 9-5; 13-10; 18-19, 18-20,
    18-25; 21-12
  eosinophils, 9-5
  eye medications, 3-15, 3-19; 17-17
  gastrointestinal procedures, 3-23; 5-11; 11-13
  gels, 12-11, 12-14
  germicides, 21-2
  and gum arabic *(see* Gum arabic, and pharma-
    ceuticals)
  and gum tragacanth *(see* Gum tragacanth, in
    pharmaceuticals)
  gynecological jellies, 11-2, 11-3, 11-9
  histological specimens, 18-24
  p-hydroxybenzoates, 8-12
  and hydroxypropylcellulose, 13-12
  intramuscular injections, 18-21
  iron arabate, 8-12
  jellies, 3-15, 3-19, 3-23; 8-12; 11-3, 11-16, 11-
    21, 11-19
  kaolin suspensions, 8-11; 17-13
  laxatives, bulk, 3-15, 3-19; 4-21; 7-8; 8-12; 10-
    8, 10-10, 10-11
  lotions, 3-15; 4-21; 12-11, 12-14; 18-16, 18-20,
    18-25; 19-16, 19-18, 19-27, *19-28F;* 24-13
  lozenges, 8-9; 11-18, 11-21
  magnesia suspensions, 8-11
  media, 7-9
  medical sundries, 18-24
  merthiolate, 8-12
  microbicide, 21-14
  microscopic diagnosis, 3-23
  mineral oil, 5-22; 8-11, 8-12; 11-3, 11-12, 11-
    16, 11-19, 11-21
  nephritic edema, 8-12, 8-16
  nonglycogenetic medicinals, 3-23
  nose drops, 3-23
  peptone complex, 8-12
  petrolatum emulsions, 8-11
  phenyl ketone therapy, 3-23
  phenylmercuric acetate, 8-12; 12-7; 13-10
  pills, 8-7, 8-12; 11-3, 11-16, 11-19, 11-21
  plasters, 8-12
  polyethylene glycols in [*see* Polyethylene gly-
    cols (PEGs), in pharmaceuticals]
  and poly(ethylene oxide), 19-16, 19-18, 19-27,
    19-28F
  and polyvinylpyrrolidone, 21-2, 21-12 to 21-
    15, 21-17, 21-20
  post-surgical drainage pouches, 10-11, 10-12
  radiology, 3-23; 7-8; 21-14
  rubbing alcohol, 19-16, 19-18, 19-27

Pharmaceuticals *(Cont.)*:
salves, 11-3, 11-19, 11-21; 18-16, 18-20, 18-25
sealing pads, 10-11, *10-12F*
silver arabate, 8-12
silver bromide, 8-12
spermicidal jellies, 11-16
steroids, 8-11
and stomach ulcers, 11-13
suppositories, 7-8; 8-12; 18-16, 18-20, 18-25, *18-26F*
and surgical casts, 3-23
surgical lubricants, 7-8
and surgical sutures, 18-24
suspensions, 3-19, 3-23; 5-18; 8-11; 18-20; 21-12
syrups, 11-3, 11-21
tablets, 3-15, 3-19, 3-23; 4-21; 7-8; 8-7, 8-12, 8-16; 11-3, 11-16, 11-19, 11-21; 13-10, 13-12, 13-13, *13-15F*; 17-17; 18-21, 18-25; 19-16, 19-21, 19-27; 21-12; 22-70
tissue cultures, 3-23
topical medicaments, 7-8
veterinarian, 18-21; 21-20
x-ray media, 3-23; 7-8; 21-14
Phenobarbitol, 18-12
Phenol, 8-8; 13-10; 12-4; 18-12; 21-11; 22-5
Phenolic resins, 19-4, 19-14; 21-2
Phenothiazine, 18-14
Phloroglucinol, 20-18
Phosphate plant tailings, 16-17
Phosphates, 12-5; 15-7
Phosphomolybdic acid, 7-4
Phosphoric acid, 20-6; 24-14
manufacture of, 16-12, 16-17
Phosphotungstic acid, 7-4
Photoconductors, 17-17
Photoelectric analysis, 9-5, 9-6
Photographic products, 7-9; 8-16; 12-6; 18-16, 18-21, 18-25; 21-16, 21-19; 24-22
Phrikolat, 4-24
Piccolyte, 21-10
Pickle products *(see under* Foods)
Pickling oils, *8-18F*
Pigments, 5-23; 9-5; 16-16, 16-17; 17-15; 20-2, 20-5, 20-9, 20-22, 20-24, 20-25; 21-2, 21-16; 22-61; 24-21
Pine oil, 13-4; 18-11
Piperazine, 18-12
Plasdone, 21-2
Plaster *(see* Building products, plaster)
Plaster casts, 21-14
Plasticizers, 2-17, 2-22, 2-24; 13-10; 20-8, 20-17
Plastic products, 18-12, 18-16, 18-21 to 18-23, 18-25
Pluracol, 18-4, 18-5, 18-30
Pluronic F-68, 22-20
Plywood *(see* Wood products)
Polar-Gel, 22-71
Polishes, 2-25, 2-26; 3-14, 3-18; 4-22; 7-9; 9-5; 11-3, 11-19, 11-21; 13-10 to 13-12; 17-11; 18-16, 18-23; 21-16; 24-21
Pollution control, 4-2, 4-19; 16-13; 18-22
Polyacrylamide, 16-1 to 16-26
additives for, 16-10
in adhesives, 16-16
in aluminum sulfate manufacture, 16-12
in ammonia manufacture, 16-16

Polyacrylamide *(Cont.)*:
applications of, 16-10, 16-17, 16-18
in asbestos-cement board, 16-12, 16-17
in barium production, 16-16
biological properties of, 16-7, 16-9
in borax manufacture, 16-12
in brine treatment, 16-16, 16-17
in building products, 16-10 to 16-12, 16-18, *16-24F*
in cement manufacture, 16-12, 16-17
chemical nature of, 16-2, 16-3
in chemicals manufacture, 16-16
in clay production, 16-3, 16-16
and coal fines, 16-12, 16-17
in coatings, 16-16
in construction, 16-10 to 16-12
and copper recovery, 16-12, 16-14
as a culture medium, 16-11
and drag reduction, 16-5, 16-7, 16-15 to 16-17
and electrorefining, 16-12
in explosives, *16-16F, 16-18F, 16-19F*
extenders for, 16-10
films from, 16-16
as a filter aid, 16-11, 16-12, 16-17
and fire fighting, 16-10
and fish, 16-9
flocculation with, 16-4, 16-5, 16-7
for flotation, 16-14
and flue-dust recovery, 16-12
formulations with, 16-16, 16-18, 16-19
and friction reduction, 16-5, 16-7, 16-15 to 16-17
and fruit washing, 16-9, 16-17
gelation of, 16-10
and gold recovery, 16-14
and green-liquor clarification, 16-13
for gypsum filtration, 16-17
handling of, 16-7, 16-10
with hydrazine, *16-18F*
hydrolysis of, 16-3, 16-4, 16-9
and iron ore concentration, 16-13, 16-17
in irrigation, 16-10
laboratory techniques for, 16-19 to 16-24
and lithium manufacture, 16-16
in lubricants, 16-16
in magnesia production, 16-12
manufacture of, 16-5, 16-6
methylolated, 16-16
and mill-tailings disposal, 16-15, 16-17, 16-18
and mine backfill, 16-15
and mine water processing, 16-13, 16-17, 16-18
in mining, 16-10, 16-13 to 16-15, 16-17, 16-18
and oil reclamation, 16-12, 16-17
for ore processing, 16-13, 16-14, 16-17
and paper products, 16-10, 16-11, 16-13, 16-17
for petroleum recovery, 16-10, 16-14, 16-15, 16-18
and pharmaceuticals, 16-16
and phosphate tailings, 16-17
physical properties of, 16-3
for pollution control, 16-13
and polyvinylpyrrolidone, 21-10
and potable water, 16-9, 16-13, 16-18
rheology of, 16-9, 16-10

Polyacrylamide *(Cont.):*
  in rocket fuels, *16-18F*
  for sewage treatment, **16**-10, **16**-13, **16**-18
  for slime removal, **16**-12, **16**-13, **16**-15, **16**-17
  in soda ash manufacture, **16**-16
  solubility of, **16**-3
  for sugar juice clarification, **16**-9, **16**-12, **16**-18
  in textile manufacture, **16**-16
  toxicological properties of, **16**-7, **16**-9
  tradenames for, **16**-25
  and Trona plant operation, **16**-18
  in uranium recovery, **16**-14, **16**-17
  for vegetable washing, **16**-9, **16**-17
  viscosity of, **16**-3 to **16**-6, **16**-9, **16**-10
  for waste removal, **16**-10, **16**-18
  and water clarification, **16**-10, **16**-13, **16**-17, **16**-18
  and water softening, **16**-9, **16**-13, **16**-17
  and white water, **16**-13
  and zinc recovery, **16**-12, **16**-15
Polyacrylates, **17**-1 to **17**-19
  in abrasive products, **17**-12, **17**-14
  acid strength of, **17**-4, **17**-5
  in adhesives, **17**-9, **17**-17
  applications of, **17**-9
  in building products, **17**-14
  in ceramic products, **17**-11, **17**-14
  chemical nature of, **17**-2 to **17**-5, **17**-7 to **17**-9
  as chemical raw materials, **17**-17
  for clarification, **17**-13
  in coatings, **17**-9, **17**-14, **17**-15
  copolymerization of, **17**-7
  in cosmetics, **17**-11, **17**-16, **17**-17
  degradation of, **17**-9
  dehydration of, **17**-9
  as dispersants, **17**-9, **17**-11 to **17**-13
  in dry cleaning solvents, **17**-13
  and electroplating, **17**-13
  esterification of, **17**-8
  and the FDA, **17**-18
  as filtration aids, **17**-13
  as flocculants, **17**-9, **17**-13, **17**-14
  in hydraulic fluids, **17**-10
  hydrolysis of, **17**-6
  in inks, **17**-12
  for ion exchange, **17**-16
  isotactic, **17**-3, **17**-4
  and leather products, **17**-15, **17**-16
  and metal products, **17**-13
  neutralization of, **17**-4, **17**-7, **17**-8
  and ore recovery, **17**-13
  in paints, **17**-10, **17**-12, **17**-15
  and paper products, **17**-15
  and petroleum recovery, **17**-10, **17**-11
  and pharmaceuticals, **17**-11 to **17**-13, **17**-17
  physical properties of, **17**-2, **17**-5
  and printing, **17**-11, **17**-15
  salts of, **17**-3
  for sewage disposal, **17**-13
  and soil conditioning, **17**-13, **17**-14
  and soil release finishes, **17**-15
  solutions of, **17**-3, **17**-4
  and solvents, **17**-3
  and starches, **22**-69
  syndiotactic, **17**-3, **17**-4
  and textile products, **17**-10, **17**-11, **17**-15
  and vinyl monomer manufacture, **17**-12

Polyacrylates *(Cont.):*
  viscosity of, **17**-3, **17**-4
  and water treatment, **17**-13
  in waxes, **17**-15
Poly(acrylic acid), **17**-1 to **17**-19
  and acids, **17**-3
  in adhesives, **17**-11, **17**-17
  in antifogging coatings, **17**-16
  in building products, **17**-12, **17**-17
  for catalyst removal, **17**-16
  in cellulose acetate manufacture, **17**-15
  in cleaners, **17**-11, **17**-13, **17**-16
  in coatings, **17**-16
  complex formation with, **17**-18
  and copper pellets, **17**-11
  and enzyme recovery, **17**-17
  and the FDA, **17**-18
  films from, **17**-8, **17**-16
  in foods, **17**-11
  with hydrazine, **17**-11
  and metal products, **17**-11, **17**-14
  and minerals processing, **17**-13
  in mold releases, **17**-16
  in optical coatings, **17**-16
  in paints, **17**-11 to **17**-13, **17**-17, **17**-18
  and paper products, **17**-16
  and photographic products, **17**-17, **17**-18
  in polishes, **17**-11
  and poly(ethylene oxide), **19**-4, **19**-5
  as a polymerization aid, **17**-12
  and polyvinylpyrrolidone, **21**-2
  and printing, **17**-14, **17**-15
  and protein recovery, **17**-17
  in release coatings, **17**-16
  in soil-release coatings, **17**-16
  and solvents, **17**-8
  and starches, **22**-66, **22**-67
  and textiles, **17**-8, **17**-14, **17**-15
  tradenames for, **17**-18
  and uremia, **17**-17
  for water treatment, **17**-12, **17**-16
  and wood products, **17**-14, **17**-17
Poly(acrylic anhydride), **17**-2, **17**-9
Polyacrylonitrile, **17**-4, **17**-14; **22**-48 to **22**-59
Polyamideamine, **22**-54
Polyamide moldings, **17**-12
Polybasic acids, **21**-2
Polychloroprene, **19**-7
Polyclar L, **2**-12
Polyesters, **18**-23
Polyethylene, **18**-12; **19**-7; **21**-10
Polyethylene glycol ether, **20**-19
Polyethylene glycols (PEGs), **18**-1 to **18**-31
  additives for, **18**-17
  in adhesives, **18**-14, **18**-16, **18**-19, **18**-23, **18**-24
  in agricultural products, **18**-14, **18**-16, **18**-19, **18**-23
  antioxidants for, **18**-14, **18**-17
  applications for, **18**-15 to **18**-17
  and archaeology, **18**-25
  and battery cases, **18**-22
  biological properties of, **18**-14
  in boiler water, **18**-14
  and butadiene recovery, **18**-22, **18**-25
  and casein adhesives, **18**-19
  and cellophane film, **18**-16, **18**-19, **18**-24
  and cellulose, **18**-21

Polyethylene glycols (PEGs) *(Cont.):*
  and ceramics, 18-14, 18-16, 18-21, 18-23
  as chemical intermediates, 18-16, 18-19, 18-23, 18-25
  chemical nature of, 18-3 to 18-5
  and citrus fruits, 18-14
  in cleaners, 18-20, 18-23
  in coatings, 18-14, 18-19, 18-20, 18-24
  compatibilities of, 18-11
  in compounded products, 18-19 to 18-21
  in cork products, 18-21
  in cosmetics, 18-16, 18-19, 18-20, 18-23, *18-27F, 18-28F*
  derivatives of, 18-17 to 18-19
  in detergents, 18-16, 18-20, 18-23
  in dialysis, 18-16, 18-21
  in drug carriers, 18-21
  in electroplating, 18-16, 18-21, 18-23
  in electropolishing, 18-21, 18-23
  in electrowinning, 18-21, 18-23
  extenders for, 18-17
  and fluxes, 18-14, 18-16
  in foods, 18-14, 18-16, 18-19, 18-21, 18-23
  formulations with, 18-19 to 18-21, 18-25 to 18-29
  in foundry cores, 18-14, 18-16
  for glass coatings, 18-20
  and halogen removal, 18-22
  handling of, 18-15
  and heat transfer baths, 18-16, 18-21, 18-25
  and histological specimens, 18-24
  in hose manufacture, 18-22, 18-25
  as humectants, 18-19, 18-24
  in hydraulic fluids, 18-21, 18-24
  and hydroxypropylcellulose, 13-4, 13-7
  hygroscopicity of, 18-12, 18-13
  in inks, 18-16, 18-20, 18-24, *18-28F*
  laboratory techniques for, 18-29
  and leather, 18-16, 18-22, 18-24
  as lubricants, 18-16, 18-21, 18-24, 18-28
  manufacture of, 18-14, 18-15
  in medical sundries, 18-14, 18-24
  and metal products, 18-16, 18-21, 18-22, 18-24
  and milk products, 18-14, 18-23
  as mold releases, 18-16, 18-22, 18-25
  molecular weights of, 18-4, 18-6, 18-7
  and nylon, 18-16, 18-23
  in packaging, 18-14, 18-16, 18-23, 18-24
  in paint removers, 18-16, 18-20, 18-24, *18-26F*
  in paints, 18-16, 18-20, 18-24, *18-26F*
  and paper products, 18-16, 18-21, 18-22, 18-24, *18-28F*
  and petroleum processing, 18-22, 18-25
  and petroleum recovery, 18-22, 18-25
  in pharmaceuticals, 18-16, 18-19 to 18-21, 18-25, *18-26F, 18-27F*
  in photographic chemicals, 18-16, 18-21, 18-25
  physical properties of, 18-4 to 18-14
  and plastic products, 18-16, 18-21 to 18-23, 18-25
  in polishes, 18-16, 18-23
  and pollution control, 18-22
  and poly(acrylic acid), 17-8
  and poly(ethylene oxide), 19-7
  and polyvinyl alcohol, 18-21; 20-4, 20-17

Polyethylene glycols (PEGs) *(Cont.):*
  and polyvinylpyrrolidone, 21-8, 21-9, 21-11
  as processing aids, 18-21
  and proteins, 18-16
  and rubber products, 18-16, 18-21, 18-22, 18-25
  and salts, 18-12, 18-14
  solubilities of, 18-8 to 18-12
  and sponges, 18-21
  and sweeteners, 18-14, 18-21
  in tablets, 18-14, 18-16, 18-21, 18-25
  in textiles, 18-16, 18-19, 18-22, 18-23, 18-25
  in tile manufacture, 18-16
  in tire manufacture, 18-16, 18-22, 18-23
  toxicological properties of, 18-14
  in veterinarian medicines, 18-21
  and viscose rayon, 18-16, 18-22, 18-25
  viscosities of, 18-7 to 18-10
  and vitamins, 18-14, 18-21
  and wood products, 18-16, 18-21, 18-23, 18-25
Poly(ethylene oxide), 19-1 to 19-33
  additives for, 19-9, 19-10
  and adhesives, 19-16
  and agricultural products, 19-25
  applications for, 19-11, 19-12
  association complexes with, 19-4, 19-14, 19-15
  biodegradability of, 19-8
  biological properties of, 19-7, 19-8
  BOD of, 19-8
  and building products, 19-26
  and ceramics, 19-25
  chemical nature of, 19-2 to 19-4
  commercial grades of, 19-3
  commercial uses of, 19-13 to 19-27
  compatibilities of, 19-7
  in cosmetics, 19-16, 19-18, 19-19, 19-26
  and drag reduction, 19-3, 19-23
  extenders for, 19-9, 19-10
  and the FDA, 19-7
  films of, 19-5, 19-9, 19-19, 19-20
  flocculation with, 19-22, 19-23
  and food additives laws, 19-12
  formulations with, 19-27 to 19-31
  grades of, 19-3
  handling of, 19-11
  hydrogels of, 19-3
  and hydroxyethylcellulose, 19-17
  and hydroxypropylcellulose, 13-4
  manufacture of, 19-6, 19-7
  melt rheology of, 19-6, 19-8
  and metal products, 19-16, 19-26
  molded products from, 19-3
  and ore processing, 19-16
  oxidative degradation of, 19-3
  in paint removers, 19-17, 19-18
  in paints, 19-17, 19-18, 19-26
  and paper products, 19-16, 19-26
  and petroleum recovery, 19-27
  and pharmaceuticals, 19-16, 19-18, 19-27
  physical properties of, 19-4 to 19-6, 19-20
  plasticizers for, 19-9
  and polyethylene glycols, 18-30
  precipitation of, 19-10
  and printing products, 19-18
  rheology of, 19-8, 19-9, 19-24
  and salts, 19-11
  shear rates of, 19-15

Poly(ethylene oxide) *(Cont.):*
   solubilities of, 19-9, 19-10
   solution properties of, 19-6
   and textiles, 19-24
   and thermoplastics manufacture, 19-24
   TOD of, 19-8
   toxicological properties of, 19-7
   tradenames of, 19-31
   uses for, 19-2
   viscosity of, 19-8, 19-13 to 19-15
Polyethylene sheet, 17-12
Polyfloc, 16-25
Poly-G, 18-4, 18-5, 18-30
Polyglycol E, 18-4, 18-5, 18-30
Polyglycols, 12-6; 16-10
Polyhall (MRL), 16-25
Poly(isopropyl acrylate), 17-6
Polymer, 12-12, 12-21; 22-71
Polymerization, 3-16, 3-19; 12-13; 13-10, 13-13;
   17-12; 19-16, 19-22, 19-26; 20-22; 21-2, 21-
   16, 21-20
Polymethacrylate, 17-3, 17-12, 17-17, 17-18
Poly(methacrylic acid), 17-1
Poly(methyl methacrylate), 17-6; 19-7
Polyox, 19-31
Polyoxyethylene, 18-30
Polyphosphates, 8-5
Polypropylene, 18-12; 19-7
Polypropylene glycol, 13-4; 19-7
Poly(sodium acrylate), 17-10, 17-13
   (See also Sodium polyacrylate)
Poly(sodium methacrylate), 17-6, 17-7
Polystyrene, 9-5, 9-6; 18-12; 19-7; 20-22; 21-10
Polyureas, 19-4, 19-5
Polyvinyl acetate, 18-11; 20-21; 22-69
Polyvinyl alcohol, 20-1 to 20-32
   in adhesives, 20-9, 20-20, 20-21, 20-24, *20-27F*
     to *20-29F*
   antifoam agents for, 20-19
   barrier properties of, 20-8
   as a binder, 20-23, 20-25, *20-29F*
   biodegradation of, 20-17
   BOD of, 20-16
   in building products, 20-25, *20-30F*
   and ceramics, 20-23, 20-25
   chemical derivatives of, 20-23
   chemical nature of, 20-3
   in coatings, 20-21, 20-24, *20-30F*
   commercial uses for, 20-2, 20-20 to 20-25
   compatibilities of, 20-13
   in cosmetics, 20-23, 20-25
   in dispersions, 20-22, 20-23
   and dyes, 20-19
   electrical properties of, 20-9
   electrolyte tolerance of, 20-6
   emulsifying properties of, 20-10, 20-11
   in emulsions, 20-22, 20-23
   extenders for, 20-18
   and the FDA, 20-16, 20-17
   films from *(see* Films, from polyvinyl alcohol)
   gelling agents for, 20-18, 20-19
   grades of, 20-5, 20-6, 20-16
   as a grease barrier, 20-21
   handling of, 20-19, 20-20
   heat effect on, 20-8, 20-9, 20-14
   hydrolysis of, 20-7
   and hydroxyethylcellulose, 12-5

Polyvinyl alcohol *(Cont.):*
   and hydroxypropylcellulose, 13-4
   insolubilizers for, 20-18
   laboratory techniques for, 20-30, 20-31
   light effect on, 20-8, 20-9, 20-14
   manufacture of, 20-15, 20-16
   mechanical properties of, 20-11 to 20-14
   modifiers for, 20-17 to 20-19
   moisture absorption of, 20-8, 20-14
   molded products from, 20-22
   molding properties of, 20-11
   oil resistance of, 20-14, 20-15
   and oils, 20-14, 20-15
   optical properties of, 20-9
   in packaging, 20-25
   and paper products, 20-2, 20-5, 20-21, 20-24,
     20-29F
   physical properties of, 20-3 to 20-15
   physiological properties of, 20-16
   and pigments, 20-19
   plasticizers for, 20-17, 20-18
   and polyethylene glycols, 18-21
   and polyvinylpyrrolidone, 21-10, 21-16
   precipitants for, 20-19
   and printing, *20-29F*
   as a quenchant, 20-23 to 20-25
   as a release coating, 20-24, *20-30F*
   safety of, 20-20
   and salts, 20-13
   shipping of, 20-19, 20-20
   and soil stabilization, 20-25
   solubility of, 20-4, 20-12
   solvent resistance of, 20-14, 20-15
   and solvents, 20-14, 20-15
   and starches *(see* Starches, and polyvinyl alco-
     hol)
   as a steel quenchant, 20-23 to 20-25
   storage of, 20-19, 20-20
   and textiles, 20-22 to 20-24, *20-25F to 20-27F*
   tradenames for, 20-31
   viscosity of, 20-3, 20-8 to 20-12
   and wetting agents, 20-19
Polyvinylbutyral, 19-7; 21-9
Polyvinyl chloride, 12-16, 12-17; 17-12; 18-12; 21-
   10
Polyvinyl formal, 21-9
Poly(vinyl isobutyl ether), 21-10
Poly(vinyl methyl ether), 21-10
Polyvinylpyrrolidone (PVP), 21-1 to 21-21
   and acids, 21-11
   and adhesives, 21-2, 21-15, 21-16, 21-18
   and agricultural products, 21-16, 21-18
   and alcohols, 21-11
   and amines, 21-11
   applications of, 21-2, 21-11 to 21-17
   and beverages, 21-2, 21-14, 21-18
   and carboxymethylcellulose, 21-8, 21-9, 21-16
   chemical nature of, 21-2
   and chlorinated hydrocarbons, 21-9, 21-11
   and coatings, 21-18
   compatibilities of, 21-8 to 21-11
   and cosmetics, 21-2, 21-8, 21-14, 21-15, 21-18
   crosslinking of, 21-2
   and detergents, 21-2
   and dyestuffs, 21-2
   and electrical products, 21-19
   and esters, 21-9, 21-11

Polyvinylpyrrolidone (PVP) *(Cont.):*
and ether-alcohols, 21-9, 21-11
and ethers, 21-9, 21-11
and ethylcellulose, 21-9
Fikentscher's K-values, 21-4
films from, 21-2, 21-8
and fluorocarbons, 21-11
and foods, 21-2
gelation of, 21-2
and glycerides, 21-9
and gum arabic, 21-9
and gum karaya, 21-9
and gums, 21-9
and gum tragacanth, 21-9
and hydrocarbons, 21-11
intrinsic viscosity of, 21-4
iodine complexes of, 21-2
and ketone-alcohols, 21-11
and ketones, 21-11
and lactams, 21-11
and lactones, 21-11
and laundry products, 21-16, 21-18
and lithography, 21-19
manufacture of, 21-2 to 21-4
and methylcellulose, 21-9
and nitroparaffins, 21-11
and paper products, 21-20
and pharmaceuticals, 21-2, 21-12 to 21-15, 21-17, 21-20
and phenols, 21-9
and photographic products, 21-16, 21-19
physical properties of, 21-2
and pigments, 21-2
plasticizers for, 21-8
and poly(acrylic acid), 17-8
and polyethylene glycols, 21-8, 21-11
as polymerization aid, 21-2, 21-16, 21-20
and polyvinyl alcohol, 21-10, 21-16
and printing, 21-2, 21-16
rheology of, 21-4, 21-5
and sodium alginate, 21-9
solubility of, 21-2
and solvents, 21-4, 21-11
and sulfonates, 21-10
and surfactants, 21-10
and textiles, 21-2, 21-19
toxicological properties of, 21-5 to 21-8
viscosity of, 21-2, 21-4 to 21-6
Porcelain enamels, 3-18
Postage stamps, 8-13
Potable water treatment, 16-9, 16-13, 16-18
Potassium alginate, 2-6 to 2-10, 2-12, 2-13, 2-17
Potassium amide, 21-3
Potassium bromide, 19-11
Potassium carbonate, 19-11
Potassium carboxymethylcellulose, 4-22
Potassium chloride, 2-22; 7-5; 19-11; 24-14
Potassium ferrocyanide, 13-7
Potassium fluoride, 19-11
Potassium halides, 19-4, 19-5
Potassium hydroxide, 9-2; 11-4; 19-11, 19-15; 22-17, 22-25, 22-26
Potassium iodide, 18-2; 19-5, 19-11; 20-6; 22-18
Potassium phosphate, dibasic, 2-22
Potassium plumbate, 23-2
Potassium polyacrylate, 17-2
Potassium salts, 22-4

Potassium silicate, 8-3
Potassium sulfate, 2-22; 19-11; 20-6; 20-19
Potassium thiocyanate, 19-4, 19-5; 20-6; 22-17
Potato starch, 22-72
and adhesives, 22-68 to 22-70
cationic, 22-55
cooked, 22-16
crosslinked, 22-62
dextrinization of, 22-36
dispersions of, 22-62
in foods, 22-29, 22-61
gelatinization of, chemical, 22-17
granule properties of, 22-7
and humidity, 22-13
and paper products, 22-22, 22-55, 22-57
pastes of, 22-19
pasting temperature of, 22-21
and phosphate groups, 22-42
solutions of, 22-18
supplies of, 22-21
swelling of, 22-14, 22-20
and tablets, 22-70
viscosity, 22-14
Poval, 20-31
Povidone, 21-11
Praestol, 16-25
Printing, 3-15; 6-13; 8-6, 8-7, 8-14, 8-15, 8-17; 11-3, 11-19, 11-21; 12-16; 14-9, 14-10, 14-14; 17-11, 17-14, 17-15; 19-16, 19-18, 19-27, *19-29F;* 21-2, 21-16, 21-19; 22-59, 22-64; 23-3, 23-7; 24-22
(See also Textiles, printing of)
Proctin, 2-41
Proger Sodio CMC, 4-24
Propanol, 7-4; 21-11
Propargyl alcohol, 21-9
Propionic acid, 21-11
Propylene alginate, 2-18, 2-20, 2-23
Propylene glycol, 2-12, 2-22; 3-12; 5-10; 13-5, 13-6, 13-10; 19-19; 21-4, 21-11; 24-15
Propylene glycol alginate:
in barbecue sauces, 2-25, 2-27
and beer, 2-25
and beverages, 2-25
BOD/COD of, 2-23
films from, 2-24
in foods, 2-18, 2-25, 2-27, 2-28, 2-31 to 2-34
and gum tragacanth, 11-12
physical properties of, 2-4
and solvents, 2-12
stability of, 2-2, 2-3
viscosity of, 2-6, 2-8, 2-9, 2-13, 22-14, 2-19
Propylene glycol linoleate, 19-7
Protacell, 2-41
Protanal, 2-41
Protatek, 2-41
Protease, 24-10
Proteins:
and agar, 7-6, 7-9
and carrageenans, 5-7, 5-18, 5-21
and glues, 12-6
and gum arabic, 8-7
and gum ghatti, 9-5
and gum karaya, 10-9
and hydroxyethylcellulose, 12-6
and pectins, 15-13
and poly(acrylic acid), 17-17

Proteins *(Cont.):*
 recovery of, 17-17
 soy, 5-21
 and starches, 22-70
Protomon, 2-39
Psyllium seeds, 10-10
Pure-Flo, 22-71
Purgol, 16-25
Purifloc, 16-25
Purity Gum, 22-71
PVP *(see* Polyvinylpyrrolidone)
Pyridine, 3-12; 13-5, 13-6; 21-11
Pyrilamine maleate, 10-9
Pyrocatechol, 8-8
Pyrodextrins, 22-34
Pyrogallol, 8-8
Pyrrolidone, 21-11

Quilon, 13-16
Quinine, 18-12
Quinine sulfate, 8-8

Rayon, 18-16, 18-22, 18-25
Redisol, 22-71
Redi-Tex Starch, 22-71
Reducing agents, 24-9
Relatin, 4-24
Resoflex, 21-8
Resole resins, 19-5
Resorcinol, 18-12; 20-18
Resorcinol-formaldehyde resins, 22-69, 22-70
Reten, 16-25
Rezista Starch, 22-71
Rhoplex, 2-21; 24-15
Rhozyme, 2-22; 24-15
Rice starch, 22-7, 22-17, 22-18, 22-21, 22-50
Rocket fuels, *16-18F*
Romano cheese dressing, 2-32
Roquefort dressing, 2-32
Rosin, 18-11
Rubber products, 3-15, 3-20; 4-26; 17-10; 18-16,
 18-21, 18-22, 18-24, 18-25; 19-23, 19-26
Rubidium halides, 19-4, 19-5
Rug backing, 17-10
Russian isinglass, 7-5
Rust removal, 17-16

Sago starch, 22-7, 22-21; 23-5
Salicylanilide, 20-18; 23-4
Salicylates, 22-17
Salol, 18-12
Salts:
 and agar, 7-4, 7-5
 and alginates, 2-17, 2-22
 and carrageenans, 5-10
 and hydroxypropylcellulose, 13-5, 13-11
 and locust bean gum, 14-6
 and poly(ethylene oxide), 19-11
 and polyvinylpyrrolidone, 21-8
 and starches, 22-69
 and tamarind gum, 23-6
 and xanthan gum, 24-3, 24-9, 24-14
 (See also specific salts)
Salt water treatment, 17-16

Sanitary napkins, 19-16, 19-19, 19-27; 22-52
Sanitizers, 3-18; 21-8
Santolite, 21-8
Saran, 21-10
Satiagel, 5-29
Satiagum, 5-29
Schweitzer reagent, 11-4
SeaGel, 5-29
SeaKem, 5-29
Sealants, 12-11, 12-14, 12-16
SeaSpen, 5-29
Secap, 22-71
Sedipur, 16-25
Select, 10-13
Sepalon, 16-25
Separan, 16-25
Sequestrants, 2-7, 2-14, 2-15, 2-17
Sewage treatment, 16-10, 16-13, 16-18; 17-13; 19-
 23; 20-17
SGP Absorbent Polymer, 22-71
Shellac, 18-20
Sherbelizer, 2-39
Ships, 19-23
Shoe heels, 18-22
Shoe polishes, 7-9; 24-21
Shur-Fil Starch, 22-71
Silica, 19-22, 19-26; 20-19
Silicotungstic acid, 7-4
Silver, 4-3
Silver nitrate, 8-8; 13-7
Silver polishes, 24-21
Sindar G-4, 2-21
Slime removal, 16-12, 16-13, 16-15, 16-17
Snow Algin, 2-41
Snow Flake Starch, 22-72
Soaps, 4-26, 4-27; 8-6, 8-7, 8-12; 16-10; 21-16, 21-
 18
Soda ash manufacture, 16-16
Sodium, 10-4
Sodium acetate, 13-7; 19-11
Sodium alginate, 12-5
 in adhesives, 2-35
 and agar, 7-5
 in bakery products, 2-30
 BOD/COD of, 2-20, 2-23
 and calcium, 2-9
 compatibilities of, 2-21, 2-22
 in dairy products, 2-29
 in desserts, 2-26, 2-28 to 2-32, 2-34, 2-35
 in dyes, 2-36
 films from, 2-24
 and fish, 2-26
 in foods, 2-17, 2-25, 2-26, 2-28 to 2-35
 in fruits, 2-31, 2-33, 2-34
 and hydroxyethylcellulose, 12-5
 and hydroxypropylcellulose, 13-4, 13-5
 manufacture of, 2-4, 2-5
 and paper products, 2-35
 physical properties of, 2-4
 and polyvinylpyrrolidone, 21-9
 and sequestrants, 2-14, 2-16
 and solvents, 2-12
 stability of, 2-10
 toxicological properties of, 4-9
 viscosity of, 2-6 to 2-9, 2-11, 2-13, 2-14, 2-18,
 2-19
 and xanthan gum, 2-7; 24-7, 24-10, 24-13

Sodium arabate, 8-3, 8-5, 8-8
Sodium benzoate, 2-16; 11-13; 13-10
Sodium bicarbonate, 23-6
Sodium bisulfate, 24-14
Sodium bisulfite, 23-7
Sodium bromide, 22-18
Sodium carbonate, 3-10, 3-11; 13-7; 19-11; 20-6, 20-19; 22-18; 23-6; 24-9, 24-14
Sodium carboxymethylcellulose, 12-5; 19-4; 22-67
 (*See also* Carboxymethylcellulose)
Sodium caseinate, 5-21; 12-5; 13-4
Sodium chloride, 2-18, 2-19, 2-22; 3-10, 3-11; 4-4, 4-11; 6-8; 8-8; 10-6; 13-7; 19-11; 22-17, 22-18; 24-7, 24-8, 24-14
Sodium chromate, 24-14
Sodium citrate, 2-22; 8-5; 24-14
Sodium fluoride, 19-4, 19-5
Sodium hexametaphosphate, 2-16, 2-17
Sodium hydroxide, 8-8; 17-3; 19-15; 20-6; 22-17, 22-18, 22-23, 22-25, 22-26, 22-70; 23-6; 24-8, 24-14
Sodium iodide, 22-17, 22-18
Sodium lauryl sulfate, 21-10; 22-20
Sodium metasilicate, 21-2; 24-9
Sodium nitrate, 13-7; 18-14; 20-6; 22-36, 22-68
Sodium Omadine, 13-16
Sodium pentachlorophenate, 23-4
Sodium pentachlorophenol, 12-5
Sodium phosphate, 2-22; 3-10, 3-11; 19-11
Sodium polyacrylate, 17-2, 17-6, 17-12, 17-16, 17-18
 and brine treatment, 17-16
 and chlorine manufacture, 17-16
 and the FDA, 17-18
 and paper sizing, 17-12
 and urea forms, 17-11
 [*See also* Poly(sodium acrylate)]
Sodium polymethacrylate, 17-18
Sodium propionate, 13-10
Sodium rosinate, 12-5
Sodium salicylate, 22-17
Sodium salts, 22-4; 23-6
Sodium silicate, 8-3; 19-11
Sodium silicofluoride, 23-4
Sodium sulfate, 2-22; 3-10, 3-11; 7-4; 13-7; 16-10; 20-6, 20-19; 22-17; 24-14
Sodium sulfite, 13-7
Sodium tetraborate, 2-22; 9-2; 23-3; 24-14
 (*See also* Borax)
Sodium thiocyanate, 20-18
Sodium thiosulfate, 13-7
Soil analysis, 9-6
Soil conditioners, 17-13, 17-14; 19-16, 19-21, 19-25; 20-24, 20-25; 23-7, 23-9
Solids recovery, 16-10, 16-13, 16-17, 16-18
Solvents *(see specific solvent or specific gum)*
Sorbitan esters, 22-21
Sorbitol, 2-24; 3-12; 10-5; 12-6; 13-10; 17-5; 19-7, 19-19; 21-8; 22-5, 22-36, 22-60, 22-68
Sorghum starch:
 amylopectin in, 22-23
 crosslinked, 22-62
 derivatives of, 22-38
 dextrinization of, 22-36
 dispersions of, 22-62
 and foods, 22-38, 22-61

Sorghum starch *(Cont.):*
 gelatinization of, 22-16
 granule properties of, 22-7
 and paper products, 22-55
 pregelatinized, 22-26
 solutions of, 22-18
 supplies of, 22-21
 swelling of, 22-20
SPAN, 22-48 to 22-51
Spectrophotometric analyses, 9-5
Speed Floc, 16-25
Sponges, 18-21
Stafome, 9-8
Sta-Lok, 22-72
Stannous salts, 18-12
Sta-O-Paque Starch, 22-72
Starches, 22-1 to 22-83
 acetates of, 22-39, 22-40, 22-57, 22-59, 22-67, 22-68, 22-70
 acetylated, 22-72
 acid converted, 22-30, 22-32, 22-33, 22-58, 22-62, 22-67, 22-68, 22-70
 acid fluidity, 22-25, 22-31
 acid hydrolyzed, 22-63
 acid modified, 22-65, 22-69, 22-72
 acrylamidomethyl ether of, 22-72
 and adhesives, 18-19; 22-22, 22-23, 22-31, 22-33, 22-45, 22-68 to 22-70, 22-72
 and agar, 7-5, 7-7
 and agricultural products, 22-47
 and alginates, 22-67
 amino derivatives of, 22-43
 ampholytic, 22-55
 amylopectin change in, 22-23
 amylopectin fraction of, 22-5, 22-6
 amylose change in, 22-23, 22-25
 amylose fraction of, 22-4 to 22-6
 anionic, 22-72
 applications of, 22-54 to 22-70
 and baking powder, 22-61
 benzyl ether of, 22-44
 and beverages, 22-22
 bleached, 22-65
 and blood plasma extenders, 22-40
 and BOD, 22-67
 buffered, 22-58, 22-72
 in building products, 22-22, 22-70, 22-72
 in candies, 22-33
 and carboxymethylcellulose, 22-67
 vs. BOD, 4-19
 carboxymethylethers of, 22-42, 22-57, 22-68, 22-70
 cationic ethers of, 22-42, 22-54 to 22-60, 22-67, 22-72
 chemical composition of, 22-2, 22-3
 chlorinated, 22-31, 22-33
 coatings (*see* Coatings, and starches)
 converted, 22-30, 22-57, 22-59, 22-64, 22-66, 22-69, 22-72
 cooked, 22-16
 cosmetics, 22-72
 crosslinked, 22-62, 22-63, 22-72
 crosslinking of, 22-27, 22-28
 cyanoethyl ether of, 22-57
 and defoamers, 22-68
 derivatives of, 22-40
 derivatization of, 22-37 to 22-47

Starches *(Cont.):*
  dextrinization of, 22-31, 22-34 to 22-36
  dextrinized, 22-32
  dextrins, 22-25, 22-59
  dialdehyde, 22-44, 22-45, 22-57
  dispersions of, 22-62
  as dusting powders, 22-43
  and encapsulation, 22-43
  enzyme converted, 22-59, 22-72
  and enzymes, 22-42
  esterification of, 22-37
  esters of, 22-41, 22-42, 22-65, 22-72
  etherification of, 22-37
  etherified, 22-65
  and FAO/WHO, 22-64
  and the FDA, 22-38, 22-43, 22-60, 22-64
  fillers for, 22-68
  films from, 22-5, 22-25, 22-26, 22-39, 22-44, 22-
    52, 22-64
  fluidity, 22-32, 22-33, 22-64, 22-67
  and foods, 22-22, 22-29, 22-31, 22-33, 22-38,
    22-60 to 22-64, 22-72
  fractionation of, 22-25
  functional groups added to, 22-40 to 22-47
  gelatinization of, 22-7, 22-12 to 22-18, 22-69
  graft polymers of, 22-72
  granule form of, 22-6, 22-7
  granule properties of, 22-7
  GRAS, 22-60, 22-64
  and gum arabic, 8-6, 8-7, 8-13
  gums from, 1-2
  and gum tragacanth, 11-12, 11-17
  humectants for, 22-68
  and humidity, 22-13
  hybrid breeding for, 22-23
  hydrophobic, 22-72
  hydroxyalkyl ethers of, 22-39
  and hydroxyethylcellulose, 12-5
  hydroxyethyl ethers of, 22-40, 22-43, 22-55, 22-
    57, 22-59, 22-60, 22-67, 22-68, 22-70, 22-
    72
  and hydroxypropylcellulose, 13-7
  hydroxypropyl ethers of, 22-40, 22-57
  inhibited, 22-25, 22-28
  laboratory techniques for, 22-70
  and laundry products, 22-33, 22-34, 22-72
  low-thermophile, 22-63
  magnesium xanthanate salt of, 22-47
  manufacture of, 22-22 to 22-24
  methacrylate esters of, 22-53
  and minerals processing, 22-22
  modification of, 22-23, 22-25 to 22-54
  modified, 22-31, 22-61, 22-65, 22-66
  molecular structure of, 22-2 to 22-6
  oxidized, 22-25, 22-30 to 22-33, 22-55, 22-57,
    22-59, 22-65, 22-67, 22-68, 22-70, 22-72
  and paper products *(see* Paper products, and
    starches)
  pastes, properties of, 22-18 to 22-21
  pasting temperature of, 22-7
  and pharmaceuticals, 22-70, 22-72
  phosphate, 22-42, 22-55, 22-57
  photomicrographs of, 22-8 to 22-13
  plasticizers for, 22-5, 22-68
  and polyethylene glycols, 18-11, 18-19
  and poly(ethylene oxide), 19-7
  polymerization of, 22-47 to 22-54

Starches *(Cont.):*
  and polyvinyl alcohol, 20-6, 20-18, 20-20, 20-
    21; 22-66, 22-67, 22-69
  powdered, 22-61
  pregelatinized, 22-26, 22-61, 22-63, 22-72
  and resins, 22-69, 22-70
  retrogradation of, 22-4
  and salts, 22-4
  saponified, 22-72
  and sodium carboxymethylcellulose, 22-67
  stabilized, 22-25, 22-72
  and stream pollution, 22-67
  structure of, 22-15
  and sugar, 22-61
  sulfates of, 22-41, 22-42
  supplies of, 22-21
  swelling of, 22-7, 22-12 to 22-18
  vs. tamarind gum, 23-1, 23-8
  tapioca, 22-26
  and textiles, 22-22, 22-23, 22-31, 22-33, 22-34,
    22-64 to 22-68, 22-72
  thermal gelatinization of, 22-13 to 22-17
  thermochemically converted, 22-59
  thin-boiling, 22-30 to 22-33
  tradenames for, 22-71, 22-72
  viscosity of, 22-14, 22-27, 22-28, 22-32, 22-69;
    23-4, 23-5
  and wastewater treatment, 22-46, 22-55, 22-
    56, 22-58
  xanthanates of, 22-45 to 22-47
  and xanthan gum, 24-9, 24-20, 24-22
  (See also British gum; Dextrins; *specific
    starches)*
Starch ethers, 23-8
Starch syrup, 23-6
Starpol, 22-71
Stearic acid, 22-20
Stearoyl fumarate, 22-21
Stearoyl-2-lactylates, 22-21
Steel quenchant, 20-23
Stein-Hall system, 22-70
Stepanol WAT, 2-22; 24-15
Sterculia caudate gum, 10-2
Sterculia gum, 10-2
Stipine, 2-41
Stoddard solvent, 21-11
Stoke's acid, 8-3; 9-2
Stream pollution, 4-19; 22-67
  (See also Water)
Strontium chloride, 22-18
Strontium hydroxide, 23-2
Strychnine sulfate, 8-8
Styrene, 20-22; 22-48, 22-53
  polymerization of, 17-12; 21-16
Sucrose, 3-10, 3-11; 12-9; 13-7; 15-7; 18-12; 22-
    18, 22-20, 22-21, 22-68; 23-6
Sugars, 5-10, 5-18; 15-3; 16-12, 16-18, 16-19; 17-
    13; 22-20, 22-36, 22-61
Sulfadiazine, 18-12
Sulfamerazine, 18-12
Sulfanilamide, 18-12
Sulfates, 12-5
Sulfathiazole, 18-12
Sulfide ore concentration, 16-17
Sulfites, 12-5
Sulfonated oils, 12-6; 22-68
Sulfonates, 21-10

Sulfur dioxide, 23-7
Sulfuric acid, *19-30F;* 20-6; 24-14
Sunflo, 2-21
Sunglasses, 20-22
Sunkist, 15-20
Sunrose, 4-24
Superfine, 10-13
Superfloc, 16-25
Super Initial, 10-13
Superloid, 2-39
Supreme, 10-13
Surfactants, 2-16, 2-17, 2-22; 18-16; 21-10; 22-20;
    24-10, 24-15
Synthetic detergents, 4-2
Syrian Gum, 11-30

Tagat, 2-41
Talc, 13-7; 22-67
Tall-oil fatty acids, 13-4
Tamarind gum, 23-1 to 23-12
    and acids, 23-6
    additives for, 23-6
    and adhesives, 23-6, 23-8
    applications of, 23-7
    biological properties of, 23-3, 23-4
    and building products, 23-6, 23-8, 23-9
    chemical nature of, 23-2, 23-3
    commercial uses for, 23-8
    derivatives of, 23-2, 23-3
    dextrinization of, 23-3
    electrochemical properties of, 23-4
    and explosives, 23-7 to 23-9
    extenders for, 23-6
    in foods, 23-4, 23-7
    formulations with, 23-8, 23-9
    vs. guar gum, 23-8
    and gum tragacanth, 23-6
    handling of, 23-7
    laboratory techniques for, 23-9, 23-10
    and latex, 23-8
    vs. locust bean gum, 23-8
    manufacture of, 23-3
    molecular weight of, 23-2
    and paper products, 23-7, 23-8
    vs. pectins, 23-8
    physical properties of, 23-3
    preservatives for, 23-4
    and printing, 23-3
    rheology of, 23-4 to 23-6
    and salts, 23-6
    and soil stabilization, 23-9
    solubility of, 23-3
    vs. starch, 23-1, 23-8
    and starch ethers, 23-8
    storage of, 23-3, 23-7
    and textiles, 23-2 to 23-4, 23-6 to 23-9
    toxicological properties of, 23-3, 23-4
    viscosity of, 23-4 to 23-6
Tamarind endosperm powder, 23-10
Tamarind kernel powder (TKP), 23-10
Tamarind seed powder, 23-10
Tannic acid, 7-4; 18-12; 21-2; 23-3
Tannin, 8-8
Tapioca starch, 22-72
    acetates of, 22-63
    and adhesives, 22-28 to 22-70

Tapioca starch *(Cont.):*
    crosslinked, 22-62, 22-63
    crosslinking of, 22-28
    derivatives of, 22-38
    dextrinization of, 22-36
    dispersions of, 22-62
    and foods, 22-29, 22-61
    granule properties, 22-7
    and humidity, 22-13
    and paper products, 22-22, 22-55, 22-57
    pastes of, 22-19
    pasting temperature of, 22-71
    and pectins, 15-7
    and polyvinyl alcohol, 20-20
    solutions of, 22-18
    stabilized, 22-63
    supplies of, 22-21
    swelling of, 22-14
    and xanthan gum, 25-14
Tenderfil, 22-72
Teratogenicity, 5-11
Tergitol, 2-22; 24-15
Terpene resin, 21-10
Terpin hydrate, 18-12
Tetrachlorethane, 20-15
Tetraethylene glycol, 20-18
Tetraethylenepentamine, 17-9
Tetrahydrofuran, 13-5, 13-6; 19-9; 21-11
Tetralin, 21-11
Tetramethylammonium hydroxide, 22-17
Tetrasodium pyrophosphate, 20-19
Tetrosan, 21-10
Textaid, 22-72
Textiles:
    adhesives for, 3-20; 10-9; 12-11, 12-14, 12-17,
        12-20F; 20-9, 20-20, 20-24
    antistats for, 18-25; 19-16, 19-24
    backings for, 16-16; 17-10
    binders for, 10-9; 12-14, 12-17; 20-22, 20-25;
        23-7
    and carboxymethylcellulose (*see* Carboxy-
        methylcellulose, in textiles)
    carding of, 18-22
    carpets, 3-15, 3-20; 6-13; 12-11, 12-14, 12-18;
        14-9, 14-14; 22-68
    coatings for, 3-20; 7-9; 12-11, 12-17; 13-10, 13-
        12, 13-13
    cotton goods, 22-67
    creaseproofing of, 16-16
    cutting of, 19-23
    double-knit, 21-15
    durable press, 17-15; 22-68
    dyeing of, 2-25, 2-26, 2-36; 3-20; 10-11; 12-11,
        12-14, 12-18; 14-9, 14-14; 19-16, 19-27; 21-
        2, 21-15; 24-21
    fiberglass, 3-18; 12-15; 19-24, 19-25; 21-15, 21-
        16; 22-25, 22-64, 22-67
        (*See also* Glass)
    finishes for, 8-15; 12-15; 18-23, 18-25; 20-2, 20-
        5; 21-15; 22-23, 22-31, 22-67, 22-68; 23-6
    flameproofing of, 22-68
    flocking of, 3-20
    fugitive tinting of, 21-15
    fugitive wefts for, 19-16, 19-24, 19-27
    and gum tragacanth, 11-20
    inks for, 8-14
    knitting of, 18-22, 18-25

Textiles *(Cont.):*
  laminating, 12-14
  latex coatings for, 3-20
  laundering of, 17-16
  and locust bean gum, 14-8, 14-9
  lubricants for, 18-16, 18-19, 18-22, 18-25
  mildew-proofed, 22-68
  nonwoven, 12-4; 20-2, 20-22, 20-25
  nylon, 17-8, 17-15; 18-16, 18-23
  pile backing for, 16-16
  polyesters, 18-23
  and polyvinylpyrrolidone, 21-15, 21-19
  printing of, 2-25, 2-26; 3-15, 3-20; 4-20; 6-10,
    6-11, 6-13, 6-18; 8-15; 10-11; 11-19, 11-21;
    12-11, 12-14, 12-18, *12-20F;* 14-9, 14-14;
    17-11, 17-14; 21-15; 22-23, 22-64, 22-68;
    23-3, 23-6, 23-8; 24-8, 24-13, 24-21
  rayon, 18-16, 18-22, 18-25
  ribbons, 20-22
  screen printing of, 12-18
  sewing thread, 22-64
  sizes for, 3-20; 4-2, 4-7, 4-13, 4-19, 4-27; 8-15,
    8-17; 11-3, 11-19, 11-21; 21-11; 17-4, 17-
    15; 18-22, 18-25; 19-16, 19-21, 19-25; 20-
    2, 20-5, 20-24, *20-25F to 20-27F;* 21-15;
    22-25, 22-31, 22-33, 22-34, 22-43, 22-64 to
    22-68, 22-72; 23-2, 23-4, 23-6, 23-7, *23-8F,*
    23-9F
  and starches *(see* Starches, and textiles)
  wash-and-wear, 22-68
  waterproofing, 22-68
  weaving of, 18-23, 18-25
  yarns, 22-66
Thin-N-Thik, 22-72
Thiocyanates, 22-4, 22-17, 22-36, 22-68, 22-69
2,2-Thiodiethanol, 21-11
Thiolignin, 19-4
Thiosulfates, 12-5
Thiourea, 19-4, 19-5
Thymol, 8-8; 18-12
Tiofloc, 16-25
Tires, 18-16, 18-22, 18-23; 19-26, *19-31F*
Titanium dioxide, 19-17, 19-22; 20-18, 20-19, 20-
  21; 21-16; 22-68
TKP (tamarind kernal powder), 23-10
Tobacco products, 3-15, 3-20, 3-22; 11-3, 11-19,
  11-21; 12-11, 12-15; 19-21
Toiletries *(see* Cosmetics/toiletries)
Toluene, 3-12; 13-5, 13-6; 18-11; 19-9; 20-15; 21-
  11
Tool making, 7-8
TOSCA (Toxic Substances Control Act), 22-69
Towels, 12-14
Toxic Substances Control Act (TOSCA), 22-69
Toys, 2-25, 2-26
Tragaya, 2-41
Tragtex, 11-30
Triazone resins, 20-18, 20-22
Tributyl phosphate, 20-19
Trichloroethylene, 18-11; 19-9, 19-15; 20-15
Triethanolamine, 2-22; 3-12; 21-4, 21-11; 24-15
Triethanolamine acetate, 20-18
Triethanolamine hydrochloride, 20-18
Triethylene glycol, 3-12; 21-11
Triethylenetetramine, 20-4
Trimethylolmelamine, 20-18
Trimethylol phenol, 19-4, 19-5

Trinitrotoluene, *16-18F, 16-19F*
Triol PEG, 18-17, 18-18
Trisodium phosphate, 21-2
Troche bases, 3-15
Trona plant operation, 16-18
Tung oil, 18-11
Turpentine, 20-15; 21-11
Tween, 2-22; 24-15
Tychem, 16-25
Tylec, 16-25
Tylose, 3-4, 3-25; 4-24

UCAR, 2-21; 24-16
Ucon Oil, 21-9
Ulceration, 5-11, 5-18
Ultra-Fast, 22-72
Ultrasonics, 8-5
Ultraviolet radiation, 8-5; 19-3
Unipectine, 15-20
Upholstery backing, 17-10
Uranium processing, 16-4, 16-7, 16-14, 16-17
Urea, 2-24; 16-3; 17-11; 18-12; 19-4, 19-5; 20-18;
  21-8; 22-5, 22-18, 22-36, 22-68, 22-69; 24-15
Urea-formaldehyde, 12-6, 12-14; 22-36, 22-59,
  22-69, 22-70; 23-6, 23-9
Urea-sodium nitrate, 2-24
Uremia, 17-17

Vancide TH, 2-21
Vanillin, 8-8; 18-12
Varnishes, 9-5; 12-17
Varnish removers, 3-17, 3-20; 19-16, 19-26, *19-*
  29F
Veterinary medicines *(see* Pharmaceuticals, vet-
  erinarian)
Vinamyl, 22-72
Vinol, 2-21; 20-31; 24-15
Vinyl acetate, 20-22; 22-53
Vinyl chloride, 20-22
Vinyl copolymers, 19-7; 21-10
Vinylidene chloride, 21-10
Vinylite VYHH, 21-10
Vinyl polymerization, 19-16, 19-22, 19-26; 21-16
Vinyl polymers, 8-13; 12-13; 17-12
N-Vinyl-2-pyrrolidone, 21-11
Viscarin, 5-29
Vitamins *(see* Foods, vitamins)
Vulca, 22-72

Wallboard cement, 16-10
Wallboard manufacture, 12-14, 12-16
Wallpaper, 3-14; 8-13; 9-5; 24-22
  (See also Adhesives, wallpaper)
Walsroder, 4-24
Water:
  boiler feed, 17-12; 18-14
  brines, 5-18; 16-16, 16-17; 17-16
  clarification of, 16-10, 16-13, 16-17, 16-18; 17-
    13; 19-23
  drilling, 16-15
  mill, 16-13, 16-17
  mine, 16-13, 16-17, 16-18
  pollution of, 22-55, 22-56, 22-58
  potable, 16-13

Water *(Cont.):*
  removal of, 16-10, 16-18
  salt, 17-16
  softening of, 16-9, 16-13, 16-17; 17-13
  and tailings recovery, 16-18
  treatment of, 12-11; 16-13; 17-13; 19-16, 19-
    22, 19-26; 20-16, 20-17; 22-46
  white, 16-13; 17-13; 19-23, 19-26
Wax products, 9-2, *9-6F;* 13-4; 17-15; 21-16; 24-
  21
Weapons, 19-23
Wheat flour, 23-4
Wheat starch:
  and acrylamide, 22-53
  and adhesives, 22-68 to 22-70
  dispersions of, 22-62
  and foods, 22-61
  granule properties of, 22-7
  and paper products, 22-22, 22-57
  pasting temperature of, 22-21
  polymerization of, 22-50
  pregelatinized, 22-26
  solutions of, 22-18
  supplies of, 22-21
  swelling of, 22-17, 22-18
Whiting, 20-19
WHO (World Health Organization), 14-8
  (See also FAO/WHO)
Windshields, 8-16; 20-22
Wood products, 3-19; *8-20F;* 12-11, 12-15, 12-16;
  17-15, 17-17; 18-16, 18-21, 18-23, 18-25; 20-
  9; 23-6 to 23-9
World Health Organization *(see* WHO)

Xanco-Frac, 24-29
Xanflood, 24-29
Xanthan gum:
  and abrasives, 24-22
  and acids, 24-8, 24-14
  and adhesives, 24-11, 24-22
  and agricultural products, 24-13, 24-20 to 24-
    22
  and alcohols, 24-9, 24-15
  and ammonium compounds, 24-8, 24-10, 24-
    14
  in animal food, 24-13, *24-27F*
  and bases, 24-8, 24-14
  and beverages, 24-13, 24-14, 24-20, 24-21
  and cellulosics, 24-9
  and ceramics, 24-13, 24-22
  chemical structure of, 24-2, 24-3
  in cleaners, 24-8, 24-22
  compatibilities of, 24-8 to 24-16
  in cosmetics, 24-22
  and dextrin, 24-9
  and dyes, 24-8, 24-11, 24-21
  and enzymes, 24-3, 24-10, 24-15
  and the EPA, 24-20
  and explosives, 24-22
  and the FAO/WHO, 24-20
  and the FDA, 24-20
  in fire fighting, 24-22
  in foods, 11-16; 24-3, 24-8, 24-13, 24-14, 24-
    16, 24-20 to 24-26, *24-22F to 24-27F*

Xanthan gum *(Cont.):*
  formulations with, 24-22 to 24-27
  and galactomannans, 24-16 to 24-19
  gelation of, 24-5, 24-6, 24-12
  and guar gum, 24-9, 24-15 to 24-17, 24-22
  and gum arabic, 24-9
  and gum tragacanth, 11-12; 24-15
  handling of, 24-16
  and inks, 24-21
  laboratory techniques for, 24-27 to 24-29
  and latex products, 24-8, 24-11, 24-16
  and locust bean gum, 14-5, 14-6; 24-9, 24-16
    to 24-19
  and metal products, 24-22
  and minerals treatment, 24-22
  moisture control of, 24-28
  and oxidizing agents, 24-9
  and paints, 24-8, 24-11, 24-13, 24-21
  and paper products, 24-8, 24-22
  and peroxides, 24-9
  in pet foods, 24-21
  and petroleum recovery, 24-13, 24-22
  in pharmaceuticals, 24-13
  in photochemical materials, 24-22
  physical properties of, 24-3, 24-4
  in polishes, 24-21
  preservatives for, 24-10, 24-15
  and printing, 24-8, 24-13, 24-21, 24-22
  and reducing agents, 24-9
  rheology of, 24-6 to 24-8, 24-12, 24-13
  safety of, 24-19, 24-20
  and salts, 24-3, 24-9, 24-10, 24-14, 24-22
  shear rates of, 24-7, 24-12
  and sodium alginate, 2-21; 24-7, 24-9, 24-10,
    24-13
  solution properties, 24-4 to 24-12
  and solvents, 24-9, 24-10, 24-15
  and starch, 24-9, 24-20, 24-22
  storage of, 24-16
  and surfactants, 24-10, 24-15
  and textiles, 24-8, 24-13, 24-21
  and thickeners, 24-9, 24-15
  toxicological properties of, 24-19, 24-20
  viscosity of, 24-4 to 24-7, 24-9 to 24-11, 24-22
  and wallpaper, 24-22
  and water-soluble resins, 24-15
  with waxes, 24-21
X-ray media, 3-23; 7-8; 21-14
X-Tra Gel, 22-72
Xylene, 13-6; 19-9; 21-11

Yarns, 22-66
Yokan, 7-18

Zein, 12-5; 13-5; 18-11
Zellin, 4-24
Zinc alginate films, 2-24
Zinc chlorides, 18-2; 20-6; 23-4, 23-6; 24-14
Zinc oxide, 20-19
Zinc recovery, 16-12, 16-15
Zinc sulfate, 19-11
Zinc sulfocarbolate, 18-12
Zirconium, 4-3